国外优秀数学著作
原 版 系 列

非线性系统及其绝妙的数学结构

Nonlinear Systems and Their Remarkable Mathematical Structures (Volume 2)

(第2卷)

(英文)

[墨]诺伯特·欧拉(Norbert Euler)
[意]玛丽亚·克拉拉·努奇(Maria Clara Nucci) 主编

HITP
哈爾濱工業大學出版社
HARBIN INSTITUTE OF TECHNOLOGY PRESS

黑版贸登字 08-2021-067 号

Nonlinear Systems and Their Remarkable Mathematical Structures: Volume 2/by Norbert Euler, Maria Clara Nucci/ISBN:978-0-367-20847-9

图书在版编目(CIP)数据

非线性系统及其绝妙的数学结构. 第 2 卷 = Nonlinear Systems and Their Remarkable Mathematical Structures: Volume 2:英文/(墨)诺伯特·欧拉(Norbert Euler),(意)玛丽亚·克拉拉·努奇(Maria Clara Nucci)主编. —哈尔滨:哈尔滨工业大学出版社,2024. 10

ISBN 978 - 7 - 5767 - 1284 - 1

Ⅰ. ①非… Ⅱ. ①诺 … ②玛… Ⅲ. ①非线性控制系统-英文 Ⅳ. ①O231. 2

中国国家版本馆 CIP 数据核字(2024)第 050331 号

FEIXIANXING XITONG JI QI JUEMIAO DE SHUXUE JIEGOU. DI 2 JUAN

策划编辑　刘培杰　杜莹雪
责任编辑　张嘉芮　李兰静
封面设计　孙茵艾
出版发行　哈尔滨工业大学出版社
社　　址　哈尔滨市南岗区复华四道街 10 号　邮编 150006
传　　真　0451 - 86414749
网　　址　http://hitpress. hit. edu. cn
印　　刷　哈尔滨市颉升高印刷有限公司
开　　本　787 mm×1 092 mm　1/16　印张 35　字数 603 千字
版　　次　2024 年 10 月第 1 版　2024 年 10 月第 1 次印刷
书　　号　ISBN 978 - 7 - 5767 - 1284 - 1
定　　价　108.00 元

(如因印装质量问题影响阅读,我社负责调换)

Contents

Preface vii

The Authors ix

A1. **Reciprocal transformations and their role in the integrability and classification of PDEs** **1**

P Albares, P G Estévez and C Sardón

1. Introduction 1
2. Fundamentals 3
3. Reciprocal transformations as a way to identify and classify PDEs 7
4. Reciprocal transformations to derive Lax pairs 17
5. A Miura-reciprocal transformation 21
6. Conclusions 25

A2. **Contact Lax pairs and associated (3+1)-dimensional integrable dispersionless systems** **29**

M Błaszak and A Sergyeyev

1. Introduction 29
2. Isospectral versus nonisospectral Lax pairs 30
3. Lax representations for dispersionless systems in (1+1)D and (2+1)D 31
4. Lax representations for dispersionless systems in (3+1)D 37
5. $\boldsymbol{R}$-matrix approach for dispersionless systems with nonisospectral Lax representations 42

A3. **Lax pairs for edge-constrained Boussinesq systems of partial difference equations** **59**

T J Bridgman and W Hereman

1. Introduction 59
2. Gauge equivalence of Lax pairs for PDEs and PΔEs 62
3. Derivation of Lax pairs for Boussinesq systems 65

4. Gauge and gauge-like equivalences of Lax pairs 72
5. Application to generalized Hietarinta systems 75
6. Summary of results 76
7. Software implementation and conclusions 85

A4. **Lie point symmetries of delay ordinary differential equations** .. **89**

V A Dorodnitsyn, R Kozlov, S V Meleshko and P Winternitz

1. Introduction 89
2. Illustrating example 91
3. Formulation of the problem for first-order DODEs 91
4. Construction of invariant first-order DODSs 93
5. First-order linear DODSs 95
6. Lie symmetry classification of first-order nonlinear DODSs 100
7. Exact solutions of the DODSs 105
8. Higher order DODSs 110
9. Traffic flow micro-model equation 112
10. Conclusions 114

A5. **The symmetry approach to integrability: recent advances** **119**

R Hernández Heredero and V Sokolov

1. Introduction 119
2. The symmetry approach to integrability 124
3. Integrable non-abelian equations 135
4. Non-evolutionary systems 143

A6. **Evolution of the concept of $\lambda-$symmetry and main applications** **158**

C Muriel and J L Romero

1. Introduction 158
2. Basic notions on Lie point symmetries and $\mathcal{C}^\infty-$ symmetries of ODEs 160
3. Analytical applications of $\mathcal{C}^\infty-$symmetries 167
4. Extensions and geometric interpretations of $\mathcal{C}^\infty-$symmetries 175

A7. **Heir-equations for partial differential equations: a 25-year review** **188**

M C Nucci

1. Introduction 188
2. Constructing the heir-equations 190
3. Symmetry solutions of heir-equations 192
4. Zhdanov's conditional Lie-Bäcklund symmetries and heir-equations 193
5. Nonclassical symmetries as special solutions of heir-equations 195
6. Final remarks 198

B1. **Coupled nonlinear Schrödinger equations: spectra and instabilities of plane waves** **206**

A Degasperis, S Lombardo and M Sommacal

1. Introduction 206
2. Spectra 211
3. Dispersion relation and instability 228
4. Conclusions 235

A. Case $r = 0$ 236

B. Polynomials: a tool box 239

B2. **Rational solutions of Painlevé systems** **249**

D Gómez-Ullate, Y Grandati and R Milson

1. Introduction 249
2. Dressing chains and Painlevé systems 252
3. Hermite τ-functions 257
4. Hermite-type rational solutions 265
5. Cyclic Maya diagrams 273
6. Examples of Hermite-type rational solutions 278

B3. **Cluster algebras and discrete integrability** **294**

A N W Hone, P Lampe and T E Kouloukas

1. Introduction 294
2. Cluster algebras: definition and examples 294

3. Cluster algebras with periodicity 302
4. Algebraic entropy and tropical dynamics 304
5. Poisson and symplectic structures 309
6. Discrete Painlevé equations from coefficient mutation 318
7. Conclusions 321

B4. **A review of elliptic difference Painlevé equations 326**

N Joshi and N Nakazono

1. Introduction 326
2. $E_8^{(1)}$-lattice 331
3. The initial-value space of the RCG equation 335
4. Cremona isometries 337
5. Birational actions of the Cremona isometries for the Jacobi setting 339
6. Special solutions of the RCG equation 342
A. $A_4^{(1)}$-lattice 344
B. General elliptic difference equations 350

B5. **Linkage mechanisms governed by integrable deformations of discrete space curves 356**

S Kaji, K Kajiwara and H Park

1. Introduction 356
2. A mathematical model of linkage 358
3. Hinged network and discrete space curve 363
4. Deformation of discrete curves 367
5. Extreme Kaleidocycles 374

B6. **The Cauchy problem of the Kadomtsev-Petviashvili hierarchy and infinite-dimensional groups 382**

J-P Magnot and E G Reyes

1. Introduction 382
2. Diffeologies, Frölicher spaces and the Ambrose-Singer theorem 383
3. Infinite-dimensional Lie groups and pseudodifferential operators 393
4. The Cauchy problem for the KP hierarchy 406
5. A non-formal KP hierarchy 410

B7. **Wronskian solutions of integrable systems** **415**

D-j Zhang

1. Introduction 415
2. Preliminary 416
3. The KdV equation 420
4. The mKdV equation 425
5. The AKNS and reductions 432
6. Discrete case: the lpKdV equation 436
7. Conclusions 440

C1. **Global gradient catastrophe in a shallow water model: evolution unfolding by stretched coordinates** **445**

R Camassa

1. Introduction 445
2. Exact solutions 446
3. Evolution beyond the gradient catastrophe 451
4. Discussion and conclusions 456

C2. **Vibrations of an elastic bar, isospectral deformations, and modified Camassa-Holm equations** **459**

X K Chang and J Szmigielski

1. Introduction 459
2. The Lax formalism: the boundary value problem 461
3. Liouville integrability 465
4. Forward map: spectrum and spectral data 469
5. Inverse problem 471
6. Multipeakons for $N = 2K$ 479
7. Multipeakons for $N = 2K + 1$ 483
8. Reductions of multipeakons 487

A. Lax pair for the 2-mCH peakon ODEs 491

B. Proof of Theorem 13 493

C. Proof of Theorem 17 495

C3. **Exactly solvable (discrete) quantum mechanics and new orthogonal polynomials** **498**

R Sasaki

1. Introduction 498
2. New orthogonal polynomials in ordinary QM 500
3. New orthogonal polynomials in discrete QM with real shifts 510
4. Summary and comments 522

编辑手记 **527**

Preface

Nonlinear systems are studied for their own sake since they describe the world from the motion of the big waves to the vibration of the smallest particle. Indeed, their mathematical structure has been found to be so rich and diverse that one needs to have very wide multiple perspectives in order to understand it. In this book the reader can find a collection of those perspectives by different authoritative authors whose work complements that of the authors of the previous book in framing nonlinear problems, a task of truly paramount importance. This book consists of 17 invited contributions written by leading experts in different aspects of nonlinear systems that include ordinary and partial differential equations, difference equations, discrete or lattice equations, non-commutative and matrix equations, and delay equations. The contents are divided into three main parts, namely **Part A**: *Integrability, Lax Pairs and Symmetry*, **Part B**: *Algebraic and Geometric Methods*, and **Part C**: *Applications*. Below we give a short description of each contribution.

Part A consists of seven contributions, numbered A1 to A7. In this part the authors mainly address the fundamental question of how to detect integrable systems, and also the use of symmetry methods for nonlinear systems. In **A1** the authors *P Albares, P G Estévez and C Sardón* discuss the role of reciprocal transformations in relating different integrable equations, certain non-integrable with integrable equations, and in constructing Lax Pairs. In **A2** the authors *M Błaszak and A Sergyeyev* review a recent approach to the construction of (3+1)-dimensional integrable dispersionless (also known as hydrodynamic-type) partial differential systems based on their contact Lax pairs, including a related R-matrix theory, and illustrate it by several examples. In **A3** the authors *T J Bridgman and W Hereman* construct Lax pairs for nonlinear partial difference systems since one of the fundamental characterizations of integrable nonlinear partial difference equations is indeed the existence of a Lax pair. In **A4** the authors *V A Dorodnitsyn, R Kozlov, S V Meleshko and P Winternitz* consider scalar first-order delay ordinary differential equations with one delay, discuss their symmetry properties and present some invariant solutions. In **A5** the authors *R Hernández Heredero and V Sokolov* review the classification of integrable partial differential equations by means of higher symmetries. In **A6** the authors *C Muriel and J L Romero* describe the main applications of the λ-symmetries theory to ordinary differential equations as well as its extension to partial differential equations. In **A7** the author *M C Nucci* gives a review of the heir-equations method and its role in the determination of extra symmetry solutions, nonclassical symmetries, and generalized symmetries of partial differential equations.

Part B consists of seven contributions, numbered B1 to B7. Here the authors describe different methods by which to obtain explicit solutions of nonlinear systems and/or describe the solution structures of the systems. In **B1** the authors *A Degasperis, S Lombardo and M Sommacal* study the linear stability problem of the continuous wave solutions of two coupled non-linear Schrödinger equations within the integrability framework. In **B2** the authors *D Gómez-Ullate, Y Grandati and R*

Milson are concerned with the classification of rational solutions to Painlevé equations and systems, and after describing a determinantal representation in terms of classical orthogonal polynomials, explicitly construct Hermite-type solutions for the Painlevé IV, Painlevé V equations and the A_4 Painlevé system. In **B3** the authors *A N W Hone, P Lampe and T E Kouloukas* give an introduction to cluster algebras, and explain how discrete integrable systems can appear in the context of cluster mutation. In **B4** the authors *N Joshi and Nakazono* review the construction of elliptic difference Painlevé equations by using Sakai's geometric way, and give rise to more examples. In **B5** the authors *S Kaji, K Kajiwara and H Park* study a family of linkages (Kaleidocycles) consisting of copies of an identical link connected by hinges, which are characterized as discrete curves with constant speed and constant torsion, and define a flow on the configuration space of a Kaleidocycle by a semi-discrete modified Korteweg-deVries equation. In **B6** the authors *J-P Magnot and E G Reyes* present proofs of the well-posedness of the Cauchy problem for the Kadomtsev-Petviashvili hierarchy by using generalized differential geometry of Frölicher and diffeological spaces. In **B7** the author *D J Zhang* provides a review of the Wronskian technique and solutions in Wronskian form, adding several examples that include the Korteweg-de Vries, the Ablowitz-Kaup-Newell-Segur hierarchy and reductions, and lattice potential Korteweg-de Vries equation.

Part C consists of three contributions, numbered C1 to C3. The authors apply different techniques in order to solve specific nonlinear problems. In **C1** the author *R Camassa* discusses the property of certain exact solutions of the Airy's shallow water system, that shed some light on the peculiar features of the dynamics when the layer thickness vanishes. In **C2** the authors *X Chang and J Szmigielski* present a general discussion of isospectral deformations of a Sturm-Liouville system that is interpreted in terms of a two-component modified Camassa-Holm equation and solved for the case of discrete measures (multipeakons). In **C3** the author *R Sasaki* provides a concise and comprehensive review of the non-classical orthogonal polynomials which satisfy second-order differential or difference equations and can be derived from the quantum mechanical reformulation of the theory of classical orthogonal polynomials.

Norbert Euler (Guangzhou, September 26, 2019)

Maria Clara Nucci (Perugia, September 26, 2019)

The Authors

1. P Albares, *University of Salamanca, Spain* [**A1**]
2. M Błaszak, *A. Mickiewicz University, Poland* [**A2**]
3. T Bridgman, *Colorado School of Mines, Golden, USA* [**A3**]
4. R Camassa, *University of North Carolina at Chapel Hill, Chapel Hill, USA* [**C1**]
5. X Chang, *Institute of Computational Mathematics and Scientific Engineering Computing, Chinese Academy of Sciences, P. R. China* [**C2**]
6. A Degasperis, *University of Rome "La Sapienza", Italy* [**B1**]
7. V A Dorodnitsyn, *Keldysh Institute of Applied Mathematics, Russian Academy of Sciences, Russia* [**A4**]
8. P G Estévez, *University of Salamanca, Spain* [**A1**]
9. D Gómez-Ullate, *Universidad Complutense de Madrid, Spain* [**B2**]
10. Y Grandati, *Universite de Lorraine-Site de Metz, France* [**B2**]
11. R Hernández Heredero, *Universidad Politécnica de Madrid, Madrid, Spain* [**A5**]
12. W Hereman, *Colorado School of Mines, Golden, USA* [**A3**]
13. A N W Hone, *University of Kent, UK* [**B3**]
14. N Joshi, *University of Sydney, Australia* [**B4**]
15. S Kaji, *Institute of Mathematics for Industry, Kyushu University, Japan* [**B5**]
16. K Kajiwara, *Institute of Mathematics for Industry, Kyushu University, Japan* [**B5**]
17. T Kouloukas, *University of Kent, UK* [**B3**]
18. R Kozlov, *Norwegian School of Economics, Bergen, Norway* [**A4**]
19. P Lampe, *University of Kent, UK* [**B3**]
20. S Lombardo, *Loughborough University, UK* [**B1**]
21. J-P Magnot, *Université d'Angers, France* [**B6**]
22. S V Meleshko, *Suranaree University of Technology, Thailand* [**A4**]

23. R Milson, *Dalhousie University, Halifax, Canada* [**B2**]

24. C Muriel, *Universidad de Cádiz, Spain* [**A6**]

25. N Nakazono, *Aoyama Gakuin University, Tokyo, Japan* [**B4**]

26. M C Nucci, *University of Perugia, Italy* [**A7**]

27. H Park, *Graduate School of Mathematics, Kyushu University, Japan* [**B5**]

28. E G Reyes, *Universidad de Santiago de Chile, Chile* [**B6**]

29. J L Romero, *Universidad de Cádiz, Spain* [**A6**]

30. C Sardón, *ICMAT, Campus de Cantoblanco, Madrid, Spain* [**A1**]

31. R Sasaki, *Shinshu University, Matsumoto, Japan* [**C3**]

32. A Sergyeyev, *Silesian University in Opava, Czech Republic* [**A2**]

33. V V Sokolov, *Landau Institute for Theoretical Physics, Chernogolovka (Moscow region), Russia* [**A5**]

34. M Sommacal, *Northumbria University, Newcastle, UK* [**B1**]

35. J Szmigielski, *University of Saskatchewan, Canada* [**C2**]

36. P Winternitz, *Centre des Recherches Mathématiques, Université de Montréal, Canada* [**A4**]

37. D J Zhang, *Shanghai University, P.R. China* [**B7**]

A1. Reciprocal transformations and their role in the integrability and classification of PDEs

P. Albares [a], *P. G. Estévez* [a] *and C. Sardón* [b]

[a] *Department of Theoretical Physics, University of Salamanca, SPAIN.*

[b] *Instituto de Ciencias Matemáticas, Madrid, SPAIN*

Abstract

Reciprocal transformations mix the role of the dependent and independent variables of (nonlinear partial) differential equations to achieve simpler versions or even linearized versions of them. These transformations help in the identification of a plethora of partial differential equations that are spread out in the physics and mathematics literature.

Two different initial equations, although seemingly unrelated at first, could be the same equation after a reciprocal transformation. In this way, the big number of integrable equations that are spread out in the literature could be greatly diminished by establishing a method to discern which equations are disguised versions of a same, common underlying equation. Then, a question arises: Is there a way to identify different differential equations that are two different versions of a same equation in disguise?

Keywords: Reciprocal transformations, integrability, differential equations, water wave, Painlevé method, Painlevé integrability

1 Introduction

On a first approximation, hodograph transformations are transformations involving the interchange of dependent and independent variables [9, 20]. When the variables are switched, the space of independent variables is called the reciprocal space. In the particular case of two variables, we refer to it as the reciprocal plane. As a physical interpretation, whereas the independent variables play the role of positions in the reciprocal space, this number is increased by turning certain fields or dependent variables into independent variables and vice versa [13]. For example, in the case of evolution equations in fluid dynamics, usually fields that represent the height of the wave or its velocity, are turned into a new set of independent variables. Reciprocal transformations share this definition with hodograph transformations, but these impose further requirements. Reciprocal transformations require the employment of conservative forms together with the fulfillment of their properties, as we shall see in forthcoming paragraphs [15, 16, 20, 41]. For example, some properties and requirements for reciprocal transformations that are not necessary for hodograph transformations are: the existence of conserved quantities for their construction [15, 16, 22, 23, 36, 37, 38], that the invariance of certain integrable hierarchies under reciprocal transformations induces auto-Bäcklund transformations [22, 23, 30, 38, 40], and these transformations map conservation laws to conservation laws and diagonalizable systems to diagonalizable systems, but act nontrivially on metrics and on Hamiltonian structures.

But finding a proper reciprocal transformation is usually a very complicated task. Notwithstanding, in fluid mechanics, a change of this type is usually reliable, specifically for systems of hydrodynamic type. Indeed, reciprocal transformations have a long story alongside the inverse scattering transform (IST) [1, 3], the two procedures gave rise to the discovery of other integrable nonlinear evolution equations similar to the KdV equations. For example, Zakharov and Shabat [45] presented the now famous nonlinear Schödinger (NLS) equation, which presents an infinite number of integrals of motion and possesses n–soliton solutions with purely elastic interaction. In 1928, the invariance of nonlinear gas dynamics, magnetogas dynamics and general hydrodynamic systems under reciprocal transformations was extensively studied [24, 39]. Stationary and moving boundary problems in soil mechanics and nonlinear heat conduction have likewise been subjects of much research [25, 35].

One of the biggest advantages of dealing with hodograph and reciprocal transformations is that many of the equations reported integrable in the bibliography of differential equations, as the mentioned hydrodynamical systems, which are considered seemingly different from one another, happen to be related via reciprocal transformations. If this were the case, two apparently unrelated equations, even two complete hierarchies of partial differential equations (PDEs) that are linked via reciprocal transformation, are tantamount versions of a unique problem. In this way, the first advantage of hodograph and reciprocal transformations is that they give rise to a procedure of relating allegedly new equations to the rest of their equivalent integrable sisters. The relation is achieved by finding simpler or linearized versions of a PDE so it becomes more tractable. For example, reciprocal transformations were proven to be a useful instrument to transform equations with peakon solutions into equations that are integrable in the Painlevé sense [14, 26]. Indeed, these transformations have also played an important role in soliton theory and providing links between hierarchies of PDEs [14, 26], as in relation to the aforementioned hydrodynamic-type systems. In this chapter we will depict straightforward reciprocal transformations that will help us identify different PDEs as different versions of a same problem, as well as slight modifications of reciprocal transformations, as it can be compositions of several transformations of this type and others. For example, the composition of a Miura transformation [2, 42] and a reciprocal transformation gives rise to the so-called Miura-reciprocal transformations that help us relate two different hierarchies of differential equations. A whole section of this chapter is devoted to illustrating Miura-reciprocal transformations.

A second significant advantage of reciprocal transformations is their utility in the identification of integrable PDEs which, a priori, are not integrable according to algebraic tests (for example, the Painlevé test is one of them) [16, 20] but they are proven indeed integrable according to Painlevé, after a reciprocal transformation. Our conjecture is that if an equation is integrable, there must be a transformation that will let us turn the initial equation into a new one in which the Painlevé test is successful. We will comment on this later in forthcoming paragraphs.

A third advantage for the use of reciprocal transformations is their role in the derivation of Lax pairs. Although it is not always possible to find a Lax pair for a given equation, a reciprocal transformation can turn it into a different one whose Lax pair is acknowledged. Therefore, by undoing the reciprocal transformation in the Lax pair of the transformed equation, we can achieve the Lax pair of the former.

These three main points describing the importance of reciprocal transformations imply the power of these transformations to classify differential equations and to sort out integrability.

2 Fundamentals

We will deal with some well-known differential equations in the literature of shallow water wave equations. In particular, we will deal with generalizations of the Camassa–Holm equation and the Qiao equation [8, 15, 16, 20, 32, 33, 34]. Such generalizations consist of a hierarchy, i.e., a set of differential equations that are related via a recursion operator. The recursive application of such operator gives members of different orders of the hierarchy, i.e., a set of different differential equations. We will understand these differential equations as submanifolds of an appropriate higher-order tangent bundle. Hence, let us introduce the necessary geometric tools for explaining PDEs as submanifolds of bundles.

2.1 PDEs and jet bundles

Let us consider a smooth k-dimensional manifold N and the following projection $\pi : (x, u) \in \mathbb{R}^n \times N \equiv N_{\mathbb{R}^n} \mapsto x \in \mathbb{R}^n$ giving rise to a trivial bundle $(N_{\mathbb{R}^n}, \mathbb{R}^n, \pi)$. Here, we choose $\{x_1, \ldots, x_n\}$ as a global coordinate system on $\mathbb{R}^n$.

We say that two sections $\sigma_1, \sigma_2 : \mathbb{R}^n \to N_{\mathbb{R}^n}$ are p–equivalent at a point $x \in \mathbb{R}^n$ or they have a contact of order p at x if they have the same Taylor expansion of order p at $x \in \mathbb{R}^n$. Equivalently,

$$\sigma_1(x) = \sigma_2(x), \qquad \frac{\partial^{|J|}(\sigma_1)_i}{\partial x_1^{j_1} \ldots \partial x_n^{j_n}}(x) = \frac{\partial^{|J|}(\sigma_2)_i}{\partial x_1^{j_1} \ldots \partial x_n^{j_n}}(x), \tag{2.1}$$

for every multi-index $J = (j_1, \ldots, j_n)$ such that $0 < |J| \equiv j_1 + \ldots + j_n \leq p$ and $i = 1, \ldots, n$. That is p-equivalent induces an equivalence relation in the space $\Gamma(\pi)$ of sections of the bundle $(N_{\mathbb{R}^n}, \mathbb{R}^n, \pi)$. Observe that if two sections have a contact of order p at a point x, then they do have a contact at that point of the same type for any other coordinate systems on $\mathbb{R}^n$ and N, i.e., this equivalence relation is geometric.

We write $j^p_x\sigma$ for the equivalence class of sections that have a contact of p-order at $x \in \mathbb{R}^n$ with a section σ. Every such an equivalence class is called a p–jet. We write $\mathrm{J}^p_x\pi$ for the space of all jets of order p of sections at x. We will denote by $\mathrm{J}^p\pi$ the space of all jets of order p. Alternatively, we will write $\mathrm{J}^p(\mathbb{R}^n, \mathbb{R}^k)$ for the jet bundle of sections of the bundle $\pi : (x, u) \in \mathbb{R}^n \times \mathbb{R}^k \mapsto x \in \mathbb{R}^n$.

Given a section $\sigma : \mathbb{R}^n \to \mathrm{J}^p\pi$, we can define the functions

$$(u_j)_J(j^p_x\sigma) = \frac{\partial^{|J|}\sigma_j}{\partial x_1^{j_1}\dots\partial x_n^{j_n}}(x), \quad \forall j, \quad |J| \leq p. \tag{2.2}$$

For $|J| = 0$, we define $u_J(x) \equiv u(x)$. Coordinate systems on $\mathbb{R}^n$ and N along with the previous functions give rise to a local coordinate system on $\mathrm{J}^p\pi$. We will also hereafter denote the n-tuple and k-tuple, respectively, by $x = (x_1, \dots, x_n)$, $u = (u_1, \dots, u_k)$, then

$$(u_j)_J = u_{x_{i_1}^{j_1}\dots x_{i_n}^{j_n}} = \frac{\partial^{|J|}u_j}{\partial x_{i_1}^{j_1}\dots\partial x_{i_n}^{j_n}}, \quad \forall j, \quad |J| \leq 0. \tag{2.3}$$

All such local coordinate systems give rise to a manifold structure on $\mathrm{J}^p\pi$. In this way, every point of $\mathrm{J}^p\pi$ can be written as

$$\left(x_i, u_j, (u_j)_{x_i}, (u_j)_{x_{i_1}^{j_1}x_{i_2}^{2-j_1}}, (u_j)_{x_{i_1}^{j_1}x_{i_2}^{j_2}x_{i_3}^{3-j_1-j_2}}, \dots, (u_j)_{x_{i_1}^{j_1}x_{i_2}^{j_2}\dots x_{i_n}^{p-\sum_{i=1}^{n-1} j_i}}\right), \tag{2.4}$$

where the numb indices run $i_1, \dots, i_p = 1, \dots, n$, $j = 1, \dots, k$, $j_1 + \dots + j_n \leq p$.

For small values of p, jet bundles have simple descriptions: $\mathrm{J}^0\pi = N_{\mathbb{R}^n}$ and $\mathrm{J}^1\pi \simeq \mathbb{R}^n \times TN$.

The projections $\pi_{p,l} : j^p_x\sigma \in \mathrm{J}^p\pi \mapsto j^l_x\sigma \in \mathrm{J}^l\pi$ with $l < p$ lead to define the smooth bundles $(\mathrm{J}^p\pi, \mathrm{J}^l\pi, \pi_{p,l})$. Conversely, for each section $\sigma : \mathbb{R}^n \to N_{\mathbb{R}^n}$, we have a natural embedding $j^p\sigma : \mathbb{R}^n \ni x \mapsto j^p_x\sigma \in \mathrm{J}^p\pi$.

The differential equations that will be appearing in the chapter will be differential equations in close connection with shallow water wave models. We will define these PDEs on a submanifold $N_{\mathbb{R}^n}$ of a higher-order bundle $J^p(\mathbb{R}^{n+1}, \mathbb{R}^{2k})$. For the reciprocal transformation, we will have to make use of conservation laws. By conservation law we will understand an expression of the form

$$\frac{\partial\psi_1}{\partial x_{i_1}} + \frac{\partial\psi_2}{\partial x_{i_2}} = 0, \tag{2.5}$$

for certain two values of the indices in between $1 \leq i_1, i_2 \leq n$ and two scalar functions $\psi_1, \psi_2 \in C^\infty(\mathrm{J}^p N_{\mathbb{R}^n})$. The scalar fields representing water wave models will generally be denoted by U or u, which depend on the independent variables x_i, and the functions ψ_1, ψ_2 will be functions of higher-order derivatives of U or u. So, let us introduce the pairs (u_j, x_i) or (U_j, X_i) as local coordinates on the product manifold $N_{\mathbb{R}^n}$ and for the further higher-order derivatives we consider the construction given in (2.4). In cases of lower dimensionality, as the 2-dim. case, we shall use upper/lower case $(X, T)/(x, t)$. In the 3-dim. case, the independent variables will be denoted by upper/lower case $(X, T, Y)/(x, t, y)$.

2.2 The Camassa–Holm hierarchy

Let us consider the well-known Camassa–Holm equation (CH equation) in $1+1$ dimensions as a submanifold of $J^3(\mathbb{R}^2, \mathbb{R})$ with local coordinates for $\mathbb{R}^2 \times \mathbb{R}$ the triple (X,T,U). It reads:

$$U_T + 2\kappa U_X - U_{XXT} + 3UU_X = 2U_X U_{XX} + UU_{XXX}. \tag{2.6}$$

We can interpret U as the fluid velocity and (X,T) as the spatial and temporal coordinates, respectively. Nonetheless, the equation (2.6) in its present form is not integrable in the strict defined Painlevé sense, but there exists a change of variables (action-angle variables) such that the evolution equation in the new variables is equivalent to a linear flow at constant speed. This change of variables is achieved by studying its associated spectral problem and it is reminiscent of the fact that integrable classical Hamiltonian systems are equivalent to linear flows on tori, [10, 11, 12]. Indeed, (2.6) is a bi-Hamiltonian model for shallow water waves propagation introduced by Roberto Camassa and Darryl Holm [8]. For κ positive, the solutions are smooth solitons and for $\kappa = 0$, it has peakon (solitons with a sharp peak, so with a discontinuity at the peak in the wave slope) solutions. A peaked solution is of the form:

$$U = ce^{-|X-cT|} + O(\kappa \log \kappa). \tag{2.7}$$

In the following, we will consider the limiting case corresponding to $\kappa = 0$.

We can show the bi–Hamiltonian character of the equation by introducing the momentum $M = U - U_{XX}$, to write the two compatible Hamiltonian descriptions of the CH equation:

$$M_T = -\mathcal{D}_1 \frac{\delta \mathcal{H}_1}{\delta M} = -\mathcal{D}_2 \frac{\delta \mathcal{H}_2}{\delta M}, \tag{2.8}$$

where

$$\begin{aligned} \mathcal{D}_1 &= M\frac{\partial}{\partial X} + \frac{\partial}{\partial X}M, & \mathcal{H}_1 &= \frac{1}{2}\int U^2 + (U_X)^2 \, \mathrm{d}X, \\ \mathcal{D}_2 &= \frac{\partial}{\partial X} - \frac{\partial^3}{\partial X^3}, & \mathcal{H}_2 &= \frac{1}{2}\int U^3 + U\,(U_X)^2 \, \mathrm{d}X. \end{aligned} \tag{2.9}$$

The CH equation (2.6) is the first member of the well-known negative Camassa-Holm hierarchy for a field $U(X,T)$ [31]. From now on, we will refer to this hierarchy by CH(1+1).

The CH(1+1) can be written in a compact form in terms of a recursion operator R, defined as follows:

$$U_T = R^{-n} U_X, \qquad R = KJ^{-1}, \tag{2.10}$$

where K and J are defined as

$$K = \partial_{XXX} - \partial_X, \qquad J = -\tfrac{1}{2}(\partial_X U + U\partial_X), \quad \partial_X = \tfrac{\partial}{\partial X}. \tag{2.11}$$

The factor $-\frac{1}{2}$ has been conveniently added for future calculations. We can include auxiliary fields $\Omega^{(i)}$ with $i = 1, \ldots, n$ when the inverse of an operator appears. These auxiliary fields are defined as follows

$$\begin{aligned} U_T &= J\Omega^{(1)}, \\ K\Omega^{(i)} &= J\Omega^{(i+1)}, \quad i = 1, \ldots, n-1, \\ U_X &= K\Omega^{(n)}. \end{aligned} \tag{2.12}$$

It is also useful to introduce the change $U = P^2$, such that the final equations read:

$$P_T = -\frac{1}{2}\left(P\Omega^{(1)}\right)_X, \tag{2.13}$$

$$\Omega^{(i)}_{XXX} - \Omega^{(i)}_X = -P\left(P\Omega^{(i+1)}\right)_X, \quad i = 1, \ldots, n-1 \tag{2.14}$$

$$P^2 = \Omega^{(n)}_{XX} - \Omega^{(n)}. \tag{2.15}$$

As we shall see in section 3, the conservative form of equation (2.13) is the key for the study of reciprocal transformations.

2.3 The Qiao hierarchy

Qiao and Liu [34] proposed an integrable equation defined as a submanifold of the bundle $J^3(\mathbb{R}^2, \mathbb{R})$. Notice that here the dependent variable is denoted by lower case u. In the future, we shall use lower cases for the dependent and independent variables related to Qiao hierarchy. The capital cases shall be used for Camassa–Holm.

$$u_t = \left(\frac{1}{2u^2}\right)_{xxx} - \left(\frac{1}{2u^2}\right)_x, \tag{2.16}$$

which also possesses peaked solutions as the CH equation, and a bi–Hamiltonian structure given by the relation

$$u_t = j\frac{\delta h_1}{\delta u} = k\frac{\delta h_2}{\delta u}, \tag{2.17}$$

where the operators j an k are

$$j = -\partial_x u \left(\partial_x\right)^{-1} u\partial_x, \qquad k = \partial_{xxx} - \partial_x, \quad \partial_x = \frac{\partial}{\partial x}, \tag{2.18}$$

and the Hamiltonian functions h_1 and h_2 correspond with

$$h_1 = -\frac{1}{2}\int \left[\frac{1}{4u^3} + \left(\frac{4}{5\,u^5} + \frac{4}{7\,u^7}\right) u_x^2\right] dx, \qquad h_2 = -\int \frac{1}{2u}\, dx. \tag{2.19}$$

We can define a recursion operator as

$$r = kj^{-1}. \tag{2.20}$$

This recursion operator was used by Qiao in [33] to construct a $1+1$ integrable hierarchy, henceforth denoted as Qiao(1+1). This hierarchy reads

$$u_t = r^{-n} u_x. \tag{2.21}$$

Equation (2.16) is the second positive member of the Qiao hierarchy. The second negative member of the hierarchy was investigated by the same author in [32]. If we introduce n additional fields $v^{(i)}$ when we encounter the inverse of an operator, the expanded equations read:

$$\begin{aligned} u_t &= j v^{(1)}, \\ k v^{(i)} &= j v^{(i+1)}, \quad i = 1, \dots, n-1, \\ u_x &= k v^{(n)}. \end{aligned} \tag{2.22}$$

If we now introduce the definition of the operators k and j, we obtain the following equations:

$$u_t = -\left(u\omega^{(1)}\right)_x, \tag{2.23}$$

$$v_{xxx}^{(i)} - v_x^{(i)} = -\left(u\omega^{(i+1)}\right)_x, \quad i = 1, \dots, n-1, \tag{2.24}$$

$$u = v_{xx}^{(n)} - v^{(n)}, \tag{2.25}$$

in which n auxiliary fields $\omega^{(i)}$ have necessarily been included to operate with the inverse term present in j. These fields have been defined as:

$$\omega_x^{(i)} = u v_x^{(i)}, \quad i = 1, \dots, n. \tag{2.26}$$

The conservative form of (2.23) allows us to define the reciprocal transformation.

3 Reciprocal transformations as a way to identify and classify PDEs

The CH(1+1) presented in the previous section is here explicitly shown to be equivalent to n copies of the Calogero-Bogoyavlenski-Schiff (CBS) equation [5, 6, 29]. This CBS equation possesses the Painlevé property and the singular manifold method can be applied to obtain its Lax pair and other relevant properties [20]. Alongside, in the previous section we have also presented another example, the Qiao(1+1) hierarchy, for which the Painlevé test is neither applicable nor constructive. Nonetheless, here we will prove that there exists a reciprocal transformation which allows us to transform this hierarchy into n copies of the modified Calogero-Bogoyavlenskii-Schiff (mCBS), which is known to have the Painlevé property [23]. We shall denote the Qiao(1+1) likewise as mCH(1+1) because it can be considered as a modified version of the CH(1+1) hierarchy introduced in [20]. Then, this subsection shows how different pairs of hierarchies and equations: CH(1+1) and CBS equation and the Qiao(1+1) or mCH(1+1) and mCBS equation are different versions of a same problem when a reciprocal transformation is performed upon them. Let us illustrate this in detail.

3.1 Hierarchies in $1+1$ dimensions

Reciprocal transformations for CH(1+1)

Given the conservative form of equation (2.13), the following transformation arises naturally:

$$dz_0 = PdX - \frac{1}{2}P\Omega^{(1)}dT, \quad dz_1 = dT. \tag{3.1}$$

We shall now propose a reciprocal transformation [23] by considering the former independent variable X as a dependent field of the new pair of independent variables $X = X(z_0, z_1)$, and therefore, $dX = X_0\, dz_0 + X_1\, dz_1$ where the subscripts zero and one refer to partial derivative of the field X with respect to z_0 and z_1, correspondingly. The inverse transformation takes the form:

$$dX = \frac{dz_0}{P} + \frac{1}{2}\Omega^{(1)}dz_1, \quad dT = dz_1, \tag{3.2}$$

which, by direct comparison with the total derivative of the field X, we obtain:

$$\partial_0 X = \frac{\partial X}{\partial z_0} = \frac{1}{P}, \qquad \partial_1 X = \frac{\partial X}{\partial z_1} = \frac{\Omega^{(1)}}{2}. \tag{3.3}$$

The important point [22, 23] is that, we can now extend the transformation (3.1) by introducing $n-1$ additional independent variables $z_2, \ldots, z_n$ which account for the transformation of the auxiliary fields $\Omega^{(i)}$ in such a way that

$$\partial_i X = \frac{\partial X}{\partial z_i} = \frac{\Omega^{(i)}}{2}, \qquad i = 2, \ldots, n. \tag{3.4}$$

Then, X is a function $X = X(z_0, z_1, z_2, \ldots, z_n)$ of $n+1$ variables. It requires some computation to transform the hierarchy (2.13)-(2.15) into the equations that $X = X(z_0, z_1, z_2, \ldots, z_n)$ should obey. For this matter, we use the symbolic calculus package Maple. Equation (2.13) is identically satisfied by the transformation, and (2.14), (2.15) lead to the following set of PDEs:

$$\partial_0\left[-\frac{\partial_{i+1}X}{\partial_0 X}\right] = \partial_i\left[\partial_0\left(\frac{\partial_{00}X}{\partial_0 X} + \partial_0 X\right) - \frac{1}{2}\left(\frac{\partial_{00}X}{\partial_0 X} + \partial_0 X\right)^2\right], \quad i = 1, \ldots, n-1, \tag{3.5}$$

which constitutes $n-1$ copies of the same system, each of which is written in three variables z_0, z_i, z_{i+1}. Considering the conservative form of (3.5), we shall introduce the change:

$$\partial_i M = \frac{1}{4}\left[-\frac{\partial_{i+1}X}{\partial_0 X}\right], \tag{3.6}$$

$$\partial_0 M = \frac{1}{4}\left[\partial_0\left(\frac{\partial_{00}X}{\partial_0 X} + \partial_0 X\right) - \frac{1}{2}\left(\frac{\partial_{00}X}{\partial_0 X} + \partial_0 X\right)^2\right], \tag{3.7}$$

with $M = M(z_0, z_i, z_{i+1})$ and $i = 1, \ldots, n-1$. The compatibility condition of $\partial_{000}X$ and $\partial_{i+1}X$ in this system gives rise to a set of equations written entirely in terms of M:

$$\partial_{0,i+1}M + \partial_{000i}M + 4\partial_i M \partial_{00}M + 8\partial_0 M \partial_{0i}M = 0, \quad i = 1, \ldots, n-1, \tag{3.8}$$

which are $n-1$ CBS equations [5, 19, 23], each one in just three variables, for the field $M = M(z_0, .., z_i, z_{i+1}, ...z_n)$.

Reciprocal transformations for mCH(1+1)

Given the conservative form of (2.23), the following reciprocal transformation [23] naturally arises:

$$dz_0 = u\,dx - u\omega^{(1)}dt, \quad dz_1 = dt. \tag{3.9}$$

We now propose a reciprocal transformation [23] by considering the initial independent variable x as a dependent field of the new independent variables such that $x = x(z_0, z_1)$, and therefore, $dx = x_0\,dz_0 + x_1\,dz_1$. The inverse transformation adopts the form:

$$dx = \frac{dz_0}{u} + \omega^{(1)}dz_1, \quad dt = dz_1. \tag{3.10}$$

By direct comparison of the inverse transform with the total derivative of x, we obtain that:

$$\partial_0 x = \frac{\partial x}{\partial z_0} = \frac{1}{u}, \qquad \partial_1 x = \frac{\partial x}{\partial z_1} = \omega^{(1)}. \tag{3.11}$$

We shall prolong this transformation in such a way that we introduce new variables $z_2, \ldots, z_n$ such that $x = x(z_0, z_1, ..., z_n)$ according to the following rule:

$$\partial_i x = \frac{\partial x}{\partial z_i} = \omega^{(i)}, \qquad i = 2, \ldots, n. \tag{3.12}$$

In this way, (2.23) is identically satisfied by the transformation, and (2.24), (2.25) are transformed into $n-1$ copies of the following equation, which is written in terms of just three variables z_0, z_i, z_{i+1}:

$$\partial_0\left[\frac{\partial_{i+1}x}{\partial_0 x} + \frac{\partial_{00i}x}{\partial_0 x}\right] = \partial_i\left[\frac{(\partial_0 x)^2}{2}\right], \quad i = 1, \ldots, n-1, \tag{3.13}$$

The conservative form of these equations allows us to write them in the form of a system as:

$$\partial_0\, m = \frac{(\partial_0 x)^2}{2}, \tag{3.14}$$

$$\partial_i\, m = \frac{\partial_{i+1}x}{\partial_0 x} + \frac{\partial_{00i}x}{\partial_0 x}, \quad i = 1, \ldots, n-1. \tag{3.15}$$

which can be considered as modified versions of the CBS equation with $m = m\,(z_0, ..z_i, z_{i+1}, ...z_n)$. The modified CBS equation has been extensively studied from the point of view of the Painlevé analysis in [19], its Lax pair was derived and hence, a version of a Lax pair for Qiao(1+1) is available in [16, 23].

3.2 Generalization to $2+1$ dimensions

Reciprocal transformations for CH(2+1)

From now on we will refer to the Camassa–Holm hierarchy in $2+1$ dimensions as CH(2+1), and we will write it in a compact form as:

$$U_T = R^{-n} U_Y, \tag{3.16}$$

where R is the recursion operator defined as:

$$R = JK^{-1}, \quad K = \partial_{XXX} - \partial_X, \quad J = -\frac{1}{2}\left(\partial_X U + U\partial_X\right), \quad \partial_X = \frac{\partial}{\partial X}. \tag{3.17}$$

This hierarchy was introduced in [20] as a generalization of the Camassa–Holm hierarchy. The recursion operator is the same as for CH(1+1). From this point of view, the spectral problem is the same [7] and the Y-variable is just another "time" variable [26, 27].

The n component of this hierarchy can also be written as a set of PDEs by introducing n dependent fields $\Omega^{[i]}, (i = 1 \dots n)$ in the following way

$$\begin{aligned} &U_Y = J\Omega^{[1]} \\ &J\Omega^{[i+1]} = K\Omega^{[i]}, \quad i = 1 \dots n-1, \\ &U_T = K\Omega^{[n]}, \end{aligned} \tag{3.18}$$

and by introducing two new fields, P and Δ, related to U as:

$$U = P^2, \qquad P_T = \Delta_X, \tag{3.19}$$

we can write the hierarchy in the form of the following set of equations

$$\begin{aligned} &P_Y = -\frac{1}{2}\left(P\Omega^{[1]}\right)_X, \\ &\Omega^{[i]}_{XXX} - \Omega^{[i]}_X = -P\left(P\Omega^{[i+1]}\right)_X, \quad i = 1 \dots n-1, \\ &P_T = \frac{\Omega^{[n]}_{XXX} - \Omega^{[n]}_X}{2P} = \Delta_X. \end{aligned} \tag{3.20}$$

The conservative form of the first and third equation allows us to define the following exact derivative

$$dz_0 = P\,dX - \frac{1}{2}P\Omega^{[1]}\,dY + \Delta\,dT. \tag{3.21}$$

A reciprocal transformation [26, 36, 37] can be introduced by considering the former independent variable X as a field depending on z_0, $z_1 = Y$ and $z_{n+1} = T$. From (3.21) we have

$$\begin{aligned} &dX = \frac{1}{P}\,dz_0 + \frac{\Omega^{[1]}}{2}\,dz_1 - \frac{\Delta}{P}\,dz_{n+1}, \\ &Y = z_1, \qquad T = z_{n+1}, \end{aligned} \tag{3.22}$$

and therefore

$$\partial_0 X = \frac{1}{P},$$
$$\partial_1 X = \frac{\Omega^{[1]}}{2},$$
$$\partial_{n+1} X = -\frac{\Delta}{P}, \tag{3.23}$$

where $\partial_i X = \frac{\partial X}{\partial z_i}$. We can now extend the transformation by introducing a new independent variable z_i for each field $\Omega^{[i]}$ by generalizing (3.23) as

$$\partial_i X = \frac{\Omega^{[i]}}{2}, \quad i = 2 \dots n. \tag{3.24}$$

Therefore, the new field $X = X(z_0, z_1, \dots z_n, z_{n+1})$ depends on $n+2$ independent variables, where each of the former dependent fields Ω_i, $(i = 1 \dots n)$ allows us to define a new dependent variable z_i through definition (3.24). It requires some calculation (see [20] for details) but it can be proved that the reciprocal transformation (3.22)-(3.24) transforms (3.20) to the following set of n PDEs:

$$\partial_0 \left[-\frac{\partial_{i+1} X}{\partial_0 X} \right] = \left[\partial_0 \left(\frac{\partial_{00} X}{\partial_0 X} + \partial_0 X \right) - \frac{1}{2} \left(\frac{\partial_{00} X}{\partial_0 X} + \partial_0 X \right)^2 \right]_i, \quad i = 1 \dots n. \tag{3.25}$$

Note that each equation depends on only three variables z_0, z_i, z_{i+1}. This result generalizes the one found in [26] for the first component of the hierarchy. The conservative form of (3.25) allows us to define a field $M(z_0, z_1, \dots z_{n+1})$ such that

$$\begin{aligned}
\partial_i M &= \frac{1}{4}\left[-\frac{\partial_{i+1} X}{\partial_0 X} \right] = -\frac{P\Omega^{[i+1]}}{8}, \qquad i = 1 \dots n-1, \\
\partial_n M &= \frac{1}{4}\left[-\frac{\partial_{n+1} X}{\partial_0 X} \right] = \frac{\Delta}{4}, \\
\partial_0 M &= \frac{1}{4}\left[\partial_0 \left(\frac{\partial_{00} X}{\partial_0 X} + \partial_0 X \right) - \frac{1}{2} \left(\frac{\partial_{00} X}{\partial_0 X} + \partial_0 X \right)^2 \right] = \\
&= \frac{1}{4P^2} \left(\frac{3P_X^2}{2P^2} - \frac{P_{XX}}{P} - \frac{1}{2} \right).
\end{aligned} \tag{3.26}$$

It is easy to prove that each M_i should satisfy the following CBS equation [7] on $J^4(\mathbb{R}^{n+2}, \mathbb{R})$,

$$\partial_{0,i+1} M + \partial_{000i} M + 4\partial_i M \partial_{00} M + 8\partial_0 M \partial_{0i} M = 0, \quad i = 1 \dots n. \tag{3.27}$$

Hence, the CH(2+1) is equivalent to n copies of a CBS equation [5, 6, 19] written in three different independent variables z_0, z_i, z_{i+1}.

Reciprocal transformation for mCH(2+1)

Another example to illustrate the role of reciprocal transformations in the identification of partial differential equations was introduced by one of us in [16], were the following $2+1$ hierarchy Qiao(2+1) or mCH(2+1) appears as follows.

$$u_t = r^{-n} u_y, \tag{3.28}$$

where r is the recursion operator, defined as:

$$r = kj^{-1}, \quad k = \partial_{xxx} - \partial_x, \quad j = -\partial_x\, u\, (\partial_x)^{-1}\, u\, \partial_x, \quad \partial_x = \frac{\partial}{\partial x}, \tag{3.29}$$

where $\partial_x = \frac{\partial}{\partial x}$. This hierarchy generalizes the one introduced by Qiao in [33]. We shall briefly summarize the results of [16] when a procedure similar to the one described above for CH(2+1) is applied to mCH(2+1).

If we introduce $2n$ auxiliary fields $v^{[i]}$, $\omega^{[i]}$ defined through

$$\begin{aligned} &u_y = jv^{[1]}, \\ &jv^{[i+1]} = kv^{[i]}, \quad \omega_x^{[i]} = uv_x^{[i]}, \quad i = 1 \ldots n-1, \\ &u_t = kv^{[n]}, \end{aligned} \tag{3.30}$$

the hierarchy can be expanded to $J^3(\mathbb{R}^3, \mathbb{R}^{2n+1})$ in the following form:

$$\begin{aligned} &u_y = -\left(u\omega^{[1]}\right)_x, \\ &v_{xxx}^{[i]} - v_x^{[i]} = -\left(u\omega^{[i+1]}\right)_x, \quad i = 1 \ldots n-1, \\ &u_t = \left(v_{xx}^{[n]} - v^{[n]}\right)_x, \end{aligned} \tag{3.31}$$

which allows us to define the exact derivative

$$dz_0 = u\, dx - u\omega^{[1]}\, dy + \left(v_{xx}^{[n]} - v^{[n]}\right)\, dt \tag{3.32}$$

and $z_1 = y, z_{n+1} = t$. We can define a reciprocal transformation such that the former independent variable x is a new field $x = x(z_0, z_1, \ldots\ldots z_{n+1})$ depending on $n+2$ variables in the form

$$\begin{aligned} &dx = \frac{1}{u} dz_0 + \omega^{[1]} dz_1 - \frac{\left(v_{xx}^{[n]} - v^{[n]}\right)}{u} dz_{n+1}, \\ &y = z_1, \qquad t = z_{n+1}, \end{aligned} \tag{3.33}$$

which implies

$$\begin{aligned} &\partial_0 x = \frac{\partial x}{\partial z_0} = \frac{1}{u}, \\ &\partial_i x = \frac{\partial x}{\partial z_i} = \omega^{[i]}, \quad i = 1 \ldots n \\ &\partial_{n+1} x = \frac{\partial x}{\partial z_{n+1}} = -\frac{\left(v_{xx}^{[n]} - v^{[n]}\right)}{u}. \end{aligned} \tag{3.34}$$

The transformation of the equations (3.31) yields the system of equations

$$\partial_0\left[\frac{\partial_{i+1}x}{\partial_0 x}+\frac{\partial_{00i}x}{\partial_0 x}\right]=\partial_i\left[\frac{x_0^2}{2}\right],\quad i=1\ldots n. \tag{3.35}$$

Note that each equation depends on only three variables: z_0, z_i, z_{i+1}.

The conservative form of (3.35) allows us to define a field $m = m(z_0, z_1, \ldots z_{n+1})$ such that

$$\begin{aligned}
\partial_0 m &= \frac{x_0^2}{2}=\frac{1}{2u^2},\\
\partial_i m &= \frac{\partial_{i+1}x}{\partial_0 x}+\frac{\partial_{00i}x}{\partial_0 x}=v^{[i]},\quad i=1\ldots n,
\end{aligned} \tag{3.36}$$

defined on $J^3(\mathbb{R}^{n+2},\mathbb{R}^2)$. Equation (3.35) has been extensively studied from the point of view of Painlevé analysis [19] and it can be considered as the modified version of the CBS equation (3.27).

Hence, we have shown again that a reciprocal transformation has proven the equivalency between two hierarchies/equations (mCH(2+1)-mCBS) that although they are unrelated at first, they are merely two different descriptions of the same common problem.

3.3 Reciprocal transformation for a fourth-order nonlinear equation

In [17, 18], we introduced a fourth-order equation in 2+1-dimensions which has the form

$$\left(H_{x_1x_1x_2}+3H_{x_2}H_{x_1}-\frac{k+1}{4}\frac{(H_{x_1x_2})^2}{H_{x_2}}\right)_{x_1}=H_{x_2x_3}. \tag{3.37}$$

The two particular cases $k=-1$ [18] and $k=2$ [17, 21] are integrable and it was possible to derive their Lax pair using the singular manifold method [44]. Based on the results in [17, 18], we proposed a spectral problem of the form:

$$\begin{aligned}
&\phi_{x_1x_1x_1}-\phi_{x_3}+3H_{x_1}\phi_{x_1}-\frac{k-5}{2}H_{x_1x_1}\phi=0,\\
&\phi_{x_1x_2}+H_{x_2}\phi+\frac{k-5}{6}\frac{H_{x_1x_2}}{H_{x_2}}\phi_{x_2}=0.
\end{aligned} \tag{3.38}$$

We can rewrite (3.37) as the system:

$$\begin{aligned}
&H_{x_1x_1x_2}+3H_{x_2}H_{x_1}-\frac{k+1}{4}\frac{H_{x_1x_2}^2}{H_{x_2}}=\Omega,\\
&\Omega_{x_1}=H_{x_2x_3}.
\end{aligned} \tag{3.39}$$

3.3.1 Reciprocal transformation I

We can perform a reciprocal transformation of equations (3.39) by proposing:

$$dx_1 = \alpha(x,t,T)[dx - \beta(x,t,T)dt - \epsilon(x,t,T)dT],$$
$$x_2 = t, \quad x_3 = T. \tag{3.40}$$

Under this reciprocal transformation the derivatives transform as

$$\frac{\partial}{\partial x_1} = \frac{1}{\alpha}\frac{\partial}{\partial x},$$
$$\frac{\partial}{\partial x_2} = \frac{\partial}{\partial t} + \beta\frac{\partial}{\partial x},$$
$$\frac{\partial}{\partial x_3} = \frac{\partial}{\partial T} + \epsilon\frac{\partial}{\partial x}. \tag{3.41}$$

The cross derivatives of (3.40) give rise to the equations:

$$\alpha_t + (\alpha\beta)_x = 0, \quad \alpha_T + (\alpha\epsilon)_x = 0, \quad \beta_T - \epsilon_t + \epsilon\beta_x - \beta\epsilon_x = 0. \tag{3.42}$$

If we select a transformation in the form for α such that

$$H_{x_2} = \alpha(x,t,T)^k, \tag{3.43}$$

this reciprocal transformation, when applied to the system (3.39), yields

$$H_{x_1} = \frac{1}{3}\left(\frac{\Omega}{\alpha^k} - k\frac{\alpha_{xx}}{\alpha^3} + (2k-1)\left(\frac{\alpha_x}{\alpha^2}\right)^2\right), \tag{3.44}$$
$$\Omega_x = -k\alpha^{(k+1)}\epsilon_x. \tag{3.45}$$

Furthermore, the compatibility condition $H_{x_2x_1} = H_{x_1x_2}$ between (3.43) and (3.44) yields

$$\Omega_t = -\beta\,\Omega_x - k\,\Omega\beta_x + \alpha^{k-2}\left[-k\beta_{xxx} + (k-2)\beta_{xx}\frac{\alpha_x}{\alpha} + 3k\alpha^k\alpha_x\right]. \tag{3.46}$$

Then, the equations (3.42), (3.45) and (3.46) constitute the transformed equations for the original system (3.39).

Still, we can find a more suitable form for the transformed equations if we introduce the following definitions:

$$A_1 = \frac{k+1}{3}, \quad A_2 = \frac{2-k}{3}, \quad M = \frac{1}{\alpha^3}. \tag{3.47}$$

In these parameters, the integrability condition $(k+1)(k-2) = 0$ is translated into

$$A_1 \cdot A_2 = 0, \quad A_1 + A_2 = 1. \tag{3.48}$$

Using the definitions above, we can finally present the reciprocally transformed system as:

$$\begin{aligned}
&A_1\left[\Omega_t+\beta\,\Omega_x+2\Omega\beta_x+2\beta_{xxx}+2\frac{M_x}{M^2}\right]+\\
&\qquad\qquad +A_2\left[\Omega_t+\beta\Omega_x-\Omega\beta_x-M\beta_{xxx}-M_x\beta_{xx}-M_x\right]=0,\\
&A_1\left(\Omega_x+2\frac{\epsilon_x}{M}\right)+A_2(\Omega_x-\epsilon_x)=0,\\
&M_t=3M\beta_x-\beta M_x,\\
&M_T=3M\epsilon_x-\epsilon M_x,\\
&\beta_T-\epsilon_t+\epsilon\beta_x-\beta\epsilon_x=0.
\end{aligned}\tag{3.49}$$

Furthermore, the reciprocal transformation can also be applied to the spectral problem (3.38). After some direct calculations, we obtain

$$\begin{aligned}
&A_1\left[\psi_{xt}+\beta\psi_{xx}-\left(\beta_{xx}-\frac{1}{M}\right)\psi\right]\\
&\qquad +A_2\left[\psi_{xt}+\beta\psi_{xx}+2\beta_x\psi_x+(\beta_{xx}+1)\,\psi\right]M^{\frac{2}{3}}=0,\\
&A_1\left[\psi_T-M\psi_{xxx}-(M\Omega-\epsilon)\,\psi_x\right]\\
&\qquad +A_2\left[\psi_T-M\psi_{xxx}-2M_x\psi_{xx}-(M_{xx}+\Omega-\epsilon)\,\psi_x\right]M^{\frac{2}{3}}=0,
\end{aligned}\tag{3.50}$$

where we have set

$$\phi(x_1,x_2,x_3)=M^{\frac{1-2k}{9}}\psi(x,t,T)\tag{3.51}$$

for convenience.

Reduction independent of T

Let us show a reduction of the set (3.49), by setting all the fields independent of T. This means that

$$\epsilon=0,\quad \Omega_x=0\Rightarrow\Omega=V(t),$$

and the system (3.49) reduces to

$$A_1\left[V_t+2\left(V\beta+\beta_{xx}-\frac{1}{M}\right)_x\right]+A_2\left[V_t-(V\beta+M\beta_{xx}+M)_x\right]=0,\tag{3.52}$$

$$M_t=3M\beta_x-\beta M_x.\tag{3.53}$$

- **Degasperis–Procesi equation**

For the case $A_1=1$ and $A_2=0$, we can integrate (3.52) as:

$$\beta_{xx}+V\beta+\frac{V_t}{2}x=\frac{1}{M}+q_0,\tag{3.54}$$

which combined with (3.53) yields

$$(\beta_{xx}+V\beta)_t+\beta\beta_{xxx}+3\beta_x\beta_{xx}+4V\beta\beta_x-3q_0\beta_x+\frac{1}{2}V_t(\beta_x+3\beta x)+\frac{x}{2}V_{tt}=0. \quad (3.55)$$

For $q_0 = 0$ and $V = -1$, this system is the well-known Degasperis-Procesi equation, [14].

- **Vakhnenko equation**

For the case $A_1 = 0, A_2 = 1$, we can integrate (3.52) as:

$$V_t x - V\beta - M\beta_{xx} - M - q_0 = 0, \quad (3.56)$$

which combined with (3.53) provides, when $V = 0$, the derivative of the Vakhnenko equation, [43],

$$\left[(\beta_t+\beta\beta_x)_x+3\beta\right]_x=0. \quad (3.57)$$

3.3.2 Reciprocal transformation II

A different reciprocal transformation can be constructed using the changes

$$\begin{aligned} dx_2 &= \eta(y,z,T)\left(dz-u(y,z,T)dy-\omega(y,z,T)dT\right), \\ x_1 &= y, \quad x_3 = T. \end{aligned} \quad (3.58)$$

The compatibility conditions for this transformation are

$$\begin{aligned} &\eta_y+(u\eta)_z=0, \\ &\eta_T+(\eta\omega)_z=0, \\ &u_T-\omega_y-u\omega_z+\omega u_z=0. \end{aligned} \quad (3.59)$$

We select the transformation by setting the field H as the new independent variable z:

$$z = H(x_1,x_2,x_3) \to dz = H_{x_1}dx_1 + H_{x_2}dx_2 + H_{x_3}dx_3. \quad (3.60)$$

By direct comparison of (3.58) and (3.60), we obtain

$$\begin{aligned} H_{x_2}(x_1,x_2,x_3) &= \frac{1}{\eta(y=x_1,z=H,T=x_3)}, \\ H_{x_1}(x_1,x_2,x_3) &= u(y=x_1,z=H,T=x_3), \\ H_{x_3}(x_1,x_2,x_3) &= \omega(y=x_1,z=H,T=x_3), \end{aligned} \quad (3.61)$$

and the transformations of the derivatives are

$$\begin{aligned} \frac{\partial}{\partial x_1} &= \frac{\partial}{\partial y}+u\frac{\partial}{\partial z}, \\ \frac{\partial}{\partial x_2} &= \frac{1}{\eta}\frac{\partial}{\partial z}, \\ \frac{\partial}{\partial x_3} &= \frac{\partial}{\partial T}+\omega\frac{\partial}{\partial z}. \end{aligned} \quad (3.62)$$

With these definitions, we get the transformation of the system (3.39), as:

$$G = (u_y + uu_z)_z + 3u - \frac{k+1}{4}u_z^2,$$
$$G_y = (\omega - uG)_z, \tag{3.63}$$

where $G(z, y, T)$ has been defined as $G = \eta\,\Omega$.

Reduction independent of T

The reduction independent of T can be obtained by setting $\omega = 0$. In this case, the system (3.63) contains the case $G = 0$,

$$(u_y + uu_z)_z + 3u - \frac{k+1}{4}u_z^2 = 0. \tag{3.64}$$

When $k = -1$, it is the Vakhnenko equation. For the other integrable case, $k = 2$, it yields a modified Vakhnenko equation if $A_2 = 0$.

4 Reciprocal transformations to derive Lax pairs

Reciprocal transformations have served us as a way to derive Lax pairs of differential equations and hierarchies of such differential equations. A differential equation in its initial form may not be Painlevé integrable as we mentioned before, but we are able to prove its integrability by transforming it into another differential equation via reciprocal transformation that makes it Painlevé integrable. In the same fashion, an initial differential equation may not have an associated Lax pair and the singular manifold method may not be applicable. Through a reciprocal transformation we can again transform such equation into another in which we can work the singular manifold method upon. We are depicting examples in the following lines.

4.1 Lax pair for the CH(2+1) hierarchy

In section 2, we have proved that the reciprocal transformations can be used to establish the equivalence between the CH(2+1) hierarchy (3.16) and $n+1$ copies of the CBS equation (3.27). This CBS equation has the Painlevé property [29] and the singular manifold method can be successfully used to derive the following Lax pair [19],

$$\partial_{00}\psi = \left(-2\partial_0 M - \frac{\lambda}{4}\right)\psi, \tag{4.1}$$
$$0 = E_i = \partial_{i+1}\psi - \lambda\partial_i\psi + 4\partial_i M\partial_0\psi - 2\partial_{0i}M\,\psi. \tag{4.2}$$

Furthermore, the compatibility condition between these two equations implies that the spectral problem is nonisospectral because λ satisfies:

$$\partial_0\lambda = 0, \qquad \partial_{i+1}\lambda - \lambda\partial_i\lambda = 0. \tag{4.3}$$

Notice that the first equation in the Lax pair is independent of the index i. Nevertheless, the second equation can be considered as a recursion relation for the derivatives of ψ with respect to each z_i.

Now, to come back to the original fields U and $\Omega^{[i]}$ as well as to the original variables X, Y, T, all we need is to perform the change

$$\psi(z_0, z_1, \ldots, z_n, z_{n+1}) = \sqrt{P}\,\phi(X, Y, T) \tag{4.4}$$

where P is defined in (3.19). Considering the reciprocal transformation (3.22), we have the following induced transformations

$$\begin{aligned}
\partial_0\psi &= \sqrt{P}\left(\frac{\phi_X}{P} + \frac{P_X}{2P^2}\phi\right),\\
\partial_{00}\psi &= \sqrt{P}\left(\frac{\phi_{XX}}{P^2} + \left[\frac{P_{XX}}{2P^3} - \frac{3}{4}\frac{P_X^2}{P^4}\right]\phi\right),\\
\partial_1\psi &= \sqrt{P}\left(\phi_Y + \frac{\Omega^{[1]}\phi_X}{2} + \left[\frac{P_Y}{2P} + \frac{P_X\Omega^{[1]}}{4P}\right]\phi\right)\\
&= \sqrt{P}\left(\phi_Y + \frac{\Omega^{[1]}\phi_X}{2} - \frac{\Omega_X^{[1]}\phi}{4}\right),\\
\partial_{n+1}\psi &= \sqrt{P}\left(\phi_T - \frac{\Delta\phi_X}{P} + \left[\frac{P_T}{2P} - \frac{P_X\Delta}{2P^2}\right]\phi\right)\\
&= \sqrt{P}\left(\phi_T - \frac{\Delta\phi_X}{P} + \left[\frac{\Delta_X}{2P} - \frac{P_X\Delta}{2P^2}\right]\phi\right).
\end{aligned} \tag{4.5}$$

With these changes, (4.1) becomes:

$$\phi_{XX} + \left(\frac{\lambda P^2}{4} - \frac{1}{4}\right)\phi = 0,$$

where equation (3.26) has been used. Finally, the combination with (3.19) yields

$$\phi_{XX} = \frac{1}{4}(1 - \lambda U)\,\phi, \tag{4.6}$$

as the spatial part of the Lax pair for the CH(2+1) hierarchy. The temporal part can be obtained from (4.2) through the following combination:

$$0 = \sum_{i=1}^{n}\lambda^{n-i}E_i = \sum_{i=1}^{n}\lambda^{n-i}\left(\partial_{i+1}\psi - \lambda\partial_i\psi\right) + \sum_{i=1}^{n}\lambda^{n-i}\left(4\partial_i M\partial_0\psi - 2\partial_{0i}M\,\psi\right). \tag{4.7}$$

It is easy to prove that

$$\sum_{i=1}^{n}\lambda^{n-i}\left(\partial_{i+1}\psi - \lambda\partial_i\psi\right) = \partial_{n+1}\psi - \lambda^n\partial_1\psi. \tag{4.8}$$

The reciprocal transformation (3.22), when applied to (4.8), and combined with (4.4) and (4.5) yields

$$\sum_{i=1}^{n} \lambda^{n-i} \left(\partial_{i+1}\psi - \lambda\partial_i\psi\right) = \sqrt{P}\left[\phi_T - \frac{\Delta\phi_X}{P} + \frac{\Delta_X\phi}{2P} - \frac{P_X\Delta}{2P^2}\phi\right]$$
$$-\lambda^n\sqrt{P}\left[\phi_Y + \frac{\Omega^{[1]}\phi_X}{2} - \frac{\Omega^{[1]}_X\phi}{4}\right]. \tag{4.9}$$

For the last sum of (4.7), we can use (3.26) and (4.5). The result is

$$\sum_{i=1}^{n} \lambda^{n-i} \left(4\partial_i M\partial_0\psi - 2\partial_{0i}M\,\psi\right) = \sqrt{P}\left[\frac{\Delta\phi_X}{P} + \frac{\Delta\, P_X}{2P^2}\phi - \frac{\Delta_X}{2P}\phi\right]$$
$$+\sqrt{P}\sum_{i=1}^{n-1} \lambda^{n-i}\left[\frac{\Omega^{[i+1]}_X}{4}\phi - \frac{\Omega^{[i+1]}}{2}\phi_X\right] = 0. \tag{4.10}$$

Substitution of (4.9) and (4.10) in (4.7) yields

$$\phi_T - \lambda^n\phi_Y + \lambda^n\left(\frac{\Omega^{[1]}_X}{4}\phi - \frac{\Omega^{[1]}}{2}\phi_X\right) + \sum_{i=1}^{n-1}\lambda^{n-i}\left[\frac{\Omega^{[i+1]}_X}{4}\phi - \frac{\Omega^{[i+1]}}{2}\phi_X\right] = 0. \tag{4.11}$$

The expression (4.11) can be written in a more compact form as

$$\phi_T - \lambda^n\phi_Y + \frac{A_X}{4}\phi - \frac{A}{2}\phi_X = 0, \tag{4.12}$$

where A is defined as

$$A = \sum_{j=1}^{n} \lambda^{n-j+1}\Omega^{[j]}, \qquad \text{with} \quad i = j-1. \tag{4.13}$$

The nonisospectral condition (4.3) reads

$$\lambda_X = 0, \qquad 0 = \sum_{i=1}^{n} \lambda^{n-i}\left(\partial_{i+1} - \lambda\partial_i\right)\lambda = \partial_{n-1}\lambda - \lambda^n\,\partial_1\lambda = \lambda_T - \lambda^n\lambda_Y = 0. \tag{4.14}$$

In sum: the Lax pair for CH(2+1) can be written as

$$\phi_{XX} + \frac{1}{4}\left(\lambda U - 1\right)\phi = 0,$$
$$\phi_T - \lambda^n\phi_Y - \frac{A}{2}\phi_X + \frac{A_X}{4}\phi = 0, \tag{4.15}$$

where

$$A = \sum_{i=1}^{n}\left[\lambda^{n-i+1}\,\Omega^{[i]}\right], \qquad \lambda_T - \lambda^n\lambda_Y = 0. \tag{4.16}$$

4.2 Lax pair for mCH(2+1)

In [19] it was proved that the CBS equation (3.27) and the mCBS equation (3.35) were linked through a Miura transformation. This is a transformation that relates the fields in the CBS and mCBS in the following form

$$\partial_0 M = -\frac{\partial_0 x^2}{8} + \frac{\partial_{00} x}{4},$$

which combined with (3.35) can be integrated as

$$4M = \partial_0 x - m. \tag{4.17}$$

The two-component Lax pair for the mCBS equation (3.35) was derived in [19]. In our variables this spectral problem reads:

$$\partial_0 \begin{pmatrix} \psi \\ \hat{\psi} \end{pmatrix} = \frac{1}{2} \begin{pmatrix} -\partial_0 x & i\sqrt{\lambda} \\ i\sqrt{\lambda} & \partial_0 x \end{pmatrix} \begin{pmatrix} \psi \\ \hat{\psi} \end{pmatrix}, \tag{4.18}$$

$$\begin{aligned} 0 = F_i = {} & \partial_{i+1} \begin{pmatrix} \psi \\ \hat{\psi} \end{pmatrix} - \lambda\, \partial_i \begin{pmatrix} \psi \\ \hat{\psi} \end{pmatrix} \\ & -\frac{1}{2} \begin{pmatrix} -\partial_{i+1} x & i\sqrt{\lambda}\, \partial_i (m - \partial_0 x) \\ i\sqrt{\lambda}\, \partial_i (m + \partial_0 x) & \partial_{i+1} x \end{pmatrix} \begin{pmatrix} \psi \\ \hat{\psi} \end{pmatrix}. \end{aligned} \tag{4.19}$$

It is easy to see that the compatibility condition of (4.18)-(4.19) yields the equation (3.35) as well as the following nonisospectral condition:

$$\partial_0 \lambda = 0, \qquad \partial_{i+1} \lambda = \lambda\, \partial_i \lambda. \tag{4.20}$$

If, from the above Lax pair, we wish to obtain the spectral problem of the mCH(2+1), we need to invert the reciprocal transformation (3.33)-(3.34), which means applying the following substitutions:

$$\begin{aligned} & \partial_0 x = \frac{1}{u}, \\ & \partial_i x = \omega^{[i]} \quad \Rightarrow \quad \partial_{0i} x = \frac{\omega_x^{[i]}}{u} = v_x^{[j]}, \qquad i = 1...n, \\ & \partial_{n+1} x = -\frac{v_{xx}^{[n]} - v^{[n]}}{u}, \\ & \partial_0 m = \frac{1}{2u^2}, \\ & \partial_i m = v^{[i]}, \end{aligned} \tag{4.21}$$

and the transformations of the derivatives are

$$\begin{aligned} & \partial_0 = \frac{1}{u}\, \partial_x\, , \\ & \partial_1 = \partial_y + \omega^{[1]}\, \partial_x\, , \\ & \partial_{n+1} = \partial_t - \frac{\left(v_{xx}^{[n]} - v^{[n]}\right)}{u}\, \partial_x. \end{aligned} \tag{4.22}$$

We can now tackle the transformation of the Lax pair (4.18)-(4.19). The spatial part (4.18) transforms trivially to:

$$\begin{pmatrix} \psi \\ \hat{\psi} \end{pmatrix}_x = \frac{1}{2}\begin{pmatrix} -1 & i\sqrt{\lambda}u \\ i\sqrt{\lambda}u & 1 \end{pmatrix}\begin{pmatrix} \psi \\ \hat{\psi} \end{pmatrix}. \tag{4.23}$$

The transformation of (4.19) is slightly more complicated. Let us compute the following sum:

$$0 = \sum_{i=1}^{n} \lambda^{n-i} F_i, \tag{4.24}$$

where F_i is defined in (4.19). It is easy to see that

$$\sum_{i=1}^{n} \lambda^{n-i}(\partial_{i+1} - \lambda\partial_i)\begin{pmatrix} \psi \\ \hat{\psi} \end{pmatrix} = (\partial_{n+1} - \lambda^n\partial_1)\begin{pmatrix} \psi \\ \hat{\psi} \end{pmatrix}, \tag{4.25}$$

and then, the inverse reciprocal transformation (4.21)-(4.22) can be applied to (4.24) in order to obtain

$$\begin{pmatrix} \psi \\ \hat{\psi} \end{pmatrix}_t - \lambda^n \begin{pmatrix} \psi \\ \hat{\psi} \end{pmatrix}_y = C\begin{pmatrix} \psi \\ \hat{\psi} \end{pmatrix}_x + \frac{i\sqrt{\lambda}}{2}\begin{pmatrix} 0 & B_{xx} - B_x \\ B_{xx} + B_x & 0 \end{pmatrix}\begin{pmatrix} \psi \\ \hat{\psi} \end{pmatrix}, \tag{4.26}$$

where

$$C = \sum_{i=1}^{n} \lambda^{n-i+1}\omega^{[i]}, \qquad B = \sum_{i=1}^{n} \lambda^{n-i} v^{[i]}. \tag{4.27}$$

The inverse reciprocal transformation, when applied to (4.20) yields

$$\lambda_x = 0, \qquad 0 = \sum_{i=1}^{n} \lambda^{n-i}\left(\partial_{i+1} - \lambda\partial_i\right)\lambda = \partial_{n-1}\lambda - \lambda^n\,\partial_1\lambda = \lambda_t - \lambda^n\lambda_y = 0. \tag{4.28}$$

Hence, we have derived a Lax pair for mCH(2+1) using the existing Miura transformation between the CBS and mCBS and the Lax pair for mCBS. This is another example of how reciprocal transformations or compositions of transformations can provide us with Lax pairs, and the implication of integrability.

5 A Miura-reciprocal transformation

Recalling the previous sections, we can summarize by saying that CH(2+1) and mCH(2+1) are related to the CBS and mCBS by reciprocal transformations, correspondingly. Aside from this property, in this section we would like to show that there exists a Miura transformation [19] relating the CBS and the mCBS equations. Hence, one wonders if mCH(2+1) is related to CH(2+1) in any way. It seems clear that the relationship between mCH(2+1) and CH(2+1) necessarily includes a composition of a Miura and a reciprocal transformation.

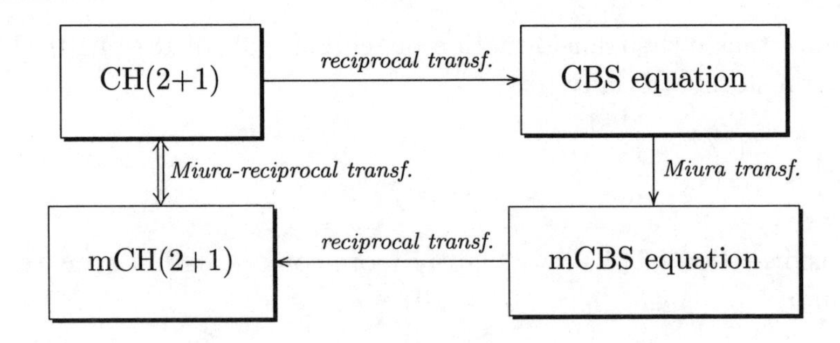

Figure 1. Miura-reciprocal transformation.

Evidently, the relationship between both hierarchies cannot be a simple Miura transformation because they are written in different variables (X, Y, T) and (x, y, t). The answer is provided by the relationship of both sets of variables with the same set (z_0, z_1, z_{n+1}). By combining (3.21) and (3.32), we have

$$\begin{aligned} &P\,dX - \frac{1}{2}P\Omega^{[1]}\,dY + \Delta\,dT = u\,dx - u\omega^{[1]}\,dy + \left(v_{xx}^{[n]} - v^{[n]}\right)\,dt, \\ &Y = y, \qquad T = t, \end{aligned} \tag{5.1}$$

which yields the required relation between the independent variables of CH(2+1) and those of mCH(2+1). The Miura transformation (4.17), combined with (3.26) and (3.36) also provides the following results

$$4\partial_0 M = \partial_{00}x - \partial_0 m \Longrightarrow \frac{\partial_{00}X}{\partial_0 X} + \partial_0 X = \partial_0 x, \tag{5.2}$$

$$4\partial_i M = \partial_{0i}x - \partial_i m \Longrightarrow -\frac{\partial_{i+1}X}{\partial_0 X} = \partial_{0i}x - \frac{\partial_{00i}x}{\partial_0 x} - \frac{\partial_{i+1}x}{\partial_0 x}, \tag{5.3}$$

with $i = 1, \ldots, n$. With the aid of (3.23), (3.24) and (3.34), the following results arise from (5.2)-(5.3)

$$\begin{aligned} &\frac{1}{u} = \left(\frac{1}{P}\right)_X + \frac{1}{P}, \\ &P\Omega^{[i+1]} = 2\left(v^{[i]} - v_x^{[i]}\right) \quad \Longrightarrow \quad \omega^{[i+1]} = \frac{\Omega_X^{[i+1]} + \Omega^{[i+1]}}{2}, \quad i = 1 \ldots n-1, \\ &\Delta = v_x^{[n]} - v^{[n]}. \end{aligned} \tag{5.4}$$

Furthermore, (5.1) can be integrated as

$$x = X - \ln P. \tag{5.5}$$

By summarizing the above conclusions, we have proven that the mCH(2+1) hierarchy

$$u_t = r^{-n}u_y, \quad u = u(x, y, t), \tag{5.6}$$

can be considered as the modified version of CH(2+1)

$$U_T = R^{-n}U_Y, \quad U = U(X, Y, T). \tag{5.7}$$

The transformation that connects the two hierarchies involves the reciprocal transformation

$$x = X - \frac{1}{2}\ln U, \tag{5.8}$$

as well as the following transformation between the fields

$$\begin{aligned}
\frac{1}{u} &= \frac{1}{\sqrt{U}}\left(1 - \frac{U_X}{2U}\right),\\
\omega^{[i]} &= \frac{\Omega_X^{[i]} + \Omega^{[i]}}{2}, \quad i = 1\dots n,\\
\frac{\delta}{u} &= \left(\frac{\Delta}{\sqrt{U}}\right)_X + \frac{\Delta}{\sqrt{U}}.
\end{aligned} \tag{5.9}$$

5.1 Particular case 1: the Qiao equation

We are now restricted to the first component of the hierarchies $n = 1$ in the case in which the field u is independent of y and U is independent of Y.

- From (3.19) and (3.20), for the restriction of CH(2+1) we have

$$\begin{aligned}
&U = P^2,\\
&U_T = \Omega_{XXX}^{[1]} - \Omega_X^{[1]},\\
&\left(P\,\Omega^{[1]}\right)_X = 0,
\end{aligned} \tag{5.10}$$

 which can be summarized as

$$\begin{aligned}
\Omega^{[1]} &= \frac{k_1}{P} = \frac{k_1}{\sqrt{U}},\\
U_T &= k_1\left[\left(\frac{1}{\sqrt{U}}\right)_{XXX} - \left(\frac{1}{\sqrt{U}}\right)_X\right],
\end{aligned} \tag{5.11}$$

 that is the Dym equation [28].

- The reduction of mCH(2+1) can be achieved from (3.31) in the form

$$\begin{aligned}
&\omega_x^{[1]} = u v_x^{[1]},\\
&u_t = v_{xxx}^{[1]} - v_x^{[1]},\\
&\left(u\omega^{[1]}\right)_x = 0,
\end{aligned} \tag{5.12}$$

 which can be written as

$$\begin{aligned}
\omega^{[1]} &= \frac{k_2}{u} \quad \Longrightarrow \quad v^{[1]} = \frac{k_2}{2u^2},\\
u_t &= k_2\left[\left(\frac{1}{2u^2}\right)_{xx} - \left(\frac{1}{2u^2}\right)\right]_x,
\end{aligned} \tag{5.13}$$

 that is the Qiao equation.

- From (5.8) and (5.9) it is easy to see that $k_1 = 2k_2$. By setting $k_2 = 1$, we can conclude that the Qiao equation

$$u_t = \left(\frac{1}{2u^2}\right)_{xxx} - \left(\frac{1}{2u^2}\right)_x, \tag{5.14}$$

is the modified version of the Dym equation

$$U_T = \left(\frac{2}{\sqrt{U}}\right)_{XXX} - \left(\frac{2}{\sqrt{U}}\right)_X. \tag{5.15}$$

- From (3.23) and (3.34), it is easy to see that the independence from y implies that $\partial_1 X = \partial_0 X$ and $\partial_1 x = \partial_0 x$, which means that the CBS and mCBS (3.27) and (3.35) reduce to the following potential versions of the KdV and modified KdV equations

$$\begin{aligned} &\partial_0\left(\partial_2 M + \partial_{000} M + 6\partial_0 M^2\right) = 0, \\ &\partial_2 x + \partial_{000} x - \frac{1}{2}\partial_0 x^3 = 0. \end{aligned} \tag{5.16}$$

5.2 Particular case 2: the Camassa–Holm equation

If we are restricted to the $n = 1$ component when $T = X$ and $t = x$, the following results hold:

- From (3.19) and (3.20), for the restriction of CH(2+1) we have

$$\begin{aligned} &\Delta = P = \sqrt{U}, \\ &U = \Omega^{[1]}_{XX} - \Omega^{[1]}, \\ &U_Y + U\Omega^{[1]}_X + \frac{1}{2}\Omega^{[1]} U_X = 0, \end{aligned} \tag{5.17}$$

which is the Camassa–Holm equation.

- The reduction of mCH(2+1) can be obtained from (3.31) in the form

$$\begin{aligned} &\delta = u = v^{[1]}_{xx} - v^{[1]}, \\ &u_y + \left(u\omega^{[1]}\right)_x = 0, \\ &\omega^{[1]}_x - uv^{[1]}_x = 0, \end{aligned} \tag{5.18}$$

which can be considered as a modified Camassa–Holm equation.

- From (3.23) and (3.34), it is easy to see that $\partial_2 X = \partial_2 x = -1$. Therefore, the reductions of (3.27) and (3.35) are

$$\partial_{0001} M + 4\partial_1 M\, \partial_{00} M + 8\partial_0 M\, \partial_{01} M = 0, \tag{5.19}$$

which is the AKNS equation, and

$$\partial_0\left(\frac{\partial_{001} x - 1}{\partial_0 x}\right) = \partial_1\left(\frac{\partial_0 x^2}{2}\right), \tag{5.20}$$

which is the modified AKNS equation.

6 Conclusions

Concerning the role of reciprocal transformations in the classification and identification of PDEs, we have shown that CH(2+1) and mCH(2+1) hierarchies can be connected with the CBS and mCBS equations via a reciprocal transformations. A big advantage of a reciprocal transformation is that it turns a whole hierarchy into a set of equations that can be studied through Painlevé analysis and other properties can afterwards be derived from this.

In this context, a reciprocal transformation has served as a way to turn a set of differential equations with multiple scalar fields and few independent variables into a unique differential equation with one scalar field depending on multiple independent variables. Furthermore, it serves to turn the initial equations into one in which the Painlevé integrability is satisfied and therefore proving the integrability of the hierarchy prior to the reciprocal transformation.

We have shown examples of higher-order by presenting a fourth-order nonlinear PDE in $2 + 1$ dimensions and investigated different reciprocal transformations for it. Reciprocal transformations have once more shown that the transformed equations (their reductions actually) in $1 + 1$ dimensions are the Vakhnenko–Parkes and Degasperis–Procesi equations.

Reciprocal transformations have been further proved to be useful for the derivation of Lax pairs. As it has been shown, the transformations of CH and mCH into CBS and mCBS, being that these equations are integrable in the algebraic Painlevé sense and their Lax pair is well known, undoing the reciprocal transformation in the Lax pairs for CBS and mCBS, we were able to retrieve Lax pairs that have not been proposed for CH and mCH in $1 + 1$ and $2 + 1$ dimensions. This verifies the importance of reciprocal transformations as a way to derive Lax pairs.

As a last instance, we have depicted Miura-reciprocal transformations, based on the composition of a Miura transformation between the CBS and mCBS and the reciprocal transformations linking CH and mCH to CBS and mCBS, correspondingly, in $1 + 1$ and $2 + 1$. Miura-reciprocal transformations verify the importance of composition of reciprocal transformations to classify hierarchies, indeed, we have successfully proven that CH and mCH in $1 + 1$ and $2 + 1$ are two different versions of a same common problem that can be reached by a transformation map that has been proposed in the last section.

The observation of all these properties shows the efficiency and importance of reciprocal transformations that we introduced at the start of the chapter, and that we here close having given proof of our arguments with remarkable examples in the physics literature of hydrodynamic systems, shallow water waves, etc.

References

[1] Ablowitz M J and Clarkson P A, *Solitons, Nonlinear Evolution Equations and Inverse Scattering*, London Mathematical Society, Lecture Notes Series 149, Cambridge University Press, Cambridge, 1991.

[2] Ablowitz M J, Kruskal M and Segur H, A note on Miura's transformation, *J. Math. Phys.* **20**, 999–1003, 1979.

[3] Ablowitz M J and Segur H, *Solitons and the Inverse Scattering Transform*, Society for Industrial and Applied Mathematics (SIAM), Philadelphia, 1981.

[4] Abraham R and Marsden J E, *Foundations of Mechanics*, 2nd edition, Addison–Wesley, 1978.

[5] Bogoyavlenskii O I, Breaking solitons in 2+1-dimensional integrable equations, *Russian Math. Surveys* **45**, 1–86, 1990.

[6] Calogero F, A method to generate solvable nonlinear evolution equations, *Lettere al Nuovo Cimento* **14**, 443–447, 1975.

[7] Calogero F, Generalized Wronskian relations, one-dimensional Schrödinger equation and non-linear partial differential equations solvable by the inverse-scattering method, *Nuovo Cimento B* **31**, 229–249, 1976.

[8] Camassa R and Holm D D, An integrable shallow water equation with peaked solitons, *Phys. Rev. Lett.* **71**(11), 1661–1664, 1993.

[9] Clarkson P A, Fokas A S and Ablowitz M J, Hodograph transformations of linearizable partial differential equations, *SIAM J. of Appl. Math.* **49**, 1188–1209, 1989.

[10] Constantin A, On the scattering problem for the Camassa Holm equation, *Proc. R. Soc. Lond. A* **457**, 953–970, 2001.

[11] Constantin A, Gerdjikov V S and Ivanov R I, Inverse scattering transform for the Camassa Holm equation, *Inverse Problems* **22**(6), 2197–2207, 2006.

[12] Constantin A and McKean H P, A shallow water equation on the circle, *Commun. Pure Appl. Math.*, **52**(8), 949–982, 1999.

[13] Conte R and Musette M, *The Painlevé Handbook*, Springer and Canopus Publishing Limited, Bristol, 2008.

[14] Degasperis A, Holm D D and Hone A N W, A new integral equation with peakon solutions, *Theor. Math. Phys.* **133**, 1463–1474, 2002.

[15] Estévez P G, Reciprocal transformations for a spectral problem in 2 + 1 dimensions, *Theor. Math. Phys.* **159**, 763–769, 2009.

[16] Estévez P G, Generalized Qiao hierarchy in 2+1 dimensions: reciprocal transformations, spectral problem and non-isospectrality, *Phys. Lett. A* **375**, 537–540, 2011.

[17] Estévez P G, Gandarias M L and Prada J, Symmetry reductions of a 2+1 Lax pair, *Phys. Lett. A* **343**, 40–47, 2005.

[18] Estévez P G and Leble S L, A wave equation in 2+1: Painlevé analysis and solutions, *Inverse Problems* **11**, 925–937, 1995.

[19] Estévez P G and Prada J, A Generalization of the Sine-Gordon Equation to 2 + 1 dimensions, *J. Nonlinear Math. Phys.* **11**, 164–179, 2004.

[20] Estévez P G and Prada J, Hodograph transformations for a Camassa–Holm hierarchy in 2+1 dimensions, *J. Phys. A: Math. Gen.* **38**, 1287–1297, 2005.

[21] Estévez P G and Prada J, Singular manifold method for an equation in $2+1$ dimensions, *J. Nonlin. Math. Phys* **12**, 266–279, 2005.

[22] Estévez P G and Sardón C, Miura reciprocal transformations for two integrable hierarchies in $1+1$ dimensions, *Proceedings GADEIS (2012)*, Protaras, Cyprus, 2012.

[23] Estévez P G and Sardón C, Miura reciprocal Transformations for hierarchies in 2+1 dimensions, *J. Nonlinear Math. Phys.* **20**, 552–564, 2013.

[24] Ferapontov E V, Reciprocal transformations and their invariants, *Differential Equations* **25**, 898–905, 1989.

[25] Ferapontov E V, Rogers C and Schief W K, Reciprocal transformations of two component hyperbolic system and their invariants, *J. Math. Anal. Appl.* **228**, 365–376, 1998.

[26] Hone A N W, Reciprocal link for $2+1$-dimensional extensions of shallow water equations, *App. Math. Letters* **13**, 37–42, 2000.

[27] Ivanov R, Equations of the Camassa–Holm hierarchy, *Theor. Math. Phys.* **160**, 953–960, 2009.

[28] Kruskal M, Nonlinear Wave Equations, In Moser J, *Dynamical Systems, Theory and Applications* **38**, 310–354, Springer, 1975.

[29] Kudryashov N and Pickering A, Rational solutions for Schwarzian integrable hierarchies, *J. Phys. A: Math. Gen.* **31**, 9505–9518, 1998.

[30] Oevel W and Rogers C, Gauge transformations and reciprocal links in $2+1$ dimensions, *Rev. Math. Phys.* **5**, 299–330, 1993.

[31] Qiao Z, The Camassa–Holm hierarchy, related N-dimensional integrable systems, and algebro–geometric solutions on a symplectic submanifold, *Commun. Math. Phys.* **239**, 309–341, 2003.

[32] Qiao Z, A new integrable equation with cuspons and W/M–shape–peaks solitons, *J. Math. Phys.* **47**, 112701, 2006.

[33] Qiao Z, New integrable hierarchy, its parametric solutions, cuspons, one–peak solitons, and M/W–shape solitons, *J. Math. Phys.* **48**, 082701, 2007.

[34] Qiao Z and Liu L, A new integrable equation with no smooth solitons, *Chaos, Solitons & Fractals* **41**, 587–593, 2009.

[35] Rogers C, Application of a reciprocal transformation to a two-phase Stefan problem, *J. Phys. A* **18**, L105–L109, 1985.

[36] Rogers C, Reciprocal transformations in (2+1) dimensions, *J. Phys. A: Math. Gen* **19**, L491–L496, 1986.

[37] Rogers C, The Harry Dym equation in 2+1 dimensions: a reciprocal link with the Kadomtsev–Petviashvili equation, *Phys. Lett. A* **120**, 15–18, 1987.

[38] Rogers C and Carillo S, On reciprocal properties of the Caudrey, Dodd–Gibbon and Kaup–Kuppersmidt hierarchies, *Phys. Scripta* **36**, 865–869, 1987.

[39] Rogers C and Kingston J G and Shadwick W F, On reciprocal type invariant transformations in magneto-gas dynamics, *J. Math. Phys.* **21**, 395–397, 1980.

[40] Rogers C and Nucci M C, On reciprocal Bäcklund transformations and the Korteweg de Vries hierarchy, *Phys. Scripta* **33**, 289–292, 1986.

[41] Rogers C and Shadwick W F, *Bäcklund transformations and their applications*, Mathematics in science and engineering, Vol 161, Academic Press, 1982.

[42] Sakovich S Y, On Miura transformations of evolution equations, *J. Phys. A: Math. Gen.* **26**, L369–L373, 1993.

[43] Vakhnenko V O, Solitons in a nonlinear model medium, *J. Phys. A: Math. Gen.* **25**, 4181–4187, 1992.

[44] Weiss J, The Painlevé property for partial differential equations. II: Bäcklund transformation, Lax pairs, and the Schwarzian derivative, *J. Math. Phys.* **24**, 1405–1413, 1983.

[45] Zakharov V E and Shabat A B, Exact theory of two dimensional self-focusing and one-dimensional self-modulation of waves in nonlinear media, *Sov. Phys. JETP* **34**, 62–69, 1972.

A2. Contact Lax pairs and associated (3+1)-dimensional integrable dispersionless systems

Maciej Błaszak [a] *and Artur Sergyeyev* [b]

[a] *Faculty of Physics, Division of Mathematical Physics, A. Mickiewicz University Umultowska 85, 61-614 Poznań, Poland*
E-mail blaszakm@amu.edu.pl

[b] *Mathematical Institute, Silesian University in Opava, Na Rybníčku 1, 74601 Opava, Czech Republic*
E-mail artur.sergyeyev@math.slu.cz

Abstract

We review the recent approach to the construction of (3+1)-dimensional integrable dispersionless partial differential systems based on their contact Lax pairs and the related R-matrix theory for the Lie algebra of functions with respect to the contact bracket. We discuss various kinds of Lax representations for such systems, in particular, linear nonisospectral contact Lax pairs and nonlinear contact Lax pairs as well as the relations among the two. Finally, we present a large number of examples with finite and infinite number of dependent variables, as well as the reductions of these examples to lower-dimensional integrable dispersionless systems.

1 Introduction

Integrable systems play an important role in modern mathematics and theoretical and mathematical physics, cf. e.g. [14, 33], and, since according to general relativity our spacetime is four-dimensional, integrable systems in four independent variables ((3+1)D for short; likewise (n+1)D is shorthand for $n+1$ independent variables) are particularly interesting. For a long time it appeared that such systems were very difficult to find but in a recent paper by one of us [37] a novel systematic and effective construction for a large new class of integrable (3+1)D systems was introduced. This construction uses Lax pairs of a new kind related to contact geometry. Moreover, later in [4] it was shown that the systems from this class are amenable to an appropriate extension of the R-matrix approach which paved the way to constructing the associated integrable hierarchies.

The overwhelming majority of integrable partial differential systems in four or more independent variables known to date, cf. e.g. [14, 15, 27, 28] and references therein, including the celebrated (anti-)self-dual Yang–Mills equations and (anti-) self-dual vacuum Einstein equations with vanishing cosmological constant, can be written as homogeneous first-order quasilinear, i.e., *dispersionless* (also known as *hydrodynamic-type*), systems, cf. e.g. [11, 14, 15, 20, 43, 46] and the discussion below for details on the latter.

Integrable (3+1)D systems from the class introduced in [37] and further studied in [4, 38, 39] also are dispersionless, and it is interesting to note that this class appears to be entirely new: it does not seem to include any of the previously known examples of integrable dispersionless (3+1)D systems with nonisospectral Lax pairs, e.g. those from [12, 13, 15, 22].

In the present paper we review the results from [37, 4] and provide some novel examples of integrable (3+1)D systems using the approach from these papers.

The rest of the text is organized as follows. After a brief review of (3+1)D dispersionless systems and their nonisospectral Lax pairs in general in Section 2, we proceed with recalling the properties of linear and nonlinear Lax pairs in (1+1)D and (2+1)D in Section 3. In Section 4 we review, following [37], the construction of linear and nonlinear contact Lax pairs and the associated integrable (3+1)D systems and illustrate it by several examples. Finally, in Section 5 we review, following [4], the version of R-matrix formalism adapted to this setting and again give a number of examples to illustrate it.

2 Isospectral versus nonisospectral Lax pairs

Dispersionless systems in four independent variables x, y, z, t by definition can be written in general form

$$A_0(\boldsymbol{u})\boldsymbol{u}_t + A_1(\boldsymbol{u})\boldsymbol{u}_x + A_2(\boldsymbol{u})\boldsymbol{u}_y + A_3(\boldsymbol{u})\boldsymbol{u}_z = 0 \tag{2.1}$$

where $\boldsymbol{u} = (u_1, \ldots, u_N)^T$ is an N-component vector of unknown functions and A_i are $M \times N$ matrices, $M \geq N$.

Integrable systems of the form (2.1) typically have scalar Lax pairs of general form

$$\begin{aligned} \chi_y &= K_1(p, \boldsymbol{u})\chi_x + K_2(p, \boldsymbol{u})\chi_z + K_3(p, \boldsymbol{u})\chi_p, \\ \chi_t &= L_1(p, \boldsymbol{u})\chi_x + L_2(p, \boldsymbol{u})\chi_z + L_3(p, \boldsymbol{u})\chi_p, \end{aligned} \tag{2.2}$$

where $\chi = \chi(x, y, z, t, p)$ and p is the (variable) spectral parameter, cf. e.g. [9, 47, 37] and references therein; we stress that $\boldsymbol{u}_p = 0$.

In general, if at least one of the quantities K_3 or L_3 is nonzero, these Lax pairs are nonisospectral as they involve χ_p. The same terminology is applied in the lower-dimension case, when e.g. the dependence on z is dropped. The isospectral case when both K_3 and L_3 are identically zero is substantially different from the nonisospectral one. In particular, it is conjectured [18] that integrable systems with isospectral Lax pairs (2.2) are linearly degenerate while those with nonisospectral Lax pairs (2.2) are not, which leads to significant differences in qualitative behavior of solutions: according to a conjecture of Majda [26], in linearly degenerate systems no shock formation for smooth initial data occurs, see also the discussion in [15]. Many examples of integrable dispersionless (3+1)D systems with Lax pairs (2.2) in the isospectral case can be found e.g. in [22, 31, 36, 40] and references therein.

On the other hand, it appears that, among dispersionless systems, only linearly degenerate systems admit recursion operators that are Bäcklund auto-transformations of linearized versions of these systems, cf. e.g. [29] and references therein for general introduction to the recursion operators of this kind, and [30, 31, 36, 40] and references therein for such operators in the context of dispersionless systems. The theory of recursion operators for integrable dispersionless systems with nonisospectral Lax pairs (2.2), if any exists, should be significantly different both from that of the recursion operators as auto-Bäcklund transformations of linearized versions of systems under study and from that of bilocal recursion operators, see e.g. [19] and references therein for the latter.

Finally, in the case of nonisospectral Lax pairs (2.2) integrability of associated nonlinear systems is intimately related to the geometry of characteristic varieties of the latter [16, 10]. On the other hand, for large classes of (1+1)D and (2+1)D dispersionless integrable systems, their nonlinear Lax representations are related to symplectic geometry, see e.g. [24, 3, 5, 6, 15, 16, 20, 17, 32, 37, 43, 47] and references therein, although there are some exceptions, cf. e.g. [27, 42] and references therein. As a consequence of this, in the (1+1)D case the systems under study can be written in the form of the Lax equations which take the form of Hamiltonian dynamics on some Poisson algebras. For the (2+1)D case, the systems under study can be written as zero-curvature-type equations on certain Poisson algebras, i.e., as Frobenius integrability conditions for some pseudopotentials or equivalently for Hamiltonian functions from the Poisson algebra under study. Moreover, thanks to some features of symplectic geometry, the original nonlinear Lax representations in (1+1)D and (2+1)D imply linear nonisospectral Lax representations written in terms of Hamiltonian vector fields of the form (2.2), as discussed in the next section.

In the view of the wealth of integrable (2+1)D dispersionless systems it is natural to look for new multidimensional integrable systems which are dispersionless, and it is indeed possible to construct in a systematic fashion such new (3+1)D systems using a *contact* geometry instead of a symplectic one in a way proposed in [37], and we review this construction below. In particular, we will show how, using this construction, one obtains a novel class of nonisospectral Lax pairs together with the associated zero-curvature-type equations in the framework of Jacobi algebras, i.e., as the Frobenius integrability condition for contact Hamiltonian functions from such algebra.

In what follows we will be interested in the class of dispersionless systems possessing nonisospectral Lax representations.

3 Lax representations for dispersionless systems in (1+1)D and (2+1)D

3.1 Nonlinear Lax pairs in (1+1)D and (2+1)D

Dispersionless systems in (2+1)D have the form (2.1) with $A_3 = 0$ and $\boldsymbol{u}_z = 0$ and these in (1+1)D have the form (2.1) with $A_3 = A_2 = 0$ and $\boldsymbol{u}_z = \boldsymbol{u}_y = 0$. For the overwhelming majority of integrable systems of this kind, see e.g. [14, 28, 47], there exists a pseudopotential ψ such that the systems under study can be written as an appropriate compatibility condition for a nonlinear (with respect to ψ) Lax pair.

The said nonlinear Lax pair takes the form (cf. e.g. [21])

$$E = \mathcal{L}(\psi_x, \boldsymbol{u}), \quad \psi_t = \mathcal{B}(\psi_x, \boldsymbol{u}), \tag{3.1}$$

where E is an arbitrary constant playing the role reminiscent of that of a spectral parameter for the linear Lax pairs, while in (2+1)D the nonlinear Lax pair takes the form [47] (cf. also e.g. [16, 17, 37] and references therein)

$$\psi_y = \mathcal{L}(\psi_x, \boldsymbol{u}), \quad \psi_t = \mathcal{B}(\psi_x, \boldsymbol{u}). \tag{3.2}$$

The compatibility relations for a Lax pair, which are necessary and sufficient conditions for the existence of a pseudopotential ψ, are equivalent to a system of PDEs for the vector $\boldsymbol{u}$ of dependent variables.

Let us illustrate this idea by a simple example.

Example 1. *Let* $\boldsymbol{u} = (v_1, v_2, u_0, u_1)^T$ *and take*

$$\mathcal{L}(\psi_x, \boldsymbol{u}) = \psi_x + u_0 + u_1\psi_x^{-1}, \quad \mathcal{B}(\psi_x, \boldsymbol{u}) = v_1\psi_x + v_2\psi_x^2. \tag{3.3}$$

Compatibility of (3.3) gives

$$0 = \frac{d\mathcal{L}}{dx} = \psi_{xx} + (u_0)_x + (u_1)_x\psi_x^{-1} - u_1\psi_{xx}\psi_x^{-2} \Rightarrow u_1\psi_{xx}\psi_x^{-2} = \psi_{xx} + (u_0)_x + (u_1)_x\psi_x^{-1}$$

and

$$\begin{aligned} 0 = \frac{d\mathcal{L}}{dt} &= \psi_{xt} + (u_0)_t + (u_1)_t\psi_x^{-1} - u_1\psi_{xt}\psi_x^{-2} \\ &= \frac{d\mathcal{B}}{dx} + (u_0)_t + (u_1)_t\psi_x^{-1} - u_1\frac{d\mathcal{B}}{dx}{}_x^{-2} \\ &= (v_2)_x\psi_x^2 + [(v_1)_x - 2v_2(u_0)_x]\,\psi_x + [(u_0)_t - 2v_2(u_1)_x - u_1(v_2)_x - v_1(u_0)_x] \\ &\quad + [(u_1)_t - u_1(v_1)_x - v_1(u_1)_x]\,\psi_x^{-1}. \end{aligned}$$

Thus, equating to zero the coefficients at the powers of ψ_x *in the above equation we obtain the following system:*

$$\begin{aligned} (v_2)_x &= 0, \\ (v_1)_x &= 2v_2(u_0)_x, \\ (u_0)_t &= 2v_2(u_1)_x + u_1(v_2)_x + v_1(u_0)_x, \\ (u_1)_t &= u_1(v_1)_x + v_1(u_1)_x. \end{aligned} \tag{3.4}$$

In particular, if we put $v_2 = \text{const} = \frac{1}{2}$ *and* $v_1 = u_0$*, we arrive at a two-component dispersionless system in 1+1 dimensions*

$$\begin{aligned} (u_0)_t &= (u_1)_x + u_0(u_0)_x, \\ (u_1)_t &= u_1(u_0)_x + u_0(u_1)_x. \end{aligned} \tag{3.5}$$

Now turn to the (2+1)D Lax pair (3.2) with (3.3). Then we have

$${}_{yt} = \psi_{xt} + (u_0)_t + (u_1)_t\psi_x^{-1} - u_1\psi_{xt}\psi_x^{-2}, \tag{3.6}$$

$$\psi_{ty} = (v_1)_y \psi_x + v_1 \psi_{xy} + (v_2)_y \psi_x^2 + 2v_2 \psi_x \psi_{xy}. \tag{3.7}$$

The compatibility of (3.2) results in

$$\begin{aligned} 0 &= \psi_{yt} - \psi_{ty} = [(u_0)_t - 2v_2(u_1)_x - u_1(v_2)_x - u_1(v_2)_x - v_1(u_0)_x] \\ &\quad + [(v_2)_x - (v_2)_y]\,\psi_x^2 + [(v_1)_x - (v_1)_y - 2v_2(u_0)_x]\,\psi_x \\ &\quad + [(u_1)_t - u_1(v_1)_x - v_1(u_1)_x]\,\psi_x^{-1}. \end{aligned}$$

Equating to zero the coefficients at the powers of ψ_x yields the system

$$\begin{aligned} (v_2)_y &= (v_2)_x, \\ (v_1)_y &= (v_1)_x - 2v_2(u_0)_x, \\ (u_0)_t &= 2v_2(u_1)_x + u_1(v_2)_x + v_1(u_0)_x, \\ (u_1)_t &= u_1(v_1)_x + v_1(u_1)_x. \end{aligned} \tag{3.8}$$

If we put $v_2 = \text{const} = \frac{1}{2}$, we arrive at a three-component dispersionless system in 2+1 dimensions:

$$\begin{aligned} &(u_0)_t = (u_1)_x + v_1(u_0)_x = (u_1)_x + v_1(v_1)_x + v_1(v_1)_y, \\ &(u_1)_t = u_1(v_1)_x + v_1(u_1)_x, \\ &(v_1)_y = (u_0)_x - (v_1)_x. \end{aligned} \tag{3.9}$$

3.2 Basics of Poisson geometry

Now we shall restate the compatibility conditions for Lax pairs (3.1) and (3.2) using the language of symplectic geometry, but first we briefly recall the setting of the latter.

Namely, consider an even-dimensional ($\dim M = 2n$) symplectic manifold (M, ω), where ω is a closed ($d\omega = 0$) differential two-form which is nondegenerate, i.e., such that the nth exterior power of ω does not vanish anywhere on M.

Then for an arbitrary smooth function H on M there exists a unique vector field $\mathfrak{X}_H$ (the Hamiltonian vector field) defined by

$$i_{\mathfrak{X}_H}\omega = dH \Longleftrightarrow \mathfrak{X}_H = \mathcal{P}dH, \tag{3.10}$$

where $i_{\mathfrak{X}_H}\omega$ is the interior product of vector field $\mathfrak{X}_H$ with ω, and $\mathcal{P}$ is the associated symplectic bivector, i.e., a nondegenerate Poisson bivector; recall that a bivector is a skew-symmetric twice contravariant tensor field. Then $\mathfrak{X}_H$ is referred to as a Hamiltonian vector field with the Hamiltonian H.

Note that a symplectic manifold is a particular case of the more general Poisson manifold. A Poisson manifold is a pair $(M, \mathcal{P})$ where $\mathcal{P}$ is a bivector (i.e., a contravariant rank two skew-symmetric tensor field) satisfying the following identity:

$$[\mathcal{P}, \mathcal{P}]_S = 0, \tag{3.11}$$

where $[\cdot, \cdot]_S$ is the Schouten bracket, cf. e.g. [44, 48].

The Poisson structure $\mathcal{P}$ induces a bilinear map

$$\{\cdot,\cdot\}_{\mathcal{P}} : \mathcal{F}(M) \times \mathcal{F}(M) \longrightarrow \mathcal{F}(M),$$

in the associative algebra $\mathcal{F}(M)$ of smooth functions on M given by

$$\{F, G\}_{\mathcal{P}} := \mathcal{P}(dF, dG), \tag{3.12}$$

which endows $\mathcal{F}(M)$ with the Lie algebra structure and also satisfies the Leibniz rule, i.e., the bracket is also a derivation with respect to multiplication in the algebra of functions. Such a bracket is called a *Poisson bracket.*

It is readily checked that once (3.11) holds we indeed have

1. $\{F, G\}_{\mathcal{P}} = -\{G, F\}_{\mathcal{P}}$, (antisymmetry),
2. $\{F, GH\}_{\mathcal{P}} = \{F, G\}_{\mathcal{P}}H + G\{F, H\}_{\mathcal{P}}$, (the Leibniz rule),
3. $\{F, \{H, G\}_{\mathcal{P}}\}_{\mathcal{P}} + \{H, \{G, F\}_{\mathcal{P}}\}_{\mathcal{P}} + \{G, \{F, H\}_{\mathcal{P}}\}_{\mathcal{P}} = 0$, (the Jacobi identity).

For a $2n$-dimensional symplectic manifold, by the Darboux theorem there exist local coordinates (x^i, p_i), $i = 1, \ldots, n$, known as the Darboux coordinates, such that $\omega = d\eta$, where $\eta = \sum_{i=1}^{n} p_i dx^i$, and hence

$$\omega = d\eta = \sum_{i=1}^{n} dp_i \wedge dx^i, \qquad P = \sum_{i=1}^{n} \partial_{x^i} \wedge \partial_{p_i},$$

$$\mathfrak{X}_H = \frac{\partial H}{\partial p_i}\frac{\partial}{\partial x^i} - \frac{\partial H}{\partial x^i}\frac{\partial}{\partial p_i}. \tag{3.13}$$

and

$$\{H, F\}_P = \mathfrak{X}_H(F) = \frac{\partial H}{\partial p_i}\frac{\partial F}{\partial x^i} - \frac{\partial H}{\partial x^i}\frac{\partial F}{\partial p_i}. \tag{3.14}$$

For any $H, F \in \mathcal{F}(M)$ we also have

$$[\mathfrak{X}_H, \mathfrak{X}_F] = \mathfrak{X}_{\{H,F\}_P}, \tag{3.15}$$

where $[\cdot,\cdot]$ is the usual Lie bracket (commutator) of vector fields.

3.3 Compatibility conditions for Lax pairs via Poisson geometry

Now let us return to the Lax pair (3.1)

$$E = \mathcal{L}(\psi_x, \boldsymbol{u}) \quad \psi_t = \mathcal{B}(\psi_x, \boldsymbol{u}) \tag{3.16}$$

for the (1+1)D case when $\boldsymbol{u} = \boldsymbol{u}(x, t)$.

We have

$$0 = \frac{d\mathcal{L}}{dx} = \frac{\partial \mathcal{L}}{\partial x} + \frac{\partial \mathcal{L}}{\partial \psi_x}\psi_{xx} \Longrightarrow \psi_{xx} = -\left(\frac{\partial \mathcal{L}}{\partial \psi_x}\right)^{-1}\frac{\partial \mathcal{L}}{\partial x} \tag{3.17}$$

and so

$$0 = \frac{d\mathcal{L}}{dt} = \frac{\partial \mathcal{L}}{\partial t} + \frac{\partial \mathcal{L}}{\partial \psi_x}\psi_{xt} = \frac{\partial \mathcal{L}}{\partial t} + \frac{\partial \mathcal{L}}{\partial \psi_x}\frac{\partial \mathcal{B}}{dx}$$

$$= \frac{\partial \mathcal{L}}{\partial t} + \frac{\partial \mathcal{L}}{\partial \psi_x}\left(\frac{\partial \mathcal{B}}{\partial x} + \frac{\partial \mathcal{B}}{\partial \psi_x}\psi_{xx}\right) \overset{(3.17)}{=} \frac{\partial \mathcal{L}}{\partial t} + \frac{\partial \mathcal{L}}{\partial \psi_x}\frac{\partial \mathcal{B}}{\partial x} - \frac{\partial \mathcal{L}}{\partial x}\frac{\partial \mathcal{B}}{\partial \psi_x}. \tag{3.18}$$

Thus, the compatibility condition for Lax pair (3.1) is equivalently expressed via the so-called Lax equation

$$L_t = \{B, L\}_P, \tag{3.19}$$

for a pair of functions $L = \mathcal{L}(p, \boldsymbol{u})$, $B = \mathcal{B}(p, \boldsymbol{u})$, where now $P = \partial_x \wedge \partial_p$ is a Poisson bivector associated to the symplectic two-form $dp \wedge dx$ on a two-dimensional symplectic manifold with global Darboux coordinates (x, p). Here p is an additional independent variable, which in the context of linear Lax pairs will be identified as a *variable spectral parameter*, see next subsection.

Now turn to the Lax pair (3.2)

$$\psi_y = \mathcal{L}(\psi_x, \boldsymbol{u}) \quad \psi_t = \mathcal{B}(\psi_x, \boldsymbol{u}) \tag{3.20}$$

for the (2+1)-dimensional case when $\boldsymbol{u} = \boldsymbol{u}(x, y, t)$.

We have

$$\psi_{yt} = \frac{\partial \mathcal{L}}{\partial t} + \frac{\partial \mathcal{L}}{\partial \psi_x}\psi_{xt}, \quad \psi_{ty} = \frac{\partial \mathcal{B}}{\partial y} + \frac{\partial \mathcal{B}}{\partial \psi_x}\psi_{xy}, \tag{3.21}$$

$$\psi_{tx} = \psi_{xt} = \frac{\partial \mathcal{B}}{\partial x} + \frac{\partial \mathcal{B}}{\partial \psi_x}\psi_{xx}, \quad \psi_{yx} = \psi_{xy} = \frac{\partial \mathcal{L}}{\partial x} + \frac{\partial \mathcal{L}}{\partial \psi_x}\psi_{xx}, \tag{3.22}$$

and thus,

$$0 = \psi_{yt} - \psi_{ty} \overset{(3.21),(3.22)}{=} \mathcal{L}_t - \mathcal{B}_y + \frac{\partial \mathcal{L}}{\partial \psi_x}\frac{\partial \mathcal{B}}{\partial x} - \frac{\partial \mathcal{L}}{\partial x}\frac{\partial \mathcal{B}}{\partial \psi_x}. \tag{3.23}$$

The compatibility condition for the Lax pair (3.2) can be now written as the so-called zero-curvature-type equation of the form [6, 3]

$$L_t - B_y + \{L, B\}_P = 0, \tag{3.24}$$

for a pair of Lax functions $L = \mathcal{L}(p, \boldsymbol{u})$, $B = \mathcal{B}(p, \boldsymbol{u})$.

For an illustration of this alternative form of the compatibility conditions for our nonlinear Lax pairs let us return to our example.

Example 1a. *Let* $L(p, \boldsymbol{u}) = p + u_0 + u_1 p^{-1}$, $B(p, \boldsymbol{u}) = v_1 p + v_2 p^2$, *where* $\boldsymbol{u} = (v_1, v_2, u_0, u_1)^T$ *and* $\boldsymbol{u} = \boldsymbol{u}(x, t)$. *Then (3.19) gives*

$$\begin{aligned} 0 &= L_t - \{B, L\}_P \\ &= (v_2)_x p^2 + [(v_1)_x - 2v_2(u_0)_x]\, p + [(u_0)_t - 2v_2(u_1)_x - v_1(u_0)_x - u_1(v_2)_x] \\ &\quad + [(u_1)_t - v_1(u_1)_x - u_1(v_1)_x]\, p^{-1}, \end{aligned}$$

and equating to zero the coefficients at the powers of p*, we again obtain the system (3.4), where we can put* $v_2 = \text{const} = \frac{1}{2}$ *and* $v_1 = u_0$*, and then again arrive at the two-component dispersionless system (3.5).*

On the other hand, in the (2+1)D case, when $\boldsymbol{u} = \boldsymbol{u}(x, y, t)$*, the zero-curvature-type equation (3.24) gives*

$$\begin{aligned}0 &= L_t - B_y + \{L, B\}_P \\ &= [(v_2)_x - (v_2)_y]\, p^2 + [(v_1)_x - 2v_2(u_0)_x - (v_1)_y]\, p \\ &+ [(u_0)_t - 2v_2(u_1)_x - v_1(u_0)_x - u_1(v_2)_x] + [(u_1)_t - v_1(u_1)_x - u_1(v_1)_x]\, p^{-1}.\end{aligned}$$

Again, equating to zero the coefficients at the powers of p *reproduces the system (3.8), and we can put* $v_2 = \text{const} = \frac{1}{2}$ *and recover the system (3.9).*

3.4 Linear nonisospectral Lax pairs in (1+1)D and (2+1)D

The relation (3.15) among the Poisson algebra of functions on M and the Lie algebra of Hamiltonian vector fields gives rise to alternative linear nonisospectral Lax pairs written in terms of Hamiltonian vector fields.

In the (1+1)D case such a linear Lax pair takes the form

$$\mathfrak{X}_L(\phi) = \{L, \phi\}_P = 0, \quad \phi_t = \mathfrak{X}_B(\phi) = \{B, \phi\}_P, \tag{3.25}$$

where $\phi = \phi(x, t, p)$, and in the (2+1)-dimensional case the form

$$\phi_y = \mathfrak{X}_L(\phi) = \{L, \phi\}_P, \quad \phi_t = \mathfrak{X}_B(\phi) = \{B, \phi\}_P, \tag{3.26}$$

where now $\phi = \phi(x, y, t, p)$.

Here p is an additional independent variable known as the *variable spectral parameter*, cf. e.g. [5, 7, 9, 14, 27] for details; recall that $\boldsymbol{u}_p \equiv 0$ by assumption.

Since the Hamiltonian vector field with a constant Hamiltonian is identically zero, the Lax equation (3.19) implies the compatibility of (3.25), and the zero-curvature-type equation (3.24) implies the compatibility of (3.26), but not vice versa.

Indeed, the compatibility condition for (3.25) reads

$$\begin{aligned}&[\partial_t - \mathfrak{X}_B, \mathfrak{X}_L]\,(\phi) = 0 \\ &\Updownarrow (3.15) \\ &\mathfrak{X}_{L_t - \{B,L\}_P}(\phi) = \{L_t - \{B, L\}_P, \phi\} = 0,\end{aligned}$$

while the compatibility condition for (3.26) takes the form

$$\begin{aligned}&[\partial_t - \mathfrak{X}_B, \partial_y - \mathfrak{X}_L]\,(\phi) = 0 \\ &\Updownarrow (3.15) \\ &\mathfrak{X}_{L_t - B_y + \{L,B\}_P}(\phi) = \{L_t - B_y + \{L, B\}_P, \phi\} = 0.\end{aligned}$$

For an explicit illustration of this we return to our example.

Example 1b. *Again let* $\boldsymbol{u} = (v_1, v_2, u_0, u_1)^T$ *and*

$$L(p, \boldsymbol{u}) = p + u_0 + u_1 p^{-1}, \quad B(p, \boldsymbol{u}) = v_1 p + v_2 p^2. \tag{3.27}$$

Then in the (1+1)D case, when $\boldsymbol{u} = \boldsymbol{u}(x, t)$, *the Lax pair (3.25) reads*

$$\begin{aligned} &(1 - u_1/p^2)\phi_x - ((u_0)_x + (u_1)_x/p)\phi_p = 0, \\ &\phi_t = (v_1 + 2pv_2)\phi_x - (p(v_1)_x + p^2(v_2)_x)\phi_p, \end{aligned} \tag{3.28}$$

which can be equivalently written as

$$\begin{aligned} \phi_x &= \frac{p}{p^2 - u_1}\left[(u_1)_x + p(u_0)_x\right]\phi_p, \\ \phi_t &= \frac{p}{p^2 - u_1}((v_1 + 2pv_2)\left[(u_1)_x + p(u_0)_x\right] - (p^2 - u_1)\left[(v_1)_x + (v_2)_x p\right])\phi_p. \end{aligned} \tag{3.29}$$

The compatibility condition for (3.29) is just $(\phi_x)_t - (\phi_t)_x = 0$ *but we cannot reproduce directly (3.4) by equating to zero the coefficients at the powers of* p. *Instead, we get a set of linear combinations of differential consequences of the latter.*

Now turn to the (2+1)D case with the same L *and* B *given by (3.27) but with* $\boldsymbol{u} = \boldsymbol{u}(x, y, t)$. *The associated linear nonisospectral Lax pair (3.26) takes the form (2.2), i.e.*

$$\begin{aligned} \phi_y &= (1 - u_1/p^2)\phi_x - ((u_0)_x + (u_1)_x/p)\phi_p, \\ \phi_t &= (v_1 + 2pv_2)\phi_x - (p(v_1)_x + p^2(v_2)_x)\phi_p, \end{aligned} \tag{3.30}$$

and, in complete analogy with the (1+1)D case, it is readily checked that its compatibility condition $(\phi_y)_t - (\phi_t)_y = 0$ *holds by virtue of the zero-curvature-type equation* $L_t - B_y + \{L, B\}_P = 0$ *but not the other way around. Besides, just as in the (1+1)D case above, equating to zero the coefficients at the powers of* p *in* $(\phi_y)_t - (\phi_t)_y = 0$, *yields a system that is a mix of algebraic and differential consequences of (3.8).*

4 Lax representations for dispersionless systems in (3+1)D

4.1 Nonlinear Lax pairs in (3+1)D

In [37] the following generalization of the (2+1)D nonlinear Lax pair (3.2) to (3+1)D was found:

$$\psi_y = \psi_z \mathcal{L}\left(-\frac{x}{z}, \boldsymbol{u}\right), \quad \psi_t = \psi_z \mathcal{B}\left(-\frac{x}{z}, \boldsymbol{u}\right), \tag{4.1}$$

where now $\psi = \psi(x, y, z, t)$. The Lax pairs of the form (4.1) are called *nonlinear contact Lax pairs.*

The above generalization leads to large new classes of integrable (3+1)D dispersionless systems for suitably chosen $\mathcal{L}$ and $\mathcal{B}$, e.g. rational functions or polynomials in ψ_x/ψ_z of certain special form.

The compatibility conditions for the Lax pair (4.1), which are necessary and sufficient conditions for the existence of a nontrivial pseudopotential ψ, are equivalent to a system of PDEs for $\boldsymbol{u}$ in (3+1)D.

Let us illustrate this idea again on our simple example.

Example 1c. *Let*

$$\begin{aligned} \psi_y &= \psi_z\mathcal{L}\left(-\frac{x}{z}, \boldsymbol{u}\right) = \psi_z\left(-\frac{x}{z} + u_0 + u_1\left(-\frac{x}{z}\right)^{-1}\right) = \psi_x + u_0\psi_z + u_1\frac{{}^2_z}{{}_x}, \\ \psi_t &= \psi_z\mathcal{B}\left(-\frac{x}{z}, \boldsymbol{u}\right) = \psi_z\left(v_1\frac{x}{\psi_z} + v_2\left(-\frac{x}{z}\right)^2\right) = v_1\psi_x + v_2\frac{{}^2_x}{{}_z}, \end{aligned} \tag{4.2}$$

where $\boldsymbol{u} = (v_1, v_2, u_0, u_1)^T$. *Then we have*

$$\begin{aligned} \psi_{yt} &= \psi_{xt} + (u_0)_t\psi_z + u_0\psi_{zt} + (u_1)_t\frac{{}^2_z}{{}_x} + 2u_1\frac{{}_z\psi_{zt}}{{}_x} - u_1\frac{\psi_z^2\psi_{xt}}{{}^2_x}, \\ \psi_{ty} &= (v_1)_y\psi_x + v_1\psi_{xy} + (v_2)_y\frac{{}^2_x}{{}_z} + 2v_2\frac{{}_x\psi_{xy}}{{}^2_z} - v_2\frac{{}^2_x\,{}_{zy}}{{}^2_z} \end{aligned} \tag{4.3}$$

and, the compatibility of (4.3) results in

$$\begin{aligned} 0 &= \psi_{yt} - \psi_{ty} = [(v_2)_x + u_0(v_2)_z - (v_2)_y + v_2(u_0)_z]\frac{{}^2_x}{{}_z} \\ &\quad + [(u_1)_t - u_1(v_1)_x - v_1(u_1)_x]\frac{{}^2_z}{{}_x} \\ &\quad + [(v_1)_x + u_0(v_1)_z + 2u_1(v_2)_z - (v_1)_y - 2v_2(u_0)_x + v_2(v_1)_z]\,\psi_x \\ &\quad + [(u_0)_t - u_1(v_2)_x + 2u_1(v_1)_z - v_1(u_0)_x - 2v_2(u_1)_x]\,\psi_z \end{aligned} \tag{4.4}$$

and we arrive at a four-component dispersionless (3+1)D integrable system

$$\begin{aligned} (u_1)_t &= u_1(v_1)_x + v_1(u_1)_x, \\ (u_0)_t &= u_1(v_2)_x - 2u_1(v_1)_z + v_1(u_0)_x + 2v_2(u_1)_x, \\ (v_1)_y &= (v_1)_x + u_0(v_1)_z + 2u_1(v_2)_z - 2v_2(u_0)_x + v_2(v_1)_z, \\ (v_2)_y &= (v_2)_x + u_0(v_2)_z + v_2(u_0)_z. \end{aligned} \tag{4.5}$$

The fields u_0 and u_1 are dynamical variables, which evolve in time, while the remaining equations can be seen as nonlocal constraints on u_0 and u_1 which define the variables v_1 and v_2. The same situation takes place in the (2+1)D case. In the (1+1)D case all fields v_i are expressible via the dynamical fields u_j.

4.2 Basics of contact geometry

Now let us restate the compatibility conditions for Lax pairs (4.1) in the language of contact geometry, and to this end recall the basics of the latter.

Consider an odd-dimensional ($\dim M = 2n+1$) *contact manifold* (M, η) with a contact one-form η such that $\eta \wedge (d\eta)^{\wedge n} \neq 0$, cf. e.g. [8] and references therein.

For a given contact form η there exists a unique vector field Y, called the *Reeb vector field*, such that

$$i_Y d\eta = 0, \quad i_Y \eta = 1. \tag{4.6}$$

For any function on M, there exists a unique vector field X_H (the *contact vector field*) defined by the formula

$$i_{X_H}\eta = H, \quad i_{X_H} d\eta = dH - i_Y dH \cdot \eta \Longleftrightarrow X_H = \mathcal{P}dH + HY, \tag{4.7}$$

where $\mathcal{P}$ is the associated bivector.

The contact manifold is a special case of the so-called Jacobi manifold. A *Jacobi manifold* [25] is a triple $(M, \mathcal{P}, Y)$ where $\mathcal{P}$ is a bivector and Y a vector field satisfying the following conditions:

$$[\mathcal{P}, \mathcal{P}]_S = 2Y \wedge \mathcal{P}, \quad [Y, \mathcal{P}]_S = 0. \tag{4.8}$$

The Jacobi structure induces a bilinear map $\{\cdot,\cdot\}_J : \mathcal{F}(M) \times \mathcal{F}(M) \longrightarrow \mathcal{F}(M)$ in the associative algebra $\mathcal{F}(M)$ of smooth functions on M through the *Jacobi bracket*

$$\{F, G\}_J := \mathcal{P}(dF, dG) + FY(G) - GY(F), \tag{4.9}$$

which turns $\mathcal{F}(M)$ into a Lie algebra and satisfies the generalized Leibniz rule, i.e., we have

1. $\{F, G\}_J = -\{G, F\}_J$ (antisymmetry),
2. $\{F, GH\}_J = \{F, G\}_J H + G\{F, H\}_J - \{F, 1\}_J GH$ (the generalized Leibniz rule),
3. $\{F, \{H, G\}_J\}_J + \{H, \{G, F\}_J\}_J + \{G, \{F, H\}_J\}_J = 0$ (the Jacobi identity).

For a $(2n+1)$-dimensional contact manifold by the Darboux theorem there exist local coordinates (x^i, p_i, z), where $i = 1, \dots, n$, known as the Darboux coordinates, such that we have

$$\eta = dz + \sum_{i=1}^{n} p_i dx^i \Rightarrow d\eta = \sum_{i=1}^{n} dp_i \wedge dx^i, \quad Y = \partial_z, \quad P = \sum_{i=1}^{n} \left(\partial_{x^i} \wedge \partial_{p_i} - p_i \partial_z \wedge \partial_{p_i}\right),$$

$$X_H = H\frac{\partial}{\partial z} + \sum_{i=1}^{n} \left(\frac{\partial H}{\partial p_i}\frac{\partial}{\partial x^i} - \frac{\partial H}{\partial x^i}\frac{\partial}{\partial p_i} - p_i \left(\frac{\partial H}{\partial p_i}\frac{\partial}{\partial z} - \frac{\partial H}{\partial z}\frac{\partial}{\partial p_i} \right) \right) \tag{4.10}$$

and the contact bracket, the relevant special case of the Jacobi bracket, reads

$$\{H, F\}_C = X_H(F) - Y(H)F = H\frac{\partial F}{\partial z} + \sum_{i=1}^{n} \left(\frac{\partial H}{\partial p_i}\frac{\partial F}{\partial x^i} - p_i \frac{\partial H}{\partial p_i}\frac{\partial F}{\partial z} \right) - (H \leftrightarrow F). \tag{4.11}$$

We also have

$$[X_H, X_F] = X_{\{H,F\}_C}. \tag{4.12}$$

4.3 Zero-curvature-type equations in (3+1)D via the contact bracket

Now return to the Lax pair (4.1) with $\boldsymbol{u} = \boldsymbol{u}(x, y, z, t)$,

$$\psi_y = \psi_z \mathcal{L}\left(-\frac{x}{z}, \boldsymbol{u}\right), \quad \psi_t = \psi_z \mathcal{B}\left(-\frac{x}{z}, \boldsymbol{u}\right). \tag{4.13}$$

Let $\theta \equiv \psi_x/\psi_z$. Then we have

$$\begin{aligned} \psi_{yt} &= \psi_{zt}\mathcal{L} + \psi_z \mathcal{L}_t + \psi_{xt}\mathcal{L}_\theta - \psi_{zt}\theta\mathcal{L}_\theta, \\ \psi_{ty} &= \psi_{zy}\mathcal{B} + \psi_z \mathcal{B}_y + \psi_{xy}\mathcal{B}_\theta - \psi_{zy}\theta\mathcal{B}_\theta. \end{aligned} \tag{4.14}$$

Again, the compatibility of (4.13) results in

$$0 = \psi_{yt} - \psi_{ty} = \psi_z[\mathcal{L}_t - \mathcal{B}_y + \mathcal{L}_\theta\mathcal{B}_x - \mathcal{L}_x\mathcal{B}_\theta - \theta\left(\mathcal{L}_\theta\mathcal{B}_z - \mathcal{L}_z\mathcal{B}_\theta\right) + \mathcal{L}\mathcal{B}_z - \mathcal{B}\mathcal{L}_z]. \tag{4.15}$$

Comparing (4.15) with (4.11) we observe that the compatibility condition for Lax pair (4.1) is equivalently given [37] by the so-called zero-curvature-type equation of the form

$$L_t - B_y + \{L, B\}_C = 0, \tag{4.16}$$

for a pair of Lax functions $L = \mathcal{L}(p, \boldsymbol{u})$, $B = \mathcal{B}(p, \boldsymbol{u})$, where the contact bracket $\{\cdot, \cdot\}_C$ now is a special case of the contact bracket (4.11) for the three-dimensional contact manifold with the (global) Darboux coordinates (x, p, z), where p is the variable spectral parameter just as in the lower-dimensional cases.

The contact bracket in this case reads

$$\{H, F\}_C = X_H(F) - Y(H)F = H\frac{\partial F}{\partial z} + \frac{\partial H}{\partial p}\frac{\partial F}{\partial x} - p\frac{\partial H}{\partial p}\frac{\partial F}{\partial z} - (H \leftrightarrow F). \tag{4.17}$$

For the illustration of that alternative Lax representation let us return to our previous example.

Example 1d. *Let*

$$L(p, \boldsymbol{u}) = p + u_0 + u_1 p^{-1}, \quad B(p, \boldsymbol{u}) = v_1 p + v_2 p^2, \tag{4.18}$$

where $\boldsymbol{u} = (v_1, v_2, u_0, u_1)^T$. *Then, for the (3+1)-dimensional case, the contact zero-curvature-type equation (4.16) reads*

$$\begin{aligned} 0 \;=\; & L_t - B_y + \{L, B\}_C = [(v_2)_x - (v_2)_y + u_0(v_2)_z + v_2(u_0)_z]\, p^2 \\ & + [(v_1)_x - 2v_2(u_0)_x - (v_1)_y + 2u_1(v_2)_z + v_2(u_1)_z + u_0(v_1)_z]\, p \\ & + [(u_0)_t - 2v_2(u_1)_x - v_1(u_0)_x - u_1(v_2)_x + 2u_1(v_1)_z] \\ & + [(u_1)_t - v_1(u_1)_x - u_1(v_1)_x]\, p^{-1}. \end{aligned}$$

and we recover the four-component (3+1)-dimensional integrable dispersionless system (4.5).

4.4 Linear nonisospectral Lax pairs in (3+1)D

Using the above results from contact geometry we readily can construct [37] two different kinds of linear nonisospectral Lax pairs in (3+1)D generalizing (3.26), that is,

$$\phi_y = \mathfrak{X}_L(\phi) = \{L, \phi\}_P, \quad \phi_t = \mathfrak{X}_B(\phi) = \{B, \phi\}_P,$$

in two different ways.

The first one replaces the Poisson bracket $\{\cdot, \cdot\}_P$ by the contact bracket (4.17) and gives us the Lax pair of the form

$$\phi_y = \{L, \phi\}_C, \quad \phi_t = \{B, \phi\}_C, \tag{4.19}$$

where now $\phi = \phi(x, y, z, t, p)$.

The second one replaces the Hamiltonian vector fields $\mathfrak{X}_H$ by their contact counterparts X_H, and we obtain

$$\chi_y = X_L(\chi), \quad \chi_t = X_B(\chi), \tag{4.20}$$

where now $\chi = \chi(x, y, z, t, p)$; here we replaced ϕ by χ in order to distinguish (4.19) from (4.20). The Lax pairs of the form (4.20) are called *linear contact Lax pairs* [37].

Recall that in our particular setting we have

$$X_H = H\frac{\partial}{\partial z} + \frac{\partial H}{\partial p}\frac{\partial}{\partial x} - \frac{\partial H}{\partial x}\frac{\partial}{\partial p} - p\left(\frac{\partial H}{\partial p}\frac{\partial}{\partial z} - \frac{\partial H}{\partial z}\frac{\partial}{\partial p}\right), \quad Y = \frac{\partial}{\partial z}. \tag{4.21}$$

In stark contrast with the (2+1)D case, the two Lax pairs (4.19) and (4.20) no longer coincide, since we have

$$X_H(F) = \{H, F\}_C + FH_z = \{H, F\}_C + FY(H) = \{H, F\}_C + F\{1, H\}_C \tag{4.22}$$

instead of

$$\mathfrak{X}_H(F) = \{H, F\}_P,$$

and the behaviour of these Lax pairs is quite different too.

As for (4.19), in complete analogy with the (2+1)D case we readily find that its compatibility condition, $(\phi_y)_t - (\phi_t)_y = 0$, can be written as

$$\{L_t - B_y + \{L, B\}_C, \phi\}_C = 0, \tag{4.23}$$

where $L = \mathcal{L}(p, \boldsymbol{u})$, $B = \mathcal{B}(p, \boldsymbol{u})$, and thus while the zero-curvature-type equation (4.16), or equivalently, the compatibility of nonlinear Lax pair (4.1), implies compatibility of the Lax pair (4.19), the converse is not true.

On the other hand, the situation for (4.20) is very different. We can, by analogy with the discussion before Example 1b, show that the compatibility condition for (4.20), that is,

$$[\partial_t - X_B, \partial_y - X_L] = 0$$

is, by virtue of (4.12), equivalent to the following:

$$X_{L_t-B_y+\{L,B\}_C} = 0. \tag{4.24}$$

Using the formula (4.21) we immediately see that, in contrast with the (2+1)-dimensional case, (4.24) implies that

$$L_t - B_y + \{L, B\}_C = 0,$$

i.e., (4.24) is *equivalent* to (4.16) rather than being just a consequence of the latter, as it is the case for (4.23).

Let us show the explicit form of the above nonisospectral Lax pairs (4.19) and (4.20) for our example.

Example 1e. *Again, let*

$$L(p, \boldsymbol{u}) = p + u_0 + u_1 p^{-1}, \quad B(p, \boldsymbol{u}) = v_1 p + v_2 p^2, \tag{4.25}$$

where $\boldsymbol{u} = (v_1, v_2, u_0, u_1)^T$. *Then the nonisospectral Lax pair (4.19) reads*

$$\begin{aligned}
\phi_y &= (1 - u_1/p^2)\phi_x + (u_0 + 2u_1/p)\phi_z + (p(u_0)_z + (u_1)_z - (u_0)_x - (u_1)_x/p)\phi_p \\
&\quad +(-(u_0)_z - (u_1)_z/p)\phi, \\
\phi_t &= (2pv_2 + v_1)\phi_x - p^2 v_2 \phi_z + ((v_2)_z p^3 + ((v_1)_z - (v_2)_x)p^2 - (v_1)_x p)\phi_p \\
&\quad -p((v_2)_z p - (v_1)_z)\phi,
\end{aligned}$$

while the nonisospectral linear contact Lax pair (4.20) has the form (2.2), i.e.

$$\begin{aligned}
\chi_y &= \left(1 - \frac{u_1}{p^2}\right)\chi_x + \left(u_0 + \frac{2u_1}{p}\right)\chi_z + \left(p(u_0)_z + (u_1)_z - (u_0)_x - \frac{(u_1)_x}{p}\right)\chi_p, \\
\chi_t &= (2pv_2 + v_1)\chi_x - p^2 v_2 \chi_z + \left((v_2)_z p^3 + ((v_1)_z - (v_2)_x)p^2 - (v_1)_x p\right)\chi_p.
\end{aligned} \tag{4.26}$$

Spelling out the compatibility condition for this Lax pair, $(\chi_y)_t - (\chi_t)_y = 0$, *and equating to zero the coefficients at* χ_x *and* χ_p *therein, we readily see that, in perfect agreement with the general discussion above, we recover (4.16) for L and B given by (4.25), and then the system (4.5).*

As for (4.19), it is readily checked that (4.16) and (4.25) imply compatibility of (4.19), that is, $(\phi_y)_t - (\phi_t)_y = 0$, *but not the other way around, i.e.,* $(\phi_y)_t - (\phi_t)_y = 0$ *gives us not the system (4.5) but merely a mix of differential and algebraic consequences thereof.*

5 *R*-matrix approach for dispersionless systems with nonisospectral Lax representations

5.1 General construction

The R-matrix approach addresses two important problems concerning the dispersionless systems under study. First, it allows for a systematic construction of consistent Lax pairs (L, B) in order to generate such systems, and second, it allows for a systematic construction of an infinite hierarchy of commuting symmetries for a given dispersionless system, proving integrability of the latter. So, let us start from some basic facts on the R-matrix formalism, see for example [7, 34, 35] and references therein.

Let $\mathfrak{g}$ be an (in general infinite-dimensional) Lie algebra. The Lie bracket $[\cdot,\cdot]$ defines the adjoint action of $\mathfrak{g}$ on $\mathfrak{g}$: $\mathrm{ad}_a b = [a, b]$.

Recall that an $R \in \mathrm{End}(\mathfrak{g})$ is called a (classical) R-matrix if the R-bracket

$$[a,b]_R := [Ra, b] + [a, Rb] \tag{5.1}$$

is a new Lie bracket on $\mathfrak{g}$. The skew symmetry of (5.1) is obvious. As for the Jacobi identity for (5.1), a sufficient condition for it to hold is the so-called classical modified Yang–Baxter equation for R,

$$[Ra, Rb] - R[a,b]_R - \alpha[a,b] = 0, \qquad \alpha \in \mathbb{R}. \tag{5.2}$$

Let $L_i \in \mathfrak{g}$, $i \in \mathbb{N}$. Consider the associated hierarchies of flows (Lax hierarchies)

$$(L_n)_{t_r} = [RL_r, L_n], \qquad r, n \in \mathbb{N}. \tag{5.3}$$

Suppose that R commutes with all derivatives ∂_{t_n}, i.e.,

$$(RL)_{t_n} = RL_{t_n}, \quad n \in \mathbb{N}, \tag{5.4}$$

and obeys the classical modified Yang–Baxter equation (5.2) for $\alpha \neq 0$. Moreover, let $L_i \in \mathfrak{g}$, $i \in \mathbb{N}$ satisfy (5.3). Then the following conditions are equivalent:

i) the zero-curvature equations

$$(RL_r)_{t_s} - (RL_s)_{t_r} + [RL_r, RL_s] = 0, \quad r, s \in \mathbb{N} \tag{5.5}$$

hold;

ii) all L_i commute in $\mathfrak{g}$:

$$[L_i, L_j] = 0, \qquad i, j \in \mathbb{N}. \tag{5.6}$$

Moreover, if one (and hence both) of the above equivalent conditions holds, then the flows (5.3) commute, i.e.,

$$((L_n)_{t_r})_{t_s} - ((L_n)_{t_s})_{t_r} = 0, \quad n, r, s \in \mathbb{N}. \tag{5.7}$$

The reader can find the proofs of the above statements for example in [7] or in [4].

Now let us present a procedure for extending the systems under study by adding an extra independent variable. This procedure bears some resemblance to that of the central extension approach, see e.g. [7, 35] and references therein. Namely, we assume that all elements of $\mathfrak{g}$ depend on an additional independent variable y not involved in the Lie bracket, so all of the above results remain valid. Consider an $L \in \mathfrak{g}$ and the associated Lax hierarchy defined by

$$L_{t_r} = [RL_r, L] + (RL_r)_y, \qquad r \in \mathbb{N}. \tag{5.8}$$

Suppose that $L_i \in \mathfrak{g}$, $i \in \mathbb{N}$ are such that the zero-curvature equations (5.5) hold for all $r, s \in \mathbb{N}$ and the R-matrix R on $\mathfrak{g}$ satisfies (5.4). Then the flows (5.8) commute, i.e.,

$$(L_{t_r})_{t_s} - (L_{t_s})_{t_r} = 0, \quad r, s \in \mathbb{N}. \tag{5.9}$$

Indeed, using the equations (5.8) and the Jacobi identity for the Lie bracket we obtain

$$\begin{aligned}(L_{t_r})_{t_s} - (L_{t_s})_{t_r} &= [(RL_r)_{t_s} - (RL_s)_{t_r} + [RL_r, RL_s], L] \\ &\quad + ((RL_r)_{t_s} - (RL_s)_{t_r} + [RL_r, RL_s])_y \\ &= 0.\end{aligned}$$

The right-hand side of the above equation vanishes by virtue of the zero curvature equations (5.5).

An important question is whether there exists a systematic procedure for constructing $R \in \mathrm{End}(\mathfrak{g})$ with the desired properties. Fortunately the answer is positive. It is well known (see e.g. [34, 35, 7]) that whenever $\mathfrak{g}$ admits a decomposition into two Lie subalgebras $\mathfrak{g}_+$ and $\mathfrak{g}_-$ such that

$$\mathfrak{g} = \mathfrak{g}_+ \oplus \mathfrak{g}_-, \qquad [\mathfrak{g}_\pm, \mathfrak{g}_\pm] \subset \mathfrak{g}_\pm, \qquad \mathfrak{g}_+ \cap \mathfrak{g}_- = \emptyset,$$

the operator

$$R = \frac{1}{2}(\Pi_+ - \Pi_-) = \Pi_+ - \frac{1}{2} \tag{5.10}$$

where $\Pi_\pm$ are projectors onto $\mathfrak{g}_\pm$, satisfies the classical modified Yang–Baxter equation (5.2) with $\alpha = \frac{1}{4}$, i.e., R is a classical R-matrix.

Next, we specify the dependence of L_j on y via the so-called Lax–Novikov equations (cf. e.g. [6] and references therein)

$$[L_j, L] + (L_j)_y = 0, \qquad j \in \mathbb{N}. \tag{5.11}$$

Then, upon applying (5.6), (5.10) and (5.11), after elementary computations, equations (5.3), (5.5) and (5.8) take the following form:

$$(L_s)_{t_r} = [B_r, L_s], \qquad r, s \in \mathbb{N}, \tag{5.12}$$

$$(B_r)_{t_s} - (B_s)_{t_r} + [B_r, B_s] = 0, \tag{5.13}$$

$$L_{t_r} = [B_r, L] + (B_r)_y, \qquad n, r \in \mathbb{N}, \tag{5.14}$$

where $B_i = \Pi_+ L_i$.

Obviously, if under the reduction to the case when all quantities are independent of y we put $L = L_n$ for some $n \in \mathbb{N}$, then the hierarchies (5.8) boil down to hierarchies (5.3) and the Lax–Novikov equations (5.11) reduce to the commutativity conditions (5.6). In particular, if the bracket $[\cdot,\cdot]$ is such that equations (5.8) give rise to integrable systems in d independent variables, then equations (5.3) yield integrable systems in $d-1$ independent variables.

A standard construction of a commutative subalgebra spanned by L_i whose existence ensures commutativity of the flows (5.3) and (5.8) is, in the case of Lie algebras which admit an additional associative multiplication $\circ$ which obeys the Leibniz rule

$$[a, b \circ c] = [a, b] \circ c + b \circ [a, c], \tag{5.15}$$

as follows: the commutative subalgebra is generated by rational powers of a given element $L \in \mathfrak{g}$, cf. e.g. [35, 7] and references therein. This is also our case for (1+1)D and (2+1)D dispersionless systems, when the Lie algebra in question is a Poisson algebra.

However, in our (3+1)D setting, when the Leibniz rule is no longer required to hold, this construction does not work anymore. In particular, it is the case of (3+1)D dispersionless systems, when the Lie algebra under study is a Jacobi algebra. In order to circumvent this difficulty, instead of an explicit construction of commuting L_i, we will *impose* the zero-curvature constraints (5.5) on chosen elements $L_i \in \mathfrak{g}$, $i \in \mathbb{N}$; it is readily seen that in the case of the Jacobi algebra that we are interested in this can be done in a consistent fashion.

Let us come back to the systems considered in the previous sections. For the (3+1)D case consider a commutative and associative algebra A of formal series in p

$$A \ni f = \sum_i u_i p^i \tag{5.16}$$

with ordinary dot multiplication

$$f_1 \cdot f_2 \equiv f_1 f_2, \qquad f_1, f_2 \in A. \tag{5.17}$$

The coefficients u_i of these series are assumed to be smooth functions of x, y, z and infinitely many times $t_1, t_2, \dots$.

The Jacobi structure on A will be induced by the contact bracket (4.11)

$$[f_1, f_2] \equiv \{f_1, f_2\}_C = \frac{\partial f_1}{\partial p}\frac{\partial f_2}{\partial x} - p\frac{\partial f_1}{\partial p}\frac{\partial f_2}{\partial z} + f_1\frac{\partial f_2}{\partial z} - (f_1 \leftrightarrow f_2). \tag{5.18}$$

Notice that this bracket is independent of y. As the unit element $e = 1$ does not belong to the center of the Jacobi algebra, the Leibniz rule (5.15) does not hold anymore, and instead we have

$$\{f_1 f_2, f_3\}_C = \{f_1, f_3\}_C f_2 + f_1\{f_2, f_3\}_C - f_1 f_2\{1, f_3\}_C. \tag{5.19}$$

For (2+1)D and (1+1)D cases, if we drop the dependence on z or on z and y, this bracket reduces to the canonical Poisson bracket (3.14) in one degree of freedom

$$\{f_1, f_2\}_P = \frac{\partial f_1}{\partial p}\frac{\partial f_2}{\partial x} - \frac{\partial f_2}{\partial p}\frac{\partial f_1}{\partial x} \tag{5.20}$$

and the Jacobi algebra $\mathfrak{g} = (A, \cdot, \{,\}_C)$ reduces to the Poisson algebra $\mathfrak{g} = (A, \cdot, \{,\}_P)$ respectively.

As for the choice of the splitting of the Jacobi algebra $\mathfrak{g} = (A, \cdot, \{,\}_C)$ into Lie subalgebras $\mathfrak{g}_\pm$ with $\Pi_\pm$ being projections onto the respective subalgebras, so that $\mathfrak{g}_\pm = \Pi_\pm(\mathfrak{g})$, it is readily checked that we have two natural choices when the R's defined by (5.10) satisfy the classical modified Yang–Baxter equation (5.2) and thus are R-matrices. These two choices are of the form

$$\Pi_+ = \Pi_{\geqslant k}, \tag{5.21}$$

where $k = 0$ or $k = 1$, and by definition

$$\Pi_{\geqslant k}\left(\sum_{j=-\infty}^{\infty} a_j p^j\right) = \sum_{j=k}^{\infty} a_j p^j.$$

Note that for (1+1)D and (2+1)D systems, associated with the Poisson algebra $\mathfrak{g} = (A, \cdot, \{,\}_P)$, the additional choice of $k = 2$ in (5.21) is also admissible [5, 6].

5.2 Integrable (3+1)D infinite-component hierarchies and their lower-dimensional reductions

We begin with the case of $k = 0$ and the nth order Lax function from A of the form

$$L \equiv L_n = u_n p^n + u_{n-1}p^{n-1} + \cdots + u_0 + u_{-1}p^{-1} + \cdots, \qquad n > 0 \tag{5.22}$$

and let

$$B_m \equiv \Pi_+ L_m = v_{m,m}p^m + v_{m,m-1}p^{m-1} + \cdots + v_{m,0}, \qquad m > 0 \tag{5.23}$$

where $u_i = u_i(x, y, z, \vec{t})$, $v_{m,j} = v_{m,j}(x, y, z, \vec{t})$, and $\vec{t} = (t_1, t_2, \ldots)$.

Substituting L and B_m into the zero-curvature-type equations

$$L_{t_m} = \{B_m, L\}_C + (B_m)_y \tag{5.24}$$

we see that one can impose a natural constraint: $u_n = c_n$, $v_{m,m} = c_{m,m}$, where $c_n, c_{m,m} \in \mathbb{R}$.

Then, if we put $c_n = c_{m,m} = 1$, we get

$$L = p^n + u_{n-1}p^{n-1} + \cdots + u_0 + u_{-1}p^{-1} + \cdots, \quad n > 0, \tag{5.25}$$

$$B_m \equiv \Pi_+ L_m = p^m + v_{m,m-1}p^{m-1} + \cdots + v_{m,0}, \qquad m > 0, \tag{5.26}$$

and equations (5.24) take the form

$$
\begin{aligned}
0 &= X_r^m[u, v_m], \qquad n < r < n+m, \\
(u_r)_{t_m} &= X_r^m[u, v_m], \qquad r \le n, \quad r \ne 0, \dots, m-1, \\
(u_r)_{t_m} &= X_r^m[u, v_m] + (v_{m,r})_y, \qquad r = 0, \dots, m-1,
\end{aligned} \tag{5.27}
$$

where $v_m = (v_{m,0}, \dots, v_{m,m} = 1)$ and

$$
\begin{aligned}
X_r^m[u, v_m] \;=\; & \sum_{s=0}^{m} [s v_{m,s} (u_{r-s+1})_x - (r-s+1) u_{r-s+1} (v_{m,s})_x \\
& -(s-1) v_{m,s} (u_{r-s})_z + (r-s-1) u_{r-s} (v_{m,s})_z],
\end{aligned} \tag{5.28}
$$

where $u_n = 1$ and $u_r = 0$ for $r > n$. The fields u_r for $r \le n$ are dynamical variables while equations for $n+m > r > n$ can be seen as nonlocal constraints on u_r which define the variables $v_{m,s}$. Observe that the additional dependent variables $v_{m,s}$ for different m are by construction related to each other through the zero-curvature equations (5.13).

There is one more constraint in the Lax pair (5.22) and (5.23). The first equation from the system (5.27), i.e., the one for $r = n+m-1$, takes the form

$$
(n-1)(v_{m,m-1})_z - (m-1)(u_{n-1})_z = 0,
$$

so the system under study for $n > 1$ admits a further constraint

$$
v_{m,m-1} = \frac{(m-1)}{(n-1)} u_{n-1}. \tag{5.29}
$$

Thus, the final Lax pair takes the form

$$
L = p^n + u_{n-1} p^{n-1} + \cdots + u_0 + u_{-1} p^{-1} + \cdots, \quad n > 0, \tag{5.30}
$$

$$
B_m = p^m + \tfrac{(m-1)}{(n-1)} u_{n-1} p^{m-1} + \cdots + v_{m,0}, \qquad m > 0. \tag{5.31}
$$

It is readily seen that for $n = 1$ the constraint (5.29) should be replaced by $u_0 = \text{const}$. Let us consider this case in more detail. Upon taking $u_0 = 0$, consider the Lax equation (5.24) for

$$
L = p + u_{-1} p^{-1} + u_{-2} p^{-2} + \cdots, \tag{5.32}
$$

$$
B_m = p^m + v_{m,m-1} p^{m-1} + \cdots + v_{m,1} p + v_{m,0}, \quad m > 0; \tag{5.33}
$$

then the related system reads

$$
\begin{aligned}
0 &= (v_{m,r})_y + X_r^m[u, v_m], \qquad r = 0, \dots, m-1, \\
(u_r)_{t_m} &= X_r^m[u, v_m], \qquad r < 0.
\end{aligned} \tag{5.34}
$$

Thus, the simplest nontrivial case is $m = 2$, so

$$
B_2 = p^2 + v_1 p + v_0
$$

and generates the following infinite-component system [4]

$$\begin{aligned}(v_1)_y &= (v_1)_x + (u_{-1})_z,\\ (v_0)_y &= (v_0)_x + (u_{-2})_z - 2(u_{-1})_x + 2u_{-1}(v_1)_z,\\ (u_r)_{t_2} &= 2(u_{r-1})_x - (u_{r-2})_z - (r+1)u_{r+1}(v_0)_x + v_0(u_r)_z\\ &\quad +(r-1)u_r(v_0)_z + v_1(u_r)_x - ru_r(v_1)_x + (r-2)u_{r-1}(v_1)_z,\end{aligned} \tag{5.35}$$

where $r < 0$ and $v_{2,s} \equiv v_s$, $s = 0, 1$.

We have a natural (2+1)D reduction of (5.35) when u_r, v_0 and v_1 are independent of y,

$$\begin{aligned}0 &= (v_1)_x + (u_{-1})_z,\\ 0 &= (v_0)_x + (u_{-2})_z - 2(u_{-1})_x + 2u_{-1}(v_1)_z,\\ (u_r)_{t_2} &= 2(u_{r-1})_x - (u_{r-2})_z - (r+1)u_{r+1}(v_0)_x + v_0(u_r)_z\\ &\quad +(r-1)u_r(v_0)_z + v_1(u_r)_x - ru_r(v_1)_x + (r-2)u_{r-1}(v_1)_z,\end{aligned} \tag{5.36}$$

another (2+1)D reduction

$$\begin{aligned}&(v_1)_y = (u_{-1})_z,\\ &(v_0)_y = (u_{-2})_z + 2u_{-1}(v_1)_z,\\ &(u_r)_{t_2} = -(u_{r-2})_z + v_0(u_r)_z + (r-1)u_r(v_0)_z + (r-2)u_{r-1}(v_1)_z,\end{aligned} \tag{5.37}$$

when u_r, v_0 and v_1 are independent of x, and yet another (2+1)D reduction

$$\begin{aligned}&(v_1)_y = (v_1)_x,\\ &(v_0)_y = (v_0)_x - 2(u_{-1})_x,\\ &(u_r)_{t_2} = 2(u_{r-1})_x - (r+1)u_{r+1}(v_0)_x + v_1(u_r)_x - ru_r(v_1)_x,\end{aligned} \tag{5.38}$$

when u_r, v_0 and v_1 are independent of z.

Moreover, system (5.38) admits a further reduction $v_1 = 0$ to the form

$$\begin{aligned}&(v_0)_y = (v_0)_x - 2(u_{-1})_x,\\ &(u_r)_{t_2} = 2(u_{r-1})_x - (r+1)u_{r+1}(v_0)_x + v_1(u_r)_x.\end{aligned} \tag{5.39}$$

The system (5.39) reduces to the $(1+1)$-dimensional system

$$(u_r)_{t_2} = 2(u_{r-1})_x - 2(r+1)u_{r+1}(u_{-1})_x, \quad r < 0, \tag{5.40}$$

when u_i are independent of y and we put $v_0 = 2u_{-1}$.

Notice that (5.40) is the well-known (1+1)D Benney system a.k.a. the Benney momentum chain [2, 23]. From that point of view, the systems (5.39) and (5.36) can be seen as natural (2+1)D extensions of the Benney chain while the system (5.35) represents a (3+1)D extension of the Benney system.

On the other hand, system (5.37) admits no reductions to $(1+1)$-dimensional systems. Note that for systems (5.35)–(5.40) there are no obvious finite-component reductions.

For systems (5.22), (5.23) and (5.25), (5.26) we have $(2+1)$-dimensional and $(1+1)$-dimensional reductions of the same types as above.

Now pass to the case of $k=1$, when $\Pi_+ = \Pi_{\geqslant 1}$, and consider the general case when

$$\begin{aligned} L &= u_n p^n + u_{n-1}p^{n-1} + \cdots + u_0 + u_{-1}p^{-1} + \ldots, \quad n>0, \\ B_m &= v_{m,m}p^m + v_{m,m-1}p^{m-1} + \cdots + v_{m,1}p, \quad m>0, \end{aligned} \tag{5.41}$$

from which we again obtain the hierarchies of infinite-component systems

$$\begin{aligned} 0 &= X_r^m[u, v_m], \qquad n < r \le n+m, \\ (u_r)_{t_m} &= X_r^m[u, v_m], \qquad r \le n, \quad r \neq 1, \ldots, m, \\ (u_r)_{t_m} &= X_r^m[u, v_m] + (v_{m,r})_y, \qquad r = 1, \ldots, m, \end{aligned} \tag{5.42}$$

where $v_m = (v_{m,1}, \ldots, v_{m,m})$ and

$$\begin{aligned} X_r^m[u, v_m] &= \sum_{s=1}^{m} [s v_{m,s}(u_{r-s+1})_x - (r-s+1)u_{r-s+1}(v_{m,s})_x \\ &\quad -(s-1)v_{m,s}(u_{r-s})_z + (r-s-1)u_{r-s}(v_{m,s})_z]. \end{aligned} \tag{5.43}$$

For $n>1, m>1$ there is an additional constraint imposed on the Lax pair (5.41). The first equation from the system (5.43), i.e., the one for $r = n+m$, takes the form

$$(n-1)u_n(v_{m,m})_z - (m-1)v_{m,m}(u_n)_z = 0,$$

and hence, for $n>1, m>1$, admits the constraint

$$v_{m,m} = (u_n)^{\frac{m-1}{n-1}}. \tag{5.44}$$

So, the final Lax pair takes the form

$$\begin{aligned} L &= u_n p^n + u_{n-1}p^{n-1} + \cdots + u_0 + u_{-1}p^{-1} + \cdots, \quad n>1, \\ B_m &= (u_n)^{\frac{m-1}{n-1}} p^m + v_{m,m-1}p^{m-1} + \cdots + v_{m,1}p, \quad m>0. \end{aligned} \tag{5.45}$$

For $n=1$ the constraint in question is replaced by $u_1 = \text{const}$. Thus, consider again in detail the simplest case when $n=1$ and $u_1 = 1$

$$\begin{aligned} L &= p + u_0 + u_{-1}p^{-1} + \cdots, \\ B_m &= v_{m,m-1}p^m + v_{m,m-2}p^{m-1} + \cdots + v_{m,1}p, \quad m>0, \end{aligned} \tag{5.46}$$

when the associated system reads

$$\begin{aligned} 0 &= (v_{m,r})_y + X_r^m[u, v_m], \qquad r = 1, \ldots, m, \\ (u_r)_{t_m} &= X_r^m[u, v_m], \qquad r<0. \end{aligned} \tag{5.47}$$

Thus, the simplest nontrivial case is $m = 2$, so $B_2 = v_2 p^2 + v_1 p$ generates the following infinite-component system [4]

$$\begin{aligned} (v_2)_y &= (v_2)_x + u_0 (v_2)_z + v_2 (u_0)_z, \\ (v_1)_y &= (v_1)_x + u_0 (v_1)_z + v_2 (u_{-1})_z + 2u_{-1}(v_2)_z - 2v_2(u_0)_x, \\ (u_r)_{t_2} &= v_1 (u_r)_x - r u_r (v_1)_x + (r-2) u_{r-1} (v_1)_z + 2 v_2 (u_{r-1})_x \\ &\quad -(r-1) u_{r-1} (v_2)_x - v_2 (u_{r-2})_z + (r-3) u_{r-2} (v_2)_z, \end{aligned} \tag{5.48}$$

where $r < 1$ and $v_{2,s} \equiv v_s$, $s = 1, 2$.

We have a natural $(2+1)$-dimensional reduction of (5.48) when u_r, v_1 and v_2 are independent of y,

$$\begin{aligned} 0 &= (v_2)_x + u_0 (v_2)_z + v_2 (u_0)_z, \\ 0 &= (v_1)_x + u_0 (v_1)_z + v_2 (u_{-1})_z + 2u_{-1}(v_2)_z - 2v_2(u_0)_x, \\ (u_r)_{t_2} &= v_1 (u_r)_x - r u_r (v_1)_x + (r-2) u_{r-1} (v_1)_z + 2 v_2 (u_{r-1})_x \\ &\quad -(r-1) u_{r-1} (v_2)_x - v_2 (u_{r-2})_z + (r-3) u_{r-2} (v_2)_z. \end{aligned} \tag{5.49}$$

On the other hand, if u_r, v_1 and v_2 are independent of x, we obtain from (5.48) another $(2+1)$-dimensional system

$$\begin{aligned} &(v_2)_y = u_0 (v_2)_z + v_2 (u_0)_z, \\ &(v_1)_y = u_0 (v_1)_z + v_2 (u_{-1})_z + 2u_{-1}(v_2)_z, \\ &(u_r)_{t_2} = (r-2) u_{r-1} (v_1)_z - v_2 (u_{r-2})_z + (r-3) u_{r-2} (v_2)_z. \end{aligned} \tag{5.50}$$

Next, if u_r, v_1 and v_2 in (5.48) are independent of z, we arrive at the third $(2+1)$-dimensional system

$$\begin{aligned} &(v_2)_y = (v_2)_x, \\ &(v_1)_y = (v_1)_x - 2v_2 (u_0)_x, \\ &(u_r)_{t_2} = v_1 (u_r)_x - r u_r (v_1)_x + 2 v_2 (u_{r-1})_x - (r-1) u_{r-1} (v_2)_x, \end{aligned} \tag{5.51}$$

whence, after the substitution $v_2 = \text{const} = 1$, we obtain

$$\begin{aligned} &(v_1)_y = (v_1)_x - 2(u_0)_x, \\ &(u_r)_{t_2} = v_1 (u_r)_x - r u_r (v_1)_x + 2(u_{r-1})_x. \end{aligned} \tag{5.52}$$

If u_r, v_1 and v_2 are independent of both y and z, we can put $v_1 = 2u_0$ and obtain

$$(u_r)_{t_2} = 2(u_{r-1})_x + 2u_0 (u_r)_x - 2r u_r (u_0)_x. \tag{5.53}$$

Finally, when u_r, v_1 and v_2 are independent of both y and x, we have

$$(u_r)_{t_2} = (r-2) u_{r-1} (v_1)_z - v_2 (u_{r-2})_z + (r-3) u_{r-2} (v_2)_z, \tag{5.54}$$

where a reduction

$$v_2 = a u_0^{-1}, \quad v_1 = -a u_{-1} u_0^{-2},$$

was performed, and $a \in \mathbb{R}$ is an arbitrary constant. Thus, in this case the system under study is rational (rather than polynomial) in u_0.

5.3 Finite-component reductions

For $k=0$, in contrast with the simplest case (5.32), we do have natural reductions to finite-component systems. Namely, they are of the form

$$\begin{aligned} L &= u_n p^n + u_{n-1}p^{n-1} + \cdots + u_r p^r, \quad r=0,1, \\ B_m &= (u_n)^{\frac{m-1}{n-1}} p^m + v_{m,m-1}p^{m-1} + \cdots + v_{m,0}, \end{aligned} \tag{5.55}$$

and

$$\begin{aligned} L &= p^n + u_{n-1}p^{n-1} + \cdots + u_r p^r, \quad r=0,1, \\ B_m &= p^m + \frac{(m-1)}{(n-1)} u_{n-1}p^{m-1} + \cdots + v_{m,0}. \end{aligned} \tag{5.56}$$

The case (5.56) for $r=0$ was considered for the first time in [37] while the remaining cases were analyzed in [4]. Notice that in (5.55) and (5.56) for $r=0$ we have $L=B_n$, and hence the variable y can be identified with t_n. Then equations (5.24) coincide with the zero-curvature equations (5.13), and the Lax–Novikov equation (5.11) reduces to equation (5.12).

Another class of natural reductions to finite-component systems arises for $k=1$ [4]. Indeed, for $n>1$ we have

$$\begin{aligned} L &= u_n p^n + u_{n-1}p^{n-1} + \cdots + u_r p^r, \quad r=1,0,-1,\ldots \\ B_m &= (u_n)^{\frac{m-1}{n-1}} p^m + v_{m,m-1}p^{m-1} + \cdots + v_{m,1}p \end{aligned} \tag{5.57}$$

while for $n=1$

$$\begin{aligned} L &= p + u_0 + u_{-1}p^{-1} + \cdots + u_r p^r, \quad r=0,1,-1,\ldots \\ B_m &= v_{m,m}p^m + v_{m,m-1}p^{m-1} + \cdots + v_{m,1}p, \qquad m>1. \end{aligned} \tag{5.58}$$

In closing we point out a large class of finite-component reductions of the hierarchy associated with (5.25) and (5.26) for $k=0$. The reductions in question for L (5.25) are given by rational Lax functions, cf. [21, 41] and references therein, for the (1+1)D case, namely,

$$L = p^n + \sum_{j=0}^{n-1} u_j p^j + \sum_{i=1}^{k} \frac{a_i}{(p-r_i)}, \qquad n>1, \quad k>0, \tag{5.59}$$

where u_j, a_i and r_i are unknown functions; in this case B_m are still given by (5.26).

Example 2. *First let us begin with the case of* $k=0$ *and the simplest Lax pair from (5.56) when* $n=2$ *and* $m=3$

$$L = p^2 + u_1 p + u_0, \quad B = p^3 + 2u_1 p^2 + v_1 p + v_0, \tag{5.60}$$

The zero-curvature-type Lax equation

$$L_t = \{B, L\}_C + (B)_y$$

generates a four-component system [37]

$$
\begin{aligned}
(u_0)_t &= (v_0)_y + v_0(u_0)_z + v_1(u_1)_x - u_0(v_0)_z - u_1(v_0)_x, \\
(u_1)_t &= (v_1)_y - 2(v_0)_x + 4u_1(u_0)_x - u_1(v_1)_x + v_1(u_1)_x + v_0(u_1)_z - u_0(v_1)_z, \\
0 &= (v_1)_z - (u_1)_x - 2(u_0)_z - 2u_1(u_1)_z, \\
0 &= (v_0)_z + 3(u_0)_x + 2(u_1)_y - 2(v_1)_x + 2u_1(u_1)_x - 2u_1(u_0)_z - 2u_0(u_1)_z,
\end{aligned}
\tag{5.61}
$$

which is a natural (3+1)D extension of the (2+1)D dispersionless Kadomtsev-Petviashvili (dKP) equation. Indeed, upon assuming that all fields are independent of z*, and that* $u_1 = 0$ *and* $v_1 = \frac{3}{2}u_0$*, and denoting* $u_0 \equiv u$ *and* $v_0 \equiv v$*, the system (5.61) reduces to the form*

$$
u_t = v_y + \tfrac{3}{2}u\,u_x, \quad 3u_y = 4v_x \Longrightarrow (u_t - \tfrac{3}{2}u\,u_x)_x = \tfrac{3}{4}u_{yy} \tag{5.62}
$$

where the last equation in (5.62) is, up to a suitable rescaling of independent variables, nothing but the celebrated dKP equation, also known as the three-dimensional Khokhlov–Zabolotskaya [45] equation.

Example 3. *Consider again our system (4.18), which is a particular case of (5.58) for* $k = 1$ *and* $r = -1$*, with notation* $u_{-1} \equiv u_1$*, being the first member of the hierarchy (5.58) generated by the Lax functions*

$$
L = p + u_0 + u_1 p^{-1}, \quad B_2 = v_2 p^2 + v_1 p,
$$

and takes the known form (4.5)

$$
\begin{aligned}
(u_1)_{t_2} &= u_1(v_1)_x + v_1(u_1)_x, \\
(u_0)_{t_2} &= -2u_1(v_1)_z + v_1(u_0)_x + u_1(v_2)_x + 2v_2(u_1)_x, \\
(v_1)_y &= (v_1)_x + 2u_1(v_2)_z + v_2(u_1)_z + u_0(v_1)_z - 2v_2(u_0)_x, \\
(v_2)_y &= (v_2)_x + u_0(v_2)_z + v_2(u_0)_z.
\end{aligned}
\tag{5.63}
$$

The second member of the hierarchy is generated by

$$
L = p + u_0 + u_1 p^{-1}, \quad B_3 = w_3 p^3 + w_2 p^2 + w_1 p
$$

and has the form

$$
\begin{aligned}
(u_1)_{t_3} &= u_1(w_1)_x + w_1(u_1)_x, \\
(u_0)_{t_3} &= w_1(u_0)_x - 2u_1(w_1)_z + u_1(w_2)_x + 2w_2(u_1)_x, \\
(w_1)_y &= (w_1)_x + w_2(u_1)_z - u_1(w_3)_x - 2w_2(u_0)_x + 2u_1(w_2)_z \\
&\quad + u_0(w_1)_z - 3w_3(u_1)_x, \\
(w_2)_y &= (w_2)_x - 3w_3(u_0)_x + 2w_3(u_1)_z + w_2(u_0)_z \\
&\quad + u_0(w_2)_z + 2u_1(w_3)_z, \\
(w_3)_y &= (w_3)_x + u_0(w_3)_z + 2w_3(u_0)_z.
\end{aligned}
\tag{5.64}
$$

Commutativity of the flows associated with t_2 *and* t_3*, i.e.*

$$
((u_i)_{t_2})_{t_3} = ((u_i)_{t_3})_{t_2}, \quad i = 0, 1,
$$

can be checked using the set of relations

$$
\begin{aligned}
(v_1)_z &= -\frac{v_2}{w_3}(w_3)_x - \frac{v_2 w_2}{4w_3^2}(w_3)_z + \frac{v_2}{2w_3}(w_2)_z + \frac{3}{2}(v_2)_x, \quad (v_2)_z = \frac{v_2}{2w_3}(w_3)_z, \\
(w_1)_{t_2} &= v_1(w_1)_x - w_1(v_1)_x + (v_1)_{t_3}, \\
(w_2)_{t_2} &= v_1(w_2)_x - w_1(v_2)_x + 2v_2(w_1)_x - 2w_2(v_1)_x + (v_2)_{t_3}, \\
(w_3)_{t_2} &= \frac{v_2 w_2}{2w_3}(w_2)_z - \frac{w_2}{2}(v_2)_x - \frac{v_2 w_2^2}{4w_3^2}(w_3)_z + \frac{(v_1 w_3 - v_2 w_2)}{w_3}(w_3)_x \\
&\quad -v_2(w_1)_z + 2v_2(w_2)_x - 3w_3(v_1)_x,
\end{aligned}
$$

which is equivalent to the zero-curvature equation

$$
(B_2)_{t_3} - (B_3)_{t_2} + \{B_2, B_3\}_C = 0. \tag{5.65}
$$

Moreover, the compatibility conditions

$$
((v_i)_y)_z = ((v_i)_z)_y, \quad i = 1, 2,
$$

are also satisfied by virtue of (5.63) and (5.65).

When u_r and v_j are independent of z, we obtain $(2+1)D$ systems with additional constraints v_2=const $= \frac{1}{2}$, w_3 = const $= \frac{1}{3}$

$$
\begin{aligned}
(u_1)_{t_2} &= u_1(v_1)_x + v_1(u_1)_x, \\
(u_0)_{t_2} &= v_1(u_0)_x + (u_1)_x, \\
(v_1)_y &= (v_1)_x - (u_0)_x,
\end{aligned}
\qquad
\begin{aligned}
(u_1)_{t_3} &= u_1(w_1)_x + w_1(u_1)_x, \\
(u_0)_{t_3} &= w_1(u_0)_x + u_1(w_2)_x + 2w_2(u_1)_x, \\
(w_1)_y &= (w_1)_x - (u_1)_x - 2w_2(u_0)_x, \\
(w_2)_y &= (w_2)_x - (u_0)_x.
\end{aligned}
\tag{5.66}
$$

see also (3.9).

When u_r and v_j are independent of x, we obtain other $(2+1)D$ systems making use of a naturally arising extra constraint $u_1 = \frac{1}{2}$, namely,

$$
\begin{aligned}
(u_0)_{t_2} &= -(v_1)_z, \\
(v_1)_y &= (v_2)_z + u_0(v_1)_z, \\
(v_2)_y &= (u_0 v_2)_z
\end{aligned}
\qquad
\begin{aligned}
(u_0)_{t_3} &= -(w_1)_z, \\
(w_1)_y &= (w_2)_z + u_0(w_1)_z, \\
(w_2)_y &= (w_3)_z + (u_0 w_2)_z, \\
(w_3)_y &= u_0(w_3)_z + 2w_3(u_0)_z.
\end{aligned}
\tag{5.67}
$$

Further reduction of (5.66) and (5.67) by assuming that u_r, v_j and w_k are independent of y leads to $(1+1)D$ systems of the form

$$
\begin{aligned}
(u_1)_{t_2} &= (u_1 u_0)_x, & (u_1)_{t_3} &= (u_1 u_0^2 + u_1^2)_x, \\
(u_0)_{t_2} &= (u_1 + \tfrac{1}{2}u_0^2)_x, & (u_0)_{t_3} &= (\tfrac{1}{3}u_0^3 + 2u_0 u_1)_x,
\end{aligned}
\tag{5.68}
$$

where we put $v_1 = u_0$, $w_2 = u_0$ and $w_1 = u_0^2 + u_1$.

Likewise, the reduction of (5.67) and (5.68) by assuming that u_r, v_j and w_k are independent of y leads to $(1+1)D$ systems of the form

$$
(u_0)_{t_2} = \tfrac{1}{2}(u_0^{-2})_z, \qquad (u_0)_{t_3} = -\tfrac{3}{4}(u_0^{-4})_z, \tag{5.69}
$$

thanks to the relations

$$
v_2 = u_0^{-1}, \quad v_1 = -u_0^{-2}, \quad w_3 = u_0^{-2}, \quad w_2 = -u_0^{-3}, \quad w_1 = \tfrac{3}{4}u_0^{-4}.
$$

Example 4. *Consider the first member of the hierarchy (5.58) for $r = 1$, generated by the Lax pair*

$$L = up^3 + wp^2 + vp, \quad B_2 = u^{\frac{1}{2}}p^2 + sp,$$

and takes the form

$$\begin{aligned} u_t &= su_x - 3us_x + ws_z + 2u^{\frac{1}{2}}w_x - u^{\frac{1}{2}}v_z - wu^{-\frac{1}{2}}u_x, \\ w_t &= sw_x - 2ws_x + 2u^{\frac{1}{2}}v_x - \tfrac{1}{2}vu^{-\frac{1}{2}}u_x + \tfrac{1}{2}u^{-\frac{1}{2}}u_y, \\ v_t &= v_x + s_y - vs_x, \\ 0 &= 2us_z - u^{\frac{1}{2}}w_z + \tfrac{1}{2}u^{\frac{1}{2}}u_x + \tfrac{1}{2}wu^{\frac{1}{2}}u_z, \end{aligned} \tag{5.70}$$

where we put $u_3 = u$, $u_2 = w$, $u_1 = v$, $v_{2,1} = s$ and $t_2 = t$. The $(2+1)D$ reduction with all fields independent of z and $u = 1$, $s = \frac{2}{3}w$ reads

$$\begin{aligned} w_t &= 2w_x - \tfrac{2}{3}ww_x, \\ v_t &= v_x - \tfrac{2}{3}vw_x + \tfrac{2}{3}w_y, \end{aligned} \tag{5.71}$$

while the case when all fields are independent of x takes the form

$$\begin{aligned} u_t &= ws_z - u^{\frac{1}{2}}v_z, \\ w_t &= (u^{\frac{1}{2}})_y, \\ v_t &= s_y, \\ 0 &= 2us_z - u^{\frac{1}{2}}w_z + w(u^{\frac{1}{2}})_z, \end{aligned} \tag{5.72}$$

or equivalently

$$\begin{aligned} &2a_ta_{tt} + a_tb_{yz} - a_yb_{tz} = 0, \\ &2a_t^2b_{tz} + a_ya_{tz} - a_ta_{yz} = 0, \end{aligned} \tag{5.73}$$

where $a_t = u^{\frac{1}{2}}$, $a_y = w$, $b_t = s$, $b_y = v$. The $(1+1)D$ reductions of (5.71) and (5.72), when all fields are additionally independent of y, take the form

$$\begin{aligned} w_t &= 2v_x - \tfrac{2}{3}ww_x, \\ v_t &= v_x - \tfrac{2}{3}vw_x, \end{aligned}$$

and

$$u_t + u^{-\frac{3}{2}}u_z = 0,$$

where $v = 0$ end $w = 2$.

Acknowledgements

The research of AS was supported in part by the Ministry of Education, Youth and Sport of the Czech Republic (MŠMT ČR) under RVO funding for IČ47813059 and the Grant Agency of the Czech Republic (GA ČR) under grant P201/12/G028.

AS gratefully acknowledges the warm hospitality extended to him in the course of his visit to the Adam Mickiewicz University where a substantial part of the present paper was written.

A number of computations in the present paper were performed using the software *Jets* for Maple [1] whose use is acknowledged with gratitude.

The authors are pleased to thank B.M. Szablikowski for helpful comments. AS also thanks A. Borowiec and R. Vitolo for stimulating discussions.

References

[1] Baran H, Marvan M, *Jets. A software for differential calculus on jet spaces and diffieties*, `http://jets.math.slu.cz`

[2] Benney D J, Some properties of long nonlinear waves, *Stud. Appl. Math.* **52**, 45–50, 1973.

[3] Błaszak M, Classical R-matrices on Poisson algebras and related dispersionless systems, *Phys. Lett. A* **297**, no. 3-4, 191–195, 2002.

[4] Błaszak M, Sergyeyev A, Dispersionless (3+1)-dimensional integrable hierarchies, *Proc. R. Soc. A.* **473**, no. 2201, 20160857, 2017, arXiv:1605.07592.

[5] Błaszak M, Szablikowski B, Classical R-matrix theory of dispersionless systems. I. (1+1) dimension theory, *J. Phys. A: Math. Gen.* **35**, no. 48, 10325–10345, 2002, arXiv:nlin/0211008.

[6] Błaszak M, Szablikowski B, Classical R-matrix theory of dispersionless systems. II. (2+1) dimension theory, *J. Phys. A: Math. Gen.* **35**, no. 48, 10345–10364, 2002, arXiv:nlin/0211018.

[7] Błaszak M, Szablikowski B, Classical R-matrix theory for bi-Hamiltonian field systems, J. Phys. A: Math. Theor. 42 (2009) 404002

[8] Bravetti A, Contact Hamiltonian dynamics: the concept and its use, *Entropy* **19**, no. 10, art. 535, 2017.

[9] Burtsev S P, Zakharov V E, Mikhailov A V, Inverse scattering method with variable spectral parameter, *Theor. Math. Phys.* **70**, no. 3, 227–240, 1987.

[10] Calderbank D M J, Kruglikov B, Integrability via geometry: dispersionless equations in three and four dimensions, preprint arXiv:1612.02753

[11] Dubrovin B A, Novikov S P, The Hamiltonian formalism of one-dimensional systems of hydrodynamic type and the Bogolyubov–Whitham averaging method, *Sov. Math. Dokl.* **270**, no. 4, 665–669, 1983.

[12] Dunajski M, Anti-self-dual four-manifolds with a parallel real spinor, *Proc. R. Soc. A* **458**, 1205–1222, 2002, arXiv:math/0102225

[13] Dunajski M, Ferapontov E V and Kruglikov B, On the Einstein–Weyl and conformal self-duality equations, *J. Math. Phys.* **56**, no. 8, art. 083501, 2015, arXiv:1406.0018.

[14] Dunajski M, *Solitons, Instantons and Twistors*, Oxford Univ. Press, Oxford, 2010.

[15] Ferapontov E V, Khusnutdinova K R, Klein C, On linear degeneracy of integrable quasilinear systems in higher dimensions, *Lett. Math. Phys.* **96**, no. 1-3, 5–35, 2011.

[16] Ferapontov E, Kruglikov B, Dispersionless integrable systems in 3D and Einstein–Weyl geometry, *J. Diff. Geom.* **97**, 215–254, 2014, arXiv:1208.2728.

[17] Ferapontov E V, Moro A, Novikov V S, Integrable equations in $2+1$ dimensions: deformations of dispersionless limits, *J. Phys. A: Math. Gen.* **42**, no. 34, art. 345205, 2009.

[18] Ferapontov E V, private communication.

[19] Fokas A S, Symmetries and integrability, *Stud. Appl. Math.* **77**, 253–299, 1987.

[20] Konopelchenko B G, Martínez Alonso L, Dispersionless scalar integrable hierarchies, Whitham hierarchy, and the quasiclassical $\bar{\partial}$-dressing method, *J. Math. Phys.* **43**, no. 7, 3807–3823, 2002, arXiv:nlin/0105071.

[21] Krichever I M, The τ-function of the universal Whitham hierarchy, matrix models and topological field theories, *Comm. Pure Appl. Math.* **47**, 437–475, 1994.

[22] Kruglikov B, Morozov O, Integrable dispersionless PDEs in 4D, their symmetry pseudogroups and deformations, *Lett. Math. Phys.* **105**, no. 12, 1703–1723, 2015, arXiv:1410.7104.

[23] Kupershmidt B A, Manin Yu I, Long wave equations with a free surface. I. Conservation laws and solutions, *Funkt. Anal. i Pril.* **11**, no. 3, 31–42, 1977; Kupershmidt B A, Manin Yu I, Long wave equations with a free surface. II. The Hamiltonian structure and the higher equations, *Funkt. Anal. i Pril.* **12**, no. 1, 25–37, 1978.

[24] Li L-C, Classical r-matrices and compatible Poisson structures for Lax equations in Poisson algebras, *Commun. Math. Phys.* **203** 573–592, 1999.

[25] Lichnerowicz A, Les variétés de Jacobi et leurs algebres de Lie associées, *J. Math. Pures Appl.* **57**, no. 4, 453–488, 1978.

[26] Majda A J, *Compressible Fluid Flows and Systems of Conservation Laws in Several Space Variables*, Springer, N.Y., 1984.

[27] Manakov S V, Santini P M, Integrable dispersionless PDEs arising as commutation condition of pairs of vector fields, *J. Phys.: Conf. Ser.* **482**, paper 012029, 2014, arXiv:1312.2740.

[28] Manakov S V, Santini P M, Solvable vector nonlinear Riemann problems, exact implicit solutions of dispersionless PDEs and wave breaking, *J. Phys. A: Math. Theor.* **44**, no. 34, paper 345203, 2011, arXiv:1011.2619.

[29] Marvan M, Another look on recursion operators, in *Differential Geometry and Applications (Brno, 1995)*, Masaryk Univ., Brno, 1996, 393–402.

[30] Marvan M, Sergyeyev A, Recursion operators for dispersionless integrable systems in any dimension, *Inverse Problems* **28**, art. 025011, 2012.

[31] Morozov O I, Sergyeyev A, The four-dimensional Martínez Alonso–Shabat equation: reductions and nonlocal symmetries, *J. Geom. Phys.* **85**, 40–45, 2014, arXiv:1401.7942

[32] Odesskii A V, Sokolov V V, Integrable pseudopotentials related to generalized hypergeometric functions, *Sel. Math. (N.S.)* **16**, 145–172, 2010, arXiv:0803.0086

[33] Olver P J, *Applications of Lie Groups to Differential Equations*, 2nd ed., Springer, New York, 2000.

[34] Semenov-Tian-Shansky M A, What is a classical r-matrix?, *Func. Anal. Appl.* **17** (1983), 259–272.

[35] Semenov-Tian-Shansky M, Integrable systems: the r-matrix approach, Preprint RIMS-1650, Kyoto, 2008.

[36] Sergyeyev A, A simple construction of recursion operators for multidimensional dispersionless integrable systems. *J. Math. Anal. Appl.* **454**, 468–480, 2017, arXiv:1501.01955.

[37] Sergyeyev A, New integrable (3+1)-dimensional systems and contact geometry, *Lett. Math. Phys.* **108**, no. 2, 359–376, 2018, arXiv:1401.2122

[38] Sergyeyev A, Integrable (3+1)-dimensional systems with rational Lax pairs, *Nonlinear Dynamics* 91 (2018), no. 3, 1677–1680, arXiv:1711.07395

[39] Sergyeyev A, Integrable (3+1)-dimensional system with an algebraic Lax pair, *Appl. Math. Lett.* 92 (2019) 196-200, arXiv:1812.02263.

[40] Sergyeyev A, Recursion operators for multidimensional integrable PDEs, arXiv:1710.05907.

[41] Szablikowski B M, Błaszak M, Meromorphic Lax representations of (1+1)-dimensional multi-Hamiltonian dispersionless systems, *J. Math. Phys.* **47**, 092701, 2006, arXiv:nlin/0510068

[42] Szablikowski B M, Hierarchies of Manakov-Santini type by means of Rota-Baxter and other identities, *SIGMA* **12**, art. 022, 2016.

[43] Takasaki K, Takebe T, Integrable hierarchies and dispersionless limit, *Rev. Math. Phys.* **7** (1995), 743–808, arXiv:hep-th/9405096.

[44] Vaisman I, *Lectures on the Geometry of Poisson Manifolds*, Birkhäuser Verlag, Basel, 1994.

[45] Zabolotskaya E A, Khokhlov R V, Quasi-plane waves in the nonlinear acoustics of confined beams, *Sov. Phys. Acoust.* **15**, 35–40, 1969.

[46] Zakharov V E, Multidimensional integrable systems, in *Proceedings of the International Congress of Mathematicians, Vol. 1, 2 (Warsaw, 1983)*, PWN, Warsaw, 1984, 1225–1243.

[47] Zakharov V E, Dispersionless limit of integrable systems in (2+1) dimensions, in *Singular Limits of Dispersive Waves*, ed. by N.M. Ercolani et al., Plenum Press, N.Y., 1994, 165–174.

[48] Zarraga J A, Perelomov A M and Perez Bueno J C, The Schouten–Nijenhuis bracket, cohomology and generalized Poisson structures, *J. Phys. A: Math. Gen.* **29**, 7993–8009, 1996.

A3. Lax pairs for edge-constrained Boussinesq systems of partial difference equations

Terry J. Bridgman and Willy Hereman

Department of Applied Mathematics and Statistics,
Colorado School of Mines, Golden, CO.
e-mail: tbridgma@mines.edu and whereman@mines.edu.

Abstract

The method due to Nijhoff and Bobenko & Suris to derive Lax pairs for partial difference equations (PΔEs) is applied to edge constrained Boussinesq systems. These systems are defined on a quadrilateral. They are consistent around the cube but they contain equations defined on the edges of the quadrilateral.

By properly incorporating the edge equations into the algorithm, it is straightforward to derive Lax matrices of minimal size. The 3 by 3 Lax matrices thus obtained are not unique but shown to be gauge-equivalent. The gauge matrices connecting the various Lax matrices are presented. It is also shown that each of the Boussinesq systems admits a 4 by 4 Lax matrix. For each system, the gauge-like transformations between Lax matrices of different sizes are explicitly given. To illustrate the analogy between continuous and lattice systems, the concept of gauge-equivalence of Lax pairs of nonlinear partial differential equations is briefly discussed.

The method to find Lax pairs of PΔEs is algorithmic and is being implemented in MATHEMATICA. The Lax pair computations for this chapter helped further improve and extend the capabilities of the software under development.

1 Introduction

As discussed by Hietarinta [16] in volume 1 of this book series, Nijhoff and Capel [27], and Bridgman [9], nonlinear partial "discrete or lattice" equations (PΔEs), arise in various contexts. They appeared early on in papers by Hirota [18] covering a soliton preserving discretization of the direct (bilinear) method for nonlinear PDEs (see, e.g., [19]). In addition, Miura [23] and Wahlquist & Estabrook [34] indirectly contributed to the development of the theory of PΔEs through their work on Bäcklund transformations. A major contribution to the study of PΔEs came from Nijhoff and colleagues [28, 30, 31]. Under the supervision of Capel, the Dutch research group used a direct linearization method and Bäcklund transformations, in connection with a discretization of the plane wave factor, to derive several PΔEs. For a detailed discussion of these methods as well as the seminal classification of scalar PΔEs by Adler *et al.* [5, 6] we refer to recent books on the subject [7, 17, 20, 21, 22].

To settle on notation, let us first consider a single *scalar* PΔE,

$$\mathcal{F}(u_{n,m}, u_{n+1,m}, u_{n,m+1}, u_{n+1,m+1}; p, q) = 0, \tag{1.1}$$

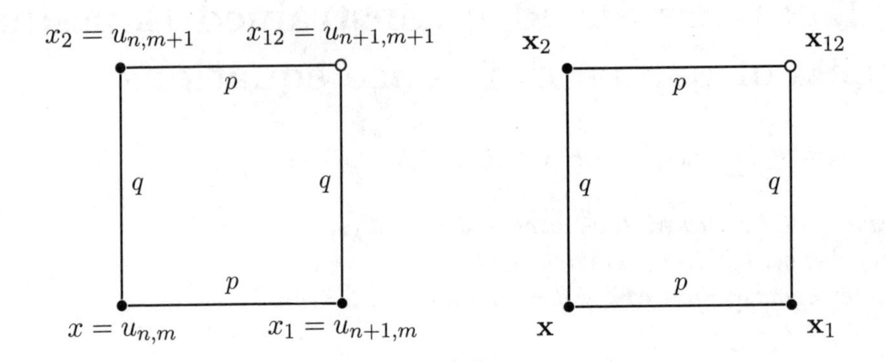

Figure 1: The PΔE is defined on the simplest quadrilateral (a square). single equation or scalar case (left), system or vector case (right).

which is defined on a 2-dimensional quad-graph as shown in Fig. 1 (left).

The one field variable $x \equiv u_{n,m}$ depends on lattice variables n and m. A shift of x in the horizontal direction (the one-direction) is denoted by $x_1 \equiv u_{n+1,m}$. A shift in the vertical or two-direction by $x_2 \equiv u_{n,m+1}$ and a shift in both directions by $x_{12} \equiv u_{n+1,m+1}$. Furthermore, the nonlinear function $\mathcal{F}$ depends on the lattice parameters p and q which correspond to the edges of the quadrilateral. In our simplified notation (1.1) is replaced by

$$\mathcal{F}(x, x_1, x_2, x_{12}; p, q) = 0. \tag{1.2}$$

Alternate notations are used in the literature. For instance, many authors denote (x, x_1, x_2, x_{12}) by $(x, \tilde{x}, \hat{x}, \hat{\tilde{x}})$ while others use $(x_{00}, x_{10}, x_{01}, x_{11})$.

As a well-studied example, consider the integrable lattice version of the potential Korteweg-de Vries equation [27], written in the various notations:

$$(p - q + u_{n,m+1} - u_{n+1,m})(p + q - u_{n+1,m+1} + u_{n,m}) = p^2 - q^2, \tag{1.3a}$$

$$(p - q + u_{01} - u_{10})(p + q - u_{11} + u_{00}) = p^2 - q^2, \tag{1.3b}$$

$$(p - q + \hat{u} - \tilde{u})(p + q - \hat{\tilde{u}} + u) = p^2 - q^2, \tag{1.3c}$$

or, in the notation used throughout this chapter,

$$(p - q + x_2 - x_1)(p + q - x_{12} + x) = p^2 - q^2. \tag{1.3d}$$

When dealing with systems of PΔEs, instead of having one field variable $u_{n,m}$, there are multiple field variables, e.g., $u_{n,m}$, $v_{n,m}$, and $w_{n,m}$, which we will denote by x, y, z. Consequently, the scalar equation (1.2) is replaced by a multi-component system involving a *vector* function $\boldsymbol{\mathcal{F}}$ which depends on field variable $\mathbf{x} \equiv (x, y, z)$ and its shifts denoted by $\mathbf{x}_1, \mathbf{x}_2$, and $\mathbf{x}_{12}$. Again, we restrict ourselves to equations defined on the quadrilateral depicted in Fig. 1 (right) which is the "vector" version of the figure on the left. We assume that the initial values (indicated by solid circles) for $\mathbf{x}, \mathbf{x}_1$ and $\mathbf{x}_2$ can be specified and that the value of $\mathbf{x}_{12}$ (indicated by an open circle) can be uniquely determined. To achieve this we require that $\mathcal{F}$ is *affine linear* (multi-linear) in the field variables. Eq. (1.3) is an example of a class [5] of scalar PΔEs which are consistent around the cube, a property that plays an important role in this chapter.

As an example of a system of PΔEs, consider the Schwarzian Boussinesq system [15, 24, 25],

$$zy_1 - x_1 + x = 0, \quad zy_2 - x_2 + x = 0, \quad \text{and} \tag{1.4a}$$

$$(z_1 - z_2)z_{12} - \frac{z}{y}(py_1z_2 - qy_2z_1) = 0, \tag{1.4b}$$

where, for simplicity of notation, p^3 and q^3 were replaced with p and q, respectively. Eq. (1.4b) is relating the four corners of the quadrilateral. Both equations in (1.4a) are not defined on the full quadrilateral. Each is restricted to a single edge of the quadrilateral. The first equation is defined on the edge connecting $\mathbf{x}$ and $\mathbf{x}_1$; the second on the edge connecting $\mathbf{x}$ and $\mathbf{x}_2$.

As we will see in Section 3.1, equations like (1.3) and (1.4) are very special. Indeed, they are multi-dimensionally consistent; a property which is inherently connected to the existence and derivation of Lax pairs.

As in the case of nonlinear partial differential equations (PDEs), one of the fundamental characterizations of *integrable* nonlinear PΔEs is the existence of a Lax pair, i.e., an associated matrix system of two *linear* difference equations for an auxiliary vector-valued function. The original nonlinear PΔE then arises by expressing the compatibility condition of that linear system via a commutative diagram.

Lax pairs for PΔEs first appeared in work by Ablowitz and Ladik [3, 4] for a discrete nonlinear Schrödinger equation, and subsequently in [30] for other equations. The existence (and construction) of a Lax pair is closely related to the so-called *consistency around the cube* (CAC) property which was (much later) proposed independently by Nijhoff [26] and Bobenko and Suris [6]. CAC is a special case of *multi-dimensional consistency* which is nowadays used as a key criterion to define integrability of PΔEs.

In contrast to the PDE case, there exists a straightforward, algorithmic approach to derive Lax pairs [6, 26] for scalar PΔEs that are consistent around the cube, i.e., 3D consistent. The algorithm was presented in [9, 12] and has been implemented in MATHEMATICA [8].

The implementation of the algorithm for systems of PΔEs [9, 12] is more subtle, in particular, when edge equations are present in the systems. In the latter case, the algorithm produces gauge equivalent Lax matrices which depend on the way the edge constraints are dealt with. As illustrated for (1.4), incorporating the edge constraints into the calculation of Lax pairs produces 3 by 3 matrices. Not using the edge constraints also leads to valid Lax pairs involving 4 by 4 matrices which are gauge-like equivalent with their 3 by 3 counterparts. This was first observed [10] when computing Lax pairs for systems of PΔEs presented in Zhang *et al.* [35]. Using the so-called direct linearization method, Zhang and collaborators have obtained 4 by 4 Lax matrices for generalizations of Boussinesq systems derived by Hietarinta

[15]. In Section 5 we show the gauge-like transformations that connect these Lax matrices with the smaller size ones presented in [9, 10].

To keep the article self-contained, in the next section we briefly discuss the concept of *gauge-equivalent Lax pairs* for nonlinear PDEs and draw the analogy between the continuous and discrete cases. The rest of the chapter is organized as follows. Section 3 has a detailed discussion of the algorithm to compute Lax pairs with its various options. The leading example is the Schwarzian Boussinesq system for which various Lax pairs are computed. The gauge and gauge-like equivalences of these Lax matrices is discussed in Section 4. In Section 5 the algorithm is applied to the generalized Hietarinta systems featured in [35]. A summary of the results is given in Section 6. The chapter ends with a brief discussion of the software implementation and conclusions in Section 7.

2 Gauge equivalence of Lax pairs for PDEs and PΔEs

In this section we show the analogy between Lax pairs for continuous equations (PDEs) and lattice equations (PΔEs). We also introduce the concept of gauge equivalence in both cases.

2.1 Lax pairs for nonlinear PDEs

A completely integrable nonlinear PDE can be associated with a system of linear PDEs in an auxiliary function Φ. The compatibility of these linear PDEs requires that the original nonlinear PDE is satisfied.

Using the matrix formalism described in [2], we can replace a given nonlinear PDE with a linear system,

$$\Phi_x = \mathbf{X}\Phi \quad \text{and} \quad \Phi_t = \mathbf{T}\Phi, \tag{2.1}$$

with vector function $\Phi(x,t)$ and unknown matrices $\mathbf{X}$ and $\mathbf{T}$. Requiring that the equations in (2.1) are compatible, that is requiring that $\Phi_{xt} = \Phi_{tx}$, readily [14] leads to the (matrix) *Lax equation* (also known as the zero curvature condition) to be satisfied by the *Lax pair* $(\mathbf{X}, \mathbf{T})$:

$$\mathbf{X}_t - \mathbf{T}_x + [\mathbf{X}, \mathbf{T}] \doteq \mathbf{0}, \tag{2.2}$$

where $[\mathbf{X}, \mathbf{T}] := \mathbf{XT} - \mathbf{TX}$ is the matrix commutator and $\doteq$ denotes that the equation holds for solutions of the given nonlinear PDE. Finding the *Lax matrices* $\mathbf{X}$ and $\mathbf{T}$ for a nonlinear PDE (or system of PDEs) is a nontrivial task for which to date no algorithm is available.

Consider, for example, the ubiquitous Korteweg-de Vries (KdV) equation [1],

$$u_t + \alpha\, u u_x + u_{xxx} = 0, \tag{2.3}$$

where α is any non-zero real constant. It is well known (see, e.g., [14]) that

$$\mathbf{X} = \begin{bmatrix} 0 & 1 \\ \lambda - \frac{1}{6}\alpha\, u & 0 \end{bmatrix} \tag{2.4a}$$

and

$$\mathbf{T} = \begin{bmatrix} \frac{1}{6}\alpha\, u_x & -4\lambda - \frac{1}{3}\alpha\, u \\ -4\lambda^2 + \frac{1}{3}\alpha\,\lambda\, u + \frac{1}{18}\alpha^2\, u^2 + \frac{1}{6}\alpha\, u_{xx} & -\frac{1}{6}\alpha\, u_x \end{bmatrix} \tag{2.4b}$$

form a Lax pair for (2.3). In this example $\Phi = \begin{bmatrix}\psi & \psi_x\end{bmatrix}^{\mathrm{T}}$, where T denotes the transpose, and $\psi(x,t)$ is the scalar eigenfunction of the Schrödinger equation,

$$\psi_{xx} - (\lambda - \tfrac{1}{6}\alpha\, u)\psi = 0, \tag{2.5}$$

with eigenvalue λ and potential proportional to $u(x,t)$.

It has been shown [13, p. 22] that if $(\mathbf{X}, \mathbf{T})$ is a Lax pair, then so is $(\tilde{\mathbf{X}}, \tilde{\mathbf{T}})$ where

$$\tilde{\mathbf{X}} = \mathbf{G}\mathbf{X}\mathbf{G}^{-1} + \mathbf{G}_x\mathbf{G}^{-1} \quad \text{and} \quad \tilde{\mathbf{T}} = \mathbf{G}\mathbf{T}\mathbf{G}^{-1} + \mathbf{G}_t\mathbf{G}^{-1}, \tag{2.6}$$

for an arbitrary invertible matrix $\mathbf{G}$ of the correct size. The above transformation comes from changing Φ in (2.1) into $\tilde{\Phi} = \mathbf{G}\Phi$ and requiring that $\tilde{\Phi}_x = \tilde{\mathbf{X}}\tilde{\Phi}$ and $\tilde{\Phi}_t = \tilde{\mathbf{T}}\tilde{\Phi}$.

In physics, transformations like (2.6) are called *gauge transformations.* Obviously, a Lax pair for a given PDE is not unique. In fact, there exists an infinite number of Lax pairs which are *gauge equivalent* through (2.6).

In the case of the KdV equation, for example using the gauge matrix

$$\mathbf{G} = \begin{bmatrix} -ik & 1 \\ -1 & 0 \end{bmatrix}, \tag{2.7}$$

we see that (2.4) is gauge equivalent to the Lax pair,

$$\tilde{\mathbf{X}} = \begin{bmatrix} -ik & \frac{1}{6}\alpha\, u \\ -1 & ik \end{bmatrix} \tag{2.8a}$$

and

$$\tilde{\mathbf{T}} = \begin{bmatrix} -4ik^3 + \frac{1}{3}i\alpha\, k\, u - \frac{1}{6}\alpha\, u_x & \frac{1}{3}\alpha\left(2k^2\, u - \frac{1}{6}\alpha\, u^2 + ik\, u_x - \frac{1}{2}u_{xx}\right) \\ -4k^2 + \frac{1}{3}\alpha\, u & 4ik^3 - \frac{1}{3}i\alpha\, k\, u + \frac{1}{6}\alpha\, u_x \end{bmatrix}, \tag{2.8b}$$

where $\lambda = -k^2$. The latter Lax matrices are complex matrices. However, in (2.8a) the eigenvalue k appears in the diagonal entries which is advantageous if one applies the Inverse Scattering Transform (IST) to solve the initial value problem for the KdV equation.

2.2 Lax pairs for nonlinear PΔEs

Analogous with the definition of Lax pairs (in matrix form) for PDEs, a Lax pair for a nonlinear PΔE is a pair of matrices, (L, M), such that the compatibility of the linear system,

$$\psi_1 = L\psi \quad \text{and} \quad \psi_2 = M\psi, \tag{2.9}$$

for an auxiliary vector function ψ, requires that the nonlinear PΔE is satisfied. The crux is to find suitable *Lax matrices* L and M so that the nonlinear PΔE can be replaced by (2.9). As shown in the (Bianchi-type) commutative diagram depicted

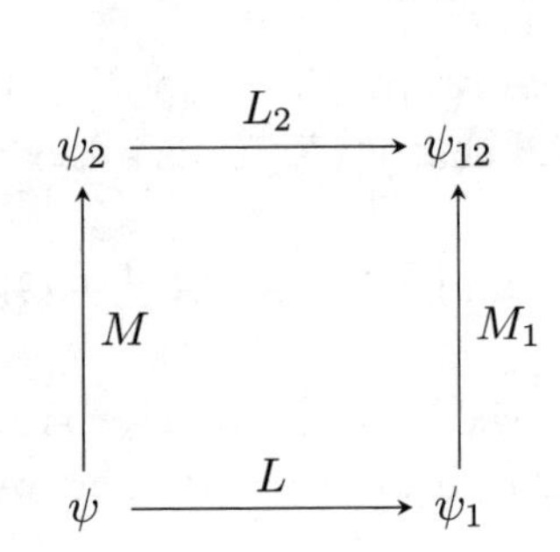

Figure 2: Commutative diagram resulting in the Lax equation.

in Fig. 2, the compatibility of (2.9) can be readily expressed by shifting both sides of $\psi_1 = L\psi$ in the two-direction, i.e., $\psi_{12} = L_2\psi_2 = L_2M\psi$, and shifting $\psi_2 = M\psi$ in the one-direction, i.e., $\psi_{21} = \psi_{12} = M_1\psi_1 = M_1L\psi$, and equating the results. Hence,

$$L_2M - M_1L \doteq 0, \tag{2.10}$$

where $\doteq$ denotes that the equation holds for *solutions* of the PΔE. In other words, the left-hand side of (2.10) should generate the PΔE and not be satisfied automatically as this would result in a "fake" Lax pair. In analogy to (2.2), equation (2.10) is called the Lax equation (or zero-curvature condition).

As in the continuous case, there is an infinite number of Lax matrices, all equivalent to each other under gauge transformations [12]. Specifically, if (L, M) is a Lax pair then so is $(\tilde{L}, \tilde{M})$ where

$$\tilde{L} = \mathcal{G}_1 L \mathcal{G}^{-1} \quad \text{and} \quad \tilde{M} = \mathcal{G}_2 M \mathcal{G}^{-1}, \tag{2.11}$$

for any arbitrary invertible matrix $\mathcal{G}$. Gauge transformation (2.11) comes from setting $\tilde{\psi} = \mathcal{G}\psi$ and requiring that $\tilde{\psi}_1 = \tilde{L}\tilde{\psi}$ and $\tilde{\psi}_2 = \tilde{M}\tilde{\psi}$.

Although (2.11) insures the existence of an infinite number of Lax matrices, it does not say how to find $\mathcal{G}$ of any two Lax matrices (which might have been derived with different methods). As we shall see, there are systems of PΔEs with Lax matrices whose gauge equivalence is presently unclear.

3 Derivation of Lax pairs for Boussinesq systems

3.1 Derivation of Lax pairs for the Schwarzian Boussinesq system

Consistency around the cube

The key idea of multi-dimensional consistency is to (i) extend the planar quadrilateral (square) to a cube by artificially introducing a third direction (with lattice parameter k) as shown in Fig. 3, (ii) impose copies of the same system, albeit with different lattice parameters, on the different faces and edges of the cube, and (iii) view the cube as a three-dimensional commutative diagram for $\mathbf{x}_{123}$. Although not explicitly shown in Fig. 3, parallel edges carry the same lattice parameter.

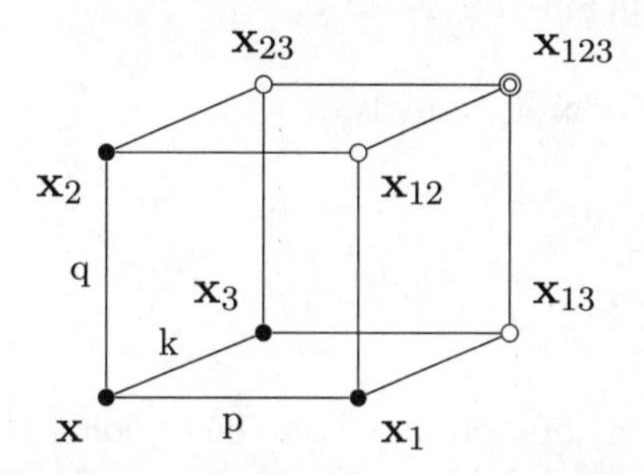

Figure 3: The system of PΔEs holds on each face of the cube.

As shown in Fig. 3, the planar quadrilateral is extended into the third dimension where k is the lattice parameter along the edge connecting $\mathbf{x}$ and $\mathbf{x}_3$. Although not explicitly shown in Fig. 3, all parallel edges carry the same lattice parameters.

With regard to (1.4), we impose that the same equations hold on all faces of the cube. The equations on the *bottom* face follow from a rotation of the *front* face along the horizontal axis connecting $\mathbf{x}$ and $\mathbf{x}_1$. Therefore, applying the substitutions $\mathbf{x}_2 \to \mathbf{x}_3$, $\mathbf{x}_{12} \to \mathbf{x}_{13}$, and $q \to k$ to (1.4), yields

$$zy_1 - x_1 + x = 0, \quad zy_3 - x_3 + x = 0, \quad \text{and} \tag{3.1a}$$

$$(z_1 - z_3)z_{13} - \frac{z}{y}(py_1 z_3 - ky_3 z_1) = 0, \tag{3.1b}$$

which visually corresponds to "folding" the front face down into the bottom face.

Likewise, the equations on the *left* face can be obtained via a rotation of the front face along the vertical axis connecting $\mathbf{x}$ and $\mathbf{x}_2$ (over ninety degrees counterclockwise from a bird's eye view). This amounts to applying the substitutions $\mathbf{x}_1 \to \mathbf{x}_3$, $\mathbf{x}_{12} \to \mathbf{x}_{23}$, and $p \to k$ to (1.4), yielding

$$zy_3 - x_3 + x = 0, \quad zy_2 - x_2 + x = 0, \quad \text{and} \tag{3.2a}$$

$$(z_3 - z_2)z_{23} - \frac{z}{y}(ky_3 z_2 - qy_2 z_3) = 0. \tag{3.2b}$$

The equations on the *back* face follow from a shift of (1.4) in the third direction letting $\mathbf{x} \to \mathbf{x}_3$, $\mathbf{x}_1 \to \mathbf{x}_{13}$, $\mathbf{x}_2 \to \mathbf{x}_{23}$, and $\mathbf{x}_{12} \to \mathbf{x}_{123}$. Likewise, the equations on the *top* and *right* faces follow from (3.1) and (3.2) by shifts in the two- and one-directions, respectively.

Gathering the equations (from all six faces) yields a system of 15 equations (after removing the three duplicates). For example, the five equations that reside on the *bottom* face (with corners $\mathbf{x}, \mathbf{x}_1, \mathbf{x}_3$, and $\mathbf{x}_{13}$) are

$$z y_1 - x_1 + x = 0, \tag{3.3a}$$
$$z y_3 - x_3 + x = 0, \tag{3.3b}$$
$$z_3 y_{13} - x_{13} + x_3 = 0, \tag{3.3c}$$
$$z_1 y_{13} - x_{13} + x_1 = 0, \tag{3.3d}$$
$$(z_1 - z_3) z_{13} - \frac{z}{y}\left(p y_1 z_3 - k y_3 z_1\right) = 0, \tag{3.3e}$$

yielding the components of $\mathbf{x}_{13}$, namely,

$$x_{13} = \frac{x_3 z_1 - x_1 z_3}{z_1 - z_3}, \quad y_{13} = \frac{x_3 - x_1}{z_1 - z_3}, \quad \text{and} \tag{3.4a}$$
$$z_{13} = \frac{z}{y}\left(\frac{p y_1 z_3 - k y_3 z_1}{z_1 - z_3}\right). \tag{3.4b}$$

Likewise, solving the equations on the front face yields the components of $\mathbf{x}_{12}$ and the equations on the left face yield the components of $\mathbf{x}_{23}$. The components of $\mathbf{x}_{123}$ can be computed using either the equations on the top face or those on the right or back faces.

Multi-dimensional consistency around the cube of the PΔE system requires that one can uniquely determine $\mathbf{x}_{123} = (x_{123}, y_{123}, z_{123})$ and that all expressions coincide, no matter which face is used (or, equivalently, no matter which path along the cube is taken to get to the corner $\mathbf{x}_{123}$). Using straightforward, yet tedious algebra, one can show [12] that (1.4) is multi-dimensionally consistent around the cube. As discussed in [7], three-dimensional consistency of a system of PΔEs establishes its complete integrability for it allows one to algorithmically compute a Lax pair.

Computation of Lax pairs

The derivation of a Lax pair for (1.4) starts with introducing *projective* variables f, g, h, F, G, and H by

$$x_3 = \frac{f}{F}, \quad y_3 = \frac{g}{G}, \quad \text{and} \quad z_3 = \frac{h}{H}. \tag{3.5}$$

Note that the numerators and denominators of (3.4) are linear in x_3, y_3, and z_3. The above fractional transformation allows one to make the top and bottom of x_{13}, y_{13}, and z_{13} linear in the projective variables (in the same vein as using fractional transformation to linearize Riccati equations).

Substitution of (3.5) into (3.4) yields

$$x_{13} = \frac{z_1 H f - x_1 F h}{F(z_1 H - h)}, \quad y_{13} = -\frac{H(x_1 F - f)}{F(z_1 H - h)}, \quad \text{and} \tag{3.6a}$$
$$z_{13} = -\frac{z}{y}\left(\frac{k z_1 H g - p y_1 G h}{G(z_1 H - h)}\right). \tag{3.6b}$$

Achieving the desired linearity requires $F = G = H$. Then, (3.6) becomes

$$x_{13} = \frac{z_1 f - x_1 h}{z_1 F - h}, \quad y_{13} = -\frac{x_1 F - f}{z_1 F - h}, \quad \text{and} \tag{3.7a}$$

$$z_{13} = -\frac{z}{y}\left(\frac{kz_1 g - py_1 h}{z_1 F - h}\right), \tag{3.7b}$$

with

$$x_3 = \frac{f}{F}, \quad y_3 = \frac{g}{F}, \quad \text{and} \quad z_3 = \frac{h}{F}. \tag{3.8}$$

How one deals with the remaining variables f, g, h, and F, leads to various alternatives for Lax matrices.

3.1.1 The first alternative

Choice 1. Note that the edge equation (3.3b) imposes an additional constraint on (3.8). Indeed, solving (3.3b) for x_3 in terms of y_3 yields

$$x_3 = zy_3 + x. \tag{3.9}$$

Using $x_3 = \frac{f}{F}$ and $y_3 = \frac{g}{F}$ one can eliminate f since $f = xF + zg$. Eqs. (3.8) and (3.7) then become

$$x_3 = \frac{xF + zg}{F}, \quad y_3 = \frac{g}{F}, \quad \text{and} \quad z_3 = \frac{h}{F}, \tag{3.10}$$

and

$$x_{13} = \frac{f_1}{F_1} = \frac{x_1 F_1 + z_1 g_1}{F_1} = \frac{xz_1 F + zz_1 g - x_1 h}{z_1 F - h}, \tag{3.11a}$$

$$y_{13} = \frac{g_1}{F_1} = \frac{(x - x_1)F + zg}{z_1 F - h}, \tag{3.11b}$$

$$z_{13} = \frac{h_1}{F_1} = -\frac{z(kz_1 g - py_1 h)}{y(z_1 F - h)}, \tag{3.11c}$$

where F, g, and h are independent (and remain undetermined).

Then, (3.11b) and (3.11c) can be split by setting

$$F_1 = t\,(z_1 F - h), \tag{3.12a}$$

$$g_1 = t\,((x - x_1)F + zg), \tag{3.12b}$$

$$h_1 = t\left(-\frac{z}{y}\,(kz_1 g - py_1 h)\right), \tag{3.12c}$$

where $t(\mathbf{x}, \mathbf{x}_1; p, k)$ is a scalar function still to be determined. One can readily verify that (3.11a) is identically satisfied.

If we define $\psi_{\rm a} := \begin{bmatrix} F & g & h \end{bmatrix}^{\rm T}$, then $(\psi_{\rm a})_1 = \begin{bmatrix} F_1 & g_1 & h_1 \end{bmatrix}^{\rm T}$, we can write (3.12) in matrix form $(\psi_{\rm a})_1 = L_{\rm a}\psi_{\rm a}$, with

$$L_{\rm a} = t\, L_{\rm a,core} := t \begin{bmatrix} z_1 & 0 & -1 \\ x - x_1 & z & 0 \\ 0 & -\frac{kzz_1}{y} & \frac{pzy_1}{y} \end{bmatrix}. \tag{3.13}$$

The partner matrix $M_{\rm a}$ of the Lax pair,

$$M_{\rm a} = s\, M_{\rm a,core} := s \begin{bmatrix} z_2 & 0 & -1 \\ x - x_2 & z & 0 \\ 0 & -\frac{kzz_2}{y} & \frac{qzy_2}{y} \end{bmatrix}, \tag{3.14}$$

comes from substituting (3.10) into the five equations (similar to (3.3)) for the left face of the cube. Formally, this amounts to replacing all indices 1 by 2 and p by q in $L_{\rm a,core}$ (see, e.g., [12] for details). In subsequent examples, the partner matrices (M) will no longer be shown.

Using the same terminology as in [12], $L_{\rm a,core}$ and $M_{\rm a,core}$ are the "core" of the Lax matrices $L_{\rm a}$ and $M_{\rm a}$, respectively. The label "a" on $\psi_{\rm a}, L_{\rm a}$, and $M_{\rm a}$ is added to differentiate the entries within each family of Lax matrices (up to trivial permutations of the components). In what follows, alternative choices will be labeled with "b," "A", "B," etc. These matrices come from alternate ways of treating the edge equations.

The functions $t(\mathbf{x}, \mathbf{x}_1; p, k)$ and $s(\mathbf{x}, \mathbf{x}_2; q, k)$ can be computed algorithmically as shown in [12] or by using the Lax equation (2.10) directly, as follows,

$$(tL_{\rm core})_2\,(sM_{\rm core})-(sM_{\rm core})_1\,(tL_{\rm core})=(s\,t_2)\,(L_{\rm core})_2\,M_{\rm core}-(t\,s_1)\,(M_{\rm core})_1\,L_{\rm core}\doteq 0, \tag{3.15}$$

which implies that

$$\frac{s\,t_2}{t\,s_1}\,(L_{\rm core})_2\,M_{\rm core} \;\doteq\; (M_{\rm core})_1\,L_{\rm core}. \tag{3.16}$$

After replacing $L_{\rm core}$ and $M_{\rm core}$ by $L_{\rm a,core}$ and $M_{\rm a,core}$ from (3.13) and (3.14), respectively, in (3.16), one gets

$$\frac{s\,t_2}{t\,s_1} \doteq \frac{z_1}{z_2}, \tag{3.17}$$

which has an infinite family of solutions. Indeed, the left-hand side of (3.17) is invariant under the change

$$t \to \frac{i_1}{i}\,t, \quad s \to \frac{i_2}{i}\,s, \tag{3.18}$$

where $i(\mathbf{x})$ is an arbitrary function and i_1 and i_2 denote the shifts of i in the one- and two-direction, respectively. One can readily verify that (3.17) is satisfied by, for example, $s = t = \frac{1}{z}$. Then, for $\psi_{\rm a} = \begin{bmatrix} F & g & h \end{bmatrix}^{\rm T}$, the Lax matrix $L_{\rm a}$ in (3.13) becomes

$$L_{\rm a} = \frac{1}{z}\begin{bmatrix} z_1 & 0 & -1 \\ x - x_1 & z & 0 \\ 0 & -\frac{kzz_1}{y} & \frac{pzy_1}{y} \end{bmatrix}. \tag{3.19}$$

Choice 2. Solving the edge equation (3.3b) for y_3,

$$y_3 = \frac{x_3 - x}{z}, \tag{3.20}$$

and using (3.8) yields $g = -\frac{xF-f}{z}$. Then (3.8) becomes

$$x_3 = \frac{f}{F}, \quad y_3 = -\frac{xF - f}{zF}, \quad \text{and} \quad z_3 = \frac{h}{F}, \tag{3.21}$$

where F, f, and h are now independent. Eqs. (3.4) now become

$$x_{13} = \frac{f_1}{F_1} = \frac{z_1 f - x_1 h}{z_1 F - h}, \tag{3.22a}$$

$$y_{13} = \frac{g_1}{F_1} = -\frac{x_1 F_1 - f_1}{z_1 F_1} = -\frac{x_1 F - f}{z_1 F - h}, \tag{3.22b}$$

$$z_{13} = \frac{h_1}{F_1} = \frac{kxz_1 F - kz_1 f + pzy_1 h}{y(z_1 F - h)}, \tag{3.22c}$$

which can be split by selecting

$$F_1 = t\,(z_1 F - h), \tag{3.23a}$$

$$f_1 = t\,(z_1 f - x_1 h), \tag{3.23b}$$

$$h_1 = t\left(\frac{1}{y}\,(kxz_1 F - kz_1 f + pzy_1 h)\right), \tag{3.23c}$$

where $t(\mathbf{x}, \mathbf{x}_1; p, k)$ is a scalar function still to be determined. Note that (3.22b) is identically satisfied.

If we define $\psi_{\rm b} := \begin{bmatrix} F & f & h \end{bmatrix}^{\rm T}$, then $(\psi_{\rm b})_1 = \begin{bmatrix} F_1 & f_1 & h_1 \end{bmatrix}^{\rm T}$, we can write (3.23) in matrix form, $(\psi_{\rm b})_1 = L_{\rm b}\psi_{\rm b}$, with

$$L_{\rm b} = t\,L_{\rm b,core} := \frac{1}{z}\begin{bmatrix} z_1 & 0 & -1 \\ 0 & z_1 & -x_1 \\ \frac{kxz_1}{y} & -\frac{kz_1}{y} & \frac{pzy_1}{y} \end{bmatrix}, \tag{3.24}$$

where we substituted $t = \frac{1}{z}$ which was computed the same way as in Choice 1.

Thus, within this first alternative, the two choices of representing the edge constraint result in minimally-sized Lax matrices which we call *representative* Lax matrices for the PΔE.

3.1.2 The second alternative

As a second alternative in the algorithm, we do not use the edge equations to replace (3.8) by (3.10) or (3.21). Instead, we incorporate the edge equations (3.1a) into *all three* equations of (3.4). Using (3.1a), we replace x_3 and x_1 in x_{13} and y_{13}, and y_1 and y_3 in z_{13}, yielding

$$\tilde{x}_{13} = x + \frac{z(y_3 z_1 - y_1 z_3)}{z_1 - z_3} = \frac{x(z_1 - z_3) + z(y_3 z_1 - y_1 z_3)}{z_1 - z_3}, \tag{3.25a}$$

$$\tilde{y}_{13} = \frac{z(y_3 - y_1)}{z_1 - z_3}, \quad \text{and} \quad \tilde{z}_{13} = \frac{kz_1(x - x_3) - pz_3(x - x_1)}{y(z_1 - z_3)}. \tag{3.25b}$$

Note that each of the above expressions has the same denominator $z_1 - z_3$ as in (3.4). Thus, in principle, any equation from (3.4) could be replaced by the matching equation from (3.25).

Choice 1. If we take x_{13}, $\tilde{y}_{13}$ and z_{13}, then substitution of (3.8) yields

$$x_{13} = \frac{f_1}{F_1} = \frac{z_1 f - x_1 h}{z_1 F - h}, \quad \tilde{y}_{13} = \frac{g_1}{F_1} = -\frac{z(y_1 F - g)}{z_1 F - h}, \quad \text{and} \tag{3.26a}$$

$$z_{13} = \frac{h_1}{F_1} = -\frac{z(kz_1 g - py_1 h)}{y(z_1 F - h)}. \tag{3.26b}$$

Hence, we set

$$F_1 = t\,(z_1 F - h), \tag{3.27a}$$

$$f_1 = t\,(z_1 f - x_1 h), \tag{3.27b}$$

$$g_1 = t\,\big(-(zy_1 F - zg)\big), \tag{3.27c}$$

$$h_1 = t\left(-\frac{z}{y}\,(kz_1 g - py_1 h)\right), \tag{3.27d}$$

where $t(\mathbf{x}, \mathbf{x}_1; p, k)$ is a scalar function still to be determined.

Defining $\psi := \begin{bmatrix} F & f & g & h \end{bmatrix}^{\mathrm{T}}$, yields $\psi_1 = \begin{bmatrix} F_1 & f_1 & g_1 & h_1 \end{bmatrix}^{\mathrm{T}}$. So, we can write (3.27) as $\psi_1 = L_A \psi$, with

$$L_{\mathrm{A}} = t\,L_{\text{core}} := \frac{1}{z}\begin{bmatrix} z_1 & 0 & 0 & -1 \\ 0 & z_1 & 0 & -x_1 \\ -zy_1 & 0 & z & 0 \\ 0 & 0 & -\frac{kzz_1}{y} & \frac{pzy_1}{y} \end{bmatrix}, \tag{3.28}$$

where we have used (3.15) to get $t = \frac{1}{z}$.

Though not obvious at first glance, L_{a} in (3.13) follows from L_{A} after removing the second row and second column and replacing zy_1 by $x_1 - x$ using (3.1a). Matrices (3.28) and M_{A} (obtained from L_{A} by replacing indices 1 by 2 and p by q in L_{A})

are a valid Lax pair despite being of larger size than $(L_{\rm a}, M_{\rm a})$. We refer to these larger-sized Lax matrices as *extended* Lax matrices of PΔEs.

Choice 2. If we work with x_{13}, $\tilde{y}_{13}$ and $\tilde{z}_{13}$ then substitution of (3.8) yields

$$x_{13} = \frac{f_1}{F_1} = \frac{z_1 f - x_1 h}{z_1 F - h}, \quad \tilde{y}_{13} = \frac{g_1}{F_1} = -\frac{z y_1 F - z g}{z_1 F - h}, \quad \text{and} \tag{3.29a}$$

$$\tilde{z}_{13} = \frac{h_1}{F_1} = \frac{k x z_1 F - k z_1 f - p(x - x_1)h}{y(z_1 F - h)}. \tag{3.29b}$$

Setting

$$F_1 = t\,(z_1 F - h), \tag{3.30a}$$

$$f_1 = t\,(z_1 f - x_1 h), \tag{3.30b}$$

$$g_1 = t\big(-(z y_1 F - z g)\big), \tag{3.30c}$$

$$h_1 = t\left(\frac{1}{y}\,(k x z_1 F - k z_1 f - p(x - x_1)h)\right), \tag{3.30d}$$

and defining ψ as in Choice 1, yields

$$L_{\rm B} = t\,L_{\rm core} := \frac{1}{z}\begin{bmatrix} z_1 & 0 & 0 & -1 \\ 0 & z_1 & 0 & -x_1 \\ -z y_1 & 0 & z & 0 \\ \frac{k x z_1}{y} & -\frac{k z_1}{y} & 0 & -\frac{p(x-x_1)}{y} \end{bmatrix}, \tag{3.31}$$

since $t = \frac{1}{z}$. Thus, $L_{\rm B}$ is another extended Lax pair for the Schwarzian Boussinesq system. Note that $L_{\rm B}$ reduces to $L_{\rm b}$ in (3.24) by removing the third row and third column and replacing $x_1 - x$ by $z y_1$ based on (3.1a).

Choices 3 and 4. Repeating the process with other combinations of solutions from (3.4) and (3.25) results in the following Lax matrices,

$$L_{\rm C} = \frac{1}{z}\begin{bmatrix} z_1 & 0 & 0 & -1 \\ x z_1 & 0 & z z_1 & -(x + z y_1) \\ -z y_1 & 0 & z & 0 \\ 0 & 0 & -\frac{k z z_1}{y} & \frac{p z y_1}{y} \end{bmatrix}, \tag{3.32}$$

and

$$L_{\rm D} = \frac{1}{z}\begin{bmatrix} z_1 & 0 & 0 & -1 \\ 0 & z_1 & 0 & -x_1 \\ -x_1 & 1 & 0 & 0 \\ \frac{k x z_1}{y} & -\frac{k z_1}{y} & 0 & \frac{p(x_1-x)}{y} \end{bmatrix}, \tag{3.33}$$

from working with $\tilde{x}_{13}, \tilde{y}_{13}, z_{13}$ and $x_{13}, y_{13}, \tilde{z}_{13}$, respectively.

Obviously L_{C} and L_{D} are trivially related to L_{a} and L_{b}, respectively. Indeed, remove the second row and second column in L_{C} and the third row and third column in L_{D} and use $zy_1 = x_1 - x$ to get L_{a} and L_{b}, respectively.

All other combinations of pieces of (3.4) and (3.25) do *not* lead to matrices that satisfy the defining equation (2.10).

The second alternative leads to extended Lax matrices but they are not always trivial extensions of the representative Lax matrices. In other words, smaller-size matrices do not necessarily follow from the larger-size matrices by simply removing rows and columns. Furthermore, if the representative Lax matrices were not known it would not be obvious which rows and columns should be removed. As shown in Section 6, some edge-constrained systems have extended Lax matrices whose "equivalence" to a representative Lax matrix is non-trivial.

In the next section we carefully investigate the connections between the Lax matrices computed with the two alternatives and various choices above.

4 Gauge and gauge-like equivalences of Lax pairs

As discussed in Section 2, if there exists *one* pair of Lax matrices for a given system of PΔEs then there is an *infinite* number of such pairs, all equivalent to each other under discrete gauge transformations of type (2.11) involving a square matrix $\mathcal{G}$.

Given two distinct Lax pairs with matrices of the same size (no matter how they were computed), it has yet to be shown if there exists a gauge transformation relating them. For edge-constrained systems, the derivation of the gauge transformation between *representative* Lax matrices is straightforward. However, the derivation of a gauge-like transformation between Lax matrices of different sizes (resulting from the application of different methods) is nontrivial. In this section we look at gauge and gauge-like transformations in more detail.

4.1 Gauge equivalence

In Section 3.1.1 we obtained (3.19) and (3.24), resulting from Choices 1 and 2 of dealing with the edge constraints.

Example 1. With regard to (2.11), computing the gauge matrix $\mathcal{G}$ such that

$$L_{\mathrm{b}} = \mathcal{G}_1 L_{\mathrm{a}} \mathcal{G}^{-1} \tag{4.1}$$

is straightforward if we consider the implications of the gauge relationship. Indeed, multiplying (4.1) by ψ_{b}, and using (2.9) yields

$$(\psi_{\mathrm{b}})_1 = L_{\mathrm{b}}\psi_{\mathrm{b}} = (\mathcal{G}_1 L_{\mathrm{a}} \mathcal{G}^{-1})\psi_{\mathrm{b}}. \tag{4.2a}$$

Hence, if we set $\psi_{\mathrm{a}} = \mathcal{G}^{-1}\psi_{\mathrm{b}}$, we obtain $(\psi_{\mathrm{b}})_1 = \mathcal{G}_1 L_{\mathrm{a}}\psi_{\mathrm{a}} = \mathcal{G}_1 (\psi_{\mathrm{a}})_1 = (\mathcal{G}\psi_{\mathrm{a}})_1$. Thus, $\psi_{\mathrm{b}} = \mathcal{G}\psi_{\mathrm{a}}$ determines $\mathcal{G}$. Not surprisingly, the gauge matrix $\mathcal{G}$ depends on how we selected the components of ψ which, in turn, depends on how the edge equation (3.3b) was treated.

Recall that $\psi_{\rm a} := \begin{bmatrix} F & g & h \end{bmatrix}^{\rm T}$ and $\psi_{\rm b} := \begin{bmatrix} F & f & h \end{bmatrix}^{\rm T}$. Using (3.10), we get

$$\mathcal{G} = \begin{bmatrix} 1 & 0 & 0 \\ x & z & 0 \\ 0 & 0 & 1 \end{bmatrix} \quad \text{and} \quad \mathcal{G}^{-1} = \begin{bmatrix} 1 & 0 & 0 \\ -\frac{x}{z} & \frac{1}{z} & 0 \\ 0 & 0 & 1 \end{bmatrix}. \tag{4.3}$$

Indeed,

$$\mathcal{G}\psi_{\rm a} = \begin{bmatrix} 1 & 0 & 0 \\ x & z & 0 \\ 0 & 0 & 1 \end{bmatrix} \begin{bmatrix} F \\ g \\ h \end{bmatrix} = \begin{bmatrix} F \\ xF + zg \\ h \end{bmatrix} = \begin{bmatrix} F \\ f \\ h \end{bmatrix} = \psi_{\rm b} \tag{4.4}$$

confirms that $\psi_{\rm b} = \mathcal{G}\psi_{\rm a}$.

Thus, the representative Lax matrices, (3.19) and (3.24), are gauge equivalent, as in (4.1), with $\mathcal{G}$ in (4.3). In essence, for Lax matrices of the same sizes, $\mathcal{G}$ "represents" the edge constraint in the system of PΔEs. Of course, (4.1) may also be represented as $L_{\rm a} = \bar{\mathcal{G}}_1 L_{\rm b} \bar{\mathcal{G}}^{-1}$, where $\bar{\mathcal{G}} = \mathcal{G}^{-1}$.

4.2 Gauge-like equivalence

In as much as gauge transformations between the representative Lax matrices of a given system of PΔEs are straightforward to derive and defined by the corresponding edge equation, the relationship between representative and extended matrices, though still dependent upon the edge equation, is not so obvious.

Consider the Schwarzian Boussinesq system (1.4) which we have shown to have representative Lax matrices, $L_{\rm a}$ and $L_{\rm b}$ (derived in Section 3.1.1) and extended Lax matrices, $L_{\rm A}$, $L_{\rm B}$, $L_{\rm C}$, and $L_{\rm D}$ derived in Section 3.1.2.

To determine a relationship between extended Lax matrices (like $L_{\rm A}$) and representative Lax matrices (like $L_{\rm a}$), (4.1) must be generalized because the matrices do not have the same sizes. We therefore introduce transformations involving non-square matrices $\mathcal{H}$ and $\bar{\mathcal{H}}$ satisfying one of the relationships,

$$L_{\rm ext} = \mathcal{H}_1 L_{\rm rep} \mathcal{H}_{\rm Left}^{-1}, \tag{4.5a}$$

$$L_{\rm rep} = \bar{\mathcal{H}}_1 L_{\rm ext} \bar{\mathcal{H}}_{\rm Right}^{-1}, \tag{4.5b}$$

where $L_{\rm rep}$ is a representative Lax pair, $L_{\rm ext}$ is an extended Lax pair, and $\mathcal{H}$ and $\bar{\mathcal{H}}$ are suitable matrices of appropriate sizes. Furthermore, the labels "Left" and "Right" refer to left and right inverses. In deriving $\mathcal{H}$ and $\bar{\mathcal{H}}$ we find that the edge equations again provide guidance.

Obviously, matrices like $\mathcal{H}$ and $\bar{\mathcal{H}}$ play the role of the gauge matrices but since they are no longer square we call them *gauge-like* matrices. Likewise, any of the transformations in (4.5) are called *gauge-like* transformations.

Example 2. To illustrate (4.5a), consider an extended Lax pair with associated vector ψ. If we consider the edge constraint expressed as in (3.21), then the linearity of $g = -\frac{xF-f}{z}$ in variables F and f allows us to express ψ in terms of ψ_{b} in a simple (unique) way

$$\psi = \begin{bmatrix} F \\ f \\ g \\ h \end{bmatrix} = \begin{bmatrix} F \\ f \\ -\frac{x}{z}F + \frac{1}{z}f \\ h \end{bmatrix} = \begin{bmatrix} 1 & 0 & 0 \\ 0 & 1 & 0 \\ -\frac{x}{z} & \frac{1}{z} & 0 \\ 0 & 0 & 1 \end{bmatrix} \begin{bmatrix} F \\ f \\ h \end{bmatrix} := \mathcal{H}\psi_{\mathrm{b}}, \tag{4.6}$$

defining the matrix $\mathcal{H}$. Since rank $\mathcal{H} = 3$, matrix $\mathcal{H}$ has a 3-parameter family of left inverses,

$$\mathcal{H}^{-1}_{\mathrm{Left},all} = \begin{bmatrix} 1-\alpha x & \alpha & -\alpha z & 0 \\ (1-\beta)x & \beta & (1-\beta)z & 0 \\ -\gamma x & \gamma & -\gamma z & 1 \end{bmatrix}, \tag{4.7}$$

where α, β and γ are free parameters (which could depend on $\mathbf{x}$).

Now we take a specific member of the family (denoted by $\mathcal{H}^{-1}_{\mathrm{Left}}$) so that

$$L_{\mathrm{C}} \doteq \mathcal{H}_1 L_{\mathrm{b}} \mathcal{H}^{-1}_{\mathrm{Left}}, \tag{4.8}$$

where, as before, $\doteq$ indicates equality when evaluated against the given PΔEs. More precisely, equality only holds when edge equation (3.3a) is used. A straightforward matrix multiplication shows that (4.8) holds if $\alpha = \beta = \gamma = 0$. Hence,

$$\mathcal{H}^{-1}_{\mathrm{Left}} = \begin{bmatrix} 1 & 0 & 0 & 0 \\ x & 0 & z & 0 \\ 0 & 0 & 0 & 1 \end{bmatrix}. \tag{4.9}$$

Instead of (4.1) we now have (4.8), i.e., a transformation of type (4.5a), which can readily be verified. Indeed, repeatedly using (4.8), (4.6), and (2.9), yields

$$\psi_1 = L_{\mathrm{C}}\psi = \mathcal{H}_1 L_{\mathrm{b}} \mathcal{H}^{-1}_{\mathrm{Left}} \mathcal{H}\psi_{\mathrm{b}} = \mathcal{H}_1 L_{\mathrm{b}} \psi_{\mathrm{b}} = \mathcal{H}_1 (\psi_{\mathrm{b}})_1 = \left(\mathcal{H}\psi_{\mathrm{b}}\right)_1, \tag{4.10}$$

confirming (4.6).

Example 3. After similar calculations involving $\psi_{\mathrm{a}} = \begin{bmatrix} F & g & h \end{bmatrix}^{\mathrm{T}}$ and ψ, and with the edge constraint expressed as in (3.10), i.e., $f = xF + zg$, we find that

$$L_{\mathrm{D}} \doteq \mathcal{H}_1 L_{\mathrm{a}} \mathcal{H}^{-1}_{\mathrm{Left}}, \tag{4.11}$$

where

$$\mathcal{H} = \begin{bmatrix} 1 & 0 & 0 \\ x & z & 0 \\ 0 & 1 & 0 \\ 0 & 0 & 1 \end{bmatrix} \quad \text{and} \quad \mathcal{H}^{-1}_{\mathrm{Left}} = \begin{bmatrix} 1 & 0 & 0 & 0 \\ -x/z & 1/z & 0 & 0 \\ 0 & 0 & 0 & 1 \end{bmatrix}. \tag{4.12}$$

Example 4. A first gauge-like relationship between L_{a} and L_{A} is simple to derive. As mentioned in Choice 1 in Section 3.1.2, removing the second row and second column from L_{A} gives L_{a}. Formally,

$$L_{\mathrm{a}} \doteq \mathcal{B} L_{\mathrm{A}} \mathcal{B}^{\mathrm{T}}, \tag{4.13}$$

with $zy_1 = x_1 - x$ and

$$\mathcal{B} = \begin{bmatrix} 1 & 0 & 0 & 0 \\ 0 & 0 & 1 & 0 \\ 0 & 0 & 0 & 1 \end{bmatrix}. \tag{4.14}$$

Continuing with L_{a} and L_{A}, we derive a second gauge-like transformation to illustrate (4.5b). To find a matrix $\bar{\mathcal{H}}$, consider the edge constraint $g = -\frac{xF-f}{z}$ expressed in (3.21). Thus,

$$\psi_{\mathrm{a}} = \begin{bmatrix} F \\ g \\ h \end{bmatrix} = \begin{bmatrix} F \\ -\frac{x}{z}F + \frac{1}{z}f \\ h \end{bmatrix} = \begin{bmatrix} 1 & 0 & 0 & 0 \\ -\frac{x}{z} & \frac{1}{z} & 0 & 0 \\ 0 & 0 & 0 & 1 \end{bmatrix} \begin{bmatrix} F \\ f \\ g \\ h \end{bmatrix} := \bar{\mathcal{H}}\psi. \tag{4.15}$$

The inverse transformation,

$$\psi = \begin{bmatrix} F \\ f \\ g \\ h \end{bmatrix} = \begin{bmatrix} F \\ xF + zg \\ g \\ h \end{bmatrix} = \begin{bmatrix} 1 & 0 & 0 \\ x & z & 0 \\ 0 & 1 & 0 \\ 0 & 0 & 1 \end{bmatrix} \begin{bmatrix} F \\ g \\ h \end{bmatrix} := \bar{\mathcal{H}}^{-1}_{\mathrm{Right}}\psi_{\mathrm{a}}, \tag{4.16}$$

determines a suitable right inverse of $\bar{\mathcal{H}}$. Thus,

$$L_{\mathrm{a}} = \bar{\mathcal{H}}_1 L_{\mathrm{A}} \bar{\mathcal{H}}^{-1}_{\mathrm{Right}} \quad \text{for} \quad \bar{\mathcal{H}} = \begin{bmatrix} 1 & 0 & 0 & 0 \\ -\frac{x}{z} & \frac{1}{z} & 0 & 0 \\ 0 & 0 & 0 & 1 \end{bmatrix}, \tag{4.17}$$

without having to use the edge equation (3.3a). To show that (4.17) is correct, use (2.9) repeatedly, together with (4.16) and (4.17), yielding

$$(\psi_{\mathrm{a}})_1 = L_{\mathrm{a}}\psi_{\mathrm{a}} = \bar{\mathcal{H}}_1 L_{\mathrm{A}} \bar{\mathcal{H}}^{-1}_{\mathrm{Right}}\psi_{\mathrm{a}} = \bar{\mathcal{H}}_1 L_{\mathrm{A}}\psi = \bar{\mathcal{H}}_1\psi_1 = \left(\bar{\mathcal{H}}\psi\right)_1, \tag{4.18}$$

confirming (4.15).

Example 5. Interestingly, gauge transformations between two distinct extended Lax matrices for a PΔE are not as straightforward. For example, a gauge transformation has not yet been found for Lax matrices L_{A} and L_{B} given in (3.28) and (3.31). Thus, even though we have shown that the corresponding representative Lax matrices are gauge equivalent, we have not been able to show the same for the corresponding extended matrices.

5 Application to generalized Hietarinta systems

In [15], Hietarinta presented the results of a search of multi-component equations which are edge-constrained and obey the property of multidimensional consistency. That search led to various generalized Boussinesq-type systems, nowadays called the Hietarinta A-2, B-2, C-3, and C-4 systems. Bridgman *et al.* [12] derived their corresponding Lax pairs using [8].

Simultaneously, Zhang *et al.* [35] showed that each of the lattice systems presented in [15] can be further generalized based on a direct linearization scheme [29] in connection with a more general dispersion law. The systems considered in [15] are then shown to be special cases. In fact, they are connected to the more general cases through point transformations.

As a by-product of the direct linearization method, Zhang *et al.* [35] obtained the Lax pairs of each of these generalized Boussinesq systems. No doubt, they are all valid Lax pairs but some of the matrices have larger than needed sizes. Using the algorithmic CAC approach discussed in Section 3, we were able to derive Lax pairs of minimal matrix sizes for these systems and unravel the connections with the Lax matrices presented in [35]. Full details of that investigation will be published elsewhere [10] but their Lax matrices are given in the next section.

6 Summary of results

6.1 Lattice Boussinesq system

The lattice Boussinesq system [32] is given by

$$\begin{aligned} &z_1 - x x_1 + y = 0, \;\; z_2 - x x_2 + y = 0, \;\; \text{and} \\ &(x_2 - x_1)(z - x x_{12} + y_{12}) - p + q = 0. \end{aligned} \tag{6.1}$$

Edge constraint $x_1 = \frac{z_1 + y}{x}$ implies that $x_3 = \frac{z_3 + y}{x}$. Here,

$$\begin{aligned} &x_{13} = \frac{y_1 - y_3}{x_1 - x_3}, \;\; y_{13} = \frac{x(y_1 - y_3) - z(x_1 - x_3) + k - p}{x_1 - x_3}, \;\; \text{and} \\ &z_{13} = \frac{x_3 y_1 - x_1 y_3}{x_1 - x_3}. \end{aligned} \tag{6.2}$$

Variants of (6.2) may be derived by incorporating edge constraints yielding

$$\begin{aligned} &\tilde{x}_{13} = \frac{x(y_1 - y_3)}{z_1 - z_3}, \;\; \tilde{y}_{13} = \frac{x\Big(k - p + x(y_1 - y_3)\Big) - z(z_1 - z_3)}{z_1 - z_3}, \;\; \text{and} \\ &\tilde{z}_{13} = \frac{y(y_1 - y_3) + (y_1 z_3 - y_3 z_1)}{z_1 - z_3}. \end{aligned} \tag{6.3}$$

Using the algorithm of Section 3, we computed 3 by 3 Lax matrices which are presented in Table 1 together with the gauge transformations. For L_{a} one has $\frac{s\,t_2}{t\,s_1} \doteq 1$, which is satisfied for $t = s = 1$. For L_{b} one obtains $\frac{s\,t_2}{t\,s_1} \doteq \frac{x_1}{x_2}$, hence, $t = s = \frac{1}{x}$.

Using only (6.2) with (3.5) yields a 4 by 4 Lax matrix (not shown) that is trivially associated with L_{a}. Similarly, using only (6.3) again with (3.5) gives a 4 by 4 Lax matrix (not shown) which is trivially associated with L_{b}.

Table 1: Boussinesq system Lax pairs and gauge matrices

Substitutions	ψ	Matrices L of Lax pair
Writing the edge constraint as $z_3 = xx_3 - y$ yields		
$x_3 = \frac{f}{F},\ y_3 = \frac{g}{F},$ $z_3 = -\frac{yF - xf}{F}.$	$\psi_{\mathrm{a}} = \begin{bmatrix} F \\ f \\ g \end{bmatrix}$	$L_{\mathrm{a}} = \begin{bmatrix} -x_1 & 1 & 0 \\ -y_1 & 0 & 1 \\ \ell_{31} & -z & x \end{bmatrix},$
where $\ell_{31} = zx_1 - xy_1 + p - k.$		
Writing the edge constraint as $x_3 = \frac{z_3+y}{x}$ yields		
$x_3 = \frac{yF + h}{xF},$ $y_3 = \frac{g}{F},\ z_3 = \frac{h}{F}.$	$\psi_{\mathrm{b}} = \begin{bmatrix} F \\ g \\ h \end{bmatrix}$	$L_{\mathrm{b}} = \frac{1}{x}\begin{bmatrix} y - xx_1 & 0 & 1 \\ \ell_{21} & x^2 & -z \\ -yy_1 & xx_1 & -y_1 \end{bmatrix},$
where $\ell_{21} = x(p - k - xy_1) - z(y - xx_1).$		
Gauge transformations for L_{a} and L_{b} are given by		
$L_{\mathrm{b}} = \mathcal{G}_1 L_{\mathrm{a}} \mathcal{G}^{-1}, \psi_{\mathrm{b}} = \mathcal{G}\psi_{\mathrm{a}},$ $L_{\mathrm{a}} = \bar{\mathcal{G}}_1 L_{\mathrm{b}} \bar{\mathcal{G}}^{-1}, \psi_{\mathrm{a}} = \bar{\mathcal{G}}\psi_{\mathrm{b}},$ where $\bar{\mathcal{G}} = \mathcal{G}^{-1}.$	$\mathcal{G} = \begin{bmatrix} 1 & 0 & 0 \\ 0 & 0 & 1 \\ -y & x & 0 \end{bmatrix}$	$\bar{\mathcal{G}} = \begin{bmatrix} 1 & 0 & 0 \\ \frac{y}{x} & 0 & \frac{1}{x} \\ 0 & 1 & 0 \end{bmatrix}.$

6.2 Schwarzian Boussinesq system

The Schwarzian Boussinesq system [15, 24, 25] is given by

$$\begin{aligned} &zy_1 - x_1 + x = 0, \quad zy_2 - x_2 + x = 0, \quad \text{and} \\ &(z_1 - z_2)z_{12} - \frac{z}{y}(py_1z_2 - qy_2z_1) = 0. \end{aligned} \tag{6.4}$$

Edge constraint $x_1 = zy_1 + x$ leads to $x_3 = zy_3 + x$. Hence,

$$\begin{aligned} &x_{13} = \frac{x_3z_1 - x_1z_3}{z_1 - z_3}, \quad y_{13} = \frac{x_3 - x_1}{z_1 - z_3}, \quad \text{and} \\ &z_{13} = \frac{z}{y}\left(\frac{py_1z_3 - ky_3z_1}{z_1 - z_3}\right). \end{aligned} \tag{6.5}$$

After incorporating edge constraints, variants of (6.5) are

$$\begin{aligned} &\tilde{x}_{13} = \frac{x(z_1 - z_3) + z(y_3z_1 - y_1z_3)}{z_1 - z_3}, \quad \tilde{y}_{13} = \frac{z(y_3 - y_1)}{z_1 - z_3}, \quad \text{and} \\ &\tilde{z}_{13} = \frac{kz_1(x - x_3) - pz_3(x - x_1)}{y(z_1 - z_3)}. \end{aligned} \tag{6.6}$$

The 3 by 3 matrices computed in Section 3 are summarized in Table 2 together with the gauge transformations that connect them. For the representative and extended Lax matrices given below we obtained $\frac{s\,t_2}{t\,s_1} \doteq \frac{z_1}{z_2}$, which holds when $t = s = \frac{1}{z}$.

Table 2: Schwarzian Boussinesq system Lax pairs and gauge matrices

Substitutions	ψ	Matrices L of Lax pair
Writing the edge constraint as $x_3 = zy_3 + x$ yields		
$x_3 = \dfrac{xF+zg}{F}$, $y_3 = \dfrac{g}{F}$, $z_3 = \dfrac{h}{F}$.	$\psi_{\mathrm{a}} = \begin{bmatrix} F \\ g \\ h \end{bmatrix}$	$L_{\mathrm{a}} = \dfrac{1}{z}\begin{bmatrix} z_1 & 0 & -1 \\ x - x_1 & z & 0 \\ 0 & -\frac{kzz_1}{y} & \frac{pzy_1}{y} \end{bmatrix}$.
Writing the edge constraint as $y_3 = \frac{x_3 - x}{z}$ yields		
$x_3 = \dfrac{f}{F}$, $z_3 = \dfrac{h}{F}$, $y_3 = -\dfrac{xF - f}{zF}$.	$\psi_{\mathrm{b}} = \begin{bmatrix} F \\ f \\ h \end{bmatrix}$	$L_{\mathrm{b}} = \dfrac{1}{z}\begin{bmatrix} z_1 & 0 & -1 \\ 0 & z_1 & -x_1 \\ \frac{kxz_1}{y} & -\frac{kz_1}{y} & \frac{pzy_1}{y} \end{bmatrix}$.
Gauge transformations for L_{a} and L_{b} are given by		
$L_{\mathrm{b}} = \mathcal{G}_1 L_{\mathrm{a}} \mathcal{G}^{-1}, \psi_{\mathrm{b}} = \mathcal{G}\psi_{\mathrm{a}}$, $L_{\mathrm{a}} = \bar{\mathcal{G}}_1 L_{\mathrm{b}} \bar{\mathcal{G}}^{-1}, \psi_{\mathrm{a}} = \bar{\mathcal{G}}\psi_{\mathrm{b}}$, where $\bar{\mathcal{G}} = \mathcal{G}^{-1}$.	$\mathcal{G} = \begin{bmatrix} 1 & 0 & 0 \\ x & z & 0 \\ 0 & 0 & 1 \end{bmatrix}$	$\bar{\mathcal{G}} = \begin{bmatrix} 1 & 0 & 0 \\ -\frac{x}{z} & \frac{1}{z} & 0 \\ 0 & 0 & 1 \end{bmatrix}$,

System (6.4) also admits the extended Lax matrices,

$$L_{\mathrm{A}} = \frac{1}{z}\begin{bmatrix} z_1 & 0 & 0 & -1 \\ 0 & z_1 & 0 & -x_1 \\ -zy_1 & 0 & z & 0 \\ 0 & 0 & -\frac{kzz_1}{y} & \frac{pzy_1}{y} \end{bmatrix} \text{ and } L_{\mathrm{B}} = \frac{1}{z}\begin{bmatrix} z_1 & 0 & 0 & -1 \\ 0 & z_1 & 0 & -x_1 \\ -zy_1 & 0 & z & 0 \\ \frac{kxz_1}{y} & -\frac{kz_1}{y} & 0 & -\frac{p(x-x_1)}{y} \end{bmatrix}, \quad (6.7)$$

when considering the edge-modified forms of y_{13}, and of y_{13} and z_{13}, respectively; and

$$L_{\mathrm{C}} = \frac{1}{z}\begin{bmatrix} z_1 & 0 & 0 & -1 \\ xz_1 & 0 & zz_1 & -(x+zy_1) \\ -zy_1 & 0 & z & 0 \\ 0 & 0 & -\frac{kzz_1}{y} & \frac{pzy_1}{y} \end{bmatrix} \text{ and } L_{\mathrm{D}} = \frac{1}{z}\begin{bmatrix} z_1 & 0 & 0 & -1 \\ 0 & z_1 & 0 & -x_1 \\ -x_1 & 1 & 0 & 0 \\ \frac{kxz_1}{y} & -\frac{kz_1}{y} & 0 & \frac{p(x_1-x)}{y} \end{bmatrix}, \quad (6.8)$$

when considering the edge-modified forms of x_{13} and y_{13}, and of z_{13}, respectively. All other combinations of (6.5) and (6.6) result in matrices which do not satisfy the defining equation (2.10).

6.3 Generalized Hietarinta systems

In [35], the authors introduced generalizations of Hietarinta's systems [15] by considering a general dispersion law,

$$\mathbb{G}(\omega,\kappa) := \omega^3 - \kappa^3 + \alpha_2(\omega^2 - \kappa^2) + \alpha_1(\omega - \kappa), \tag{6.9}$$

where α_1 and α_2 are constant parameters. For example, for the special case $a = \alpha_1 = \alpha_2 = 0$, one gets $\mathbb{G}(-p,-a) = -p^3$ and $\mathbb{G}(-q,-a) = -q^3$. Then (6.11) (below) reduces to Hietarinta's original A-2 system in [17, p. 95]. The term with coefficient b_0 could be removed by a simple transformation [15]. We will keep it to cover the most general case. In [12, 15], p^3 and q^3 are identified with p and q, respectively.

The explicit form of $\mathbb{G}(\omega,\kappa)$ in (6.9) is not needed [10] to compute Lax pairs. However, for the B-2 system the condition

$$\mathbb{G}(-p,-k) + \mathbb{G}(-k,-q) = \mathbb{G}(-p,-q) \tag{6.10}$$

must hold for 3D consistency and, consequently, for the computation of Lax pairs.

Zhang *et al.* [35] computed 4 by 4 Lax matrices for these generalized systems with the direct linearization method. By incorporating the edge equations (as shown in Section 3), we were able to find 3 by 3 matrices which are presented in this section. Computational details will appear in a forthcoming paper [10].

6.3.1 Generalized Hietarinta A-2 system

The generalized Hietarinta A-2 system [35] is given by

$$\begin{aligned} &z x_1 - y_1 - x = 0, \quad z x_2 - y_2 - x = 0, \quad \text{and} \\ &y - x z_{12} + b_0 x + \frac{\mathbb{G}(-p,-a)x_1 - \mathbb{G}(-q,-a)x_2}{z_2 - z_1} = 0. \end{aligned} \tag{6.11}$$

From edge constraint $x_1 = \frac{x + y_1}{z}$ one gets $x_3 = \frac{x + y_3}{z}$. Here,

$$\begin{aligned} &x_{13} = \frac{x_1 - x_3}{z_1 - z_3}, \quad y_{13} = \frac{x_1 z_3 - x_3 z_1}{z_1 - z_3}, \quad \text{and} \\ &z_{13} = \frac{(y + b_0 x)(z_1 - z_3) + \mathbb{G}(-k,-a)x_3 - \mathbb{G}(-p,-a)x_1}{x(z_1 - z_3)}. \end{aligned} \tag{6.12}$$

The 3 by 3 Lax matrices are presented in Table 3, together with the gauge transformations that connect them. For L_{a} one has $\frac{s\,t_2}{t\,s_1} \doteq \frac{z_1}{z_2}$, hence, $t = s = \frac{1}{z}$. For L_{b} we set $t = s = 1$ since $\frac{s\,t_2}{t\,s_1} \doteq 1$.

Alternative forms of (6.12) (after incorporating edge constraints) are

$$\begin{aligned} &\tilde{x}_{13} = \frac{y_1 - y_3}{z(z_1 - z_3)}, \quad \tilde{y}_{13} = -\left(\frac{x}{z} + \frac{y_3 z_1 - y_1 z_3}{z(z_1 - z_3)}\right), \quad \text{and} \\ &\tilde{z}_{13} = \frac{y + b_0 x}{x} + \frac{\mathbb{G}(-k,-a)(x + y_3) - \mathbb{G}(-p,-a)(x + y_1)}{x z(z_1 - z_3)}. \end{aligned} \tag{6.13}$$

Using only (6.12) with (3.5) leads to a 4 by 4 Lax matrix (not shown) that is trivially associated with L_{a}. Similarly, as shown in [35], using only (6.13) with (3.5) results in a 4 by 4 Lax matrix (not shown) which is trivially associated with L_{b}.

Table 3: Generalized Hietarinta A-2 system Lax pairs and gauge matrices

Substitutions	ψ	Matrices L of Lax pair
Writing the edge constraint as $x_3 = \frac{y_3+x}{z}$ yields		
$x_3 = \dfrac{xF+g}{zF}$, $y_3 = \dfrac{g}{F}$, $z_3 = \dfrac{h}{F}$.	$\psi_{\mathrm{a}} = \begin{bmatrix} F \\ g \\ h \end{bmatrix}$	$L_{\mathrm{a}} = \dfrac{1}{z}\begin{bmatrix} -zz_1 & 0 & z \\ xz_1 & z_1 & -zx_1 \\ \ell_{31} & -\frac{\mathbb{G}(-k,-a)}{x} & \frac{(y+b_0x)z}{x} \end{bmatrix}$,
where $\ell_{31} = \frac{1}{x}\big(\mathbb{G}(-p,-a)zx_1 - \mathbb{G}(-k,-a)x - (y+b_0x)zz_1\big)$.		
Writing the edge constraint as $y_3 = zx_3 - x$ yields		
$x_3 = \dfrac{f}{F}$, $z_3 = \dfrac{h}{F}$, $y_3 = -\dfrac{xF-zf}{F}$.	$\psi_{\mathrm{b}} = \begin{bmatrix} F \\ f \\ h \end{bmatrix}$	$L_{\mathrm{b}} = \begin{bmatrix} -z_1 & 0 & 1 \\ -x_1 & 1 & 0 \\ \tilde{\ell}_{31} & -\frac{\mathbb{G}(-k,-a)}{x} & \frac{y+b_0x}{x} \end{bmatrix}$,
where $\tilde{\ell}_{31} = \frac{1}{x}\big(\mathbb{G}(-p,-a)x_1 - (y+b_0x)z_1\big)$.		
Gauge transformations for L_{a} and L_{b} are given by		
$L_{\mathrm{b}} = \mathcal{G}_1 L_{\mathrm{a}} \mathcal{G}^{-1}, \psi_{\mathrm{b}} = \mathcal{G}\psi_{\mathrm{a}}$, $L_{\mathrm{a}} = \bar{\mathcal{G}}_1 L_{\mathrm{b}} \bar{\mathcal{G}}^{-1}, \psi_{\mathrm{a}} = \bar{\mathcal{G}}\psi_{\mathrm{b}}$, where $\bar{\mathcal{G}} = \mathcal{G}^{-1}$.	$\mathcal{G} = \begin{bmatrix} 1 & 0 & 0 \\ \frac{x}{z} & \frac{1}{z} & 0 \\ 0 & 0 & 1 \end{bmatrix}$	$\bar{\mathcal{G}} = \begin{bmatrix} 1 & 0 & 0 \\ -x & z & 0 \\ 0 & 0 & 1 \end{bmatrix}$,

6.3.2 Generalized Hietarinta B-2 system

The generalized Hietarinta B-2 system [35],

$$
\begin{aligned}
&xx_1 - z_1 - y = 0, \quad xx_2 - z_2 - y = 0, \quad \text{and} \\
&y_{12} + \alpha_1 + z + \alpha_2(x_{12} - x) - xx_{12} + \frac{\mathbb{G}(-p,-q)}{x_2 - x_1} = 0,
\end{aligned}
\tag{6.14}
$$

has edge constraint $x_1 = \dfrac{z_1 + y}{x}$ which yields $x_3 = \dfrac{z_3 + y}{x}$. Here,

$$
\begin{aligned}
&x_{13} = \frac{y_1 - y_3}{x_1 - x_3}, \quad z_{13} = \frac{x_3 y_1 - x_1 y_3}{x_1 - x_3}, \quad \text{and} \\
&y_{13} = (\alpha_2 x - \alpha_1 - z) + \frac{(x - \alpha_2)(y_1 - y_3) + \mathbb{G}(-p,-k)}{x_1 - x_3}.
\end{aligned}
\tag{6.15}
$$

The 3 by 3 Lax matrices with their gauge transformations are listed in Table 4. For L_{a} one has $\frac{s\,t_2}{t\,s_1} \doteq \frac{x_1}{x_2}$, hence, $t = s = \frac{1}{x}$. For L_{b} we set $t = s = 1$ since $\frac{s\,t_2}{t\,s_1} \doteq 1$.

Other forms of (6.15) by incorporating edge constraints are

$$
\begin{aligned}
&\tilde{x}_{13} = \frac{x(y_1 - y_3)}{z_1 - z_3}, \quad \tilde{z}_{13} = -\frac{y(y_3 - y_1) + y_3 z_1 - y_1 z_3}{z_1 - z_3}, \quad \text{and} \\
&\tilde{y}_{13} = (\alpha_2 x - \alpha_1 - z) + \frac{x\Big((x - \alpha_2)(y_1 - y_3) + \mathbb{G}(-p,-k)\Big)}{z_1 - z_3}.
\end{aligned}
\tag{6.16}
$$

As shown in [35], using only (6.16) with (3.5) leads to a 4 by 4 Lax matrix (not shown) that is trivially associated with L_{b}. Using only (6.15) with (3.5) results in a 4 by 4 Lax matrix (not shown) which is trivially associated with L_{a} when evaluated against the given system.

Table 4: Generalized Hietarinta B-2 system Lax pairs and gauge matrices

Substitutions	ψ	Matrices L of Lax pair
Writing the edge constraint as $x_3 = \frac{y+z_3}{x}$ yields		
$x_3 = \dfrac{yF+h}{xF}$, $y_3 = \dfrac{g}{F}$, $z_3 = \dfrac{h}{F}$.	$\psi_{\mathrm{a}} = \begin{bmatrix} F \\ g \\ h \end{bmatrix}$	$L_{\mathrm{a}} = \dfrac{1}{x}\begin{bmatrix} y - xx_1 & 0 & 1 \\ \ell_{21} & x(x-\alpha_2) & \ell_{23} \\ -yy_1 & xx_1 & -y_1 \end{bmatrix}$,
where $\ell_{21} = (\alpha_2 x - \alpha_1 - z)(y - xx_1) + x\big((\alpha_2 - x)y_1 - \mathbb{G}(-p,-k)\big)$ and $\ell_{23} = \alpha_2 x - \alpha_1 - z$.		
Writing the edge constraint as $z_3 = xx_3 - y$ yields		
$x_3 = \dfrac{f}{F}$, $y_3 = \dfrac{g}{F}$, $z_3 = -\dfrac{yF - xf}{F}$	$\psi_{\mathrm{b}} = \begin{bmatrix} F \\ f \\ g \end{bmatrix}$	$L_{\mathrm{b}} = \begin{bmatrix} -x_1 & 1 & 0 \\ -y_1 & 0 & 1 \\ \ell_{31} & \alpha_2 x - \alpha_1 - z & x - \alpha_2 \end{bmatrix}$,
where $\ell_{31} = -(\alpha_2 x - \alpha_1 - z)x_1 + (\alpha_2 - x)y_1 - \mathbb{G}(-p,-k)$.		
Gauge transformations for L_{a} and L_{b} are given by		
$L_{\mathrm{b}} = \mathcal{G}_1 L_{\mathrm{a}} \mathcal{G}^{-1}, \psi_{\mathrm{b}} = \mathcal{G}\psi_{\mathrm{a}}$, $L_{\mathrm{a}} = \bar{\mathcal{G}}_1 L_{\mathrm{b}} \bar{\mathcal{G}}^{-1}, \psi_{\mathrm{a}} = \bar{\mathcal{G}}\psi_{\mathrm{b}}$, where $\bar{\mathcal{G}} = \mathcal{G}^{-1}$.	$\mathcal{G} = \begin{bmatrix} 1 & 0 & 0 \\ \frac{y}{x} & 0 & \frac{1}{x} \\ 0 & 1 & 0 \end{bmatrix}$	$\bar{\mathcal{G}} = \begin{bmatrix} 1 & 0 & 0 \\ 0 & 0 & 1 \\ -y & x & 1 \end{bmatrix}$,

6.3.3 Generalized Hietarinta C-3 system

The generalized Hietarinta C-3 system [35],

$$
\begin{aligned}
&zy_1 + x_1 - x = 0, \quad zy_2 + x_2 - x = 0, \quad \text{and} \\
&\mathbb{G}(-a,-b)x_{12} - yz_{12} + z\left(\frac{\mathbb{G}(-q,-b)y_2 z_1 - \mathbb{G}(-p,-b)y_1 z_2}{z_1 - z_2}\right) = 0,
\end{aligned} \tag{6.17}
$$

has edge constraint $x_1 = x - zy_1$ leading to $x_3 = x - zy_3$. Here,

$$
\begin{aligned}
x_{13} &= \frac{x_3 z_1 - x_1 z_3}{z_1 - z_3}, \quad y_{13} = \frac{x_1 - x_3}{z_1 - z_3}, \quad \text{and} \\
z_{13} &= \frac{\mathbb{G}(-a,-b)(x_3 z_1 - x_1 z_3) + z\Big(\mathbb{G}(-k,-b)y_3 z_1 - \mathbb{G}(-p,-b)y_1 z_3\Big)}{y(z_1 - z_3)}.
\end{aligned} \tag{6.18}
$$

The 3 by 3 Lax matrices with their gauge transformations are given in Table 5. For $L_{\rm a}$ and $L_{\rm b}$ we set $t = s = \frac{1}{z}$ since $\frac{s\,t_2}{t\,s_1} \doteq \frac{z_1}{z_2}$.

Table 5: Generalized Hietarinta C-3 system Lax pairs and gauge matrices

Substitutions	ψ	Matrices L of Lax pair
Writing the edge constraint as $x_3 = x - zy_3$ yields		
$x_3 = \dfrac{xF - zg}{F}$, $y_3 = \dfrac{g}{F}$, $z_3 = \dfrac{h}{F}$.	$\psi_{\rm a} = \begin{bmatrix} F \\ g \\ h \end{bmatrix}$	$L_{\rm a} = \dfrac{1}{z}\begin{bmatrix} -z_1 & 0 & 1 \\ x - x_1 & -z & 0 \\ -\mathbb{G}(-a,-b)\frac{xz_1}{y} & \ell_{32} & \ell_{33} \end{bmatrix}$,
with $\ell_{32} = \big(\mathbb{G}(-a,-b) - \mathbb{G}(-k,-b)\big)\frac{xz_1}{y}$, $\ell_{33} = \big(\mathbb{G}(-a,-b)x_1 + \mathbb{G}(-p,-b)zy_1\big)\frac{1}{y}$.		
Writing the edge constraint as $y_3 = \frac{x - x_3}{z}$ yields		
$x_3 = \dfrac{f}{F}$, $z_3 = \dfrac{h}{F}$, $y_3 = \dfrac{xF - f}{zF}$.	$\psi_{\rm b} = \begin{bmatrix} F \\ f \\ h \end{bmatrix}$	$L_{\rm b} = \dfrac{1}{z}\begin{bmatrix} -z_1 & 0 & 1 \\ 0 & -z_1 & x_1 \\ -\mathbb{G}(-k,-b)\frac{xz_1}{y} & \tilde{\ell}_{32} & \tilde{\ell}_{33} \end{bmatrix}$,
with $\tilde{\ell}_{32} = \big(\mathbb{G}(-k,-b) - \mathbb{G}(-a,-b)\big)\frac{z_1}{y}$, $\tilde{\ell}_{33} = \big(\mathbb{G}(-a,-b)x_1 + \mathbb{G}(-p,-b)zy_1\big)\frac{1}{y}$.		
Gauge transformations for $L_{\rm a}$ and $L_{\rm b}$ are given by		
$L_{\rm b} = \mathcal{G}_1 L_{\rm a} \mathcal{G}^{-1}$, $\psi_{\rm b} = \mathcal{G}\psi_{\rm a}$, $L_{\rm a} = \bar{\mathcal{G}}_1 L_{\rm b} \bar{\mathcal{G}}^{-1}$, $\psi_{\rm a} = \bar{\mathcal{G}}\psi_{\rm b}$, where $\bar{\mathcal{G}} = \mathcal{G}^{-1}$.	$\mathcal{G} = \begin{bmatrix} 1 & 0 & 0 \\ x & -z & 0 \\ 0 & 0 & 1 \end{bmatrix}$	$\bar{\mathcal{G}} = \begin{bmatrix} 1 & 0 & 0 \\ \frac{x}{z} & -\frac{1}{z} & 0 \\ 0 & 0 & 1 \end{bmatrix}$.

Incorporating edge constraints into (6.18) yields

$$\tilde{x}_{13} = x + \frac{z(y_1 z_3 - y_3 z_1)}{z_1 - z_3}, \quad \tilde{y}_{13} = \frac{z(y_3 - y_1)}{z_1 - z_3}, \quad \text{and}$$
$$\tilde{z}_{13} = \frac{\mathbb{G}(-a,-b)\big(x(z_1 - z_3) + z(y_1 z_3 - y_3 z_1)\big)}{y(z_1 - z_3)} + \frac{z\Big(\mathbb{G}(-k,-b)y_3 z_1 - \mathbb{G}(-p,-b)y_1 z_3\Big)}{y(z_1 - z_3)}. \tag{6.19}$$

System (6.17) also admits extended Lax matrices:

$$L_{\rm A} = \frac{1}{z}\begin{bmatrix} -z_1 & 0 & 0 & 1 \\ 0 & -z_1 & 0 & x_1 \\ zy_1 & 0 & -z & 0 \\ -\mathbb{G}(-a,-b)\frac{xz_1}{y} & 0 & \big(\mathbb{G}(-a,-b) - \mathbb{G}(-k,-b)\big)\frac{zz_1}{y} & \ell_{44} \end{bmatrix}, \tag{6.20}$$

when considering the edge-modified solutions for y_{13} and z_{13} and where

$\ell_{44} = \big(\mathbb{G}(-a,-b)x - \big(\mathbb{G}(-a,-b) - \mathbb{G}(-p,-b)\big)zy_1\big)\frac{1}{y}$; and

$$L_{\mathrm{B}} = \frac{1}{z}\begin{bmatrix} -z_1 & 0 & 0 & 1 \\ -xz_1 & 0 & zz_1 & x - zy_1 \\ zy_1 & 0 & -z & 0 \\ -\mathbb{G}(-a,-b)\frac{xz_1}{y} & 0 & \big(\mathbb{G}(-a,-b) - \mathbb{G}(-k,-b)\big)\frac{zz_1}{y} & \ell_{44} \end{bmatrix}, \tag{6.21}$$

when considering the edge-modified solutions (6.19) and with ℓ_{44} as above; and

$$L_{\mathrm{C}} = \frac{1}{z}\begin{bmatrix} -z_1 & 0 & 0 & 1 \\ 0 & -z_1 & 0 & x_1 \\ zy_1 & 0 & -z & 0 \\ 0 & -\mathbb{G}(-a,-b)\frac{z_1}{y} & -\mathbb{G}(-k,-b)\frac{zz_1}{y} & \tilde{\ell}_{44} \end{bmatrix}, \tag{6.22}$$

when considering the edge-modified solutions for y_{13} and where $\tilde{\ell}_{44} = (\mathbb{G}(-a,-b)x_1 + \mathbb{G}(-p,-b)zy_1)\frac{1}{y}$. The matrix L_{C} was derived in [35, eq. (95)] using $\tilde{y}_{13}$. All other combinations of (6.18) and (6.19) result in matrices which do not satisfy the defining equation (2.10).

6.3.4 Generalized Hietarinta C-4 system

The generalized Hietarinta C-4 system [35] is given by

$$\begin{gathered} zy_1 + x_1 - x = 0, \quad zy_2 + x_2 - x = 0, \quad \text{and} \\ yz_{12} - z\left(\frac{\mathcal{G}(p)y_1z_2 - \mathcal{G}(q)y_2z_1}{z_1 - z_2}\right) - xx_{12} + \frac{1}{4}\mathbb{G}(-a,-b)^2 = 0, \end{gathered} \tag{6.23}$$

where

$$\mathcal{G}(\tau) := -\frac{1}{2}\big(\mathbb{G}(-\tau,-a) + \mathbb{G}(-\tau,-b)\big). \tag{6.24}$$

Edge constraint $x_1 = x - zy_1$ yields $x_3 = x - zy_3$. Here,

$$\begin{gathered} x_{13} = \frac{x_3z_1 - x_1z_3}{z_1 - z_3}, \quad y_{13} = \frac{x_1 - x_3}{z_1 - z_3}, \quad \text{and} \\ z_{13} = \frac{x(x_3z_1 - x_1z_3) - z\big(\mathcal{G}(k)y_3z_1 - \mathcal{G}(p)y_1z_3\big)}{y(z_1 - z_3)} - \frac{\mathbb{G}(-a,-b)^2}{4y}. \end{gathered} \tag{6.25}$$

Variants of (6.25) obtained by incorporating edge constraints are

$$\begin{gathered} \tilde{x}_{13} = \frac{x(z_1 - z_3) - z(y_3z_1 - y_1z_3)}{z_1 - z_3}, \quad \tilde{y}_{13} = -\frac{z(y_1 - y_3)}{z_1 - z_3}, \quad \text{and} \\ \tilde{z}_{13} = \frac{xz(y_1z_3 - y_3z_1) - z\big(\mathcal{G}(k)y_3z_1 - \mathcal{G}(p)y_1z_3\big)}{y(z_1 - z_3)} + \frac{4x^2 - \mathbb{G}(-a,-b)^2}{4y}. \end{gathered} \tag{6.26}$$

The 3 by 3 Lax matrices and the gauge transformations are given in Table 6. For L_{a} and L_{b} we take $t = s = \frac{1}{z}$ since $\frac{s\,t_2}{t\,s_1} \doteq \frac{z_1}{z_2}$.

Table 6: Generalized Hietarinta C-4 system Lax pairs and gauge matrices

Substitutions	ψ	Matrices L of Lax pair
Writing the edge constraint as $x_3 = x - zy_3$ yields		
$x_3 = \dfrac{xF - zg}{F}$, $y_3 = \dfrac{g}{F}$, $z_3 = \dfrac{h}{F}$.	$\psi_{\mathrm{a}} = \begin{bmatrix} F \\ g \\ h \end{bmatrix}$	$L_{\mathrm{a}} = \dfrac{1}{z}\begin{bmatrix} -z_1 & 0 & 1 \\ x - x_1 & -z & 0 \\ \ell_{31} & \frac{zz_1}{y}\left(x + \mathcal{G}(k)\right) & \ell_{33} \end{bmatrix}$,
with $\ell_{31} = -\frac{z_1}{y}\left(x^2 - \frac{1}{4}\mathbb{G}(-a,-b)^2\right)$, $\ell_{33} = \frac{1}{y}\left(xx_1 - \frac{1}{4}\mathbb{G}(-a,-b)^2 - \mathcal{G}(p)zy_1\right)$.		
Writing the edge constraint as $y_3 = \frac{x - x_3}{z}$ yields		
$x_3 = \dfrac{f}{F}$, $z_3 = \dfrac{h}{F}$, $y_3 = \dfrac{xF - f}{zF}$.	$\psi_{\mathrm{b}} = \begin{bmatrix} F \\ f \\ h \end{bmatrix}$	$L_{\mathrm{b}} = \dfrac{1}{z}\begin{bmatrix} -z_1 & 0 & 1 \\ 0 & -z_1 & x_1 \\ \tilde{\ell}_{31} & -\frac{z_1}{y}\left(x + \mathcal{G}(k)\right) & \tilde{\ell}_{33} \end{bmatrix}$.
with $\tilde{\ell}_{31} = \frac{z_1}{y}\left(\frac{1}{4}\mathbb{G}(-a,-b)^2 + \mathcal{G}(k)x\right)$, $\tilde{\ell}_{33} = \frac{1}{y}\left(xx_1 - \frac{1}{4}\mathbb{G}(-a,-b)^2 - \mathcal{G}(p)zy_1\right)$.		
Gauge transformations for L_{a} and L_{b} are given by		
$L_{\mathrm{b}} = \mathcal{G}_1 L_{\mathrm{a}} \mathcal{G}^{-1}, \psi_{\mathrm{b}} = \mathcal{G}\psi_{\mathrm{a}}$, $L_{\mathrm{a}} = \bar{\mathcal{G}}_1 L_{\mathrm{b}} \bar{\mathcal{G}}^{-1}, \psi_{\mathrm{a}} = \bar{\mathcal{G}}\psi_{\mathrm{b}}$, where $\bar{\mathcal{G}} = \mathcal{G}^{-1}$.	$\mathcal{G} = \begin{bmatrix} 1 & 0 & 0 \\ x & -z & 0 \\ 0 & 0 & 1 \end{bmatrix}$	$\bar{\mathcal{G}} = \begin{bmatrix} 1 & 0 & 0 \\ \frac{x}{z} & -\frac{1}{z} & 0 \\ 0 & 0 & 1 \end{bmatrix}$.

System (6.23) has the following extended Lax matrices:

$$L_{\mathrm{A}} = \frac{1}{z}\begin{bmatrix} -z_1 & 0 & 0 & 1 \\ 0 & -z_1 & 0 & x_1 \\ zy_1 & 0 & -z & 0 \\ \ell_{41} & 0 & \frac{zz_1}{y}\left(x + \mathcal{G}(k)\right) & \ell_{44} \end{bmatrix}, \tag{6.27}$$

when considering edge-modified y_{13} and z_{13}, and where $\ell_{41} = -\frac{z_1}{y}\left(x^2 - \frac{1}{4}\mathbb{G}(-a,-b)^2\right)$ and $\ell_{44} = \frac{1}{y}\left(x^2 - \frac{1}{4}\mathbb{G}(-a,-b)^2 - zy_1(x + \mathcal{G}(p))\right)$;

$$L_{\mathrm{B}} = \frac{1}{z}\begin{bmatrix} -z_1 & 0 & 0 & 1 \\ -xz_1 & 0 & zz_1 & x - zy_1 \\ zy_1 & 0 & -z & 0 \\ \ell_{41} & 0 & \frac{zz_1}{y}\left(x + \mathcal{G}(k)\right) & \ell_{44} \end{bmatrix}, \tag{6.28}$$

by taking the edge-modified expression of (6.26), with ℓ_{41} and ℓ_{44} as above; and

$$L_{\mathrm{C}} = \frac{1}{z}\begin{bmatrix} -z_1 & 0 & 0 & 1 \\ 0 & -z_1 & 0 & x_1 \\ zy_1 & 0 & -z & 0 \\ \frac{z_1\mathbb{G}(-a,-b)^2}{4y} & -\frac{xz_1}{y} & \frac{zz_1}{y}\mathcal{G}(k) & \tilde{\ell}_{44} \end{bmatrix} \tag{6.29}$$

when using the edge-modified expression for y_{13} and with $\tilde{\ell}_{44} = \frac{1}{y}\big(xx_1 - \frac{1}{4}\mathbb{G}(-a,-b)^2 - \mathcal{G}(p)zy_1\big)$. All other combinations of (6.25) and (6.26) result in matrices that fail to satisfy the defining equation (2.10).

7 Software implementation and conclusions

The method to find Lax pairs of PΔEs based on multi-dimensional consistency is being implemented in MATHEMATICA. Using our prototype MATHEMATICA package [8] we derived Lax matrices of minimal sizes for various Boussinesq-type equations. In turn, the research done for this chapter helped us improve and extend the capabilities of the software under development [11].

The way we symbolically compute (and verify) Lax pairs might slightly differ from the procedure used by other authors (by hand or interactively with a computer algebra system). Indeed, for a system of PΔEs, the software generates all equations (and solutions) necessary to define a full face of the quadrilateral. That is, for a system of PΔEs including full-face expressions (involving at least 3 corners of the quadrilateral) and edge equations (involving two adjacent corners of the quadrilateral), the software will first augment the given system with the additional edge equations necessary to complete the set of equations for a particular face of the cube. For example, the Schwarzian Boussinesq system (1.4) discussed in Section 1.4 is augmented with two additional edge equations,

$$z_2 y_{12} - x_{12} + x_2 = 0, \quad z_1 y_{12} - x_{12} + x_1 = 0, \tag{7.1}$$

to generate the full set of equations for the front face of the cube. Then, using lexicographical ordering ($x \prec y \prec z$) and an index ordering (double-subscripts $\prec$ no-subscripts $\prec$ single-subscripts), the software solves (1.4) and (7.1) yielding

$$x_{12} = \frac{x_2 z_1 - x_1 z_2}{z_1 - z_2}, \quad y_{12} = \frac{x_2 - x_1}{z_1 - z_2}, \quad z_{12} = \frac{z}{y}\left(\frac{p y_1 z_2 - q y_2 z_1}{z_1 - z_2}\right), \quad \text{with} \tag{7.2a}$$

$$x = x_1 - z y_1, \quad \text{and} \quad z = \frac{x_1 - x_2}{y_1 - y_2}. \tag{7.2b}$$

This process is then repeated for the left and bottom faces of the cube, always substituting solutions such as (7.2) to enforce consistency and remove redundancies.

The complete front-corner system (i.e., front, left and bottom faces connected at corner $\mathbf{x}$, see Fig. 3) is used to simplify the Lax equation when a Lax pair is finally tested. If all works as planned, the evaluation of the Lax equation then automatically results in a zero matrix.

For verification of consistency about the cube, the equations for the faces connected at the back-corner (where $\mathbf{x}_{123}$ is located as shown in Fig. 3) are computed and then solved (adhering to the above ordering). Next, these solutions are then checked for consistency with the front-corner system. Finally, if the system is 3D consistent, the multiple expressions obtained for $\mathbf{x}_{123}$ should be equal when reduced using (7.2) augmented with like equations for $\mathbf{x}_{13}$ and $\mathbf{x}_{23}$.

Why would one care about different Lax matrices, in particular, if they are gauge equivalent? In the PDE case, application of the IST is easier if one selects a Lax pair of a specific form (i.e., the eigenvalues should appear in the diagonal entries), chosen from the infinite number of gauge equivalent pairs. Thus, for the KdV equation one may prefer to work with (2.8) instead of (2.4). Similar issues arise for PΔEs. Among the family of gauge-equivalent Lax matrices for PΔEs, which one should be selected so that, for example, the IST or staircase method [33] could be applied? (The latter method is used to find first integrals for periodic reductions of integrable PΔEs). In addition, one has to select an appropriate (separation) factor $t(\mathbf{x}, \mathbf{x}_1; p, k)$ (see Sec. 3.1.1). These issues are not addressed in this chapter for they require further study.

Acknowledgments

This material is based in part upon research supported by the National Science Foundation (NSF) under Grant No. CCF-0830783. Any opinions, findings, and conclusions or recommendations expressed in this material are those of the authors and do not necessarily reflect the views of the NSF.

References

[1] Ablowitz M J and Clarkson P A, *Solitons, Nonlinear Evolution Equations and Inverse Scattering*, (London Math. Soc. Lect. Note Ser. **149**) Cambridge Univ. Press, Cambridge, UK, 1991.

[2] Ablowitz M J, Kaup D J, Newell A C, and Segur H, The inverse scattering transform–Fourier analysis for nonlinear problems, *Stud. Appl. Math.* **53**(4), 249–315, 1974.

[3] Ablowitz M J and Ladik F J, A nonlinear difference scheme and inverse scattering, *Stud. Appl. Math.* **55**(3), 213–229, 1976.

[4] Ablowitz M J and Ladik F J, On the solution of a class of nonlinear partial difference equations, *Stud. Appl. Math.* **57**(1), 1–12, 1977.

[5] Adler V E, Bobenko A I, and Suris Yu B, Classification of integrable equations on quad-graphs. The consistency approach, *Commun. Math. Phys.* **233**(3), 513–543, 2003.

[6] Bobenko A I and Suris Yu B, Integrable systems on quad-graphs, *Int. Math. Res. Not.* **2002**(11), 573–611, 2002.

[7] Bobenko A I and Suris Yu B, *Discrete Differential Geometry: Integrable Structure*, (Grad. Stud. Math. **98**) AMS, Philadelphia, PA, 2008.

[8] Bridgman T J, `LaxPairPartialDifferenceEquations.m`: a Mathematica package for the symbolic computation of Lax pairs of systems of nonlinear partial difference equations defined on quadrilaterals,

`http://inside.mines.edu/~whereman/software/LaxPairPartialDifferenceEquations`, 2012-2019.

[9] Bridgman T J, *Symbolic Computation of Lax Pairs of Nonlinear Partial Difference Equations*, Ph.D Thesis, Dept. Appl. Maths. Stats., Colorado School of Mines, 2018.

[10] Bridgman T and Hereman W, Gauge-equivalent Lax pairs for Boussinesq-type systems of partial difference equations, Report, Dept. Appl. Maths. Stats., Colorado School of Mines, 26 pages, submitted, 2018.

[11] Bridgman T and Hereman W, Symbolic software for the computation of Lax pairs of nonlinear partial difference equations, in preparation, 2019.

[12] Bridgman T, Hereman W, Quispel G R W, and van der Kamp P H, Symbolic computation of Lax pairs of partial difference equations using consistency around the cube, *Found. Comput. Math.* **13**(4), 517–544, 2013.

[13] Faddeev L D and Takhtajan L A, *Hamiltonian Methods in the Theory of Solitons*, (Springer Ser. Sov. Math.) Springer-Verlag, Berlin, Germany, 1987.

[14] Hickman M, Hereman W, Larue J, and Göktaş Ü, Scaling invariant Lax pairs of nonlinear evolution equations, *Applicable Analysis* **91**(2), 381–402, 2012.

[15] Hietarinta J, Boussinesq-like multi-component lattice equations and multi-dimensional consistency, *J. Phys. A: Math. Theor.* **44**(16), 165204, 22 pages, 2011.

[16] Hietarinta J, Elementary introduction to discrete soliton equations. In: Euler N (Ed), *Nonlinear Systems and Their Remarkable Mathematical Structures* **1**, Chapter A4, 74–93, CRC Press, Boca Raton, Florida, 2018.

[17] Hietarinta J, Joshi N, and Nijhoff F W, *Discrete Systems and Integrability*, (Cambridge Texts Appl. Math.) Cambridge Univ. Press, Cambridge, UK, 2016.

[18] Hirota R, Nonlinear partial difference equations. I - A difference analogue of the Korteweg-de Vries equation, *J. Phys. Soc. Jpn.* **43**(4), 1424–1433, 1977.

[19] Hirota R, *The Direct Methods in Soliton Theory*, (Cambridge Tracts Math.) Cambridge Univ. Press, Cambridge, UK, 2004.

[20] Hydon P E, *Difference Equations by Differential Equation Methods*, (Cambridge Monographs Appl. Comp. Math. **27**) Cambridge Univ. Press, Cambridge, UK, 2014.

[21] Levi D, Olver P, Thomova Z, and Winternitz P (Eds), *Symmetries and Integrability of Difference Equations*, (London Math. Soc. Lect. Note Ser. **381**) Cambridge Univ. Press, Cambridge, UK, 2011.

[22] Levi D, Verge-Rebelo R, and Winternitz P (Eds), *Symmetries and Integrability of Difference Equations*, (CRM Ser. Math. Phys.) Springer Int. Publ., New York, 2017.

[23] Miura R M, Korteweg-de Vries equation and generalizations. I. A remarkable explicit nonlinear transformation, *J. Math. Phys.* **9**(8), 1202–1204, 1968.

[24] Nijhoff F W, On some "Schwarzian" equations and their discrete analogues. In: Fokas A S and Gelfand I M (Eds), *Algebraic Aspects of Integrable Systems: In memory of Irene Dorfman*, 237–260, Birkhäuser Verlag, Boston, MA, 1997.

[25] Nijhoff F W, Discrete Painlevé equations and symmetry reduction on the lattice. In: Bobenko A I and Seiler R (Eds), *Discrete Integrable Geometry and Physics*, 209–234, (Oxford Lect. Ser. Math. Appls.) Oxford Univ. Press, New York, 1999.

[26] Nijhoff F W, Lax pair for the Adler (lattice Krichever-Novikov) system, *Phys. Lett. A* **297**(1-2), 49–58, 2002.

[27] Nijhoff F and Capel H, The discrete Korteweg-de Vries equation, *Acta Appl. Math.* **39**(1-3), 133–158, 1995.

[28] Nijhoff F W and Papageorgiou V G, Similarity reductions of integrable lattices and discrete analogues of the Painlevé II equation, *Phys. Lett. A* **153**(6-7), 337–344, 1991.

[29] Nijhoff F W, Papageorgiou V G, Capel H W, and Quispel G R W, The lattice Gel'fand-Dikii hierarchy, *Inv. Probl.* **8**(4), 597–621, 1992.

[30] Nijhoff F W, Quispel G R W, and Capel H W, Direct linearization of nonlinear difference-difference equations, *Phys. Lett. A* **97**(4), 125–128, 1983.

[31] Nijhoff F W, van der Linden J, Quispel G R W, Capel H W, and Velthuizen J, Linearization of the nonlinear Schrödinger equation and the isotropic Heisenberg spin chain, *Physica A* **116**(1-2), 1–33, 1982.

[32] Tongas A and Nijhoff F W, The Boussinesq integrable system: compatible lattice and continuum structures, *Glasgow Math. J.* **47**(A), 205–219, 2005.

[33] van der Kamp P H and Quispel G R W, The staircase method: integrals for periodic reductions of integrable lattice equations, *J. Phys. A: Math. Theor.* **43**(46), 465207, 34 pages, 2010.

[34] Wahlquist H D and Estabrook F B, Bäcklund transformation for solutions of the Korteweg-de Vries equation, *Phys. Rev. Lett.* **31**(23), 1386–1390, 1973.

[35] Zhang D-J, Zhao S-L, and Nijhoff F W, Direct linearization of extended lattice BSQ systems, *Stud. Appl. Math.* **129**(2), 220–248, 2012.

A4. Lie point symmetries of delay ordinary differential equations

Vladimir A. Dorodnitsyn [a], *Roman Kozlov* [b], *Sergey V. Meleshko* [c] *and Pavel Winternitz* [d]

[a] *Keldysh Institute of Applied Mathematics, Russian Academy of Science, Miusskaya Pl. 4, Moscow, 125047, Russia;*
e-mail: Dorodnitsyn@Keldysh.ru
[b] *Department of Business and Management Science, Norwegian School of Economics, Helleveien 30, 5045, Bergen, Norway;*
e-mail: Roman.Kozlov@nhh.no
[c] *School of Mathematics, Institute of Science, Suranaree University of Technology, 30000, Thailand;*
e-mail: sergey@math.sut.ac.th
[d] *Centre de Recherches Mathématiques and Département de mathématiques et de statistique, Université de Montréal, Montréal, QC, H3C 3J7, Canada;*
e-mail: wintern@crm.umontreal.ca

Abstract

Lie point symmetries of delay ordinary differential equations (DODEs) accompanied by an equation for the delay parameter (delay relation) are considered. A subset of such systems (delay ordinary differential systems or DODSs) which consists of linear DODEs and solution independent delay relations have infinite-dimensional symmetry algebras, as do nonlinear ones that are linearizable by an invertible transformation of variables. Moreover, the symmetry algebras of these linear or linearizable DODSs of order N contain a subalgebra of dimension $\dim L = 2N$ realized by linearly connected vector fields. Genuinely nonlinear DODSs of order N have symmetry algebras of dimension n, $0 \leq n \leq 2N + 2$. It is shown how exact analytical solutions of invariant DODSs can be obtained using symmetry reduction. In particular we present invariant solutions of a DODS originating in a study of traffic flow.

1 Introduction

Lie groups have provided efficient tools for studying ordinary and partial differential equations since their introduction in the seminal work of Sophus Lie in the late 19th century [28, 29, 30, 31]. The symmetry group of a differential equation transforms solutions into solutions while leaving the set of all solutions invariant. The symmetry group can be used to obtain new solutions from known ones and to classify equations into equivalence classes according to their symmetry groups. It can also be used to obtain exact analytic solutions that are invariant under some subgroup of the symmetry group, so called "group invariant solutions". Most solutions of the nonlinear differential equations occurring in physics and in other applications were obtained in this manner. Applications of Lie group theory to differential equations, known as group analysis, is the subject of many books and review articles [41, 22, 40, 2, 23, 17].

More recently, applications of Lie group analysis have been extended to discrete equations [34, 35, 4, 26, 5, 6, 49, 16, 8, 9, 27, 52, 7, 21], both difference and differential-difference ones. The applications are the same as in the case of differential equations. One can classify discrete equations into symmetry classes, obtain invariant solutions and do everything that one does for differential equations.

The present paper reviews extension of group analysis applications to another type of equations, namely delay differential equations. These are equations involving a field $y(x)$ and its derivatives, all evaluated not only at the point x, but also at some preceding points x_-. In general, the independent and dependent variables x and y can be vectors. We restrict to the case when both are scalars. Thus we have just one equation and one independent variable, the case of a delay ordinary differential equation (DODE). For simplicity we first consider only first-order DODEs with one delay. However, one can treat more complicated DODEs in a similar manner and we present results on N-th order DODEs in Section 8 below.

Differential delay equations play an important role in the description of many phenomena in physics, chemistry, engineering, biology, economics, etc. They occur whenever the state of a system depends on previous states. For instance, a patient's blood pressure $y(t)$ may depend on the blood pressure $y(t_-)$ at the moment t_- when medication had been delivered. The state of a bank account may depend on transactions made at previous times. Traffic flow on a road may depend on the driver's reaction time.

A sizable literature exists on delay differential equations and their applications [14, 20, 36, 13, 37, 53, 24, 50, 15]. In particular for applications in many fields and for a useful list of references we turn to the book [15] by T. Erneux. Studies have been devoted to the existence and stability of solutions, to the general properties of solutions and to questions of periodicity and oscillations.

Nearly all known solutions to delay differential equations are numerical ones. Analytical solutions have been obtained by postulating a specific parameterized form and then adjusting the parameters to satisfy the equation [44, 45, 46, 43]. For specific applications in biology, see [1, 3, 32, 33]. For earlier work on symmetry analysis for delay-differential equations see [48].

We consider scalar first-order delay ordinary differential equations (DODEs) with one delay, discuss their symmetry properties and present some invariant solutions. The chapter is organized as follows. Section 2 gives an illustrating example of DODE symmetries and shows how one can use them to find particular solutions. In Section 3 we recall some basic facts concerning DODEs. We define a DODS, that is a DODE together with a delay equation that specifies the position of the delay point x_-. In Section 4 we provide the general theory and outline the method for constructing invariant DODSs. Section 5 is devoted to linear DODEs with solution independent delay relations. They all admit infinite-dimensional symmetry groups. The Lie group classification of nonlinear first-order DODSs is discussed in Section 6. It will be given in Table 1. In Section 7 we show how symmetries can be used to find particular (namely, group invariant) solutions of DODEs. We discuss higher order

DODSs in Section 8. In Section 9 we use symmetries to find particular solutions of a traffic micro-model equation. Finally, the conclusions are presented in Section 10.

2 Illustrating example

Let us consider a DODE with a constant delay parameter

$$\dot{y} = \frac{\Delta y}{\Delta x} + C_1 e^x, \qquad \Delta x = C_2, \qquad C_2 > 0. \tag{2.1}$$

These equations admit symmetries

$$X_1 = \frac{\partial}{\partial y}, \quad X_2 = x\frac{\partial}{\partial y}, \quad X_3 = \frac{\partial}{\partial x} + y\frac{\partial}{\partial y}. \tag{2.2}$$

One can easily see that the corresponding Lie group transformations

$$\begin{aligned}
X_1 : \quad & \bar{x} = e^{\varepsilon X_1}(x) = x, & \bar{y} = e^{\varepsilon X_1}(y) = y + \varepsilon, \\
X_2 : \quad & \bar{x} = e^{\varepsilon X_2}(x) = x, & \bar{y} = e^{\varepsilon X_2}(y) = y + \varepsilon x, \\
X_3 : \quad & \bar{x} = e^{\varepsilon X_3}(x) = x + \varepsilon, & \bar{y} = e^{\varepsilon X_3}(y) = e^{\varepsilon} y
\end{aligned}$$

do not change the form of the DODE, nor the delay relation.

We can find a particular solution of (2.1), which is an invariant solution for the symmetry X_3. It has the form

$$y = Ae^x, \qquad x_- = x - B, \tag{2.3}$$

where A and B are constants. This form will be explained in Section 7. For now we just mention that the solution form is invariant with respect to the group transformation generated by the symmetry X_3. Substituting the form (2.3) into the DODS (2.1), we obtain the conditions on the constants

$$A = \frac{C_1 C_2}{C_2 - 1 + e^{-C_2}}, \qquad B = C_2.$$

3 Formulation of the problem for first-order DODEs

We consider first-order delay ordinary differential equations (DODEs)

$$\dot{y} = f(x, y, y_-), \qquad \frac{\partial f}{\partial y_-} \not\equiv 0, \qquad x \in I, \tag{3.1}$$

where $I \subset \mathbb{R}$ is some finite or semifinite interval. We will be interested in symmetry properties of these DODEs, which are considered locally, independently of the initial conditions. For equation (3.1) we have to specify the delayed point x_- where the delayed function value $y_- = y(x_-)$ is taken, otherwise the problem is not fully determined. Therefore, we supplement the DODE with a delay relation

$$x_- = g(x, y, y_-), \qquad g(x, y, y_-) \not\equiv \text{const}, \qquad x_- < x. \tag{3.2}$$

The two equations (3.1) and (3.2) together will be called a *delay ordinary differential system* (DODS). Here f and g are arbitrary smooth functions. Sometimes for convenience we shall write (3.2) in the equivalent form

$$\Delta x = x - x_- = \tilde{g}(x, y, y_-), \qquad \tilde{g}(x, y, y_-) = x - g(x, y, y_-).$$

In most of the existing literature, the delay parameter Δx is considered to be constant

$$\Delta x = \tau > 0, \qquad \tau = \text{const}. \tag{3.3}$$

An alternative is to impose a specific form of the function (3.2) to include some physical features of the delay Δx.

In the classification that we are performing we will find all special cases of functions f and g, when the DODS under consideration will possess a nontrivial symmetry group. For all such cases of f and g the corresponding group will be presented.

We will be interested in group transformations, leaving the Eqs. (3.1),(3.2) invariant. That means that the transformation will transform solutions of the DODS into solutions. They leave the set of all solutions invariant. Let us stress that we need to consider these two equations together. This makes our approach similar to one of the approaches for considering symmetries of discrete equations (see, for example, [7, 8]), where the invariance is required for both the discrete equation and the equation for the lattice on which the discrete equation is considered. Here, we have the DODE (3.1) instead of a discrete equation and the delay relation (3.2) instead of a lattice equation.

In order to solve a DODE on some interval I, one must add initial conditions to the DODS (3.1),(3.2). Contrary to the case of ordinary differential equations, the initial condition must be given by a function $\varphi(x)$ on an initial interval $I_0 \subset \mathbb{R}$, e.g.

$$y(x) = \varphi(x), \qquad x \in [x_{-1}, x_0]. \tag{3.4}$$

For $x - x_- = \tau = x_0 - x_{-1}$ constant this leads to the *method of steps* [37] for solving the DODE either analytically or numerically. Thus for $x_0 \leq x \leq x_1 = x_0 + \tau$ we replace

$$y_- = y(x_-) = \varphi(x - \tau)$$

and this reduces the DODE (3.1) to an ODE

$$\dot{y}(x) = f(x, y(x), \varphi(x - \tau)), \qquad x_0 \leq x \leq x_1 = x_0 + \tau \tag{3.5}$$

which is solved with initial condition $y(x_0) = \varphi(x_0)$.

On the second step we consider the same procedure: we solve the ODE

$$\dot{y}(x) = f(x, y(x), y(x - \tau)), \qquad x_1 \leq x \leq x_2 = x_1 + \tau, \tag{3.6}$$

where $y(x - \tau)$ and the initial condition are known from the first step. Thus we continue until we cover the entire interval I, at each step solving the ODE with input from the previous step. This procedure provides a solution that is in general continuous in the points $x_n = x_0 + n\tau$, $n = 0, 1, 2, ...$ but not smooth.

The condition (3.3) is very restrictive and rules out most of the symmetries that could be present for ODEs. Hence we consider the more general case of the DODS (3.1),(3.2) and we must adapt the method of steps to this case. We again use the initial condition (3.4), which is required to satisfy the condition

$$x_{-1} = g(x_0, \varphi(x_0), \varphi(x_{-1})).$$

We will also assume that

$$x_{-1} \leq x_{-} < x. \tag{3.7}$$

On the first step we solve the system

$$\dot{y}(x) = f(x, y(x), \varphi(x_{-})), \tag{3.8a}$$
$$x_{-} = g(x, y(x), \varphi(x_{-})), \tag{3.8b}$$

for $y(x)$ and $x_{-}(x)$ from the point x_0 with initial condition $y(x_0) = \varphi(x_0)$. The system is solved forward till a point x_1 such that $x_0 = g(x_1, y(x_1), y(x_0))$. Thus, we obtain the solution $y(x)$ on the interval $[x_0, x_1]$. In the general case of delay relation (3.2) point x_1 can depend on the particular solution $y(x)$ generated by the initial values (3.4).

Once we know the solution on the interval $[x_0, x_1]$, we can proceed to the next interval $[x_1, x_2]$ such that $x_1 = g(x_2, y(x_2), y(x_1))$ and so on. Thus the method of steps for the solution of the initial value problem introduces a natural sequence of intervals

$$[x_n, x_{n+1}], \qquad n = -1, 0, 1, 2, ... \tag{3.9}$$

We emphasize that the introduction of the intervals (3.9) is not a discretization. The variable x varies continuously over the entire region where Eqs. (3.1),(3.2) are defined as does the dependent variable y.

4 Construction of invariant first-order DODSs

In this section we describe the method to be used to classify all first-order delay ordinary differential systems (3.1),(3.2) that are invariant under some nontrivial Lie group of point transformations into conjugacy classes under local diffeomorphisms. For each class we propose a representative DODS.

We realize the Lie algebra L of the point symmetry group G of the system (3.1), (3.2) by vector fields of the same form as in the case of ordinary differential equations, namely

$$X_\alpha = \xi_\alpha(x, y)\frac{\partial}{\partial x} + \eta_\alpha(x, y)\frac{\partial}{\partial y}, \qquad \alpha = 1, ..., n. \tag{4.1}$$

The prolongation of these vector fields acting on the system (3.1),(3.2) will have the form

$$\mathbf{pr}X_\alpha = \xi_\alpha \frac{\partial}{\partial x} + \eta_\alpha \frac{\partial}{\partial y} + \xi_\alpha^- \frac{\partial}{\partial x_-} + \eta_\alpha^- \frac{\partial}{\partial y_-} + \zeta_\alpha \frac{\partial}{\partial \dot{y}} \tag{4.2}$$

with

$$\xi_\alpha = \xi_\alpha(x, y), \qquad \eta_\alpha = \eta_\alpha(x, y),$$

$$\xi_\alpha^- = \xi_\alpha(x_-, y_-), \qquad \eta_\alpha^- = \eta_\alpha(x_-, y_-),$$

$$\zeta_\alpha(x, y, \dot{y}) = D(\eta_\alpha) - \dot{y} D(\xi_\alpha),$$

where D is the total derivative operator. Let us note that Eq. (4.2) combines prolongation for shifted discrete variables $\{x_-, y_-\}$ [7, 8] with standard prolongation for the continuous derivative $\dot{y}$ [41, 40].

All finite-dimensional complex Lie algebras of vector fields of the form (4.1) were classified by S. Lie [28, 31]. The real ones were classified more recently in [19] (see also [47, 38]). The classification is performed under the local group of diffeomorphisms

$$\bar{x} = \bar{x}(x, y), \qquad \bar{y} = \bar{y}(x, y).$$

Let us assume that a Lie group G is given and that its Lie algebra L is realized by the vector fields of the form (4.1). If we wish to construct a first-order DODE with a delay relation that are invariant under this group, we proceed as follows. We choose a basis of the Lie algebra, namely $\{X_\alpha, \alpha = 1, ..., n\}$, and impose the equations

$$\mathbf{pr}X_\alpha \Phi(x, y, x_-, y_-, \dot{y}) = 0, \qquad \alpha = 1, ..., n, \tag{4.3}$$

with $\mathbf{pr}X_\alpha$ given by Eq. (4.2). Using the method of characteristics, we obtain a set of elementary invariants $I_1, ..., I_k$. Their number is

$$k = \dim M - (\dim G - \dim G_0), \tag{4.4}$$

where M is the manifold that G acts on and G_0 is the stabilizer of a generic point on M. In our case we have $M \sim (x, y, x_-, y_-, \dot{y})$ and hence $\dim M = 5$.

Practically, it is convenient to express the number of invariants as

$$k = \dim M - \text{rank } Z, \qquad k \geq 0, \tag{4.5}$$

where Z is the matrix

$$Z = \begin{pmatrix} \xi_1 & \eta_1 & \xi_1^- & \eta_1^- & \zeta_1 \\ \vdots & & & & \\ \xi_n & \eta_n & \xi_n^- & \eta_n^- & \zeta_n \end{pmatrix}. \tag{4.6}$$

The rank of Z is calculated at a generic point of M.

The invariant DODE and delay relation are written as

$$F(I_1, ..., I_k) = 0, \tag{4.7a}$$

$$G(I_1, ..., I_k) = 0, \tag{4.7b}$$

where F and G satisfy

$$\det\left(\frac{\partial(F,G)}{\partial(\dot{y},x_-)}\right) \neq 0.$$

Note that Eqs. (4.7) can be rewritten in the form (3.1),(3.2). Equation (4.7a) with delay relation (4.7b) obtained in this manner are "strongly invariant", i.e. $\mathbf{pr}X_\alpha F = 0$ and $\mathbf{pr}X_\alpha G = 0$ are satisfied identically.

Further invariant equations are obtained if the rank of Z is less than maximal on some manifold described by the equations

$$F(x, y, x_-, y_-, \dot{y}) = 0, \tag{4.8a}$$

$$G(x, y, x_-, y_-, \dot{y}) = 0, \tag{4.8b}$$

$$\det\left(\frac{\partial(F,G)}{\partial(\dot{y},x_-)}\right) \neq 0,$$

which satisfy the conditions

$$\mathbf{pr}X_\alpha F|_{F=0,\ G=0} = 0, \qquad \alpha = 1, ..., n, \tag{4.9a}$$

$$\mathbf{pr}X_\alpha G|_{F=0,\ G=0} = 0, \qquad \alpha = 1, ..., n. \tag{4.9b}$$

Thus we obtain "weakly invariant" DODS (4.8), i.e. equations (4.9) are satisfied on the solutions of the system $F = 0$, $G = 0$.

Note that we must discard trivial cases when the obtained equations (3.1),(3.2) do not satisfy the conditions

$$\frac{\partial f}{\partial y_-} \neq 0, \qquad g(x, y, y_-) \not\equiv \text{const} \qquad \text{or} \qquad x_- < x \tag{4.10}$$

as expected for a DODS.

If convenient, we will use the following notations

$$\Delta x = x - x_-, \qquad \Delta y = y - y_-.$$

In the "no delay" limit we have $\Delta x \to 0$ and $\Delta y \to 0$.

5 First-order linear DODSs

Two symmetry generators of the form (4.1) are called "linearly connected" if their coefficients are proportional with a nonconstant coefficient of proportionality. Otherwise they are called "linearly nonconnected".

It was observed in [10] that first-order DODSs which admit two linearly connected symmetries have more restricted form than the general form (3.1),(3.2). They consist of a linear DODE and a solution independent delay relation. Let us show how this result was obtained.

There are two possibilities for a two-dimensional algebra with linearly connected generators (4.1):

1. The non-Abelian Lie algebra with basis elements

$$X_1 = \frac{\partial}{\partial y}, \qquad X_2 = y\frac{\partial}{\partial y} \tag{5.1}$$

provides us with the invariants

$$I_1 = x, \qquad I_2 = x_-, \qquad I_3 = \frac{\dot{y}}{\Delta y}.$$

The most general invariant DODS can be written as

$$\dot{y} = f(x)\frac{\Delta y}{\Delta x}, \qquad x_- = g(x). \tag{5.2}$$

Here and below we will assume that conditions (4.10) hold. For the equations (5.2) they imply

$$f(x) \not\equiv 0, \qquad g(x) \not\equiv \text{const}, \qquad g(x) < x.$$

2. The Abelian Lie algebra with basis elements

$$X_1 = \frac{\partial}{\partial y}, \qquad X_2 = x\frac{\partial}{\partial y}. \tag{5.3}$$

A basis for the invariants of the group corresponding to the Lie algebra is

$$I_1 = x, \qquad I_2 = x_-, \qquad I_3 = \dot{y} - \frac{\Delta y}{\Delta x}$$

and the invariant DODE with the equation for the delay can be presented as

$$\dot{y} = \frac{\Delta y}{\Delta x} + f(x), \qquad x_- = g(x). \tag{5.4}$$

We shall assume that $f(x) \not\equiv 0$; otherwise (5.4) is a special case of (5.2).

We remark that the *method of steps* allows us to solve the systems (5.2) and (5.4) using successive quadratures.

Thus we arrive at the following general results.

Theorem 1. *([10]) Let the DODE (3.1) with the delay relation (3.2) admit two linearly connected symmetries. Then it can be transformed into the homogeneous linear equation (5.2) if the symmetries do not commute; and into the inhomogeneous linear equation (5.4) if these symmetries commute.*

In both cases the DODE is supplemented by the delay relation $x_- = g(x)$, which does not depend on the solutions of the considered DODE.

Thus, first-order DODEs which admit two linearly connected symmetries are linearizable. From now on we shall use the term *linear* DODS precisely for the systems (5.2) and (5.4).

Corollary 1. Any Lie algebra containing a two-dimensional subalgebra realized by linearly connected vector fields will provide an invariant DODS that can be transformed into the form (5.2) or (5.4) (i.e. a linear DODE with a solution independent delay relation). The larger Lie algebra will at most put constraints on the functions $f(x)$ and $g(x)$.

Let us show that linear DODEs with solution independent delays actually admit infinite-dimensional symmetry algebras due to the linear superposition principle. It is convenient to consider the most general linear DODS in the considered class, namely

$$\dot{y} = \alpha(x)y + \beta(x)y_- + \gamma(x), \qquad x_- = g(x), \tag{5.5}$$

where $\alpha(x)$, $\beta(x)$, $\gamma(x)$ and $g(x)$ are arbitrary real functions, smooth in some interval $x \in I$, satisfying

$$\beta(x) \not\equiv 0, \qquad g(x) \not\equiv \text{const}, \qquad g(x) < x. \tag{5.6}$$

We will also need its homogeneous counterpart

$$\dot{y} = \alpha(x)y + \beta(x)y_-, \qquad x_- = g(x). \tag{5.7}$$

Note that equation (5.5) includes (5.2) and (5.4) as particular cases and can be transformed into one of these two.

Proposition 1. *The change of variables*

$$\bar{x} = x, \qquad \bar{y} = y - \sigma(x), \tag{5.8}$$

where $\sigma(x)$ is an arbitrary solution of the inhomogeneous DODS (5.5), transforms the inhomogeneous DODS (5.5) into its homogeneous counterpart (5.7).

Theorem 2. *([10]) Consider the linear DODS (5.5). For all functions $\alpha(x)$, $\beta(x)$, $\gamma(x)$ and $g(x)$ the DODS admits an infinite-dimensional symmetry algebra represented by the vector fields*

$$X(\rho) = \rho(x)\frac{\partial}{\partial y}, \qquad Y(\sigma) = (y - \sigma(x))\frac{\partial}{\partial y}, \tag{5.9}$$

where $\rho(x)$ is the general solution of the homogeneous DODS (5.7) and $\sigma(x)$ is any one particular solution of the DODS (5.5).

Remark 1. For the homogeneous linear DODS (5.7) the theorem gets simplified since we can use $\sigma(x) = 0$. We get an infinite-dimensional symmetry algebra represented by the vector fields

$$X(\rho) = \rho(x)\frac{\partial}{\partial y}, \qquad Y = y\frac{\partial}{\partial y}, \tag{5.10}$$

where $\rho(x)$ is the general solution of the homogeneous DODS (5.7).

The admitted symmetry algebra (5.9) can be supplemented by at most one additional symmetry. We present this result as the following theorem.

Theorem 3. *([10]) Consider the linear DODS (5.5). For specific choices of the arbitrary functions $\alpha(x)$, $\beta(x)$ and $g(x)$, namely for functions satisfying the compatibility condition*

$$K(g(x))(\dot{g}(x))^2 = \ddot{g}(x) + K(x)\dot{g}(x), \tag{5.11}$$

where

$$K(x) = \alpha(x) - \dot{g}(x)\alpha(g(x)) - \frac{\dot{\beta}(x)}{\beta(x)}, \tag{5.12}$$

the symmetry algebra is larger. It contains one additional basis element of the form

$$Z = \xi(x)\frac{\partial}{\partial x} + (A(x)y + B(x))\frac{\partial}{\partial y}, \qquad \xi(x)\not\equiv 0, \tag{5.13}$$

where

$$\xi(x) = e^{\int K(x)dx}, \qquad A(x) = \xi(x)\alpha(x) \tag{5.14}$$

and the function $B(x)$ is a particular solution of the DODS

$$\dot{B}(x) = \alpha(x)B(x) + \beta(x)B(x_-) + \gamma(x)\dot{\xi}(x) + \dot{\gamma}(x)\xi(x) - \gamma(x)A(x), \qquad x_- = g(x).$$

Remark 2. For the homogeneous linear DODS (5.7) the theorem gets simplified and we can present the admitted symmetries as

$$X(\rho) = \rho(x)\frac{\partial}{\partial y}, \qquad Y = y\frac{\partial}{\partial y}, \qquad Z = \xi(x)\frac{\partial}{\partial x} + \xi(x)\alpha(x)y\frac{\partial}{\partial y}, \tag{5.15}$$

where $\rho(x)$ is the general solution of the homogeneous DODS (5.7) and $\xi(x)\not\equiv 0$ is given in (5.14).

Using the results of the previous theorems, we can present all symmetries of the homogeneous DODS (5.2) as

$$X(\rho) = \rho(x)\frac{\partial}{\partial y}, \qquad Y = y\frac{\partial}{\partial y}, \tag{5.16}$$

where $\rho(x)$ is an arbitrary solution of the DODS (5.2). The symmetries of the inhomogeneous DODS (5.4) can be presented as

$$X(\rho) = \rho(x)\frac{\partial}{\partial y}, \qquad Y(\sigma) = (y - \sigma(x))\frac{\partial}{\partial y}, \tag{5.17}$$

where $\sigma(x)$ is a particular solution of the original inhomogeneous DODS (5.4) and $\rho(x)$ is the general solution of its homogeneous counterpart.

We can use the results established above to provide simplifications of the linear DODS.

Theorem 4. *([10]) If the linear DODS (5.5) has functions $\alpha(x)$, $\beta(x)$ and $g(x)$ satisfying the compatibility condition (5.11), the DODS can be transformed into the representative form*

$$\dot{y} = y_- + h(x), \qquad x_- = x - C, \qquad C > 0, \tag{5.18}$$

which admits symmetries

$$X(\rho) = \rho(x)\frac{\partial}{\partial y}, \qquad Y(\sigma) = (y - \sigma(x))\frac{\partial}{\partial y}, \qquad Z = \frac{\partial}{\partial x} + B(x)\frac{\partial}{\partial y}, \tag{5.19}$$

where $\sigma(x)$ is any one particular solution of the inhomogeneous DODS (5.18), $\rho(x)$ is the general solution of the corresponding homogeneous DODS and $B(x)$ is a particular solution of the DODS

$$\dot{B}(x) = B(x_-) + \dot{h}(x), \qquad x_- = x - C, \qquad C > 0.$$

Further simplification is possible if we know at least one particular solution of the DODS and can bring the DODE into the homogeneous form.

Corollary 2. If the linear homogeneous DODS (5.7) has functions $\alpha(x)$, $\beta(x)$ and $g(x)$ satisfying the compatibility condition (5.11), the DODS can be transformed into the representative form

$$\dot{y} = y_-, \qquad x_- = x - C, \qquad C > 0, \tag{5.20}$$

which admits symmetries

$$X(\rho) = \rho(x)\frac{\partial}{\partial y}, \qquad Y = y\frac{\partial}{\partial y}, \qquad Z = \frac{\partial}{\partial x}, \tag{5.21}$$

where $\rho(x)$ is the general solution of the DODS (5.20).

Let us note that we could choose the representative form

$$\dot{y} = C_1 y + C_2 y_-, \qquad x_- = x - C, \qquad C_2 \neq 0, \qquad C > 0. \tag{5.22}$$

instead of the form (5.20).

Theorem 5. *([10]) If the linear DODS (5.5) has functions $\alpha(x)$, $\beta(x)$ and $g(x)$ which do not satisfy the compatibility condition (5.11), the DODS can be transformed into the representative form*

$$\dot{y} = f(x)y_- + h(x), \qquad x_- = x - C, \qquad \dot{f}(x) \neq 0, \qquad C > 0 \tag{5.23}$$

or into the form

$$\dot{y} = y_- + h(x), \qquad x_- = g(x), \qquad \ddot{g}(x) \neq 0, \qquad g(x) < x. \tag{5.24}$$

Corollary 3. If the linear homogeneous DODS (5.7) has functions $\alpha(x)$, $\beta(x)$ and $g(x)$ which do not satisfy the compatibility condition (5.11), the DODS can be transformed into the representative form

$$\dot{y} = f(x)y_-, \qquad x_- = x - C, \qquad \dot{f}(x) \neq 0, \qquad C > 0 \tag{5.25}$$

or into the form

$$\dot{y} = y_-, \qquad x_- = g(x), \qquad \ddot{g}(x) \neq 0, \qquad g(x) < x. \tag{5.26}$$

Linear DODSs can be fully characterized by a symmetry algebra of dimension $2 \leq n \leq 4$ [11]. Such algebra always contains the non-Abelian algebra (5.1) or the Abelian Lie algebra (5.3), or possibly both.

6 Lie symmetry classification of first-order nonlinear DODSs

In the previous section it was shown that linear DODEs with solution independent delays admit infinite-dimensional symmetry algebras. Here we examine the other cases: the genuinely nonlinear DODEs and/or the solution dependent delays. We will refer to such DODSs as *nonlinear*.

Their Lie group classification was performed in [10]. Proceeding by dimension, all realizations of Lie algebras by vector fields of the form (4.1) we considered. The cases containing two linearly connected operators were discarded. For the other cases we constructed invariant DODEs with invariant delay relations as described in Section 4.

6.1 Dimension 1

We start with the simplest case of a symmetry group, namely a one-dimensional one. Its Lie algebra is generated by one vector field of the form (4.1). By an appropriate change of variables we take this vector field into its rectified form (locally in a nonsingular point (x, y)). Thus we have

$$X_1 = \frac{\partial}{\partial y}. \tag{6.1}$$

In order to write a first-order DODE and a delay relation invariant under this group we need the invariants annihilated by the prolongation (4.2) of X_1 to the prolonged space $(x, y, x_-, y_-, \dot{y})$. A basis for the invariants is

$$I_1 = x, \qquad I_2 = x_-, \qquad I_3 = \Delta y, \qquad I_4 = \dot{y}.$$

The most general first-order DODE with delay relation invariant under the corresponding group can be written as

$$\dot{y} = f\left(x, \frac{\Delta y}{\Delta x}\right), \qquad \Delta x = g(x, \Delta y), \tag{6.2}$$

where f and g are arbitrary functions. Writing $\dot{y} = h(x, \Delta y)$, $\Delta x = g(x, \Delta y)$ would obviously be equivalent.

6.2 Dimension 2

The non-Abelian Lie algebra with nonconnected elements

$$X_1 = \frac{\partial}{\partial y}, \qquad X_2 = x\frac{\partial}{\partial x} + y\frac{\partial}{\partial y} \tag{6.3}$$

yields a convenient basis for the invariants in the form

$$I_1 = \frac{\Delta x}{x}, \qquad I_2 = \frac{\Delta y}{\Delta x}, \qquad I_3 = \dot{y}.$$

The general invariant DODE with delay can be written as

$$\dot{y} = f\left(\frac{\Delta y}{\Delta x}\right), \qquad x_- = xg\left(\frac{\Delta y}{\Delta x}\right). \tag{6.4}$$

The Abelian Lie algebra with nonconnected basis elements is represented by the generators

$$X_1 = \frac{\partial}{\partial x}, \qquad X_2 = \frac{\partial}{\partial y}. \tag{6.5}$$

A convenient set of invariants is

$$I_1 = \Delta x, \qquad I_2 = \Delta y, \qquad I_3 = \dot{y}.$$

The most general invariant DODE and delay relation can be written as

$$\dot{y} = f\left(\frac{\Delta y}{\Delta x}\right), \qquad \Delta x = g(\Delta y). \tag{6.6}$$

6.3 Dimension 3

The results are given in Table 1. In Column 1 we give the isomorphism class using the notations of [51]. Thus $\mathfrak{s}_{i,k}$ is the k-th solvable Lie algebra of dimension i in the list. For solvable Lie algebras the nilradical precedes a semicolon. The only nilpotent algebra is $\mathfrak{n}_{1,1}$. Simple Lie algebras are identified by their usual names ($\mathfrak{sl}(2,\mathbb{R})$, $\mathfrak{o}(3,\mathbb{R})$). In Column 2 $\mathbf{A}_{i,k}$ gives algebras in the list of subalgebras of diff$(2,\mathbb{R})$ according to Ref. [10] and i is again the dimension of the algebra. The numbers in brackets correspond to notations used in the list of Ref. [19]. In Columns 3, 4 and 5 we give vector fields spanning each representative algebra, the invariant DODE and the invariant delay relation, respectively.

6.4 Dimensions $n \geq 4$

In [10] it was shown that Lie algebras of dimension $n \geq 4$ can only lead to linear DODEs with y-independent delay relations.

The main idea was to take all representations of finite Lie algebras of dimension $n \geq 4$ by vector fields of the form (4.1), exclude the cases containing two linearly connected operators and show that the remaining cases do not lead to invariant DODSs (3.1),(3.2). It leads to the following theorem.

Theorem 6. *The only DODSs of the form (3.1),(3.2) that have symmetry algebras of dimension $n \geq 4$ are equivalent to linear DODSs with solution independent delay equations.*

Table 1. Lie group classification of invariant nonlinear first-order DODSs

Dimensions 1 and 2

Lie algebra	Case	Operators	DODE	Delay relation
$\mathfrak{n}_{1,1}$	$\mathbf{A}_{1,1}(9)$	$X_1 = \partial_y$	$\dot{y} = f\left(x, \frac{\Delta y}{\Delta x}\right)$	$\Delta x = g(x, \Delta y)$
$\mathfrak{s}_{2,1}$	$\mathbf{A}_{2,2}(22)$	$X_1 = \partial_y$; $X_2 = x\partial_x + y\partial_y$	$\dot{y} = f\left(\frac{\Delta y}{\Delta x}\right)$	$x_- = xg\left(\frac{\Delta y}{\Delta x}\right)$
$2\mathfrak{n}_{1,1}$	$\mathbf{A}_{2,4}(22)$	$X_1 = \partial_x$, $X_2 = \partial_y$	$\dot{y} = f\left(\frac{\Delta y}{\Delta x}\right)$	$\Delta x = g(\Delta y)$

Dimension 3

Lie algebra	Case	Operators		DODE	Delay relation
$\mathfrak{s}_{3,1}$	$\mathbf{A}^a_{3,2}(12)$	$X_1 = \partial_x$, $X_2 = \partial_y$; $X_3 = x\partial_x + ay\partial_y$, $0 < \|a\| \leq 1$	i) $a \neq 1$	$\dot{y} = C_1 \dfrac{\Delta y}{\Delta x}$	$\Delta x = \|\Delta y\|^{1/a}$
			ii) $a = 1$	No DODE	or no delay relation
$\mathfrak{s}_{3,2}$	$\mathbf{A}_{3,4}(25)$	$X_1 = \partial_x$, $X_2 = \partial_y$; $X_3 = x\partial_x + (x+y)\partial_y$		$\dot{y} = \dfrac{\Delta y}{\Delta x} + C_1$	$\Delta x = C_2 e^{\frac{\Delta y}{\Delta x}}$
$\mathfrak{s}_{3,3}$	$\mathbf{A}^b_{3,6}(1)$	$X_1 = \partial_x$, $X_2 = \partial_y$; $X_3 = (bx+y)\partial_x + (by-x)\partial_y$,	$b \geq 0$	$\dot{y} = \dfrac{\dfrac{\Delta y}{\Delta x} + C_1}{1 - C_1 \dfrac{\Delta y}{\Delta x}}$	$\Delta x e^{b \arctan \frac{\Delta y}{\Delta x}} \sqrt{1 + \left(\dfrac{\Delta y}{\Delta x}\right)^2} = C_2$

Lie algebra	Case	Operators	DODE	Delay relation
$\mathfrak{sl}(2,\mathbb{R})$	$\mathbf{A}_{3,8}(18)$	$X_1=\partial_y,$ $X_2=x\partial_x+y\partial_y,$ $X_3=2xy\partial_x+y^2\partial_y$	$\dot{y}=\dfrac{\Delta y}{2x+C_1\Delta y}$	$xx_-=C_2(\Delta y)^2$
	$\mathbf{A}_{3,9}(2)$	$X_1=\partial_y,$ $X_2=x\partial_x+y\partial_y,$ $X_3=2xy\partial_x+(y^2-x^2)\partial_y$	$\dot{y}=\dfrac{(\Delta y)^2+x_-^2-x^2+2C_1x\Delta y}{2x\Delta y-C_1((\Delta y)^2+x_-^2-x^2)}$	$(\Delta y)^2+(x-x_-)^2=C_2xx_-$
	$\mathbf{A}_{3,10}(17)$	$X_1=\partial_y,$ $X_2=x\partial_x+y\partial_y,$ $X_3=2xy\partial_x+(y^2+x^2)\partial_y$	$\dot{y}=\dfrac{(\Delta y)^2+x^2-x_-^2+2C_1x\Delta y}{2x\Delta y+C_1((\Delta y)^2+x^2-x_-^2)}$	$(\Delta y)^2-(x-x_-)^2=C_2xx_-$
$\mathfrak{o}(3,\mathbb{R})$	$\mathbf{A}_{3,12}(3)$	$X_1=(1+x^2)\partial_x+xy\partial_y,$ $X_2=xy\partial_x+(1+y^2)\partial_y,$ $X_3=y\partial_x-x\partial_y$	$\dfrac{\dot{y}-\dfrac{y-y_-}{x-x_-}}{\sqrt{1+\dot{y}^2+(y-x\dot{y})^2}}=C_1\dfrac{\sqrt{1+x_-^2+y_-^2}}{x-x_-}$	$(x-x_-)^2=\dfrac{C_2(1+x^2+y^2)(1+x_-^2+y_-^2)}{1+\left(\frac{y-y_-}{x-x_-}\right)^2+\left(y-x\frac{y-y_-}{x-x_-}\right)^2}$

7 Exact solutions of the DODSs

7.1 Find the symmetry algebra L

We are given a general DODS of the form (3.1),(3.2) where f and g are given functions of the indicated arguments. The symmetry algebra of the system can be found as follows. We put

$$E_1 = \dot{y} - f(x, y, y_-), \qquad E_2 = x_- - g(x, y, y_-) \tag{7.1}$$

and X_α as in (4.1), $\mathbf{pr}X_\alpha$ as in (4.2) and request

$$\mathbf{pr}X_\alpha E_1|_{E_1=0,\ E_2=0} = 0, \qquad \mathbf{pr}X_\alpha E_2|_{E_1=0,\ E_2=0} = 0. \tag{7.2}$$

Eqs. (7.2) provide us with the determining equations for the coefficients ξ_α and η_α in (4.1). From the analysis in Sections 5 and 6 it follows that the possible dimensions of the symmetry algebra L are $n = 0,\ 1,\ 2,\ 3$ and $n = \infty$.

7.2 Invariant solutions

If the DODS (3.1),(3.2) has a symmetry algebra L of dimension $n = \dim L \geq 1$, it can, at least in principle, be used to construct explicit analytical solutions. These particular solutions will satisfy very specific initial conditions. Essentially the method consists of constructing solutions that are invariant under a subgroup of the symmetry group of the DODS.

In the case of the DODS (3.1),(3.2) it is sufficient to consider one-dimensional subalgebras. They will all have the form

$$X = \sum_{\alpha=1}^{n} c_\alpha X_\alpha, \qquad c_\alpha \in \mathbb{R}, \tag{7.3}$$

where c_α are constants and X_α are of the form (4.1) and are elements of the symmetry algebra L.

The method consists of several steps.

1. Construct a representative list of one-dimensional subalgebras of L_i of the symmetry algebra L of the DODS. The subalgebras are classified under the group of inner automorphisms $G = \exp L$.

2. For each subalgebra in the list, calculate the invariants of the subgroups $G_i = \exp L_i$ in the four-dimensional space with local coordinates $\{x, y, x_-, y_-\}$. There will be three functionally independent invariants. For the method to be applicable, two of the invariants must depend on two variables only, namely (x, y) and (x_-, y_-) respectively. We set them equal to constants as follows:

$$J_1(x, y) = A, \qquad J_2(x, y, x_-, y_-) = B. \tag{7.4}$$

They must satisfy the Jacobian condition

$$\det\left(\frac{\partial(J_1,J_2)}{\partial(y,x_-)}\right)\neq 0 \tag{7.5}$$

(we have $J_3(x_-,y_-) = J_1(x_-,y_-)$, i.e., $J_3(x_-,y_-)$ is obtained by shifting $J_1(x,y)$ to (x_-,y_-)).

All elements of the Lie algebra have the form (4.1). A **necessary condition** for invariants of the form (7.4) to exist is that at least one of the vector fields in L_i satisfies

$$\xi(x,y)\not\equiv 0. \tag{7.6}$$

3. Solve equations (7.4) to obtain the reduction formulas

$$y = h(x,A), \qquad x_- = k(x,A,B) \tag{7.7}$$

(we also have $y_- = h(x_-,A)$).

4. Substitute the reduction formulas (7.7) into the DODS (3.1),(3.2) and request that the equations should be satisfied identically. This will provide relations which define the constants and therefore determine the functions h and k. It may also impose constraints on the functions $f(x,y,y_-)$ and $g(x,y,y_-)$ in (3.1) and (3.2). Once the relations are satisfied the invariant solution is given by (7.7).

5. Apply the entire group $G_i = \exp L_i$ to the obtained invariant solutions. This can provide a more general solution depending on up to (dim $L-1$) parameters (corresponding to the factor algebra L/L_i). Check whether the DODS imposes further constraints involving the new parameters.

Following Olver's [40] definition for ODEs and PDEs, we shall call solutions obtained in this manner *group invariant solutions* (for DODSs). These solutions are solutions of DODSs supplemented by very specific initial conditions (dictated by the symmetries).

7.3 Examples

Let us show how the general method for finding invariant solutions is applied. To obtain a representative list of all subalgebras of the symmetry algebras we can use the tables given in [42] (after appropriately identifying basis elements). Alternatively one can calculate them in each case using the method presented in the same article.

Example. Algebra $\mathbf{A}_{2,2}$ with symmetries (6.3) and invariant DODS (6.4):
Step 1. The representative list of one-dimensional subalgebras [42] is

$$\{X_1\}, \qquad \{X_2\}.$$

Since X_1 does not satisfy condition $\xi(x,y)\not\equiv 0$ we proceed only for X_2.

Step 2. Invariants for X_2 are

$$J_1 = \frac{y}{x} = A, \qquad J_2 = \frac{x_-}{x} = B. \tag{7.8}$$

Step 3. We obtain reduction formulas

$$y = Ax, \qquad x_- = Bx. \tag{7.9}$$

Step 4. Substitution in (6.4) gives

$$A = f(A), \qquad B = g(A). \tag{7.10}$$

Step 5. Applying group transformation exp (εX_1) to (7.9), we obtain a more general solution

$$y = Ax + \alpha, \qquad x_- = Bx. \tag{7.11}$$

In (7.11) α is arbitrary, A and B are solutions of (7.10). Thus, the number of different invariant solutions is determined by the number of a solutions of the equations (7.10) (the functions $f(z)$ and $g(z)$ in (6.4) are given).

Example. Algebra $\mathbf{A}_{2,4}$ with symmetries (6.5) and invariant DODS (6.6):

The representative list is

$$\{\cos(\varphi)X_1 + \sin(\varphi)X_2, \qquad 0 \leq \varphi < \pi\}.$$

Invariant solutions can be obtained for $\cos(\varphi) \neq 0$. Let $a = \tan(\varphi) < \infty$, then we can rewrite the symmetry as $X_1 + aX_2$ and find invariant solutions

$$y = ax + A, \qquad x_- = x - B, \tag{7.12}$$

where a and B satisfy

$$a = f(a), \qquad B = g(Ba) \tag{7.13}$$

and A is arbitrary. Note that the invariant solutions exist only for $a = \tan(\varphi)$ which satisfies the first equation in (7.13).

Application of the group transformations $\exp(\varepsilon(\cos(\varphi)X_1 + \sin(\varphi)X_2))$ does not give more solutions (it only changes the value of A).

Example. Algebra $\mathbf{A}_{3,4}$ with symmetries

$$X_1 = \frac{\partial}{\partial x}, \qquad X_2 = \frac{\partial}{\partial y}; \qquad X_3 = x\frac{\partial}{\partial x} + (x+y)\frac{\partial}{\partial y} \tag{7.14}$$

and invariant DODS

$$\dot{y} = \frac{\Delta y}{\Delta x} + C_1, \qquad \Delta x = C_2 e^{\frac{\Delta y}{\Delta x}}. \tag{7.15}$$

The representative subalgebras are

$$\{X_1\}, \qquad \{X_2\}, \qquad \{X_3\}.$$

Note that X_2 does not satisfy $\xi(x,y)\not\equiv 0$.

1. For X_1 the invariant solution has the form

$$y = A, \qquad x_- = x - B. \tag{7.16}$$

It exists under conditions

$$0 = C_1, \qquad B = C_2. \tag{7.17}$$

Applying the group, we generate more general solutions

$$y = A + \alpha x, \qquad x_- = x - B, \tag{7.18}$$

which satisfy the DODE under the same conditions (7.17).

2. For symmetry X_3 we get the invariant solution form

$$y = x \ln |x| + Ax, \qquad x_- = Bx. \tag{7.19}$$

Group transformations extend this form to more general solutions

$$y - \alpha = (x - \beta) \ln |x - \beta| + A(x - \beta), \qquad x_- - \beta = B(x - \beta), \tag{7.20}$$

which satisfy the DODS provided that

$$\frac{B \ln |B|}{1 - B} = C_1 - A - 1, \qquad \operatorname{sgn}(x - \beta)(1 - B)|B|^{\frac{B}{1-B}} = C_2 e^A, \tag{7.21}$$

where

$$\operatorname{sgn}(x) = \begin{cases} 1, & x > 0; \\ 0, & x = 0; \\ -1 & x < 0. \end{cases}$$

Example. Lie algebra $\mathbf{A}_{3,8}$ is often (and equivalently to the realization given in Table 1) realized by the vector fields

$$X_1 = \frac{\partial}{\partial x}, \qquad X_2 = 2x\frac{\partial}{\partial x} + y\frac{\partial}{\partial y}, \qquad X_3 = x^2\frac{\partial}{\partial x} + xy\frac{\partial}{\partial y}. \tag{7.22}$$

There are two invariants

$$I_1 = \frac{\Delta x}{yy_-}, \qquad I_2 = y\left(\dot{y} - \frac{\Delta y}{\Delta x}\right),$$

which provide us with the invariant DODS

$$\dot{y} = \frac{\Delta y}{\Delta x} + \frac{C_1}{y}, \qquad \Delta x = C_2 yy_-, \qquad C_2 \neq 0. \tag{7.23}$$

The representative list of one-dimensional subalgebras consists of algebras

$$\{X_1\}, \qquad \{X_2\}, \qquad \{X_1 + X_3\}.$$

We obtain the following invariant solutions:

1. The subalgebra corresponding to the algebra X_1 suggests the invariant solution in the form

$$y = A, \qquad x_- = x - B \tag{7.24}$$

provided that

$$\frac{C_1}{A} = 0, \qquad B = C_2 A^2.$$

We find only constant solutions $y = A$, which exist if the DODE in (7.23) satisfies $C_1 = 0$. There is one relation which gives B for any value of A.

2. The operator X_2 implies

$$y = A\sqrt{|x|}, \qquad x_- = Bx \tag{7.25}$$

and we obtain the conditions

$$\frac{A}{2}\operatorname{sgn}(x) = A\frac{1-\sqrt{|B|}}{\operatorname{sgn}(x) - \operatorname{sgn}(x_-)|B|} + \frac{C_1}{A}, \qquad \frac{1}{\sqrt{|B|}} - \sqrt{|B|} = C_2 A^2$$

for the constants.

3. The subalgebra corresponding to the operator $X_1 + X_3$ gives the invariant solution form

$$y = A\sqrt{x^2+1}, \qquad x_- = \frac{x-B}{1+Bx}. \tag{7.26}$$

Substitution into the DODS leads to the requirements

$$\frac{A}{B}\left(1 - \operatorname{sgn}(1+Bx)\sqrt{B^2+1}\right) + \frac{C_1}{A} = 0, \qquad B = C_2 A^2 \operatorname{sgn}(1+Bx)\sqrt{B^2+1},$$

which define the constants A and B.

Example. For Lie algebra $\mathbf{A}_{3,10}$ we can use realization

$$X_1 = \frac{\partial}{\partial x} + \frac{\partial}{\partial y}, \qquad X_2 = x\frac{\partial}{\partial x} + y\frac{\partial}{\partial y}, \qquad X_3 = x^2\frac{\partial}{\partial x} + y^2\frac{\partial}{\partial y}. \tag{7.27}$$

We obtain invariants

$$I_1 = \frac{(y-y_-)(x-x_-)}{(y-x)(y_- - x_-)}, \qquad I_2 = \left(\frac{x-x_-}{y-x_-}\right)^2 \dot{y}$$

and the invariant DODS

$$\dot{y} = C_1\left(\frac{y-x_-}{x-x_-}\right)^2, \qquad \frac{(y-y_-)(x-x_-)}{(y-x)(y_- - x_-)} = C_2, \qquad C_1 C_2 \neq 0. \tag{7.28}$$

The representative list is

$$\{X_1\}, \qquad \{X_2\}, \qquad \{X_1 + X_3\}$$

as for the previous realization of $\mathfrak{sl}(2,\mathbb{R})$. We obtain invariant solutions for all three elements.

1. Invariant solutions for X_1 have the form

$$y = x + A, \qquad x_- = x - B \tag{7.29}$$

provided that

$$1 = C_1 \left(\frac{A+B}{B}\right)^2, \qquad \frac{B^2}{A^2} = C_2.$$

2. For X_2 we look for invariant solution in the form

$$y = Ax, \qquad x_- = Bx \tag{7.30}$$

and obtain the system

$$A = C_1 \left(\frac{A-B}{1-B}\right)^2, \qquad \frac{A(1-B)^2}{(A-1)^2 B} = C_2,$$

which defines the constants A and B.

3. For $X_1 + X_3$ we look for invariant solutions in the form

$$y = \frac{x+A}{1-Ax}, \qquad x_- = \frac{x-B}{1+Bx} \tag{7.31}$$

and get restrictions

$$1 + A^2 = C_1 \left(1 + \frac{A}{B}\right)^2, \qquad \frac{(1+A^2)B^2}{A^2(1+B^2)} = C_2.$$

More examples of invariant solutions can be found in [10, 11].

8 Higher order DODSs

Let us now consider a class of equations that generalizes the class (3.1),(3.2) of first-order DODSs considered above and the second-order DODSs considered in [12]. We have in mind N-th order DODSs

$$y^{(N)} = f(x, y, y_-, \dot{y}, \dot{y}_-, ..., y^{(N-1)}, y_-^{(N-1)}), \tag{8.1a}$$

$$x_- = g(x, y, y_-, \dot{y}, \dot{y}_-, ..., y^{(N-1)}, y_-^{(N-1)}) \tag{8.1b}$$

with

$$y^{(k)} = y^{(k)}(x) = \frac{d^k y}{dx^k}(x), \qquad y_-^{(k)} = y^{(k)}(x_-) = \frac{d^k y}{dx_-^k}(x_-), \qquad 1 \le k \le N.$$

The smooth functions f and g are essentially arbitrary, but satisfy some nontriviality conditions:

$$\frac{\partial f}{\partial y_-^{(k)}} \not\equiv 0 \qquad \text{for at least one value of } k, \qquad 1 \le k \le N;$$

$$g \not\equiv \text{const}, \qquad x_- < x.$$

The vector fields realizing the Lie point symmetry algebras of the system (8.1) have the same form for all N, namely (4.1). They must however be prolonged to order N. Eq. (4.2) must be replaced by

$$\mathbf{pr}X = \xi\frac{\partial}{\partial x} + \eta\frac{\partial}{\partial y} + \xi^{-}\frac{\partial}{\partial x_{-}} + \eta^{-}\frac{\partial}{\partial y_{-}} + \sum_{k=1}^{N}\zeta^{k}\frac{\partial}{\partial y^{(k)}} + \sum_{k=1}^{N}\zeta^{k,-}\frac{\partial}{\partial y_{-}^{(k)}}, \quad (8.3a)$$

$$\zeta^{k}(x,y,\dot{y},...,y^{(k)}) = D(\zeta^{k-1}) - y^{(k)}D(\xi), \qquad \zeta^{0} = \eta(x,y), \quad (8.3b)$$

$$\zeta^{k}_{-} = \zeta^{k}(x_{-},y_{-},\dot{y}_{-},...,y_{-}^{(k)}). \quad (8.3c)$$

Functions $\xi(x,y)$, $\eta(x,y)$, $\xi^{-}(x_{-},y_{-})$ and $\eta^{-}(x_{-},y_{-})$ are the same as in (4.2) and the functions ζ^{k} and $\zeta^{k,-}$ are given recursively.

The functions ξ, η, ζ^{k} should also have a label α, with $\alpha = 1,...,n$ where $n = \dim L$ is the dimension of the Lie algebra and the independent vector fields X_{α} form a basis of L.

So far we have classified and constructed representatives of all DODSs of the form (8.1) for $N = 1$ and $N = 2$ [10, 11, 12]. The pattern that is emerging for all orders N has the following features.

1. Consider the linear DODS

$$y^{(N)} = \sum_{i=0}^{N-1}\alpha_{i}(x)y^{(i)} + \sum_{i=0}^{N-1}\beta_{i}(x)y_{-}^{(i)} + \gamma(x), \quad (8.4a)$$

$$x_{-} = g(x). \quad (8.4b)$$

The symmetry algebra of this system is infinite-dimensional and its Lie point symmetry algebra is represented by the vector fields

$$X(\rho) = \rho(x)\frac{\partial}{\partial y}, \qquad Y(\sigma) = (y - \sigma(x))\frac{\partial}{\partial y}. \quad (8.5)$$

The functions $\rho(x)$ are linearly independent solutions of the homogeneous DODS obtained from (8.4) by setting $\gamma(x) = 0$. The function $\sigma(x)$ is any one particular solution of the DODS (8.4) (for $\gamma(x) \neq 0$). Eqs. (8.5) apply for homogeneous equations as well: we simply put $\sigma(x) = 0$.

2. The vector fields in (8.5) are all linearly connected (since $\partial/\partial x$ does not figure in any of them). The algebra is solvable with the subalgebra $\{X(\rho)\}$ as its nilradical. The nilradical is Abelian.

The DODS (8.1) is linearizable by a point transformation if it allows a Lie point symmetry algebra of dimension $\dim L = 2N$ realized by linearly connected vector fields (it then automatically allows an infinite-dimensional symmetry algebra).

A transformation

$$(x,y) \to (t,z): \qquad t = t(x,y), \qquad z = z(x,y) \quad (8.6)$$

that transforms the vector fields $\{X_{1},...,X_{2N}\}$ of the form (4.1) in (x,y) into the form

$$\bar{Y} = z\frac{\partial}{\partial z},\ \bar{X}_{1} = \frac{\partial}{\partial z},\ \bar{X}_{2} = t\frac{\partial}{\partial z},\ \bar{X}_{3} = X_{4}(t)\frac{\partial}{\partial z},\ ...,\ \bar{X}_{2N-1} = X_{2N}(t)\frac{\partial}{\partial z} \quad (8.7)$$

or

$$\bar{X}_1 = \frac{\partial}{\partial z}, \ \bar{X}_2 = t\frac{\partial}{\partial z}, \ \bar{X}_3 = X_3(t)\frac{\partial}{\partial z}, \ ... \ \bar{X}_{2N} = X_{2N}(t)\frac{\partial}{\partial z} \tag{8.8}$$

exists.

This transformation will take (8.1) into a homogeneous linear DODS (Eq. (8.4) with $\gamma(x) = 0$) in the case (8.7) and into an inhomogeneous linear DODS in the case (8.8). Notice that (8.7) is solvable but not Abelian, (8.8) is Abelian.

3. Nonlinear and nonlinearizable DODSs have finite dimensional Lie point symmetry algebras for any N. Indeed, any Lie algebra of real vector fields in two dimensions can be transformed by a point transformation into one of the 28 classes of algebras in the list provided in Ref. [19]. The list contains 19 finite-dimensional algebras of dimension n with $1 \le n \le 8$ (two of these algebras depend on a parameter α).

The list also contains 9 infinite series of finite-dimensional Lie algebras L of dimension n that depends on the integer r , satisfying $1 \le r \le \infty$. Each of these algebras contain a subalgebra $L_0 \subseteq L$ of linearly connected vector fields of the form $f_k(x)\ \partial/\partial y$ or $y\ \partial/\partial y$. From Table 1 of Ref. [19] we see that $n_0 = \dim L_0$ satisfies $n_0 = r$, $n_0 = r+1$ or $n_0 = r+2$. On the other hand it was shown above that $n_0 \ge 2N$ implies that the DODS is linearizable (or linear). Hence, for a genuinely nonlinear DODS of order N we must have

$$n_0 = \dim L_0 \le 2N - 1, \qquad n = \dim L \le 2N + 2. \tag{8.9}$$

These bounds also hold for the 19 finite-dimensional algebras of dimension n with $1 \le n \le 8$ provided in Ref. [19]. The cases analyzed in detail, namely $N = 1, 2$, satisfy $\dim L = 3$ and $\dim L = 6$, respectively.

More details on the symmetry algebras of N-th order DODSs, together with rigorous proofs of the above statements, will be published in a forthcoming article.

9 Traffic flow micro-model equation

Traffic flow is one of the areas in which time delay plays an important role. The time delay τ corresponds to the time it takes a driver of a car to react to changes of velocity of a preceding vehicle in a row of cars. One of the standard nonlinear "follow-the-leader" models is given by the system of delay ordinary differential equations [18, 39]

$$\ddot{x}_{n+1}(t) = \alpha\,[\dot{x}_{n+1}(t)]^{n_1}\,\frac{\dot{x}_n(t-\tau) - \dot{x}_{n+1}(t-\tau)}{[x_n(t-\tau) - x_{n+1}(t-\tau)]^{n_2}}, \qquad \alpha \ne 0, \tag{9.1}$$

where α, n_1, n_2 are adjustable constant parameters and τ is the constant time delay $t - t_- = \tau > 0$. The function $x_n(t)$ is the position of the n-th vehicle at time t, the dots denote time derivatives. Equations (9.1) can be viewed as a finite $(0 \le n \le N < \infty)$ or infinite $(0 \le n < \infty)$ set of DODEs, or as a delay ordinary differential-difference equation (since $x_n(t)$ is a function of a continuous variable t

and a discrete one n). The leading car's position is $x_0(t)$ and must be given as part of the input in the model.

In order to apply the Lie theory formalism developed above we restrict to the case of two cars: the leading one with position $x_0(t)$ and a following one with position $x_1(t) \equiv x(t)$. The system (9.1) then simplifies to the DODE

$$\ddot{x} = \alpha \dot{x}^{n_1} \frac{\dot{x}_{0,-} - \dot{x}_-}{(x_{0,-} - x_-)^{n_2}}, \qquad t_- = g(t, x, x_-, \dot{x}, \dot{x}_-), \tag{9.2}$$

where we replaced (x, y) by (t, x) (as usual in traffic flow studies) and allowed t_- to be a function of the indicated variables, instead of imposing $t_- = t - \tau$ with constant τ. We have put $x_- = x(t_-)$, $x_{0,-} = x_0(t_-)$, etc.

A group analysis of equations (9.2) and also (9.1) is in progress. Here we will just present an example showing how group analysis can lead to exact solutions invariant under some subgroup of the symmetry group.

Example. Consider the equation (9.2) with

$$n_1(n_1 - 1) \neq 0, \qquad n_2 = 0, \qquad x_0(t) = kt^n,$$

where $n = 1 - n_1^{-1}$. The DODS

$$\ddot{x} = \alpha \dot{x}^{n_1}(knt_-^{n-1} - \dot{x}_-), \qquad t_- = qt, \qquad 0 < q < 1 \tag{9.3}$$

admits the symmetry

$$X = t\frac{\partial}{\partial t} + n(x - \beta)\frac{\partial}{\partial x}. \tag{9.4}$$

An invariant solution has the form

$$x(t) = \beta + At^n, \qquad t_- = Bt. \tag{9.5}$$

Substituting it into the DODS, we get restrictions for the constants

$$\alpha \frac{(n_1 - 1)^{n_1}}{n_1^{n_1 - 1}} q^{-\frac{1}{n_1}} (k - A) A^{n_1 - 1} = -1, \qquad B = q. \tag{9.6}$$

The model and the solution (9.5) make sense. The leading car moves with acceleration $\ddot{x}_0 = kn(n-1)t^{n-2}$. The second car tries to adapt with solution (9.5). The distance between the two cars would be constant for $A = k$. This is however not possible since $A = k$ contradicts the constraint (9.6). To avoid a collision we must have $A < k$. For example, for $\alpha > 0$, $n_1 > 1$ a collision is inevitable.

The example shows the usefulness of simple exact solutions of DODSs. They help identify useful limits on the parameters of the model like α, n_1 and n_2 in (9.2).

10 Conclusions

We examined Lie symmetry properties of delay ordinary differential systems, which consist of DODEs of order N and delay relations. This approach has some common features with the Lie group classification of ordinary difference equations supplemented by lattice equations (ordinary difference systems). However solutions of delay differential systems are continuous while solutions of ordinary difference systems exist only in lattice points.

The dimension of a symmetry algebra admitted by a DODS of order N can satisfy $0 \leq n \leq 2N + 2$ or $n = \infty$. If $n = \infty$, the DODS consists of a linear DODE and a solution independent delay relation, or it can be transformed into such a form by an invertible transformation. Genuinely nonlinear DODS can admit symmetry algebras of dimension $0 \leq n \leq 2N + 2$. The Lie group classification of nonlinear DODSs is given in Table 1.

For invariant DODSs we obtained several theoretical results. Namely, if the symmetry algebra has $2N$ linearly connected symmetries, it provides a delay differential system which can be transformed into a linear DODE supplemented by a solution independent delay relation (see Theorem 1 for $N = 1$). Such linear delay differential systems admit infinite-dimensional symmetry groups since they allow linear superposition of solutions (Theorem 2). In this respect DODSs are similar to PDEs rather than ODEs [25, 2]. The reason is that initial conditions for ODEs are given in one point and hence consist of a finite number of constants. For DODSs (and PDEs) initial conditions must be given on an interval and hence involve arbitrary functions.

Linear DODSs (8.4) split into two classes. In one of them there are DODSs which, in addition to the infinite-dimensional symmetry algebra corresponding to the superposition principle, admit one further element (5.13). The superposition group acts only on the dependent variable y whereas the additional symmetry also acts on the independent variable x.

In Section 7, as an application of the symmetries admitted by DODS, we presented a procedure for calculating particular solutions, which are invariant solutions with respect to one-dimensional subgroups of the symmetry group. Several examples were given. In order to provide invariant solutions the symmetry algebra must contain at least one element with $\xi(x, y)\not\equiv 0$ in (4.1). New previously unpublished results are concentrated in Sections 8 and 9.

Finally, we remark that one can approach more complicated DODEs (for example, DODEs with several delays, etc.) in a similar manner.

Acknowledgments

The research of VD was partly supported by research grant No. 18-01-00890 of the Russian Fund for Base Research. The research of PW was partially supported by a research grant from NSERC of Canada.

References

[1] Baker C T H, Bocharov G A, Ford J M, Lumb P M, Norton S J, Paul C A H, Junt T, Krebs P and Ludewig B, Computational approaches to parameter estimation and model selection in immunology, *J. Computat. Appl. Math.* **184**, 50–76, 2005.

[2] Bluman G and Kumei S, *Symmetries and Differential Equations*, Springer-Verlag, New York, 1989.

[3] Craig M, Humphries A R and Mackey M C, A mathematical model of granulopoiesis incorporating the negative feedback dynamics and kinetics of G-CSF/neutrophil binding and internalization, *Bull. Math. Biol.* **78**, 2304–2357, 2016.

[4] Dorodnitsyn V A, Transformation groups in net spaces, *J. Sov. Math.* **55**, 1490–1517, 1991.

[5] Dorodnitsyn V A, Finite-difference models entirely inheriting symmetry of original differential equations, in Ibragimov N H et al (Ed), *Modern Group Analysis: Advanced Analytical and Computational Methods in Mathematical Physics*, Springer, Dordrecht, p 191, 1993.

[6] Dorodnitsyn V A, The finite-difference analogy of Noethers theorem, *Dokl. RAN* **328**, 678–682, 1993 (Russian); *Phys. Dokl.* **38**, 66–68, 1993 (translation in).

[7] Dorodnitsyn V, *Applications of Lie Groups to Difference Equations*, Chapman and Hall/CRC Differential and Integral Equations Series, Chapman and Hall, London, 2011.

[8] Dorodnitsyn V, Kozlov R and Winternitz P, Lie group classification of second-order ordinary difference equations, *J. Math. Phys.* **41**, 480–504, 2000.

[9] Dorodnitsyn V, Kozlov R and Winternitz P, Continuous symmetries of Lagrangians and exact solutions of discrete equations, *J. Math. Phys.* **45**, 336–359, 2004.

[10] Dorodnitsyn V A, Kozlov R, Meleshko S V and Winternitz P, Lie group classification of first-order delay ordinary differential equations, *J. Phys. A: Math. Theor.* **51**, 205202, 2018.

[11] Dorodnitsyn V A, Kozlov R, Meleshko S V and Winternitz P, Linear or linearizable first-order delay ordinary differential equations and their Lie point symmetries, *J. Phys. A: Math. Theor.* **51**, 205203, 2018.

[12] Dorodnitsyn V A, Kozlov R, Meleshko S V and Winternitz P, Lie group classification of second-order delay ordinary differential equations, arXiv:1901.06251.

[13] Driver R D, *Ordinary and Delay Differential Equations*, Springer-Verlag, New York, 1977.

[14] El'sgol'c L E, *Qualitative Methods in Mathematical Analysis*, GITTL, Moscow, 1955 (Russian); American Mathematical Society, Providence, RI, 1964 (Translations of Mathematical Monographs).

[15] Erneux T, *Applied Delay Differential Equations*, Springer, New York, 2009.

[16] Floreanini R and Vinet L, Lie symmetries of finite-difference equations, *J. Math. Phys.* **36**, 7024–7042, 1995.

[17] Gaeta G, *Nonlinear Symmetries and Nonlinear Equations*, Kluwer, Dordrecht, 1994.

[18] Gazis D C, Herman R and Rothery R W, Nonlinear follow-the-leader models of traffic flows, *Operations Research* **9** (4) 545–567, 1961.

[19] Gonzalez-Lopez A, Kamran N and Olver P J, Lie algebras of vector fields in the real plane, *Proc. Lond. Math. Soc.* **s3–64**, 339–368, 1992.

[20] Hale J, *Functional Differential Equations*, Springer, New York, 1971.

[21] Hydon P E, *Difference Equations by Differential Equation Methods*, Cambridge University Press, Cambridge, 2014.

[22] Ibragimov N H, *Transformation Groups Applied to Mathematical Physics*, Reidel, Boston, 1985.

[23] Ibragimov N H (Ed), *CRC Handbook of Lie Group Analysis of Differential Equations*, vol 1, 2, 3, CRC Press, Boca Raton, 1994–1996.

[24] Kolmanovskii V and Myshkis A, *Applied Theory of Functional Differential Equations*, Kluwer, Dordrecht, 1992.

[25] Kumei S and Bluman G, When nonlinear differential equations are equivalent to linear differential equations, *SIAM J. Appl. Math.* **42**, 1157–1173, 1982.

[26] Levi D and Winternitz P, Continuous symmetries of discrete equations, *Phys. Lett.* **152**, 335–338, 1991.

[27] Levi D and Winternitz P, Continuous symmetries of difference equations, *J. Phys. A: Math. Gen.* **39**, R1–R63, 2006.

[28] Lie S, Theorie der Transformationsgruppen, *Math. Ann.* **16**, 441–528, 1880.
Lie S, *Gesammelte Abhandlungen*, vol 6, BG Teubner, Leipzig, pp 1–94, 1927.

[29] Lie S, *Theorie der Transformationsgruppen*, vol I, BG Teubner, Leipzig, 1888.
Lie S, *Theorie der Transformationsgruppen*, vol II, BG Teubner, Leipzig, 1890.
Lie S, *Theorie der Transformationsgruppen*, vol III, BG Teubner, Leipzig, 1893.

[30] Lie S, Vorlesungen uber Differentialgleichungen mit bekannten infinitesimalen Transformationen, Bearbeitet und herausgegehen von Dr. G Scheffers, BG Teubner, Leipzig, 1891, (Reprinted as Differentialgleichungen, Chelsea, New York, 1967).

[31] Lie S, Gruppenregister, *Gesammelte Abhandlungen*, vol 5, BG Teubner, Leipzig, pp 767–773, 1924.

[32] Luzyanina T, Engelborghs K, Ehl S, Klenerman P and Bocharov G, Low level viral persistence after infection with LCMV: a quantitative insight through numerical bifurcation analysis, *Math. Biosci.* **173**, 1–23, 2001.

[33] Luzyanina T, Roosea D and Bocharov G, Numerical bifurcation analysis of immunological models with time delays, *J. Comput. Appl. Math.* **184**, 165–176, 2005.

[34] Maeda S, Extension of discrete Noether theorem, *Math. Japonica* **26**, 85–90, 1981.

[35] Maeda S, The similarity method for difference equations, *J. Inst. Math. Appl.* **38**, 129–134, 1987.

[36] Myshkis A D, *Linear Differential Equations with Retarded Argument*, Nauka, Moscow, 1972.

[37] Myshkis A D, *Differential Equations, Ordinary with Distributed Arguments* (Encyclopaedia of Mathematics vol 3), pp 144–147, Kluwer, Boston, 1989.

[38] Nesterenko M O, Transformation groups on real plane and their differential invariants, *Int. J. Math. Math. Sci.* **2006**, 1–17, 2006.

[39] Nagel K, Wagner P and Woesler R, Still flowing: approaches to traffic flow and traffic jam modeling *Operations Research* **51** (5) 681–710, 2003.

[40] Olver P J, *Applications of Lie Groups to Differential Equations*, Springer, New York, 1986.

[41] Ovsiannikov L V, *Group Analysis of Differential Equations*, Academic, New York, 1982.

[42] Patera J and Winternitz P, Subalgebras of real three- and four-dimensional Lie algebras, *J. Math. Phys.* **18**, 1449–1455, 1977.

[43] Polyanin A D and Sorokin V G, Nonlinear delay reaction-diffusion equations: traveling-wave solutions in elementary functions, *Appl. Math. Lett.* **46**, 38–43, 2015.

[44] Polyanin A D and Zhurov A I, The functional constraints method: application to non-linear delay reaction-diffusion equations with varying transfer coefficients, *Int. J. Non-Linear Mech.* **67**, 267–277, 2014.

[45] Polyanin A D and Zhurov A I, Generalized and functional separable solutions to nonlinear delay Klein-Gordon equations, *Commun. Nonlinear Sci. Numer. Simul.* **19**, 2676–2689, 2014.

[46] Polyanin A D and Zhurov A I, New generalized and functional separable solutions to nonlinear delay reaction-diffusion equations, *Int. J. Non-Linear Mech.* **59**, 16–22, 2014.

[47] Popovych R O, Boyko V M, Nesterenko M O and Lutfullin M W, Realizations of real low-dimensional Lie algebras, *J. Phys. A: Math. Gen.* **36**, 7337–7360, 2003.

[48] Pue-on P and Meleshko S V, Group classification of second-order delay ordinary differential equation, *Commun. Nonlinear Sci. Numer. Simul.* **15**, 1444–1453, 2010.

[49] Quispel G R W and Sahadevan R, Lie symmetries and integration of difference equations, *Phys. Lett. A* **184**, 64–70, 1993.

[50] Richard J P, Time-delay systems: an overview of some recent advances and open problems, *Automatica* **39**, 1667–1694, 2003.

[51] Šnobl L and Winternitz P, *Classification and Identification of Lie Algebras (CRM Monograph Series vol 33)*, American Mathematical Society, Providence, RI, 2014.

[52] Winternitz P, Symmetry preserving discretization of differential equations and Lie point symmetries of differential-difference equations, in Levi D et al (Ed) *Symmetries and Integrability of Difference Equations*, Cambridge University Press, Cambridge, ch I, pp 292–341, 2011.

[53] Wu J, *Theory and Applications of Partial Functional Differential Equations*, Springer, New York, 1996.

A5. The symmetry approach to integrability: recent advances

Rafael Hernández Heredero [a] *and Vladimir Sokolov* [b,c]

[a] *Depto. de Matemática Aplicada a las TIC, Universidad Politécnica de Madrid, C. Nikola Tesla s/n. 28031 Madrid. Spain*

[b] *Landau Institute for Theoretical Physics, 142432 Chernogolovka (Moscow region), Russia*

[c] *Universidade Federal do ABC, 09210-580 Sao Paulo, Brazil*

Abstract

We provide a concise introduction to the symmetry approach to integrability. Some results on integrable evolution and systems of evolution equations are reviewed. Quasi-local recursion and Hamiltonian operators are discussed. We further describe non-abelian integrable equations, especially matrix (ODE and PDE) systems. Some non-evolutionary integrable equations are studied using a formulation of formal recursion operators that allows to study non-diagonalizable systems of evolution equations.

1 Introduction

The symmetry approach to the classification of integrable PDEs has been developed since 1979 by: A. Shabat, A. Zhiber, N. Ibragimov, A. Fokas, V. Sokolov, S. Svinolupov, A. Mikhailov, R. Yamilov, V. Adler, P. Olver, J. Sanders, J.P. Wang, V. Novikov, A. Meshkov, D. Demskoy, H. Chen, Y. Lee, C. Liu, I. Khabibullin, B. Magadeev, R. H. Heredero, V. Marikhin, M. Foursov, S. Startcev, M. Balakhnev, and others. It is very efficient for PDEs with two independent variables and, under additional assumptions, it can be applied for ODEs.

The basic definition of the symmetry approach is the following.

Definition 1. A differential equation is integrable if it possesses infinitely many higher infinitesimal symmetries.

This definition is a priori reasonable: linear equations have infinitely many higher symmetries and all known integrable equations are related to linear equations by some transformations. These transformations produce symmetries on the integrable nonlinear equations coming from symmetries of the corresponding linear equations.

Requiring the existence of higher symmetries is a powerful method to find all integrable equations from a prescribed class of equations. The first classification result in the frame of the symmetry approach was the following:

Theorem 1 ([59]). *A nonlinear hyperbolic equation of the form*

$$u_{xy} = F(u)$$

possesses higher symmetries iff (up to scalings and shifts)

$$F(u) = e^u, \quad F(u) = e^u + e^{-u}, \quad or \quad F(u) = e^u + e^{-2u}.$$

There are several reviews [4,5,35,36,39,50] devoted to the symmetry approach (see also [13,19,23,42]). In this report we mostly concentrate on results not covered in those papers and books. Due to lack of space, we prefer to illustrate some significant ideas by examples and to give informal but constructive definitions, and refer to the cited reviews (e.g. [35,39,42]) for more detailed and rigorous developments. With respect to citations of original sources, we usually cite general reviews where the references can be found.

1.1 Infinitesimal symmetries

Consider a dynamical system of ODEs

$$\frac{dy^i}{dt} = F_i(y^1, \ldots, y^n), \qquad i = 1, \ldots, n. \tag{1.1}$$

Definition 2. The dynamical system

$$\frac{dy^i}{d\tau} = G_i(y^1, \ldots, y^n), \qquad i = 1, \ldots, n \tag{1.2}$$

is called an infinitesimal symmetry of (1.1) iff (1.1) and (1.2) are compatible.

Compatibility means that $XY - YX = 0$, where

$$X = \sum F_i \frac{\partial}{\partial y^i}, \quad \text{and} \quad Y = \sum G_i \frac{\partial}{\partial y^i}. \tag{1.3}$$

Consider now an evolution equation

$$u_t = F(u, u_x, u_{xx}, \ldots, u_n), \qquad u_i = \frac{\partial^i u}{\partial x^i}. \tag{1.4}$$

A higher (or generalized) infinitesimal symmetry of (1.4) is an evolution equation

$$u_\tau = G(u, u_x, u_{xx}, \ldots, u_m), \qquad m > 1 \tag{1.5}$$

that is compatible with (1.4).

Remark. Infinitesimal symmetries (1.5) with $m \leq 1$ correspond to one-parameter groups of point or contact transformations [42], the so called *classical symmetries.*

Compatibility of (1.4) and (1.5) means that

$$\frac{\partial}{\partial t}\frac{\partial u}{\partial \tau} = \frac{\partial}{\partial \tau}\frac{\partial u}{\partial t},$$

where the partial derivatives are calculated in virtue of (1.4) and (1.5). The compatibility condition can be rewritten as $G_*(F) = F_*(G)$ or

$$D_t(G) - F_*(G) = 0. \tag{1.6}$$

Here and below for any function $a(u, u_1, \dots)$ we denote

$$a_* \stackrel{\text{def}}{=} \sum_k \frac{\partial a}{\partial u_k} D^k,$$

where

$$D = \sum_{i=0}^{\infty} u_{i+1} \frac{\partial}{\partial u_i}$$

is the total x-derivative.

The concept of infinitesimal symmetry is connected with the procedure of linearization of a differential equation. For an arbitrary partial differential equation

$$Q(u, u_x, u_t, \dots) = 0$$

consider a linear equation

$$0 = \frac{\partial}{\partial \varepsilon}\Big(Q(u + \varepsilon\varphi, u_x + \varepsilon\varphi_x, u_t + \varepsilon\varphi_t, \dots)\Big)\Big|_{\varepsilon=0} \stackrel{\text{def}}{=} \mathcal{L}_Q(\varphi),$$

where the linear differential operator

$$\mathcal{L}_Q = \frac{\partial Q}{\partial u} + \frac{\partial Q}{\partial u_x} D + \frac{\partial Q}{\partial u_t} D_t + \dots$$

is called the *linearization operator* for the equation $Q = 0$. Here D and D_t are the total derivatives [42] with respect to x and t, respectively.

In the case of an evolution equation (1.4), we have $Q = u_t - F$ and

$$\mathcal{L}_Q = D_t - F_*.$$

Formula (1.6) means that the symmetry generator, G, is an element of the kernel of the linearization operator. This is a definition of infinitesimal symmetry which is directly applicable to the case of non-evolution PDEs.

For a more rigorous definition of symmetries of evolution equations in terms of evolution vector fields, see [39, 42].

1.2 Examples

In this section we present several examples of polynomial evolution equations and their symmetries, and also how the existence of a symmetry allows us to find integrable equations.

Example 1. Any equation of the form (1.4) has the classical symmetry $u_\tau = u_x$, which corresponds to the group of displacement parameters $x \to x + \lambda$.

Example 2. For all m and n, the equation $u_\tau = u_m$ is a symmetry of the linear equation $u_t = u_n$. Symmetries with different m are compatible with each other. Thus, we have an infinite *hierarchy* of evolution linear equations such that each of the equations is a symmetry for all the others.

Example 3. The Burgers equation

$$u_t = u_{xx} + 2uu_x \tag{1.7}$$

has a third-order symmetry

$$u_\tau = u_{xxx} + 3uu_{xx} + 3u_x^2 + 3u^2 u_x.$$

Example 4. The simplest higher symmetry of the Korteweg-de Vries (KdV) equation

$$u_t = u_{xxx} + 6\,uu_x \tag{1.8}$$

is

$$u_\tau = u_5 + 10\,uu_3 + 20\,u_1u_2 + 30\,u^2u_1. \tag{1.9}$$

Consider equations of the form

$$u_t = u_5 + a_1\,uu_3 + a_2\,u_1u_2 + a_3\,u^2u_1, \tag{1.10}$$

where a_i are constants. Let us find all equations (1.10) admitting a symmetry

$$u_\tau = u_7 + c_1uu_5 + c_2u_1u_4 + c_3u_2u_3 + c_4u^2u_3 + c_5uu_1u_2 + c_6u_1^3 + c_7u^3u_1.$$

The left-hand side of the compatibility condition (1.6) is a polynomial P in the variables $u_1, \ldots, u_{10}$. There are no linear terms in P. Equating to zero the coefficients of quadratic terms, we find,

$$c_1 = \frac{7}{5}a_1, \qquad c_2 = \frac{7}{5}(a_1 + a_2), \qquad c_3 = \frac{7}{5}(a_1 + 2a_2).$$

The vanishing conditions for the cubic part allow us to express c_4, c_5 and c_6 in terms of a_1, a_2, a_3. In addition, it turns out that

$$a_3 = -\frac{3}{10}a_1^2 + \frac{7}{10}a_1a_2 - \frac{1}{5}a_2^2.$$

Fourth degree terms lead to a formula expressing c_7 in terms of a_1, a_2 and to the basic algebraic relation

$$(a_2 - a_1)(a_2 - 2a_1)(2a_2 - 5a_1) = 0$$

between the coefficients a_1 and a_2. Solving this equation, we find that up to a scaling $u \to \lambda u$ there are only four integrable cases: the linear equation $u_t = u_5$, equations

$$u_t = u_5 + 5uu_3 + 5u_1u_2 + 5u^2u_1, \tag{1.11}$$

$$u_t = u_5 + 10uu_3 + 25u_1u_2 + 20u^2u_1, \tag{1.12}$$

and (1.9). In each of these cases, the terms of the fifth and sixth degrees in the defining equation are canceled automatically. The equations (1.11) and (1.12) are well known [22, 47].

Subsection 2.2 deals with an advanced version of the symmetry test, where there is no requirement for the polynomiality of the right-hand side of equation (1.4), as well as fixing the order of symmetry. Only the existence of an infinite hierarchy of symmetries is required. A similar approach is being developed for equations with local higher conservation laws.

1.3 First integrals and local conservation laws

In the ODE case the concept of a first integral (or integral of motion) is one of the basic notions. A function $f(y^1, \dots, y^n)$ is a *first integral* for the system (1.1) if its value does not depend on t for any solution $\{y^1(t), \dots, y^n(t)\}$ of (1.1). Since

$$\frac{d}{dt}\Big(f(y^1(t), \dots, y^n(t))\Big) = X(f),$$

where the vector field X is defined by (1.3), from the algebraic point of view a first integral is a solution of the first order PDE

$$X\Big(f(y^1, \dots, y^n)\Big) = 0.$$

In the case of evolution equations of the form (1.4) the concept of integral of motion is substituted by that of *local conservation law*: a pair of functions $\rho(u, u_x, ...)$ and $\sigma(u, u_x, ...)$ such that

$$\frac{\partial}{\partial t}\Big(\rho(u, u_x, \dots, u_p)\Big) = \frac{\partial}{\partial x}\Big(\sigma(u, u_x, \dots, u_q)\Big) \tag{1.13}$$

for any solution $u(x, t)$ of (1.4). The functions ρ and σ are called *density* and *flow* of the conservation law (1.13), respectively.

For soliton-type solutions, which are decreasing with derivatives as $x \to \pm\infty$, we obtain

$$\frac{\partial}{\partial t} \int_{-\infty}^{+\infty} \rho \, dx = 0$$

for any polynomial density ρ with a constant free term. This justifies the name *conserved density* for the function ρ. Similarly, if $u(x, t)$ is periodic in x with a period L, then the value of the functional $\int_0^L \rho \, dx$ on the solution u does not depend on time, so it is an integral of motion for equation (1.4).

Example 5. The functions

$$\rho_1 = u, \qquad \rho_2 = u^2, \qquad \rho_3 = -u_x^2 + 2u^3$$

are conserved densities for the Korteweg-de Vries equation (1.8).

Example 6. For any n the function $\rho_n = u_n^2$ is a conserved density for the linear equation

$$u_t = u_3. \tag{1.14}$$

2 The symmetry approach to integrability

The symmetry approach to the classification of integrable PDEs with two independent variables is based on the existence of higher symmetries and/or local conservation laws.

In the terminology by F. Calogero, an equation is *S-integrable* if it has infinitely many higher symmetries *and* conservation laws. And it is *C-integrable* if it has infinitely many higher symmetries but only a finite number of higher conservation laws.

Proposition 1 ([1], Theorem 29 in [39]). *Scalar evolution equations* (1.4) *of even order $n = 2k$ cannot possess infinitely many higher local conservation laws.*

Typical examples of *S-integrable* and *C-integrable* are the KdV-equation (1.8) and the Burgers equation (1.7), respectively.

Remark. Usually, the inverse scattering method can be applied to S-integrable equations while C-integrable equations can be reduced to linear equations by differential substitutions. However, to eliminate obvious exceptions like the linear equation (1.14), we need to refine the definition of S-integrable equations (see Definition 7).

There are two types of classifications obtained with the symmetry approach: a "weak" version, with equations admitting conservation laws and symmetries (S-integrable equations), and a "strong" version related only to symmetries, containing both S-integrable and C-integrable equations.

2.1 Description of some classification results

2.1.1 Hyperbolic equations

The first classification result using the symmetry approach was formulated in Theorem (1), and concerned hyperbolic equations. Nevertheless, to fully classify more general integrable hyperbolic equations remains an open problem. Some partial results were obtained in [30, 58].

Example 7. The following equation

$$u_{xy} = S(u)\sqrt{u_x^2+1}\sqrt{u_y^2+1}, \qquad \text{where} \qquad S'' - 2S^3 + c\,S = 0,$$

is integrable.

For hyperbolic equations $u_{xy} = \Psi(u, u_x, u_y)$ the symmetry approach assumes the existence of both x-symmetries of the form $u_t = A(u, u_x, u_{xx}, \dots,)$ and y-symmetries of the form $u_\tau = B(u, u_y, u_{yy}, \dots,)$, as it happens with the famous integrable sin-Gordon equation $u_{xy} = \sin u$. In [30] a classification was given assuming that both x and y-symmetries are integrable evolution equations of third order. In the survey [60] the reader can find further results on integrable hyperbolic equations.

2.1.2 Evolution equations

For evolution equations of the form (1.4), some necessary conditions for the existence of higher symmetries not depending on symmetry order were found in [18,50] (cf. Subsection 2.2). It was proved in [54] that the same conditions hold if the equation (1.4) admits infinitely many local conservation laws. In fact, the conditions for conservation laws are stronger (see Theorem 7) than the conditions for symmetries.

Second order equations. All nonlinear integrable equations of the form

$$u_t = F(x,\, t,\, u,\, u_1,\, u_2)$$

were listed in [53] and [51]. The answer is:

$$\begin{aligned} u_t &= u_2 + 2uu_x + h(x), \\ u_t &= u^2 u_2 - \lambda x u_1 + \lambda u, \\ u_t &= u^2 u_2 + \lambda u^2, \\ u_t &= u^2 u_2 - \lambda x^2 u_1 + 3\lambda x u. \end{aligned}$$

This list is complete up to contact transformations of the form

$$\hat{t} = \chi(t), \qquad \hat{x} = \varphi(x, u, u_1), \qquad \hat{u} = \psi(x, u, u_1),$$

$$\hat{u}_i = \left(\frac{1}{D(\varphi)} D\right)^i (\psi), \qquad D(\varphi)\frac{\partial \psi}{\partial u_1} = D(\psi)\frac{\partial \varphi}{\partial u_1}.$$

The first three equations of the list possess local symmetries and form a list obtained in [53]. The latter equation has so-called weakly non-local symmetries (see [51]). According to Proposition 1 all these equations are C-integrable. They are related to the heat equation $v_t = v_{xx}$ by differential substitutions of Cole-Hopf type [56].

Third order equations. A first result of the "weak" type for equations (1.4) is the following:

Theorem 2 ([54]). *A complete list (up to "almost invertible" transformations* [56]) *of equations of the form*

$$u_t = u_{xxx} + f(u, u_x, u_{xx}) \tag{2.1}$$

with an infinite sequence of conservation laws can be written as:

$$\begin{aligned} u_t &= u_{xxx} + 6u\, u_x, \\ u_t &= u_{xxx} + u^2 u_x, \\ u_t &= u_{xxx} - \frac{1}{2} u_x^3 + (\alpha e^{2u} + \beta e^{-2u}) u_x, \\ u_t &= u_{xxx} - \frac{1}{2} Q''\, u_x + \frac{3}{8} \frac{(Q - u_x^2)_x^2}{u_x\,(Q - u_x^2)}, \\ u_t &= u_{xxx} - \frac{3}{2} \frac{u_{xx}^2 + Q(u)}{u_x}, \qquad \textit{with } Q^{(\mathrm{v})}(u) = 0. \end{aligned} \tag{2.2}$$

For the "strong" version of this Theorem see [31, 55].

More general integrable third order equations of the form $u_t = F(u, u_x, u_{xx}, u_{xxx})$ admit three possible types of u_{xxx}-dependence [35]:

$$1)\ u_t = a\, u_{xxx} + b, \quad 2)\ u_t = \frac{a}{(u_{xxx}+b)^2} + c, \quad \text{and } 3)\ u_t = \frac{2a\, u_{xxx} + b}{\sqrt{a\, u_{xxx}^2 + b\, u_{xxx} + c}} + d,$$

where the functions a, b, c and d depend on u, u_x, u_{xx}. A complete classification of integrable equations of such type is not finished yet [15, 17], but there is the following insight.

Conjecture. All integrable third order equations are related to the KdV equation or to the Krichever-Novikov equation (2.2) by differential substitutions of Cole-Hopf and Miura type [57].

Fifth order equations. All equations of the form

$$u_t = u_5 + F(u, u_x, u_2, u_3, u_4),$$

possessing higher conservation laws were found in [11].

The list of integrable cases contains well-known equations and several new equations like

$$\begin{aligned} u_t &= u_5 + 5(u_2 - u_1^2 + \lambda_1 e^{2u} - \lambda_2^2 e^{-4u})\, u_3 - 5u_1 u_2^2 \\ &+ 15(\lambda_1 e^{2u}\, u_3 + 4\lambda_2^2 e^{-4u})\, u_1 u_2 + u_1^5 - 90\lambda_2^2 e^{-4u}\, u_1^3 + 5(\lambda_1 e^{2u} - \lambda_2^2 e^{-4u})^2\, u_1. \end{aligned}$$

The "strong" version of this classification result appears in [31].

The problem of classifying integrable equations

$$u_t = u_n + F(u,\, u_x,\, u_{xx}, \ldots, u_{n-1}), \qquad u_i = \frac{\partial^i u}{\partial x^i} \tag{2.3}$$

with arbitrary n seems to be far from being solved. Nevertheless, there are some clues to conclude that the only relevant classifications are those with $n = 2, 3, 5$. Each integrable equation together with all its symmetries form a hierarchy of integrable equations. As a rule, the members of a hierarchy also commute between themselves, i.e. each equation of the hierarchy is a higher symmetry for all others (cf. [49] for more details). In the case of equations (2.3) polynomial and homogeneous, it was proved in [44, 46] that the corresponding hierarchy contains an equation of second, third, or fifth order. It seems quite plausible that this fact could be extended to the general, non-polynomial case (2.3).

Further references on the classification of scalar evolution equations are the reviews [31, 35, 36, 39, 50] and papers [2, 7, 12, 21].

2.1.3 Systems of two equations

In [33, 34] necessary conditions of integrability were generalized to the case of systems of evolution equations. Computations become involved and the most general classification problem solved [33, 34, 36] is that of all S-integrable systems of the form

$$u_t = u_2 + F(u, v, u_1, v_1), \qquad v_t = -v_2 + G(u, v, u_1, v_1). \tag{2.4}$$

Besides the well-known NLS equation written as a system of two equations

$$u_t = -u_{xx} + 2u^2 v, \qquad v_t = v_{xx} - 2v^2 u, \tag{2.5}$$

basic integrable models from a long list of such integrable models are:

- a version of the Boussinesq equation

$$u_t = u_2 + (u+v)^2, \qquad v_t = -v_2 + (u+v)^2;$$

- and the two-component form of the Landau-Lifshitz equation

$$\begin{cases} u_t = u_2 - \dfrac{2u_1^2}{u+v} - \dfrac{4\,(p(u,v)\,u_1 + r(u)\,v_1)}{(u+v)^2}, \\ v_t = -v_2 + \dfrac{2v_1^2}{u+v} - \dfrac{4\,(p(u,v)\,v_1 + r(-v)\,u_1)}{(u+v)^2}, \end{cases}$$

where $r(y) = c_4 y^4 + c_3 y^3 + c_2 y^2 + c_1 y + c_0$ and

$$p(u,v) = 2c_4 u^2 v^2 + c_3(uv^2 - vu^2) - 2c_2 uv + c_1(u-v) + 2c_0.$$

A complete list of integrable systems (2.4) up to transformations

$$u \to \Phi(u), \qquad v \to \Psi(v)$$

should contain more than 100 systems. Such a list has never been published. In [36] appears a list complete up to "almost invertible" transformations [37]. All of these equations have a fourth order symmetry of the form

$$\begin{cases} u_\tau = \quad u_{xxxx} + f(u, v, u_x, v_x, u_{xx}, v_{xx}, u_{xxx}, v_{xxx}), \\ v_\tau = -v_{xxxx} + g(u, v, u_x, v_x, u_{xx}, v_{xx}, u_{xxx}, v_{xxx}) \end{cases} . \tag{2.6}$$

Reference [52] contains a classification of integrable equations of the form

$$\begin{cases} u_t = \quad u_{xx} + A_1(u,v)\,u_x + A_2(u,v)\,v_x + A_0(u,v), \\ v_t = -v_{xx} + B_1(u,v)\,v_x + B_2(u,v)\,u_x + B_0(u,v) \end{cases}$$

that includes C-integrable equations. The integrability requirement is that the system admits a fourth order symmetry (2.6) and triangular systems like $u_t = u_{xx} + 2uv_x$, $v_t = -v_{xx} - 2vv_x$ are disregarded.

2.2 Integrability conditions

We denote by $\mathcal{F}$ a field of functions depending on a finite number of variables $u, u_1, \ldots$. The field of constants is $\mathbb{C}$.

2.2.1 Pseudo-differential series

Consider a skew field of (non-commutative) pseudo-differential series of the form

$$A = a_m D^m + a_{m-1} D^{m-1} + \ldots + a_0 + a_{-1} D^{-1} + a_{-2} D^{-2} + \ldots, \qquad a_k \in \mathcal{F}. \quad (2.7)$$

The number $\mathrm{ord}(A) = m \in \mathbb{Z}$ is called the *order* of A. If $a_i = 0$ for $i < 0$, then A is a *differential operator.*

The product of two pseudo-differential series is defined over monomials by

$$D^k \circ bD^m = bD^{m+k} + C_k^1 D(b) D^{k+m-1} + C_k^2 D^2(b) D^{k+m-2} + \cdots,$$

where $k, m \in \mathbb{Z}$ and $C_n^j = \binom{n}{j}$ is the binomial coefficient. The formally conjugated pseudo-differential series A^+ is defined as

$$A^+ = (-1)^m D^m \circ a_m + (-1)^{m-1} D^{m-1} \circ a_{m-1} + \cdots + a_0 - D^{-1} \circ a_{-1} + D^{-2} \circ a_{-2} + \cdots.$$

For any series (2.7) there is a unique inverse series B such that $A \circ B = B \circ A = 1$, and there are m-th roots C such that $C^m = A$, unique up to a numeric factor ε with $\varepsilon^m = 1$. Notice that $\mathrm{ord}(B) = -m$ and $\mathrm{ord}(C) = 1$.

Definition 3. The *residue* of series (2.7) is the coefficient of D^{-1}: $\mathrm{res}(A) = a_{-1}$. The *logarithmic residue* of A is defined as $\mathrm{res} \log A = a_{n-1}/a_n$.

Theorem 3 ([3]). *For any two series A, B the residue of the commutator belongs to* $\mathrm{Im}\, D$*:*

$$\mathrm{res}[A, B] = D(\sigma(A, B)),$$

$$\text{where} \quad \sigma(A, B) = \sum_{p \le \mathrm{ord}(B),\; q \le \mathrm{ord}(A)}^{p+q+1>0} C_q^{p+q+1} \times \sum_{s=0}^{p+q} (-1)^s D^s(a_q) D^{p+q-s}(b_q).$$

2.2.2 Formal symmetry

Definition 4. A pseudo-differential series R that satisfies

$$D_t(R) = [F_*, R], \qquad \text{where} \qquad F_* = \sum_{i=0}^{n} \frac{\partial F}{\partial u_i} D^i \quad (2.8)$$

is called *formal symmetry*, or *formal recursion operator*,[1] of eq. (1.4).

[1] Relation (2.8) can be rewritten as $[D_t - F_*^+, R] = 0$. Therefore any genuine operator that satisfies (2.8) maps higher symmetries of the equation (1.4) to higher symmetries.

Proposition 2 ([50]).

1) If R_1 *and* R_2 *are formal symmetries, then* $R_1 \circ R_2$ *is a formal symmetry too;*

2) If R *is a formal symmetry of order* k*, so is* $R^{i/k}$ *for any* $i \in \mathbb{Z}$*;*

3) Let Λ *be a formal symmetry of order 1*

$$\Lambda = l_1 D + l_0 + l_{-1} D^{-1} + \cdots . \tag{2.9}$$

Then R *can be written in the form*

$$R = \sum_{-\infty}^{k} a_i \Lambda^i, \qquad k = \mathrm{ord}(R), \quad a_i \in \mathbb{C};$$

4) In particular, any formal symmetry $\bar{\Lambda}$ *of order 1 has the form*

$$\bar{\Lambda} = \sum_{-\infty}^{1} c_i \Lambda^i, \qquad c_i \in \mathbb{C}.$$

In Sections 2 and 3 we will only consider a formal symmetry Λ of order 1, without loss of generality (see Item 2) of Proposition 2).

Example 8. Consider finding formal symmetries of equations of KdV type

$$u_t = u_3 + f(u, u_1). \tag{2.10}$$

With

$$F_* = D^3 + \frac{\partial f}{\partial u_1} D + \frac{\partial f}{\partial u}, \qquad \Lambda = l_1 D + l_0 + l_{-1} D^{-1} + \cdots .$$

equation (2.8) becomes an infinite system of differential equations whose members are the coefficients of $D^3, D^2, \ldots$ equaled to zero. The first equations are

$$D^3: \quad 3D(l_1) = 0; \qquad D^2: \quad 3D^2(l_1) + 3D(l_0) = 0;$$

$$D: \quad D^3(l_1) + 3D^2(l_0) + 3D(l_{-1}) + \frac{\partial f}{\partial u_1} D(l_1) = D_t(l_1) + l_1 D\left(\frac{\partial f}{\partial u_1}\right).$$

The first three equations above imply that $l_1 = 1$, $l_0 = 0$ and $l_{-1} = \frac{1}{3}\frac{\partial f}{\partial u_1}$, i.e.

$$\Lambda = D + \frac{1}{3}\frac{\partial f}{\partial u_1} D^{-1} + \cdots$$

The first obstacle for the existence of Λ appears in the coefficient of D^{-1}, requiring that $\partial^4 f(u, u_1)/\partial u_1^4 = 0$. Thus, there are no formal symmetries for an arbitrary function $f(u, u_1)$ in (2.10).

Remark 1. In general, for any k the equation for the coefficient l_k of Λ can be written as $D(l_k) = S_k$, where $S_k \in \mathcal{F}$ is an already known function. The equation is solvable only if $S_k \in \operatorname{Im} D$. Thus there are infinitely many obstacles for the existence of a formal symmetry. The integration constant appearing in l_k can be taken as 0 (except for $k = 1$) because of Proposition 2.

Theorem 4 ([18, 50]). *If an equation $u_t = F$ possesses an infinite sequence of higher symmetries*

$$u_{\tau_i} = G_i(u, \ldots, u_{m_i}), \qquad m_i \to \infty$$

then it has a formal symmetry.

2.2.3 Formal symplectic operator

It is known [35, p. 122] that for any conserved density ρ the variational derivative

$$X = \frac{\delta \rho}{\delta u} = \sum_k (-1)^k D^k \left(\frac{\partial \rho}{\partial u_k} \right),$$

satisfies the equation conjugate to (1.6):

$$D_t\,(X) + F_*^+\,(X) = 0\,. \tag{2.11}$$

Any solution $X \in \mathcal{F}$ of equation (2.11) is called *cosymmetry.*

The conjugate concept of a formal symmetry is the following.

Definition 5. A pseudo-differential series

$$S = s_m D^m + s_{m-1} D^{m-1} + \cdots + s_0 + s_{-1} D^{-1} + \cdots, \qquad s_m \neq 0, \quad s_i \in \mathcal{F}$$

is called a *formal symplectic operator*[2] of order m for equation (1.4) if it satisfies

$$D_t(S) + S\,F_* + F_*^+\,S = 0\,. \tag{2.12}$$

It follows from (2.12) that equations of the form (1.4) of even order n have no formal symplectic operators.

Lemma 1. *The ratio $S_1^{-1} S_2$ of any two formal symplectic operators S_1 and S_2 satisfies the equation* (2.8) *of a formal symmetry.*

Theorem 5 ([18, 50]). *If equation $u_t = F$ possesses an infinite sequence of local conservation laws, then the equation has:*

1) a formal recursion operator Λ and

2) a formal symplectic operator S of first order.

Remark (see [50]). Without loss of generality one can assume that

$$S^+ = -S, \qquad \Lambda^+ = -S^{-1} \Lambda S.$$

[2]Relation (2.12) can be rewritten as $(D_t + F_*^+) \circ S = S(D_t - F_*)$. This means that a genuine operator $S : \mathcal{F} \to \mathcal{F}$ maps symmetries to cosymmetries. If equation (1.4) is Hamiltonian, then the symplectic operator, which is inverse to the Hamiltonian operator, satisfies equation (2.12) [10].

2.2.4 Canonical densities and necessary integrability conditions

In this section we formulate the necessary conditions over equations (1.4) to admit infinite higher symmetries or conservation laws. According to Theorems 4 and 5, such equations possess a formal symmetry. The obstructions in Remark 1 to the existence of a formal symmetry, are thus integrability conditions. It tuns out that these conditions can be written in the form of conservation laws.

Definition 6. For equations (1.4) possessing a formal symmetry Λ, the functions

$$\rho_i = \mathrm{res}\,(\Lambda^i), \qquad i = -1, 1, 2, \ldots, \qquad \text{and} \qquad \rho_0 = \mathrm{res}\,\log(\Lambda) \tag{2.13}$$

are called *canonical densities* for equation (1.4) .

Adler's theorem 3 implies the following result.

Theorem 6. *If an equation* (1.4) *has a formal symmetry* Λ, *then the canonical densities* (2.13) *define corresponding local conservation laws*

$$D_t(\rho_i) = D(\sigma_i), \quad \sigma_i \in \mathcal{F}, \qquad i = -1, 0, 1, 2, \ldots. \tag{2.14}$$

Theorem 7 ([50])**.** *Under the assumptions of Theorem 5, all even canonical densities* ρ_{2j} *belong to* $\mathrm{Im}\, D$.

Example 9. The differential operator $\Lambda = D$ is a formal symmetry for any linear equation of the form $u_t = u_n$. Therefore all canonical densities are equal to zero.

Example 10. The KdV equation (1.8) has a recursion operator

$$\hat{\Lambda} = D^2 + 4u + 2u_1 D^{-1}\,, \tag{2.15}$$

which satisfies equation (2.8). A corresponding formal symmetry of order 1 for the KdV equation is $\Lambda = \hat{\Lambda}^{1/2}$. The infinite commutative hierarchy of symmetries for the KdV equation is generated by the recursion operator:

$$G_{2k+1} = \hat{\Lambda}^k(u_1)\,.$$

The first five canonical densities for the KdV equation are

$$\rho_{-1} = 1, \qquad \rho_0 = 0, \qquad \rho_1 = 2u, \qquad \rho_2 = 2u_1, \qquad \rho_3 = 2u_2 + u^2.$$

Example 11. The Burgers equation (1.7) has the recursion operator

$$\Lambda = D + u + u_1 D^{-1}\,.$$

Functions $G_n = \Lambda^n(u_1)$ are generators of symmetries for the Burgers equation. The canonical densities for the Burgers equation are

$$\rho_{-1} = 1, \qquad \rho_0 = u, \qquad \rho_1 = u_1, \qquad \rho_2 = u_2 + 2uu_1, \ldots.$$

Although ρ_0 is not trivial (i.e. $\rho_0 \notin \mathrm{Im}\, D$), all other canonical densities are trivial.

Now we can refine the definition of C-integrability such that linear equations become C-integrable.

Definition 7. Equation (1.4) is called *S-integrable* if it has a formal symmetry that provides infinitely many non-trivial canonical densities. An equation is called *C-integrable* if it has a formal symmetry such that only finite number of canonical densities are non-trivial.

Remark. It follows from Theorems 4, 5, 6 and 7 that if we are going to find equations (1.4) with higher symmetries, we have to use conditions (2.14) only, while for equations with higher conservation laws we may additionally assume that $\rho_{2j} = D(\theta_j) + c_j$, where $\theta_j \in \mathcal{F}$ and $c_j \in \mathbb{C}$. Thus the necessary conditions, which we employ for conservation laws are stronger than ones for symmetries.

Using the ideas of [7, 29], a recursive formula for the whole infinite chain of canonical conserved densities can be derived. For equations of the form (2.1) such a formula was obtained in [31]:

$$\rho_0 = -\frac{1}{3}f_2, \qquad \rho_1 = \frac{1}{9}f_2^2 - \frac{1}{3}f_1 + \frac{1}{3}D(f_2),$$

$$\begin{aligned}\rho_{n+2} = \frac{1}{3}\Bigg[&\sigma_n - \delta_{n,0}f_0 - f_1\rho_n - f_2\Big(D(\rho_n) + 2\rho_{n+1} + \sum_{s=0}^{n}\rho_s\,\rho_{n-s}\Big)\Bigg] \\ &- \sum_{s=0}^{n+1}\rho_s\,\rho_{n+1-s} - \frac{1}{3}\sum_{0\le s+k\le n}\rho_s\,\rho_k\,\rho_{n-s-k} \\ &- D\Bigg[\rho_{n+1} + \frac{1}{2}\sum_{s=0}^{n}\rho_s\,\rho_{n-s} + \frac{1}{3}D(\rho_n)\Bigg], \qquad n \ge 0. \end{aligned} \tag{2.16}$$

Here, $\delta_{i,j}$ is the Kronecker delta and $f_i = \partial f/\partial u_i$, for $i = 0, 1, 2$.

Using the integrability conditions, one can find all equations of the prescribed type, which have a formal symmetry. A full classification result includes:

1) a complete[3] list of integrable equations that satisfy the necessary integrability conditions and

2) a confirmation of integrability for each equation from the list.

For Item 2) one can find a Lax representation or a transformation that links the equation with an equation known to be integrable. The existence of an auto-Bäcklud transformation with an arbitrary parameter is also a proper justification of integrability.

Remark. From a proof of a classification result one can derive a constructive description of transformations that bring a given integrable equation to one from the list and the number of necessary conditions, which should be verified for a given equation to establish its integrability.

The classifications of integrable evolution equations discussed in Section 2.1 have been performed using the theory described in this section.

[3] usually complete up to a class of admissible transformations.

2.3 Recursion and Hamiltonian quasi-local operators

Recursion and Hamiltonian operators establish additional relations between higher symmetries and conserved densities.

Proposition 3. *If the operator* $\mathcal{R}: \mathcal{F} \to \mathcal{F}$ *satisfies the equation*[4]

$$D_t(\mathcal{R}) = F_* \mathcal{R} - \mathcal{R} F_*, \tag{2.17}$$

then, for any symmetry[5] G *of equation* (1.4), $\mathcal{R}(G)$ *is also a symmetry of* (1.4).

Definition 8. An operator $\mathcal{R}: \mathcal{F} \to \mathcal{F}$ satisfying (2.17) is called a *recursion operator* for equation (1.4).

The set of all recursion operators forms an associative algebra over $\mathbb{C}$.

The simplest symmetry for any equation (1.4) is $u_\tau = u_x$. Acting with a recursion operator over u_x usually yields the generators of all the other symmetries.

A recursion operator is usually non-local (see, for instance, (2.15)) so it can only be applied to a very special subset of $\mathcal{F}$ to get a function in $\mathcal{F}$.

2.3.1 Quasi-local recursion operators

Most of the known recursion operators have the following special non-local structure:

$$\mathcal{R} = R + \sum_{i=1}^{k} G_i\, D^{-1} \circ g_i, \qquad g_i, G_i \in \mathcal{F}, \tag{2.18}$$

where R is a differential operator. Such operators are called *quasi-local* or *weakly nonlocal.*

Definition 9. An operator $\mathcal{R}$ of the form (2.18) is called a *quasi-local recursion operator* for equation (1.4) if

1) $\mathcal{R}$, considered as a pseudo-differential series, satisfies (2.17);

2) the functions G_i are generators of some symmetries for (1.4);

3) the functions g_i are variational derivatives of conserved densities.[6]

Example 12. The recursion operator (2.15) for the KdV equation is quasi-local with $k = 1$, $G_1 = u_x/2$ and $g_1 = \delta u/\delta u = 1$.

The first reference we know of where a quasi-local ansatz for finding a recursion operator was used, is [48].

It can be proved that the set of all quasi-local recursion operators for the KdV equation form a commutative associative algebra $A_{\rm rec}$ over $\mathbb{C}$ generated by the

[4]In the language of differential geometry, this relation means that the Lie derivative of the operator $\mathcal{R}$, by virtue of equation (1.4), is zero.

[5]For brevity, we often refer to the symmetry by its generator G.

[6]It might be reasonable to add the hereditary property [14] of the operator $\mathcal{R}$ to the properties 1)–3).

operator (2.15), i.e. $A_{\rm rec}$ is isomorphic to the algebra of all polynomials in one variable.

It turns out that this is not true for integrable models such as the Krichever-Novikov and the Landau-Lifshitz equations. In particular, the Krichever-Novikov equation (2.2) has two quasi-local recursion operators $\mathcal{R}_1$ and $\mathcal{R}_2$ such that $\mathcal{R}_2^2 = \mathcal{R}_1^3 - \phi\mathcal{R}_1 - \theta$, where the constants ϕ, θ are polynomial in the coefficients of Q.

Remark. In the case of the Krichever-Novikov equation (2.2), the ratio $\mathcal{R}_3 = \mathcal{R}_2\mathcal{R}_1^{-1}$ satisfies equation (2.17). It belongs to the skew field of differential operator fractions [45]. However, this operator is not quasi-local and it is unclear how to apply it even to the simplest symmetry generator u_x.

Further information about this matter can be found in [9, 48].

2.3.2 Hamiltonian operators

Most of the known integrable equations (1.4) can be written in a Hamiltonian form as

$$u_t = \mathcal{H}\left(\frac{\delta\rho}{\delta u}\right),$$

where ρ is a conserved density and $\mathcal{H}$ is a Hamiltonian operator. The analog of the operator identity (2.17) for Hamiltonian operators is given by

$$(D_t - F_*)\,\mathcal{H} = \mathcal{H}(D_t + F_*^+), \tag{2.19}$$

which means that $\mathcal{H}$ maps cosymmetries to symmetries. This formula, justified by the general theory [10], states that Hamiltonian operators play an inverse role of that of symplectic operators (see (2.12)).

The Poisson bracket corresponding to a Hamiltonian operator $\mathcal{H}$ is defined by

$$\{f,\,g\} = \frac{\delta f}{\delta u}\,\mathcal{H}\Big(\frac{\delta g}{\delta u}\Big). \tag{2.20}$$

Skew-symmetricity and the Jacobi identity for (2.20) are required. Namely,

$$\{f, g\} + \{g, f\} \in \operatorname{Im} D, \tag{2.21}$$

$$\{\{f, g\}, h\} + \{\{g, h\}, f\} + \{\{h, f\}, g\} \in \operatorname{Im} D \tag{2.22}$$

for $f, g \in \mathcal{F}$. Thus, Hamiltonian operators should satisfy, besides (2.19), some identities (see for example [10, 42]) equivalent to (2.21), (2.22). Lemma 1 suggests the next result.

Lemma 2. *If operators $\mathcal{H}_1$ and $\mathcal{H}_2$ satisfy* (2.19), *then $\mathcal{R} = \mathcal{H}_2\mathcal{H}_1^{-1}$ satisfies* (2.8).

As a rule, Hamiltonian operators are local (differential) or quasi-local operators

$$\mathcal{H} = H + \sum_{i=1}^{m} G_i D^{-1} \bar{G}_i,$$

where H is a differential operator and $G_i, \bar{G}_i$ are symmetries [26, 40].

The KdV equation possesses two local Hamiltonian operators

$$\mathcal{H}_1 = D, \qquad \mathcal{H}_2 = D^3 + 4uD + 2u_x.$$

Their ratio gives the recursion operator (2.15).

The first example

$$\mathcal{H}_0 = u_x D^{-1} u_x$$

of a quasi-local Hamiltonian operator was found in paper [48], where the Krichever-Novikov equation (2.2) was studied. In [9] it was shown that, besides $\mathcal{H}_0$, equation (2.2) possesses two more quasi-local Hamiltonian operators.

3 Integrable non-abelian equations

3.1 ODEs on free associative algebras

We consider ODE systems of the form

$$\frac{dx_\alpha}{dt} = F_\alpha(\mathbf{x}), \qquad \mathbf{x} = (x_1, ..., x_N), \qquad \alpha = 1, \ldots, N, \tag{3.1}$$

where $x_i(t)$ are $m \times m$ matrices and F_α are (non-commutative) polynomials with constant scalar coefficients. As usual, a symmetry is defined as an equation

$$\frac{dx_\alpha}{d\tau} = G_\alpha(\mathbf{x}), \tag{3.2}$$

compatible with (3.1).

In the case $N = 2$ we denote $x_1 = u$, $x_2 = v$.

3.1.1 Manakov top

The system

$$u_t = u^2\, v - v\, u^2, \qquad v_t = 0 \tag{3.3}$$

has infinitely many symmetries for any size of the matrices u and v. Many important *multi-component* integrable systems can be obtained as reductions of (3.3). For instance, if u is $m \times m$ matrix such that $u^t = -u$, and v is a constant diagonal matrix, then (3.3) is equivalent to the m-dimensional Euler top. The integrability of this model by the inverse scattering method was established by S.V. Manakov in [27].

Consider the cyclic reduction

$$u = \begin{pmatrix} 0 & u_1 & 0 & 0 & \cdot & 0 \\ 0 & 0 & u_2 & 0 & \cdot & 0 \\ \cdot & \cdot & \cdot & \cdot & \cdot & \cdot \\ 0 & 0 & 0 & 0 & \cdot & u_{m-1} \\ u_m & 0 & 0 & 0 & \cdot & 0 \end{pmatrix}, \qquad v = \begin{pmatrix} 0 & 0 & 0 & \cdot & 0 & J_m \\ J_1 & 0 & 0 & \cdot & 0 & 0 \\ 0 & J_2 & 0 & \cdot & 0 & 0 \\ \cdot & \cdot & \cdot & \cdot & \cdot & \cdot \\ 0 & 0 & 0 & \cdot & J_{m-1} & 0 \end{pmatrix},$$

where u_k and J_k are matrices of smaller size. Then (3.3) is equivalent to the non-abelian Volterra chain

$$\frac{d}{dt}u_k = u_k u_{k+1} J_{k+1} - J_{k-1} u_{k-1} u_k, \qquad k = 1, \ldots, m.$$

If we assume $m = 3$, $J_1 = J_2 = J_3 = \mathrm{Id}$ and $u_3 = -u_1 - u_2$ the system becomes

$$u_t = u^2 + uv + vu\,, \qquad v_t = -v^2 - uv - vu\,.$$

3.1.2 Matrix generalization of a flow on an elliptic curve

The system

$$\begin{cases} u_t = v^2 + cu + a\,\mathrm{I}, \\ v_t = u^2 - cv + b\,\mathrm{I}, \end{cases} \qquad a, b, c \in \mathbb{C}, \tag{3.4}$$

where u and v are $m \times m$-matrices and I is the identity matrix, is integrable for any m. It has a Lax pair [61] and possesses an infinite sequence of polynomial symmetries. If $m = 1$, this system can be written in the Hamiltonian form

$$u_t = -\frac{\partial H}{\partial v}, \qquad v_t = \frac{\partial H}{\partial u}$$

with Hamiltonian

$$H = \frac{1}{3}u^3 - \frac{1}{3}v^3 - cuv + bu - av.$$

For generic a, b, c the relation $H =$ constant defines an elliptic curve, and equations (3.4) describe the motion of a point along this curve.

In the homogeneous case the system (3.4) has the form [38]

$$u_t = v^2, \qquad v_t = u^2. \tag{3.5}$$

F. Calogero has observed that in the matrix case the functions $x_i = \lambda_i^{1/2}$, where λ_i are the eigenvalues of the matrix $u - v$, satisfy the following integrable system:

$$x_i'' = -x_i^5 + \sum_{j \neq i} \Big[(x_i - x_j)^{-3} + (x_i + x_j)^{-3} \Big].$$

3.1.3 Non-abelian systems

The variables $x_1, \ldots, x_N$ in (3.1) can be regarded as generators of a free associative algebra $\mathcal{A}$. We call systems on $\mathcal{A}$ *non-abelian systems.* In order to understand what compatibility of equations (3.1) and (3.2) means, we use the following definition.

Definition 10. A linear map $d : \mathcal{A} \to \mathcal{A}$ is called a derivation if it satisfies the Leibniz rule: $d(xy) = xd(y) + d(x)y$.

Fixing $d(x_i) = F_i(\mathbf{x})$ over all generators x_i of $\mathcal{A}$ uniquely determines $d(z)$ for any $z \in \mathcal{A}$, through the Leibnitz rule. The polynomials F_i can be taken arbitrarily. Instead of dynamical system (3.1) one considers the derivation $D_t : \mathcal{A} \to \mathcal{A}$ such that $D_t(x_i) = F_i$. Compatibility of (3.1) and (3.2) means that the corresponding derivations D_t and D_τ commute: $D_t D_\tau - D_\tau D_t = 0$.

From the symmetry approach point of view, system (3.1) is integrable if it possesses infinitely many linearly independent symmetries.

Two-component non-abelian systems. Consider non-abelian systems

$$u_t = P(u, v), \qquad v_t = Q(u, v), \qquad P, Q \in \mathcal{A}$$

on the free associative algebra $\mathcal{A}$ over $\mathbb{C}$ with generators u and v. Define an involution $\star$ on $\mathcal{A}$ by the formulas

$$u^\star = u, \quad v^\star = v, \quad (a\, b)^\star = b^\star\, a^\star, \quad a, b \in \mathcal{A}. \tag{3.6}$$

Two systems related to each other by a linear transformation of the form

$$\hat{u} = \alpha u + \beta v, \qquad \hat{v} = \gamma u + \delta v, \qquad \alpha\delta - \beta\gamma \neq 0 \tag{3.7}$$

and involutions (3.6) are defined as *equivalent.*

The simplest class is that of quadratic systems of the form

$$\begin{cases} u_t = \alpha_1 u\, u + \alpha_2 u\, v + \alpha_3 v\, u + \alpha_4 v\, v, \\ v_t = \beta_1 v\, v + \beta_2 v\, u + \beta_3 u\, v + \beta_4 u\, u. \end{cases} \tag{3.8}$$

The problem is to describe all non-equivalent systems (3.8) which possess infinitely many symmetries. Some preliminary results were obtained in [38]. Here we follow the paper [61].

It is reasonable to assume that the corresponding scalar system

$$\begin{cases} u_t = a_1 u^2 + a_2 uv + a_3 v^2, \\ v_t = b_1 v^2 + b_2 uv + b_3 u^2, \end{cases} \tag{3.9}$$

where $a_1 = \alpha_1,\ a_2 = \alpha_2 + \alpha_3,\ a_3 = \alpha_4,\ b_1 = \beta_1,\ b_2 = \beta_2 + \beta_3,\ b_3 = \beta_4$ should be integrable. The main feature of integrable systems of the form (3.9) is the existence of infinitesimal polynomial symmetries and first integrals. Another evidence of integrability is the absence of movable singularities in solutions for complex t. The so-called Painlevé approach is based on this assumption. Our first requirement is that system (3.9) should possess a polynomial first integral I.

Lemma 3. *Suppose that a system* (3.9) *has a homogeneous polynomial integral $I(u, v)$. Then it has an infinite sequence of polynomial symmetries of the form*

$$\begin{cases} u_\tau = I^N\,(\alpha_1 u^2 + \alpha_2 uv + \alpha_3 v^2), \\ v_\tau = I^N\,(\beta_1 v^2 + \beta_2 uv + \beta_3 u^2) \end{cases} \qquad N \in \mathbb{N}. \tag{3.10}$$

Writing I in factorized form:

$$I=\prod_{i=1}^{k}(u-\kappa_i v)^{n_i}, \qquad n_i\in\mathbb{N}, \qquad \kappa_i\neq\kappa_j \text{ if } i\neq j.$$

Theorem 8. *Suppose that at least one of the coefficients of a system (3.9) is not equal to zero. Then $k\leq 3$.*

Consider the case $k=3$.[7] Transformations (3.7) reduce I to the form

$$I=u^{k_1}(u-v)^{k_2}v^{k_3}, \tag{3.11}$$

where k_i are natural numbers which are defined up to permutations. Without loss of generality we assume that

$$k_1\leq k_2\leq k_3$$

and that k_1,k_2,k_3 have no non-trivial common divisor.

Lemma 4. *A system (3.9) has an integral (3.11) iff up to a scaling $u\to\mu u$, $v\to\mu v$ it has the following form:*

$$\begin{cases} u_t=-k_3\,u^2+(k_3+k_2)\,uv\\ v_t=-k_1\,v^2+(k_1+k_2)\,uv.\end{cases} \tag{3.12}$$

Proposition 4. *A system (3.12) satisfies the Painlevé test in the following three cases:*

Case 1. $k_1=k_2=k_3=1$;

Case 2. $k_1=k_3=1, \quad k_2=2$;

Case 3. $k_1=1, \quad k_2=2, \quad k_3=3$.

Any non-abelian system which coincides with (3.12) in the scalar case, has the form

$$\begin{cases} u_t=-k_3\,u^2+(k_2+k_3)\,uv+\alpha(uv-vu)\\ v_t=-k_1\,v^2+(k_1+k_2)\,vu+\beta(vu-uv).\end{cases} \tag{3.13}$$

Let us find the parameters α and β such that (3.13) has infinitely many symmetries that reduce to (3.10) in the scalar case.

Consider Case 1: $k_1=k_2=k_3=1$. The integral I is of degree 3 and (3.10) implies that the simplest symmetry is supposed to be of fifth degree.

Theorem 9. *In the case $k_1=k_2=k_3=1$ there exist only 5 non-equivalent non-abelian systems of the form (3.13) that have a fifth degree symmetry. They correspond to the following pairs α,β in (3.13):*

[7] For the cases $k=1,2$ see [61].

1. $\alpha = -1, \quad \beta = -1,$

2. $\alpha = \ \ 0, \quad \beta = -1,$

3. $\alpha = \ \ 0, \quad \beta = -2,$

4. $\alpha = \ \ 0, \quad \beta = \ \ 0,$

5. $\alpha = \ \ 0, \quad \beta = -3.$

System (3.5) is equivalent to the system in Item 1.

Consider now Case 2: $k_1 = k_3 = 1$, $k_2 = 2$. We thus suppose a simplest symmetry of degree 6.

Theorem 10. *In the case* $k_1 = k_3 = 1$, $k_2 = 2$ *there exist only 4 non-equivalent systems* (3.13) *that have the symmetry of degree six. They correspond to:*

1. $\alpha = -1, \quad \beta = -1,$

2. $\alpha = \ \ 0, \quad \beta = -2,$

3. $\alpha = \ \ 0, \quad \beta = \ \ 0,$

4. $\alpha = \ \ 0, \quad \beta = -4.$

The classification of integrable non-abelian systems (3.13) ends with Case 3:

Theorem 11. *In the case* $k_1 = 1$, $k_2 = 2$, $k_3 = 3$ *there exist only 5 non-equivalent systems* (3.13) *with the symmetry of degree 8. They correspond to:*

1. $\alpha = -2, \quad \beta = \ \ 0,$

2. $\alpha = -4, \quad \beta = \ \ 0,$

3. $\alpha = -6, \quad \beta = \ \ 0,$

4. $\alpha = \ \ 0, \quad \beta = -6,$

5. $\alpha = \ \ 0, \quad \beta = \ \ 0.$

The integrable systems found in [61] contain all examples from [38] as well as new integrable non-abelian systems of the form (3.8). Moreover, all integrable inhomogeneous generalizations of these systems were found in [61]. System (3.4) is one of them.

There are interesting integrable non-abelian Laurent systems. In this case we extend the free associative algebra $\mathfrak{A}$ with generators u and v by new symbols u^{-1} and v^{-1} such that $uu^{-1} = u^{-1}u = vv^{-1} = v^{-1}v = \mathrm{I}$.

Example 13. In the paper [62] the following integrable non-abelian Laurent system

$$u_t = uv - uv^{-1} - v^{-1}, \qquad v_t = -vu + vu^{-1} + u^{-1},$$

proposed by M. Kontsevich, was investigated. It can be regarded as a non-trivial deformation of the integrable system[8]

$$u_t = uv, \qquad v_t = -vu$$

by Laurent terms of smaller degree.

[8]In the scalar case this system has a first integral of first degree.

3.2 PDEs on free associative algebra

In this subsection we consider the so-called non-abelian evolution equations, which are natural generalizations of evolution matrix equations.

3.2.1 Matrix integrable equations

The matrix KdV equation has the following form

$$\mathbf{U}_t = \mathbf{U}_{xxx} + 3\,(\mathbf{U}\mathbf{U}_x + \mathbf{U}_x\mathbf{U}), \tag{3.14}$$

where $\mathbf{U}(x,t)$ is an unknown $m \times m$-matrix. It is known that this equation has infinitely many higher symmetries for arbitrary m. All of them can be written in matrix form. The simplest higher symmetry of (3.14) is given by

$$\mathbf{U}_\tau = \mathbf{U}_{xxxxx} + 5\,(\mathbf{U}\mathbf{U}_{xxx} + \mathbf{U}_{xxx}\mathbf{U}) + 10\,(\mathbf{U}_x\mathbf{U}_{xx} + \mathbf{U}_{xx}\mathbf{U}_x) \\ + 10\,(\mathbf{U}^2\mathbf{U}_x + \mathbf{U}\mathbf{U}_x\mathbf{U} + \mathbf{U}_x\mathbf{U}^2).$$

For $m = 1$ this matrix hierarchy of symmetries coincides with the usual KdV hierarchy.

In general [43], we may consider matrix equations of the form

$$\mathbf{U}_t = F(\mathbf{U},\, \mathbf{U}_1,\, \ldots,\, \mathbf{U}_n), \qquad \mathbf{U}_i = \frac{\partial^i \mathbf{U}}{\partial x^i},$$

where F is a (non-commutative) polynomial with constant scalar coefficients. The criterion of integrability is the existence of matrix higher symmetries

$$\mathbf{U}_\tau = G(\mathbf{U},\, \mathbf{U}_1,\, \ldots,\, \mathbf{U}_m).$$

The matrix KdV equation is not an isolated example. Many known integrable models have matrix generalizations [24, 28, 43]. In particular, the mKdV equation $u_t = u_{xxx} + u^2u_x$ has two different matrix generalizations:

$$\mathbf{U}_t = \mathbf{U}_{xxx} + 3\mathbf{U}^2\mathbf{U}_x + 3\mathbf{U}_x\mathbf{U}^2,$$

and (see [24])

$$\mathbf{U}_t = \mathbf{U}_{xxx} + 3[\mathbf{U}, \mathbf{U}_{xx}] - 6\mathbf{U}\mathbf{U}_x\mathbf{U}.$$

The matrix generalization of the NLS equation (2.5) is given by

$$\mathbf{U}_t = \mathbf{U}_{xx} - 2\,\mathbf{U}\mathbf{V}\mathbf{U}, \qquad \mathbf{V}_t = -\mathbf{V}_{xx} + 2\,\mathbf{V}\mathbf{U}\mathbf{V}.$$

The Krichever-Novikov equation (2.2) with $Q = 0$ is called the *Schwartz KdV equation.* Its matrix generalization is given by

$$\mathbf{U}_t = \mathbf{U}_{xxx} - \frac{3}{2}\,\mathbf{U}_{xx}\mathbf{U}_x^{-1}\mathbf{U}_{xx}.$$

The Krichever-Novikov equation with generic Q probably has no matrix generalizations.

The matrix Heisenberg equation has the form

$$\mathbf{U}_t = \mathbf{U}_{xx} - 2\,\mathbf{U}_x(\mathbf{U}+\mathbf{V})^{-1}\mathbf{U}_x, \qquad \mathbf{V}_t = -\mathbf{V}_{xx} + 2\,\mathbf{V}_x(\mathbf{U}+\mathbf{V})^{-1}\mathbf{V}_x.$$

One of the most renowned hyperbolic matrix integrable equations is the principal chiral σ-model

$$\mathbf{U}_{xy} = \frac{1}{2}\,(\mathbf{U}_x\mathbf{U}^{-1}\mathbf{U}_y + \mathbf{U}_y\mathbf{U}^{-1}\mathbf{U}_x).$$

The system

$$\begin{aligned} \mathbf{U}_t &= \lambda_1\mathbf{U}_x + (\lambda_2-\lambda_3)\mathbf{W}^t\mathbf{V}^t,\\ \mathbf{V}_t &= \lambda_2\mathbf{V}_x + (\lambda_3-\lambda_1)\mathbf{U}^t\mathbf{W}^t,\\ \mathbf{W}_t &= \lambda_3\mathbf{W}_x + (\lambda_1-\lambda_2)\mathbf{V}^t\mathbf{U}^t \end{aligned}$$

is a matrix generalization of the 3-wave model. In contrast with the previous equations, it contains matrix transpositions, denoted by ^{t}.

Let $\mathbf{e}_1,\ldots,\mathbf{e}_N$ be a basis of some associative algebra $\mathcal{B}$ and

$$U = \sum_{i=1}^{N} u_i\,\mathbf{e}_i. \tag{3.15}$$

Then, all the matrix equations presented above give rise to corresponding integrable systems in the unknown functions $u_1,\ldots,u_N$ in (3.15). Indeed, just the associativity of the product in $\mathcal{B}$ is enough to ensure that the symmetries of a matrix equation remain as symmetries of the corresponding system for u_i.

In the matrix case $\mathcal{B} = \mathfrak{gl}_m$ interesting examples of integrable multi-component systems are produced by Clifford algebras and by group algebras of associative rings.

The most fundamental setting for the non-abelian equations is the formalism of free associative algebras, leading to a generalization of matrix equations.

3.2.2 Non-abelian evolution equations over free associative algebras

Let us consider evolution equations on an infinitely generated free associative algebra $\mathcal{A}$. In the case of one-field non-abelian equations the generators of $\mathcal{A}$ are denoted by

$$U, \quad U_1 = U_x, \quad \ldots, \quad U_k, \quad \ldots. \tag{3.16}$$

As $\mathcal{A}$ is free, there are no algebraic relations between the generators. All definitions can be easily generalized to the case of several non-abelian variables.

The formula

$$U_t = F(U, U_1, \ldots, U_n), \qquad F \in \mathcal{A} \tag{3.17}$$

defines a derivation D_t of $\mathcal{A}$ which commutes with the basic derivation

$$D = \sum_0^\infty U_{i+1} \frac{\partial}{\partial U_i}.$$

It is easy to check that D_t is defined by the vector field

$$D_t = \sum_0^\infty D^i(F) \frac{\partial}{\partial U_i}.$$

The concepts of symmetry, conservation law, the operation *, and formal symmetry have to be specified for differential equations on free associative algebras.

As in the scalar case, a symmetry is an evolution equation

$$U_\tau = G(U, U_1, \ldots, U_m),$$

such that the vector field

$$D_G = \sum_0^\infty D^i(G) \frac{\partial}{\partial u_i}$$

commutes with D_t. The polynomial G is called *the symmetry generator*.

The condition $[D_t, D_G] = 0$ is equivalent to $D_t(G) = D_G(F)$. The latter relation can be rewritten as

$$G_*(F) - F_*(G) = 0,$$

where the differential operator H_* for any $H \in \mathcal{A}$ can be defined as follows.

For any $a \in \mathcal{A}$ we denote by L_a and R_a the operators of left and right multiplication by a:

$$L_a(X) = a\,X, \qquad R_a(X) = X\,a, \qquad X \in \mathcal{A}.$$

The associativity of $\mathcal{A}$ is equivalent to the identity $[L_a, R_b] = 0$ for any a and b. Moreover,

$$L_{ab} = L_a\,L_b, \qquad R_{ab} = R_b\,R_a, \qquad L_{a+b} = L_a + L_b, \qquad R_{a+b} = R_a + R_b.$$

Definition 11. We denote by $\mathcal{O}$ the associative algebra generated by all operators of left and right multiplication by any element (3.16). This algebra is called the *algebra of local operators.*

Extending the set of generators with an additional non-commutative symbol V_0 and prolonged symbols $V_{i+1} = D(V_i)$, one can define, given $H(U, U_1, \ldots, U_k) \in \mathcal{A}$,

$$H_*(V_0) = \frac{\partial}{\partial \varepsilon} H(U + \varepsilon V_0,\ U_1 + \varepsilon V_1,\ U_2 + \varepsilon V_2, \ldots)\big|_{\varepsilon=0}.$$

Here H_* is a linear differential operator of order k with coefficients in $\mathcal{O}$. For example, $(U_2 + UU_1)_* = D^2 + L_U D + R_{U_1}$.

The definition of conserved density has to be modified. In the scalar case [42] conserved densities are defined up to total x-derivatives, i.e. two conserved densities are equivalent if $\rho_1 - \rho_2 \in \mathcal{F}/\operatorname{Im} D$. In the matrix case the conserved densities are traces of some matrix polynomials defined up to total derivatives.

In the non-abelian case $\rho_1 \sim \rho_2$ iff $\rho_1 - \rho_2 \in \mathcal{A}/(\operatorname{Im} D + [\mathcal{A}, \mathcal{A}])$. The equivalence class of an element ρ is called the *trace of* ρ and is denoted by $\operatorname{tr} \rho$.

Definition 12. The equivalence class of an element $\rho \in \mathcal{A}$ is called conserved density for equation (3.17) if $D_t(\rho) \sim 0$.

Poisson brackets (2.20) are defined on the vector space $\mathcal{A}/(\operatorname{Im} D + [\mathcal{A}, \mathcal{A}])$. A general theory of Poisson and double Poisson brackets on algebras of differential functions was developed in [8]. An algebra of (non-commutative) differential functions is defined as a unital associative algebra $\mathcal{D}$ with a derivation D and commuting derivations ∂_i, $i \in \mathbb{Z}_+$ such that the following two properties hold:

1) For each $f \in \mathcal{D}$, $\partial_i(f) = 0$ for all but finitely many i;
2) $[\partial_i, D] = \partial_{i-1}$.

Formal symmetry. At least for non-abelian equations of the form (3.17), where

$$F = U_n + f(U, U_1, \dots, U_{n-1}) \tag{3.18}$$

all definitions and results concerning formal symmetries (as in Subsection 2.2) can be easily generalized.

Definition 13. A formal series

$$\Lambda = D + l_0 + l_{-1}D^{-1} + \cdots, \qquad l_k \in \mathcal{O}$$

is a *formal symmetry* of order 1 for an equation (3.18) if it satisfies the equation

$$D_t(\Lambda) - [F_*, \Lambda] = 0\,.$$

For example, for the non-abelian Korteweg-de Vries equation (3.14) one can take $\Lambda = \mathcal{R}^{1/2}$, where $\mathcal{R}$ is the following recursion operator for (3.14) (see [43]):

$$\mathcal{R} = D^2 + 2(L_U + R_U) + (L_{U_x} + R_{U_x})\,D^{-1} + (L_U - R_U)\,D^{-1}\,(L_U - R_U)\,D^{-1}.$$

There are analogues to Theorems 4, 5 in the non-commutative case.

4 Non-evolutionary systems

Consider non-evolutionary equations of the form

$$u_{tt} = F(u, u_1, \dots, u_n; u_t, u_{1t}, \dots, u_{mt}) \tag{4.1}$$

where $u_{mt} = \partial^{m+1}u/\partial x^m \partial t$. In this section $\mathcal{F}$ denotes the field of functions of variables $u, u_1, u_2, \ldots$ and $u_t, u_{1t}, u_{2t}, \ldots$. We will say that equation (4.1) is of order (n, m). The total x-derivative D and t-derivative D_t are

$$D = \sum_{i=0}^{\infty} u_{i+1} \frac{\partial}{\partial u_i} + \sum_{j=0}^{\infty} u_{j+1t} \frac{\partial}{\partial u_{jt}}, \qquad D_t = \sum_{i=0}^{\infty} u_{it} \frac{\partial}{\partial u_i} + \sum_{j=0}^{\infty} D^j(F) \frac{\partial}{\partial u_{jt}}$$

and commute. It is easy to see that $\operatorname{Ker} D = \mathbb{C}$ in $\mathcal{F}$.

Evolutionary vector fields that commute with D, can be written as

$$D_H = \sum_{i=0}^{\infty} D^i(H) \frac{\partial}{\partial u_i} + \sum_{j=0}^{\infty} D^j(D_t(H)) \frac{\partial}{\partial u_{jt}}.$$

For any function $H(u, u_1, \ldots, u_t, u_{1t}, \ldots) \in \mathcal{F}$ the differential operator H_* is defined as

$$H_* = \sum_{i=0}^{\infty} \frac{\partial H}{\partial u_i} D^i + \sum_{j=0}^{\infty} \frac{\partial H}{\partial u_{jt}} D^j D_t.$$

The linearization operator (see Section 1.1) for (4.1) has the form

$$\mathcal{L} = D_t^2 - F_* = D_t^2 - U - V D_t,$$

where the differential operators U and V

$$\begin{aligned} U &= \mathfrak{u}_n D^n + \mathfrak{u}_{n-1} D^{n-1} + \cdots + \mathfrak{u}_0, \\ V &= \mathfrak{v}_m D^m + \mathfrak{v}_{m-1} D^{m-1} + \cdots + \mathfrak{v}_0 \end{aligned} \quad \text{with} \quad \mathfrak{u}_i = \frac{\partial F}{\partial u_i}, \quad \mathfrak{v}_j = \frac{\partial F}{\partial u_{jt}} \tag{4.2}$$

are defined by the rhs of (4.1).

Remark 2. We can rewrite (4.1) in an evolutionary form as

$$u_t = v, \qquad v_t = F(u, u_1, \ldots, u_r, v, v_x, \ldots, v_s). \tag{4.3}$$

The matrix linearization operator for (4.3) is $D_t - \mathbf{F}$, where

$$\mathbf{F} = \begin{pmatrix} 0 & 1 \\ U & V \end{pmatrix}.$$

The approach in [33, 34] is not applicable in the classification of integrable systems (4.3), since it requires the diagonalizability of the matrix differential operator $\mathbf{F}$, and here it is not diagonalizable. Assuming polynomiality of the equations, a powerful symbolic technique has been developed and applied to perform classifications of integrable systems (4.1) in [32, 41].

In this chapter we develop the approach proposed in [16], which does not assume the polynomiality of the equations.

4.1 Formal recursion operators

Definition 14. The pseudo-differential operator $\mathcal{R} = X + Y\,D_t$ with components

$$X = \sum_{-\infty}^{p} x_i D^i, \qquad Y = \sum_{-\infty}^{q} y_i D^i$$

is called a *formal recursion operator of order* (p, q) for equation (4.1) if it satisfies the relation

$$\mathcal{L}(X + Y\,D_t) = (\bar{X} + \bar{Y}\,D_t)\mathcal{L} \tag{4.4}$$

for some formal series $\bar{X}$, $\bar{Y}$.

It follows from (4.4) that $\bar{Y} = Y$ and $\bar{X} = X + 2Y_t + [Y, V]$. If X and Y are differential operators (or ratios of differential operators), condition (4.4) implies the fact that operator $\mathcal{R}$ maps symmetries of equation (4.1) to symmetries.

Lemma 5. *Relation* (4.4) *is equivalent to the identities*

$$X_{tt} - VX_t + [X, U] + (2Y_t + [Y, V])U + YU_t = 0, \tag{4.5}$$

$$Y_{tt} + 2X_t + [Y, U] + [X, V] + ([Y, V] + 2Y_t)V + YV_t - VY_t = 0. \tag{4.6}$$

Let $\mathcal{R}_1 = X_1 + Y_1\,D_t$ and $\mathcal{R}_2 = X_2 + Y_2\,D_t$ be two formal recursion operators. Then the product $\mathcal{R}_3 = \mathcal{R}_1\mathcal{R}_2$, in which D_t^2 is replaced by $(U + VD_t)$ is also a formal recursion operator whose components are given by

$$X_3 = X_1X_2 + Y_1Y_2U + Y_1(X_2)_t, \qquad Y_3 = X_1Y_2 + Y_1X_2 + Y_1Y_2V + Y_1(Y_2)_t.$$

Thus the set of all formal recursion operators forms an associative algebra A_{frec}. In the evolution case this algebra is generated by one generator of the form (2.9). For equations of the form (4.1) the structure of A_{frec} essentially depends on the numbers n and m.

Definition 15. The pseudo-differential operator $\mathcal{S} = P + Q\,D_t$ is called a *formal symplectic operator* for equation (4.1) if it satisfies the relation[9]

$$\mathcal{L}^{+}(P + Q\,D_t) + (\bar{P} + \bar{Q}\,D_t)\mathcal{L} = 0 \tag{4.7}$$

for some formal series $\bar{P}, \bar{Q}$.

The operator equations for the components of a formal symplectic operator $\mathcal{S}$ have the following form

$$P_{tt} + V^*P_t + 2Q_tU + QU_t = U^*P - PU - (QV + V^*Q)\,U - V_t^*P, \tag{4.8}$$

$$\begin{aligned} Q_{tt} + 2P_t + 2Q_tV + V^*Q_t = U^*Q - QU - (QV + V^*Q)\,V \\ - (PV + V^*P) - (V_t^*Q + QV_t) \end{aligned} \tag{4.9}$$

and

$$\bar{P} = -P - 2Q_t - V^{+}Q - QV, \qquad \bar{Q} = -Q.$$

[9] In the evolution case, $\mathcal{L} = D_t - F_*$ and (4.7) coincides with (2.12).

Definition 16. We call equation (4.1) formally integrable if it possesses a formal recursion operator $\mathcal{R}$ of some order (p,q) with a complete set of arbitrary integration constants.

Since $\mathcal{R}$ depends on integration constants linearly, we actually have an infinite-dimensional vector space of formal recursion operators.

4.2 Examples

Example of order (3,1).

Example 14. The simplest example of an integrable equation of the form (4.1) is given by [25]:

$$u_{tt} = u_{xxx} + 3u_x u_{xt} + \left(u_t - 3u_x^2\right) u_{xx}. \tag{4.10}$$

The formal recursion operator of order $(n+1,n)$[10] for this equation has the following structure on higher order terms

$$\begin{aligned}&(l_{n+1} + m_n n u_x)\, D^{n+1} + \left[l_n + m_{n-1}(n-1)u_x + l_{n+1}(n+1)\left(\tfrac{1}{2}(n-4)u_x^2 + u_t\right)\right.\\ &\left.+m_n\left(\tfrac{1}{6}(n+1)\left(n^2-7n-6\right)u_x^3 + n(n+1)u_t u_x + \tfrac{1}{2}\left(n^2-n+2\right)u_{xx}\right)\right] D^n + \cdots\\ &+\left\{m_n D^n + \left[m_{n-1} + l_{n+1}(n+1)u_x + m_n(n+1)\left(\tfrac{n}{2}u_x^2 + u_t\right)\right] D^{n-1} + \cdots\right\} D_t\end{aligned}$$

where m_i, l_i are the integration constants. Denote by $\mathcal{Y}_i$ the recursion operator corresponding to $m_i = 1$, $m_j = 0$ when $j \neq i$, and $l_j = 0$, and denote by $\mathcal{X}_i$ the recursion operator corresponding to $l_i = 1$, $l_j = 0$ if $j \neq i$ and $m_j = 0$. The operator $\mathcal{Y}_{-1}$ can be written in an explicit form as

$$\mathcal{Y}_{-1} = -u_x + D^{-1}(D_t + 2u_{xx}).$$

We have that

$$\mathcal{Y}_{-1}^{-1} = \mathcal{Y}_{-2}, \quad \mathcal{Y}_{-1}^2 = \mathcal{X}_1, \quad \mathcal{X}_1^{-1} = \mathcal{X}_{-1}, \quad \mathcal{X}_1^i = \mathcal{X}_i.$$

Therefore any recursion operator $\mathcal{R}$ can be uniquely represented in the form

$$\mathcal{R} = \sum_{-\infty}^{k} c_i \mathcal{Y}_{-1}^i, \qquad c_i \in \mathbb{C}.$$

This equation does not admit any formal symplectic operator (see Lemma 6).

[10] We do not assume that the leading coefficients are non-zeros and therefore we may postulate this order without loss of generality.

The potential Boussinesq equation.

Example 15. An example of an integrable equation of order (4,1) is the potential Boussinesq equation

$$u_{tt} = u_{xxxx} - 3u_x u_{xx}. \tag{4.11}$$

Similar to Example 14, we can find two different types of formal recursion operators:

$$\begin{aligned}\mathcal{Y}_i = &-\tfrac{3}{8}(i+2)u_t\,D^{-1+i} - \tfrac{3}{16}(i-2)(i+1)u_{xt}\,D^{-2+i}\\ &+\tfrac{1}{32}\left[9(i-1)(i+2)u_x u_t - 2i\left(i^2-3i+8\right)u_{xxt}\right]D^{-3+i}+\cdots\\ &+\{D^i - \tfrac{3}{4}iu_x\,D^{-2+i} - \tfrac{3}{8}(i-2)iu_{xx}\,D^{-3+i}\\ &\quad+\left[\tfrac{9}{32}(i^2-3i+2)u_x^2 - \tfrac{1}{16}(i-3)(i-2)(2i+1)u_{xxx}\right]D^{-4+i}+\cdots\}\,D_t \end{aligned} \tag{4.12}$$

and

$$\begin{aligned}\mathcal{X}_i = &\,D^i - \tfrac{3}{4}iu_x\,D^{-2+i} - \tfrac{3}{8}(i-2)iu_{xx}\,D^{-3+i}\\ &+\tfrac{1}{32}\left[9(i-3)iu_x^2 - 2(2i-7)(i-1)iu_{xxx}\right]D^{-4+i}+\cdots\\ &+\{-\tfrac{3}{8}iu_t\,D^{-5+i} - \tfrac{3}{16}(i-5)iu_{xt}\,D^{-6+i}\\ &\quad+\left[\tfrac{9}{32}(i-5)iu_x u_t - \tfrac{1}{16}(i-5)(i-4)iu_{xxt}\right]D^{-7+i}+\cdots\}\,D_t. \end{aligned} \tag{4.13}$$

However, algebraic relations between them are completely different:

$$\mathcal{X}_i = \mathcal{X}_1^i, \qquad \mathcal{Y}_i = \mathcal{Y}_1\mathcal{X}_1^{i-1}.$$

We also have that

$$\mathcal{Y}_1^2 = \mathcal{X}_6, \quad \mathcal{Y}_1^{-1} = \mathcal{Y}_{-5}.$$

so, notably, $\mathcal{Y}_{-2}^2 = 1$. We see that the algebra of all formal recursion operators is generated by $\mathcal{X}_1$ and $\mathcal{Y}_1$. The generators commute with each other[11] and are related by the algebraic curve

$$\mathcal{X}_1^2 = \mathcal{Y}_1^6.$$

The operator $\mathcal{Y}_1$ can be written in closed form as

$$\begin{aligned}\mathcal{Y}_1 &= -\tfrac{9}{8}u_t + \tfrac{3}{8}D^{-1}\cdot u_{xt} + \left[D - \tfrac{3}{8}u_x D^{-1} - \tfrac{3}{8}D^{-1}\cdot u_x\right]D_t\\ &= -\tfrac{9}{8}u_t + DD_t + \tfrac{3}{8}D^{-1}\cdot(u_{xt} - u_x D_t) - \tfrac{3}{8}u_x D^{-1}D_t. \end{aligned} \tag{4.14}$$

The independent classical symmetries of (4.11) are 1, $s_0^a = u_x$ and $s_0^b = u_t$. Applying the recursion operator $\mathcal{Y}_1$ to them, we obtain a chain of symmetries:

$$s_1^{\mathrm{b}} = \mathcal{Y}_1(u_x) = u_{xxt} - \tfrac{3}{2}u_x u_t,$$

[11] A proof that these statements are true uses the homogeneity of (4.11) and (4.8), (4.9).

$$s_2^{\mathrm{a}} = \mathcal{Y}_1(u_t) = u_{xxxxx} - \tfrac{15}{4}u_x u_{xxx} - \tfrac{15}{16}u_t^2 - \tfrac{45}{16}u_{xx}^2 + \tfrac{15}{16}u_x^3.$$

In general, $\mathcal{Y}_1(s_i^{\mathrm{a}}) = s_{i+1}^{\mathrm{b}}$ and $\mathcal{Y}_1(s_i^{\mathrm{b}}) = s_{i+2}^{\mathrm{a}}$, $i = 0, 1, \dots$ where the notation indicates that there are two types of symmetries

$$\begin{aligned} s_i^{\mathrm{a}} &= c_i\, u_{2i+1} + \text{lower order terms}, \\ s_i^{\mathrm{b}} &= d_i\, u_{2it} + \text{lower order terms}, \end{aligned} \qquad i = 0, 1, \dots$$

Notice that symmetries of type s_{1+3i}^{a} and s_{2+3i}^{b} with $i = 0, 1, \dots$ are not produced by this scheme.

Remark. The potential Boussinesq equation (4.11) is Lagrangian (see Subsection 4.4.2) and therefore its linearization operator is self-adjoint: $\mathcal{L}^+ = \mathcal{L}$. For such equations, recursion and symplectic operators coincide. In particular, the recursion operator (4.14) is also a symplectic operator for equation (4.11).

4.3 Integrability conditions

For any equation (4.1) with $2m < n$, the terms with highest order on D in (4.5)–(4.6) are in $[X, U]$ and $[Y, U]$ and produce equations where, using integrating factors of the form u_4^{α}, the unknowns x_i, y_j can be isolated in two equations of the form

$$D\left(\mathrm{u}_4^{-i/4} x_i\right) = A_i, \qquad D\left(\mathrm{u}_4^{-i/4} y_i\right) = B_i. \tag{4.15}$$

The right-hand sides A_i, B_i depend only on variables x_j and y_j of lower indexes. Therefore the coefficients of the series X and Y can be found recursively from relations (4.5) and (4.6) but there appear an infinite number of integrability obstructions because equations (4.15) have no solutions for arbitrary A_i and B_i.[12] Since $\operatorname{Ker} D = \mathbb{C}$, some integration constants arise from (4.15) of each step (cf. Remark 1).

Proposition 2 implies that in the evolution case the obstructions to formal integrability do not depend on the choice of the integration constants and on the order of formal recursion operator.

Conjecture. This is also true for equations (4.1) with $2m < n$.

To find the coefficients of a formal symplectic operator we have to use the relations (4.8), (4.9). An analysis of the terms with highest order on D leads (cf. Subsection 2.2.3) to the following result:

Lemma 6. *If $2m < n$ and n is odd, then no non-trivial formal symplectic operator exists.*

Here we consider two classes of equations (4.1), of order (3,1) and of order (4,1). The computations justify the conjecture in these two cases.

[12] These functions must be total derivatives.

4.3.1 Equations of order (3,1)

In the paper [16] equations of the form

$$u_{tt} = f(u_3, u_{1t}, u_2, u_t, u_1, u), \qquad \frac{\partial f}{\partial u_3} \neq 0 \tag{4.16}$$

were considered. According to Lemma 6, such equations have no formal symplectic operator.

The first integrability conditions for the existence of a formal recursion operator have the form of conservation laws $D_t(\rho_i) = D(\sigma_i)$, $i = 0, 1, \ldots$ (see [6,16]) and the conserved densities can be written as

$$\rho_0 = \frac{1}{\sqrt[3]{\mathfrak{u}_3}}, \qquad \rho_1 = \frac{3\mathfrak{u}_2}{\mathfrak{u}_3} + \frac{2\sigma_1}{\sqrt[3]{\mathfrak{u}_3}} + \frac{\sigma_0\mathfrak{v}_1}{\mathfrak{u}_3^{2/3}} + \frac{\sigma_0^2}{\sqrt[3]{\mathfrak{u}_3}}, \qquad \rho_2 = \frac{\mathfrak{v}_1}{\mathfrak{u}_3^{2/3}} - \frac{2\sigma_0}{\sqrt[3]{\mathfrak{u}_3}},$$

$$\rho_3 = \frac{\mathfrak{v}_1^3}{\mathfrak{u}_3^{4/3}} - \frac{27\mathfrak{v}_0}{\sqrt[3]{\mathfrak{u}_3}} + \frac{9\mathfrak{u}_2\mathfrak{v}_1}{\mathfrak{u}_3^{4/3}} - \frac{9\mathfrak{v}_1 D\mathfrak{u}_3}{\mathfrak{u}_3^{4/3}} - \frac{6\sigma_2}{\sqrt[3]{\mathfrak{u}_3}} + \frac{3\sigma_1\mathfrak{v}_1}{\mathfrak{u}_3^{2/3}} + \frac{6\sigma_1\sigma_0}{\sqrt[3]{\mathfrak{u}_3}} + \frac{3\sigma_0^2\mathfrak{v}_1}{\mathfrak{u}_3^{2/3}} + \frac{4\sigma_0^3}{\sqrt[3]{\mathfrak{u}_3}},$$

$$\begin{aligned}\rho_4 = &-\frac{81\mathfrak{u}_1}{\mathfrak{u}_3^{2/3}} + \frac{27\mathfrak{u}_2^2}{\mathfrak{u}_3^{5/3}} + \frac{\mathfrak{v}_1^4}{\mathfrak{u}_3^{5/3}} + \frac{9\mathfrak{u}_2\mathfrak{v}_1^2}{\mathfrak{u}_3^{5/3}} - \frac{27\mathfrak{v}_0\mathfrak{v}_1}{\mathfrak{u}_3^{2/3}} - \frac{27\mathfrak{u}_2 D\mathfrak{u}_3}{\mathfrak{u}_3^{5/3}} - \frac{9\mathfrak{v}_1^2 D\mathfrak{u}_3}{\mathfrak{u}_3^{5/3}} + \frac{9\,(D\mathfrak{u}_3)^2}{\mathfrak{u}_3^{5/3}}\\ &+\frac{2\sigma_3}{\sqrt[3]{\mathfrak{u}_3}} + \frac{3\sigma_2\mathfrak{v}_1}{\mathfrak{u}_3^{2/3}} + \frac{6\sigma_2\sigma_0}{\sqrt[3]{\mathfrak{u}_3}} - \frac{3\sigma_1^2}{\sqrt[3]{\mathfrak{u}_3}} - \frac{6\sigma_1\sigma_0\mathfrak{v}_1}{\mathfrak{u}_3^{2/3}} - \frac{12\sigma_1\sigma_0^2}{\sqrt[3]{\mathfrak{u}_3}} - \frac{7\sigma_0^4}{\sqrt[3]{\mathfrak{u}_3}} - \frac{5\sigma_0^3\mathfrak{v}_1}{\mathfrak{u}_3^{2/3}}\\ &-\frac{\sigma_0\mathfrak{v}_1^3}{\mathfrak{u}_3^{4/3}} + \frac{27\sigma_0\mathfrak{v}_0}{\sqrt[3]{\mathfrak{u}_3}} - \frac{9\sigma_0\mathfrak{u}_2\mathfrak{v}_1}{\mathfrak{u}_3^{4/3}} + \frac{9\sigma_0\mathfrak{v}_1 D\mathfrak{u}_3}{\mathfrak{u}_3^{4/3}} - 27D_t\left(\frac{\sigma_0}{\sqrt[3]{\mathfrak{u}_3}}\right) - \frac{54\mathfrak{v}_1 D\sigma_0}{\sqrt[3]{\mathfrak{u}_3}}.\end{aligned}$$

4.3.2 Equations of order (4,1)

Similar conditions for integrable equations of the form

$$u_{tt} = f(u_4, u_3, u_{1t}, u_2, u_t, u_1, u), \qquad \frac{\partial f}{\partial u_4} \neq 0 \tag{4.17}$$

are given [6] by the conserved densities:

$$\rho_0 = \frac{1}{\sqrt[4]{\mathfrak{u}_4}}, \qquad \rho_1 = \frac{\mathfrak{u}_3}{\mathfrak{u}_4}, \qquad \rho_2 = \frac{\mathfrak{v}_1}{\sqrt{\mathfrak{u}_4}} - 2\frac{\sigma_0}{\sqrt[4]{\mathfrak{u}_4}},$$

$$\rho_3 = -\frac{4\mathfrak{v}_0}{\sqrt[4]{\mathfrak{u}_4}} + \frac{\mathfrak{u}_3\mathfrak{v}_1}{\mathfrak{u}_4^{5/4}} - \frac{3\mathfrak{v}_1 D\mathfrak{u}_4}{2\mathfrak{u}_4^{5/4}} - \frac{2\sigma_1}{\sqrt[4]{\mathfrak{u}_4}},$$

$$\rho_4 = -\frac{32\mathfrak{u}_2}{\mathfrak{u}_4^{3/4}} - \frac{4\mathfrak{v}_1^2}{\mathfrak{u}_4^{3/4}} + \frac{12\mathfrak{u}_3^2}{\mathfrak{u}_4^{7/4}} + \frac{5\,(D\mathfrak{u}_4)^2}{\mathfrak{u}_4^{7/4}} - \frac{12\mathfrak{u}_3 D\mathfrak{u}_4}{\mathfrak{u}_4^{7/4}} - \frac{16\sigma_2}{\sqrt[4]{\mathfrak{u}_4}} - \frac{16\sigma_0^2}{\sqrt[4]{\mathfrak{u}_4}},$$

$$\begin{aligned}\rho_5 = &\frac{16\mathfrak{u}_1}{\sqrt{\mathfrak{u}_4}} + \frac{3\mathfrak{v}_1^2 D\mathfrak{u}_4}{\mathfrak{u}_4^{3/2}} - \frac{3\mathfrak{u}_3^2 D\mathfrak{u}_4}{\mathfrak{u}_4^{5/2}} - \frac{2\mathfrak{u}_3\,(D\mathfrak{u}_4)^2}{\mathfrak{u}_4^{5/2}} + \frac{4\mathfrak{u}_3 D^2\mathfrak{u}_4}{\mathfrak{u}_4^{3/2}}\\ &+\frac{8\mathfrak{u}_2 D\mathfrak{u}_4}{\mathfrak{u}_4^{3/2}} - \frac{8\mathfrak{u}_2\mathfrak{u}_3}{\mathfrak{u}_4^{3/2}} + \frac{8\mathfrak{v}_0\mathfrak{v}_1}{\sqrt{\mathfrak{u}_4}} + \frac{2\mathfrak{u}_3^3}{\mathfrak{u}_4^{5/2}} - \frac{2\mathfrak{u}_3\mathfrak{v}_1^2}{\mathfrak{u}_4^{3/2}} - \frac{2\sigma_3}{\sqrt[4]{\mathfrak{u}_4}} - \frac{4\sigma_0\sigma_1}{\sqrt[4]{\mathfrak{u}_4}}\end{aligned}$$

$$+ 8D_t\left(\frac{\sigma_0}{\sqrt[4]{\mathfrak{u}_4}}\right) + \frac{16\mathfrak{v}_1(D\sigma_0)}{\sqrt[4]{\mathfrak{u}_4}} - \frac{3\sigma_0\mathfrak{v}_1(D\mathfrak{u}_4)}{\mathfrak{u}_4^{5/4}} + \frac{2\sigma_0\mathfrak{u}_3\mathfrak{v}_1}{\mathfrak{u}_4^{5/4}} - \frac{8\sigma_0\mathfrak{v}_0}{\sqrt[4]{\mathfrak{u}_4}}.$$

Remark. Supposing X of order $n+1$ and Y of order n in (4.5), (4.6), the density ρ_0 comes from the coefficient of D^{n+4} in (4.5), ρ_1 from the coefficient of D^{n+3} in (4.5), ρ_2 from D^{n+1} in (4.6), ρ_3 from D^n in (4.6), ρ_4 from D^{n+2} in (4.5), ρ_5 from D^{n+1} in (4.5), ρ_6 from D^{n-1} in (4.6), ρ_7 from D^{n-2} in (4.6), etc.

It would be interesting to find a recurrent formula for these conserved densities similar to (2.16).

The existence of a formal symplectic operator imposes additional conditions (cf. Theorem 7). The first calculated conditions are of a similar nature to the evolution case: the densities ρ_i must be conserved and, additionally, some of them must be trivial, i.e. total derivatives. Concretely, at the time of writing this report, we know that $\rho_1 = D\omega_1$, $\rho_3 = D\omega_3$, $\rho_5 = D\omega_5$, and $\rho_7 = D\omega_7$, where $\omega_i \in \mathcal{F}$ are local functions.

The general formulas for the first coefficients of formal recursion operators for a general equation (4.17), written in terms of coefficients of the linearization operator (see (4.2)), are given by

$$\mathcal{X}_i = \mathfrak{u}_4^{\frac{i}{4}} D^i + \left[\tfrac{1}{4} i\mathfrak{u}_3\mathfrak{u}_4^{\frac{i}{4}-1} + \tfrac{1}{8}(i-4)i\,(D\mathfrak{u}_4)\,\mathfrak{u}_4^{\frac{i}{4}-1}\right] D^{-1+i} + \cdots$$
$$+ \left\{\left[\tfrac{1}{4} i\mathfrak{u}_4^{\frac{i}{4}-1}\mathfrak{v}_1 - \tfrac{1}{2} i\sigma_0\mathfrak{u}_4^{\frac{i}{4}-\frac{3}{4}}\right] D^{-3+i} + \cdots\right\} D_t \quad (4.18)$$

$$\mathcal{Y}_i = \left[\tfrac{1}{4} i\mathfrak{u}_4^{\frac{i}{4}}\mathfrak{v}_1 - \tfrac{1}{2}(i+2)\sigma_0\mathfrak{u}_4^{\frac{i}{4}+\frac{1}{4}}\right] D^{1+i} + \cdots$$
$$+ \left\{\mathfrak{u}_4^{\frac{i}{4}} D^i + \left[\tfrac{1}{4} i\mathfrak{u}_3\mathfrak{u}_4^{\frac{i}{4}-1} + \tfrac{1}{8}(i-4)i\,(D\mathfrak{u}_4)\,\mathfrak{u}_4^{\frac{i}{4}-1}\right] D^{-1+i} + \cdots\right\} D_t. \quad (4.19)$$

The next coefficients depend on the fluxes σ_i, which correspond to the canonical conserved densities ρ_i shown above. The integration constants are hidden in the fluxes.

The formal symplectic operators have the following structure:

$$\mathcal{P}_i = \kappa\mathfrak{u}_4^{\frac{i}{4}-1} D^i + \kappa\left[\tfrac{1}{4} i\mathfrak{u}_3\mathfrak{u}_4^{\frac{i}{4}-2} + \tfrac{1}{8}(i-4)i(D\mathfrak{u}_4)\mathfrak{u}_4^{\frac{i}{4}-2}\right] D^{-1+i} + \cdots$$
$$+ \left\{\kappa\left[\tfrac{1}{4}(i-1)\mathfrak{v}_1\mathfrak{u}_4^{\frac{i}{4}-2} - \tfrac{1}{8}\omega_3\mathfrak{u}_4^{\frac{i}{4}-\frac{7}{4}} - \tfrac{1}{2}(i-4)\sigma_0\mathfrak{u}_4^{\frac{i}{4}-\frac{7}{4}}\right] D^{-3+i} + \cdots\right\} D_t \quad (4.20)$$

$$\mathcal{Q}_i = \kappa\left[\tfrac{1}{4}(i-1)\mathfrak{v}_1\mathfrak{u}_4^{\frac{i}{4}-1} - \tfrac{1}{8}\omega_3\mathfrak{u}_4^{\frac{i}{4}-\frac{3}{4}} - \tfrac{1}{2}(i-2)\sigma_0\mathfrak{u}_4^{\frac{i}{4}-\frac{3}{4}}\right] D^{1+i} + \cdots$$
$$+ \left\{\kappa\mathfrak{u}_4^{\frac{i}{4}-1} D^i + \kappa\left[\tfrac{1}{4} i\mathfrak{u}_3\mathfrak{u}_4^{\frac{i}{4}-2} + \tfrac{1}{8}(i-4)i\,(D\mathfrak{u}_4)\,\mathfrak{u}_4^{\frac{i}{4}-2}\right] D^{-1+i} + \cdots\right\} D_t \quad (4.21)$$

where the function κ is defined by the formula $D\kappa = 2\mathfrak{u}_3\kappa/\mathfrak{u}_4$. Since the density ρ_1 is trivial, $\kappa \in \mathcal{F}$.

4.4 Lists of integrable equations

4.4.1 Equations of order (3, 1)

In [16] we find that the only integrable equations of type

$$u_{tt} = u_3 + f(u_{1t}, u_2, u_t, u_1, u) \tag{4.22}$$

up to certain point transformations, are

$$\begin{aligned} u_{tt} = u_3 &+ (3u_1 + k)u_{1t} + (u_t - u_1^2 - 2ku_1 + 6\wp)u_2 - 2\wp' u_t + 6\wp' u_1^2 \\ &+ (\wp'' + k\wp')u_1, \end{aligned} \tag{4.23}$$

$$\begin{aligned} u_{tt} = u_3 &+ \left[3\frac{u_t}{u_1} + \tfrac{3}{2}X(u)\right] u_{1t} - \frac{u_2^2}{u_1} - \left[3\frac{u_t^2}{u_1^2} + \tfrac{3}{2}X(u)\frac{u_t}{u_1}\right] u_2 \\ &+ c_2\left[u_1 u_t + \tfrac{3}{2}X(u)u_1^2\right], \end{aligned} \tag{4.24}$$

where $\wp = \wp(u)$ is a solution of $(\wp')^2 = 8\wp^3 + k^2\wp^2 + k_1\wp + k_0$, $X(u) = c_2 u + c_1$ and k_1, k_2, c_1, c_2 are arbitrary constants. Both are linearizable.

4.4.2 Lagrangian equations of order (4, 1)

In [6], a classification of integrable Lagrangian systems with Lagrangians of the form

$$L = \frac{1}{2}L_2(u_2, u_1, u)\, u_t^2 + L_1(u_2, u_1, u)\, u_t + L_0(u_2, u_1, u)$$

was performed. The corresponding Euler-Lagrange equations are of the form (4.17), and a total of eight systems were found, with Lagrangians:

$$L_1 = \frac{u_t^2}{2} + \epsilon\, u_1 u_t + \frac{u_2^2}{2} + \delta_2\frac{u_1^2}{2} + \delta_1\frac{q^2}{2} + \delta_0 q, \tag{4.25}$$

$$L_2 = \frac{u_t^2}{2} - u_1^2 u_t + \frac{u_2^2}{2} + \frac{u_1^4}{2}, \tag{4.26}$$

$$L_3 = \frac{u_t^2}{2} + \frac{u_2^2}{2} + \frac{u_1^3}{2}, \tag{4.27}$$

$$L_4 = \frac{u_t^2}{2} + a(q)\, u_1 u_t + \frac{u_2^2}{2u_1^4} + a'(q)u_1 \log u_1 + \frac{a^2(q)}{2}u_1^2 + d(q), \tag{4.28}$$

$$L_5 = \frac{u_t^2}{2} + \left(\frac{\gamma}{u_1} + \epsilon\, u_1\right) u_t + \frac{u_2^2}{2u_1^4} + \frac{\epsilon^2}{2}u_1^2 + \frac{\gamma^2}{2u_1^2} + \frac{\delta}{u_1}, \quad |\gamma| + |\delta| \neq 0, \tag{4.29}$$

$$L_6 = \frac{u_1}{2}u_t^2 + (\epsilon u_1 + \beta)u_1 u_t + \frac{u_2^2}{2u_1^3} + \frac{\epsilon^2}{2}u_1^3 + \epsilon\beta\, u_1^2 + \frac{\delta}{u_1}, \tag{4.30}$$

$$L_7 = \frac{u_t^2}{2u_1} + \frac{b(q)}{u_1}\, u_t + \frac{u_2^2}{2a(q)^4 u_1^5} + \frac{d_2(q)}{u_1}, \tag{4.31}$$

$$L_8 = \frac{u_t^2}{2(u_1^2 - 1)} + \frac{u_2^2}{2(u_1^2 - 1)} + d(q)u_1^2 - \frac{d(q)}{3}, \quad d'''(q) - 8d(q)d'(q) = 0. \tag{4.32}$$

The Greek letters are arbitrary constants, and the Lagrangians are nonequivalent through contact transformations and total derivatives. The integrability of L_4 was proved only if $a'(q) = 0$, while a non-constant $a(q)$ leads probably to a non-integrable equation. All the other Lagrangians were proved integrable by providing explicit quasi-local recursion operators $\mathcal{R}$, all of them being of type $\mathcal{Y}_i$ or $\mathcal{X}_i$.

4.5 Quasi-local recursion operators

The linear case (4.25) admits two recursion operators $\mathcal{L}_1 = D$ and $\mathcal{X}_{\prime} = D_t$ that create a double chain of symmetries starting from u.

The search of explicit recursion operators is again facilitated by the quasi-local ansatz (2.18) adapted to the non-evolutionary case:

$$\mathcal{R} = \mathcal{D} + \sum_k s_k D^{-1} \cdot \mathcal{C}_k, \tag{4.33}$$

where $\mathcal{D}$ is a differential operator, the coefficients s_k are symmetries and $\mathcal{C}_k$ the variational derivatives of conserved densities. The only difference with the evolution case is that the variational derivative is not a function but a differential operator of the form $pD_t + q,\ p, q \in \mathcal{F}$. Namely, if $\rho_* = P + QD_t$ then

$$\frac{\delta \rho}{\delta u} = P^+(1) + Q^+(1)D_t.$$

Example 16. Equation (4.23) has essentially only one non-trivial conserved density given by

$$\rho = u_t - u_x^2 + 2\wp(u). \tag{4.34}$$

The explicit recursion operator for this equation can be written in the form

$$\mathcal{R} = D + (u_t - 2u_x^2 - ku_x + 2\wp) + u_x D^{-1}(D_t + 2u_{xx} + 2\wp').$$

In the non-local term the factor u_x is a symmetry and the operator $D_t + 2u_{xx} + 2\wp'$ is the variational derivative of the conserved density. Thus this recursion operator fits into the quasi-local ansatz (4.33). Symmetries of equation (4.23) can be constructed by applying this recursion operator to the seed symmetries u_x and u_t.

Another quasi-local recursion operator for (4.23) has the form

$$\bar{\mathcal{R}} = D_t + (u_{xx} - u_x^3 - ku_x^2 + 6\wp u_x + k\wp + \wp') + u_t D^{-1}(D_t + 2u_{xx} + 2\wp').$$

One can verify that

$$\bar{\mathcal{R}}^2 = \mathcal{R}^3 + c_2\mathcal{R}^2 + c_1\mathcal{R} + c_0$$

for some constants c_i. Recall that the situation is similar for the Krichever-Novikov equation (see Subsection 2.3).

The recursion operator (4.14) for the potential Boussinesq equation (4.11) has the form (4.33), where the two non-local terms are generated by the symmetries $s_1 = 1, s_2 = u_x$ and by the conserved densities $\rho_1 = u_x u_t$ and $\rho_2 = u_t$.

Another example is system (4.31) that admits for all the arbitrary functions involved, notably, a local differential recursion operator

$$\mathcal{Y}_0 = -D \cdot \frac{u_t}{u_x} + D_t.$$

This operator generates a chain of symmetries starting from u_t. Usually, the existence of a local recursion operator shows that the equation is linearizable.

Acknowledgments

V. Sokolov acknowledges support from state assignment No 0033-2019-0006 and of the State Programme of the Ministry of Education and Science of the Russian Federation, project No 1.12873.2018/12.1. R. Hernández Heredero acknowledges partial support from Ministerio de Economía y Competitividad (MINECO, Spain) under grant MTM2016-79639-P (AEI/FEDER, EU).

References

[1] Abellanas L and Galindo A, Conserved densities for nonlinear evolution equations I. Even order Case, *J. Math. Phys.*, **20**(6), 1239–1243, 1979.

[2] Abellanas L and Galindo A, Evolution equations with high order conservation laws. *J. Math. Phys.*, **24**(3), 504–509, 1983.

[3] Adler M, On a trace functional for formal pseudo differential operators and the symplectic structure of the Korteweg de Vries type equations, *Invent. Math.*, **50**(3), 219-248, 1979.

[4] Adler V E, Marikhin V G and Shabat A B, Lagrangian chains and canonical Backlund transformations, *Theoret. and Math. Phys.*, **129**(2), 1448–1465, 2001.

[5] Adler V E, Shabat A B and Yamilov R I, Symmetry approach to the integrability problem, *Theoret. and Math. Phys.*, **125**(3), 355-424, 2000.

[6] Caparrós Quintero A and Hernández Heredero R, Formal recursion operators of integrable nonevolutionary equations and Lagrangian Systems. *J. Phys. A: Math. Theor.*, **51**, 385201, 2018.

[7] Chen H H, Lee Y C and Liu C S, Integrability of nonlinear Hamiltonian systems by inverse scattering method, *Phys. Scr.*, **20**(3-4), 490–492, 1979.

[8] De Sole A, Kac V G, and Valeri D, Double Poisson vertex algebras and noncommutative Hamiltonian equations, *Adv. Math.*, **281**, 1025–1099, 2015.

[9] Demskoy D K and Sokolov V V, On recursion operators for elliptic models, *Nonlinearity*, **21**, 1253–1264, 2008.

[10] Dorfman I Ya, *Dirac Structures and Integrability of Nonlinear Evolution Equations*, John Wiley & Sons, Chichester, 1993.

[11] Drinfel'd V G, Svinolupov S I and Sokolov V V, Classification of fifth order evolution equations possessing infinite series of conservation laws, *Dokl. AN USSR,*, **A10**, 7–10, 1985 (in Russian).

[12] Fokas A S, A symmetry approach to exactly solvable evolution equations, *J. Math. Phys.*. **21**(6), 1318–1325, 1980.

[13] Fokas A S, Symmetries and integrability, *Studies in Applied Mathematics*, 77(3), 253–299, 1987.

[14] Fuchssteiner B, Application of hereditary symmetries to nonlinear evolution equations, *Nonlinear Anal. Theory Meth. Appl.*, **3**, 849–862, 1979.

[15] Hernández Heredero R, Sokolov V V, and Svinolupov S I, Toward the classification of third order integrable evolution equations, *J. Phys. A.* **13**, 4557–4568, 1994.

[16] Hernández Heredero R, Sokolov V V, and Shabat A B, A new class of linearizable equations. *J. Phys. A: Math. Gen.*, **36**, L605–L614, 2003.

[17] Hernández Heredero R, Classification of fully nonlinear integrable evolution equations of third order, *J. Nonlin. Math. Phys.*, **12**(4), 567–585, 2005.

[18] Ibragimov N Kh and Shabat A B, Infinite Lie-Bäklund algebras, *Funct. anal. appl.*, **14**(4), 313–315, 1980.

[19] Ibragimov N H, *Transformation Groups Applied to Mathematical Physics Mathematics*, Dordrecht: D. Reidel, 1985.

[20] Kaplansky I, *An Introduction to Differential Algebra*, Paris, Hermann, 1957.

[21] Kaptsov, O V, Classification of evolution equations with respect to conservation laws, *Func. analiz i pril.*, **16**(1), 73–73, 1982.

[22] Kaup D J, On the inverse scattering problem for the cubic eigenvalue problem of the class $\varphi_{xxx} + 6Q\phi_x + 6R\phi = \lambda\phi$, *Stud. Appl. Math.*, **62**, 189–216, 1980.

[23] Krasilchchik I S, Lychagin V V and Vinogradov A M, *Geometry of Jet Spaces and Nonlinear Partial Differential Equations*, Series (Adv. Stud. Contemp. Math., **1**, Gordon and Breach, New York, 1996.

[24] Kupershmidt B A, *KP or mKP*, Providence, RI: American Mathematical Society, 600 pp., 2000.

[25] Martínez Alonso L and Shabat A B, Towards a theory of differential constraints of a hydrodynamic hierarchy, *J. Non. Math. Phys.*, **10**, 229–242, 2003.

[26] Maltsev A Ya and Novikov S P, On the local systems Hamiltonian in the weakly non-local Poisson brackets, *Phys. D*, **156**(1-2), 53–80, 2001.

[27] Manakov S V., Note on the integration of Euler's equations of the dynamics of an n-dimensional rigid body, *Funct. Anal. Appl.*, **10**(4), 93–94, 1976.

[28] Marchenko V A, *Nonlinear equations and operator algebras*, Naukova dumka, Kiev, 152 pp., 1986.

[29] Meshkov A G, Necessary conditions of the integrability, *Inverse Problems*, **10**, 635–653, 1994.

[30] Meshkov A G and Sokolov V V, Hyperbolic equations with symmetries of third order, *Theoret. and Math. Phys.*, **166**(1), 43–57, 2011.

[31] Meshkov A G and Sokolov V V, Integrable evolution equations with constant separant, *Ufa Mathematical Journal*, **4**(3), 104–154, 2012.

[32] Mikhailov A V, Novikov V S and Wang J P. On Classification of Integrable Nonevolutionary Equations. *Studies in Applied Mathematics*, **118**(4), 419–457, 2007.

[33] Mikhailov A V and Shabat A B, Integrability conditions for systems of two equations $u_t = A(u)u_{xx}+B(u, u_x)$ I, *Theor. Math. Phys.*, **62**(2), 163–185, 1985.

[34] Mikhailov A V and Shabat A B, Integrability conditions for systems of two equations $u_t = A(u)u_{xx} + B(u, u_x)$ II, *Theor. Math. Phys.*, **66**(1), 47–65, 1986.

[35] Mikhailov A V, Shabat A B and Sokolov V V, Symmetry Approach to Classification of Integrable Equations, in *What Is Integrability?* Ed. V.E. Zakharov, Springer Series in Nonlinear Dynamics, Springer-Verlag, 115–184, 1991.

[36] Mikhailov A V, Shabat A B and Yamilov R I, Symmetry approach to classification of nonlinear equations. Complete lists of integrable systems. *Russian Math. Surveys*, **42**(4), 1–63, 1987.

[37] Mikhailov A V, Shabat A B and Yamilov R I, Extension of the module of invertible transformations. Classification of integrable systems. *Commun. Math. Phys.*, **115**(1), 1–19, 1988.

[38] Mikhailov A V and Sokolov V V, Integrable ODEs on associative algebras, *Comm. in Math. Phys.*, **211**(1), 231–251, 2000.

[39] Mikhailov A V and Sokolov V V, Symmetries of differential equations and the problem of Integrability, in *Integrability*, Ed. Mikhailov A V, *Lecture Notes in Physics*, Springer, **767**, 19–88, 2009.

[40] Mokhov O I and Ferapontov E V, Nonlocal Hamiltonian operators of hydrodynamic type associated with constant curvature metrics, *Russian Math. Surveys*, **45**(3), 218–219, 1990.

[41] Novikov V S and Wang J P, Symmetry structure of integrable nonevolutionary equations. *Studies in Applied Mathematics*, **119**(4), 393–428, 2007.

[42] Olver P J, *Applications of Lie Groups to Differential Equations*, (2nd edn), Graduate Texts in Mathematics, **107**, Springer-Verlag, New York, 1993.

[43] Olver P J and Sokolov V V, Integrable evolution equations on associative algebras, *Commun. Math. Phys.*, **193**, 245–268, 1998.

[44] Olver P J and Wang J P, Classification of integrable one-component systems on associative algebras, *Proc. London Math. Soc.*, **81**(3), 566–586, 2000.

[45] Ore O, Theory of non-commutative polynomials, *Ann. Math.*, **34**, 480–508, 1933.

[46] Sanders J A and Wang J P, On the integrability of homogeneous scalar evolution equations, *J. Differential Equations*, **147**, 410–434, 1998.

[47] Sawada S and Kotera T, A method for finding N-soliton solutions of the KdV and KdV-like equation, *Prog. Theor. Phys.*, **51**, 1355–1367, 1974.

[48] Sokolov V V, Hamiltonian property of the Krichever-Novikov equation, *Sov. Math. Dokl.*, **30**, 44–47, 1984.

[49] Sokolov V V, On the symmetries of evolution equations, *Russ. Math. Surv.*, **43**(5), 165–204, 1988.

[50] Sokolov V V and Shabat A B, Classification of integrable evolution equations, *Soviet Scientific Reviews*, Section C, **4**, 221–280, 1984.

[51] Sokolov V V and Svinolupov S I, Weak nonlocalities in evolution equations, *Math. Notes* **48**(6), 1234–1239, 1990.

[52] Sokolov V V and Wolf T, A symmetry test for quasilinear coupled systems, *Inverse Problems*, **15**, L5–L11, 1999.

[53] Svinolupov S I, Second-order evolution equations with symmetries, *Russian Mathematical Surveys*, **40**(5), 241–242, 1985.

[54] Svinolupov S I and Sokolov V V, Evolution equations with nontrivial conservative laws, *Funct. Anal. Appl.*, **16**(4), 317–319, 1982.

[55] Svinolupov S I and Sokolov V V, On conservations laws for the equations possessing nontrivial Lie-Bäcklund algebra, in *Integrable Systems: Collection of the Papers*. Ed. Shabat A B, BB AS USSR, Ufa., 53–67, (in Russian), 1982.

[56] Svinolupov S I and Sokolov V V, Factorization of evolution equations, *Russian Math. Surveys*, **47**(3), 127–162, 1992.

[57] Svinolupov S I, Sokolov V V and Yamilov R I, On Bäcklund transformations for integrable evolution equations, *Sov. Math., Dokl.*, **28**, 165–168, 1983.

[58] Zhiber A V, Quasilinear hyperbolic equations with an infinite-dimensional symmetry algebra, *Russ AC SC Izv. Math.*, **45**(1), 33–54, 1995.

[59] Zhiber A V and Shabat A B, Klein-Gordon equations with a nontrivial group, *Sov. Phys. Dokl.*, **24**(8), 608–609, 1979.

[60] Zhiber A V and Sokolov V V, Exactly integrable hyperbolic equations of Liouville type, *Russ. Math. Surv.* **56**(1), 61–101, 2001.

[61] Sokolov V V and Wolf T, On non-abelization of integrable polynomial ODEs, to appear, arXiv nlin. 1809.03030, 1807.05583.

[62] Wolf T and Efimovskaya O, On integrability of the Kontsevich non-abelian ODE system, Lett. Math. Phys., **100**(2), 161–170, 2012, arXiv:1108.4208v1 [nlin.SI]

A6. Evolution of the concept of $\lambda-$symmetry and main applications

C. Muriel [a] *and J.L. Romero* [a]

[a] *Departamento de Matemáticas. Universidad de Cádiz.*
11510 Puerto Real. Spain

Abstract

For ordinary differential equations, the genesis of the concept of $\mathcal{C}^{\infty}-$symmetry (or $\lambda-$symmetry), several of its equivalent formulations and some of its relationships with other concepts of symmetry are considered. A description of the main applications of the $\lambda-$symmetries theory, including its role in linearisation problems and the search of integrating factors, first integrals and Jacobi last multipliers, is provided. Some connections with other integration or reduction procedures are also explored.

A review of the generalisations of the concept of $\lambda-$prolongation of a vector field and the extensions of the concept of $\lambda-$symmetry outside the framework of ordinary differential equations has also been performed. Several extensions and open problems of interest in this theory are discussed.

1 Introduction

Symmetry analysis of differential equations has its origin in the investigations of Sophus Lie on continuous groups of transformations leaving a given ordinary differential equation (ODE) invariant. Lie's initial aim was to extend the works of Lagrange, Cauchy, Abel and (especially) Galois on groups of permutations to the world of differential equations in order to set a general theory for the integration of ODEs [109]. Galois' methods to the study of algebraic equations by means of finite groups served to Lie as a model to use continuous groups of infinitesimal transformations for studying a given differential equation and finally to solve it. One of Lie's main results is the proof that it is always possible to assign a (Lie) algebra to a given continuous (Lie) group and, conversely, from each (Lie) algebra the construction of an associated (Lie) group is always possible.

This theory soon became widely known (see the book by Jordan [61], for instance) and, gradually, Lie theory was understood as one of the most powerful methods to search for the solutions of differential equations. Several (apparently) unrelated methods of integration were unified in a single procedure. However, for half a century the applications of Lie's approach to concrete differential equations were scarcely exploited and only the abstract theory of Lie groups grew, primarily due to its applications to particle physics.

Although the application of Lie's theory to particular differential equations is completely algorithmic, it commonly requires long and tedious calculations. Maybe, this is the main reason why the applications of Lie theory to differential equations remained a bit on hold until the 1950s; two exceptions were the books [15, 103], that use similarity methods. The interest in the theory reappeared in the 1960s thanks to the work of the Russian school (specially the work of L. V. Ovsiannikov and his group [94]), who exploited systematically the methods of symmetry analysis of differential equations for the study of a large amount of problems of mathematical physics.

In the course of the 1970s and 1980s, modern group analysis applied to differential equations became a worldwide standard tool thanks to the availability of affordable and powerful computers (and fundamental progress on Computer Algebra Systems) and the books [18, 94], the first edition of [93], and others. These books were the support of many studies about the existence of symmetries (maybe in a wider sense than Lie point symmetries) and integrability by quadratures of the given ODEs. However, the existence of many instances in which the theory was of no help led to extensions and generalisations of the original Lie method (which classically employs only point and contact transformations) in order to broaden the class of equations solvable by this method.

During the last three decades of the 20th century, many different occurrences and some counterintuitive examples were reported in the literature about nonlocal symmetries [3, 7, 93]; see also [62] and the references therein quoted. Some examples are: (1) equations that lack Lie point symmetries but can be integrated [45, 93], (2) equations that lack Lie point symmetries but can be integrated by using a nonlocal change of variables, (3) equations that gain a symmetry after the reduction of order by using a symmetry (hidden symmetry of Type II), and (4) equations that lost a symmetry in addition to the one used to reduce the order of the ODE (Type I hidden symmetry).

Exponential vector fields [93] were the origin of the concept of nonlocal symmetry; i.e., a symmetry with one or more of the coefficient functions containing an integral [62]. Due to the presence of these nonlocal terms, exponential vector fields and, in general, nonlocal symmetries are not well-defined vector fields in the space of the variables of the equation and are hard to be found. They usually appear as gained or lost symmetries in order reduction procedures [3, 4, 5, 6]. As it is explained in Section 2, exponential vector fields were the origin of $\mathcal{C}^\infty-$symmetries of ODEs. One of the key points of $\mathcal{C}^\infty-$symmetries (also called $\lambda-$symmetries) is the use of a different type of prolongations of vector fields. This raised some interest on the geometric theory of ordinary and partial differential equations, and several interpretations, extensions and generalisations of the initial concept have been developed in the last two decades. Because there are several authors that have participated in that development, it seems obvious that the notations and nomenclature differ from one author to another. Since this work is devoted to the evolution and applications of the initial concept of $\lambda-$symmetry, in order to provide a comprehensive/coherent report on the subject we have had to adopt a uniform nomenclature and notations to make clear the similarities and differences between the several approaches.

A full understanding of the concepts that are considered in this work might require a very long introduction, for which some parts of excellent books like [93, 107] would be enough. Nevertheless, in Section 2, we have tried to do a very straightforward introduction to the main ideas involved in the processes of order reduction of an ODE by using Lie point symmetries, with the primary objective of fixing notations and making clear the essential points involved in these processes. In Section 2 we have also introduced a subsection about the genesis of $\mathcal{C}^\infty$−symmetries and the basic notions on this last concept.

For Section 3, we have selected some of the main generic applications of $\mathcal{C}^\infty$−symmetries to problems of order reductions of equations lacking Lie point symmetries, linearisation of 2nd-order ODEs, search of integrating factors, integration of chains of ODEs, use of Jacobi last multipliers, reductions of Euler-Lagrange equations and some specific equations that appear in the applied sciences, etc.

On the other hand, from a geometrical point of view, two very complete reviews on the interpretations and the generalisations of λ−symmetries (μ−symmetries, Λ−symmetries, σ−symmetries, etc.) have appeared in [39, 41]. Therefore, in Section 4 of this work we will briefly focus on some analytical aspects of the corresponding prolongations of vector fields.

2 Basic notions on Lie point symmetries and $\mathcal{C}^\infty$− symmetries of ODEs

2.1 Terminology and basic notions on Lie point symmetries

In this section, with the aim of fixing notations and terminology, we first recall the standard notion of prolongation of a vector field, as it is usually considered for the study of symmetries of an ODE. In order to avoid the introduction of too much machinery (differential manifolds, Lie groups, Lie algebras, etc.), we work exclusively in Euclidean spaces here, and we will concentrate on a few topics which together may serve as a basis of the concept of $\mathcal{C}^\infty$−symmetry of an ODE.

Throughout this chapter a scalar nth-order ODE is considered:

$$u_n = \phi(x, u, u_1, \ldots, u_{n-1}), \tag{2.1}$$

where x (resp. u) denotes the independent (resp. dependent) variable and u_i denotes the derivative of order i of u with respect to x. Eq. (2.1) is defined for points of the corresponding nth-order jet space whose projections to the $(n-1)$th–order jet space belong to the domain of ϕ. Let $M \subset \mathbb{R}^2$ be an open set of the projection of this domain to the zero-order jet space. To simplify the notation we will write $u^{(i)} = (u, u_1, \ldots, u_i)$, for $i \in \mathbb{N}$. Then, with this notation Eq. (2.1) can be also written as $u_n = \phi\big(x, u^{(n-1)}\big)$.

In the following, when dealing with vector fields, we will use the shorthand notation $\partial_x := \frac{\partial}{\partial x}$ and similarly ∂_u, ∂_{u_1}, etc. The total derivative operator with respect to x is then defined by

$$\mathbf{D}_x = \partial_x + u_1\partial_u + u_2\partial_{u_1} + \cdots . \tag{2.2}$$

The restriction of $\mathbf{D}_x$ to the $(n+1)th$−dimensional manifold defined by Eq. (2.1) is denoted by

$$\mathbf{A} = \partial_x + u_1\partial_u + \cdots + \phi\big(x, u^{(n-1)}\big)\partial_{u_{n-1}} \tag{2.3}$$

and will be called the **vector field associated to** Eq. (2.1).

For any vector field $\mathbf{v} = \xi(x,u)\partial_x + \eta(x,u)\partial_u$ on M and $(\bar{x},\bar{u}) \in M$ we can consider the autonomous system $\dot{z} = \mathbf{v}_{|z}$, where $z = (x,u) \in M$, with initial condition $z(0) = (\bar{x},\bar{u})$. For any $(\bar{x},\bar{u}) \in M$, the maximal solution of this initial value problem will be denoted by $\psi(\cdot;(\bar{x},\bar{u}))$ and the corresponding domain will be denoted by $I_{(\bar{x},\bar{u})}$, where $0 \in I_{(\bar{x},\bar{u})}$. This means that, for $(\bar{x},\bar{u}) \in M$, if $\gamma :]a,b[\to M$ satisfies the differential equation $\dot{\gamma}(t) = \mathbf{v}_{|\gamma(t)}$ with the initial condition $\gamma(0) = (\bar{x},\bar{u})$, then $]a,b[\subset I_{(\bar{x},\bar{u})}$ and $\gamma(t) = \psi(t;(\bar{x},\bar{u}))$ for $t \in]a,b[$. Hence $\psi(\cdot;(\bar{x},\bar{u}))$ defines the maximal integral curve passing through $(\bar{x},\bar{u})$.

If we denote $\mathcal{U} = \{(t;(x,u)) : t \in I_{(x,u)}, (x,u) \in M\}$ then we can use former maximal solutions to define a function $\psi : \mathcal{U} \to M$ that assigns to $(t;(x,u)) \in \mathcal{U}$ the point $\psi(t;(x,u)) \in M$. This map is known as the **flow of the equation** $\dot{z} = \mathbf{v}_{|z}$ (or the flow generated by $\mathbf{v}$). From the properties of the solutions of any autonomous system (see [56, Ch. 8]), it follows that the function ψ has, among others, the following properties: (1) $\psi(0;(x,u)) = (x,u)$ for $(x,y) \in M$, (2) $\partial_t\psi(t;(x,u)) = \mathbf{v}_{|\psi(t;(x,u))}$ for $(t;(x,u)) \in \mathcal{U}$, (3) $\psi(s;\psi(t;(x,u))) = \psi(s+t;(x,u))$ for $(x,u) \in M$ whenever $(t;(x,u)) \in \mathcal{U}$, $(s;\psi(t;(x,u))) \in \mathcal{U}$ and $\psi(s+t;(x,u)) \in \mathcal{U}$, and (4) if $(t;(x_0,u_0)) \in \mathcal{U}$ then (x_0,u_0) has a neighbourhood $U \subset M$ with $t \times U \subset \mathcal{U}$ and the function $(x,u) \mapsto \psi(t;(x,u))$ defines a homeomorphism $\psi_t : U \to V \subset M$, where $V \subset M$ is an open set. As a consequence, for $(x,u) \in M$, if $(t;(x,u)) \in \mathcal{U}$ then $\psi(-t;\psi(t;(x,u))) = (x,u)$. In the literature, these are the properties that define a **local group of transformations**, which will be denoted by $G_{\mathbf{v}}$, the vector field $\mathbf{v}$ will be referred as the **infinitesimal generator** of $G_{\mathbf{v}}$, and ξ,η are the **infinitesimals** of $\mathbf{v}$. In fact, if $\psi : \mathcal{U} \to M$ is any smooth map with former properties, the corresponding infinitesimal generator is given by $\mathbf{v}_{|(x,u)} = \partial_t\psi(0;(x,u))$ for $(x,u) \in M$.

If $U \subset \mathbb{R}$ is open and $f : U \to \mathbb{R}$ is a smooth function whose graph $\mathcal{G}(f) = \{(x,u) : x \in U, u = f(x)\}$ is contained in M, then we could think of considering the set $\psi(t;\mathcal{G}(f)) = \{\psi(t;(x,f(x))) : x \in U\}$; however, for a given $t \in \mathbb{R}$ and $x \in U$ it may happen that $(t;(x,f(x))) \notin \mathcal{U}$. Nevertheless [93], since $\psi(0;(x,u)) = (x,u)$ for $(x,u) \in M$, we may shrink the domain U of f to an open set $V \subset U$ so that $\psi(t;\mathcal{G}(f))$ contains the graph of a certain function $\tilde{f}_t$, for t in some ball $B(0,r) \subset \mathbb{R}$.

For an equation of the form (2.1), the basic idea of a Lie point symmetry is related to a vector field $\mathbf{v}$ with the property that whenever $u = f(x)$ is a solution of the equation, defined in some open set $U \subset \mathbb{R}$, and $t \in \mathbb{R}$ is such that $u = \tilde{f}_t$ is defined then $u = \tilde{f}_t$ is a solution of the same equation. To deal with this idea, it becomes necessary to relate the successive derivatives of a solution and the successive derivatives of the transformed solution. This leads to extend the definition of $\mathbf{v}$ to the jth–order jet space $M^{(j)}$, for $j \in \mathbb{N}$, in such a way that the jth–order prolongation of a function f is transformed, by the local group of transformations defined by the extension, into the jth–order prolongation of the transformed function $\tilde{f}_t$, for $t \in \mathbb{R}$ small enough. For $j \in \mathbb{N}$, this extension of $\mathbf{v}$ to $M^{(j)}$ is denoted by $\mathbf{v}^{(j)}$ and can be easily constructed [93, 107]:

$$\mathbf{v}^{(j)} = \mathbf{v} + \eta^{(1)}(x,u^{(1)})\partial_{u_1} + \cdots + \eta^{(j)}(x,u^{(j)})\partial_{u_j}, \tag{2.4}$$

where

$$\eta^{(1)} = \mathbf{D}_x \eta - u_1 \mathbf{D}_x(\xi), \qquad \eta^{(j)} = \mathbf{D}_x \eta^{(j-1)} - u_j \mathbf{D}_x(\xi), \quad (j > 1). \tag{2.5}$$

It can be checked that any (formal) differential operator

$$\mathbf{v}_\infty = \mathbf{v} + \sum_{j=1}^{\infty} \eta^j \big(x, u^{(j)}\big) \partial_{u_j} \tag{2.6}$$

satisfies (formally) the equation

$$[\mathbf{v}_\infty, \mathbf{D}_x] = -\mathbf{D}_x(\xi)\mathbf{D}_x \tag{2.7}$$

if and only if $\eta^j\big(x, u^{(j)}\big) = \eta^{(j)}\big(x, u^{(j)}\big)$ for $j \geq 1$, where $\eta^{(j)}\big(x, u^{(j)}\big)$ is given in (2.5) [107]. For $1 \leq j \leq n$, the flow associated to the vector field $\mathbf{v}^{(j)}$ defines a local group of transformations on the jth–order jet space $M^{(j)}$, called the **jth–order prolongation of the action of $G_{\mathbf{v}}$** on M. This new local group of transformations will be denoted by $G_{\mathbf{v}}^{(j)}$.

An important property of the family of prolongations (2.4) is the so-called **invariants by derivation (IBD) property** ([76],[93, Prop. 2.53, Cor. 2.54]): *if $f = f\big(x, u^{(j)}\big)$ and $g = g\big(x, u^{(j)}\big)$ are two invariants of the jth–prolongation $\mathbf{v}^{(j)}$ $\big($i.e., $\mathbf{v}^{(j)}(f) = \mathbf{v}^{(j)}(g) = 0\big)$ then $\frac{df}{dg} := \frac{\mathbf{D}_x(f)}{\mathbf{D}_x(g)}$ is an invariant of $\mathbf{v}^{(j+1)}$*, for $j \geq 1$. As a consequence, if $y = y(x, u)$ and $w = w(x, u, u_1)$ define a complete set of invariants of $\mathbf{v}^{(1)}$ then $\left\{y, w, \frac{dw}{dy}, \ldots, \frac{d^{n-1}w}{dy^{n-1}}\right\}$ is a complete set of invariants of $\mathbf{v}^{(n)}$, where $\frac{d^{j+1}f}{dg^{j+1}} := \frac{\mathbf{D}_x(h)}{\mathbf{D}_x(g)}$ being $h = \frac{d^j f}{dg^j}$, for $j = 1, \ldots, n-2$.

The use of the prolongations of the vector field $\mathbf{v}$ allows the formulation of the concept of Lie point symmetry for Eq. (2.1) in terms of the equation and the prolongation $\mathbf{v}^{(n)}$ of $\mathbf{v}$, instead of the solutions of the equation [93, 107]. *Eq. (2.1) admits a Lie point symmetry with generator $\mathbf{v}$ if and only if*

$$\mathbf{v}^{(n)}\big(u_n - \phi\big(x, u^{(n-1)}\big)\big) = 0 \quad \text{when} \quad u_n = \phi\big(x, u^{(n-1)}\big). \tag{2.8}$$

For the objectives of this work, it is convenient to use a third formulation of the concept of symmetry in terms of the vector field $\mathbf{A}$ given in (2.3) and the Lie bracket of two differential operators in $M^{(n-1)}$: *Eq. (2.1) admits a Lie point symmetry with generator $\mathbf{v} = \xi(x, u)\partial_x + \eta(x, u)\partial_u$ if and only if*

$$\big[\mathbf{v}^{(n-1)}, \mathbf{A}\big] = -\mathbf{A}(\xi)\mathbf{A}. \tag{2.9}$$

We now describe how a Lie point symmetry can be used to reduce the order of an equation. If $\mathbf{v} = \xi(x, u)\partial_x + \eta(x, u)\partial_u$ is the generator of a symmetry for Eq. (2.1) and $y = y(x, u)$, $z = z(x, u)$ are two functions such that $\mathbf{v}(y) = 0$, $\mathbf{v}(z) = 1$, then $(y, z) = \big(y(x, u), z(x, u)\big)$ defines a change of variables for which, in the new variables and with the pertinent prolongations [107, Pag. 42], $\mathbf{v} = \partial_z$ and $\mathbf{v}^{(n)} = \partial_z$. The action of $\mathbf{v}^{(n)}$ on $M^{(n)}$ changes Eq. (2.1) into an equation of the form $z_n = \tilde{\phi}(y, z_1, \ldots, z_{n-1})$, where $z_i = \frac{d^i z}{dy^i}$ for $1 \leq i \leq n$, because the condition

(2.8) for the transformed equation and $\mathbf{v}^{(n)} = \partial_z$ imply that $\tilde{\phi}$ does not depend on z. Therefore the solutions to Eq (2.1) can be obtained by solving the equation $z_n = \tilde{\phi}(y, z_1, \ldots, z_{n-1})$, which can be considered as an ODE of order $n-1$ for the function $z_1 = z'$, and the quadrature $z = \int z'(y)dy$. In other words, by setting $\tilde{z} = z'$, the solutions of Eq. (2.1) can be obtained from the solutions $\tilde{z} = h(y)$ of the **reduced equation** $\tilde{z}_{n-1} = \tilde{\phi}(y, \tilde{z}, \ldots, \tilde{z}_{n-2})$, and the **auxiliary equation** $\tilde{z}(x, u, u_x) = h\big(y(x, u)\big)$, which can be integrated by a quadrature.

2.2 Exponential vector fields and the genesis of the concept of $\mathcal{C}^{\infty}$–symmetry

If $\mathbf{v}_1$ and $\mathbf{v}_2$ are two symmetry generators for Eq. (2.1) then, in principle, there is no direct relationship between the corresponding reduced equations. However, it remains the question of whether and how $\mathbf{v}_2$ can be used in the integration of the reduced equation obtained with $\mathbf{v}_1$. To this respect, a crucial remark is that the Lie bracket $[\mathbf{v}_1, \mathbf{v}_2]$ is a vector field which is also the generator of a symmetry of Eq. (2.1). This is an easy consequence of the Jacobi identity for vector fields and the condition (2.9) applied both to $\mathbf{v}_1$ and $\mathbf{v}_2$. Therefore, the set of the infinitesimal generators of the Lie point symmetries of Eq. (2.1) constitutes an algebra, with the Lie bracket as the product in the algebra. This algebra is known as the Lie algebra of symmetry of Eq (2.1) and will be denoted by $\mathcal{L}$. The answer to the former question depends strongly on the structure of the algebra $\mathcal{L}$.

Let us assume that $\{\mathbf{v}_1, \ldots, \mathbf{v}_r\}$, $1 \le r \le n$, is the set of generators of an r-dimensional subalgebra $\mathcal{L}_r \subset \mathcal{L}$. Then, by constructing two common invariants $y = y\big(x, u^{(r)}\big)$ and $z = z\big(x, u^{(r)}\big)$ for the vector fields $\mathbf{v}_1^{(r)}, \ldots, \mathbf{v}_r^{(r)}$ and by using the IBD property, it can be checked that Eq. (2.1) is equivalent [93] to an $(n-r)$th–order equation

$$\tilde{\Delta}\big(y, z^{(n-r)}\big) = 0, \qquad \left(z_j = \frac{d^j z}{dy^j}, \qquad j = 1, \ldots, n-r \right), \tag{2.10}$$

called a **reduced equation**, in such a way that if $z = h(y)$ is the general solution of (2.10) then the general solution of (2.1) can be obtained by solving the rth–order **auxiliary equation**

$$z\big(x, u^{(r)}\big) = h\big(y\big(x, u^{(r)}\big)\big). \tag{2.11}$$

However, if $r > 1$ then Eq. (2.11) cannot, in general, be integrated by quadratures. If the subalgebra $\mathcal{L}_r$ is *solvable* [93] then there exists a general procedure to solve (2.11) by means of r successive quadratures. This is related to the use in a specific order of the generators of $\mathcal{L}_r$. If the appropriate order is not followed then some "undesirable" situations may arise. As an example of these situations, we consider the following well-known example, which appears in [93, Example 2.62, Exercise 2.31] in relation with the concept of an *exponential vector field.*

Example. The equation

$$x^2 u_2 = H(xu_1 - u) \tag{2.12}$$

admits the Lie point symmetries defined by the generators $\mathbf{v}_1 = x\partial_u$ and $\mathbf{v}_2 = x\partial_x$. Since $[\mathbf{v}_1, \mathbf{v}_2] = -\mathbf{v}_1$, then $\mathbf{v}_1$ and $\mathbf{v}_2$ generate a solvable symmetry algebra $\mathcal{L}_2$. By following the general procedure, Eq. (2.12) can be integrated by quadratures if we first use $\mathbf{v}_1$ and then $\mathbf{v}_2$. However, if one tries to integrate (2.12) by using the symmetry generators in the reverse order, the reduced equation obtained by using the invariants $y = u$, and $z = xu_1$ of $\mathbf{v}_2^{(1)}$ becomes

$$z(z_y - 1) = H(z - y), \tag{2.13}$$

because $\frac{dz}{dy} = x\frac{u_2}{u_1} + 1$. Eq. (2.13) does not inherit a Lie point symmetry from $\mathbf{v}_1$ because $\mathbf{v}_1^{(1)} = x\partial_u + \partial_{u_1} = x(\partial_y + \partial_z)$ is not a well-defined vector field in the (y, z)–coordinates; in fact, it can be checked that it takes the form

$$\mathbf{w} = e^{\int z^{-1} dy}(\partial_y + \partial_z), \tag{2.14}$$

where $e^{\int z^{-1} dy}$ is, formally, the integral of z^{-1}, once a function $z = f(y)$ has been chosen. These types of formal expressions are called *exponential vector fields* in [93, Exer. 2.31] and they can be prolonged to $M^{(j)}$ by following the standard procedure provided by (2.4) and (2.5). A very interesting property of this family of prolongations is that it has the IBD property [76]; consequently, if $y = y(x, u)$ and $w = w(x, u, u_1)$ define a complete set of invariants of $\mathbf{w}^{(1)}$ then $\left\{y, w, \frac{dw}{dy}, \ldots, \frac{d^{k-1}w}{dy^{k-1}}\right\}$ is a complete set of invariants of $\mathbf{w}^{(k)}$, for $k \geq 1$. The main consequence of having the IBD property is that the order of an equation which is invariant under an exponential vector field (2.14) can be reduced by one, as in the case of standard symmetries.

Nonlocal terms of exponential vector fields are easy to avoid: for instance, $\mathbf{w}^* = e^{-\int z^{-1} dy}\mathbf{w} = \partial_y + \partial_z$ is a well-defined vector field in the (y, z)–coordinates and we also have that $\mathbf{w}_k^* := e^{-\int z^{-1} dy}\mathbf{w}^{(k)}$ is a true vector field on $M^{(k)}$, for $k \geq 1$. We now consider the (formal) vector field $\mathbf{w}_\infty^*$ on M^∞ defined in such a way that its restriction to variables $(y, z^{(k)})$ is $\mathbf{w}_k^*$; i.e.,

$$\mathbf{w}_\infty^* = \partial_y + \partial_z + \left(\frac{1 - z_1}{z}\right)\partial_{z_1} + \left(\left(\frac{z_1 - 1}{z}\right)^2 - \frac{2z_2}{z}\right)\partial_{z_2} + \cdots. \tag{2.15}$$

It is clear that $\mathbf{w}_k^*$ is a well-defined vector field that, in some sense, is a prolongation of $\mathbf{w}^* = \partial_y + \partial_z$, but $\mathbf{w}_k^*$ is not the standard kth– prolongation of $\mathbf{w}^*$ (which coincides with itself). This shows that standard prolongations are not the only prolongations with the IBD property and that prolongations with this last property can be used in order reductions for ODEs.

The former example motivates two questions:

(a) Is there a systematic procedure to directly generate the infinitesimals of $\mathbf{w}_\infty^*$ from the infinitesimals of $\mathbf{w}^*$?

(b) What are the prolongations that have the IBD property?

(a) Recalling that relation (2.7) determines completely the infinitesimals of standard prolongations, it is natural to think that $[\mathbf{w}_\infty^*, \mathbf{D}_y]$ could help to answer the first question. In fact, it is easy to check that (formally)

$$[\mathbf{w}_\infty^*, \mathbf{D}_y] = \lambda \mathbf{w}_\infty^* + \mu\, \mathbf{D}_y, \tag{2.16}$$

where $\lambda = z^{-1}$ and $\mu = -(\mathbf{D}_y + \lambda)\big(\mathbf{w}_\infty^*(y)\big) = -z^{-1}$. Equation (2.16) can be used to generate all the infinitesimals of $\mathbf{w}_\infty^*$ in terms of the infinitesimals of $\mathbf{w}$ and the function λ. So, for instance, the application of both sides of (2.16) to the coordinate function z provides

$$\mathbf{w}_\infty^*(z_1) = (\mathbf{D}_y + \lambda)\big(\mathbf{w}_\infty^*(z)\big) - (\mathbf{D}_y + \lambda)\big(\mathbf{w}_\infty^*(y)\big) z_1 = \lambda - \lambda\, z_1. \tag{2.17}$$

Similarly, by successively applying both sides of (2.16) to $z_1, \ldots, z_k, \ldots$, it can be checked that for $k \geq 1$,

$$\mathbf{w}_\infty^*(z_k) = (\mathbf{D}_y + \lambda)\big(\mathbf{w}_\infty^*(z_{k-1})\big) - (\mathbf{D}_y + \lambda)\big(\mathbf{w}_\infty^*(y)\big) z_k. \tag{2.18}$$

Let us observe that (2.18) is similar to (2.5), by considering the operator $(\mathbf{D}_y + \lambda)$ instead of $\mathbf{D}_x$.

(b) In order to answer the second question, we observe that if f is any smooth function such that $\mathbf{w}_\infty^*(f) = 0$ then the application of both sides of (2.16) to f gives $\mathbf{w}_\infty^*(\mathbf{D}_y(f)) = \mu\, \mathbf{D}_y(f)$. Consequently, if f and g satisfy $\mathbf{w}_\infty^*(f) = \mathbf{w}_\infty^*(g) = 0$ then a straightforward calculation gives:

$$\mathbf{w}_\infty^* \left(\frac{\mathbf{D}_y(f)}{\mathbf{D}_y(g)} \right) = 0. \tag{2.19}$$

By proceeding in this way, it can be proved [76] that the prolongations $\mathbf{w}_k^*$, $k \geq 1$, are the unique prolongations of $\mathbf{w}^*$ that satisfy the IBD property, although these are not the standard prolongations.

2.3 The concept of $\mathcal{C}^\infty$−symmetry

The previous discussion suggests to consider (2.18) to construct prolongations of any vector field $\mathbf{v} = \xi(x,u)\partial_x + \eta(x,u)\partial_u$ by using a given function $\lambda = \lambda(x,u,u_1) \in \mathcal{C}^\infty(M^{(1)})$. This led to the primary concept of $\mathcal{C}^\infty$−prolongation of the pair $(\mathbf{v}, \lambda)$ or the λ−prolongations of $\mathbf{v}$ [74]. The jth–order λ−prolongation of $\mathbf{v}$ will be denoted by $\mathbf{v}^{[\lambda,(j)]}$ and is defined as

$$\mathbf{v}^{[\lambda,(j)]} := \mathbf{v} + \eta^{[\lambda,(1)]}(x, u^{(1)})\partial_{u_1} + \cdots + \eta^{[\lambda,(j)]}(x, u^{(j)})\partial_{u_j}, \tag{2.20}$$

where

$$\eta^{[\lambda,(0)]} = \eta, \qquad \eta^{[\lambda,(i)]} = (\mathbf{D}_x + \lambda)\big(\eta^{[\lambda,(i-1)]}\big) - u_i(\mathbf{D}_x + \lambda)\big(\xi\big), \tag{2.21}$$

for $i = 1, \ldots, j$. If we consider the (formal) differential operator

$$\mathbf{v}_\infty^\lambda = \mathbf{v} + \sum_{j=1}^{\infty} \eta^{[\lambda,(j)]}(x, u^{(j)})\partial_{u_j}, \tag{2.22}$$

then (formally)

$$[\mathbf{v}_\infty^\lambda, \mathbf{D}_x] = \lambda\, \mathbf{v}_\infty^\lambda - (\mathbf{D}_x + \lambda)(\xi)\mathbf{D}_x. \tag{2.23}$$

As a consequence of (2.23), it can be checked as before that the family of prolongations $\mathbf{v}^{[\lambda,(j)]}$, $j \geq 1$, satisfies the IBD property.

The concept of $\mathcal{C}^\infty$−symmetry arises naturally from the concept of Lie point symmetry when, instead of standard prolongations, λ−prolongations are considered. The pair $(\mathbf{v}, \lambda)$ defines a $\mathcal{C}^\infty$−symmetry (or $\mathbf{v}$ is a λ−symmetry) of Eq. (2.1) if and only if [74]

$$\mathbf{v}^{[\lambda,(n)]}\big(u_n - \phi(x, u^{(n-1)})\big) = 0 \quad \text{when} \quad u_n = \phi\big(x, u^{(n-1)}\big). \tag{2.24}$$

Equivalently, $(\mathbf{v}, \lambda)$ is a $\mathcal{C}^\infty$−symmetry of (2.1) if and only if [74]

$$[\mathbf{v}^{[\lambda,(n-1)]}, \mathbf{A}] = \lambda \mathbf{v}^{[\lambda,(n-1)]} - (\mathbf{A} + \lambda)(\xi)\mathbf{A}. \tag{2.25}$$

Geometrically, this condition means that $\mathbf{v}^{[\lambda,(n)]}$ is tangent to the $(n+1)$-dimensional manifold defined by Eq. (2.1).

Since $\mathcal{C}^\infty$−prolongations keep the IBD property, the order of any nth-order ODE invariant by the nth-order λ−prolongation of a vector field $\mathbf{v}$ can be reduced by one, by following a procedure similar to the standard reduction associated to a Lie point symmetry. In some sense, λ−symmetries can be considered as perturbations of classical Lie point symmetries, that correspond to $\lambda = 0$.

Let us observe that although equations (2.24) and (2.25) are very similar to equations (2.8) and (2.9) respectively, in general the local group of transformations defined by $\mathbf{v}^{[\lambda,(n)]}$ on $M^{(n)}$ does not transform solutions of (2.1) into solutions of the equation.

It might be wondered whether there are other prolongations leading to a family of vector fields having the IBD property. It has been demonstrated [76] that λ−prolongations are the unique possible prolongations of vector fields on M with this property. Indeed, relation (2.16) completely characterizes the prolongations that have the IBD property (see [76, Th. 2] for details).

The primitive concept of $\mathcal{C}^\infty$−symmetry was later extended to permit the function λ or the infinitesimals ξ, η of $\mathbf{v}$ belong to the space $\mathcal{F}$ of smooth functions on x, u and the derivatives of u with respect to x up to some finite but unspecified order [77, Def. 2.1], [84, Sect. 2.3]. In these cases the pair $(\mathbf{v}, \lambda)$ is called a generalised λ−symmetry (or a generalised $\mathcal{C}^\infty$−symmetry).

The former discussion proves that $\mathcal{C}^\infty$−symmetries provide a suitable theoretical framework for dealing with exponential vector fields [74, Th. 5.1] and nonlocal symmetries generated by type I hidden symmetries [78]. In the next section we discuss some applications of $\mathcal{C}^\infty$−symmetries. In particular, we show how they are very effective to reduce or integrate equations lacking Lie point symmetries, as well as to determine integrating factors and first integrals.

3 Analytical applications of $\mathcal{C}^\infty$−symmetries

3.1 Classical integration methods and $\mathcal{C}^\infty$−symmetries

One of the most immediate applications of $\mathcal{C}^\infty$−symmetries is that they can be used to reduce or to integrate ODEs for which the classical Lie method is not applicable.

(a) In the papers [44, 45], the authors consider several examples of equations with a trivial symmetry algebra which are integrable by quadratures. These papers raised many questions about the relationships between the existence of symmetries and the integrability by quadratures. In [74] it is proved that these equations admit $\mathcal{C}^\infty$−symmetries that explain their integrability.

(b) It is well known that there exist examples of equations lacking Lie point symmetries whose order can be trivially reduced [93, Pag. 182]; this is the case of an equation of the form $\mathbf{D}_x\big(\Delta(x, u^{(n-1)})\big) = 0$. In [74, Th. 4.1] and [77, Th. 3.2]), it has been demonstrated that any equation of that form admits a (possibly generalised) $\mathcal{C}^\infty$−symmetry. Moreover, the trivial reduction $\Delta\big(x, u^{(n-1)}\big) = C$, $C \in \mathbb{R}$, admitted by that equation is a consequence of the reduction procedure associated to that $\mathcal{C}^\infty$−symmetry.

(c) There are many order-reduction methods that can also be explained by the existence of $\mathcal{C}^\infty$−symmetries. In fact, if Eq. (2.1) can be reduced to an $(n-1)$th–order equation by means of a transformation $y = y(x, u)$, $w = w(x, u, u_1)$ then [77, Th. 3.1] the mentioned reduced equation is also the reduced equation that corresponds to a certain $\mathcal{C}^\infty$−symmetry of (2.1).

    It has been also proved that a class of potential symmetries [19], called *super-potential symmetries*, can be recovered as $\mathcal{C}^\infty$−symmetries [77], avoiding the determination of the Bäcklund transformations needed to write the equation in conserved form, that in practice are difficult to find.

Several methods derived from the existence of $\mathcal{C}^\infty$−symmetries have been used for the analytical study of relevant equations from both physical and mathematical points of view. Next we list some of them, as an illustration of the large variety of possible applications:

(d) The existence of a common $\mathcal{C}^\infty$−symmetry for all the equations in certain hierarchies of autonomous ODEs permits us to connect these chains of equations to the well-studied Riccati chain [85]. As a consequence, a closed-form solution for any equation in the Riccati chain has been obtained. This can be used to derive a unified procedure to solve any equation in any of the considered hierarchies.

(e) For some second-order equations of the Painlevé-Gambier classification [60] several $\mathcal{C}^\infty$−symmetries have been obtained in the literature [34, 49, 110]. In particular, two $\mathcal{C}^\infty$−symmetries of a family of equations included in the XXVII case of the mentioned classification can be found in [49]. Although they were initially used to obtain two reduced equations of Riccati-type, the general solution of the equation was later obtained by using a different procedure [88] that uses simultaneously both $\mathcal{C}^\infty$−symmetries; this general solution can be expressed in terms of a fundamental set of solutions for a second-order linear equation related to the mentioned Riccati–type reductions.

(f) The Liénard type I equation $u_2 + a_1(u)u_1 + a_0(u) = 0$ models a wide class of physically important nonlinear oscillators. One of the most well-known equations in this family is the modified Emden equation, which has been intensively studied [12, 13, 14, 53, 104, 106] by using methods based on $\mathcal{C}^\infty$−symmetries. New solutions, conservation laws and classification properties of the equations in the family have been obtained in [52] by using partial Noether operators and λ−symmetry approaches. It has been recently proved that any equation in this family of equations admits a specific $\mathcal{C}^\infty$−symmetry which can be used to determine directly an integrating factor [99, Th. 1]. The associated first integral can be calculated by a quadrature and it is functionally independent of the first integral associated to the (in general) unique Lie point symmetry ∂_x admitted by the Liénard type I equation.

(g) The λ−symmetry reduction method has been successfully applied to obtain some steady solutions of a generalised lubrication equation that models the evolution of the free surface of a thin film spreading on a solid substrate [1], as well as to obtain the exact implicit solution of a second-order nonlinear ODE governing heat transfer in a rectangular fin [2]. First integrals of nonlinear equations involving arbitrary functions have also been obtained by using $\mathcal{C}^\infty$−symmetries (some of them come from Lie point symmetries): this is the case of the studies on the path equation describing the minimum drag work [51] and the equation describing the nonlinear heat conduction equation with fins considered in [50].

3.2 First integrals, integrating factors and λ−symmetries

Let us recall some simple facts about first integrals: a function $I = I\big(x, u^{(n-1)}\big)$ is a first integral of (2.1) if $\mathbf{A}(I) = 0$. In this case an immediate reduced equation of Eq. (2.1) is $I\big(x, u^{(n-1)}\big) = C$, where $C \in \mathbb{R}$. Moreover, if n functionally independent first integrals of $\mathbf{A}$ are known then the general solution of Eq. (2.1) can be locally determined in implicit form, by the Implicit Function Theorem.

If $(\mathbf{v}, \lambda)$ is a $\mathcal{C}^\infty$−symmetry of Eq. (2.1) then, by (2.25), the system of vector fields $\big\{\mathbf{v}^{[\lambda,(n-1)]}, \mathbf{A}\big\}$ on $M^{(n-1)}$ are in involution [93, Def. 1.39]. Consequently, $\big\{\mathbf{v}^{[\lambda,(n-1)]}, \mathbf{A}\big\}$ is integrable and, by the Frobenius theorem, there exists (locally) a common first integral $I = I\big(x, u^{(n-1)}\big)$ for the vector fields $\mathbf{A}$ and $\mathbf{v}^{[\lambda,(n-1)]}$. In this case we will say that I is a first integral of $\mathbf{A}$ associated to the $\mathcal{C}^\infty$−symmetry $(\mathbf{v}, \lambda)$.

For 2nd-order equations, some useful information on the functional dependence between two first integrals I_1 and I_2 of $\mathbf{A}$ that are associated respectively to two $\mathcal{C}^\infty$−symmetries $(\mathbf{v}_1, \lambda_1)$ and $(\mathbf{v}_2, \lambda_2)$ can be obtained by using the concept of

$\mathbf{A}$−equivalence: two $\mathcal{C}^\infty$−symmetries $(\mathbf{v}_1, \lambda_1)$ and $(\mathbf{v}_2, \lambda_2)$ are $\mathbf{A}$−equivalent if the following condition holds

$$\frac{\mathbf{A}(Q_1)}{Q_1} + \lambda_1 = \frac{\mathbf{A}(Q_2)}{Q_2} + \lambda_2, \tag{3.1}$$

where $Q_1 = \eta_1 - \xi_1 u_1$ and $Q_2 = \eta_2 - \xi_2 u_1$ are the characteristics of $\mathbf{v}_1$ and $\mathbf{v}_2$, respectively. In this case we will write $(\mathbf{v}_1, \lambda_1) \overset{\mathbf{A}}{\sim} (\mathbf{v}_2, \lambda_2)$ and it is easy to check that this is an equivalence relationship in the set of $\mathcal{C}^\infty$−symmetries of (2.1). Further details on this concept can be found in [88]. In case of $\mathbf{A}$−equivalence, if I is a common first integral for $\mathbf{v}_1^{[\lambda_1,(1)]}$ and $\mathbf{A}$, then I is also a first integral for $\mathbf{v}_2^{[\lambda_2,(1)]}$.

As a consequence of (3.1), any $\mathcal{C}^\infty$−symmetry $(\mathbf{v}_1, \lambda_1)$ is $\mathbf{A}$−equivalent to a pair of the form (∂_u, λ) where $\lambda = \dfrac{\mathbf{A}(Q_1)}{Q_1} + \lambda_1$. In any equivalence class there exists a unique pair of that form, which is called its *canonical representative* and can be used to calculate the first integrals of $\mathbf{A}$ associated to any of its $\mathbf{A}$−equivalent $\mathcal{C}^\infty$−symmetries. If $\mathbf{v}_1$ and $\mathbf{v}_2$ are two $\mathbf{A}$−equivalent Lie point symmetries whose characteristics are Q_1 and Q_2 then Q_1/Q_2 is a first integral of (2.1).

On the other hand, let us recall that an integrating factor [17] for (2.1) is a function $\mu = \mu(x, u^{(l)})$, with $0 \le l \le n-1$, such that $\mu(x, u^{(l)})\big(u_n - \phi(x, u^{(n-1)})\big) = \mathbf{D}_x\big(I(x, u^{(n-1)})\big)$, for some first integral I of $\mathbf{A}$.

For second-order ODEs, several relationships among $\mathcal{C}^\infty$−symmetries, first integrals and integrating factors have been thoroughly investigated in [79]; we now mention some of these relationships. If $I = I\big(x, u^{(1)}\big)$ is a first integral then $\mu = I_{u_1}$ is an integrating factor and (∂_u, λ) is a $\mathcal{C}^\infty$−symmetry of the equation, for $\lambda = -I_u/I_{u_1}$. Conversely, if (∂_u, λ) is a $\mathcal{C}^\infty$− symmetry of the equation then a well-defined procedure for determining a first integral and an integrating factor is also provided.

If $\mu = \mu(x, u^{(1)})$ is an integrating factor and (∂_u, λ) is a $\mathcal{C}^\infty$−symmetry of the 2nd-order equation then a system of the form

$$I_x = \mu(\lambda u_1 - \phi), \quad I_u = -\mu\,\lambda, \quad I_{u_1} = \mu, \tag{3.2}$$

is compatible and any of its solutions is a first integral. Conversely, if for some smooth functions $\mu = \mu(x, u^{(1)})$ and $\lambda = \lambda(x, u^{(1)})$ the corresponding system (3.2) is compatible the μ is an integrating factor and (∂_u, λ) is a $\mathcal{C}^\infty$−symmetry.

Integrating factors of a 2nd-order equation can be characterised as the functions $\mu = \mu(x, u^{(1)})$ such that $\mu_u + [A(\mu) + \mu\phi_{u_1}]_{u_1} = 0$ and (∂_u, λ) is a $\mathcal{C}^\infty$−symmetry of the equation. Consequently, if (∂_u, λ), with $\lambda = \mathbf{A}(\mu)/\mu + \phi_{u_1}$, is a $\mathcal{C}^\infty$−symmetry of the equation then any solution of the first-order system

$$\mathbf{A}(\mu) + (\phi_{u_1} - \lambda)\mu = 0, \qquad \mu_u + (\lambda\mu)_{u_1} = 0 \tag{3.3}$$

becomes an integrating factor of the equation. At this point, it should be mentioned that the classical determining equations of the integrating factors of a second-order equation constitute a system of two second-order PDEs [17, 57], whereas the equations in (3.3) are of first order.

These $\mathcal{C}^\infty$−symmetry methods for integrating factors have been related to the Prelle-Singer method, firstly introduced in [95] for first-order ODEs and later extended for second-order ODEs [36] and nth-order ODEs [24]. For a second-order equation, the method tries to add to the differential one-form $\phi dx - du$ a *ghost* differential one-form $S(x, u, u_1)(u_1 dx - du)$ in order for the resulting differential form to admit an integrating factor. It has been demonstrated that this happens if and only if (∂_u, λ) is a $\mathcal{C}^\infty$−symmetry of the equation for $\lambda = -S$. The $\mathcal{C}^\infty$−symmetry approach also covers and complements the method for integrating factors based on variational symmetries [17, 57].

For third-order ODEs, some relationships between first integrals, integrating factors and λ−symmetries have been studied in [91].

3.3 $\mathcal{C}^\infty$−symmetries and some classes of first integrals

The above-described relationships between first integrals and $\mathcal{C}^\infty$−symmetries suggest that the classification problem of ODEs in terms of $\mathcal{C}^\infty$−symmetries goes in parallel to the study of equations admitting first integrals of an specific form. This problem has been addressed by considering first integrals that are linear in the first-order derivative: $I = A(x, u)u_1 + B(x, u)$.

The equations in the class $\mathcal{A}$ of second-order ODEs admitting a first integral of the form $I = A(x, u)u_1 + B(x, u)$ have been thoroughly studied in [80, 81]. One of the main results in this regard identifies $\mathcal{A}$ with the class or equations that are linearisable by a generalised Sundman transformation [37] of the form $U = F(x, u), dX = G(x, u)dx$.

It has been proved that an equation in $\mathcal{A}$ is necessarily of the form

$$u_2 + a_2(x, u)u_1^2 + a_1(x, u)u_1 + a_0(x, u) = 0, \tag{3.4}$$

whose coefficients satisfy either

$$\begin{aligned} S_1 &:= a_{1u} - 2a_{2x} = 0, \\ S_2 &:= (a_0 a_2 + a_{0u})_u + (a_{2x} - a_{1u})_x + (a_{2x} - a_{1u})a_1 = 0 \end{aligned} \tag{3.5}$$

or, if $S_1 \neq 0$,

$$\begin{aligned} S_3 &:= \left(\frac{S_2}{S_1}\right)_u - (a_{2x} - a_{1u}) = 0, \\ S_4 &:= \left(\frac{S_2}{S_1}\right)_x + \left(\frac{S_2}{S_1}\right)^2 + a_1\left(\frac{S_2}{S_1}\right) + a_0 a_2 + a_{0u} = 0. \end{aligned} \tag{3.6}$$

The properties of the equations whose coefficients satisfy (3.5) (class $\mathcal{A}_1$) are different from those that satisfy (3.6) (class $\mathcal{A}_2$):

1. Any equation in $\mathcal{A}_1$ admits two functionally independent first integrals of the form $I = A(x, u)u_1 + B(x, u)$, whereas each equation in $\mathcal{A}_2$ admits a unique (up to constant) first integral of this form.

2. The equations in class $\mathcal{A}$ are the unique equations of the form (3.4) that admit a $\mathcal{C}^\infty$−symmetry (∂_u, λ) where λ is of the form $\lambda = -a_2(x,u)u_1 + \beta(x,u)$, for some function β. Whereas each equation in class $\mathcal{A}_1$ admits an infinite number of λ−symmetries of this form, any equation in $\mathcal{A}_2$ admits only one λ−symmetry of this type, which is defined by $\lambda = -a_2(x,u)u_1 + S_2/S_1$, where S_1 and S_2 have been defined in (3.5).

3. The Lie symmetry algebra of any equation in $\mathcal{A}_1$ is maximal (of dimension eight) and hence all the equations in $\mathcal{A}_1$ are linearisable by point transformations. In contrast, none of the equations in $\mathcal{A}_2$ passes the Lie test of linearisation [58, 57, 66]. Nevertheless, as it has been said before, the equations in $\mathcal{A}$ are precisely the equations (3.4) that can be linearised by a generalised Sundman transformation of the form mentioned above. Constructive methods to derive the corresponding linearising generalised Sundman transformations have been derived; moreover, these procedures have been adapted to linearise any equation in $\mathcal{A}_1$ by invertible point transformations [81].

   As far as we know all the examples of equations in the class $\mathcal{A}_2$ reported in the literature admit either one Lie point symmetry or none, but whether this fact holds in general has not been proved yet. Several examples of equations in $\mathcal{A}_2$ lacking Lie point symmetries appear in [68], and correspond to a family of equations that admit Riccati-type first integrals. In this case, the general solution can be explicitly expressed in terms of a fundamental set of solutions of a related second-order linear equation, although none of the equations passes Lie's test of linearisation.

The study of the equations in the class $\mathcal{A}$ was later extended to consider the larger class $\mathcal{B}$ of the second-order equations (3.4) that admit a $\mathcal{C}^\infty$−symmetry of the form (∂_u, λ) with $\lambda = \alpha(x,u)u_1 + \beta(x,u)$, where the function α is not necessarily equal to $-a_2$, as for the class $\mathcal{A}$. Due to the correspondence between $\mathcal{C}^\infty$−symmetries and first integrals, this class coincides [83] with the class of second-order ODEs that admit a first integral of the form $I = C(x) + 1/\big(A(x,u)u_1 + B(x,u)\big)$. The characterisation of equations in $\mathcal{B}$ in terms of their coefficients is a quite complicated problem, but a surprising consequence of the procedure followed in [82] to deal with this issue is that the functions A, B and C that define the first integral $I = C(x) + 1/\big(A(x,u)u_1 + B(x,u)\big)$ can be expressed in terms of the coefficients of Eq. (3.4) and their derivatives. These functions can be also used to calculate explicitly the functions α and β that define the admitted $\mathcal{C}^\infty$−symmetry $(\partial_u, \lambda) = (\partial_u, \alpha u_1 + \beta)$. Moreover, it has been demonstrated that the class $\mathcal{B}$ coincides with the class of the equations that can be linearised through some nonlocal transformation of the form $U = F(x,u)$, $dX = \big(G_1(x,u)u_1 + G(x,u)\big)dx$, which are more general than the generalised Sundman transformations obtained in the study of the equations in the class $\mathcal{A}$. Procedures to construct these linearising nonlocal transformations by using the first integral $I = C(x) + 1/\big(Au_1 + B\big)$ have also been derived. Since any equation (3.4) that passes the Lie test of linearisation has been proved to be contained in class $\mathcal{B}$, those linearising procedures can be applied to explicitly construct the corresponding invertible linearising point transformations. Some of these results were later extended to other families of second-order ODEs [67].

The theory of $\mathcal{C}^\infty$−symmetries has been used in [8] to provide an alternative proof of Lie's approach for the linearisation of scalar second-order ODEs, which provides a new approach for constructing the linearisation transformations with lower complexity. The procedure has been applied to examples of linearisable nonlinear ordinary differential equations which are quadratic or cubic in the first derivative.

3.4 $\mathcal{C}^\infty$−symmetries and solvable structures

In this subsection we describe the role of $\mathcal{C}^\infty$−symmetries in the integrability by quadratures when the symmetry algebra $\mathcal{L}$ of Eq. (2.1) is nonsolvable. We recall that in this case, the integrability by quadratures, or the recovery of solutions by means of quadratures, cannot be assured if $\dim(\mathcal{L}) < n$. The lowest dimension of a nonsolvable symmetry algebra is three and it corresponds to the case of symmetry algebras isomorphic to either $\mathfrak{sl}(2,\mathbb{R})$ or $\mathfrak{so}(3,\mathbb{R})$. The case of $\mathfrak{sl}(2,\mathbb{C})$ has been intensively studied in the literature by using different approaches. In [35], the three inequivalent realisations of $\mathfrak{sl}(2,\mathbb{C})$ were connected via the standard prolongation process, in order to derive integration methods of cases 2 and 3 from the basic unimodular action of case 1. In [59] a two-dimensional subalgebra of $\mathfrak{sl}(2,\mathbb{C})$ was used to reduce the given ODE to a first-order equation, which cannot be integrated by quadrature, but can be transformed into a Riccati equation by using a nonlocal symmetry [7, 47, 48].

The theory of $\mathcal{C}^\infty$−symmetries can be used to provide a unified study of ODEs admitting three-dimensional nonsolvable symmetry algebras, including the fourth additional realisation of $\mathfrak{sl}(2,\mathbb{R})$ on a two-dimensional real manifold [46] and the case of $\mathfrak{so}(3,\mathbb{R})$, which had been scarcely investigated before. It was first proved [73, 75] that the symmetry generators that are lost when the Lie method of reduction is applied, can be recovered as $\mathcal{C}^\infty$−symmetries of the reduced equations. Consequently, these $\mathcal{C}^\infty$−symmetries can be used to reduce the order step-by-step, as in the case of solvable symmetry algebras. Additionally, in the case $n = 3$, a classification of the first-order reduced equations that appear in the last stage of the reduction process was also performed.

These preliminary studies have been recently extended by considering solvable structures, introduced in [10]. In the context of ODEs, solvable structures include, as a particular case, the solvable symmetry algebras for differential equations [9, 23, 55, 105]. In [10] it has been proved the equivalence between the existence of a solvable structure and the integrability by quadratures of an involutive system of vector fields. However, it is a very complicated task to determine solvable structures in practice.

For any third-order equation with a symmetry algebra isomorphic to $\mathfrak{sl}(2,\mathbb{R})$ a solvable structure can be constructed by using the symmetry generators [97]. Once the solvable structure has been determined, different strategies can be followed to determine a complete set of first integrals of the equation under study. Remarkably, such first integrals are obtainable by quadratures. Indeed, it was later proved [98] that these three first integrals can be directly expressed in terms of a fundamental set of solutions to a related second-order linear equation, for each one of the four realisations of $\mathfrak{sl}(2,\mathbb{R})$. For the first realisation, the general solution can be easily obtained from the implicit solutions defined by the level sets of the first integrals, but this is quite complicated for the three remaining cases. This problem was finally solved by expressing the independent variable in terms of a suitable parameter, which permitted expressing the general solution in parametric form and in terms of a fundamental set of solutions to second-order linear ODEs. These results have been applied to a number of examples, covering the four realisations, including a generalised Chazy equation [25], whose first integrals and general solution can be expressed in terms of well-known special functions. These methods and results have been also extended to the case of symmetry algebras isomorphic to $\mathfrak{so}(3,\mathbb{R})$ [100].

For higher-order equations the presence of the symmetry algebra $\mathfrak{sl}(2,\mathbb{R})$ can be exploited to reduce the order by three and to reconstruct the solution in terms of a complete set of solutions of a second-order linear equations. This is the case, for instance, of the fourth-order equation that arises in the static case of a Euler-Bernoulli beam equation [65], for which only particular solutions had been reported in the literature. The joint differential invariants of its symmetry algebra (isomorphic to $\mathfrak{sl}(2,\mathbb{R})$) can be used to reduce the equation to a first-order equation that can be integrated by a quadrature. The reconstruction of solutions requires us to solve an auxiliary third-order equation which keeps $\mathfrak{sl}(2,\mathbb{R})$ as a symmetry algebra. The above-described results on these types of equations permit us to obtain an expression of the general solution of the Euler-Bernoulli beam equation in parametric form and in terms of a fundamental set of solutions to a linear second-order Lamé equation [101]. Although this procedure could be applied (in principle) to equations of arbitrary order n, it might happen that the reduced equation, of order $n-3$, cannot be solved. In this situation, generalised solvable structures, introduced in [71], can be very useful and their determination is definitely easier than finding standard symmetries or solvable structures. Once a generalised solvable structure has been determined, the general solution is obtained by quadratures and can be expressed in terms of a fundamental set of solutions of an associated second-order linear ODE, as in the case of order three.

The results collected in this subsection enlarged the classes of vector fields that can be used to integrate or reduce the order of the equations, which opens new alternatives for future integrability studies.

3.5 $\mathcal{C}^\infty$–symmetries and variational problems

For differential equations that can be derived from a variational principle

$$\mathcal{L}[u] = \int L\big(x, u^{(n)}\big)\,dx, \tag{3.7}$$

the existence of special types of symmetries (variational symmetries) *doubles* the power of Lie's method of reduction. Due to the special structure of the Euler-Lagrange equation $E[L] = 0$ derived from (3.7), the knowledge of a variational symmetry allows us to reduce the order by two [93].

The concept of variational C^∞−symmetry was introduced in [87] as a generalisation of the concept of variational symmetry based on λ−prolongations. A pair $(\mathbf{v}, \lambda)$, where $\mathbf{v}$ is a generalised vector field and $\lambda \in \mathcal{F}$, will be called a *generalised variational $\mathcal{C}^\infty$−symmetry* of the functional (3.7) if there exists some $B \in \mathcal{F}$, such that

$$\mathbf{v}^{[\lambda,(n)]}(L) + L(\mathbf{D}_x + \lambda)\big(\mathbf{v}(x)\big) = (\mathbf{D}_x + \lambda)(B). \tag{3.8}$$

The order of any Euler-Lagrange equation that admits a variational C^∞−symmetry can be reduced by two. This is a "partial" reduction, because, in general, a one-parameter family of solutions is lost (in general) when the reduced equation is considered.

The correspondence between variational symmetries and conservation laws for Euler-Lagrange equations is completely determined by the celebrated Theorem of Noether [92]. This is a one-to-one correspondence when divergence variational symmetries are considered. Therefore we cannot expect the existence of a conservation law associated to any variational C^∞−symmetry. Nevertheless, the corresponding version of the Theorem of Noether for variational C^∞−symmetries [87, Th. 3] connects, from a new perspective, the original Euler-Lagrange equation and the reduced one. Its main consequence is the construction of a conservation law for the one-parameter family of solutions that has been lost in the reduction process, which can be done by relating the variational C^∞−symmetry to a special pseudo-variational symmetry.

The extension of the concept of C^∞−symmetry to partial differential equations led to the notion of μ−symmetry (see [30, 43] and Section 4.3 below), whose application to the variational framework has also been studied in the recent literature [29, 90].

Variational C^∞−symmetries have been recently applied to the problem of conservation of symmetries through successive order reductions [102]. It has been demonstrated in [102] that two variational C^∞−symmetries of the functional (3.7), satisfying a certain solvability condition, can be used to reduce the order of the associated Euler-Lagrange equation by four. It should be noted that such a result cannot hold in the case of two-standard non-commuting symmetries [93, Exer. 4.11]. The solvability condition involves the Lie bracket of two λ−prolonged vector fields for which, as far as we know, no convenient characterisation has been given yet. When the λ−prolonged vector fields are in evolutionary form, its commutator involves a new type of symmetry that remains to be investigated in detail. The extension of the described results to higher-order reductions associated with more than two C^∞−symmetries also needs to be further investigated.

3.6 Jacobi last multipliers

There are several objects related to the differential operator $\mathbf{A}$ that play an important role in the integrability of Eq. (2.1). Perhaps, Jacobi last multipliers (JLMs) are among the most known ones. JLMs were introduced by C. G. Jacobi around 1844. They are commonly used to obtain the complete integral of a given system of first-order ODEs with $n+1$ variables and $n-1$ known first integrals [11, 20, 108]. With regard to the role of JLMs in the $\mathcal{C}^\infty$-symmetries reduction method, in [86] it has been shown that any JLM is inherited by the auxiliary equation that appears after the elimination of the last $n-k$ derivatives from $n-k$ known first integrals. Conversely, any JLM for this auxiliary equation provides a JLM for the original one. For the reduction derived from a $\mathcal{C}^\infty$-symmetry, JLMs are inherited by the auxiliary equations; in fact they are integrating factors of those first–order auxiliary equations. Several examples illustrate that the combination of these two tools, JLMs and $\mathcal{C}^\infty$-symmetries, leads to the complete solution of ODEs, even if they lack Lie point symmetries [86].

3.7 Equations with two $\mathcal{C}^\infty$-symmetries

Two non-equivalent $\mathcal{C}^\infty$-symmetries of a given a second-order ODE can be used separately to obtain two functionally independent first integrals. This method can be improved by using simultaneously both $\mathcal{C}^\infty$-symmetries, in order to calculate the two functionally independent first integrals by quadratures. The procedure is based on the existence of two generalised commuting symmetries, which can be calculated with the help of the two known $\mathcal{C}^\infty$-symmetries [88]. The functions used in the construction of the commuting symmetries are closely related with integrating factors of the reduced and the auxiliary equations associated to the $\mathcal{C}^\infty$-symmetries. These functions can also be used to calculate two integrating factors and a Jacobi last multiplier of the given equation. New relationships between $\mathcal{C}^\infty$-symmetries and generalised symmetries have been recently established [89], by interrelating the respective determining equations. Some solutions of these determining equations can be derived by using several equations related to the differential operator $\mathbf{A}$, that can be combined in various ways and can provide, in particular, first integrals and JLMs without any kind of integration.

4 Extensions and geometric interpretations of $\mathcal{C}^\infty$-symmetries

In this section we consider, from an analytical point of view, several interpretations and generalisations of $\mathcal{C}^\infty$-symmetries that have appeared in the last two decades. From a geometric and rather theoretical point of view, two very complete reviews of these issues appear in [39] and [41]. The idea that links both points of view is the way vector fields on M or $M^{(1)}$ are prolonged to $M^{(n)}$, which is essential in the geometric theory of differential equations and for the IBD property, that is fundamental for the reductions of ODEs by symmetry methods.

4.1 Telescopic vector fields

The most general class of transformations that can be used as standard symmetries on the reduction of a scalar ODE has been identified in [96], by considering the idea of the IBD property. This led the authors to the so-called *telescopic vector fields* and to establish some relationships between telescopic vector fields and $\lambda-$prolongations (which also have the IBD property). The study of these relationships was completed in [84], where it is proved that a vector field on $M^{(k)}$ of the form

$$\tau^{(k)} = \alpha(x,u,u_1)\partial_x + \beta(x,u,u_1)\partial_u + \sum_{i=1}^{k} \gamma^{(i)}(x,u^{(i)})\partial_{u_i} \tag{4.1}$$

is telescopic if and only if

$$[\tau^{(k)}, \mathbf{D}_x] = \lambda\tau^{(k)} - (\mathbf{D}_x + \lambda)(\alpha)\mathbf{D}_x \qquad (k \geq 1) \tag{4.2}$$

where $\lambda = \dfrac{\gamma^{(1)} + u_1\mathbf{D}_x\alpha - \mathbf{D}_x\beta}{\beta - u_1\alpha}$. The similitude between (4.2) and (2.23) should be observed. Consequently, $\tau^{(k)} = (\alpha\partial_x + \beta\partial_u)^{[\lambda,(k)]}$ and, by (2.21) and (2.7), a telescopic vector field can be considered as a $\lambda-$prolongation of a vector field on $M^{(1)}$ whose first two infinitesimals can depend on the first derivative of the dependent variable, and λ may depend on u_2. In other words, the pair $(\alpha\partial_x + \beta\partial_u, \lambda)$ is a generalised $\mathcal{C}^\infty-$symmetry.

4.2 $\lambda-$coverings

We have shown before that an exponential vector field is a nonlocal symmetry that can be associated to a specific $\mathcal{C}^\infty-$symmetry. By using the same ideas that relate the classical symmetries with the potential symmetries [16], in [21] it was shown that a $\mathcal{C}^\infty-$symmetry can always be interpreted as a nonlocal symmetry of the given equation (see also [22]). In fact, any $\mathcal{C}^\infty-$symmetry of Eq. (2.1) corresponds to a local standard but generalised symmetry of the system ($\lambda-$covering) defined by (2.1) together with the additional equation $w_1 = \lambda(x,u,u_1)$. That local symmetry of the $\lambda-$covering corresponds, in the original variables, to a nonlocal symmetry of the equation and, on the other hand, the reduction method derived from the $\lambda-$symmetry corresponds to the standard method of reduction by using invariants applied to that system. So, $\lambda-$symmetries of Eq. (2.1) can be interpreted as *shadows* of some nonlocal symmetries [21]. In practice this means that by embedding the equation in a suitable system determined by the function $\lambda(x,u,u_1)$, any $\lambda-$symmetry can be recovered as a local (generalised) symmetry of the system. In this way, the reduction method derived from a $\lambda-$symmetry can be interpreted as a particular case of the standard symmetry reduction [23, Pag. 2]. On the other hand, the idea of considering the $\lambda-$covering for the given equation was used in [63, 64] for extending $\lambda-$symmetries to ordinary difference equations.

The existence of a $\lambda-$symmetry associated to a nonlocal symmetry of this type is proved in [84], where it has also been proved that for some special cases such nonlocal symmetries (also called semi-classical in [21]) correspond to the exponential vector fields considered above. An *ansatz* to search nonlocal symmetries useful to reduce the order does also appear in [84, Th. 5].

4.3 Geometric interpretation of λ–symmetries and the introduction of μ–symmetries

A further geometric interpretation of λ–symmetries, motivated by [96], appears in [43]: if λ is a smooth function on $M^{(1)}$, and $\mathbf{Y}$ is a vector field on $M^{(n)}$ which projects to a vector field $\mathbf{v}$ on M, then $\mathbf{Y}$ is the λ–prolongation of $\mathbf{v}$ to $M^{(n)}$ if and only if $\mathbf{Y}$ preserves the standard contact structure $\mathcal{E}$ of $M^{(n)}$ (this means that for any contact 1-form ω the relation $\mathcal{L}_{\mathbf{Y}}(\omega) + (\mathbf{Y} \lrcorner \omega)\lambda dx = \tilde{\omega}$ holds, for some 1-form $\tilde{\omega} \in \mathcal{E}$).

The basic ideas of λ–prolongations and $\mathcal{C}^\infty$–symmetries were extended to PDEs in [43] (see also ([69]). We assume that there are p independent variables $x = (x^1, \ldots, x^p)$ and q dependent variables $u = (u^1, \ldots, u^q)$; we denote, as before, the corresponding kth–order jet space by $M^{(k)}$. In this subsection we will adopt the standard notations that are commonly used for the reduction of PDEs [93, 107] and the Einstein summation convention over repeated multi-indices. A vector field on M can be written as $\mathbf{v} = \xi^i(x,u)\partial_{x^i} + \psi^j(x,u)\partial_{u^j}$. The general prolongation formula [93, Th. 2.36] is given by $\mathbf{v}^{(n)} = \mathbf{v} + \Psi^j_J \partial_{u^j_J}$, with

$$\Psi^j_{J,k} = \mathbf{D}_{x^k}\Psi^j_J - \left(\mathbf{D}_{x^k}\xi^m\right) u^j_{J,m},$$

where $|J| = 0, \ldots, k-1$, $\Psi^j_0 = \psi^j$, and $\mathbf{D}_{x^k}$ denotes the total derivative operator respect to the independent variable x^k, $k = 1, \ldots, p$.

For the generalisation of this type of prolongation, the authors of [43, 69] use a horizontal 1-form $\mu = \Lambda_i dx^i$ where $\Lambda_i = \Lambda_i\big(x, u^{(n)}\big)$, $1 \le i \le p$, is a $q \times q$ matrix whose coefficients are smooth functions of $\big(x, u^{(n)}\big)$, for $i = 1, \ldots, n$, and a certain compatibility condition is satisfied. This condition, which is given below, is related to the Euler-Young theorem on mixed derivatives and to the condition that $d\mu$ must be in the exterior ideal generated by the contact forms. The kth–order μ–prolongation of a vector field $\mathbf{v}$ on M is a vector field $\mathbf{v}^{(k)}_\mu$ on $M^{(k)}$ given by $\mathbf{v}^{(k)}_\mu = \mathbf{v} + \tilde{\Psi}^j_J \partial_{u^j_J}$, with the coefficients of the $\partial_{u^j_J}$ satisfying the prolongation formula

$$\tilde{\Psi}^j_{J,k} = \big(\widehat{\mathbf{D}}_{x^k} + \Lambda_k\big)^j_s \tilde{\Psi}^s_J - u^s_{J,m}\big(\widehat{\mathbf{D}}_{x^k} + \Lambda_k\big)^j_s \xi^m$$

where $\widehat{\mathbf{D}}_{x^k}$ is the (matrix) differential operator $\widehat{\mathbf{D}}_{x^k} := I\mathbf{D}_{x^k}$ and I denotes the $q \times q$ identity matrix. The mentioned compatibility condition for the matrices Λ_i can be written as $[\nabla_i, \nabla_j] = 0$, $1 \le i, j \le p$, where $\nabla_i = \widehat{\mathbf{D}}_{x^i} - \Lambda_i$. By using these prolongations, the concept of μ–symmetry goes as usual. The geometric interpretation of μ–prolongations and the relationships between μ–symmetries and standard symmetries for PDEs is studied in [26]. From a geometric point of view and by using generalised gauge transformations, in [38] it is shown how μ–prolongations and μ–symmetries arise naturally, which might suggest directions for further developments.

4.4 Λ-symmetries

For systems of ODEs of an arbitrary order, with one independent variable x and q dependent variables $u^1, \ldots, u^q$, a first generalisation of the λ−prolongation for vector fields of the form $\mathbf{v} = \xi(x,u)\partial_x + \eta_j(x,u)\partial_{u^j}$, where $u = (u^1, \ldots, u^q)$, was considered in [72]. For a real function $\lambda = \lambda(x, u, u^{(1)})$, the λ−prolongation of order n of $\mathbf{v}$, denoted by $\mathbf{v}^{[\lambda,(n)]}$, is the vector field on $M^{(n)}$ defined by $\mathbf{v}^{[\lambda,(n)]} = \mathbf{v} + \eta_j^{[\lambda,(i)]}(x, u^{(i)})\partial_{u_i^j}$ where, for $j = 1, \ldots, q$, $\eta_j^{[\lambda,(0)]} = \eta_j$, and

$$\eta_j^{[\lambda,(i)]} = (\mathbf{D}_x + \lambda)\big(\eta_j^{[\lambda,(i-1)]}\big) - (\mathbf{D}_x + \lambda)(\xi)(u_i^j), \tag{4.3}$$

where u_i^j denotes $u_i^j = \frac{d^i u^j}{dx^i}$.

A different generalisation of the λ−prolongation formula was considered in [27], by using a $q \times q$ matrix Λ whose coefficients are smooth functions of $(x, u^{(1)})$. For $n = 1$, the first-order Λ−prolongation of the vector field $\mathbf{v}$ is given by $\mathbf{v}_\Lambda^{(1)} = \mathbf{v} + \eta_{(1)}^j(x, u^{(1)})\partial_{u_1^j}$ where the coefficients $\eta_{(1)}^j$ of $\partial_{u_1^j}$, $1 \le j \le q$, are given by

$$\eta_{(1)}^j = \big(\widehat{\mathbf{D}}_x + \Lambda\big)_s^j \eta^s - u_1^s\big(\widehat{\mathbf{D}}_x + \Lambda\big)_s^j \xi. \tag{4.4}$$

Let us observe that $\mathbf{v}_\Lambda^{(1)}$ can also be expressed as $\mathbf{v}_\Lambda^{(1)} = \mathbf{v}^{(1)} + (\Lambda Q)^i \partial_{u_1^i}$, where $\mathbf{v}^{(1)}$ is the standard prolongation of $\mathbf{v}$ and Q is the vector function whose jth-component is $Q^j = \eta^j - \xi u_1^j$.

By considering Λ−prolongations, the notion of Λ−symmetry (also called ρ-symmetry) follows as for μ−symmetries, by considering $p = 1$. Therefore, the class of Λ−prolongations is a specific class of μ−prolongations which allows reductions of systems of first-order ODEs. This is achieved by passing to suitable (symmetry-adapted) coordinates. For a n-dimensional dynamical system, the existence of a ρ-symmetry makes the system split into an $(n-1)$-dimensional reduced system and an auxiliary scalar ODE.

Some applications of Λ−symmetries to Hamiltonian equations and, in particular, to Hamiltonian problems derived from a Lagrangian which is invariant under $\mathbf{v}^{[\Lambda,(1)]}$ have been considered in [28].

4.5 σ-symmetries

A series of three papers [31, 32, 33] developed a combined generalisation of λ−symmetries. For systems with one independent variable and q dependent variables $u = (u^1, \ldots, u^q)$ the authors consider in [31] a family $\mathcal{X} = \{\mathbf{X}_1, \ldots, \mathbf{X}_r\}$ of vector fields on M (written in coordinates as $\mathbf{X}_i = \xi_i\partial_x + \psi_i^j\partial_{u^j}$) that are in involution (i.e. $[\mathbf{X}_i, \mathbf{X}_j] = \mu_{ij}^k\mathbf{X}_k$, where $\mu_{ij}^k = \mu_{ji}^k$ are smooth functions on M). A joint λ−prolongation is considered in these papers to obtain a family $\mathcal{Y} = \{\mathbf{Y}_i : 1 \le i \le r\}$ of vector fields on $M^{(k)}$ that can be written as

$$\mathbf{Y}_i = \xi_i\partial_x + \psi_{i,(k)}^j\partial_{u_{(k)}^j},$$

where the index (k) refers to derivatives of order k. In this case the prolongation to obtain the set $\mathcal{Y}$ is done through an $r \times r$ matrix σ whose components are smooth functions on $M^{(1)}$. It is said that the family $\mathcal{Y}$ is jointly σ-prolonged from $\mathcal{X}$ if and only if $\psi^j_{i,(0)} = \psi^j_i$ and, for $k \geq 0$,

$$\psi^j_{i,(k+1)} = \left(\mathbf{D}_x(\psi^j_{i,(k)}) - u^j_{(k+1)}\mathbf{D}_x\xi_i\right) + \sigma^i_h\left(\psi^j_{h,(k)} - u^j_{(k+1)}\xi_h\right). \tag{4.5}$$

In this expression the components of several vector fields appear combined. In vector form, (4.5) can be written as $\psi^j_{(k+1)} = \left(\mathbf{D}_x + \sigma\right)(\psi^j_{(k)}) - u^j_{(k+1)}\left(\mathbf{D}_x + \sigma\right)(\xi)$, where ξ and ψ^j are vectors of dimension r whose components are ξ_i and ψ^j_i, respectively, with $1 \leq i \leq r$ and $1 \leq j \leq q$.

Although the vector fields in $\mathcal{X}$ are in involution, the elements of $\mathcal{Y}$ do not necessarily have that property. However, some sufficient conditions for them to be in involution are provided in [31], in terms of the coefficients σ^i_j and μ^k_{ij}. If $\mathcal{Y}$ is a family of σ-prolonged vector fields, then from two common invariants of order k a common invariant of order $k+1$ can be obtained. In other words, these types of joint prolongations maintain the IBD property and therefore can be used for the reduction of ODEs. In [31] the relationships between $\sigma-$ and $\mu-$prolongations are also discussed.

A concept of reducibility of autonomous systems that generalises the reductions derived from symmetries was introduced in [54]. In [33] the concept of orbital reducibility is considered (two equations are called orbit-equivalent if they admit the same first integrals near a non-stationary point) and, among several issues, some relationships between orbital reducibility and a construction similar to the joint $\sigma-$prolongations is established.

Reference [32] considers the application of $\sigma-$symmetries to dynamical systems where the direct application of the IBD property cannot be used and some modifications must be introduced. As a consequence, in case the system $\mathcal{Y}$ satisfies the same involution properties as $\mathcal{X}$ and the system $\dot{x} = f(x)$ is invariant under $\mathcal{Y}$, the autonomous system can be reduced by using suitable symmetry adapted coordinates, to an autonomous system of dimension r and an auxiliary (or reconstruction) system of dimension $n - r$.

4.6 Twisted symmetries

By the end of the first decade of the 21st century, the term "twisted symmetry" started to be used to include several generalisations of the concept of Lie point symmetry ($\lambda-$symmetries, $\mu-$symmetries, $\Lambda-$symmetries, $\sigma-$symmetries, etc.) that involve similar prolongation operations [69]; therefore that term corresponds to a collective name for some types of symmetries with similar geometric properties. In [69] it is observed that $\lambda-$prolongations, $\mu-$prolongations and Λ-prolongations can be described in terms of certain deformations of the Lie derivative and the exterior derivative. This formalism is very useful for several generalisations and extensions. So, for instance, a vector field $\mathbf{Y}$ on $M^{(n)}$ that leaves M invariant and which reduces to $\mathbf{v}$ when restricted to M is the $\mu-$prolongation of $\mathbf{v}$ if the deformed Lie derivative associated to Y and μ, denoted by $\mathcal{L}_Y^\mu$, preserves the contact ideal $\mathcal{E}$. Another geometric interpretation of $\lambda-$prolongations (and twisted prolongations, in general) is given in [40], where their relationships with certain gauge transformation on vector fields on jet bundles are considered.

The geometrical aspects of twisted symmetries were reviewed in detail in [39] and [41]. In these reviews, the author provides a unifying geometrical description of the different types of twisted symmetries: this is based on the Frobenius distributions generated by the sets of vector fields. This variation of focus allows a sound understanding of the geometrical aspects, explaining the reason for the appearance of gauge transformations in the study of these prolongations.

The effectiveness of twisted symmetries in the context of Lie-Frobenius reductions has been studied in [42] from the point of view of the distribution generated by the prolonged (symmetry) vector fields, not directly from the pertinent prolongations. In fact (see [70]) the $\lambda-$symmetry and $\sigma-$symmetry reductions can be recovered as particular cases of the Frobenius reduction theorem for distributions of vector fields, which might be useful as a starting point for the reconstruction problem.

Acknowledgments

This research is supported in part by Junta de Andalucía (Research Group FQM377) and FEDER/MICINN-AEI Project PGC2018-101514-B-I00.

References

[1] Abdel Kader A, Latif M A and Nour N, Exact solutions of a third-order ODE from thin film flow using $\lambda-$symmetry method, *International Journal of Non-Linear Mechanics* **55**, 147–152, 2013.

[2] Abdel Latif M S, Kader A H A and Nour H M, Exact implicit solution of nonlinear heat transfer in rectangular straight fin using symmetry reduction methods, *Appl. Appl. Math.* **10**, 864–877, 2015.

[3] Abraham-Shrauner B, Hidden symmetries, first integrals and reduction of order of nonlinear ordinary differential equations, *J. Nonlinear Math. Phys.* **9**(2), 1–9, 2002.

[4] Abraham-Shrauner B and Guo A, Hidden symmetries associated with the projective group of nonlinear first-order ordinary differential equations, *J. Phys. A: Math. Gen.* **25**, 5597–08, 1992.

[5] Abraham-Shrauner B and Guo A, Hidden and nonlocal symmetries of nonlinear differential equations, in *Modern Goup Analysis: Advanced Analytical and Computational Methods in Mathematical Physics*, Ibragimov N H, Torrisi M, and Valenti A (Eds), Springer, Dordrecht, 1993.

[6] Abraham-Shrauner B and Leach P G L, Hidden symmetries of nonlinear ordinary differential equations, in *Exploiting Symmetry in Applied and Numerical*

Analysis (Fort Collins, CO, 1992), in *Lectures in Appl. Math. (N. 29)*, 1–10. Amer. Math. Soc., Providence, RI, 1993.

[7] Adam A and Mahomed F, Integration of ordinary differential equations via nonlocal symmetries, *Nonlinear Dynamics* **30**(3), 267–275, 2002.

[8] Al-Dweik A Y, Mustafa M, Mara'Beh R A and Mahomed F M, An alternative proof of Lie's linearization theorem using a new $\lambda-$symmetry criterion, *Commun. Nonlinear. Sci. Numer. Simulat.* **26**(1), 45–51, 2015.

[9] Barco M and Prince G, Solvable symmetry structures in differential form applications, *Acta Applicandae Mathematicae* **66**(1), 89–121, 2001.

[10] Basarab-Horwath P, Integrability by quadratures for systems of involutive vector fields, *Ukrainian Math. Zh.* **43**, 1330–1337, 1991.

[11] Berrone L R and Giacomini H, Inverse Jacobi multipliers, *Rend. Circ. Mat. Palermo (Ser 2)* **52**(1), 77–130, 2003.

[12] Bhuvaneswari A, Chandrasekar V, Senthilvelan M and Lakshmanan M, On the complete integrability of a nonlinear oscillator from group theoretical perspective, *J. Math. Phys.* **53**(7), 073504, 2012.

[13] Bhuvaneswari A, Kraenkel R and Senthilvelan M, Lie point symmetries and the time-independent integral of the damped harmonic oscillator, *Physica Scripta* **83**(5), 055005, 2011.

[14] Bhuvaneswari A, Kraenkel R and Senthilvelan M, Application of the $\lambda-$symmetries approach and time independent integral of the modified Emden equation, *Nonlinear Anal.-Real World Appl.* **13**(3), 1102–1114, 2012.

[15] Birkhoff G, *Hydrodynamics – A study in Logic, Fact and Similitude,* Princeton University Press, Princeton, NJ, USA, 1950.

[16] Bluman G W and Kumei S, *Symmetries and Differential Equations* (2nd Ed.), Springer-Verlag, New York, 1989.

[17] Bluman G W and Anco S C, *Symmetry and Integration Methods for Differential Equations*, Springer-Verlag, New York, 2002.

[18] Bluman G W and Cole J D, *Similarity Methods for Differential Equations,* Springer-Verlag, New York, 1974.

[19] Bluman G W and Reid G J, New symmetries for ordinary differential equations, *IMA J. Appl. Math.*, 40, 87–94, 1988.

[20] Boole G, *A Treatise on Differential Equations* (5th ed.), Chelsea, New York, 1959, (First ed. 1859).

[21] Catalano-Ferraioli D, Nonlocal aspects of $\lambda-$symmetries and ODEs reduction, *J. Phys. A: Math. Theor.* **40**(21), 5479, 2007.

[22] Catalano-Ferraioli D and Morando P, Applications of solvable structures to the nonlocal symmetry-reduction of ODEs, *J. Nonlinear Math. Phys.* **16**(1), 27–42, 2009.

[23] Catalano-Ferraioli D and Morando P, Local and nonlocal solvable structures in the reduction of ODEs, *J. Phys. A: Math. Theor.* **42**(3), 035210, 15, 2009.

[24] Chandrasekar V K, Senthilvelan M and Lakshmanan M, Extended Prelle-Singer method and integrability/solvability of a class of nonlinear nth order ordinary differential equations, *J. Nonlinear Math. Phys.* **12**(1), 184–201, 2005.

[25] Chazy J, Sur les équations differentielles du troisiéme ordre et d'ordre supérier dont l'intégrale générale a ses points critiques fixes, *Acta Math.* **34**, 317–385, 1911.

[26] Cicogna G, On the relation between standard and $\mu-$symmetries for PDEs, *J. Phys. A: Math. Gen.* **37**(40), 9467, 2004.

[27] Cicogna G, Reduction of systems of first-order differential equations via Λ-symmetries, *Phys. Lett. A* **372**(20), 3672–3677, 2008.

[28] Cicogna G, Symmetries of Hamiltonian equations and $\Lambda-$constants of motion, *J. Nonlinear Math. Phys.* **16**, 43–60, 2009.

[29] Cicogna G. and Gaeta G, Noether theorem for μ-symmetries. *J. Phys. A: Math. Theor.* **40**(39), 11899–11921, 2007.

[30] Cicogna G, Gaeta G and Morando P, On the relation between standard and μ-symmetries for PDEs, *J. Phys. A: Math. Gen.* **37**(40), 9467–9486, 2004.

[31] Cicogna G, Gaeta G and Walcher S, A generalization of $\lambda-$symmetry reduction for systems of ODEs: $\sigma-$symmetries, *J. Phys. A: Math. Theor.* **45**(35), 355205 , 2012.

[32] Cicogna G, Gaeta G and Walcher S, Dynamical systems and σ-symmetries. *J. Phys. A: Math. Theor.* **46**(23), 235204, 2013.

[33] Cicogna G, Gaeta G and Walcher S, Orbital reducibility and a generalization of $\lambda-$symmetries. *J. Lie Theory* **23**(23), 357–381, 2013.

[34] Cimpoiasu R and Cimpoiasu V M, $\lambda-$symmetry reduction for nonlinear ODEs without Lie symmetries, *Physics AUC* **25**, 22–26, 2015.

[35] Clarkson P A and Olver P J, Symmetry and the Chazy equation, *J. Diff. Eq.* **124**, 225–246, 1996.

[36] Duarte L G S, Duarte S E S, da Mota L A C P and Skea J E F, Solving second-order ordinary differential equations by extending the Prelle-Singer method, *J. Phys. A: Math. Gen.* **34**(14), 3015–3024, 2001.

[37] Duarte L G S, Moreira I C and Santos F C, Linearization under non-point transformations, *J. Phys. A: Math. Gen.* **27**, 739–743, 1994.

[38] Gaeta G, A gauge-theoretic description of μ–prolongations, and μ–symmetries of differential equations, *J. Geom. Phys.* **59**, 519, 2009.

[39] Gaeta G, Twisted symmetries of differential equations, *J. Nonlinear Math. Phys.* **16**, 107–136, 2009.

[40] Gaeta G, Gauge fixing and twisted prolongations, *J. Phys. A: Math. Theor.* **44**(32), 325203, 2011.

[41] Gaeta G, Simple and collective twisted symmetries. *J. Nonlinear Math. Phys.* **21**, 593, 2014.

[42] Gaeta G, Symmetry and Lie-Frobenius reduction of differential equations, *J. Phys. A: Math. Theor.* **48**(1), 015202, 2015.

[43] Gaeta G and Morando P, On the geometry of λ–symmetries and PDE reduction, *J. Phys. A: Math. Gen.* **37**(27), 6955–6975, 2004.

[44] González-Gascón F and González-López A, Newtonian systems of differential equations, integrable via quadratures, with trivial group of point symmetries, *Phys. Lett. A* **129**(3), 153–156, 1988.

[45] González-López A, Symmetry and integrability by quadratures of ordinary differential equations, *Phys. Lett. A* **133**(4-5), 190–194, 1988.

[46] González-López A, Kamram N and Olver P J, Lie algebras of vector fields in the real plane, *Proc. London Math. Soc. 2* **64**, 339–368, 1992.

[47] Govinder K S and Leach P G L, On the determination of non-local symmetries, *J. Phys. A: Math. Gen.* **28**(18), 5349–5359, 1995.

[48] Govinder K S and Leach P G L, A group-theoretic approach to a class of second-order ordinary differential equations not possessing Lie point symmetries, *J. Phys. A: Math. Gen.* **30**(6), 2055–2068, 1997.

[49] Guha P, Choudhury A and Khanra B, λ–Symmetries, isochronicity and integrating factors of nonlinear ordinary differential equations, *J. Eng. Math.* **82**(1), 85–99, 2013.

[50] Gün G, Orhan Ö and Özer T, On new conservation laws of fin equation, *Advances in Mathematical Physics* **2014**, Article ID 695408 (16 pp), 2014.

[51] Gün G and Özer T, First integrals, integrating factors, and invariant solutions of the path equation based on Noether and λ–symmetries, In *Abstr. Appl. Anal.* **2013**, Article ID 284653 (15 pp), 2013.

[52] Gün Polat G and Özer T, On conservation forms and invariant solutions for classical mechanics problems of Liénard type, *Advances in Mathematical Physics* **2014**, Article ID 107895 (16 pp), 2014.

[53] Gün Polat G and Özer T, New conservation laws, Lagrangian forms, and exact solutions of modified Emden equation, *ASME J. Comput. Nonlinear Dynam.* **12**(4), 41001–1–041001–15, 2017.

[54] Hadeler K P and Walcher S, Reducible ordinary differential equations, *J. Nonlinear Sci.* **16**, 583–613, 2006.

[55] Hartl T and Athorne C, Solvable structures and hidden symmetries, *J. Phys. A: Math. Gen.* **27**(10), 3463, 1994.

[56] Hirsch M W and Smale S, *Differential Equations, Dynamical Systems and Linear Algebra*, Academic Press, San Diego, 1974.

[57] Ibragimov N H, *A Practical Course in Differential Equations and Mathematical Modelling: Classical and New Methods, Nonlinear Mathematical Models, Symmetry and Invariance Principles*, World Scientific, Beijing, 2010.

[58] Ibragimov N H and Magri F, Geometric proof of Lie's linearization theorem, *Nonlinear Dynamics* **36**, 41–46, 2004.

[59] Ibragimov N H and Nucci M C, Integration of third order ordinary differential equations by Lie's method: equations admitting three-dimensional Lie algebras, *Lie Groups Appl.* **1**(2), 49–64, 1994.

[60] Ince E L, *Ordinary Differential Equations*, Dover, New York, 1956.

[61] Jordan C, *Cours d'analyse de l'École polytechnique. Tome troisième.* Gauthiers-Villars, Paris, 1896.

[62] Leach P G L and Andriopoulos K, Nonlocal symmetries past, present and future, *Appl. Anal. Discrete Math.* **1**(1), 150–171, 2007.

[63] Levi D, Nucci M C and Rodríguez M, $\lambda-$symmetries for the reduction of continuous and discrete equations, *Acta Appl. Math.* **122**(1), 311–321, 2012.

[64] Levi D and Rodríguez M A, λ-symmetries for discrete equations. *J. Phys. A: Math. Theor.* **43**(29), 292001, 9, 2010.

[65] Love A, *A Treatise on the Mathematical Theory of Elasticity*, Dover, New York, 1944.

[66] Meleshko S V, On linearization of third-order ordinary differential equations, *J. Phys. A: Math. Gen.* **39**, 15135–15145, 2006.

[67] Meleshko S V, Moyo S, Muriel C, Romero J L, Guha P, and Choudhury A G, On first integrals of second-order ordinary differential equations, *J. Eng. Math.* **82**(1), 17–30, 2013.

[68] Mendoza J and Muriel C, Exact solutions and Riccati-type first integrals, *J. Nonlinear Math. Phys.* **24**(1), 75–89, 2017.

[69] Morando P, Deformation of Lie derivative and μ−symmetries, *J. Phys. A: Math. Theor.* **40**, 11547, 2007.

[70] Morando P, Reduction by λ−symmetries and σ−symmetries: a Frobenius approach, *J. Nonlinear Math. Phys.* **22**(1), 47–59, 2015.

[71] Morando P, Ruiz A and Muriel C, Generalized solvable structures and first integrals for ODEs admitting an $\mathfrak{sl}(2,\mathbb{R})$ symmetry algebra, *J. Nonlinear Math. Phys.* **26**(2), 188–201, 2019.

[72] Muriel C, Conservación de simetrías por reducción de orden de una ecuación diferencial ordinaria, in *Proceedings of the XV Congreso de Ecuaciones Diferenciales y Aplicaciones, V Congreso de Matemática Aplicada,* Vol. 1, (Vigo, 1997), 1998, 23–26.

[73] Muriel C and Romero J L, C^∞−Symmetries and non-solvable symmetry algebras, *IMA J. Appl. Math.* **66**(5), 477–498, 2001.

[74] Muriel C and Romero J L, New methods of reduction for ordinary differential equations, *IMA J. Appl. Math.* **66**(2), 111–125, 2001.

[75] Muriel C and Romero J L, Integrability of equations admitting the nonsolvable symmetry algebra so$(3,\mathbb{R})$, *Stud. Appl. Math.* **109**(4), 337–352, 2002.

[76] Muriel C and Romero J L, Prolongations of vector fields and the property of the existence of invariants obtained by differentiation, *Teoret. Mat. Fiz.* **133**(2), 289–300, 2002.

[77] Muriel C and Romero J L, C^∞−Symmetries and reduction of equations without Lie point symmetries, *J. Lie Theory,* **13**(1), 167–188, 2003.

[78] Muriel C and Romero J L, $\mathcal{C}^\infty$−Symmetries and nonlocal symmetries of exponential type, *IMA J. Appl. Math.* **72**(2), 191–205, 2007.

[79] Muriel C and Romero J L, First integrals, integrating factors and λ-symmetries of second-order differential equations, *J. Phys. A: Math. Theor.* **42**(36), 365207 (17pp), 2009.

[80] Muriel C and Romero J L, Second-order ordinary differential equations and first integrals of the form $A(t,x)\dot{x} + B(t,x)$, *J. Nonlinear Math. Phys.* **16**(1), 209–222, 2009.

[81] Muriel C and Romero J L, Nonlocal transformations and linearization of second-order ordinary differential equations, *J. Phys. A: Math. Theor.* **43**(43), 434025 (13pp), 2010.

[82] Muriel C and Romero J L, A $\lambda-$symmetry-based method for the linearization and determination of first integrals of a family of second-order ordinary differential equations, *J. Phys. A: Math. Theor.* **44**(24), 245201, 2011.

[83] Muriel C and Romero J L, Second-order ordinary differential equations and first integrals of the form $C(t) + 1/(A(t,x)\dot{x} + B(t,x))$, *J. Nonlinear Math. Phys.* **6**, 237–250, 2011.

[84] Muriel C and Romero J L, Nonlocal symmetries, telescopic vector fields and $\lambda-$symmetries of ordinary differential equations, *SIGMA* **8**, 106 (21 pp), 2012.

[85] Muriel C and Romero J L, $\lambda-$Symmetries of some chains of ordinary differential equations, *Nonlinear Anal.-Real World Appl.* **16**, 191–201, 2014.

[86] Muriel C and Romero J L, The $\lambda-$symmetry reduction method and Jacobi last multipliers, *Commun. Nonlinear Sci. Numer. Simul.* **19**(4), 807–820, 2014.

[87] Muriel C, Romero J L and Olver P J, Variational C^{∞}-symmetries and Euler-Lagrange equations, *J. Diff. Eq.* **222**(1), 164–184, 2006.

[88] Muriel C, Romero J L and Ruiz A, $\lambda-$symmetries and integrability by quadratures, *IMA J. Appl. Math.* **82**(5), 1061–1087, 2017.

[89] Muriel C, Romero J L and Ruiz A, The calculation and use of generalized symmetries for second-order ordinary differential equations. In *Symmetries, Differential Equations and Applications*, Kac V G, Olver P J, Winternitz P, and Özer T (Eds), 137-158, Springer, New York, 2018.

[90] Nadjafikhah M, Dodangeh S and Kabi-Nejad P, On the variational problems without having desired variational symmetries, *Journal of Mathematics* **2013**, Article ID 685212 (4 pp), 2013.

[91] Nadjafikhah N and Goodarzi K, Integrating factor and $\lambda-$symmetry for third-order differential equations, in *Proceedings of the 7th Seminar on Geometry and Topology,* (Tehran, 2014), 2014, 73–81.

[92] Noether E, Invariante variationsprobleme, *Nachr. König. Gessell. Wissen. Göttingen, Math-phys. Kl.*, 235–257, 1918.

[93] Olver P J, *Applications of Lie Groups to Differential Equations*, Springer-Verlag, New York, second edition, 1993.

[94] Ovsiannikov L V, *Group Analysis of Differential Equations*, Academic Press Inc., New York, 1982. Translated from the Russian by Y. Chapovsky, Translation edited by William F. Ames.

[95] Prelle M and Singer M, Elementary first integrals of differential equations, *Trans. Amer. Math. Soc.* **279**, 215–229, 1983.

[96] Pucci E and Saccomandi G, On the reduction methods for ordinary differential equations, *J. Phys. A: Math. Gen.* **35**(29), 6145–6155, 2002.

[97] Ruiz A and Muriel C, Solvable structures associated to the non-solvable symmetry algebra $\mathfrak{sl}(2,\mathbb{R})$, *SIGMA* **12**, 77 (18 pp), 2016.

[98] Ruiz A and Muriel C, First integrals and parametric solutions of third-order ODEs admitting $\mathfrak{sl}(2,\mathbb{R})$, *J. Phys. A: Math. Theor.* **50**, 205201 (21 pp), 2017.

[99] Ruiz A and Muriel C, On the integrability of Liénard I-type equations via $\lambda-$symmetries and solvable structures, *Appl. Math. Comput.* **339**, 888–898, 2018.

[100] Ruiz A and Muriel C, Construction of Solvable Structures from $\mathfrak{so}(3,\mathbb{R})$. In *Symmetries, Differential Equations and Applications*, Kac V G, Olver P J, Winternitz P, and Özer T (Eds), 53–65, Springer, New York, 2018.

[101] Ruiz A and Muriel C, Exact general solution and first integrals of a remarkable static Euler-Bernoulli beam equation, *Commun. Nonlinear Sci. Numer. Simul.* **69**, 261–69, 2019.

[102] Ruiz A, Muriel C and Olver P J, On the commutator of $\mathcal{C}^{\infty}-$symmetries and the reduction of Euler-Lagrange equations, *J. Phys. A: Math. Theor.* **51**, 145202, 2018.

[103] Sedov L I, *Similarity and Dimensional Methods in Mechanics,* Academic Press, New York, 1959.

[104] Senthilvelan M and Chandrasekar V and Mohanasubha M, Symmetries of nonlinear ordinary differential equations: The modified Emden equation as a case study, *Pramana-J. Phys.* **85**(5), 755–787, 2015.

[105] Sherring J and Prince G, Geometric aspects of reduction of order, *Trans. Amer. Math. Soc.* **334**(1), 433–453, 1992.

[106] Sinelshchikov D I and Kudryashov N A, On the Jacobi last multipliers and Lagrangians for a family of Liénard-type equations, *Appl. Math. Comput.* **307**, 257–264, 2017.

[107] Stephani H, *Differential Equations: Their Solution Using Symmetries,* Cambridge University Press, Cambridge, 1989.

[108] Whittaker E, *A Treatise on the Analytical Dynamics of Particles and Rigid Bodies,* Cambridge University Press, Cambridge, 1988 (First ed. 1908).

[109] Yaglom I, *Felix Klein and Sophus Lie: Evolution of the Idea of Symmetry in the Nineteenth Century,* Birkhäuser, Boston, 1988.

[110] Yaşar E, $\lambda-$symmetries, nonlocal transformations and first integrals to a class of Painlevé-Gambier equations, *Math. Meth. Appl. Sci* **35**(6), 684–692, 2012.

A7. Heir-equations for partial differential equations: a 25-year review

M.C. Nucci

Dipartimento di Matematica e Informatica
Università degli Studi di Perugia, 06123 Perugia, Italy

Abstract

Heir-equations were found by iterating the nonclassical symmetry method. Apart from inheriting the same Lie symmetry algebra of the original partial differential equation, and thus yielding more (and different) symmetry solutions than expected, the heir-equations are connected to conditional Lie-Bäcklund symmetries, and generalized conditional symmetries; moreover they solve the inverse problem, namely a special solution corresponds to the nonclassical symmetry. A 25-year review of work is presented, and open problems are brought forward.

1 Introduction

The most famous and established method for finding exact solutions of differential equations is the classical symmetry method, also called group analysis, which originated in 1881 from the pioneering work of Sophus Lie [54]. Many textbooks have been dedicated to this subject and its generalizations, e.g., [4], [13], [75], [71], [14], [78], [83], [47], [35], [52], [49], [21], [11], [5].

The nonclassical symmetry method was introduced fifty years ago in a seminal paper by Bluman and Cole [12] to obtain new exact solutions of the linear heat equation, i.e. solutions not deducible from the classical symmetry method. The nonclassical symmetry method consists of adding the invariant surface condition to the given equation, and then applying the classical symmetry method on the system consisting of the given differential equation and the invariant surface condition. The main difficulty of this approach is that the determining equations are no longer linear. On the other hand, the nonclassical symmetry method may give more solutions than the classical symmetry method.

After twenty years and few occasional papers, e.g. [74], [15], in the early 1990s there was a sudden spur of interest and several papers began to appear, e.g. [53], [30], [63], [70], [77], [56], [39], [27], [29], [6], [69], [7], [37], [32]. Since then the nonclassical symmetry method has been applied to various equations and systems in hundreds of published papers, e.g., [41], [57], [25], [42], [26], [82], [22], [18], [20], [24] [76], [8], [23], [48], [85], [16], the latest being [86], [9], [17].

One should be aware that some authors call nonclassical symmetries Q-conditional symmetries[1] of the second type, e.g. [34] and [22], while others call them reduction operators, e.g. [76].

[1]In [38] this name was introduced for the first time.

The nonclassical symmetry method can be viewed as a particular instance of the more general differential constraint method that, as stated by Kruglikov [51], *dates back at least to the time of Lagrange... and was introduced into practice by Yanenko* [91]. The method was set forth in detail in Yanenko's monograph [81] that was not published until after his death [31]. A more recent account and generalization of Yanenko's work can be found in [59].

Less than thirty years ago in [40] and [50], solutions were found, which apparently did not seem to follow either the classical or nonclassical symmetries method. Twenty-five years ago, we showed [65] how these solutions could be obtained by iterating the nonclassical symmetries method. A special case of the nonclassical symmetries method generates a new nonlinear equation (the so-called G-equation [64]), which inherits the prolonged symmetry algebra of the original equation. Another special case of the nonclassical symmetries method is then applied to this heir-equation to generate another heir-equation, and so on. Invariant solutions of these heir-equations are just the solutions derived in [40] and [50].

The heir-equations can also yield nonclassical symmetries (as well as classical symmetries) as shown in [68]. The difficulty in applying the method of nonclassical symmetries consists in solving nonlinear determining equations in contrast to the linear determining equations in the case of classical symmetries. The concept of the Gröbner basis has been used [29] for this purpose.

In [68] it was shown that one can find the nonclassical symmetries of any evolution equation of any order by using a suitable heir-equation and searching for a given particular solution among all its solutions, thus avoiding any complicated calculations.

Fokas and Liu [36] and Zhdanov [92] independently introduced the method of generalised conditional symmetries, i.e., conditional Lie-Bäcklund symmetries. In [66] it was shown that the heir-equations can retrieve all the conditional Lie-Bäcklund symmetries found by Zhdanov.

In [43] Goard has shown that Nucci's method of constructing heir-equations by iterating the nonclassical symmetry method is equivalent to the generalized conditional symmetries method.

In [10] Bîlă and Niesen presented another method that reduces the partial differential equation to an ordinary differential equation by using the invariant surface condition and then applying the Lie classical symmetry method in order to find nonclassical symmetries of the original partial differential equation. Recently, Goard in [44] has shown that Bîlă and Niesen's method, and its extension by Bruzón and Gandarias in [20], are equivalent to Nucci's method [68].

The use of a symbolic manipulator became imperative, because the heir-equations can be quite long: one more independent variable is added at each iteration. We employed our own interactive REDUCE programs [67] to calculate both the classical and the nonclassical symmetries, while we use MAPLE in order to generate the heir-equations.

In the next sections, after recalling what heir-equations are and how to construct them, we show some illustrative examples and applications that have been drawn from our publications of the last 25 years, and include some open problems.

2 Constructing the heir-equations

Let us consider an evolution equation in two independent variables and one dependent variable:

$$u_t = H(t, x, u, u_x, u_{xx}, u_{xxx}, \ldots). \tag{2.1}$$

The invariant surface condition is given by:

$$V_1(t, x, u)u_t + V_2(t, x, u)u_x = G(t, x, u). \tag{2.2}$$

Let us take the case with $V_1 = 0$ and $V_2 = 1$, so that (2.2) becomes:

$$u_x = G(t, x, u). \tag{2.3}$$

Applying the nonclassical symmetry method leads to an equation for G. We call this the G-equation [64], which has the following invariant surface condition:

$$\xi_1(t, x, u, G)G_t + \xi_2(t, x, u, G)G_x + \xi_3(t, x, u, G)G_u = \eta(t, x, u, G). \tag{2.4}$$

Let us consider the case $\xi_1 = 0$, $\xi_2 = 1$, and $\xi_3 = G$, so that (2.4) becomes:

$$G_x + GG_u = \eta(t, x, u, G). \tag{2.5}$$

Applying the nonclassical symmetry method leads to an equation for η called the η-equation. Clearly:

$$G_x + GG_u \equiv u_{xx} \equiv \eta. \tag{2.6}$$

We could keep iterating to obtain the Ω-equation, which corresponds to:

$$\eta_x + G\eta_u + \eta\eta_G \equiv u_{xxx} \equiv \Omega(t, x, u, G, \eta), \tag{2.7}$$

the ρ-equation, which corresponds to:

$$\Omega_x + G\Omega_u + \eta\Omega_G + \Omega\Omega_\eta \equiv u_{xxxx} \equiv \rho(t, x, u, G, \eta, \Omega), \tag{2.8}$$

and so on. Each of these equations inherits the symmetry algebra of the original equation, with the correct prolongation: first prolongation for the G-equation, second prolongation for the η-equation, and so on. Therefore, these equations were named heir-equations in [65]. This implies that even in the case of few Lie point symmetries, many more Lie symmetry reductions can be performed by using the invariant symmetry solution of any of the possible heir-equations, as was shown in [65], [3] and [58].

Also, it should be noticed that the $u\underbrace{xx\cdots}_{r}$-equation of (2.1) is just one of many possible r-extended equations as defined by Guthrie in [45].

We point out that the above described iteration method is strongly connected to the definition of partial symmetries given by Vorobev in [87]. To exemplify, we consider the heat equation:

$$u_t = u_{xx}. \tag{2.9}$$

Its G-equation is:

$$2GG_{xu} + G^2G_{uu} - G_t + G_{xx} = 0. \tag{2.10}$$

Its η-equation is:

$$2\eta\eta_{xG} + 2G\eta\eta_{uG} + \eta^2\eta_{GG} + 2G\eta_{xu} + \eta_{xx} - \eta_t + G^2\eta_{uu} = 0. \tag{2.11}$$

The G-equation corresponds to the zeroth-order differential constraint as given by Vorobev [88] on p. 76, formula (3.5), while the η-equation gives the partial symmetry of the heat equation as in [87] on p. 324, formula (14), and in [88] on p. 83, formula (4.10).

Now, let us consider a hyperbolic equation in two independent variables and one dependent variable:

$$u_{tt} = u_{xx} + h(t, x, u, u_x, u_t), \tag{2.12}$$

and take $V_1 = 1$, and $V_2 = 1$, so that (2.2) becomes[2]:

$$u_t + u_x = G(t, x, u). \tag{2.13}$$

Applying the nonclassical symmetry method leads to an equation for G. We call this equation the G-equation. If we take $\xi_1 = 1$, $\xi_2 = 1$, and $\xi_3 = G$, then (2.4) becomes[3]:

$$G_t + G_x + GG_u = \eta(t, x, u, G). \tag{2.14}$$

Applying the nonclassical symmetry method leads to an equation for η. We call this equation the η-equation. Clearly:

$$G_t + G_x + GG_u \equiv u_{tt} + 2u_{tx} + u_{xx} \equiv \eta. \tag{2.15}$$

We could keep iterating to obtain the Ω-equation, which corresponds to:

$$\eta_t + \eta_x + G\eta_u + \eta\eta_G \equiv u_{ttt} + 3u_{ttx} + 3u_{txx} + u_{xxx} \equiv \Omega(t, x, u, G, \eta), \tag{2.16}$$

and so on.

[2]There exists another case with $V_1 = 1$, and $V_2 = -1$, which leads to $u_t - u_x = G(t, x, u)$.

[3]There exists another case with $\xi_1 = 1$, $\xi_2 = -1$, and $\xi_3 = G$, which leads to $G_t - G_x + GG_u = \eta(t, x, u, G)$.

Let us consider an elliptic equation in two independent variables and one dependent variable:

$$u_{tt} + u_{xx} = h(t, x, u, u_x, u_t), \tag{2.17}$$

and take $V_1 = 1$, and $V_2 = i$, so that (2.2) becomes[4]:

$$u_t + iu_x = G(t, x, u). \tag{2.18}$$

Then the η-equation will be given through:

$$G_t + iG_x + GG_u \equiv u_{tt} + 2iu_{tx} - u_{xx} \equiv \eta(t, x, u, G), \tag{2.19}$$

the Ω-equation will be given through:

$$\eta_t + i\eta_x + G\eta_u + \eta\eta_G \equiv u_{ttt} + 3iu_{ttx} - 3u_{txx} - iu_{xxx} \equiv \Omega(t, x, u, G, \eta), \tag{2.20}$$

and so on.

3 Symmetry solutions of heir-equations

In [65] we have shown that solutions obtained in [40] are just invariant solutions of the $u_{xx} \equiv \eta$–equation.

We seek t-independent invariant solutions, which have x as the similarity variable of the heir-equations. In this way, we obtain ordinary differential equations of order two. Their general solution depends on arbitrary functions of t. Substituting into the original equation yields ordinary differential equations to be satisfied by these t-dependent functions.

We recall Galaktionov's equation:

$$u_t = u_{xx} + u_x^2 + u^2. \tag{3.1}$$

Its G-equation is:

$$2GG_{xu} + G^2G_{uu} + G^2G_u - u^2G_u - G_t + G_{xx} + 2GG_x + 2uG = 0. \tag{3.2}$$

Its η-equation is:

$$\begin{aligned} 2\eta\eta_{xG} + 2G\eta\eta_{uG} + \eta^2\eta_{GG} - 2uG\eta_G + 2G\eta_{xu} + \eta_{xx} \\ +2G\eta_x - \eta_t + G^2\eta_{uu} + G^2\eta_u - u^2\eta_u + 2\eta^2 + 2u\eta + 2G^2 = 0. \end{aligned} \tag{3.3}$$

The symmetry algebra of (3.1) is spanned by the two vector fields $X_1 = \partial_t$, and $X_2 = \partial_x$. Therefore, t-independent invariant solutions of (3.3) are given in the form $\eta = \eta(x, u, G)$. A particular case is $\eta_u = 0$, which implies $\eta = L(x, G)$. Substituting this expression for η into (3.3) leads to $L = f(x)G$ with[5]:

$$f(x) = \frac{-c_1 \sin x + c_2 \cos x}{c_2 \sin x + c_1 \cos x}. \tag{3.4}$$

[4]There exists another case with $V_1 = 1$, and $V_2 = -i$, which leads to $u_t - iu_x = G(t, x, u)$.
[5]c_n $(n = 1, 2, 3)$ are arbitrary constants.

If we let $c_1 = 0$, then:

$$\eta = \cot(x)G, \tag{3.5}$$

which is just the differential constraint for (3.1) given by Olver in [72], i.e.:

$$u_{xx} = \cot(x)u_x. \tag{3.6}$$

Integrating (3.6) with respect to x gives rise to[6]:

$$u = w_1(t)\cos(x) + w_2(t). \tag{3.7}$$

Finally, the substitution of (3.7) into (3.1) leads to:

$$\dot{w}_1 = w_1^2 + w_2^2, \qquad \dot{w}_2 = 2w_1w_2 - w_2 \tag{3.8}$$

This is the solution derived by Galaktionov for (3.1).

4 Zhdanov's conditional Lie-Bäcklund symmetries and heir-equations

We recall Zhdanov's conditional Lie-Bäcklund symmetries [92] and their relationship with heir-equations as determined in [66].

In [92], Zhdanov introduced the concept of conditional Lie-Bäcklund symmetry, i.e., given an evolution-type equation

$$u_t = H(t, x, u, u_x, u_{xx}, u_{xxx}, \ldots) \tag{4.1}$$

and some smooth Lie-Bäcklund vector field (LBVF)

$$Q = S\partial_u + (D_tS)\partial_{u_t} + (D_xS)\partial_{u_x} + \ldots \tag{4.2}$$

with $S = S(t, x, u, u_t, u_x, \ldots)$, equation (4.1) is said to be conditionally invariant under LBVF (4.2) if the condition

$$Q(u_t - H)|_{M \cap L_x} = 0 \tag{4.3}$$

holds. Here M is a set of all differential consequences of the equation (4.1), and L_x is the set of all x-differential consequences of the equation $S = 0$. Zhdanov claimed that this definition can be applied to construct new exact solutions of (4.1), which cannot be obtained by either Lie point or Lie-Bäcklund symmetries.

However, $S = 0$ is just a particular invariant solution of a suitable heir-equation. Of course, we assume that $S = 0$ can be written in explicit form with respect to the highest derivative of u.

[6] w_n $(n = 1, 2)$ are arbitrary functions of t.

We present here one example from [92], which we have also discussed in [66]. Zhdanov introduced the following nonlinear heat conductivity equation with a logarithmic-type nonlinearity

$$u_t = u_{xx} + (\alpha + \beta \log(u) - \gamma^2 \log(u)^2)u, \tag{4.4}$$

and obtained new solutions by showing that (4.4) is conditionally invariant with respect to LBVF (4.2) with

$$S = u_{xx} - \gamma u_x - u_x^2/u. \tag{4.5}$$

It can easily be shown that equation

$$S \equiv u_{xx} - \gamma u_x - u_x^2/u = 0 \tag{4.6}$$

admits an eight-dimensional Lie point symmetry algebra and therefore is linearizable.[7] In fact, the change of dependent variable $u = \exp(v)$ transforms (4.6) into $v_{xx} - \gamma v_x = 0$. Therefore, the following general solution of (4.6) can be obtained [92]

$$u(t,x) = \exp\left(\phi_1(t) + \phi_2(t)\exp(\gamma x)\right),$$

which, substituted into (4.4), gives rise to the following system of two ordinary differential equations

$$\dot{\phi}_1 = \alpha + \beta\phi_1 - \gamma^2\phi_1^2, \qquad \dot{\phi}_2 = (\beta + \gamma^2 - 2\gamma^2\phi_1)\phi_2,$$

and its general solution can easily be derived [92].

Now, let us apply the heir-equation method to equation (4.4). Its G-equation is

$$\begin{aligned} &2G_{xu}G + G_{uu}G^2 + G_u \log(u)^2\gamma^2 u - G_u \log(u)\beta u - G_u \alpha u - G_t \\ &+G_{xx} - \log(u)^2 G\gamma^2 + \log(u)\beta G - 2\log(u)G\gamma^2 + \alpha G + \beta G = 0. \end{aligned} \tag{4.7}$$

Its η-equation is

$$\begin{aligned} &2\eta_{uG}\eta G u + \eta_{GG}\eta^2 u + \eta_G \log(u)^2\gamma^2 G u - \eta_G \log(u)\beta G u - \eta_G \alpha G u \\ &\quad +2\eta_G \log(u)\gamma^2 G u - \eta_G \beta G u + \eta_{uu} G^2 u + \eta_u \log(u)^2\gamma^2 u^2 \\ &\quad -\eta_u \log(u)\beta u^2 - \eta_u \alpha u^2 - \log(u)^2\gamma^2\eta u + \log(u)\beta\eta u \\ &-2\log(u)\gamma^2\eta u - 2\log(u)\gamma^2 G^2 + \alpha\eta u + \beta\eta u + \beta G^2 - 2\gamma^2 G^2 = 0. \end{aligned} \tag{4.8}$$

The Lie point symmetry algebra of (4.4) is spanned by the two vector fields $X_1 = \partial_t$, and $X_2 = \partial_x$. Therefore, (x,t)-independent invariant solutions of (4.8) are given in the form $\eta = \eta(u, G)$. A particular case is $\eta = r_1(u)G^2 + r_2(u)G + r_3(u)$, i.e., a polynomial of second degree in G. Substituting into (4.8) and assuming $r_3 = 0$ gives rise to

$$\eta = \frac{G^2}{u} \pm \gamma G. \tag{4.9}$$

Finally, substituting $\eta = u_{xx}$, and $G = u_x$ into (4.9) yields (4.6).

All Zhdanov's examples in [92] were similarly framed within the heir-equation method in [66].

[7] Zhdanov integrated equation (4.6) without any mention of this property.

5 Nonclassical symmetries as special solutions of heir-equations

We recall the method that allows one to find nonclassical symmetries of an evolution equation by using a suitable heir-equation [68].

For the sake of simplicity, let us assume that the highest order x-derivative in the equation is two, i.e.:

$$u_t = H(t, x, u, u_x, u_{xx}). \tag{5.1}$$

First, we use (5.1) to replace u_t in (2.2), with the condition $V_1 = 1$, i.e.:

$$H(t, x, u, u_x, u_{xx}) + V_2(t, x, u)u_x = F(t, x, u). \tag{5.2}$$

Then we generate the η-equation with $\eta = \eta(x, t, u, G)$, and replace $u_x = G$, $u_{xx} = \eta$ in (5.2), i.e.:

$$H(t, x, u, G, \eta) = F(t, x, u) - V_2(t, x, u)G. \tag{5.3}$$

By the implicit function theorem, we can isolate η in (5.3), e.g.:

$$\eta = [h_1(t, x, u, G) + F(t, x, u) - V_2(t, x, u)G]\, h_2(t, x, u, G), \tag{5.4}$$

where $h_i(t, x, u, G)(i = 1, 2)$ are known functions. Thus, we have obtained a particular solution of η which must yield an identity if substituted in the η-equation. The only unknown functions are $V_2 = V_2(t, x, u)$ and $F = F(t, x, u)$. We remind the reader that there are two kinds of nonclassical symmetries, namely those with $V_1 \neq 0$ in (2.2) or those with $V_1 = 0$ in (2.2) [29]. In the first case, we can assume without loss of generality that $V_1 = 1$, while in the second case we can assume $V_2 = 1$, which generates the G-equation. If there does exist a nonclassical symmetry[8], our method will recover it. Otherwise, only the classical symmetries will be found. If we are only interested in finding nonclassical symmetries, impose F and V_2 to be functions only of the dependent variable u. Moreover, any such solution should be singular, i.e. should not form a group.

If we are dealing with a third order equation, then we need to construct the heir-equation of order three, i.e. the Ω-equation. Then, a similar procedure will yield a particular solution of the Ω-equation given by a formula of the form:

$$\Omega = [h_1(t, x, u, G, \eta) + F(t, x, u) - V_2(t, x, u)G]\, h_2(t, x, u, G, \eta) \tag{5.5}$$

where $h_i(t, x, u, G, \eta)(i = 1, 2)$ are known functions.

In the case of a fourth order equation, we need to construct the heir-equation of order four, i.e. the ρ-equation. Then, a similar procedure will yield a particular solution of the ρ-equation given by a formula of the form:

$$\rho = [h_1(t, x, u, G, \eta, \Omega) + F(t, x, u) - V_2(t, x, u)G]\, h_2(t, x, u, G, \eta, \Omega) \tag{5.6}$$

[8]Of course, we mean one such that $V_1 \neq 0$, i.e. $V_1 = 1$.

where $h_i(t, x, u, G, \eta, \Omega)(i = 1, 2)$ are known functions.

And so on.

We would like to underline how easy this method is in comparison with the nonclassical symmetry method itself since one has just to check if a particular solution is admitted by the same-order heir-equation instead of solving nonlinear determining equations. The only difficulty consists of deriving the heir-equations, which become longer and longer. However, they can be automatically determined by using any computer algebra system.

We now present some examples to show how the method works.

In [68], the following family of second order evolution equations:

$$u_t = u_{xx} + R(u, u_x), \tag{5.7}$$

with $R(u, u_x)$ a known function of u and u_x, was considered. Several well-known equations that possess nonclassical symmetries belong to (5.7). In particular, Burgers' equation [4], Fisher's equation [28], real Newell-Whitehead's equation [62], Fitzhugh-Nagumo's equation [70], and Huxley's equation [28], [7].
The G-equation of (5.7) is:

$$R_G(GG_u + G_x) + GR_u + 2G_{xu}G + G_{uu}G^2 - G_uR - G_t + G_{xx} = 0. \tag{5.8}$$

The η-equation of (5.7) is:

$$\begin{aligned} &2R_{uG}\eta G + R_{GG}\eta^2 + R_G\eta_x + GR_G\eta_u + R_{uu}G^2 - GR_u\eta_G + R_u\eta \\ &+2\eta_{xG}\eta + 2\eta_{uG}\eta G + \eta_{GG}\eta^2 - \eta_t + 2\eta_{xu}G + \eta_{xx} + \eta_{uu}G^2 - R\eta_u = 0. \end{aligned} \tag{5.9}$$

The particular solution (5.4) that we are looking for is:

$$\eta = -R(u, G) + F(t, x, u) - V_2(t, x, u)G \tag{5.10}$$

which, substituted in (5.9), yields an overdetermined system in the unknowns functions F and V_2, whereby $R(u, G)$ is given explicitly. Otherwise, after solving a first-order linear partial differential equation in $R(u, G)$, we obtain that equation (5.7) may possess a nonclassical symmetry (2.2) with $V_1 = 1, V_2 = v(u), F = f(u)$ if $R(u, u_x)$ has the following form

$$R(u, u_x) = \frac{u_x}{f^2}\left(\left(-\frac{df}{du}fu_x + \frac{dv}{du}\right)fu_x^2 + \Psi(\xi)u_x^2 + 2f^2v - 3fu_xv^2 + u_x^2v^3\right) \tag{5.11}$$

with f, v arbitrary functions of u, and Ψ arbitrary function of

$$\xi = \frac{f(u)}{u_x} - v(u). \tag{5.12}$$

This means that infinitely many cases can be found. For example, equation (5.7) with $R(u, u_x)$ given by

$$R(u, u_x) = (2u_x + u^4)\frac{u_x}{u} \tag{5.13}$$

i.e.

$$u_t = u_{xx} + (2u_x + u^4)\frac{u_x}{u}, \tag{5.14}$$

admits a nonclassical symmetry[9] with $v = u^3/2$ and $f = -u^7/12$. It is interesting to note that the corresponding reduction leads to the solution of the following ordinary differential equation in $u(x)$:

$$u_{xx} = -2\frac{u_x^2}{u} - \frac{3}{2}u^3 u_x - \frac{u^7}{12}, \tag{5.15}$$

which is linearizable. In fact, it admits a Lie symmetry algebra of dimension eight [55], and consequently the two-dimensional abelian intransitive subalgebra generated by the following operators:

$$\frac{1}{6}u^3\left(6\partial_x - u^4\partial_u\right), \quad \frac{1}{12}(2 - xu^3)\left(6\partial_x - u^4\partial_u\right)$$

which yields the transformation:

$$\tilde{x} = \frac{2u^3}{2 - xu^3}, \quad \tilde{u} = \frac{x(4 - xu^3)}{2(2 - xu^3)}$$

that takes equation (5.15) into

$$\frac{d\tilde{u}}{d\tilde{x}} = 0.$$

Thus, the general solution of (5.15) is

$$\frac{x^2u^3}{2(2 - xu^3)} + x - \frac{2s_1u^3}{2 - xu^3} = s_2,$$

with s_1, s_2 arbitrary functions of t. Maple 16 solves this third degree polynomial in u,

$$u = \left(\frac{4(x - s_2)}{4s_1 - 2s_2x + x^2}\right)^{1/3},$$

which, substituted into (5.14), yields the following solution:

$$u = \left(\frac{4(x - a_2)}{4a_1 + 2t - 2a_2x + x^2}\right)^{1/3}.$$

[9]We remark that equation (5.14) admits a five-dimensional Lie symmetry algebra, generated by the following operators:

$$\Gamma_1 = t^2\partial_t + tx\partial_x - \frac{x + tu^3}{3u^2}\partial_u, \; \Gamma_2 = t\partial_t + 2x\partial_x + \partial_u, \; \Gamma_3 = \partial_t, \; \Gamma_4 = t\partial_x - \frac{1}{3u^2}\partial_u, \; \Gamma_5 = \partial_x.$$

6 Final remarks

As stated in [68], we have determined an algorithm which is easier to implement than the usual method to find nonclassical symmetries admitted by an evolution equation in two independent variables. Moreover, one can retrieve both classical and nonclassical symmetries with the same method.

While one can apply the heir-equations and their properties to a system of evolution equations [3], [46], [19], it is still an open problem to determine suitable heir-equations for more than two independent variables.

Moreover, the heir-equation method raises many other intriguing questions [68]:

- Could an a priori knowledge of the existence of nonclassical symmetries apart from classical symmetries be achieved by looking at the properties of the right-order heir-equation? We have shown that our method leads to both classical and nonclassical symmetries. Nonclassical symmetries could exist if we impose F and V_2 to be functions only of the dependent variable u in either (5.4), or (5.5), or (5.6), etc. Of course, any such solution of F and V_2 does not yield a nonclassical symmetry, unless it is isolated, i.e. does not form a group.

- What is integrability? The existence of infinitely many higher order symmetries is one of the criteria [60], [73]. In [66], we have shown that invariant solutions of the heir-equations yield Zhdanov's conditional Lie-Bäcklund symmetries [92]. Higher order symmetries may be interpreted as special solutions of heir-equations (up to which order? see [80], [73]). Another criterion for integrability consists of looking for Bäcklund transformations [2], [79]. In [64], we have found that a nonclassical symmetry of the G-equation for the modified Korteweg-deVries equation gives the known Bäcklund transformation between the modified Korteweg-deVries and Korteweg-deVries equations [61]. Another integrability test is the Painlevé property [89] which when satisfied leads to Lax pairs (hence, inverse scattering transform) [2], Bäcklund transformations, and Hirota bilinear formalism [84]. In [33], the singularity manifold of the modified Korteweg-deVries equation was found to be connected to an equation which is exactly the G-equation for the modified Korteweg-deVries, and the same was done for five other equations. Could heir-equations be the common link among all the integrability methods?

- In order to reduce a partial differential equation to ordinary differential equations, one of the first steps is to find the admitted Lie point symmetry algebra. In most instances, it is very small, and therefore not many reductions can be obtained. However, if heir-equations are considered, then many more ordinary differential equations can be derived using the same Lie algebra [65], [3], [58]. Of course, the classification of all subalgebras [90] becomes imperative [58]. In the case of known integrable equations, it would be interesting to investigate which ordinary differential equations result from using the admitted Lie point symmetry algebra and the corresponding heir-equations. Do all these ordinary differential equations possess the Painlevé property (see the Painlevé conjecture as stated in [1])?

- Researchers often find solutions of partial differential equations which apparently do not follow from any symmetry reduction. Are the heir-equations the ultimate method which keeps Lie symmetries at center stage?

References

[1] Ablowitz M J, Ramani A, and Segur H, Nonlinear evolution equations and ordinary differential equations of Painlevé type, *Lett. Nuovo Cim.* **23**, 333–338, 1978.

[2] Ablowitz M J and Segur H, *Solitons and the Inverse Scattering Transform*, SIAM, Philadelphia, 1975.

[3] Allassia F and Nucci M C, Symmetries and heir equations for the laminar boundary layer model, *J. Math. An. Appl.* **201**, 911–942, 1996.

[4] Ames W F, *Nonlinear Partial Differential Equations in Engineering, Vol. 2.*, Academic Press, New York, 1972.

[5] Arrigo D J, *Symmetry Analysis of Differential Equations: An Introduction*, John Wiley & Sons, Chichester, 2015.

[6] Arrigo D J, Broadbridge P, Hill J M, Nonclassical symmetry solutions and the methods of Bluman-Cole and Clarkson-Kruskal, *J. Math. Phys.* **34**, 4692–4703, 1993.

[7] Arrigo D J, Broadbridge P, Hill J M, Nonclassical symmetry reductions of the linear diffusion equation with a nonlinear source, *IMA J. Appl. Math.* **52**, 1–24, 1994.

[8] Arrigo D J, Ekrut D A, Fliss J R, Long Le, Nonclassical symmetries of a class of Burgers systems, *J. Math. Anal. Appl.* **371**, 813–820, 2010.

[9] Barannyk T A, Nonclassical Symmetries of a System of Nonlinear Reaction-Diffusion Equations, *J. Math. Sci.* **238**, 207–214, 2019.

[10] Bîlă N and Niesen J, On a new procedure for finding nonclassical symmetries, *J. Symbol. Comp.* **38**, 1523–1533, 2004.

[11] Bluman G W and Anco S C, *Symmetry and integration methods for differential equations*, Springer, New York, 2002.

[12] Bluman G W and Cole J D, The general similarity solution of the heat equation, *J. Math. Mech.* **18**, 1025–1042, 1969.

[13] Bluman G W and Cole J D, *Similarity Methods for Differential Equations*, Springer-Verlag, Berlin, 1974.

[14] Bluman G W and Kumei S, *Symmetries and Differential Equations*, Springer-Verlag, Berlin, 1989.

[15] Bluman G W, Reid G J, Kumei S, New classes of symmetries for partial differential equations, *J. Math. Phys.* **29**, 806-811, 1988.

[16] Bluman G W, Shou-fu Tian, Zhengzheng Yang, Nonclassical analysis of the nonlinear Kompaneets equation, *J. Eng. Math.* **84**, 87–97, 2014.

[17] Bradshaw-Hajek B H, Nonclassical symmetry solutions for non-autonomous reaction-diffusion equations, *Symmetry* **11**, 208, 2019.

[18] Bradshaw-Hajek B H, Edwards M P, Broadbridge P, Williams G H, Nonclassical symmetry solutions for reaction-diffusion equations with explicit spatial dependence, *Nonlin. Anal.* **67**, 2541–2552, 2007.

[19] Bradshaw-Hajek B H and Nucci M C, Heir-equations for systems of evolution equations, 2019, in preparation.

[20] Bruzón M S and Gandarias M L, Applying a new algorithm to derive nonclassical symmetries, *Commun. Nonlinear Sci. Numer. Simul.* **13**, 517–523, 2008.

[21] Cantwell B J, *Introduction to Symmetry Analysis*, Cambridge University Press, Cambridge, 2002.

[22] Cherniha R, New Q-conditional symmetries and exact solutions of some reaction-diffusion-convection equations arising in mathematical biology, *J. Math. Anal. Appl.* **326**, 783–799, 2007.

[23] Cherniha R and Davydovych V, Conditional symmetries and exact solutions of nonlinear reaction-diffusion systems with non-constant diffusivities, *Commun. Nonlinear Sci. Numer. Simulat.* **17**, 3177–3188, 2012.

[24] Cherniha R and Pliukhin O, New conditional symmetries and exact solutions of reaction-systems with power diffusivities, *J. Phys. A: Math. Gen.* **41**, 185208–185222, 2008.

[25] Cherniha R M and Serov M I, Symmetries, Ansätze and exact solutions of nonlinear second-order evolution equations with convection term, *European J. Appl. Math.* **9**, 527–542, 1998.

[26] Cherniha R and Serov M, Nonlinear systems of the Burgers-type equations: Lie and Q-conditional symmetries, Ansätze and solutions, *J. Math. Anal. Appl.* **282**, 305–328, 2003.

[27] Clarkson P A, Nonclassical symmetry reductions of nonlinear partial differential equations, *Math. Comput. Modell.* **18**, 45–68, 1993.

[28] Clarkson P A and Mansfield E L, Nonclassical symmetry reductions and exact solutions of nonlinear reaction-diffusion equations, in *Applications of Analytic and Geometric Methods to Nonlinear Differential Equations*, edited by P. A. Clarkson, Kluwer, Dordrecht, 375–389, 1993.

[29] Clarkson P A and Mansfield E L, Symmetry reductions and exact solutions of a class of nonlinear heat equations, *Physica D* **70**, 250–288, 1994.

[30] Clarkson P A and Winternitz P, Nonclassical symmetry reductions for the Kadomtsev-Petviashvili equation, *Physica D* **49**, 257–272, 1991.

[31] Dulov V G, Novikov S P, Ovsyannikov L V, Rozhdestvenskii B L, Samarskii A A, Shokin Yu I, Nikolai Nikolaevich Yanenko (obituary), *Russ. Math. Surv.* **39**, 99–110, 1984.

[32] Estevez P G and Gordoa P R, Nonclassical symmetries and the singular manifold method: the Burgers equation, *Theor. Math. Phys.* **99**, 562–566, 1994.

[33] Estevez P G and Gordoa P R, Nonclassical symmetries and the singular manifold method: theory and six examples, *Stud. Appl. Math.* **95**, 73–113, 1995.

[34] Euler N, Köhler A and Fushchich W I, Q-symmetry generators and exact solutions for nonlinear heat conduction, *Physics Scripta* **49**, 518 - 524, 1994.

[35] Euler N and Steeb W-H, *Continuous Symmetries, Lie Algebras and Differential Equations*, BI Wissenschaftsverlag, Mannheim, 1992.

[36] Fokas A S and Liu Q M, Generalised conditional symmetries and exact solutions of nonintegrable equations, *Theor. Math. Phys.* **99**, 263–277, 1994.

[37] Fushchych W I, Serov M I, Tulupova L A, The conditional invariance and exact solutions of the nonlinear diffusion equation, *Proc. Acad. of Sci. Ukraine* **4**, 37–40, 1993.

[38] Fushchych W I, Shtelen W M, Serov M I, *Symmetry Analysis and Exact Solutions of Equations of Nonlinear Mathematical Physics*, Kluwer, Dordrecht, 1993.

[39] Fushchych W I, Shtelen W M, Serov M I, Popovych R O, Q-conditional symmetry of the linear heat equation, *Proc. Acad. of Sci. Ukraine* **12**, 28–33, 1992.

[40] Galaktionov V A, On new exact blow-up solutions for nonlinear heat conduction equations with source and applications, *Diff. Int. Eqns.* **3**, 863–874, 1990.

[41] Gandarias M L, Nonclassical symmetries of a porous medium equation with absorption, *J. Phys. A: Math. Gen.* **30**, 6081–6091, 1997.

[42] Gandarias M L, Romero J L, Díaz J M, Nonclassical symmetry reductions of a porous medium equation with convection, *J. Phys. A: Math. Gen.* **32**, 1461–1473, 1999.

[43] Goard J, Generalised conditional symmetries and Nucci's method of iterating the nonclassical symmetries method, *Appl. Math. Lett.* **16**, 481–486, 2003.

[44] Goard J, A note on the equivalence of methods to finding nonclassical determining equations, *J. Nonlinear Math. Phys.*, **26**, 1–6, 2019.

[45] Guthrie G, *Constructing Miura transformations using symmetry groups*, Research report No. 85, 1993. (https://core.ac.uk/download/pdf/35472859.pdf)

[46] Hashemi M S and Nucci M C, Nonclassical symmetries for a class of reaction-diffusion equations: the method of heir-equations, *J. Nonlinear Math. Phys.* **20**, 44–60, 2013.

[47] Hill J M, *Differential Equations and Group Methods for Scientists and Engineers*, CRC Press, Boca Raton, 1992.

[48] Huang D -J and Zhou S, Group-theoretical analysis of variable coefficient nonlinear telegraph equations, *Acta Appl. Math.* **117**, 135–183, 2012.

[49] Hydon P E, *Symmetry Methods for Differential Equations: a Beginner's Guide*, Cambridge University Press, Cambridge, 2000.

[50] King J R, Exact polynomial solutions to some nonlinear diffusion equations, *Physica D* **64**, 35–65, 1993.

[51] Kruglikov B, Symmetry approaches for reductions of PDEs, differential constraints and Lagrange-Charpit method, *Acta Appl. Math.* **101**, 145–161, 2008.

[52] Ibragimov N H, *Elementary Lie Group Analysis and Ordinary Differential Equations*, John Wiley & Sons, Chichester, 1999.

[53] Levi D and Winternitz P, Nonclassical symmetry reduction: Example of the Boussinesq equation, *J. Phys. A: Math. Gen.* **22**, 2915–2924, 1989.

[54] Lie S, Über die Integration durch bestimmte Integrale von Einer Classe linearer partieller Differentialgleichungen, *Arch. Math.* **6**, 328–368, 1881.

[55] Lie S, *Vorlesungen über Differentialgleichungen mit Bekannten Infinitesimalen Transformationen*, B. G. Teubner, Leipzig, 1912.

[56] Lou S -Y, Nonclassical symmetry reductions for the dispersive wave equations in shallow water, *J. Math. Phys.* **33**, 4300–4305, 1992.

[57] Ludlow D K, Clarkson P A, Bassom A P, Nonclassical symmetry reductions of the three-dimensional incompressible Navier-Stokes equations, *J. Phys. A: Math. Gen.* **31**, 7965–7980, 1998.

[58] Martini S, Ciccoli N, Nucci M C, Group analysis and heir-equations of a mathematical model for thin liquid films, *J. Nonlinear Math. Phys.* **16**, 77–92, 2009.

[59] Meleshko S V, *Methods for Constructing Exact Solutions of Partial Differential Equations: Mathematical and Analytical Techniques with Applications to Engineering*, Springer, New York, 2005.

[60] Mikhailov A V, Shabat A B, and Sokolov V V, The symmetry approach to classification of integrable equations, in *What Is Integrability?*, edited by V E Zakharov, Springer-Verlag, Berlin, 115–184, 1991.

[61] Miura R M, Korteweg-de Vries equation and generalizations. I. A remarkable explicit nonlinear transformation, *J. Math. Phys.* **9**, 1202–1204, 1968.

[62] Newell A C and Whitehead J A, Finite bandwidth, finite amplitude convection, *J. Fluid Mech.* **38**, 279–303, 1969.

[63] Nucci M C, Symmetries of linear, C-integrable, S-integrable, and non-integrable equations, in *Nonlinear Evolution Equations and Dynamical Systems. Proceedings NEEDS '91*, M. Boiti, L. Martina, and F. Pempinelli, Eds., World Scientific, Singapore, 374–381, 1992.

[64] Nucci M C, Nonclassical symmetries and Bäcklund transformations, *J. Math. An. Appl.* **178**, 294–300, 1993.

[65] Nucci M C, Iterating the nonclassical symmetries method, *Physica D* **78**, 124–134, 1994.

[66] Nucci M C, Iterations of the nonclassical symmetries method and conditional Lie-Bäcklund symmetries, *J. Phys. A: Math. Gen.* **29**, 8117–8122, 1996.

[67] Nucci M C, Interactive REDUCE programs for calculating Lie point, nonclassical, Lie-Backlund, and approximate symmetries of differential equations: manual and floppy disk, in *CRC Handbook of Lie Group Analysis of Differential Equations. Vol. 3: New Trends in Theoretical Developments and Computational Methods*, ed. N. H. Ibragimov, CRC Press, Boca Raton, 415–481, 1996.

[68] Nucci M C, Nonclassical symmetries as special solutions of heir-equations, *J. Math. Anal. Appl.* **279**, 168–179, 2003.

[69] Nucci M C and Ames W F, Classical and nonclassical symmetries of the Helmholtz equation, *J. Math. Anal. Appl.* **178**, 584–591, 1993.

[70] Nucci M C and Clarkson P A, The nonclassical method is more general than the direct method for symmetry reduction. An example of the Fitzhugh-Nagumo equation, *Phys. Lett. A* **184**, 49–56, 1992.

[71] Olver P J, *Applications of Lie Groups to Differential Equations*, Springer-Verlag, Berlin, 1986 and 1993.

[72] Olver P J, Direct reduction and differential constraints, *Proc. R. Soc. Lond. A* **444**, 509–523, 1994.

[73] Olver P J, Sanders J, and Wang J P, Classification of symmetry-integrable evolution equations, in *Bäcklund and Darboux Transformations. The Geometry of Solitons (Halifax, NS, 1999)*, A.M.S., Providence, 363–372, 2001.

[74] Oron P and Rosenau P, Some symmetries of the nonlinear heat and wave equations, *Phys. Lett. A* **118**, 172–176, 1986.

[75] Ovsjannikov L V, *Group Analysis of Differential Equations*, Academic Press, New York, 1982.

[76] Popovych R O, Reduction operators of linear second-order parabolic equations, *J. Phys. A: Math. Theor.* **41**, 185202, 2008.

[77] Pucci E, Similarity reductions of partial differential equations, *J. Phys. A: Math. Gen.* **25** 2631–2640, 1992.

[78] Rogers C and Ames W F, *Nonlinear Boundary Value Problems in Science and Engineering*, Academic Press, New York, 1989.

[79] Rogers C and Shadwick W F, *Bäcklund Transformations and Their Applications*, Academic Press, New York, 1982.

[80] Sanders J A and Jing Ping Wang, On the integrability of homogeneous scalar evolution equations, *J. Differential Equations* **147**, 410–434, 1998.

[81] Sidorov A F, Shapeev V P, Yanenko N N, *The Method of Differential Constraints and Its Applications in Gas Dynamics*, (in Russian) Nauka, Novosibirsk, 1984.

[82] Sophocleous C, Transformation properties of a variable-coefficient Burgers equation, *Chaos, Solitons and Fractals* **20**, 1047–1057, 2004.

[83] Stephani H, *Differential Equations. Their Solution Using Symmetries*, Cambridge University Press, Cambridge, 1989.

[84] Tabor M and Gibbon J D, Aspects of the Painlevé property for partial differential equations, *Phyisica D* **18**, 180–189, 1986.

[85] Vaneeva O O, Popovych R O, Sophocleous C, Extended group analysis of variable coefficient reaction-diffusion equations with exponential nonlinearities, *J. Math. Anal. Appl.* **396**, 225–242, 2012.

[86] Vaneeva O, Boyko V, Zhalij A, Sophocleous C, Classification of reduction operators and exact solutions of variable coefficient Newell-Whitehead-Segel equations, *J. Math. Anal. Appl.* **474**, 264–275, 2019.

[87] Vorob'ev E M, Partial symmetries and integrable multidimensional differential equations, *Differ. Eqs.* **25**, 322–325, 1989.

[88] Vorob'ev E M, Symmetries of compatibility conditions for systems of differential equations, *Acta Appl. Math.* **26**, 61–86, 1992.

[89] Weiss J, Tabor M, and Carnevale G, The Painlevé property for partial differential equations, *J. Math. Phys.* **24**, 522–526, 1983.

[90] Winternitz P, Lie groups and solutions of nonlinear partial differential equations, in *Integrable Systems, Quantum Groups, and Quantum Field Theories (Salamanca, 1992)*, edited by L. A. Ibort and M. A. Rodriguez, Kluwer, Dordrecht, 429–495, 1999.

[91] Yanenko N N, Compatibility theory and integration methods for systems of nonlinear partial differential equations, in *Proc. Fourth All-Union Math. Congr. in Leningrad 1961, Vol. II* (in Russian), 247–252, Nauka, Leningrad, 1964.

[92] Zhdanov R Z, Conditional Lie-Bäcklund symmetry and reduction of evolution equations, *J. Phys. A: Math. Gen.* **28**, 3841–3850, 1995.

B1. Coupled nonlinear Schrödinger equations: spectra and instabilities of plane waves

Antonio Degasperis [a], *Sara Lombardo* [b] *and Matteo Sommacal* [c]

[a] *Dipartimento di Fisica, "Sapienza" Università di Roma, Italy*
E-mail: antonio.degasperis@uniroma1.it

[b] *Department of Mathematical Sciences, School of Science*
Loughborough University, UK
E-mail: s.lombardo@lboro.ac.uk

[c] *Department of Mathematics, Physics and Electrical Engineering*
Northumbria University, Newcastle upon Tyne, UK
E-mail: matteo.sommacal@northumbria.ac.uk

Abstract

The coherent coupling of two regular wave trains is considered as a solution of a system of a pair of nonlinear Schrödinger equations with cubic self- and cross-interactions. The purpose is the study of the linear stability of this plane wave solution against small and localised perturbations. The model wave equation considered here is integrable in three different regimes, the focusing, defocusing and mixed regime, according to the values of the coupling constants. The approach we take is based on the construction of the *eigenfunctions* of the linearised wave equations via the so-called *squared eigenfunctions*, obtained from the Lax pair. The notion of *stability spectrum* is introduced as associated in particular to the plane wave solution. This is an algebraic curve in the complex plane of the spectral variable of the Lax pair. By means of the geometric features of the spectrum, we completely classify spectra in the parameter space of amplitudes and coupling constants. The stability is then assessed by computing the eigenfrequencies of the eigenfunctions. The instability band is shown to be related to the properties of the non-real part of the spectrum. This analysis confirms that instabilities exist also in the defocusing regime. Different types of spectra are displayed.

1 Introduction

The quest for stable solutions is central in applied mathematics and physics. *Stability* is a fundamental concept in nonlinear dynamics, and, in particular, in the field of wave propagation. Wave stability is also an interdisciplinary subject because of its relevance in many contexts, such as optics, fluids, plasma physics, solid state and Bose-Einstein condensates, to name a few. Indeed, after the first observations of wave instability in several nonlinear wave phenomena, such as radio waves in transmission lines [1], wave beams in nonlinear media [2], plasma waves [3], optical fibres [4] and water waves [5], and after further experimental evidence [6, 7] (see also *e.g.* [8]), the research on this subject has grown very rapidly as similar phenomena appear also in different contexts (see *e.g.* [9] plasma physics, [10] in water waves, [11, 12] in optics and [13] in Bose-Einstein condensation). As the literature on the subject is nowadays vast, we refer the interested reader to *e.g.* [14, 15], and references therein. Theoretical and computational works followed these experimental findings. In particular, predictions regarding the short time evolution of small perturbations of the initial profile is the obvious way to obtain informations on the wave stability. It can be followed by standard linear methods, see *e.g.* [16] and [17] and references therein. Here we are specifically concerned with the early stage of amplitude modulation instabilities due to cubic nonlinearity combined with a quadratic dispersive propagation in a one-dimensional space. In a sketchy presentation of the approach we have in mind, suppose we wish to assess the linear stability of the particular known solution $u(x,t)$ of a nonlinear wave equation, we then consider the perturbed solution $u+\delta u$ of the same equation. At the first order of approximation, δu satisfies a linear equation, whose coefficients depend on the solution $u(x,t)$ itself, and therefore they are generally non-constant. Consequently, solving the initial value problem $\delta u(x,t_0) \to \delta u(x,t)$ in general is not tractable by analytical methods and it requires heavy numerical work. Only for very special solutions $u(x,t)$, such as nonlinear plane waves or solitary localised waves, see *e.g.* [16], this initial value problem can be transformed in an eigenvalue problem for an ordinary differential operator in the space variable x. In this way the computational task reduces to constructing the eigenmodes, *i.e.* the eigenfunctions of an ordinary differential operator. In this lucky case, the corresponding eigenvalues are simply related to the proper frequencies whose computation is the main goal of the method. For very special solutions $u(x,t)$, this procedure exceptionally leads to a linear evolution equation for $\delta u(x,t)$ with constant coefficients which can be therefore solved via Fourier analysis. A simple and well-known example of this case is the linearisation of the focusing Non-Linear Schrödinger (NLS) equation $iu_t + u_{xx} + 2|u|^2u = 0$ around its Continuous Wave (CW) solution $u(x,t) = e^{2it}$ (hereafter subscripts x and t denote partial differentiation), which leads to the well-known modulation (or modulational) instability of a regular wave train.

Employing either analytical or numerical methods, the computation of all the complex eigenfrequencies yields the relevant information about the linear stability of $u(x,t)$, provided the set of eigenmodes is complete. In fact, the computed eigenfrequencies may be real, as for stable modes, or may be complex, say with strictly non-vanishing imaginary part, as for unstable modes. It turns out that this computational step of the method requires finding the *spectrum* of a differential operator. Because of this, the stability property of the solution $u(x,t)$ is also referred to as *spectral stability*. It is clear that this approach successfully works for a limited class of solutions of the wave equation. Here we are concerned with this linear stability only. However, a number of different approaches to nonlinear wave stability have been introduced which use different mathematical techniques and aim at various physical applications. For instance, variational methods to investigate orbital stability have been applied to solitary waves and standing waves, *e.g.*, see [18] [19].

An alternative and powerful approach to linear stability originated from [20], shortly after the discovery of the complete integrability and of the spectral method to solve the Kortewg-de Vries (KdV) and NLS equations (*e.g.*, see the textbooks [21, 22, 23]). With regard to stability, a convenient starting point, the one we adopt here, is to associate to an integrable wave equation a Lax pair of ordinary differential equations. By means of this pair of equations, one is not only able to construct particular solutions $u(x,t)$ of the integrable wave equation, and to investigate the Cauchy initial value problem for appropriate boundary conditions, but also to construct, by *algebraic* operations on solutions of the Lax pair itself, a complete set of solutions (eigenmodes) of the linearised integrable wave equation for the small perturbation $\delta u(x,t)$. These solutions of the linearised equation have been termed *squared eigenfunctions* (see Section 2) and have been mainly used to investigate first order effects of small perturbations of the wave equation itself [24], see also [25] and references quoted there. Here we make use of these solutions of the linearised equation to control instead the effect of small changes of the initial condition $\delta u(x,t_0)$. We do this by considering a spectral representation of the perturbation $\delta u(x,t)$ of the solution $u(x,t)$ as linear combination of such squared eigenfunctions to the purpose of investigating linear stability. With respect to other approaches to linear stability in use, this method shows its power by formally applying also to solutions $u(x,t)$, where standard computations fail. Moreover, it proves to be applicable (see [26] and below) to a very large family of matrix Lax pairs and, therefore, to many integrable systems other than KdV and NLS equations, (*e.g.* to Sine-Gordon, mKdV, derivative NLS, coupled NLS (see below), three-wave resonant interaction, massive Thirring model, and other equations of interest in applications). The evident drawback of the squared eigenfunctions technique is that its applicability is limited to the very special class of integrable wave equations. Notwithstanding this restriction, it remains of important practical interest because several integrable partial differential equations have been derived in various physical contexts as reliable, though approximate, models [21, 27, 28]. A further argument which motivates the interest in integrable wave equations is that the stability properties of a given solution of these equations provide a strong insight on similar solutions of a close enough non-integrable equation.

The main observation is that the construction of the squared eigenfunctions yields the corresponding eigenfrequencies, which gives the (necessary and sufficient) information to assess linear stability. Explicit expressions of such eigenmodes have been obtained if the unperturbed wave solution $u(x,t)$ is a cnoidal wave (*e.g.* see [29, 30, 31] for the KdV equation [30] and for the NLS equation [32]), or if it is a soliton solution [33] and, although only formally, an arbitrary solution [34] with vanishing boundary values. The computational strategy amounts to constructing the set of eigenmodes together with their eigenfrequencies. It should be pointed out that, once the functional space of the wave fields $u(x,t)$ has been fixed, the integrability methods possibly provide also a way of deriving the closure and completeness relations of the eigenmodes, see *e.g.* [24, 33, 34] for solutions which vanish sufficiently fast as $|x| \to \infty$. In this respect, a word of warning is appropriate. The boundary conditions imposed on the solutions $u(x,t)$ are an essential ingredient of the proof that the wave equation is indeed integrable. Thus, in particular for the NLS equation, integrability methods have been applied so far to linear stability of wave solutions which, as $|x| \to \infty$, either vanish as a localised soliton [33, 35], or go to a CW solution [36], or else are periodic, $u(x,t) = u(x+L,t)$ [37, 38]. We also note that the connection between the linearised equation and the squared eigenfunctions follows from the inverse spectral transform machinery. However, this property is local [26] since it follows directly from the Lax pair without any need of the spectral transform method. This method however allows to go beyond the linear stage of the evolution of small perturbations. In fact the spectral method of solving the initial value problem for the perturbed solution $u + \delta u$ yields the long time evolution of δu beyond the linear approximation, see for instance [39, 40]. This important problem falls outside the scope of the present work and will not be considered here (for the initial value problem and unstable solutions of the NLS equation see [41, 42, 43, 44]).

The stability properties of a given solution $u(x,t)$ may depend on parameters. These parameters come from the coefficients of the wave equation, and from the parameters (if any) which characterise the solution $u(x,t)$ itself. This obvious observation implies that one may expect the parameter space to be divided into regions where the solution $u(x,t)$ features different behaviours in terms of linear stability. Indeed, this is generically the case, and crossing the border of one of these regions by varying the parameters, for instance a wave amplitude, may correspond to the opening of a gap in the instability frequency band, so that a *threshold* occurs at that amplitude value which corresponds to crossing. The investigation of such thresholds is rather simple for scalar (one-component) waves. For instance, by playing with symmetries, the KdV equation $u_t + u_{xxx} + uu_x = 0$ has no free coefficients, while the NLS equation comes with a sign in front of the cubic term, which distinguishes between defocusing and focusing self-interaction. These two different versions of the NLS equation lead to different phenomena such as modulation stability and instability of the continuous wave solution. Much less simple is the stability analysis of model equations which have more structural coefficients whose values cannot be simultaneously, and independently, rescaled. This is the case when two or more waves resonate and propagate while coupling to each other. In this case, the wave equations do not happen to be integrable for all choices of the coefficients. A well-known example, which is our main focus, is that of two interacting fields, u_j, $j = 1, 2$, which evolve according to the coupled system of NLS equations

$$iu_{jt} + u_{jxx} - 2\,(s_{1j}|u_1|^2 + s_{2j}|u_2|^2)\,u_j = 0\,, \quad j = 1,\, 2\,,$$

where $(s_{1j}|u_1|^2 + s_{2j}|u_2|^2)$ are the self- and cross-interaction terms. This is integrable only if the real coupling constants s_{1j} and s_{2j} are independent on j. With this restriction, and dropping then the index j, this system is integrable in three cases [45], namely (after appropriate rescaling): $s_1 = s_2 = 1$ (defocusing Manakov model), $s_1 = s_2 = -1$ (focusing Manakov model) [46], and the mixed case $s_1 = -s_2 = 1$. These three integrable systems of two Coupled NLS (CNLS) equations are of interest in few special applications in optics [47, 48, 49] and in fluid dynamics [50], while, in various contexts (*e.g.* in optics [11] and in fluid dynamics [10, 51]), the coupling constants s_{1j}, s_{2j} take different values and the CNLS system happens to be non-integrable. Yet the analysis of the three integrable cases is still relevant in the study of non-integrable equations if they are sufficiently close to those which are integrable [52]. The linear stability of CW solutions, $|u_j(x,t)|$ = constant, of integrable CNLS systems is of special interest not only because of its experimental observability, but also because it can be analysed via both standard methods and the squared eigenfunctions approach. As far as the standard methods are concerned, the linear stability of CW solutions has been investigated by means of the Fourier transform [53] only in the integrable focusing and defocusing regimes, but not for the mixed one. Conversely, say regarding the squared eigenfunctions method, it has been partially discussed in [54] to mainly show that instability may occur also in defocusing media, in contrast with the single NLS equation where waves are unstable only in the focusing case.

Here we study the linear stability problem of the CW solutions of the CNLS equations within the integrability framework to prove that the main object to be computed is a *spectrum* (to be defined below) as a curve in the complex plane of the spectral variable, together with the eigenmodes wave numbers and frequencies of the eigenfunctions of the linearised system. In particular, we show that the spectrum which is relevant to our analysis is related to, but *not coincident* with, the spectrum of the Lax equation. In addition, if λ is the spectral variable, the computational outcome is the wave number $k(\lambda)$ and frequency $\omega(\lambda)$, so that the dispersion relation, and also the instability band, are implicitly defined over the spectrum through their dependence on λ. Since spectrum and eigenmodes depend on parameters, we explore the entire parameter space of the two amplitudes and coupling constants to arrive at a complete classification of spectra by means of numerically-assisted, algebraic techniques. Our investigation in Section 2 illustrates how the linear stability analysis works within the theory of integrability. Our focus is on x- and t-dependent CWs, a case which is both of relevance to physics and is computationally doable. It is worth observing again that the linear stability of the CW solutions can indeed be discussed also by standard Fourier analysis, *e.g.*, see [53] for the CNLS systems in the focussing and defocussing regimes. However, such way is of no help to investigate the stability of other solutions. On the contrary, at least for the integrable CNLS system, our method relies only on the existence of a Lax pair and as such it has the advantage of being applicable also to other solutions as well. In particular, it can be applied to the CW solutions in all regimes (as we do it here), as well as to solutions such as, for instance, dark-dark, bright-dark and higher order soliton solutions traveling on a CW background, to which the standard methods are not applicable.

This chapter is organised as follows. In the next section (Section 2) we specialise the general approach [26] together with the expression of the eigenmodes (aka squared eigenfunctions) of the linearised equation to the CNLS equations. There we define the x-spectrum in the complex plane of the spectral variable, together with wave numbers and frequencies as functions defined on the spectrum. In the Subsection 2.1 we introduce the geometric properties which are distinctive of each spectrum and characterise spectra according to their topological features. In Subsection 2.2 we divide the entire parameter space according to five distinct classes of spectra. Section 3 is devoted to discussing the dispersion relation of the eigenfunctions of the linearised equation. There we introduce the *gain function* for each type of spectrum, and we show the difference between instability basebands and passbands. Conclusions with open problems and perspectives of future work are the content of Section 4. Details regarding a limit case and numerical aspects are confined in appendices.

2 Spectra

This section is meant to make contact with our paper [26], and to set the notation we introduced there. We do so because some of those results we established in that paper serve our present discussion and are therefore reported here for convenience. Details, proofs and further results can be found there. Our main focus here is on the linear stability of continuous wave (CW) solutions of the following two coupled nonlinear Schrödinger (CNLS) equations

$$iu_{jt} + u_{jxx} - 2\,(s_1|u_1|^2 + s_2|u_2|^2)\,u_j = 0\,, \quad u_j = u_j(x,t) \quad ,j = 1, 2\,. \tag{2.1}$$

Here and in the following, a subscripted continuous variable (as x and t) stands for partial differentiation with respect to that variable. The dependent variables u_j, $j = 1, 2$, are the complex envelope amplitudes of two quasi monochromatic waves that are weakly resonant and propagate with self and cross interactions which are modeled by cubic terms with real j-independent coupling constants s_1 and s_2. Among other model equations of resonantly coupled waves, this system (2.1) has the peculiar property of being integrable for any choice of s_1, s_2. With no loss of generality, we adopt the rescaled values of the coupling constants $s_1 = \pm 1$, $s_2 = \pm 1$. Thus the system (2.1) describes three different physical settings: the defocusing Manakov model for $s_1 = s_2 = 1$, the focusing Manakov model for $s_1 = s_2 = -1$ and the mixed case $s_1 = -s_2$. This system is therefore a natural integrable vector generalization of the NLS scalar equation which is obtained from (2.1) by setting $u_2 = 0$. Its integrality stems from the existence of its associated Lax pair of matrix ordinary linear differential equations

$$\Psi_x = X\Psi\,, \quad \Psi_t = T\Psi\,, \tag{2.2}$$

which are compatible with each other, namely

$$X_t - T_x + [X\,,\,T] = 0\,. \tag{2.3}$$

Both the matrix solution $\Psi(x,t,\lambda)$ and the matrix coefficients $X(x,t,\lambda)$, $T(x,t,\lambda)$ are 3×3, and these last two matrices are assumed to have the simple polynomial dependence on the complex *spectral variable* λ

$$X(\lambda) = i\lambda\Sigma + Q\,, \quad T(\lambda) = 2i\lambda^2\Sigma + 2\lambda Q + i\Sigma(Q^2 - Q_x)\,, \tag{2.4}$$

where

$$\Sigma = \mathrm{diag}\{1\,, -1\,, -1\} \tag{2.5}$$

and

$$Q = \begin{pmatrix} 0 & s_1 u_1^* & s_2 u_2^* \\ u_1 & 0 & 0 \\ u_2 & 0 & 0 \end{pmatrix}, \tag{2.6}$$

the upper scripted asterisk meaning complex conjugation. This choice of the matrices X, T implies that the compatibility condition (2.3) is equivalent to the CNLS system (2.1).
Once a solution $u_1(x,t)$, $u_2(x,t)$ has been fixed, then $u_1(x,t) + \delta u_1(x,t)$, $u_2(x,t) + \delta u_2(x,t)$ is an approximate solution of the CNLS system provided $\delta u_1(x,t)$, $\delta u_2(x,t)$ solve the linearized equations

$$\begin{aligned} \delta u_{1t} &= i\{\delta u_{1xx} - 2[(2s_1|u_1|^2 + s_2|u_2|^2)\delta u_1 + s_1 u_1^2 \delta u_1^* + s_2 u_1 u_2^* \delta u_2 + s_2 u_1 u_2 \delta u_2^*]\} \\ \delta u_{2t} &= i\{\delta u_{2xx} - 2[(s_1|u_1|^2 + 2s_2|u_2|^2)\delta u_2 + s_2 u_2^2 \delta u_2^* + s_1 u_2 u_1^* \delta u_1 + s_1 u_2 u_1 \delta u_1^*]\}\,. \end{aligned} \tag{2.7}$$

The linear stability analysis of a given bounded solution u_1, u_2 demands solving these two linear equations; in the following we will investigate perturbations $\delta u_1(x,t)$, $\delta u_2(x,t)$ whose initial data are assumed to be *localized and bounded* in x. Proving that the solution $u_1(x,t)$, $u_2(x,t)$ is linearly stable requires showing that the initially small data $\delta u_1(x,t_0)$, $\delta u_2(x,t_0)$ remains small at any time t. A standard way to tackle this problem is by constructing a (complete) set of solutions, i. e. eigenmodes, $F_1(x,t,\lambda)$, $F_2(x,t,\lambda)$ of the linear equation (2.7), say

$$\begin{aligned} F_{1t} &= i\{F_{1xx} - 2[(2s_1|u_1|^2 + s_2|u_2|^2)F_1 + s_1 u_1^2 F_1^* + s_2 u_1 u_2^* F_2 + s_2 u_1 u_2 F_2^*]\} \\ F_{2t} &= i\{F_{2xx} - 2[(s_1|u_1|^2 + 2s_2|u_2|^2)F_2 + s_2 u_2^2 F_2^* + s_1 u_2 u_1^* F_1 + s_1 u_2 u_1 F_1^*]\}\,, \end{aligned} \tag{2.8}$$

which depend on the spectral parameter λ in such a way that the solution of (2.7) can be formally represented as the linear combination

$$\delta u_j(x,t) = \int \mathrm{d}\lambda\, F_j(x,t,\lambda)\,, \quad j = 1\,,\,2\,. \tag{2.9}$$

This task, which is only very exceptionally achievable since the equations (2.8) have x,t-dependent coefficients, becomes feasible if the wave dynamics is integrable, say, associated to a Lax pair, and if the matrix solution $\Psi(x,t,\lambda)$ of the Lax pair (2.2) corresponding to the given solution u_1, u_2, see (2.6), is known [26]. Here is where integrability plays a role in linear stability. In fact, as it is known for a number of integrable systems and as it has been recently proved for a fairly larger class of integrable equations [26], the eigenmode solutions F_j of the linearized equations (2.8) (also known as squared eigenfunctions [25]), are constructed by the following transformation:

$$\Psi \mapsto F = [\Sigma\,,\,\Psi M \Psi^{-1}] = \begin{pmatrix} 0 & s_1 F_1^* & s_2 F_2^* \\ F_1 & 0 & 0 \\ F_2 & 0 & 0 \end{pmatrix}, \tag{2.10}$$

where the matrix $M(\lambda)$ is arbitrary and x, t-independent. As it will be shown below, we point out here that this construction of the eigenfunctions F_1, F_2 is sufficient to prove stability while there is no need to actually compute the integral (2.9).

Next we apply this stability analysis to the CW solution

$$u_1(x,t) = a_1 e^{i(qx-\nu t)}\,, \quad u_2(x,t) = a_2 e^{-i(qx+\nu t)}\,, \quad \nu = q^2 + 2p \tag{2.11}$$

of the CNLS equations (2.1). Here the real parameter p,

$$p = s_1 a_1^2 + s_2 a_2^2\,, \tag{2.12}$$

is given in terms of the coupling constants and of the amplitudes a_1, a_2 which, with no loss of generality, are real and positive. The parameter q, which has a relevant role in the following analysis, measures the half-difference of the wave numbers of the two continuous wave components (2.11). According to the map (2.10), the first step is constructing the matrix solution $\Psi(x,t,\lambda)$ of the Lax pair (2.2). This is given by the expression

$$\Psi(x,t,\lambda) = R(x,t)e^{i(xW(\lambda)-tZ(\lambda))}\,, \tag{2.13a}$$

$$R(x,t) = \mathrm{diag}\{e^{ipt},\, e^{i(qx-q^2t-pt)},\, e^{-i(qx+q^2t+pt)}\}, \tag{2.13b}$$

where the x, t-independent matrices W and Z are found to be

$$W(\lambda) = \begin{pmatrix} \lambda & -is_1a_1 & -is_2a_2 \\ -ia_1 & -\lambda - q & 0 \\ -ia_2 & 0 & -\lambda + q \end{pmatrix}, \tag{2.14}$$

$$Z(\lambda) = \lambda^2 - 2\lambda W(\lambda) - W^2(\lambda) - p\,, \tag{2.15}$$

with the property that they commute, $[W\,,\,Z] = 0$, consistently with the compatibility condition (2.3). We consider here the eigenvalues $w_j(\lambda)$ and $z_j(\lambda)$, $j = 1,2,3$, of $W(\lambda)$ and, respectively, of $Z(\lambda)$ as simple, as indeed they are for generic values of λ. In this case both $W(\lambda)$ and $Z(\lambda)$ are diagonalized by the same matrix $U(\lambda)$, namely

$$W(\lambda) = U(\lambda)W_D(\lambda)U^{-1}(\lambda)\,, \quad W_D = \mathrm{diag}\{w_1,\, w_2,\, w_3\}\,, \tag{2.16}$$

$$Z(\lambda) = U(\lambda)Z_D(\lambda)U^{-1}(\lambda)\,, \quad Z_D = \mathrm{diag}\{z_1,\, z_2,\, z_3\}\,. \tag{2.17}$$

Next we construct the matrix $F(x,t,\lambda)$ via its definition (2.10) which, because of the explicit expression (2.13), reads

$$F(x,t,\lambda) = R(x,t)\left[\Sigma\,,\, e^{i(xW(\lambda)-tZ(\lambda))}M(\lambda)e^{-i(xW(\lambda)-tZ(\lambda))}\right]R^{-1}(x,t)\,. \tag{2.18}$$

As for the matrix $M(\lambda)$, it lies in a nine-dimensional linear space whose standard basis is given by the matrices $B^{(jm)}$, whose entries are

$$B^{(jm)}_{kn} = \delta_{jk}\delta_{mn}\,, \tag{2.19}$$

where δ_{jk} is the Kronecker symbol ($\delta_{jk} = 1$ if $j = k$ and $\delta_{jk} = 0$ otherwise). However the alternative basis $V^{(jm)}$, which is obtained via the similarity transformation

$$V^{(jm)}(\lambda) = U(\lambda)B^{(jm)}U^{-1}(\lambda)\,, \tag{2.20}$$

where $U(\lambda)$ diagonalizes W and Z (see (2.16)), is more convenient to our purpose. Indeed, expanding the generic matrix $M(\lambda)$ in this basis as

$$M(\lambda) = \sum_{j,m=1}^{3} \mu_{jm}(\lambda)V^{(jm)}(\lambda)\,, \tag{2.21}$$

the scalar functions μ_{jm} being its components, and inserting this decomposition into the expression (2.18), leads to the following representation of F

$$F(x,t,\lambda) = R(x,t)\sum_{j,m=1}^{3} \mu_{jm}(\lambda)e^{i[(x(w_j-w_m)-t(z_j-z_m)]}F^{(jm)}(\lambda)R^{-1}(x,t)\,, \tag{2.22}$$

where we have introduced the x, t-independent matrices

$$F^{(jm)}(\lambda) = \left[\Sigma\,,\, V^{(jm)}(\lambda)\right]\,. \tag{2.23}$$

The advantage of such expression (2.22) is to explicitly show the x, t-dependence of the matrix F, and therefore of its entries F_1, F_2, see (2.10), through the six exponentials $e^{i[(x(w_j-w_m)-t(z_j-z_m)]}$. Moreover, F_1, F_2 are solutions of the linearized equation (2.8)) for any choice of the functions $\mu_{jm}(\lambda)$. On the other hand, the requirement that the solution (2.9) be localized in the variable x implies the necessary condition that the functions $\mu_{jm}(\lambda)$ be vanishing for $j = m$, $\mu_{jj} = 0$, $j = 1, 2, 3$. The further condition that the solution $\delta u_j(x, t)$ be bounded in x at any fixed time t imposes integrating with respect to the variable λ, see (2.9), over an appropriate subset $\mathbf{S}_x$ of the complex λ-plane :

$$\delta u_j(x,t) = \int_{\mathbf{S}_x} \mathrm{d}\lambda\, F_j(x,t,\lambda)\,. \tag{2.24}$$

In order to define this subset $\mathbf{S}_x$, we have to introduce first the three "wave-numbers", see (2.22),

$$k_1(\lambda) = w_2(\lambda) - w_3(\lambda)\,,\quad k_2(\lambda) = w_3(\lambda) - w_1(\lambda)\,,\quad k_3(\lambda) = w_1(\lambda) - w_2(\lambda)\,, \tag{2.25}$$

as complex functions of the spectral variable λ. These are therefore the differences of the three eigenvalues of the matrix $W(\lambda)$, see (2.16). It should be also noted that the labeling by the index j of the eigenvalues w_j, and therefore of the k_j's, is arbitrary and it does not imply any order. Then $\mathbf{S}_x$, namely the spectrum which is relevant to linear stability, can be defined as follows:

Definition 1. *The spectrum* $\mathbf{S}_x$ *is the subset of values of the spectral variable* λ *such that at least one of the three wave-numbers* $k_j(\lambda), j = 1, 2, 3$*, see (2.25), is real.*

As already noticed, to the purpose of establishing the stability properties of the continuous wave solution (2.11), we do not need to compute the integral representation (2.24) of the solution δu_j of (2.7). Indeed, it is sufficient to compute the "eigenfrequecies", whose definition

$$\omega_1(\lambda) = z_2(\lambda) - z_3(\lambda), \quad \omega_2(\lambda) = z_3(\lambda) - z_1(\lambda), \quad \omega_3(\lambda) = z_1(\lambda) - z_2(\lambda), \quad (2.26)$$

is suggested by the exponentials which appear in (2.22). Their expression follows from the matrix relation (2.15) between Z and W, which implies

$$z_j = \lambda^2 - 2\lambda w_j - w_j^2 - p, \quad (2.27)$$

and reads

$$\omega_j = -k_j(2\lambda + w_{j+1} + w_{j+2}), \quad j = 1, 2, 3\,(\text{MOD}\,3). \quad (2.28)$$

This formula looks even simpler by using the relation $w_1 + w_2 + w_3 = -\lambda$ implied by the trace of the matrix $W(\lambda)$ (see (2.14)), and gets the expression

$$\omega_j = k_j(w_j - \lambda), \quad j = 1,\, 2,\, 3\,. \quad (2.29)$$

However, before looking at the instability of the CW solution (2.11), namely at the possible exponential growth with time of the eigenmodes, we investigate the x-spectrum $\mathbf{S}_x$ and the way it depends on parameters. As evident from its definition, and from the expression (2.25) of the wave numbers k_j, the starting point is the characteristic polynomial of the matrix $W(\lambda)$ (2.14),

$$P_W(w, \lambda) = \det[w\mathbb{1} - W(\lambda)] = w^3 + \lambda w^2 + (p - q^2 - \lambda^2)w - \lambda^3 + (p + q^2)\lambda - qr\,, \quad (2.30)$$

whose coefficients depend not only on λ but also on the real parameters p defined by (2.12), on q, see (2.11), and on the additional parameter

$$r = s_1 a_1^2 - s_2 a_2^2\ . \quad (2.31)$$

As for the three parameters p, r, q, it is worth observing that the case $q = 0$ can be explicitly treated [26], and it leads to defocusing and focusing phenomena which are similar to those associated to the NLS equation with corresponding modulational stability and, respectively, instability effects. Thus we consider only non-vanishing values of q, say $q \neq 0$; moreover, even if this parameter can be rescaled to $q = 1$, and disappear from all formulae, we find it convenient to keep it explicit. We also observe that the symmetry $P_W(w, \lambda; p, r, q)) = -P_W(-w, -\lambda; p, -r, q))$ allows us to consider, with no loss of generality, only non-negative values of r. In conclusion, the parameter space we explore in the following analysis is the half (r, p) plane with $r \geq 0$.

We first note that the wave numbers k_j (2.25) are invariant with respect to translations in the complex w-plane. Then we make the following preliminary

Remark 1. *No complex translation* $w \mapsto w + \alpha$ *can take the polynomial* $P_W(w;\lambda)$ *(2.30) into the new polynomial* $P_W(w+\alpha,\lambda)$ *whose coefficients differ from the coefficients of* $P_W(w,\lambda)$ *for just a different value of* λ.

This implies that the spectrum $\mathbf{S}_x$ has to be independently investigated for real λ and for $\lambda \neq \lambda^*$.

The following two propositions point out two general properties of the spectrum. The first concerns an important distinction between the *real part* and the *non-real part* of the spectrum $\mathbf{S}_x$:

Proposition 1. *The three wave numbers* k_j, $j = 1,2,3$ *can be all real if and only if* λ *is real. If* λ *is not real* only one *of the wave numbers* k_j*'s can be real.*

Proof. Indeed, if λ is real, $\lambda = \lambda^*$, then the roots w_j are either all real or one is real and two are complex conjugate of each other. In the first case the k_j's (2.25) are all real as well, and in the other case no one of them is real. Vice versa, if all k_j are real then, because of the remark 1 above, also the roots of the polynomial $P_W(w;\lambda)$ are all real and therefore, since $w_1 + w_2 + w_3 = -\lambda$, also λ is real. The case in which only two of the three wave numbers k_j are real is excluded because $k_1 + k_2 + k_3 = 0$. In the alternative case in which λ has a non-real value, $\lambda \neq \lambda^*$, only one wave number can be real, since only two of the roots w_j, see (2.25), may have the same imaginary part. Here and in the following we adopt the convention that this particular real wave number be k_3. Thus, on the non-real part of the spectrum $\mathbf{S}_x$, the physical wave number is $k_3(\lambda)$ while k_1 and k_2 are not real. ■

The second proposition states the following symmetry property of the spectrum:

Proposition 2. *At any point* $(r\,,\,p)$ *of the parameter space, the spectrum* $\mathbf{S}_x$ *is symmetric with respect to the real axis of the* λ*-plane.*

This is an obvious consequence of the reality property $P_W(w;\lambda) = P_W^*(w^*;\lambda^*)$ which guarantees that if $\lambda \in \mathbf{S}_x$, say $\mathrm{Im}w_1(\lambda) = \mathrm{Im}w_2(\lambda)$, then also $\mathrm{Im}w_1(\lambda^*) = \mathrm{Im}w_2(\lambda^*)$, and therefore $\lambda^* \in \mathbf{S}_x$.

Our task below is twofold, first we show in the next subsection the way to construct both the real and non-real parts of the spectrum $\mathbf{S}_x$ for a *fixed* value of the parameters r, p, as defined by (2.11, 2.12, 2.31). In the second subsection below we let these parameters vary in the entire parameter space to the aim of classifying all possible types of spectra.

2.1 Gaps, branches and loops

Here we specifically show that spectra are geometrically characterized by their gaps, branches and loops. We start by warning the reader that in different contexts we freely label roots of polynomials by arbitrarily appending an index as far as this does not cause misunderstandings or lead to invalid statements. A general property of the spectrum $\mathbf{S}_x$ is well described by the following proposition.

Proposition 3. *A real part of the spectrum $\mathbf{S}_x$ is always present around the point at ∞, namely for $|\lambda| > \lambda_0$ for some positive λ_0.*

This is so because the three roots of the characteristic polynomial $P_W(w;\lambda)$ have the following asymptotic behavior as $|\lambda| \to \infty$ (free root-indexing)

$$\begin{cases} w_1(\lambda) &= -\lambda + q + (p-r)/(4\lambda) + O(1/\lambda^2)\,, \\ w_2(\lambda) &= -\lambda - q + (p+r)/(4\lambda) + O(1/\lambda^2)\,, \\ w_3(\lambda) &= \lambda - p/(2\lambda) + O(1/\lambda^2)\,. \end{cases} \tag{2.32}$$

This behavior implies that, if λ is sufficiently large and real, then the three roots w_j are also real, and therefore the three wave-numbers k_j, see (2.25), are as well real. This real part of $\mathbf{S}_x$ may be the entire real axis or the real axis with *gaps*, namely finite intervals in which $\lambda \notin \mathbf{S}_x$. If instead λ is not real, a complex component of the spectrum, namely one which lies off the real axis of the λ-plane, may occur. This important part of the spectrum, where the eigenmodes $F_j(x,t,\lambda)$ are exponentially growing with time (see [26] and below), may feature open curves, which will be referred to as *branches*, and/or closed curves which will be termed *loops*.

Let us consider first the definition and construction of gaps. These are finite intervals of the real λ-axis, which may or may not exist. A real value of the spectral variable λ belongs to a gap if it is not in $\mathbf{S}_x$. According to definition 1, this happens if, and only if, the characteristic polynomial $P_W(w,\lambda)$ (2.30) has two complex conjugate roots, or, equivalently, if its discriminant $\Delta_w(P_W)$ (see (B.5)) with respect to w,

$$\Delta_w(P_W) = 64q^2\lambda^4 - 32qr\lambda^3 + 4(p^2-20q^2p-8q^4)\lambda^2 + 36qr(p+2q^2)\lambda - 4(p-q^2)^3 - 27q^2r^2 \tag{2.33}$$

is negative (see Appendix B). Thus, if $\lambda \in \mathbf{S}_x$ then $\Delta_w(P_W) > 0$, while if $\lambda \notin \mathbf{S}_x$ then $\Delta_w(P_W) < 0$. This implies that at the end points of a gap $\Delta_w(P_W)$ changes its sign, or, in other words, these end points are simple zeros of the discriminant (2.33). Therefore, since the discriminant (2.33) is a fourth degree polynomial with real coefficients and assuming that $\Delta_w(P_W)$ has no double real roots, there may occur only one, two or no gaps if $\Delta_w(P_W)$ has, respectively, two, four or no real zeros. In the following, the values of r and p such that $\Delta_w(P_W)$ has no double real roots, namely $\Delta_\lambda\Delta_w(P_W) \neq 0$, will be referred to as *non-generic*; in Section 2.2 we will see that the condition of non-genericity of r and p play a crucial role in the classification of the spectra, see 2.50 and Proposition 11. Counting the number of gaps and computing the location of their end points amounts therefore to finding the roots of the polynomial $\Delta_w(P_W)$. In this context, and to our present and future purposes, rather than looking at the characteristic polynomial $P_W(w,\lambda)$, we find it far more convenient to introduce the polynomial $\mathcal{P}(\zeta,\lambda)$ whose roots ζ_j, $j=1,2,3$, are the squares of the differences of the roots of P_W, namely the squares of the wave numbers k_j, see (2.25),

$$\zeta_1 = k_1^2 = (w_2-w_3)^2\,, \quad \zeta_2 = k_2^2 = (w_3-w_1)^2\,, \quad \zeta_3 = k_3^2 = (w_1-w_2)^2\,. \tag{2.34}$$

The map $P_W(w,\lambda) \mapsto \mathcal{P}(\zeta,\lambda)$ with respect to the variable change $w \mapsto \zeta$ is specified in Appendix B to which we refer the reader for its construction and properties. The outcome of this mapping is summarised by the formulae (B.19) which in the present instance

$$\mathcal{P}(\zeta,\lambda) = \prod_{j=1}^{3} (\zeta - \zeta_j) \, , \tag{2.35}$$

takes the compact expression

$$\mathcal{P}(\zeta,\lambda) = \begin{pmatrix} 1 & \lambda & \lambda^2 & \lambda^3 & \lambda^4 \end{pmatrix} C \begin{pmatrix} 1 \\ \zeta \\ \zeta^2 \\ \zeta^3 \end{pmatrix} , \tag{2.36a}$$

where the 5×4 rectangular coefficient matrix $C = C(r,p)$ has the following dependence on the parameters r, p

$$C = \begin{pmatrix} 4(p-q^2)^3 + 27q^2r^2 & 9(p-q^2)^2 & 6(p-q^2) & 1 \\ -36qr(p+2q^2) & 0 & 0 & 0 \\ 4[8q^4 - p(p-20q^2)] & -24(p-q^2) & -8 & 0 \\ 32qr & 0 & 0 & 0 \\ -64q^2 & 16 & 0 & 0 \end{pmatrix} , \tag{2.36b}$$

or, more explicitly

$$\begin{aligned} \mathcal{P}(\zeta,\lambda) &= 16\left(\zeta - 4q^2\right)\lambda^4 + 32qr\lambda^3 \\ &\quad - 4\left[2\zeta^2 + 6\left(p-q^2\right)\zeta + p^2 - 20q^2p - 8q^4\right]\lambda^2 - 36qr\left(p+2q^2\right)\lambda \\ &\quad + \zeta^3 + 6\left(p-q^2\right)\zeta^2 + 9\left(p-q^2\right)^2\zeta + 4\left(p-q^2\right)^3 + 27q^2r^2 \, . \end{aligned} \tag{2.36c}$$

Here and in the following we call ζ-roots the solutions of $\mathcal{P}(\zeta,\lambda) = 0$ with respect to ζ for a given value of λ, and λ-roots the solutions of $\mathcal{P}(\zeta,\lambda) = 0$ with respect to λ for a given value of ζ. In the following we investigate the x-spectrum $\mathbf{S}_x$ by means of the polynomial $\mathcal{P}(\zeta,\lambda)$. Definition 1 of the spectrum can be so rephrased as follows:

Definition 2. *The subset $\mathbf{S}_x$ of the complex λ-plane is the set of values of the spectral variable λ such that at least one of the three ζ-roots ζ_j, $j = 1,2,3$, see (2.34), be non-negative.*

First we note that the existence of the real part of the spectrum for any value of p, r as $\lambda \to \infty$ we have established with Proposition 3 can be restated as

Proposition 4. *For any value of the parameters p, r the three ζ-roots of the polynomial $\mathcal{P}(\zeta,\lambda)$ have the following asymptotic behavior as $|\lambda| \to \infty$*

$$\begin{cases} \zeta_1(\lambda) &= (2\lambda+q)^2 - (3p+r) + O(1/\lambda) \, , \\ \zeta_2(\lambda) &= (2\lambda-q)^2 - (3p-r) + O(1/\lambda) \, , \\ \zeta_3(\lambda) &= 4q^2 - 2qr/\lambda + O(1/\lambda^2) \, . \end{cases} \tag{2.37}$$

This proposition implies that only if the spectral variable λ is real and sufficiently large, the three wave numbers k_j, see (2.34), are real. Moreover in this case, say if λ is large and real, the eigenmode wave number k_3 remains finite and equal to $\pm 2q$. We also note that our observations above, regarding the gap end-points as zeros of the discriminant of the characteristic polynomial $P_W(w, \lambda)$, can be differently restated as follows:

Proposition 5. *For all p, r the endpoints of the gaps correspond to real λ-roots of $\mathcal{P}(0, \lambda)$.*

This is a straight consequence of the relation (B.20), as proved in Appendix B, which is presently reading

$$\Delta_w(P_W) = -\mathcal{P}(0, \lambda)\,. \tag{2.38}$$

We use the polynomial $\mathcal{P}(\zeta, \lambda)$ also to investigate branches and loops. Indeed, according to Definition 2, the spectrum $\mathbf{S}_x$ is the locus of all the λ-roots $\lambda_j(\zeta)$ of the equation $\mathcal{P}(\zeta, \lambda) = 0$ for $1 \leq j \leq 4$ and for all values of the variable ζ which are non-negative, $0 \leq \zeta < +\infty$. In this regard, we start with noticing that the assumption that ζ be real makes $\mathcal{P}(\zeta, \lambda)$, see (2.36a), a fourth degree polynomial of the complex variable λ with *real* coefficients for any value of the parameters r, p. This implies that its four zeros λ_j are either real or complex conjugate pairs, consistently with the spectrum symmetry proved by proposition 2. Most importantly, the non-real part of the spectrum originates from a finite interval $0 \leq \zeta \leq \zeta_o$ for some finite value ζ_o. This is so because the λ-roots of $\mathcal{P}(\zeta, \lambda)$ are necessarily all real for sufficiently large positive values of ζ. This property comes from the following asymptotic behaviour:

Proposition 6. *For any value of the parameters r, p the four λ-roots of the polynomial $\mathcal{P}(\zeta, \lambda)$ have the following asymptotic behaviour as $\zeta \to +\infty$*

$$\begin{cases} \lambda_1(\zeta) &=& \frac{1}{2}\sqrt{\zeta} + \frac{1}{2}q + \frac{1}{4}(3p - r)/\sqrt{\zeta} + O(\zeta^{-3/2})\,, \\ \lambda_2(\zeta) &=& \frac{1}{2}\sqrt{\zeta} - \frac{1}{2}q + \frac{1}{4}(3p + r)/\sqrt{\zeta} + O(\zeta^{-3/2})\,, \\ \lambda_3(\zeta) &=& -\frac{1}{2}\sqrt{\zeta} + \frac{1}{2}q - \frac{1}{4}(3p - r)/\sqrt{\zeta} + O(\zeta^{-3/2})\,, \\ \lambda_4(\zeta) &=& -\frac{1}{2}\sqrt{\zeta} - \frac{1}{2}q - \frac{1}{4}(3p + r)/\sqrt{\zeta} + O(\zeta^{-3/2})\,. \end{cases} \tag{2.39}$$

These formulae, which are derived by direct calculation, are the counterpart of (2.32) and of Proposition 3 for the positive variable ζ.

A simple, and appealing, way of understanding the occurrence of gaps, branches and loops comes from the recognition that the spectrum is the union set of the four trajectories $\lambda_j(\zeta)$ of the λ-roots of the polynomial $\mathcal{P}(\zeta, \lambda)$ for all values of j, $j = 1, \ldots, 4$, and of the *evolution* variable ζ over the semi-line of non-negative values, $0 \leq \zeta < +\infty$. En passant, we note that viewing the λ-roots of a polynomial $\mathcal{P}(\zeta, \lambda)$, whose coefficients depend on a real parameter ζ (say the time), as particle coordinates which move in a plane is a known technique to discover and investigate many-body dynamical systems on the plane (see *e.g.* [55]). In our present case, the trajectories $\lambda_j(\zeta)$ obey few rules dictated by the polynomial $\mathcal{P}(\zeta, \lambda)$.

From now on, we assume $r > 0$; the case $r = 0$ will be treated in Appendix A. The following list of rules can be used to trace the trajectories backwards in ζ, namely from $\zeta = +\infty$, where we have the explicit expressions (2.39), to their dead-stop at $\zeta = 0$. We remind the reader that while ζ is a point of the positive real line, the λ-roots belong to the Riemann sphere, namely $\mathbb{C} \cup \{\infty\}$.

1. For positive and sufficiently large ζ, two trajectories, λ_1 and λ_2 according to the labelling in (2.39), come along the real axis from $\lambda = +\infty$ and the other two of them, λ_3 and λ_4, come as well on the real axis but from $\lambda = -\infty$. This is proved by proposition 6. For positive and sufficiently large ζ, two trajectories, λ_1 and λ_2 according to the labelling in (2.39), come from the point at ∞ along the positive part of the real line, whereas the other two of them, λ_3 and λ_4, come as well on the real axis but from $\lambda = -\infty$. This is proven in Proposition 6.

2. One trajectory, for instance λ_4, has a simple pole at $\zeta = 4q^2$ with the asymptotic behaviour

$$\lambda_4 = -2qr/(\zeta - 4q^2) + O(1)\,, \quad \zeta \to 4q^2\,. \tag{2.40}$$

   Namely, for $\zeta > 4q^2$, it goes to the point at ∞ along the negative part of the real axis, whereas, for $\zeta < 4q^2$, it goes to the point at ∞ along the positive part of the real axis. The other three trajectories are instead regular and finite at this singularity. This statement is a consequence of the explicit expression (2.36c).

3. Two trajectories may cross each other, say at $\zeta = \zeta_c > 0$. At the crossing point the polynomial $\mathcal{P}(\zeta, \lambda)$ has a double root. This implies that its discriminant with respect to λ, $\Delta_\lambda(\mathcal{P}) \equiv \mathcal{Q}(\zeta)$, vanishes at ζ_c. This discriminant turns out to be a ten degree polynomial of ζ. It is however factorized as

$$\Delta_\lambda(\mathcal{P}(\zeta, \lambda)) = \mathcal{Q}(\zeta) = 2^{16}\mathcal{Q}_1^2(\zeta)\mathcal{Q}_2(\zeta)\,, \tag{2.41}$$

   where $\mathcal{Q}_1(\zeta)$ and $\mathcal{Q}_2(\zeta)$ have degree 3 and, respectively, 4, and their expression reads

$$\mathcal{Q}_1(\zeta) = 4\zeta^3 - 36\,q^2\,\zeta^2 - 3\,(p - 4q^2)\,(p + 8q^2)\,\zeta + (p - q^2)\,(p + 8q^2)^2 - 27q^2\,r^2\,, \tag{2.42}$$

   and

$$\mathcal{Q}_2(\zeta) = \sum_{j=0}^{4} \mathcal{Q}_j\,\zeta^j\,, \tag{2.43a}$$

   with

$$\mathcal{Q}_0 = -16\,q^2\,(p-r)\,(p+r)\,\left[(p-q^2)\,(p+8q^2)^2 - 27\,q^2\,r^2\right]\,, \tag{2.43b}$$

$$\mathcal{Q}_1 = 4\,(p-4q^2)\,\left[p\,(p+8q^2)\,(p^2+10q^2\,p-8\,q^4)\right.$$
$$\left.-6\,q^2\,(5\,p+4\,q^2)\,r^2\right]\,, \tag{2.43c}$$

$$\mathcal{Q}_2 = p\left[p\,(p^2+48\,q^2\,p+256\,q^4) - 96\,q^2\,(r^2+8\,q^4)\right]$$
$$+\,32\,q^4\,(3\,r^2+8\,q^4)\,, \tag{2.43d}$$

$$\mathcal{Q}_3 = 8\,q^2\left[p\,(p+16\,q^2) - 2\,(r^2+8\,q^4)\right]\,, \tag{2.43e}$$

$$\mathcal{Q}_4 = 16\,q^4\,, \tag{2.43f}$$

4. All trajectories stop at $\zeta = 0$. The four stop points $\lambda_j(0)$, which may be either real or complex conjugate pairs, are therefore the four roots of the polynomial $\mathcal{P}(0,\lambda)$. Note that $\lambda_j(0)$ are also the roots of the discriminant $\Delta_w(P_W)$ (2.33) because of the relation (2.38).

Let us consider first the values ζ_c at which two trajectories meet each other. These can be either zeros of the polynomial $\mathcal{Q}_1(\zeta)$ or zeros of the polynomial $\mathcal{Q}_2(\zeta)$ (see(2.41)). As it is plain because of the square $\mathcal{Q}_1^2(\zeta)$ which appears in the expression (2.41), in the first case no deviation of the two crossing trajectories occurs. Thus these zeros of the discriminant $\mathcal{Q}(\zeta)$ have no scattering effect. On the contrary, at the zeros ζ_c of the polynomial $\mathcal{Q}_2(\zeta)$, if the two trajectories are, for $\zeta > \zeta_c$, on the real axis they scatter off the real axis and go in the complex plane as complex conjugate of each other. If these trajectories do not collide again for $\zeta < \zeta_c$, they eventually stop at $\zeta = 0$ and *form a branch* of the spectrum $\mathbf{S}_x$. Therefore, in order to count the number of branches of the spectrum, it is sufficient to count the number of pairs of complex conjugate roots of the polynomial $\mathcal{P}(0,\lambda)$. Combining this result with that given above on gaps yields the following

Proposition 7. *If G is the number of gaps and B is the number of branches, then for any generic value of the parameters r, p the spectrum satisfies the condition*

$$G + B = 2\,. \tag{2.44}$$

Indeed each branch stops at a pair of complex conjugate roots of $\mathcal{P}(0,\lambda)$ and each real trajectory stops at the end point of a gap. Therefore the condition (2.44) is merely a consequence of the degree of the polynomial $\mathcal{P}(0,\lambda)$ that is 4. Collisions of more than two trajectories, if any, are non-generic, and can be considered as limit cases. We point out that a trajectory coming from the point at ∞ along the negative part of the real axis may well end up at a right-hand endpoint of a gap (if any), without entering the gap. Indeed, this is possible by going back to the point at ∞ thanks to the pole singularity at $\zeta = 4q^2$ (as in item 2 of the above list), and then reappearing on the positive part of the real axis. A quite different way for two real trajectories to go from one side of a gap to its opposite side, without going through it, is to collide at a zero ζ_{c1} of $\mathcal{Q}_2(\zeta)$. In this way they leave the real axis by scattering into the complex plane as a pair of complex conjugate roots. Then, at a different positive real zero $\zeta_{c2} < \zeta_{c1}$ of $\mathcal{Q}_2(\zeta)$, these two trajectories collide again against each other to scatter back on the real axis.

This mechanism for a pair of trajectories to avoid entering a gap leads to the formation of a *loop* in the spectrum $\mathbf{S}_x$. Thus, the appearing of one loop requires the existence of two real positive roots, ζ_{c1} and ζ_{c2} of $\mathcal{Q}_2(\zeta)$. As for the number of loops in a spectrum, we have the following proposition.

Proposition 8. *For any generic value of the parameters r, p the spectrum S_x may have either one loop or none.*

This result is a consequence of the classification of the spectra, as described in the next subsection. Regarding the roots ζ_c of $\mathcal{Q}_2(\zeta)$, it is plain that those which are real and positive, $\zeta_c > 0$, play a special role since they correspond to λ-root collisions. A straight consequence of this correspondence may be stated as follows:

Proposition 9. *For any generic value of the parameter r, p, the number of real positive roots of $\mathcal{Q}_2(\zeta)$ equals $B + 2L$, B is the number of branches and L is the number of loops.*

Spectra are classified according to the number and nature of their components, namely gaps, branches and loops. Combining the algebraic properties of the polynomial $\mathcal{P}(\zeta, \lambda)$ with the geometry of the trajectories of its four λ-roots $\lambda_j(\zeta)$ leads finally to the following proposition (see [26]):

Proposition 10. *If all spectra are characterized by their number of gaps, branches and loops, then, for any generic value of the parameters r, p, the spectrum can be only one out of the following five different types of spectra:*

$$2G\,0B\,1L\,,\quad 1G\,1B\,0L\,,\quad 1G\,1B\,1L\,,\quad 0G\,2B\,0L\,,\quad 0G\,2B\,1L\,,$$

where the notation nX stands for n components of type X, with X being either G, or B, or L.

Proof. Propositions 7 and 8 are a priori compatible with the existence of six types of spectra, namely the three cases with $G + B = 2$ times the two possible cases $L = 0$ and $L = 1$. However, the type $2G\,0B\,0L$ is excluded by the argument that no trajectory of the four λ-roots can be associated to this spectrum. Indeed, this follows from Proposition 9, for it would imply that there would be no positive root, and therefore no collisions. This would entail that all trajectories would move along the real axis. But then, there is no way for four trajectories on the real axis starting, for $\zeta = \infty$, at the point at ∞ to reach, for $\zeta = 0$, the right endpoint of the left gap and the left endpoint of the right gap, without entering the gaps themselves. For future reference, we note that the exclusion of the type $2G\,0B\,0L$ can be also stated as the condition $B + L > 0$, which is equivalent to the property of the polynomial $\mathcal{Q}_2(\zeta)$ (2.43a) to have at least one positive root (see also Proposition 9). ∎

Spectra are algebraic varieties on the complex λ plane. Alternatively, branches and loops of the spectrum $\mathbf{S}_x$ can be regarded as components of the real algebraic curve $\mathcal{C}(\mu,\rho)=0$, where $\lambda=\mu+i\rho$, so that their computation requires finding the real zeros of the polynomial $\mathcal{C}(\mu,\rho)$ of two variables. We now consider the parameters r, p as fixed, and look for the algebraic expression of $\mathcal{C}(\mu,\rho)$. The starting point is the third degree polynomial $\mathcal{P}(\zeta,\lambda)$ in the variable ζ with complex coefficients depending on the complex λ, as explicitly shown by

$$\mathcal{P}(\zeta,\lambda)=\zeta^3+d_2\zeta^2+d_1\zeta+d_0\,, \tag{2.45a}$$

with

$$d_2=2\left[-4\lambda^2+3(p-q^2)\right]\,,\quad d_1=\frac{1}{4}d_2^2\,,\quad d_0=-\Delta_w(P_W)\,. \tag{2.45b}$$

We now have to look for a real zero, $\zeta=\zeta_r\in\mathbb{R}$, of $\mathcal{P}(\zeta,\lambda)$, which, if it exists, is unique (none of the other two zeros can be real for $\rho\neq 0$), as stated in Proposition 1. The problem of finding this zero has been solved in Appendix B for the generic cubic polynomial (B.9) with complex coefficients (B.10). In particular, the relevant results, which are given in Proposition 16, are the explicit expression of the real zero (B.12) and the necessary and sufficient conditions (B.11) on the coefficients for its existence. By applying that technique as in Appendix B, we introduce the real and imaginary parts of the coefficients of the polynomial $\mathcal{P}(\zeta,\lambda)$, see (2.45b), which are defined by the expressions

$$\begin{aligned} &d_2=\alpha+i\beta\,,\\ &d_0=-\Delta_w(P_W)=-(c_4\lambda^4+c_3\lambda^3+c_2\lambda^2+c_1\lambda+c_0)=-(\mathcal{A}+i\rho\mathcal{B})\,. \end{aligned} \tag{2.46a}$$

Here α and β are

$$\alpha=2[4(\rho^2-\mu^2)+3(p-q^2)]\,,\quad \beta=-16\mu\rho\,; \tag{2.46b}$$

the real coefficients c_j of the discriminant $\Delta_w(P_W)$ are given by (2.33)

$$\begin{aligned} &c_0=-4(p-q^2)^3-27q^2r^2\,,\quad c_1=36qr(p+2q^2)\,,\\ &c_2=4(p^2-20q^2p-8q^4)\,,\quad c_3=-32qr\,,\quad c_4=64q^2\,; \end{aligned} \tag{2.46c}$$

and $\mathcal{A}(\mu,\rho)$ and $\mathcal{B}(\mu,\rho)$ are two auxiliary functions whose expressions read

$$\begin{aligned} \mathcal{A}&=c_4(\mu^4+\rho^4-6\mu^2\rho^2)+c_3\mu(\mu^2-3\rho^2)+c_2(\mu^2-\rho^2)+c_1\mu+c_0\,,\\ \mathcal{B}&=4c_4\mu(\mu^2-\rho^2)+c_3(3\mu^2-\rho^2)+2c_2\mu+c_1\,. \end{aligned} \tag{2.46d}$$

Splitting the equation $\mathcal{P}(\zeta,\mu+i\rho)=0$ itself into two real equations for real ζ, μ and $\rho\neq 0$, one obtains the system

$$\zeta^3+\alpha\zeta^2+\frac{1}{4}(\alpha^2-\beta^2)\zeta-\mathcal{A}=0\,,\quad \beta\zeta^2+\frac{1}{2}\alpha\beta\zeta-\rho\mathcal{B}=0\,. \tag{2.47}$$

These two equations readily imply a linear equation for the unknown ζ_r whose unique solution has the expression

$$\zeta_r=\frac{4\mathcal{A}\beta-2\rho\alpha\mathcal{B}}{4\rho\mathcal{B}-\beta^3}\,,\quad \beta^3-4\rho\mathcal{B}\neq 0\,. \tag{2.48}$$

A polynomial implicit relation between μ and ρ is obtained by replacing ζ_r with its expression (2.48) in the second one of equations (2.47). The graph in the (μ, ρ)-plane of this polynomial $\mathcal{C}(\mu, \rho)$, which takes the expression

$$\begin{aligned}\mathcal{C}(\mu,\rho) = 256\mu^2[-16\mu\mathcal{A}^2 + 64\mu^2\rho^2\mathcal{B}(4\mu^2 - 4\rho^2 - 3p + 3q^2)^2 \\ +\mathcal{A}(\mathcal{B} - 1024\mu^3\rho^2)(4\mu^2 - 4\rho^2 - 3p + 3q^2)] - \mathcal{B}(\mathcal{B} + 1024\mu^3\rho^2)^2\,,\end{aligned} \tag{2.49a}$$

together with the constraint $\zeta_r \geq 0$ (see Definition 2),

$$\frac{4\mathcal{A}\beta - 2\rho\alpha\mathcal{B}}{4\rho\mathcal{B} - \beta^3} \geq 0\,, \quad \mathcal{B} + 1024\rho^2\mu^3 \neq 0\,, \tag{2.49b}$$

provides the non-real part of the spectrum, namely its branches and loops. We note that the polynomial $\mathcal{C}(\mu, \rho)$ is of the 10-th degree in the variable μ (the coefficient of the μ^{11} vanishes) and of the 8-th degree in the variable ρ. Examples of such curves are given in 2 -8 and in [26].

2.2 Classification of spectra

The five types of spectra, as established by Proposition 10, are here related to the values of the parameters r, p so as to complete our classification scheme. This requires that five subsets of the (r, p) half-plane $r \geq 0$ be found such that, at each point (r, p) of the same subset, the spectrum features the same number of gaps, branches and loops. This involves the identification of those boundary (threshold) curves $p = p(r)$ whose crossing implies changing the type of spectrum.

Consider first how the number of gaps changes by varying the parameters r, p. Proposition 5 shows that the extremal points of the gaps are the real zeros of the polynomial $\mathcal{P}(0, \lambda)$. As the gaps move by changing r and/or p, their number may change due to collisions of these zeros. Thus, for instance, two gaps may coalesce into one, or a gap may disappear or appear. In any case, where two zeros of $\mathcal{P}(0, \lambda)$ meet each other, at that point a double root of $\mathcal{P}(0, \lambda)$ occurs, corresponding to a zero of its own discriminant $\Delta_\lambda \mathcal{P}(0, \lambda)$, namely when (see 2.41)

$$\begin{aligned}\Delta_\lambda \mathcal{P}(0,\lambda) &= 2^{16}\, Q_1^2(0)\, Q_2(0) = \\ &= -2^{20}\, q^2\, (p-r)\,(p+r)\, \left[(p-q^2)\,(p+8\,q^2)^2 \; - 27q^2\, r^2\right]^3 = 0\,.\end{aligned} \tag{2.50}$$

This implies the following proposition.

Proposition 11. *Let $p_+ \equiv p_+(r)$, $p_- \equiv p_-(r)$, and $p_S \equiv p_S(r)$ be the three curves of the (r, p)-plane defined implicitly by the vanishing of the three polynomial factors appearing in (2.50), namely,*

$$p_+ = \left\{(p,r)\,|\,(p-r) = 0\,,\; r \geq 0\right\}, \tag{2.51a}$$

$$p_- = \left\{(p,r)\,|\,(p+r) = 0\,,\; r \geq 0\right\}, \tag{2.51b}$$

$$p_S = \left\{(p,r)\,|\,(p-q^2)\,(p+8\,q^2)^2 \; - 27q^2\, r^2 = 0\,,\; r \geq 0\right\}, \tag{2.51c}$$

or, in explicit form,

$$p_+(r) = r\,, \tag{2.51d}$$

$$p_-(r) = -r\,, \tag{2.51e}$$

$$p_S(r) = -5q^2 + 3q^2\left[\eta^{-\frac{1}{3}} + \eta^{\frac{1}{3}}\right]\,,\quad \eta = \frac{2q^4 + r\left(r + \sqrt{4q^4 + r^2}\right)}{2q^4}\,. \tag{2.51f}$$

Then, the curves $p_+(r)$, $p_-(r)$, and $p_S(r)$ bound the regions of the (r,p)-plane characterized by different numbers of gaps and branches.

Proof. Note that the third degree equation (2.51c) for the unknown p has only one real solution, namely the explicit expression (2.51f). Moreover we observe that, for $r \geq 0$, the curve $p_S(r)$ does not intersect $p_-(r)$ and intersects $p_+(r)$ at $r = 4q^2$, $p_+(4q^2) = p_S(4q^2) = 4q^2$.

It should be pointed out here that crossing one of the three threshold curves $p_+(r)$, $p_-(r)$, and $p_S(r)$ implies as well changing the number of branches. This is due to the relation $G + B = 2$ (Proposition 7) which shows that gap and branch endpoints are real and, respectively, complex conjugate zeros of $\mathcal{P}(0,\lambda)$. Thus the spectrum $\mathbf{S}_x$ has i) no gap and 2 branches ($0G\ 2B$) ii) one gap and one branch ($1G\ 1B$), iii) 2 gaps and no branch ($2G\ 0B$), if the polynomial $\mathcal{P}(0,\lambda)$ has i) two pairs of complex conjugate roots, ii) one pair of complex conjugate roots and two real roots, iii) four real roots. As reported in Appendix B, see (B.8), one can assess the reality of the zeros of a fourth degree polynomial $f(x)$ from its coefficients by checking the sign of its discriminant $\Delta_x(f)$ (B.6) and of the two functions $\mathcal{D}^{(1)}(f)$ and $\mathcal{D}^{(2)}(f)$ (B.7). However, the application of these results to our present polynomial $\mathcal{P}(0,\lambda)$ does not yield further threshold curves in addition to those (2.51) given in proposition (11). ■

Let us proceed now to identify the threshold curve that bounds regions of the parameter half-plane (r,p), $r \geq 0$, with different number of loops only. Since one loop requires the existence of two real and positive zeros of the 4-th degree, real polynomial $\mathcal{Q}_2(\zeta)$ (2.43), see Proposition 9, one may start with checking the reality of the roots of $\mathcal{Q}_2(\zeta)$ by applying standard methods for polynomials (see e.g. Appendix B, (B.6), (B.7) and (B.8)). We first compute the discriminant

$$\Delta_\zeta(\mathcal{Q}_2) = -2^{12}\,q^4\,r^2\,(p-r)\,(p+r)\left[\left(p^2 - 16q^4\right)^3 + 432\,q^4\,r^2\,\left(p^2 - r^2\right)\right]^3\,, \tag{2.52}$$

and the two auxiliary functions

$$\mathcal{D}^{(1)}(\mathcal{Q}_2) = -2^6\,q^4\,\left[p^4 + 160\,q^4\,p^2 + 256\,q^8 - 12\,r^2\,\left(p^2 - r^2\right)\right]\,, \tag{2.53a}$$

$$\begin{aligned}\mathcal{D}^{(2)}(\mathcal{Q}_2) = 2^{15}\,q^8\Big[&p^6\,\left(r^2 - 16\,q^4\right) - p^4\,\left(7\,r^4 - 16\,q^4\,r^2 + 1024\,q^8\right)\\ &+ 4\,p^2\,\left(3\,r^6 - 4\,q^4\,r^4 + 640\,q^8\,r^2 - 1024\,q^{12}\right)\\ &- 2\,r^4\,\left(3\,r^4 + 1280\,q^8\right)\Big]\,.\end{aligned} \tag{2.53b}$$

Then, we take into account both the inequalities (B.8) and the existence of at least one real positive root of $\mathcal{Q}_2(\zeta)$, see Proposition 10. Finally, we arrive at the following proposition.

Proposition 12. *According to the sign of the functions (2.52) and (2.53), the roots of the discriminant of $\mathcal{Q}_2(\zeta)$ satisfy the reality conditions*

$$\Delta_\zeta(\mathcal{Q}_2) < 0\,, \quad \text{entailing 2 real roots and 1 pair of cc roots,} \tag{2.54a}$$

$$\Delta_\zeta(\mathcal{Q}_2) > 0\,, \quad \mathcal{D}^{(1)}(\mathcal{Q}_2) < 0\,, \quad \mathcal{D}^{(2)}(\mathcal{Q}_2) < 0\,, \quad \text{entailing 4 real roots.} \tag{2.54b}$$

Moreover, the novel threshold curve $p_C \equiv p_C(r)$, which is defined implicitly as

$$p_C = \left\{(p,r)\,|\,(p^2 - 16\,q^4)^3 + 432\,q^4\,r^2\,(p^2 - r^2) = 0\,,\ r \geq 0\,,\ p \leq 0\right\}, \tag{2.55a}$$

or, in explicit form, as

$$p_C(r) = -\sqrt{16\,q^4 - 12\,q^2\,(4\,q^2\,r^2)^{1/3} + 3\,(4\,q^2\,r^2)^{2/3}}\,, \tag{2.55b}$$

is such that, if crossed by changing r and/or p, only the number of loops changes, but not the number of gaps and branches. Then, the 4 curves $p_+(r)$, $p_-(r)$, $p_S(r)$, and $p_C(r)$ bound the regions of the (r,p)-plane characterized by different numbers of positive, negative and complex conjugate roots of $\mathcal{Q}_2(\zeta)$. In other words, for any choice of p and r inside a single region, $\mathcal{Q}_2(\zeta)$ has the same number of positive, negative and complex conjugate roots.

Proof. The validity of this statement has been verified by sampling the parameter space in each region and numerically computing the roots of $\mathcal{Q}_2(\zeta)$ and the corresponding spectra. As for the expression (2.55b), we point out that it comes from the vanishing of one of the factors of the discriminant (2.52), namely $(p^2 - 16\,q^4)^3 + 432\,q^4\,r^2\,(p^2 - r^2) = 0$, which yields a third degree equation for the unknown p^2. For each choice of r, this equation has one real solution, namely $p^2 = p_C^2$. This entails $p = \pm p_C$. However, the sampling of points in each region and the numerical computation of the corresponding spectra, show that only the negative expression (2.55b) is indeed a threshold curve for the topology of the spectra. ∎

The classification of spectra is illustrated in Figure 1, and it is summarised by the following theorem.

Theorem 1. *For generic choices of r and p with $r > 0$, the roots of $\mathcal{P}(0,\lambda)$ and of $\mathcal{Q}_2(\zeta)$, as well as the types of spectra, are displayed in the following table:*

for $0 < r < 4q^2$

p range	*real roots of $\mathcal{P}(0,\lambda)$*	*roots of $\mathcal{Q}_2(\zeta)$*	*spectrum*
$-\infty < p < p_C(r)$	0	*2 positive, 0 negative, 2 c.c.*	*0G 2B 0L*
$p_C(r) < p < -r$	0	*4 positive, 0 negative, 0 c.c.*	*0G 2B 1L*
$-r < p < r$	2	*1 positive, 1 negative, 0 c.c.*	*1G 1B 0L*
$r < p < p_S(r)$	4	*2 positive, 2 negative, 0 c.c.*	*2G 0B 1L*
$p_S(r) < p < -p_C(r)$	2	*1 positive, 3 negative, 0 c.c.*	*1G 1B 0L*
$-p_C(r) < p < +\infty$	2	*1 positive, 1 negative, 2 c.c.*	*1G 1B 0L*

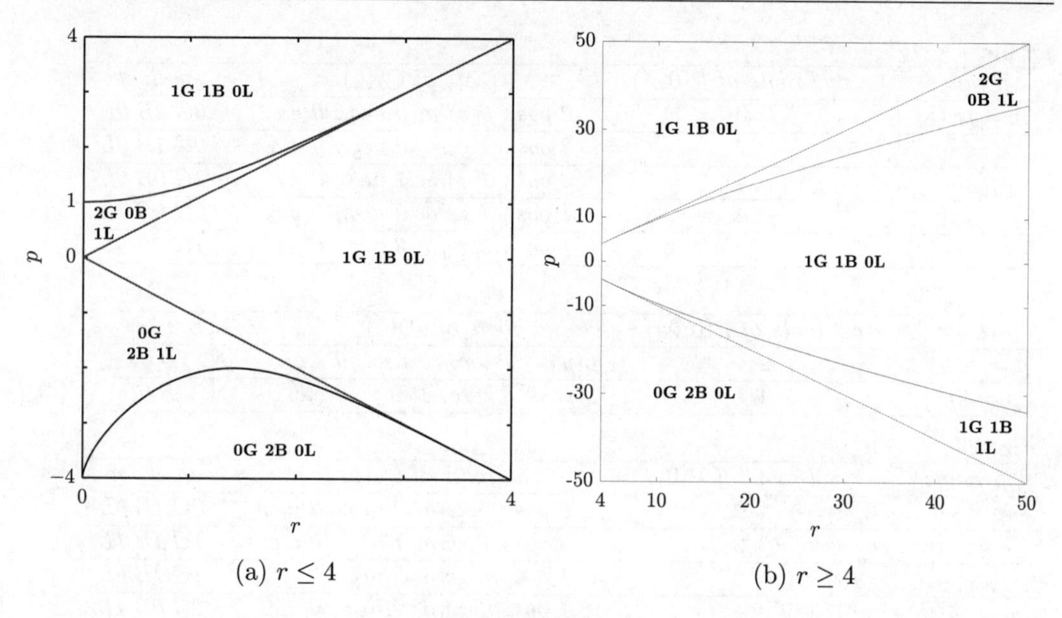

(a) $r \leq 4$ (b) $r \geq 4$

Figure 1: The structure of the spectrum $\mathbf{S}_x$ in the (r, p) plane (see Theorem(1).

for $r = 4q^2$

p range	*real roots of* $\mathcal{P}(0, \lambda)$	*roots of* $\mathcal{Q}_2(\zeta)$	*spectrum*
$-\infty < p < -4q^2$	0	*2 positive, 0 negative, 2 c.c.*	*0G 2B 0L*
$-4q^2 < p < 4q^2$	0	*1 positive, 1 negative, 0 c.c.*	*1G 1B 0L*
$4q^2 < p < +\infty$	2	*1 positive, 1 negative, 2 c.c.*	*1G 1B 0L*

for $r > 4q^2$

p range	*real roots of* $\mathcal{P}(0, \lambda)$	*roots of* $\mathcal{Q}_2(\zeta)$	*spectrum*
$-\infty < p < -r$	0	*2 positive, 0 negative, 2 c.c.*	*0G 2B 0L*
$-r < p < p_{\mathrm{C}}(r)$	2	*3 positive, 1 negative, 0 c.c.*	*1G 1B 1L*
$p_{\mathrm{C}}(r) < p < -p_{\mathrm{C}}(r)$	2	*1 positive, 1 negative, 2 c.c.*	*1G 1B 0L*
$-p_{\mathrm{C}}(r) < p < p_{\mathrm{S}}(r)$	2	*1 positive, 3 negative, 0 c.c.*	*1G 1B 0L*
$p_{\mathrm{S}}(r) < p < r$	4	*2 positive, 2 negative, 0 c.c.*	*2G 0B 1L*
$r < p < +\infty$	2	*1 positive, 1 negative, 2 c.c.*	*1G 1B 0L*

The validity of these results follows from the propositions proven in this section.

Observe that, as already stated, crossing the curve $-p_{\mathrm{C}}(r)$ does not result in a change of the topology of the spectrum.

Threshold values for $p = \pm r$ and $p = p_{\mathrm{S}}(r)$ correspond to non-generic choices of r and p. More in general, all threshold values are such that either $\mathcal{Q}_2(\zeta)$ features a double root, or a simple root passes through the origin. In particular, for these values, Propositions 7, 9 and 10 are not valid. These cases can be treated on a case-by-case basis, and are summarized in the following theorem.

Theorem 2. *For the non-generic choices* $p = \pm r$, $p = p_{\mathrm{S}}(r)$, *and for the choices* $p = \pm p_{\mathrm{C}}(r)$, *for* $r > 0$, *one has:*

for $0 < r < 4q^2$

p range	*real roots of* $\mathcal{P}(0,\lambda)$	*roots of* $\mathcal{Q}_2(\zeta)$	*spectrum*
$p = p_C(r)$	0	*2 pos., 0 zero, 0 neg., 0 c.c.*	*0G 2B 0L*
$p = -r$	2	*3 pos., 1 zero, 0 neg., 0 c.c.*	*0G 1B 0L*
$p = r$	4	*2 pos., 1 zero, 1 neg., 0 c.c.*	*1G 0B 0L*
$p = p_S(r)$	4	*1 pos., 1 zero, 2 neg., 0 c.c.*	*2G 0B 1L*
$p = -p_C(r)$	2	*1 pos., 0 zero, 3 neg., 0 c.c.*	*1G 1B 0L*

for $r = 4q^2$

p range	*real roots of* $\mathcal{P}(0,\lambda)$	*roots of* $\mathcal{Q}_2(\zeta)$	*spectrum*
$p = -4q^2$	2	*3 pos., 1 zero, 0 neg., 0 c.c.*	*1G 0B 0L*
$p = 4q^2$	4	*0 pos., 3 zero, 1 neg., 0 c.c.*	*1G 0B 0L*

for $r > 4q^2$

p range	*real roots of* $\mathcal{P}(0,\lambda)$	*roots of* $\mathcal{Q}_2(\zeta)$	*spectrum*
$p = -r$	2	*3 pos., 1 zero, 0 neg., 0 c.c.*	*0G 1B 0L*
$p = p_C(r)$	2	*3 pos., 0 zero, 1 neg., 0 c.c.*	*1G 1B 1L*
$p = -p_C(r)$	2	*1 pos., 0 zero, 3 neg., 0 c.c.*	*1G 1B 0L*
$p = p_S(r)$	4	*1 pos., 1 zero, 2 neg., 0 c.c.*	*2G 0B 1L*
$p = r$	4	*2 pos., 1 zero, 1 neg., 0 c.c.*	*1G 0B 0L*

The remaining case $r = 0$ follows within the description of the case $0 < r < 4q^2$ as given in Theorems 1 and 2, however its understanding requires some care and thus it is treated for completeness in Appendix A.

3 Dispersion relation and instability

In the previous section we have confined our analysis to the wave numbers k_j of the eigenmodes of the linearized equation (2.7). There we arrived at the definition of the spectrum $\mathbf{S}_x$ by requiring that these wave numbers be real. As an important upshot, we realized that three wave numbers, $k_j(\lambda)$, $j = 1, 2, 3$, are functions of the real spectral variable λ as they are defined on the real part of the spectrum. In contrast, we also found that just one wave number function, $k_3(\lambda)$, exists on the complex (i.e. non-real) part of the spectrum, namely on branches and loops. As a consequence of these findings the formal representation (2.24) can be better specialized as

$$\delta u_1(x,t) = e^{i[qx-(q^2+2p)t]} \left\{ \int_{\mathbb{R}\setminus G} d\lambda \sum_{j=1}^{3} \left[f_{1+}^{(j)}(\lambda) e^{i(k_j x - \omega_j t)} + f_{1-}^{(j)}(\lambda) e^{-i(k_j x - \omega_j t)} \right] \right.$$
$$\left. + \int_{B \cup L} d\lambda \left[f_{1+}^{(3)}(\lambda) e^{i(k_3 x - \omega_3 t)} + f_{1-}^{(3)}(\lambda) e^{-i(k_3 x - \omega_3 t)} \right] \right\}, \tag{3.1a}$$

$$\delta u_2(x,t) = e^{-i[qx+(q^2+2p)t]} \left\{ \int_{\mathbb{R}\setminus G} d\lambda \sum_{j=1}^{3} \left[f_{2+}^{(j)}(\lambda) e^{i(k_j x - \omega_j t)} + f_{2-}^{(j)}(\lambda) e^{-i(k_j x - \omega_j t)} \right] \right.$$
$$\left. + \int_{B \cup L} d\lambda \left[f_{2+}^{(3)}(\lambda) e^{i(k_3 x - \omega_3 t)} + f_{2-}^{(3)}(\lambda) e^{-i(k_3 x - \omega_3 t)} \right] \right\}, \tag{3.1b}$$

where the sets G, B, L in the integration operations are, respectively, gaps, branches and loops of the spectrum $\mathbf{S}_x$ and $f_{1\pm}^{(j)}$, $f_{2\pm}^{(j)}$ can be obtain by expanding (2.22). Here there is no interest in finding the expressions of the amplitudes. We rather turn our attention to the time domain, namely to the frequencies $\omega_j(\lambda)$, $j = 1, 2, 3$, as defined by (2.26), which correspond to the wave numbers $k_j(\lambda)$.

Proposition 13. *If λ is on the real part of the spectrum $\mathbf{S}_x$ then the three frequencies $\omega_j(\lambda)$, $j = 1, 2, 3$ are real. If instead λ is on the complex part of the spectrum, namely if it belongs to branches or loops, the corresponding frequency $\omega_3(\lambda)$ is not real,*

$$\omega_3(\lambda) = \Omega(\lambda) + i\Gamma(\lambda) \ , \ \lambda = \mu + i\rho \ . \tag{3.2}$$

Proof. The expression (2.29) of the frequency ω_j, together with proposition (1) we derived in the previous section, proves that if λ i real ($\rho = 0$), then so is ω_3 ($\Gamma(\lambda) = 0$); in fact all ω_j, $j = 1, 2, 3$ are real. In order to prove the converse, namely, that if λ is not real ($\rho \neq 0$), so is not ω_3, we proceed as follows. Let us assume that λ is complex, $\lambda = \mu + i\,\rho$, with non-vanishing imaginary part, $\rho \neq 0$. Let $w_j = w_j^{(R)} + i\,w_j^{(I)}$ be the (generically complex) roots of $P_W(w; \lambda)$. With this notation, we have that one of the wave-numbers, say k_3, will be real only if $w_1^{(I)} = w_2^{(I)}$. Then, from (2.29), we have that ω_3 will also be real only if $w_3^{(I)} = \rho$. Writing the polynomial $P_W(w; \lambda)$ as $P_W(w; \lambda) = \prod_{j=1}^{3}(w - \alpha_j - i\,\beta_j)$, and expanding and comparing the real and imaginary parts of the coefficients of same powers of w with those appearing in (2.30), we get a system of six polynomial equations for the six unknowns $w_1^{(R)}$, $w_2^{(R)}$, $w_3^{(R)}$, $w_1^{(I)}$, μ, and ρ. The explicit form of this system can be found in [26]. It is straightforward to see that such systems admit real solutions only if $\rho = w_1^{(I)} = 0$ or if $p = r = 0$. The former contradicts the original assumption $\rho \neq 0$; the latter can be excluded on physical grounds. ■

In this case the time scale of the exponential growth of the wave amplitude is measured by $1/|\Gamma|$, and $\Gamma(\lambda)$ will be referred to as *gain function* corresponding to the wave number $k_3(\lambda)$. Moreover, we notice that computing the integrals (3.1) over branch and loop curves, requires knowing the functions $\rho(\mu)$ by solving the equation $\mathcal{C}(\mu, \rho) = 0$, see (2.49a).

Regarding dispersion phenomena of wave packets, (3.1) shows also that one has to integrate over the spectral variable λ, and therefore one has to compute wave numbers together with their corresponding frequencies, as functions of λ over the entire spectrum $\mathbf{S}_x$. Regarding frequencies, the expression (2.29) of ω_j is however not quite convenient to this purpose because that expression, $\omega_j = k_j(w_j - \lambda)$, depends not only on k_j and on λ itself but also on the root w_j of $P_W(w, \lambda)$. Thus, we first aim to derive an alternative expression of ω_j which does not have an explicit dependence on w_j. We do so analytically by algebraic arguments which do not depend on whether λ belongs to the spectrum. So we arrive at the following proposition.

Proposition 14. *For each $j = 1, 2, 3$, for any complex value of the spectral variable λ and for any value of the parameters r, p, the expression (2.29) of ω_j may be given the alternative and equivalent form*

$$\omega_j = k_j \left[\frac{2\lambda(2k_j^2 - p - 8q^2) + 9qr}{4\lambda^2 - 3(k_j^2 + p - q^2)} \right] . \tag{3.3}$$

Proof. The linear Vieta formula $w_{j+1} + w_{j+2} = -\lambda - w_j$, $j = 1, 2, 3 \,(\text{MOD}\, 3)$, as applied to the roots of $P_W(w, \lambda)$ (see (2.30)), and the definition (2.25) of the wave-numbers, $w_{j+1} - w_{j+2} = k_j$ for $j = 1, 2, 3 \;(\text{MOD}\, 3)$, yield the following expressions

$$w_{j+1} = \frac{1}{2}(k_j - w_j - \lambda) , \quad w_{j+2} = -\frac{1}{2}(k_j + w_j + \lambda) , \quad j = 1, 2, 3 \;(\text{MOD}\, 3) . \tag{3.4}$$

On the other hand, replacing w_{j+1} and w_{j+2} with these expressions (3.4) in the other two Vieta formulae $w_j\, w_{j+1} + w_{j+1}\, w_{j+2} + w_{j+2}\, w_j = p - q^2 - \lambda^2$ and $w_j\, w_{j+1}\, w_{j+2} = \lambda^3 - \lambda(p + q^2) + qr$ leads to the couple of equations

$$\begin{cases} w_j = \frac{4\lambda^3 + \lambda(k_j^2 - 5p - 13q^2) + 9qr}{4\lambda^2 - 3k_j^2 - 3(p - q^2)} , \\ 3w_j^2 + 2\lambda w_j + k_j^2 + 4(p - q^2) - 5\lambda^2 = 0 . \end{cases} \tag{3.5}$$

These two expressions, as expected, are just equivalent to the equations $P(w_j, \lambda) = 0$, and $\mathcal{P}(\zeta_j, \lambda) = 0$, with $\zeta_j = k_j^2$. This shows that w_j and, respectively, ζ_j are the roots of the polynomial (2.30) and, respectively, (2.36a). Inserting the expression of w_j, as provided by the first of the equations (3.5), into the formula (2.29) finally proves this proposition. ■

As a consequence of this last proposition, the expression of the real frequency Ω and of the gain function Γ, see (3.2), reads

$$\begin{aligned} \Omega = -\, k_3 \Big\{ & 2\mu(2\zeta_3 - p - 8q^2)[3(\zeta_3 + p - q^2) - 4(\mu^2 + \rho^2)] + \\ & + 9qr[3(\zeta_3 + p - q^2) - 4(\mu^2 - \rho^2)] \Big\} / \Theta, \end{aligned} \tag{3.6a}$$

and

$$\begin{aligned} &\Gamma = -2\rho k_3 \left\{ (2\zeta_3 - p - 8q^2)[3(\zeta_3 + p - q^2) + 4(\mu^2 + \rho^2)] + 36qr\mu \right\} / \Theta , \\ &\zeta_3 = k_3^2 , \end{aligned} \tag{3.6b}$$

with

$$\Theta = \left[3(\zeta_3 + p - q^2) - 4(\mu^2 - \rho^2) \right]^2 + 64\mu^2\rho^2 , \tag{3.6c}$$

where μ and ρ, i.e. the real and, respectively, imaginary parts of the spectral variable, $\lambda = \mu + i\,\rho$, are related to each other by the function $\rho(\mu)$ implicitly defined by the equation $\mathcal{C}(\mu, \rho) = 0$, see (2.49a).

Let us consider now the computation of wave numbers and frequencies for a given value of λ over the spectrum, as required by the integral expression (3.1).

We find it convenient to discuss this matter separately for real λ and for complex λ. If λ is fixed real (not in a gap), one may first look for the three roots ζ_j, $j = 1, 2, 3$ of the polynomial $\mathcal{P}(\zeta, \lambda)$ (2.45a) which are real and non-negative according to Propositions 1 and 2.34. The corresponding three wave numbers k_j are then solutions of the two equations: $k_j^2 = \zeta_j$ and $k_1 + k_2 + k_3 = 0$ (see (2.25)). There are only two solutions of this problem. Let us assume the ordering $0 \leq \zeta_1 < \zeta_2 < \zeta_3$, then one solution is $k_1 = \sqrt{\zeta_1}$, $k_2 = \sqrt{\zeta_2}$, $k_3 = -\sqrt{\zeta_1} - \sqrt{\zeta_2}$. Note that the condition $\zeta_3 = \zeta_1 + \zeta_2 + 2\sqrt{\zeta_1}\sqrt{\zeta_2}$ is valid by construction of the polynomial $\mathcal{P}(\zeta, \lambda)$ (see Appendix B). Here all squared roots have been taken positive, $\sqrt{a^2} = |a| > 0$. The second solution is merely the one with the opposite sign $-k_j$. Once these wave numbers have been computed, the three corresponding frequencies ω_j are also real and have the expression (3.3). The obvious conclusion is that the contributions to the perturbation $\delta u_j(x, t)$ that come from the eigenmodes with real λ do not lead to instabilities of the CW solution (2.11).

If $\lambda = \mu + i\rho \in \mathbf{S}_x$ is instead off the real axis, say on a branch or on a loop, there exists only one root ζ_3 of $\mathcal{P}(\zeta, \lambda)$ real and positive. The two corresponding wave numbers are therefore $k_3 = \pm\sqrt{\zeta_3}$. If the algebraic curve $\rho(\mu)$, which represents this component of the spectrum, is known (e.g. by solving the equation $\mathcal{C}(\mu, \rho) = 0$ (see Section 2.1)), then the function $\zeta_3(\mu)$ on this part of the spectrum is explicitly given by the expression (2.48) of the real ζ-solution ζ_r with $\lambda = \mu + i\rho(\mu)$ and the condition $\zeta_r > 0$. The expressions of the frequency Ω and of the gain function Γ, which correspond to this wave number k_3, are consequently also explicit by using (3.6). Since frequencies and gains are relevant to physical applications, particularly to wave stability, we add here some remarks on branches and loops.

As proven in the previous section, the variable λ moves on the spectrum $\mathbf{S}_x$ by changing ζ according to the equation $\mathcal{P}(\zeta, \lambda) = 0$. We also showed that branches and loops intersect the real axis of the complex λ-plane at the zeros of the function $Q_2(\zeta)$. Thus a branch crosses the real axis for $\zeta = \zeta_B > 0$, $Q_2(\zeta_B) = 0$, and stops for $\zeta = 0$. Similarly, a loop crosses the real axis in two points, say for $\zeta = \zeta_{L1} > 0$, $Q_2(\zeta_{L1}) = 0$, and for $\zeta = \zeta_{L2} > 0$, $Q_2(\zeta_{L2}) = 0$, see Theorem 1. If, for instance, the non-real part of the spectrum has only one branch, i.e. the spectrum is $1G\ 1B\ 0L$ type (see Proposition 10), the wave number $k_3 = \pm\sqrt{\zeta_3}$ is defined on the interval $0 \leq |k_3| \leq \sqrt{\zeta_B}$, and therefore, on this same interval, also the frequency Ω and the gain function Γ, see (3.6), are defined. Most importantly, it is clear that only in this interval of values of k_3 the gain is different from zero. This implies that, in this interval of wave numbers, the eigenmodes of the linearized equation (2.7) are unstable. By similar reasonings, if λ is on a loop, the corresponding instability interval of wave numbers is $\sqrt{\zeta_{L1}} \leq |k_3| \leq \sqrt{\zeta_{L2}}$. Thus, the instability band that is associated to a branch differs from that associated to a loop. This difference is of practical interest since the first one is around $k_3 = 0$, and it is commonly known as *baseband instability* as it involves very long wavelengths. The one associated instead to a loop is known as *passband instability*. We also note that, when branches and loops are present in the spectrum, instability bands may overlap and lead to a rather complicate overall gain function. These new and distinctive features of the instability of two coupled background plane waves are quite evident in our numerical results as plotted in the following Figures. Thus in Figure 2c the two baseband instability curves of the gain function overlap while the contribution due to the loop remains in a separated band. The overlapping of two branch-generated instability bands is also shown in Figure 7c where the spectrum has no loop. An instability over a single passband which is rather peculiar of the wave coupling is presented in Figure 6c since it corresponds to a spectrum with no branches. The same feature in the defocussing regime is evident in Figure 8c. Gain functions which are similar to the one due to the modulation instability predicted by the single NLS equation are present in spectra that have no loops, as in Figures 3c and 4c.

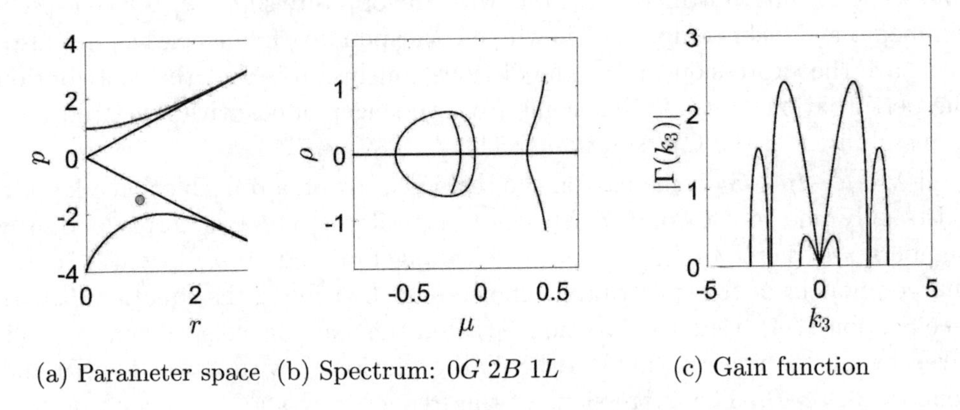

(a) Parameter space (b) Spectrum: $0G\ 2B\ 1L$ (c) Gain function

Figure 2: (a) $r = 1$, $p = -1.5$ (focussing); (b) spectrum; (c) gain Γ.

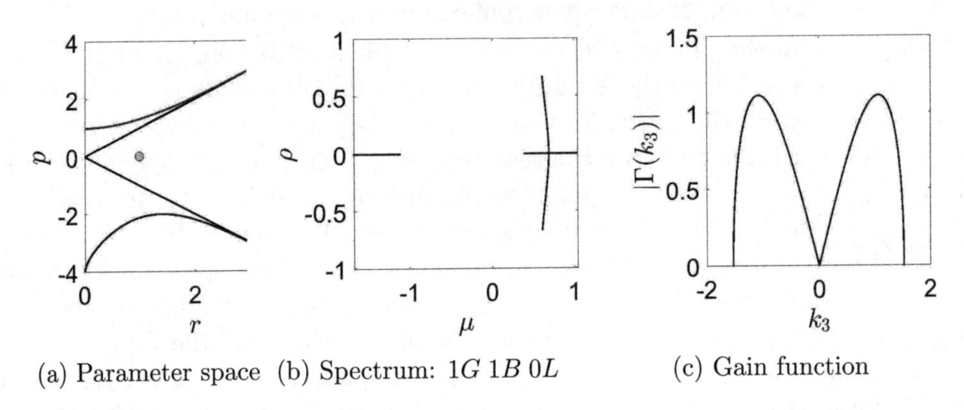

(a) Parameter space (b) Spectrum: $1G\ 1B\ 0L$ (c) Gain function

Figure 3: (a) $r = 1$, $p = 0$ (mix); (b) spectrum; (c) gain Γ.

We conclude this section by going back to the algebraic side of the dispersion relation between wave numbers $k_j(\lambda)$ and frequencies $\omega_j(\lambda)$. With respect to the preceding analysis, we now change perspective by eliminating λ. To this aim, we first drop the index j appended to the wave number k and to the frequency ω as unnecessary. Then we note that the pair of algebraic equations (see (3.3) and (2.36a))

$$\begin{cases} \omega\left[4\lambda^2 - 3k^2 - 3(p - q^2)\right] - k\left[2\lambda(2k^2 - p - 8q^2) + 9qr\right] = 0\,, \\ \mathcal{P}(k^2, \lambda) = 0\,, \end{cases} \tag{3.7}$$

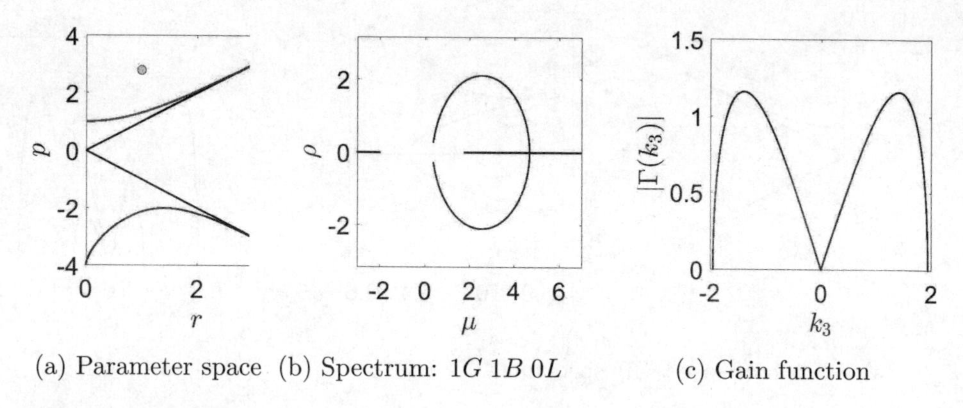

(a) Parameter space (b) Spectrum: $1G\ 1B\ 0L$ (c) Gain function

Figure 4: (a) $r = 1$, $p = 2.8$ (defocussing); (b) spectrum; (c) gain Γ.

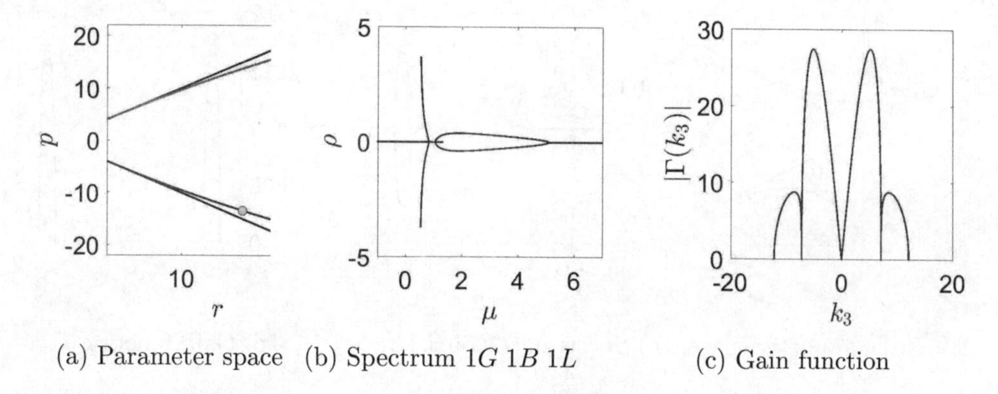

(a) Parameter space (b) Spectrum $1G\ 1B\ 1L$ (c) Gain function

Figure 5: (a) $r = 15$, $p = -13.45$ (mix); (b) spectrum; (c) gain function Γ.

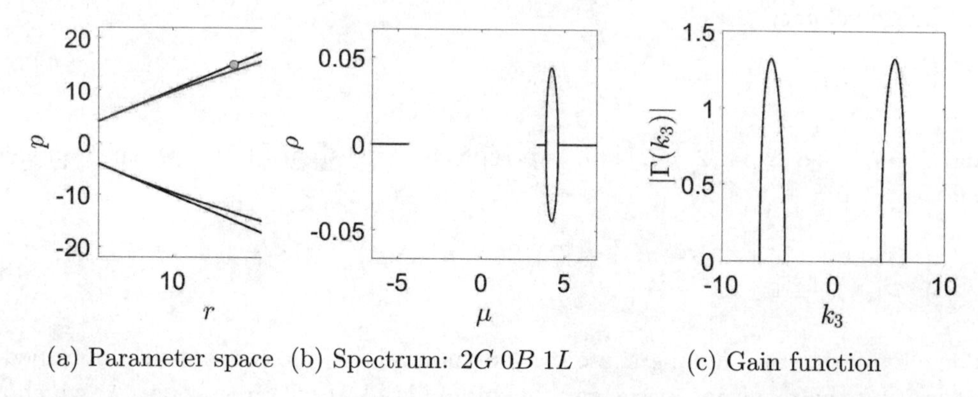

(a) Parameter space (b) Spectrum: $2G\ 0B\ 1L$ (c) Gain function

Figure 6: (a) $r = 15$, $p = 14.9$ (mix); (b) spectrum; (c) gain function Γ.

can be viewed as a *dispersion relation* in implicit-parametric form, the parameter being the spectral variable λ. The strategy therefore is eliminating λ among these two equations (3.7) to end up with an algebraic relation involving ω and k only. This computational task looks neater if these two equations (3.7) are rewritten for

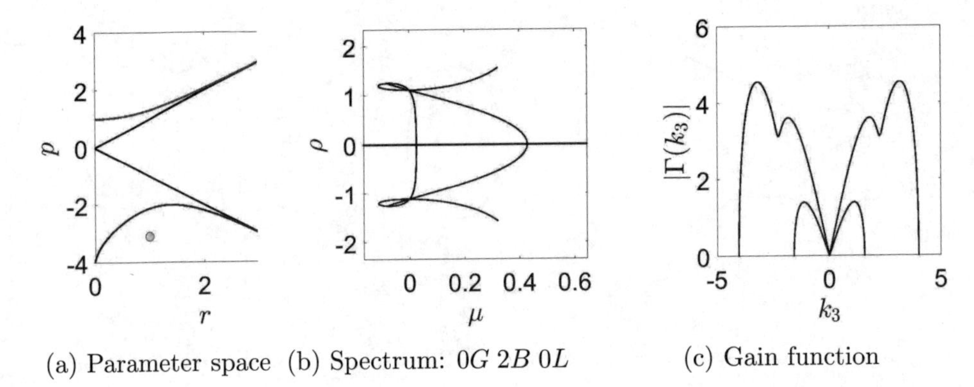

(a) Parameter space (b) Spectrum: $0G\ 2B\ 0L$ (c) Gain function

Figure 7: (a) $r = 1$, $p = -3.1$ (focussing); (b), spectrum; (c) gain function Γ.

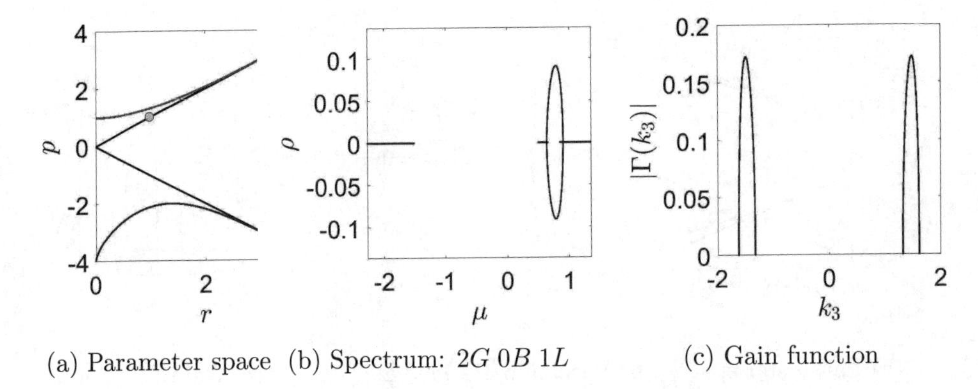

(a) Parameter space (b) Spectrum: $2G\ 0B\ 1L$ (c) Gain function

Figure 8: (a) $r = 1$, $p = 1.025$ (defocussing); (b) spectrum; (c) gain function Γ.

the phase velocity

$$\phi = \omega/k \tag{3.8}$$

and $\zeta = k^2$, see (2.34, 2.35). Thus we replace the system (3.7) with the equivalent one

$$\begin{cases} \phi\left[4\lambda^2 - 3\zeta - 3(p - q^2)\right] - \left[2\lambda(2\zeta - p - 8q^2) + 9qr\right] = 0\,, \quad \phi = \omega/k\,, \\ \mathcal{P}(\zeta, \lambda) = 0\,, \quad \zeta = k^2\,. \end{cases} \tag{3.9}$$

The elimination of λ among these two polynomials can be conveniently obtained by computing the resultant of the two polynomials (3.9) with respect to λ, which can be factorised as

$$256\zeta^2[4(\zeta - 4q^2)^2(\zeta - q^2) - 12pq^2(\zeta - 4q^2) - 3p^2(\zeta - 5q^2) + p^3 - 27q^2r^2]^2\, H(\zeta, \phi)\,, \tag{3.10a}$$

with

$$H(\zeta, \phi) = \phi^4 - 2(\zeta + 2p + 4q^2)\phi^2 - 16qr\phi + (\zeta - 4q^2)(\zeta + 4p - 4q^2)\,. \tag{3.10b}$$

This implies that the relation between ω and k takes the implicit polynomial expression

$$H(k^2, \omega/k) = 0\,. \tag{3.11}$$

To each ϕ-root $\phi(\zeta)$ of $H(\zeta, \phi)$, there corresponds the complex frequency $\omega(k) = k\phi(k^2)$. One should also note that, at each zero ζ_0 of the discriminant $\Delta_\phi(H)$ of the polynomial $H(\zeta, \phi)$ with respect to ϕ, two of the four ϕ-roots of $H(\zeta_0, \phi)$ coincide with each other. On the other hand, in Section 2 we showed that the spectrum $\mathbf{S}_x$ itself is the union set of all λ-roots of the polynomial $\mathcal{P}(\zeta, \lambda)$ for all values of ζ on the real non-negative semi-axis $0 \leq \zeta < +\infty$. In that case, at each zero of the function $Q_2(\zeta)$, see (2.41, 2.43), two λ-roots coincide by marking the crossing of a branch, or a loop, with the real λ-axis. At this crossing point we know as well that three real frequencies ω_j reduce to one complex frequency $\omega_3 = \Omega + i\Gamma$ when moving from the real to the complex part of the spectrum. This change of the frequencies, and therefore of the phase velocities ϕ_j, see (3.8), should also occur by the same way with the polynomial $H(\zeta, \phi)$ (3.10b). To this end, we have the following proposition.

Proposition 15. *For any value of the parameters (r, p), the remarkable equality*

$$\Delta_\phi(H) = (2)^{12} Q_2(\zeta) \tag{3.12}$$

holds true, where $Q_2(\zeta)$ is the same polynomial (2.43) which appears in the factorization (2.41) of the discriminant $\Delta_\lambda(\mathcal{P})$ of $\mathcal{P}(\zeta, \lambda)$ with respect to λ.

Proof. A direct calculation yields the proof. ∎

4 Conclusions

We have generalised the well-known approach to the linear instability of the plane wave solution of the NLS equation to the case of two weakly resonant plane waves. To this purpose, we have considered two coupled NLS equations with interaction terms satisfying the integrability condition. In contrast with the NLS case, where this instability emerges only if the interaction is self-focusing, the coupled plane waves are unstable for any type of interaction, both focusing and defocusing, provided their wave numbers have not the same value. Our approach to linear stability makes use of the construction of the eigenmodes (aka squared eigenfunctions) of the linearised coupled NLS equations by just solving the Lax pair equations, with no need of dealing with the inverse spectral problem technique. To this end, we introduce and compute the *stability spectrum* associated with the two plane waves solution. This spectrum, which is a piecewise algebraic curve in the complex plane of the spectral variable λ, delivers information which is relevant to physics. In fact, the spectrum turns out to be one of five different types, according to its geometric components. These five types provide a complete classification of spectra. The distinctive features are the number of gaps on the real axis, and the number of open and closed curves, respectively termed *branches* and *loops*, which spring off the real axis. Spectra change by changing physical parameters. These are the self- and cross-interaction coupling constants s_1, s_2, and the plane wave amplitudes a_1, a_2 (or rather, their combinations $r = s_1 a_1^2 - s_2 a_2^2$ and $p = s_1 a_1^2 + s_2 a_2^2$). Therefore the parameter space is divided into regions in such a way that crossing their boundary corresponds to changing the spectrum type. The boundary curves have an explicit expression.

Important functions defined over the spectrum are the wave numbers $k(\lambda)$, and the corresponding frequencies $\omega(\lambda)$, of the small perturbations propagating over the background plane waves. These eigenfrequencies turn out to be complex on branches and loops, while their imaginary part, known as *gain function*, is related to the temporal growth of the initially small amplitudes. As a novel feature of these instabilities, we find that the gain function associated with loops (which are not present in the NLS spectrum) are non-vanishing only on a finite interval of wave lengths, say on a passband. On the contrary, the instability corresponding to branches have the same kind of band as the known modulational instability, which includes very long wave lengths.

The findings reported here open up a number of interesting problems. The approach we have used can be extended to solutions of the coupled NLS system which differ from pure plane waves by adding one or more solitons of dark or bright type. Interesting cases include those associated to real eigenvalues in a gap or to complex eigenvalues, *i.e.* in analogy with Peregrine-type solutions (see [56, 36] and [57]). In these directions one would shed further light on the life-time of rogue waves.

A Case $r = 0$

In this appendix we investigate the case $r = 0$, corresponding to the case $s_1 = s_2$, $a_1 = a_2$. Since the parameter q is assumed to be non-vanishing, for the sake of simplicity and with no loss of generality, we set $q = 1$ in all following formulae. Note that the parameter q can be easily reinserted in all expressions by dimensional arguments. If $r = 0$, it is immediate to observe that the polynomial $Q_2(\zeta)$ factorises into

$$Q_2(\zeta) = (\zeta - 4)\,(\zeta + 4p - 4)\left[\zeta + \frac{p\,(p+8)}{4}\right]^2 .$$

Consequently, when $r = 0$, we have that $Q_2(\zeta)$ has 1 real, positive root if $p > 1$; 2 real, positive roots if $0 < p < 1$; 3 real, positive roots if $-8 < p < 0$; and 2 real, positive roots if $p < -8$. Moreover, we observe that the curve $p_C(r)$, see 2.55b, intersects the p-axis at $p = -4$; however, this does not result in a change of the topology of the spectrum S_x.

When $r = 0$, the spectrum S_x can be given explicitly in parametric form with respect to ζ, since the four λ-roots of the polynomial $\mathcal{P}(\zeta;\lambda)$, see 2.36c, have the explicit expressions:

$$\lambda = \pm\sqrt{\frac{p^2 + p\,(6\,\zeta - 20) + 2\,(\zeta - 4)\,(\zeta + 1) \pm |8 + p - 2\,\zeta|\,\sqrt{p\,(p+8) + 4\,\zeta}}{8\,(\zeta - 4)}},\ \zeta \geq 0. \tag{A.1}$$

From the above formula, it is clear that, generically, λ approaches complex infinity as $\zeta \to 4$ and as $\zeta \to \infty$. However, when $r = 0$, $\zeta = 4$ is also a zero of $Q_2(\zeta)$, thus in this case we expect that the point at infinity on the λ-plane may possibly intersect branches and/or loops of S_x.

A thorough analysis of the geometry of S_x in the λ-plane for $r \approx 0$ has been carried out based on the explicit expression in (A.1). However, because of space limitation, in the following, we only present a few examples of the spectrum S_x as $r \to 0$, illustrating how the case $r = 0$ fits within the spectra classification provided in Section 2.2. This is true for all the choices of the parameter p. Observe that, as r approaches zero, branches and loops may contain intervals of the imaginary axis, thus the identification of the equivalence classes within the classification may not appear evident. Hence, for each choice of p and r, the strategy is to project the resulting λ-plane onto the unit sphere (applying an inverse stereographic sphere projection), allowing a visualization of the whole curve, which may include the point at infinity.

Consider as an example the case $p > 1$. If $r \approx 0$ and $p > 1$, the classification predicts 1 gap and 1 branch (see Figures 9a and 9b). The branch $\widehat{B'AB''}$ has endpoints B' and B'', intersecting the real axis at A. Observe that, when $r > 0$, neither B', nor B'' are purely imaginary. The gap extends over the interval $\overline{C'C''}$. As r approaches 0, the point A approaches ∞ and the endpoints B' and B'' move towards the imaginary axis. When $r = 0$, the gap is symmetric around the origin ($C' = -C''$), the endpoints B' and B'' are purely imaginary, and the branch intersects the real axis at infinity (see Figures 9c and 9d). Therefore, when $r = 0$ and $p > 1$, as predicted by the classification scheme, we have 1 gap and 1 branch, which describes the imaginary axis, except for the (imaginary) gap $\overline{B'B''}$.

Similarly, if $r \approx 0$ and $r < p < p_S(r)$, the classification scheme predicts 1 loop and 2 gaps (see Figures 10a and 10b). The loop $\widehat{BAB}$ intersects the real axis in two points, A and B, with the "big" gap $\overline{C'C''}$ lying, in the λ-plane, outside the loop, and the "small" gap $\overline{D'D''}$ lying, in the λ-plane, inside the loop (the notion of "inside" and "outside" is meaningless for the projection onto the stereographic sphere). Observe that, when $r > 0$, the point B does not coincide with the origin. As r approaches 0, the point A approaches infinity and the point B moves towards the origin. When $r = 0$, the two gaps are symmetric around the origin ($C' = -D''$ and $C'' = -D'$), the point B coincides with the origin, and the loop intersects the real axis at infinity, covering the whole imaginary axis (see Figures 10c and 10d). Therefore, when $r = 0$ and $r < p < p_S(r)$, as predicted by the classification scheme, we have 2 gaps on the real axis and 1 loop, which describes the whole imaginary axis.

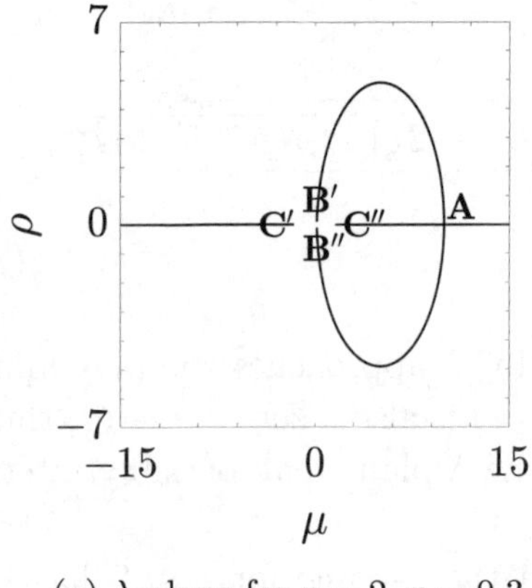

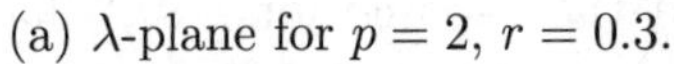

(a) λ-plane for $p = 2$, $r = 0.3$.

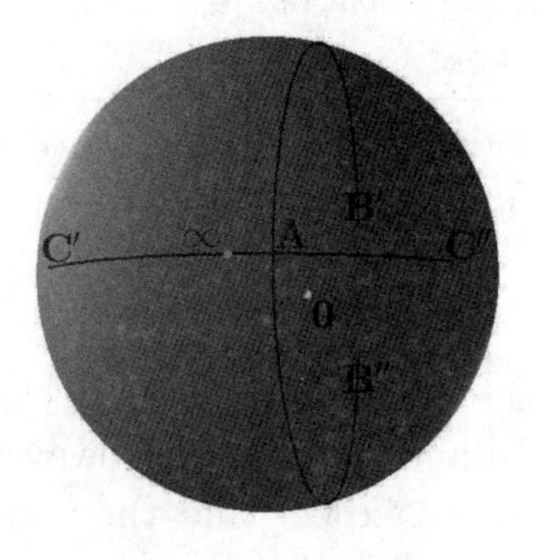

(b) Sphere projection for $p = 2$, $r = 0.3$.

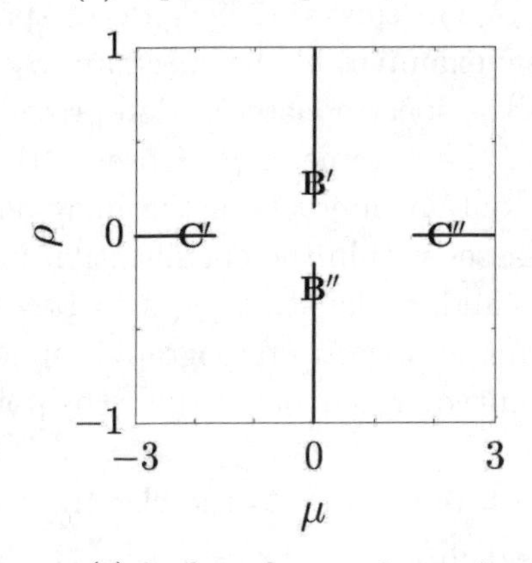

(c) λ-plane for $p = 2$, $r = 0$.

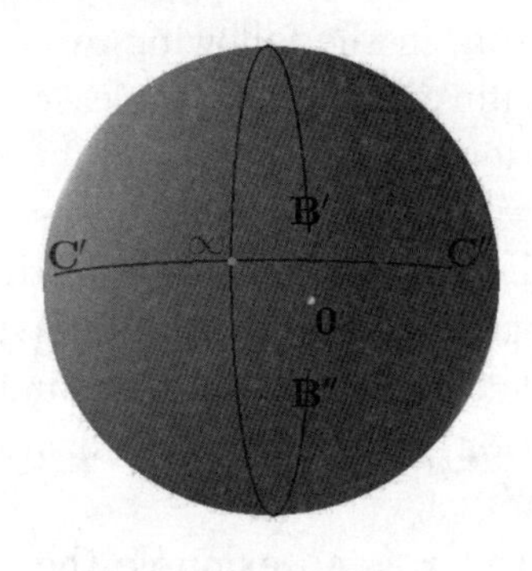

(d) Sphere projection for $p = 2$, $r = 0$.

Figure 9: Transition from $r = 0.3$ to $r = 0$ at $p = 2$, illustrating the change of S_x in the λ-plane as $r \to 0$ and $p > 1$.

In the focussing region ($p < 0$) the transitions are more involved; for example, in order to understand what happens when $r = 0$ for $-4 < p < 0$, one has first to understand what happens to the $\mathbf{S}_x$ curve in the λ-plane when, for a fixed value of p in the interval $(-4, 0)$, r is moved from a value larger than $p_C^{-1}(p)$ to exactly this latter value (see Figures 11a and 11b). If $r > p_C^{-1}(p)$, the classification scheme predicts two branches (no gaps and no loops). The first branch, $\widehat{A'\,A\,A''}$, starts in A', intersects once with the other branch in the upper half-plane, intersects the real axis in the point A, intersects again with the other branch in the lower half-plane, and reaches the endpoint A''. The second branch, $\widehat{B'\,B\,B''}$, starts in B', intersects twice with the other branch, forming a loop (or noose) around the point A' in the upper half-plane, descends almost vertically intersecting the real axis in the point B, intersects twice with the other branch, forming another loop (or noose) around the point A'' in the lower half-plane, and closes into the endpoint B''. As r approaches the value $p_C^{-1}(p)$, two symmetric points of the nooses formed by the branch $\widehat{B'\,B\,B''}$ around the endpoints A' and A'' move towards each other, and, when $r = p_C^{-1}(p)$, the two nooses touch each other in the point B_0 on the real axis (see Figures 11c

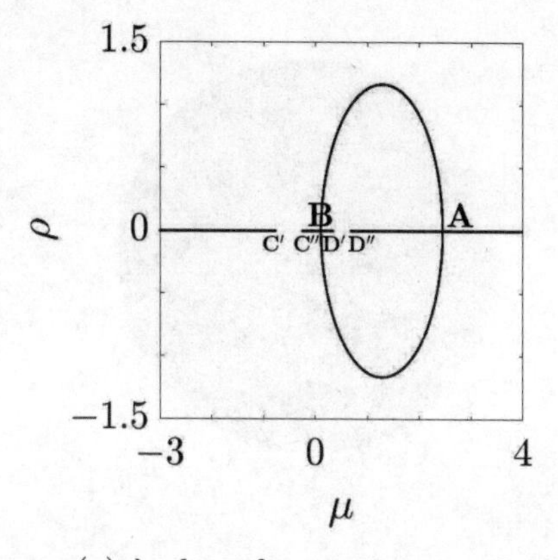

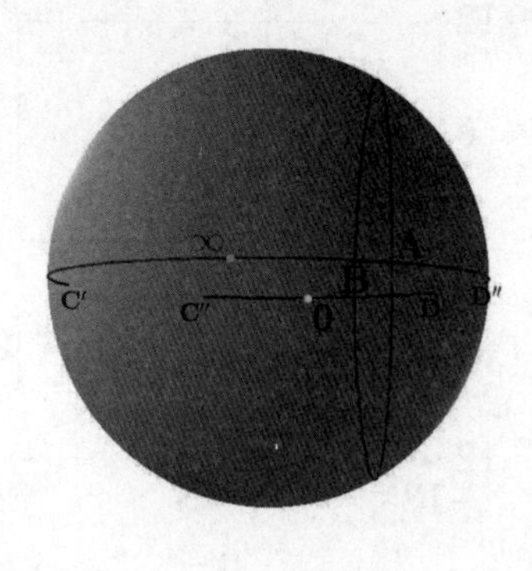

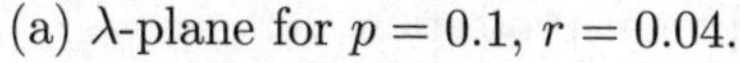

(a) λ-plane for $p = 0.1$, $r = 0.04$.

(b) Sphere projection for $p = 0.1$, $r = 0.04$.

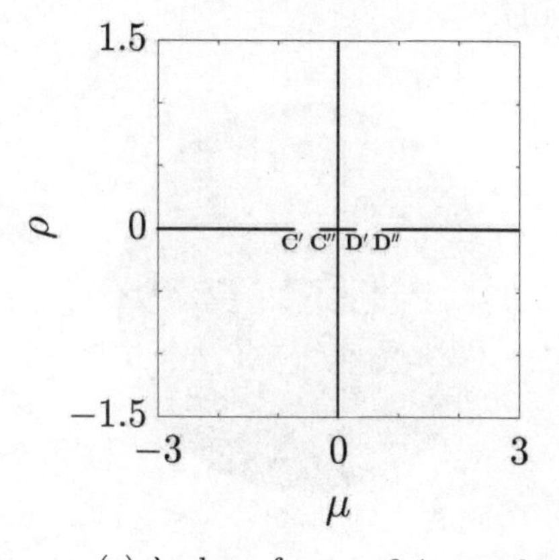

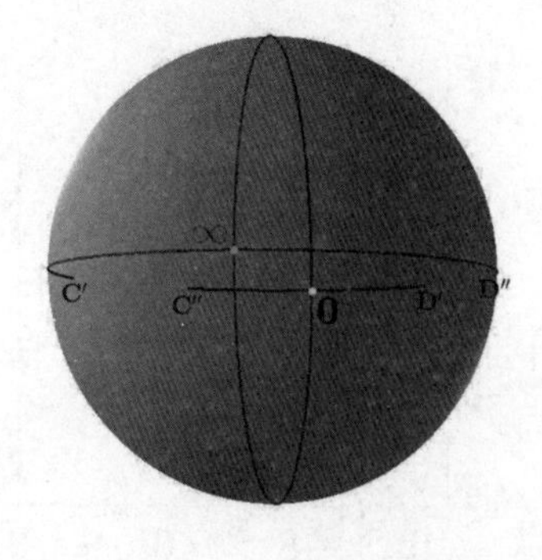

(c) λ-plane for $p = 0.1$, $r = 0$.

(d) Sphere projection for $p = 0.1$, $r = 0$.

Figure 10: Transition from $r = 0.04$ to $r = 0$ at $p = 0.1$, illustrating the change of S_x in the λ-plane as $r \to 0$ and $r < p < p_S(r)$.

and 11d).

This analysis shows the advantage of projecting the λ-plane onto the unit sphere (applying an inverse stereographic sphere projection), for transitions which are not clear as seen in the λ-plane, resolve on the sphere. All other transitions can be similarly analyized.

B Polynomials: a tool box

Our investigation of the spectra in the λ-plane, as in Section 2, makes use of few properties of polynomials. Though some of them can be found in textbooks, we deem it useful to collect them here, with restriction to those results which are in use in our computations. The expression of a generic polynomial of degree N, either in terms of its coefficients f_j,

$$f(x) = f_N x^N + \cdots + f_0 \,, \quad f_N \neq 0 \,, \tag{B.1}$$

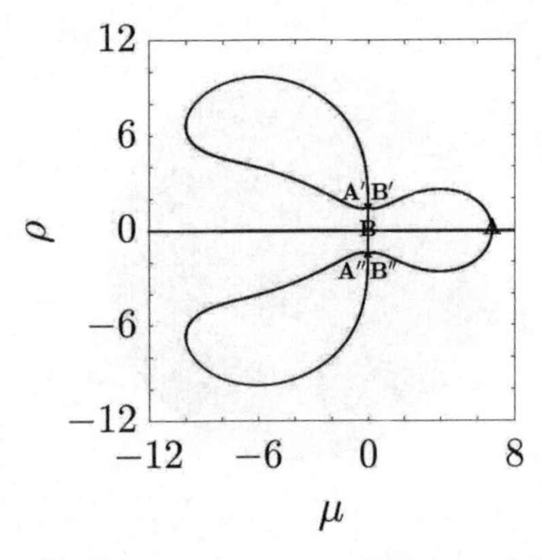

(a) λ-plane for $p = -3.97$, $r = 0.0017$.

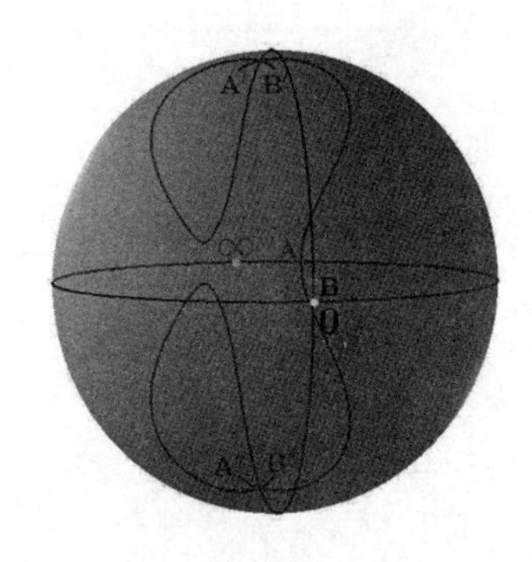

(b) Sphere projection for $p = -3.97$, $r = 0.0017$.

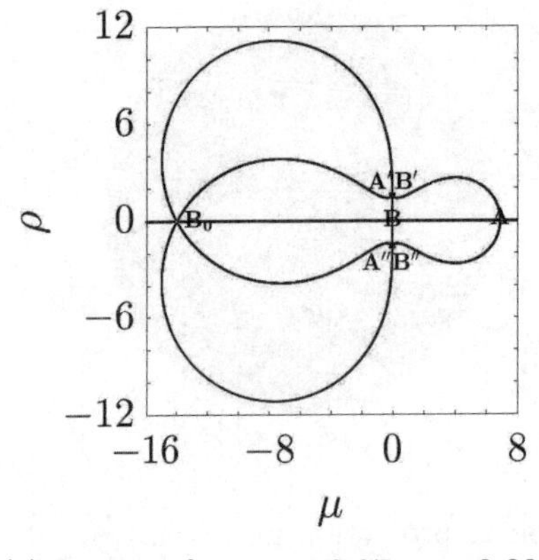

(c) λ-plane for $p = -3.97$, $r = 0.0014169$.

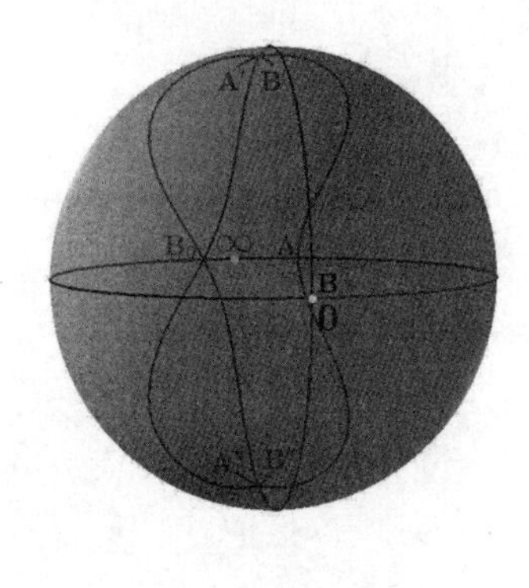

(d) Sphere projection for $p = -3.97$, $r = 0.0014169$.

Figure 11: Transition from $r = 0.0017$ to $r = 0.0014169 \approx p_C^{-1}(-3.97)$ at $p = -3.97$, illustrating the change of S_x in the λ-plane as $r \to p_C^{-1}(p)$ and $p < p_C(r)$.

or in terms of its roots x_j,

$$f(x) = f_N \Pi_{j=1}^N (x - x_j)\,, \tag{B.2}$$

can be used to express its discriminant $\Delta_x(f)$. If in terms of roots, the formula for the discriminant of $f(x)$ reads

$$\Delta_x(f) = f_N^{2N-2} \Pi_{j<m} (x_j - x_m)^2\,. \tag{B.3}$$

Instead, if in terms of coefficients, this discriminant can be expressed as a homogeneous polynomial of the f_j's. However, since this formula is particularly complicated if N is an arbitrary integer and we deal here only with polynomials of degree 2, 3 and 4, we content ourselves with reporting only these three cases. Thus, for $N = 2$, it reads

$$\Delta_x(f) = f_1^2 - 4 f_0 f_2\,, \tag{B.4}$$

for $N = 3$

$$\Delta_x(f) = f_1^2 f_2^2 - 4f_0 f_2^3 - 4f_1^3 f_3 - 27f_0^2 f_3^2 + 18f_0 f_1 f_2 f_3 \,, \tag{B.5}$$

and, for $N = 4$,

$$\begin{aligned} \Delta_x(f) \;=\; & f_1^2 f_2^2 f_3^2 - 4f_1^3 f_3^3 - 27f_1^4 f_4^2 + 256f_0^3 f_4^3 - 27f_0^2 f_3^4 - 4f_1^2 f_2^3 f_4 - 4f_0 f_2^3 f_3^2 \\ & +16f_0 f_2^4 f_4 - 128f_0^2 f_2^2 f_4^2 + 18f_1^3 f_2 f_3 f_4 + 18f_0 f_1 f_2 f_3^3 - 6f_0 f_1^2 f_3^2 f_4 \\ & +144f_0 f_1^2 f_2 f_4^2 + 144f_0^2 f_2 f_3^2 f_4 - 192f_0^2 f_1 f_3 f_4^2 - 80f_0 f_1 f_2^2 f_3 f_4 \,. \end{aligned} \tag{B.6}$$

If the coefficients of the polynomial (B.1) are all real, the roots x_j are either real themselves, or they lie off the real axis as pair of complex conjugate numbers. The number of real roots of $f_N(x)$ can be extracted from the coefficients f_j. Thus, if we assume that the discriminant is non-vanishing, $\Delta_x(f) \neq 0$, so that all roots are simple, for $N = 2$ and $N = 3$ knowing the sign of the discriminant $\Delta_x(f)$ is sufficient to distinguish between the only two possible cases: i) two real and ii) two complex conjugate roots for $N = 2$, and i) three real and ii) one real + two complex conjugate roots for $N = 3$. This is not so if the degree of $f_N(x)$ is greater than three, say for $N \geq 4$. In particular, for $N = 4$ which is the only case we consider here as of interest to us, it is necessary to compute also the following two functions $\mathcal{D}^{(1)}(f)$ and $\mathcal{D}^{(2)}(f)$, which are defined in terms of the coefficients of $f(x)$, see (B.1), by the expressions

$$\mathcal{D}^{(1)}(f) = 8\, f_2\, f_4 - 3\, f_3^2 \;\; , \mathcal{D}^{(2)}(f) = 64\, f_0\, f_4^3 - 16\, f_2^2\, f_4^2 + 16\, f_2\, f_3^2\, f_4 - 16\, f_1\, f_3\, f_4^2 - 3\, f_3^4 \,. \tag{B.7}$$

Then the reality properties of the four simple roots, say $\{x_1\,,\, x_2\,,\, x_3\,,\, x_4\}$, in terms of the coefficients reads

$$\Delta_x(f) < 0 \quad \text{2 real+1 pair of cc roots } \{x_1\,,\, x_2\,,\, x_3\,,\, x_3^*\}\,, \tag{B.8a}$$

$$\Delta_x(f) > 0\,, \quad \mathcal{D}^{(1)}(f) > 0 \quad \text{2 pair of cc roots } \{x_1\,,\, x_1^*\,,\, x_3\,,\, x_3^*\}\,, \tag{B.8b}$$

$$\Delta_x(f) > 0\,, \quad \mathcal{D}^{(1)}(f) < 0 \quad \begin{cases} \mathcal{D}^{(2)}(f) > 0 & \text{2 pair of cc roots } \{x_1\,,\, x_1^*\,,\, x_3\,,\, x_3^*\}\,, \\ \mathcal{D}^{(2)}(f) < 0 & \text{4 real roots } \{x_1\,,\, x_2\,,\, x_3\,,\, x_4\}\,. \end{cases} \tag{B.8c}$$

If instead the discriminant vanishes, $\Delta_x(f) = 0$, then one or more roots of $f(x)$ are multiple. This happens in six different ways, namely if the roots are: 2 simple real and 1 double real, 2 double real, 1 simple real and 1 triple real, 1 quadruple real, 1 double real and 2 simple complex conjugate, 2 double complex conjugate. For the conditions on the coefficients f_j which correspond to each of these six different cases of multiple roots see *e.g.* [58].

Let us consider now the different case in which we assume that the coefficients f_j of the polynomial (B.1) are complex, $f_j^* \neq f_j$. To our aims, and for the sake of simplicity, we confine our present discussion to cubic monic polynomials, $N = 3$ and $f_3 = 1$, namely

$$f(x) = x^3 + f_2 x^2 + f_1 x + f_0 \,. \tag{B.9}$$

We intend to show the conditions on the coefficients,

$$f_j = a_j + i b_j \,, \quad b_j \neq 0 \,, \quad j = 0, 1, 2 \,, \tag{B.10}$$

a_j and b_j being their real, and respectively, imaginary parts, which are necessary and sufficient for this polynomial (B.9) to possess either one or two real roots. In addition we give also the explicit expression of the three roots of $f(x)$. This information is summarised as follows:

Proposition 16. *The polynomial (B.9) has one real root ξ_r and two complex roots ξ_j, $j = 1, 2$ if, and only if, the two conditions*

$$b_1^2 - 4b_0 b_2 > 0 \,, \quad b_0 b_2 - b_1^2 - a_1 b_2^2 + a_2 b_1 b_2 \neq 0 \,, \tag{B.11}$$

are satisfied. In this case, the expression of the real root reads

$$\xi_r = (a_0 b_2^2 - a_2 b_0 b_2 + b_0 b_1)/(b_0 b_2 - b_1^2 - a_1 b_2^2 + a_2 b_1 b_2) \,, \tag{B.12}$$

and that of the two complex roots is

$$\xi_j = -\frac{1}{2} \left[(\xi_r + f_2) + (-1)^j \sqrt{(\xi_r + f_2)^2 + 4(f_0/\xi_r)} \right] \,, \quad j = 1, 2 \,. \tag{B.13}$$

Proposition 17. *The polynomial (B.9) has two real roots $\xi_\pm$ and one complex root ξ_c if, and only if, the three conditions*

$$b_1^2 - 4b_0 b_2 > 0 \,, \quad a_0 b_2^2 - a_2 b_0 b_2 + b_0 b_1 = 0 \,, \quad b_0 b_2 - b_1^2 - a_1 b_2^2 + a_2 b_1 b_2 = 0 \,, \tag{B.14}$$

are met. In this case, the expression of the real roots reads

$$\xi_\pm = -\frac{1}{2b_2} \left(b_1 \pm \sqrt{b_1^2 - 4 b_0 b_2} \right) \,, \tag{B.15}$$

and that of the complex root is

$$\xi_c = b_1 / b_2 - f_2 \,. \tag{B.16}$$

The proof of these propositions is elementary and follows from the observation that, if ξ is a real zero of the polynomial (B.9), then it is a solution of the system

$$\xi^3 + a_2 \xi^2 + a_1 \xi + a_0 = 0 \,, \quad b_2 \xi^2 + b_1 \xi + b_0 = 0 \,, \tag{B.17}$$

which has one solution, as in Proposition 16, and two solutions in the other case of Proposition 17. Moreover the explicit expressions of the roots are a straight consequence of Vieta' s formulae.

A further issue we take up here is the mapping which associates to a generic monic polynomial $f(x)$ of degree N, see (B.1) with $f_N = 1$, a novel monic polynomial $g(y)$ of degree $N(N-1)/2$ whose roots are the squared differences of the N roots x_j of $f(x)$, namely

$$g(y) = \Pi_{j<m}[y - (x_j - x_m)^2]\,. \tag{B.18}$$

Since the roots of $g(y)$ are by construction quadratic symmetric functions of the roots x_j of $f(x)$, the coefficients of $g(y)$ have to be polynomials of the coefficients of $f(x)$. Deriving this connection among coefficients for arbitrary degree N is unnecessary to our purposes. Thus we give here only the following explicit formulae for $N = 3$, which have been obtained via Vieta relations.

$$f(x) = x^3 + f_2 x^2 + f_1 x + f_0 = \Pi_{j=1}^3 (x - x_j)\,, \tag{B.19a}$$

$$g(y) = y^3 + g_2 y^2 + g_1 y + g_0 = \Pi_{j=1}^3 (y - y_j)\,, \tag{B.19b}$$

$$y_1 = (x_2 - x_3)^2 \;\; , \;\; y_2 = (x_3 - x_1)^2 \;\; , \;\; y_3 = (x_1 - x_2)^2\,, \tag{B.19c}$$

$$\begin{aligned} g_0 &= 27 f_0^2 + 4 f_1^3 - 18 f_0 f_1 f_2 + 4 f_0 f_2^3 - f_1^2 f_2^2\,, \\ g_1 &= (3 f_1 - f_2^2)^2\,, \\ g_2 &= 2(3 f_1 - f_2^2)\,. \end{aligned} \tag{B.19d}$$

We note that, while the polynomial $f(x)$ has generic coefficients, this map (B.19) yields the resulting polynomial $g(y)$ in the non-generic form $g(y) = y(y + g_2/2)^2 + g_0$. Three simple consequences of these equations (B.19), which deserve particular attention as relevant to our computations, are the following. One reads

$$\Delta_x(f) = -g(0)\,, \tag{B.20}$$

and comes from (B.3), with $f_3 = 1$, and (B.18). The second one is clearly implied by (B.19c) which tells that the polynomial $g(y)$ is invariant with respect to x-translations of the polynomial $f(x)$, namely $g(y)$ does not change if $f(x)$ is replaced by $f(x + A)$ for any constant A,

$$\text{if } \; f(x) \mapsto g(y)\,, \quad \text{then also } \; f(x + A) \mapsto g(y)\,. \tag{B.21}$$

The third implication of the map (B.19) is the following relation between the roots x_j of the polynomial $f(x)$ (B.9) and those y_j of its companion $g(y)$ (B.18), see (B.19c). This is specified by:

Proposition 18. *The six roots x_j and y_j of $f(x)$ and, respectively, $g(y)$ which satisfy the system of equations*

$$f(x_j) = 0\,, \quad g(y_j) = 0\,, \quad j = 1, 2, 3\,, \tag{B.22}$$

do also satisfy the following equivalent system

$$x_j = -\frac{f_2 y_j + 4f_2 f_1 - f_2^3 - 9f_0}{3y_j + 3f_1 - f_2^2}\,, \quad 3x_j^2 + 2f_2 x_j + y_j + 4f_1 - f_2^2 = 0\ . \tag{B.23}$$

The equivalence of these two systems (B.22) and (B.23) of six equations follows from Vieta relations as applied to both the polynomials $f(x)$ and $g(x)$, and its validity is readily verified by direct computation.

References

[1] Zagryadskaya L I and Ostrovskii L A, Observed self-influence of modulated waves in a nonlinear line, *Radiophysics and Quantum Electronics*, **11**, Issue 6, 548–550, 1928.

[2] Talanov V, Self-focusing of wave beams in nonlinear media, *JETP Lett.*, **2**, 138, 1965.

[3] Hasegawa A, Stimulated modulational instabilities of plasma waves, *Phys. Rev. A*, **1**, 1746, 1970.

[4] Tai K, Hasegawa A and Tomita A, Observation of modulation instability in optical fibres, *Phys. Rev. Lett.*, **56**, 135, 1986.

[5] Benjamin T B and Feir J E, The disintegration of wave trains on deep water Part 1. Theory, *Journal of Fluid Mechanics*, **27**, 417–430, 1967.

[6] Rothenberg J E, Modulational instability for normal dispersion, *Phys. Rev. A*, **42**, 682–685, 1990.

[7] Rothenberg J E, Observation of the buildup of modulational instability from wave breaking, *Opt. Lett.*, **16**(1), 18–20, 1991.

[8] Zakharov, V and Ostrovsky L, Modulation instability: the beginning, *Physica D Nonlinear Phenomena*, **238**, 540–548, 2009.

[9] Kuznetsov E A, Solitons in a parametrically unstable plasma, *Dokl. Akad. Nauk SSSR*, **236**(9), 575–577, 1977.

[10] Yuen H C and Lake B M, Instabilities of waves on deep water, *Annual Review of Fluid Mechanics*, **12**(1), 303–334, 1980.

[11] Agrawal G, *Nonlinear Fiber Optics*, Academic Press, 1995.

[12] Dudley J, Gent G, Dias F, Kibler B, and Akhmediev N, Modulation instability, Akhmediev Breathers and continuous wave supercontinuum generation, *Opt. Express*, **17**, 21497–21508, 2009.

[13] Kevrekidis P G, Frantzeskakis D J, and Carretero-González R, *Emergent Nonlinear Phenomena in Bose-Einstein Condensates: Theory and Experiment*, Springer Series on Atomic, Optical, and Plasma Physics. Springer, 2007.

[14] Ostrovskii L A, Potapov A I, Modulated waves, in *Theory and Applications*, John Hopkins University Press, Baltimore, London, 1999.

[15] Tajiri M, Wave Stability and Instability, in Scott A (Ed.), *Encyclopedia of Nonlinear Science*, Routledge, NY, London, 2005.

[16] Skryabin D V and Firth W J, Modulational instability of bright solitary waves in incoherently coupled nonlinear Schrödinger equations, *Phys. Rev. E*, **60**(1), 1019–1029, 1999.

[17] Zakharov V E and Kuznetsov E A, Solitons and collapses: two evolution scenarios of nonlinear wave systems, *Physics-Uspekhi*, **55**(6), 535–556, 2012.

[18] Maddocks J and Sachs R, On the stability of KdV multi-solitons, *Comm. Pure Appl. Math.*, **46**, 867–901, 1993.

[19] Georgiev V and Ohta M, Nonlinear instability of linearly unstable standing waves for nonlinear Schrödinger equations, *Journal of the Mathematical Society of Japan*, **64**(2), 533–548, 2012.

[20] Ablowitz M J, Kaup D, Newell A, and Segur H, The inverse scattering transform-Fourier analysis for nonlinear problems, *Studies in Applied Mathematics*, vol. , **53**, 4, 249–315, 1974.

[21] Ablowitz M J and Segur H, *Solitons and the Inverse Scattering Transform*, Studies in Applied Mathematics. SIAM, 1981.

[22] Calogero F and Degasperis A, *Spectral Transform and Solitons: Tools to Solve and Investigate Nonlinear Evolution Equations. V. 1*, North-Holland, 1982.

[23] Novikov S, Manakov S, Pitaevskii L, and Zakharov V, *Theory of Solitons: The Inverse Scattering Method*, Contemporary Soviet Mathematics, Plenum, 1984.

[24] Kaup D J, Closure of the squared Zakharov-Shabat eigenstates, *J. Math. Anal. Appl.*, **54**, 849–864, 1976.

[25] Kaup D J, and Van Gorder R A, Squared eigenfunctions and the perturbation theory for the nondegenerate $N \times N$ operator: a general outline. *Journal of Physics A Mathematical and Theoretical*, **43**(43), 2010.

[26] Degasperis A, Lombardo S and Sommacal M, Integrability and linear stability of nonlinear waves, *M. J Nonlinear Sci*, **28**, 1251, 2018.

[27] Dodd R, Eilbeck J, Gibbon J, and Morris H, *Solitons and Nonlinear Wave Equations*, Academic Press, 1982.

[28] Dauxois T and Peyrard M, *Physics of Solitons.* Cambridge University Press, 2006.

[29] Kuznetsov E A and Mikhailov A V, Stability of stationary waves in nonlinear weakly dispersive media, *Zh. Eksp. Teor. Fiz.*, **67**, 1717–1727, 1974.

[30] Sachs R L, Completeness of derivatives of squared Schrödinger eigenfunctions and explicit solutions of the linearized KdV equation, *SIAM J. Math. Anal.*, **14**, 674–683, 1983.

[31] Kuznetsov E A, Spector M D, and Fal'kovich G E, On the stability of nonlinear waves in integrable models, *Physica D*, **10**, 379–386, 1984.

[32] Kuznetsov E A and Spector M D, Modulation instability of soliton trains in fiber communication systems, *Theor. Math. Phys.*, **120**(2), 997–1008, 1999.

[33] Yang J, Complete eigenfunctions of linearized integrable equations expanded around a soliton solution, *J. Math. Phys.*, **41**(9), 6614–6638, 2000.

[34] Yang J, Eigenfunctions of linearized integrable equations expanded around an arbitrary solution, *Studies in Applied Mathematics*, **108**, 145–159, 2002.

[35] Yang J and Kaup D J, Squared eigenfunctions for the Sasa-Satsuma equation, *J. Math. Phys.*, **50**, 023504, 2009.

[36] Degasperis A and Lombardo S, Integrability in action: solitons, instability and rogue waves. In M. Onorato, S. Resitori, and F. Baronio, editors, *Rogue and Shock Waves in Nonlinear Dispersive Media*, Springer, 2016.

[37] Bottman N, Deconinck D, KdV cnoidal waves are spectrally stable, *Discrete and Continuous*, Dynamical Systems-Series A (DCDS-A) **25**(4), 1163, 2009.

[38] Bottman N, Deconinck D, and Nivala M, Elliptic solutions of the defocusing NLS equation are stable, *J. Phys. A: Math. Theor.*, **44**, 285201, 2009.

[39] Zakharov V E and Gelash A A, Superregular solitonic solutions: a novel scenario of the nonlinear stage of Modulation Instability, *Theor. Math. Phys.*, **120**(2), 997–1008, 2013.

[40] Biondini G and Mantzavinos D, Universal nature of the nonlinear stage of modulational instability, *Phys. Rev. Lett.*, **116**(4), 043902, 2016.

[41] Biondini G, Kovacic G, Inverse scattering transform for the focusing nonlinear Schrödinger equation with nonzero boundary conditions, *Journal of Mathematical Physics*, **55** (3), 031506, 2014.

[42] Grinevich P G and Santini P M, Numerical instability of the Akhmediev breather and a finite-gap model of it. In Buchstaber V., Konstantinou-Rizos S., Mikhailov A. (eds), *Recent Developments in Integrable Systems and Related Topics of Mathematical Physics.* Springer, 2016.

[43] Grinevich P G and Santini P M, The finite gap method and the analytic description of the exact rogue wave recurrence in the periodic NLS Cauchy problem, *Nonlinearity*, **31** (11), 5258, 2018.

[44] Grinevich P G and Santini P M, The exact rogue wave recurrence in the NLS periodic setting via matched asymptotic expansions, for 1 and 2 unstable modes *Physics Letters A*, **382** (14), 973-979, 2018.

[45] Zakharov V E and Shulmann E I, To the integrability of the system of two coupled nonlinear Schrödinger equations, *Physica D*, **4**(2), 270–274, 1982.

[46] Manakov S V, On the theory of two-dimensional stationary self-focusing of electromagnetic waves, *Soviet Journal of Experimental and Theoretical Physics*, **38**, 1974.

[47] Menyuk C R, Nonlinear pulse propagation in birefringent optical fibers, *IEEE Journal of Quantum Electronics*, **23**(2), 174–176, Feb 1987.

[48] Evangelides G S, Mollenauer L F, Gordon J P, and Bergano N S, Polarization multiplexing with solitons, *Lightwave Technology, Journal of*, **10**, 1, 28–35, 1992.

[49] Wang D and Menyuk C R, Polarization evolution due to the Kerr nonlinearity and chromatic dispersion, *J. Lightwave Technol.*, **17**(12), 2520, 1999.

[50] Onorato M, Proment D, and Toffoli A, Freak waves in crossing seas, *The European Physical Journal Special Topics*, **185**, 1, 45–55, 2010.

[51] Ablowitz M J and Horikis T P, Interacting nonlinear wave envelopes and rogue wave formation in deep water, *Physics of Fluids*, **27**(1), 012107, 2015.

[52] Yang J and Benney D J, Some properties of nonlinear wave systems, *Studies in Applied Mathematics*, **96**(1), 111–139, 1996.

[53] Forest M G, McLaughlin D W, Muraki D J, and Wright O C, Nonfocusing instabilities in coupled, integrable nonlinear Schrödinger PDEs, *Journal of Nonlinear Science*, **10**, 3, 291–331, 2000.

[54] Ling L and Zhao L C, Modulational instability and homoclinic orbit solutions in vector nonlinear Schrödinger equation, *Communications in Nonlinear Science and Numerical Simulation*, **72**, 2017.

[55] Calogero F, *Zeros of Polynomials and Solvable Nonlinear Evolution Equations*, Cambridge University Press, 2018.

[56] Degasperis A and Lombardo S, Rational solitons of wave resonant-interaction models, *Physical Review. E*, **88**, 5, 052914, 2013.

[57] Degasperis A, Lombardo S and Sommacal M, Rogues waves type solutions and spectra of coupled nonlinear Schrödinger equations, *Fluids*, **4**(1), 57, 2019.

[58] Rees E L, Graphical discussion of the roots of a quartic equation, *The American Mathematical Monthly*, **29**(2), 51–55, 1922.

B2. Rational solutions of Painlevé systems.

David Gómez-Ullate [a,b], *Yves Grandati* [c] *and Robert Milson* [d]

[a] *Escuela Superior de Ingeniería, Universidad de Cádiz,*
Avda. Universidad de Cádiz,
Campus Universitario de Puerto Real, 11519, Spain.
david.gomezullate@uca.es

[b] *Departamento de Física Teórica,*
Universidad Complutense de Madrid,
28040 Madrid, Spain.

[c] *Laboratoire de Physique et Chimie Théoriques,*
Université de Lorraine, 1 Bd Arago,
57078 Metz, Cedex 3, France.
yves.grandati@univ-lorraine.fr

[d] *Department of Mathematics and Statistics,*
Dalhousie University,
Halifax, NS, B3H 3J5, Canada.
rmilson@dal.ca

Abstract

Although the solutions of Painlevé equations are transcendental in the sense that they cannot be expressed in terms of known elementary functions, there do exist rational solutions for specialized values of the equation parameters. A very successful approach in the study of rational solutions to Painlevé equations involves the reformulation of these scalar equations into a symmetric system of coupled, Riccati-like equations known as dressing chains. Periodic dressing chains are known to be equivalent to the A_N-Painlevé system, first described by Noumi and Yamada. The Noumi-Yamada system, in turn, can be linearized as using bilinear equations and τ-functions; the corresponding rational solutions can then be given as specializations of rational solutions of the KP hierarchy.

The classification of rational solutions to Painlevé equations and systems may now be reduced to an analysis of combinatorial objects known as Maya diagrams. The upshot of this analysis is an explicit determinantal representation for rational solutions in terms of classical orthogonal polynomials. We illustrate this approach by describing Hermite-type rational solutions of Painlevé of the Noumi-Yamada system in terms of cyclic Maya diagrams. By way of example we explicitly construct Hermite-type solutions for the P_{IV}, P_V equations and the A_4 Painlevé system.

1 Introduction

The defining property of the six nonlinear second order Painlevé equations is that their solutions have fixed monodromy; that is all movable singularities are poles.

The resulting Painlevé transcendents are now considered to be the nonlinear analogues of special functions [7, 11]. Although these functions are transcendental in the sense that they cannot be expressed in terms of known elementary functions, Painlevé equations also possess special families of solutions that, for special values of the parameters, can be expressed via known special functions such hypergeometric functions or even rational functions [2].

Rational solutions of P_{II} were studied by Yablonskii [41] and Vorob'ev [39], in terms of a special class of polynomials that are now named after them. For P_{III}, classical solutions have been considered in [26]. Okamoto [32] obtained special polynomials associated with some of the rational solutions of the fourth Painlevé equation (P_{IV})

$$y'' = \frac{1}{2y}(y')^2 + \frac{3}{2}y^3 + 4ty^2 + 2(t^2 - a)y + \frac{b}{y}, \quad y = y(t), \tag{1.1}$$

with a and b constants, which are analogous to the Yablonskii–Vorob'ev polynomials. Noumi and Yamada [29] generalized Okamoto's results and expressed all rational solutions of P_{IV} in terms of two types of special polynomials, now known as the *generalized Hermite polynomials* and *generalized Okamoto polynomials*, both of which may be given as determinants of sequences of Hermite polynomials.

A very successful approach in the study of rational solutions to Painlevé equations has been the set of geometric methods developed by the Japanese school, most notably by Noumi and Yamada, [30]. The core idea is to write the scalar equations as a set of first order coupled nonlinear system of equations. For instance, the fourth Painlevé (1.1) equation P_{IV} is equivalent to the following autonomous system of three first order equations

$$\begin{aligned} f_0' &= f_0(f_1 - f_2) + \alpha_0, \\ f_1' &= f_1(f_2 - f_0) + \alpha_1, \\ f_2' &= f_2(f_0 - f_1) + \alpha_2, \end{aligned} \tag{1.2}$$

subject to the condition

$$(f_0 + f_1 + f_2)' = \alpha_0 + \alpha_1 + \alpha_2 = 1. \tag{1.3}$$

Once this equivalence is shown, it is clear that the symmetric form of P_{IV} (1.2), sometimes referred to as sP_{IV}, is easier to analyze. In particular, [29] showed that the system (1.2) possesses a symmetry group of Bäcklund transformations acting on the tuple of solutions and parameters $(f_0, f_1, f_2|\alpha_0, \alpha_1, \alpha_2)$. This symmetry group is the affine Weyl group $A_2^{(1)}$, generated by the operators $\{\pi, \mathbf{s}_0, \mathbf{s}_1, \mathbf{s}_2\}$ whose action on the tuple $(f_0, f_1, f_2|\alpha_0, \alpha_1, \alpha_2)$ is given by:

$$\begin{aligned} \mathbf{s}_k(f_j) &= f_j - \frac{\alpha_k \delta_{k+1,j}}{f_k} + \frac{\alpha_k \delta_{k-1,j}}{f_k}, \\ \mathbf{s}_k(\alpha_j) &= \alpha_j - 2\alpha_j \delta_{k,j} + \alpha_k(\delta_{k+1,j} + \delta_{k-1,j}), \\ \pi(f_j) &= f_{j+1}, \qquad \pi(\alpha_j) = \alpha_{j+1} \end{aligned} \tag{1.4}$$

where $\delta_{k,j}$ is the Kronecker delta and $j, k = 0, 1, 2 \mod (3)$. The technique to generate rational solutions is to first identify a number of very simple rational *seed solutions*, and then successively apply the Bäcklund transformations (1.4) to generate families of rational solutions.

This is a beautiful approach which makes use of the hidden group theoretic structure of transformations of the equations, but the solutions built by dressing seed solutions are not very explicit, in the sense that one needs to iterate a number of Bäcklund transformations (1.4) on the functions and parameters in order to obtain the desired solutions. Questions such as determining the number of zeros or poles of a given solution constructed in this manner seem very difficult to address. For this reason, alternative representations of the rational solutions have also been investigated, most notably the determinantal representations [20, 21].

The system of first order equations (1.2) admits a natural generalization to any number of equations, and it is known as the A_N-Painlevé or the Noumi-Yamada system. The higher order Painlevé system, exhibited below in (2.16) and (2.17), is considerably simpler (for reasons that will be explained later), and it is the one we will focus on in this paper. The symmetry group of this higher order system is the affine Weyl group $A_N^{(1)}$, acting by Bäcklund transformations as in (1.4). The system has the Painlevé property, and thus can be considered a proper higher order generalization of $\mathrm{sP_{IV}}$ (1.2), which corresponds to $N = 2$.

The next higher order system belonging to this hierarchy is the A_3-Painlevé system, which is known to be equivalent to the scalar $\mathrm{P_V}$equation. Rational solutions of $\mathrm{P_V}$ were classified using direct analysis by Kitaev, Law and McLeod [22]. These rational solutions can be described in terms of generalized Umemura polynomials [36, 23], which admit a description in terms of Schur functions [28]. A more general determinantal representation based on universal characters was given by Tsuda [35]. In general, these determinants are constructed Laguerre polynomials, but for some particular values these degenerate to Hermite polynomials and fit into the framework of the present paper.

The A_4-Painlevé system cannot be reduced to a scalar equation, and so represents a genuine generalization of the classic Painlevé equations. Special solutions have been studied by Clarkson and Filipuk [10], who provide several classes of rational solutions via an explicit Wronskian representation, and by Matsuda [24], who uses the classical approach to identify the set of parameters that lead to rational solutions. However, a complete classification and explicit description of the rational solutions of A_{2n}-Painlevé for $n \geq 2$ is, to the best of our knowledge, still not available in the literature.

Of particular interest are the special polynomials associated to these rational solutions, whose zeros and poles structure shows extremely regular patterns in the complex plane, and have received a considerable amount of study [32, 29, 37, 12, 9]. Some of these polynomial families are known as generalized Hermite, Okamoto or Umemura polynomials, and they can be described as Wronskian determinants of given sequences of Hermite polynomials. We will show that all these polynomial families are only particular cases of a larger one.

Our approach for describing rational solutions to the Noumi-Yamada system differs from the one used by the Japanese school in that it makes no use of symmetry groups of Bäcklund transformations. Instead, we will be influenced by the approach of Darboux dressing chains introduced by the Russian school [1, 38], which has received comparatively less attention in connection to Painlevé systems, and which makes use of the notion of trivial monodromy [31]. Our interest in rational solutions of Painlevé equations follows from the recent advances in the theory of exceptional polynomials [14, 15, 18], and especially exceptional Hermite polynomials [17]. Nonetheless we strive to maintain the connection to the theory of integrable systems by employing the concepts of a Maya diagram and of bilinear relations [25].

The chapter is organized as follows: in Section 2 we introduce the equations for a dressing chain of Darboux transformations of Schrödinger operators and prove that they are equivalent to the Noumi-Yamada system. These results are well known [1] but recalling them is useful to fix notation and make the chapter self-contained. In Section 3 we introduce the class of Hermite-type τ functions and their representations via Maya diagrams and pseudo-Wronskian determinants. We introduce the key notion of *cyclic Maya diagrams* and reformulate the problem of classifying rational solutions of the Noumi-Yamada system as that of classifying cyclic Maya diagrams. In Section 5 we introduce the notion of genus and interlacing for Maya diagrams which allows us to achieve a complete classification of p-cyclic Maya diagrams for any period p. Finally, we apply the theory to exhibit rational solutions of the A_2, A_3, A_4 systems in Section 6 and we write out explicitly these solutions using the representation developed in the previous sections.

While the Japanese school has built a beautiful framework around Painlevé equations, including reductions of the KP hierarchy in Sato theory, for the particular task of describing rational solutions of higher order Painlevé equations, we find the approach of cyclic Maya diagrams to be more direct, simple and explicit.

2 Dressing chains and Painlevé systems

A factorization chain is a sequence of Schrödinger operators connected by Darboux transformations. By replacing the second-order Schrödinger equations with first order Riccati equations, one obtains a closely related system called a dressing chain. The theory of dressing chains was developed by Adler [1], Veselov and Shabat [38]. The connection to Painlevé equations was already noted by the just-mentioned authors, and further developed by others [34, 5].

Let us recall the well-known connection between Riccati and Schrödinger equations. An elementary calculation shows that a function $w(z)$ that satisfies a Riccati equation

$$w' + w^2 + \lambda = U \tag{2.1}$$

is the log-derivative of a solution $\psi(z)$ of the corresponding Schrödinger equation:

$$-\psi'' + U\psi = \lambda\psi, \qquad w = \frac{\psi'}{\psi}. \tag{2.2}$$

The Riccati equation (2.1) is equivalent to the factorization relation

$$-D^2 + U = (D + w)(-D + w) + \lambda. \tag{2.3}$$

It follows that a Schrodinger operator $-D^2 + U$ admits a factorization (2.3) if and only if w is the log-derivative of a formal eigenfunction of L with eigenvalue λ.

A Darboux transformation is the transformation $U \mapsto \hat{U}$ where

$$-D^2 + \hat{U} = (D - w)(-D - w) + \lambda$$

is a second-order operator obtained by interchanging the factors in (2.3). Equivalently, the correspondence $U \mapsto \hat{U}$ may be engendered by the transformation $w \mapsto -w$ in (2.3).

Consider a doubly infinite sequence of Schrödinger operators $-D^2 + U_i, i \in \mathbb{Z}$ where neighbouring operators are related by a Darboux transformation

$$\begin{aligned} -D^2 + U_i &= (D + w_i)(-D + w_i) + \lambda_i, \\ -D^2 + U_{i+1} &= (-D + w_i)(D + w_i) + \lambda_i. \end{aligned} \tag{2.4}$$

Since functions w_i are solutions of the Riccati equations

$$w_i' + w_i^2 + \lambda_i = U_i, \quad -w_i' + w_i^2 + \lambda_i = U_{i+1}, \tag{2.5}$$

the above potentials are related by

$$U_{i+1} = U_i - 2w_i', \tag{2.6}$$

$$U_{i+n} = U_i - 2\left(w_i' + \cdots + w_{i+n-1}'\right), \quad n \geq 2. \tag{2.7}$$

If we eliminate the potentials in (2.5) and set

$$a_i = \lambda_i - \lambda_{i+1} \tag{2.8}$$

we obtain a system of coupled differential equations called the doubly infinite dressing chain:

$$(w_i + w_{i+1})' + w_{i+1}^2 - w_i^2 = a_i, \quad i \in \mathbb{Z} \tag{2.9}$$

If we impose a cyclic condition

$$U_{i+p} = U_i + \Delta, \quad i \in \mathbb{Z}, \qquad p \in \mathbb{N}, \Delta \in \mathbb{C} \tag{2.10}$$

on the potentials of the above chain, we obtain a finite-dimensional system of ordinary differential equations. If this holds, then necessarily $w_{i+p} = w_i$, $\alpha_{i+p} = \alpha_i$, and

$$\Delta = -(a_0 + \cdots + a_{p-1}). \tag{2.11}$$

Going forward, we impose the non-degeneracy assumption that

$$\Delta \neq 0.$$

Degenerate dressing chains with $\Delta = 0$ are more closely related to elliptic functions [38] and will not be considered here.

Definition 1. A solution to the p-cyclic dressing chain with shift Δ is a sequence of p functions $w_0, \dots, w_{p-1}$ and complex numbers $a_0, \dots, a_{p-1}$ that satisfy the following coupled Riccati-like equations:

$$(w_i + w_{i+1})' + w_{i+1}^2 - w_i^2 = a_i, \qquad i = 0, 1, \dots, p-1 \mod p \tag{2.12}$$

subject to the condition (2.11).

The cyclic chain has a number of evident symmetries: the reversal symmetry

$$\hat{w}_i = -w_{-i}, \quad \hat{a}_i = -a_{-i}; \tag{2.13}$$

the cyclic symmetry

$$\hat{w}_i = w_{i+1}, \quad \hat{a}_i = a_{i+1}; \tag{2.14}$$

and the scaling symmetry

$$\hat{w}_i(z) = k w_i(kz), \quad \hat{a}_i = k^2 a_i, \quad k \neq 0. \tag{2.15}$$

In the classification of solutions to (2.12) it will be convenient to regard solutions related by reversal, cyclic, and scaling symmetries as being equivalent.

The p-cyclic dressing chain is closely related to higher order Painlevé systems of type A_N, $N = p - 1$ introduced by Noumi and Yamada in [27]. In the even case of $N = 2n$, the Noumi-Yamada system has the form

$$f_i' = \sum_{j=1}^{p-1} (-1)^{j+1} f_i f_{i+j} + \alpha_i, \quad i = 0 \dots, 2n \mod 2n+1 \tag{2.16}$$

In the odd case of $N = 2n - 1$, the Noumi-Yamada system has a more complicated form:

$$x f_i' = f_i \left(1 - 2 \sum_{k=1}^{n-1} \alpha_{i+2k} + 2 \sum_{j=1}^{n} \sum_{k=1}^{n-1} \operatorname{sgn}(2j - 1 - 2k) f_{2j+i-1} f_{2k+i} \right) + 2\alpha_i \sum_{k=1}^{n-1} f_{i+2k}, \tag{2.17}$$

where $i = 0, \dots, 2n - 1 \mod 2n$ and $f_i = f_i(x)$. In both cases, the parameters $\alpha_0, \dots, \alpha_N$ are subject to the constraint

$$\alpha_0 + \cdots \alpha_N = 1.$$

Proposition 1. *The A_{2n} and A_{2n-1} Noumi-Yamada systems* (2.16) (2.17) *are related to the p-cyclic dressing chain* (2.12) *by the following change of variables:*

$$-\sqrt{\Delta} f_i(x) = w_i(z) + w_{i+1}(z), \quad \alpha_i = -\frac{a_i}{\Delta}, \qquad z = \frac{x}{\sqrt{\Delta}}, \tag{2.18}$$

where $i = 0, \dots, p-1 \mod p$ and where $p = 2n + 1$ in the first case, and $p = 2n$ in the second case.

Proof. The proof for the case $p = 2n + 1$ is quite direct. Set

$$d_i(x) = K(w_i(Kx) - w_{i+1}(Kx)), \quad K = \frac{1}{\sqrt{\Delta}},$$

which allows us to rewrite relation (2.9) as

$$f_i' = d_i f_i + \alpha_i. \tag{2.19}$$

Then, observe that because p is odd,

$$d_i = \sum_{j=1}^{p-1} (-1)^{j+1} f_{i+j}. \tag{2.20}$$

This transforms (2.19) into (2.16). ■

If $p = 2n$ is even, the linear relation (2.20) no longer holds. Rather we have the following quadratic relation.

Lemma 1. *For each* $i = 1, \ldots, 2n \mod 2n$ *we have*

$$\begin{aligned}(w_{i+1} - w_i)(w_1 + \cdots + w_{2n}) + (w_i^2 - w_{i+1}^2 + \cdots - w_{i+2n-1}^2) = \\ = \sum_{j=1}^{n} \sum_{k=1}^{n-1} \operatorname{sgn}(2k + 1 - 2j)(w_{2j+i-1} + w_{2j+i})(w_{2k+i} + w_{2k+i+1}).\end{aligned} \tag{2.21}$$

Proof. The left side of (2.21) expands to

$$w_{i+1} \sum_{a=2}^{2n-1} w_{i+a} - w_i \sum_{a=2}^{2n-1} w_{i+a} + \sum_{a=2}^{n-1} (-1)^a w_{i+a}^2 . \tag{2.22}$$

The right side of (2.21) expands to

$$\sum_{k=1}^{n-1}\sum_{j=1}^{k}(w_{i+2j-1}+w_{i+2j})(w_{i+2k}+w_{i+2k+1})$$
$$-\sum_{j=2}^{n}\sum_{k=1}^{j-1}(w_{i+2j-1}+w_{i+2j})(w_{i+2k}+w_{i+2k+1})$$
$$=\sum_{k=2}^{n}\sum_{j=1}^{k-1}(w_{i+2j-1}+w_{i+2j})(w_{i+2k-2}+w_{i+2k-1})$$
$$-\sum_{j=2}^{n}\sum_{k=1}^{j-1}(w_{i+2j-1}+w_{i+2j})(w_{i+2k}+w_{i+2k+1})$$
$$=\sum_{k=2}^{n}\sum_{j=1}^{k-1}(w_{i+2j-1}+w_{i+2j})(w_{i+2k-2}+w_{i+2k-1})$$
$$-(w_{i+2k-1}+w_{i+2k})(w_{i+2j}+w_{i+2j+1})$$
$$=\sum_{k=2}^{n}\sum_{j=1}^{k-1}(w_{i+2j-1}-w_{i+2j+1})w_{i+2k-1}+w_{i+2j}(w_{i+2k-2}-w_{i+2k})+$$
$$+w_{i+2j-1}w_{i+2k-2}-w_{i+2j+1}w_{i+2k}$$
$$=\sum_{k=2}^{n}(w_{i+1}-w_{i+2k-1})w_{i+2k-1}+\sum_{j=1}^{n-1}w_{i+2j}(w_{i+2j}-w_{i+2n})$$
$$+\sum_{k=1}^{n-1}w_{i+1}w_{i+2k}-\sum_{j=1}^{n-1}w_{i+2j+1}w_{i+2n}$$
$$=w_{i+1}\sum_{a=2}^{2n-1}w_{i+a}-w_i\sum_{a=2}^{2n-1}w_{i+a}+\sum_{a=2}^{2n-1}(-1)^a w_{i+a}^2\,,$$

which matches (2.22). ■

Proof of Proposition 1 continued. Every dressing chain has an obvious first integral, obtained by summing (2.12):

$$w_1(z)+\cdots+w_{2n}(z)=-\frac{1}{2}\Delta\, z\,. \tag{2.23}$$

For $p=2n$, the even-cyclic dressing chain also has an additional first integral — obtained by taking an alternating sum of (2.12):

$$2(w_1^2-w_2^2+\cdots-w_{2n}^2)=-a_1+a_2-\cdots+a_{2n}. \tag{2.24}$$

Using (2.21) and (2.23) we obtain

$$x\,d_i(x)=2\sum_{j=1}^{n}\sum_{k=1}^{n-1}\operatorname{sgn}(2k+1-2j)f_{2j+i-1}(x)f_{2k+i}(x)+2\sum_{j=0}^{2n-1}(-1)^j\alpha_{i+j}\,.$$

Relation (2.19) may now be rewritten as

$$\begin{aligned} xf_i'(x) = 2f_i(x)\sum_{j=1}^{n}\sum_{k=1}^{n-1}\operatorname{sgn}(2j-1+2k)f_{2j+i-1}(x)f_{2k+i}(x) \\ - 2f_i(x)\sum_{j=0}^{2n-1}(-1)^j\alpha_{i+j} + \alpha_i x\,. \end{aligned} \tag{2.25}$$

Since

$$\sum_{j=1}^{n} f_{2j-1}(x) = \sum_{j=1}^{n} f_{2j}(x) = \frac{1}{2}x,$$

and since

$$\sum_{j=1}^{n}\alpha_{2j-1} + \sum_{j=1}^{n}\alpha_{2j} = 1,$$

relation (2.25) may be rewritten as (2.17). ■

The problem now becomes that of finding and classifying cyclic dressing chains, sequences of Darboux transformations that reproduce the initial potential up to an additive shift Δ after a fixed given number of transformations. The theory of exceptional polynomials is intimately related with families of Schrödinger operators connected by Darboux transformations [16, 13]. Each of these exceptional operators admits a bilinear formulation in terms of τ-functions which suggests a strong connection with integrable systems theory, and which will be the basis of the development here. Each τ-function in this class can be indexed by a finite set of integers, or equivalently by a Maya diagram, which becomes a very useful representation to capture a notion of equivalence and relations of the type (2.10).

3 Hermite τ-functions

In this section we introduce Hermite-type τ-functions, their bilinear relations, and 3 determinantal representations of these objects: pseudo-Wronskians, Jacobi-Trudi formula, and a Boson-Fermion correspondence formula. We also introduce Maya diagrams, partitions, and indicate the relation between these two types of objects. In a nutshell, the Hermite-type τ function is a specialization of the more general Schur function. Various instances of this observation can be found in [8]. Schur functions arise in integrable systems theory as polynomial solutions of the KP hierarchy [40]. As shown by Tsuda [35], the Painlevé systems are actually reductions of the KP hierarchy. Theorem 1 is an indication of how this reduction manifests at the level of solutions.

Following Noumi [30], we introduce the following.

Definition 2. A Maya diagram is a set of integers $M \subset \mathbb{Z}$ that contains a finite number of positive integers, and excludes a finite number of negative integers. We will use $\mathcal{M}$ to denote the set of all Maya diagrams.

Let $k_1 > k_2 > \cdots$ be the decreasing enumeration of a Maya diagram $M \subset \mathbb{Z}$. The condition that M be a Maya diagram is equivalent to the condition that $k_{i+1} = k_i - 1$ for i sufficiently large. Thus, there exists a unique integer $\sigma_M \in \mathbb{Z}$ such that $k_i = -i + \sigma_M$ for all i sufficiently large. We call σ_M to the index of M.

We visualize a Maya diagram as a horizontally extended sequence of ◼ and □ symbols with the filled symbol ◼ in position i indicating membership $i \in M$. The defining assumption now manifests as the condition that a Maya diagram begins with an infinite filled ◼ segment and terminates with an infinite empty □ segment.

Definition 3. Let M be a Maya diagram, and

$$M_- = \{-m - 1 \colon m \notin M, m < 0\}, \qquad M_+ = \{m \colon m \in M\,, m \geq 0\}.$$

Let $s_1 > s_2 > \cdots > s_r$ and $t_1 > t_2 > \cdots > t_q$ be the elements of M_- and M_+ arranged in descending order. We call the double list $(s_1, \ldots, s_r \mid t_q, \ldots, t_1)$ the *Frobenius symbol* of M and use $M(s_1, \ldots, s_r \mid t_q, \ldots, t_1)$ to denote the Maya diagram with the indicated Frobenius symbol.

It is not hard to show that $\sigma_M = q - r$ is the index of M. The classical Frobenius symbol [4, 33, 3] corresponds to the zero index case where $q = r$.

If M is a Maya diagram, then for any $k \in \mathbb{Z}$ so is

$$M + k = \{m + k \colon m \in M\}.$$

The behaviour of the index σ_M under translation of k is given by

$$M' = M + k \quad \Rightarrow \quad \sigma_{M'} = \sigma_M + k. \tag{3.1}$$

We will refer to an equivalence class of Maya diagrams related by such shifts as an *unlabelled Maya diagram.* One can visualize the passage from an unlabelled to a labelled Maya diagram as the choice of placement of the origin.

Definition 4. A Maya diagram $M \subset \mathbb{Z}$ is said to be in standard form if $p = 0$ and $t_q > 0$. Equivalently, M is in standard form if the index $\sigma_M = q$ is the number of positive elements of M. Visually, a Maya diagram in standard form has only filled boxes ◼ to the left of the origin and one empty box □ just to the right of the origin. Every unlabelled Maya diagram permits a unique placement of the origin so as to obtain a Maya diagram in standard form.

In [18] it was shown that to every Maya diagram we can associate a polynomial called a Hermite pseudo-Wronskian. For $n \geq 0$, let

$$H_n(x) = (-1)^n e^{x^2} \left(\frac{d}{dx}\right)^n e^{-x^2} \tag{3.2}$$

denote the degree n Hermite polynomial, and

$$\tilde{H}_n(x) = \mathrm{i}^{-n} H_n(\mathrm{i}x) \tag{3.3}$$

the conjugate Hermite polynomial. A number of equivalent definition of H_n are available. One is that $y = H_n$ is the polynomial solution of the Hermite differential equation

$$y''(z) - 2zy'(z) + 2ny(z) = 0 \tag{3.4}$$

subject to the normalization condition

$$y(z) \sim 2^n z^n \quad z \to \infty.$$

Setting

$$\hat{H}_n(z) = e^{z^2} \tilde{H}_n(z)$$

we also note that $\hat{H}_{-n-1}$ is a solution of (3.4) for negative integers $n < 0$.

A third definition involves the 3-term recurrence relation:

$$H_{n+1}(z) = 2zH_n(z) - 2nH_{n-1}(z), \qquad H_0(z) = 1,\ H_1(z) = 2z.$$

A fourth definition involves the generating function

$$\sum_{n=0}^{\infty} H_n(z) \frac{t^n}{n!} = e^{2zt - t^2}. \tag{3.5}$$

Definition 5. For $s_1, \ldots, s_r, t_q, \ldots, t_1 \in \mathbb{Z}$ set

$$\tau(s_1, \ldots, s_r | t_q, \ldots, t_1) = e^{-rz^2} \operatorname{Wr}[\hat{H}_{s_1}(z), \ldots, \hat{H}_{s_r}(z), H_{t_q}(z), \ldots H_{t_1}(z)] \tag{3.6}$$

where Wr denotes the Wronskian determinant of the indicated functions. For a Maya diagram $M(s_1, \ldots, s_r \mid t_q, \ldots, t_1)$ we let

$$\tau_M(z) = \tau(s_1, \ldots, s_r | t_q, \ldots, t_1). \tag{3.7}$$

Note: when $r = 0$ it will be convenient to simply write

$$\tau(t_q, \ldots, t_1) = \operatorname{Wr}[H_{t_q}, \ldots, H_{t_1}] \tag{3.8}$$

to indicate a Wronskian of Hermite polynomials.

The polynomial nature of $\tau_M(z)$ becomes evident once we represent it using a slightly different determinant.

Proposition 2. *The Wronskian in* (3.6) *admits the following alternative* pseudo-Wronskian *representation*

$$\tau(s_1, \ldots, s_r | t_q, \ldots, t_1) = \begin{vmatrix} \tilde{H}_{s_1} & \tilde{H}_{s_1+1} & \cdots & \tilde{H}_{s_1+r+q-1} \\ \vdots & \vdots & \ddots & \vdots \\ \tilde{H}_{s_r} & \tilde{H}_{s_r+1} & \cdots & \tilde{H}_{s_r+r+q-1} \\ H_{t_q} & H'_{t_q} & \cdots & H^{(r+q-1)}_{t_q} \\ \vdots & \vdots & \ddots & \vdots \\ H_{t_1} & H'_{t_1} & \cdots & H^{(r+q-1)}_{t_1} \end{vmatrix}. \tag{3.9}$$

The proof of the above result can be found in [18]. The term Hermite pseudo-Wronskian was also introduced in that paper, because (3.9) is a mix of a Casoratian and a Wronskian determinant. The just mentioned article also demonstrated that the pseudo-Wronskians of two Maya diagrams related by a translation are proportional.

Proposition 3. *Let $\hat{\tau}_M$ be the normalized pseudo-Wronskian*

$$\hat{\tau}_M = \frac{(-1)^{rq}\tau(s_1,\ldots,s_r|t_q,\ldots,t_1)}{\prod_{1\le i<j\le r}(2s_j-2s_i)\prod_{1\le i<j\le q}(2t_i-2t_j)}. \tag{3.10}$$

Then for any Maya diagram M and $k\in\mathbb{Z}$ we have

$$\hat{\tau}_M = \hat{\tau}_{M+k}. \tag{3.11}$$

Observe that the identity in (3.11) involves determinants of different sizes, and a Wronskian of Hermite polynomials will not, in general, be the smallest determinant in the equivalence class. The question of which determinant has the smallest size was solved in [18].

We define a partition to be a non-increasing sequence of natural numbers $\lambda_1 \ge \lambda_2 \ge \cdots$ such that

$$|\lambda| := \sum_{i=1}^{\infty}\lambda_i < \infty.$$

Implicit in this definition is the assumption that $\lambda_i = 0$ for i sufficiently large. We define $\ell(\lambda)$, the length of λ, to be the smallest $q\in\mathbb{N}$ such that $\lambda_{q+1}=0$.

To a partition λ of length $q=\ell(\lambda)$ we associate the Maya diagram M_λ consisting of

$$t_i = \lambda_i + q - i, \quad i = 1,2,\ldots. \tag{3.12}$$

By construction, we have

$$t_q > 0, \quad \text{and} \quad t_{i+1}+1 = t_i < 0, \qquad i > q.$$

Therefore M_λ is a Maya diagram in standard form. Indeed, (3.12) defines a bijection between the set of partitions and the set of Maya diagrams in standard form. Going forward, let

$$\tau_\lambda = \mathrm{Wr}[H_{t_q},\ldots,H_{t_1}]. \tag{3.13}$$

For $n\in\mathbb{Z}$ and λ a partition, let

$$M_\lambda^{(n)} = M_\lambda + n - \ell(\lambda),$$

and let $t_1 > t_2 > \cdots$ be the decreasing enumeration of $M_\lambda^{(n)}$. Equivalently,

$$t_i = \lambda_i + n - i, \quad i = 1,2,\ldots. \tag{3.14}$$

Note that the condition $n \geq \ell(\lambda)$ holds if and only if $M_\lambda^{(n)}$ contains all negative integers and exactly n non-negative integers, that is if

$$M_\lambda^{(n)} = M(|\, t_n, \ldots, t_1).$$

Given univariate polynomials $p_1(z), \ldots, p_n(z)$, define the multivariate functions

$$\Delta[p_1, \ldots, p_n](z_1, \ldots, z_n) = \begin{vmatrix} p_1(z_1) & p_1(z_2) & \ldots & p_1(z_n) \\ p_2(z_1) & p_2(z_2) & \ldots & p_2(z_n) \\ \vdots & \vdots & \ddots & \vdots \\ p_n(z_1) & p_n(z_2) & \ldots & p_n(z_n) \end{vmatrix} \tag{3.15}$$

$$S[p_1, \ldots, p_n] = \frac{\Delta[p_1, \ldots, p_n]}{\Delta[\mathfrak{m}_{n-1}, \ldots, \mathfrak{m}_1, \mathfrak{m}_0]} \tag{3.16}$$

where

$$\mathfrak{m}_k(z) = z^k \tag{3.17}$$

is the kth degree monomial function. Thus,

$$\Delta[\mathfrak{m}_{n-1}, \ldots, \mathfrak{m}_0](z_1, \ldots, z_n) = \begin{vmatrix} z_1^{n-1} & \ldots & z_n^{n-1} \\ \vdots & \ddots & \vdots \\ 1 & \ldots & 1 \end{vmatrix} = \prod_{1 \leq i < j \leq n} (z_i - z_j)$$

is the usual Vandermonde determinant, while $S[p_1, \ldots, p_n]$ is a symmetric polynomial in $z_1, \ldots, z_n$.

Let λ be a partition. For $n \geq \ell(\lambda)$, let $t_1 > t_2 > \cdots$ be the decreasing enumeration of $M_\lambda^{(n)}$ as per (3.14). The n-variate Schur polynomial is the symmetric polynomial

$$\mathfrak{s}_\lambda^{(n)} = S[\mathfrak{m}_{t_1}, \ldots, \mathfrak{m}_{t_n}].$$

The Schur polynomial $\mathfrak{s}_\lambda^{(n)}$ is the character of the irreducible representation of the general linear group GL_n corresponding to partition λ. Moreover, the Weyl dimension formula asserts that

$$\mathfrak{s}_\lambda^{(n)}(1, \ldots, 1) = \prod_{1 \leq i < j \leq n} \frac{\lambda_i - \lambda_j + j - i}{j - i} = \left(\prod_{j=1}^{n-1} j! \right)^{-1} \prod_{1 \leq i < j \leq n} (t_i - t_j) \tag{3.18}$$

is the dimension of the representation in question.

For $n \geq 1$, let

$$\mathfrak{h}_k^{(n)}(z_1, \ldots, z_n) = \sum_{1 \leq i_1 \leq i_2 \leq \cdots \leq i_k \leq n} z_{i_1} z_{i_2} \cdots z_{i_k}, \quad k = 1, 2, \ldots$$

denote the complete symmetric polynomial of degree k in n variables. These polynomials may also be defined by means of the generating function

$$\sum_{k=0}^{\infty} \mathfrak{h}_k^{(n)}(z_1, \ldots, z_n) u^k = \prod_{i=1}^{n} \frac{1}{1 - z_i u}. \tag{3.19}$$

The classical Jacobi-Trudi identity is a determinantal representation of the Schur polynomials in terms of complete symmetric polynomials.

Proposition 4. *Let λ be a partition and $n \geq \ell(\lambda)$ we have*

$$\mathfrak{s}_\lambda^{(n)} = \det \left(\mathfrak{h}_{\lambda_i + j - i}^{(n)} \right)_{i,j=1}^{\ell(\lambda)}. \tag{3.20}$$

We now describe a closely related identity based on symmetric power functions. Define the ordinary Bell polynomials $\mathfrak{B}_k(t_1, \ldots, t_k)$, $k = 0, 1, 2, \ldots$ by means of the power generating function

$$\exp \left(\sum_{k=0}^{\infty} t_k u^k \right) = \sum_{k=0}^{\infty} \mathfrak{B}_k(t_1, \ldots, t_k) u^k. \tag{3.21}$$

Since

$$\exp \left(\sum_{j=0}^{\infty} t_k u^k \right) = \sum_{j=0}^{\infty} \frac{1}{j!} \left(\sum_{k=0}^{\infty} t_k u^k \right)^j,$$

the multinomial formula implies that,

$$\begin{aligned} \mathfrak{B}_k(t_1, \ldots, t_k) &= \sum_{\substack{j_1, \ldots, j_\ell \geq 0 \\ \|j\| = n}} \frac{t_1^{j_1}}{j_1!} \frac{t_2^{j_1}}{j_2!} \cdots \frac{t_\ell^{j_\ell}}{j_\ell!}, \qquad \|j\| = j_1 + 2j_2 + \cdots + \ell j_\ell \\ &= \frac{t_1^k}{k!} + \frac{t_1^{k-2} t_2}{(k-2)!} + \cdots + t_{k-1} t_1 + t_k. \end{aligned}$$

The Bell polynomials are instrumental in describing the relation between complete homogeneous polynomials and symmetric power polynomials. For a given $n \geq 1$, let

$$\mathfrak{p}_k^{(n)}(z_1, \ldots, z_n) = \sum_{j=1}^{n} z_j^k$$

denote the symmetric k^{th} power polynomial in n variables. These polynomials admit the following generating function

$$-\log \prod_{j=1}^{n} (1 - z_i u) = \sum_{j=1}^{\infty} \mathfrak{p}_j^{(n)}(z_1, \ldots, z_n) \frac{u^j}{j}. \tag{3.22}$$

Comparing the generating functions (3.19) (3.21) (3.22) yields the following identity

$$\mathfrak{h}_k^{(n)} = \mathfrak{B}_k\left(\mathfrak{p}_1^{(n)}, \frac{1}{2}\mathfrak{p}_2^{(n)}, \ldots, \frac{1}{k}\mathfrak{p}_k^{(n)}\right). \tag{3.23}$$

With these preliminaries out of the way we can present the following alternative version of the Jacobi-Trudi formula. For a partition λ, we define the Schur function $\mathfrak{S}_\lambda$ to be the multivariate polynomial

$$\mathfrak{S}_\lambda = \det(\mathfrak{B}_{\lambda_i+j-i})_{i,j=1}^{\ell(\lambda)}. \tag{3.24}$$

Relation (3.20) may now be restated as

$$\mathfrak{s}_\lambda^{(n)} = \mathfrak{S}_\lambda(\mathfrak{p}_1^{(n)}, \frac{1}{2}\mathfrak{p}_2^{(n)}, \ldots, \frac{1}{k}\mathfrak{p}_k^{(n)}, \ldots). \tag{3.25}$$

Relation (3.25) will be instrumental in the proof of the following.

Theorem 1. *Let λ be a partition. Then,*

$$\tau_\lambda(z) = C_\lambda \mathfrak{S}_\lambda(2z, -1),$$

where

$$C_\lambda = 2^{n(n-1)/2} \prod_{j=1}^{n} (\lambda_i + n - i)!, \qquad n = \ell(\lambda).$$

Proof. Specializing (3.21) and using (3.5), we observe that

$$\sum_{k=0}^{\infty} \mathfrak{B}(2z, -1)u^k = \exp(2zu - u^2/2) = \sum_{k=0}^{\infty} H_n(z)\frac{u^k}{n!}$$

Hence,

$$\mathfrak{B}_n(2z, -1) = \frac{H_n(z)}{n!}, \quad n = 0, 1, 2, \ldots,$$

Hence, by (3.24),

$$\mathfrak{S}_\lambda(2z, -1) = \det\left(\frac{H_{\lambda_i+j-i}(z)}{(\lambda_i + j - i)!}\right)_{i,j=1}^{\ell(\lambda)}.$$

Hence, by the identity

$$H_n' = 2nH_{n-1}, \quad n = 1, 2, \ldots$$

we have

$$H_{t_i}^{(n-j)} = \frac{t_i!}{(\lambda_i + j - i)!} 2^{n-j} H_{\lambda_i+j-i}, \quad i, j = 1, \ldots, n,$$

where as above,

$$t_i = \lambda_i + n - i, \qquad n = \ell(\lambda).$$

Hence, from the definition (3.13), we have

$$\tau_\lambda = \det\left(2^{n-j} t_i! \frac{H_{\lambda_i+j-i}(z)}{(\lambda_i + j - i)!}\right) = C_\lambda \mathfrak{S}_\lambda(2z, -1).$$

■

The following results will lead to yet another description of the Hermite-type τ-function, one that is related to the Boson-Fermion correspondence [19] (although we do not discuss this here).

Proposition 5. *Let $p_1(z), \ldots, p_n(z)$ be polynomials. Then,*

$$W[p_1, \ldots, p_n](z) = \left(\prod_{j=1}^{n-1} j!\right) S[p_1, \ldots, p_n](z, \ldots, z). \tag{3.26}$$

Proof. Express the given polynomials as

$$p_i = \sum_{j=0}^{\infty} p_{ij} \mathfrak{m}_j, \quad p_{ij} \in \mathbb{C}.$$

Note: The above sum is actually finite, because $p_{ij} = 0$ for j sufficiently large. Hence,

$$\begin{aligned}
\mathrm{Wr}[p_1, \ldots, p_n] &= \sum_{t_1, \ldots, t_n = 0}^{\infty} \left(\prod_{i=1}^{n} p_{it_i}\right) \mathrm{Wr}[\mathfrak{m}_{t_1}, \ldots, \mathfrak{m}_{t_n}] \\
&= \sum_{t_1 > \cdots > t_n \geq 0} \sum_{\pi \in \mathcal{S}_n} \mathrm{sgn}(\pi) \left(\prod_{i=1}^{n} p_{it_{\pi_i}}\right) \mathrm{Wr}[\mathfrak{m}_{t_1}, \ldots, \mathfrak{m}_{t_n}]
\end{aligned}$$

where $\mathcal{S}_n$ is the group of permutations of $\{1, \ldots, n\}$.

By an elementary calculation,

$$\mathrm{Wr}[\mathfrak{m}_{t_1}, \ldots, \mathfrak{m}_{t_n}] = \prod_{1 \leq i < j \leq n} (t_j - t_i)\, \mathfrak{m}_{|\lambda|}.$$

Thus, by introducing the abbreviation

$$p_\lambda = \sum_{\pi \in \mathcal{S}_n} \mathrm{sgn}(\pi) \prod_{i=1}^{n} p_{it_{\pi_i}}$$

and making use of (3.18) we may write

$$\mathrm{Wr}[p_1, \ldots, p_n] = \prod_{j=1}^{n-1} j! \sum_{\ell(\lambda) \leq n} \mathfrak{s}_\lambda(1, \ldots, 1) p_\lambda \mathfrak{m}_{|\lambda|}. \tag{3.27}$$

Of course, the sum is actually finite because $p_\lambda = 0$ if $\lambda_1 > \max\{\deg p_1, \ldots, \deg p_n\}$.

By the multi-linearity and skew-symmetry of the determinant (3.16),

$$\begin{aligned} S[p_1, \ldots, p_n] &= \sum_{t_1, \ldots, t_n = 0}^{\infty} \left(\prod_{i=1}^{n} p_{it_i} \right) S[\mathfrak{m}_{t_1}, \ldots, \mathfrak{m}_{t_n}] \\ &= \sum_{t_1 > \cdots > t_n \geq 0} \sum_{\pi \in \mathcal{S}_n} \operatorname{sgn}(\pi) \left(\prod_{i=1}^{n} p_{it_{\pi_i}} \right) S[\mathfrak{m}_{t_1}, \ldots, \mathfrak{m}_{t_n}] \\ &= \sum_{\ell(\lambda) \leq n} p_\lambda \mathfrak{s}_\lambda . \end{aligned}$$

Since $\mathfrak{s}_\lambda$ is a homogeneous polynomial whose total degree is equal to $|\lambda|$, we have

$$\mathfrak{s}_\lambda(z, \ldots, z) = \mathfrak{s}_\lambda(1, \ldots, 1) z^{|\lambda|}. \tag{3.28}$$

Hence,

$$S[p_1, \ldots, p_n](z, \ldots, z) = \sum_{\lambda} p_\lambda \mathfrak{s}_\lambda(1, \ldots, 1) z^{|\lambda|}.$$

The desired conclusion now follows directly by (3.27). ■

Corollary 1. For $0 \leq t_q < \cdots < t_1$ we have

$$\tau(t_q, \ldots, t_1)(z) = \left(\prod_{j=1}^{n-1} j! \right) S[H_{t_q}, \ldots, H_{t_1}](z, \ldots, z).$$

4 Hermite-type rational solutions

In this section we develop the relationship between cyclic dressing chains and Hermite-type τ-functions. Every element of the chain will be represented by a Maya diagram with successive elements related by flip operations. With this representation, the construction of rational solutions to a cyclic dressing chain is reduced to a combinatorial question regarding cyclic Maya diagrams.

In what follows we make use of the Hirota bilinear notation:

$$Df \cdot g = f'g - g'f \tag{4.1}$$

$$D^2 f \cdot g = f''g - 2f'g' + g''f \tag{4.2}$$

A direct calculation then establishes the following.

Proposition 6. *Let $f = f(z), g = g(z)$ be rational functions, and let*

$$w = -z - \frac{f'}{f} + \frac{g'}{g}, \tag{4.3}$$

$$U = z^2 - 2(\log f)'', \tag{4.4}$$

$$V = z^2 - 2(\log g)'', \tag{4.5}$$

where $w = w(z), U = U(z), V = V(z)$. *Then,*

$$\begin{aligned}(D^2 - 2zD)f \cdot g &= (w^2 + w' + 1 - U)fg \\ &= (w^2 - w' - 1 - V)fg.\end{aligned} \tag{4.6}$$

We are now able to exhibit a bilinear formulation for the dressing chain (2.12).

Proposition 7. *Suppose that* $\tau_i = \tau_i(z), \epsilon_i, \sigma_i \in \{-1, 1\},\ i = 0, 1, \ldots, p-1$ *is a sequence of functions and constants that satisfies*

$$(D^2 + 2\sigma_i zD + \epsilon_i)\tau_i \cdot \tau_{i+1} = 0, \qquad i = 0, 1, \ldots, p-1 \mod p. \tag{4.7}$$

Then,

$$\begin{aligned}w_i &= \sigma_i z - \frac{\tau_i'}{\tau_i} + \frac{\tau_{i+1}'}{\tau_{i+1}}, \qquad i = 0, 1, \ldots, p-1 \mod p \\ a_i &= \epsilon_i - \epsilon_{i+1} + \sigma_i + \sigma_{i+1}\end{aligned} \tag{4.8}$$

satisfy the p-cyclic dressing chain (2.12) *with*

$$\Delta = -2\sum_{j=0}^{p-1} \sigma_i. \tag{4.9}$$

Proof. Set

$$U_i(z) = z^2 - 2(\log \tau_i)''(z) - 2\sum_{j=0}^{i-1} \sigma_j, \quad i = 0, 1, 2, \ldots, p$$

$$\lambda_i = \epsilon_i - 2\sum_{j=0}^{i-1} \sigma_j - \sigma_i.$$

Observe that

$$(D^2 - 2\sigma_i zD + \lambda)\tau_{i+1} \cdot \tau_i = (D^2 + 2\sigma_i zD + \lambda)\tau_i \cdot \tau_{i+1}.$$

Hence, by (4.6),

$$\begin{aligned}w_i^2 + w_i' + \epsilon_i &= z^2 - 2(\log \tau_i)'' + \sigma_i \\ w_i^2 - w_i' + \epsilon_i &= z^2 - 2(\log \tau_{i+1})'' - \sigma_i.\end{aligned}$$

The above relations are equivalent to (2.5), and hence the $U_i(z), \lambda_i$ constitute a periodic factorization chain (2.4) with $\Delta = U_p - U_0$ given by (4.9). Applying (2.8), we obtain

$$a_i = \lambda_i - \lambda_{i+1} = \epsilon_i - \epsilon_{i+1} + \sigma_i + \sigma_{i+1}.$$

Hence, with (4.8) as definition of $w_i(z)$ and a_i, relation (2.12) is satisfied. ∎

Note: This proposition should not be taken as a claim that *all* rational solutions of (2.12) may be obtained in this fashion.

In order to obtain polynomial solutions of (4.7) we now introduce the following.

Definition 6. A flip at $m \in \mathbb{Z}$ is the involution $\phi_m : \mathcal{M} \to \mathcal{M}$ defined by

$$\phi_m : M \mapsto \begin{cases} M \cup \{m\} & \text{if } m \notin M \\ M \setminus \{m\} & \text{if } m \in M \end{cases}, \qquad M \in \mathcal{M}. \tag{4.10}$$

In the first case, we say that ϕ_m acts on M by a state-deleting transformation (□ → ▣). In the second case, we say that ϕ_m acts by a state-adding transformation (▣ → □). We define the *flip group* $\mathcal{F}$ to be the group of transformations of $\mathcal{M}$ generated by flips ϕ_m, $m \in \mathbb{Z}$. A *multi-flip* is an element of $\mathcal{F}$.

Theorem 2. *Let $M_1 \subset \mathbb{Z}$ be a Maya diagram, and $M_2 = M_1 \cup \{m\}$, $m \notin M_1$ another Maya diagram obtained by a state-deleting transformation. Then, the corresponding pseudo-Wronskians satisfy the bilinear relation*

$$(D^2 - 2zD + \epsilon)\tau_{M_1} \cdot \tau_{M_2} = 0, \quad \tau_M = \tau_M(z) \tag{4.11}$$

where

$$\epsilon = 2(\deg \tau_{M_2} - \deg \tau_{M_1}). \tag{4.12}$$

Conversely, suppose that (4.11) *holds for some Maya diagrams $M_1, M_2 \subset \mathbb{Z}$ and some $\epsilon \in \mathbb{C}$. Then, necessarily M_2 is obtained from M_1 by a state-deleting transformation and ϵ takes the value shown above.*

The proof of this theorem requires a number of intermediate results.

Lemma 2. *Let $\tau_0(z), \tau_1(z), \tau_2(z), \tau_3(z)$ be rational functions such that*

$$\tau_0\tau_2 = \mathrm{Wr}[\tau_1, \tau_3]. \tag{4.13}$$

Associate the edges of the following diagram to the bilinear relations displayed below:

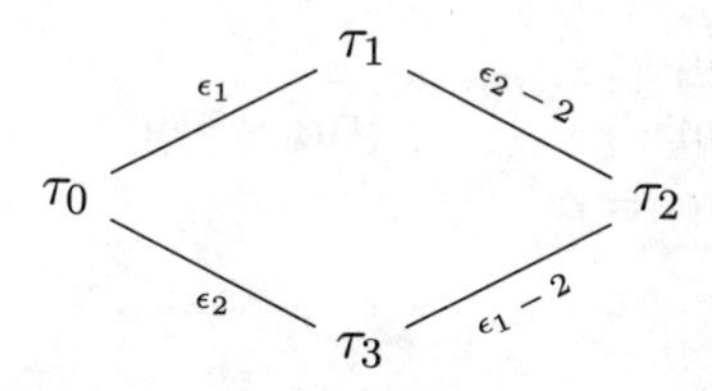

$$(D^2 - 2zD + \epsilon_1)\tau_0 \cdot \tau_1 = 0, \tag{4.14}$$
$$(D^2 - 2zD + \epsilon_2)\tau_0 \cdot \tau_3 = 0, \tag{4.15}$$
$$(D^2 - 2zD + \epsilon_2 - 2)\tau_1 \cdot \tau_2 = 0, \tag{4.16}$$
$$(D^2 - 2zD + \epsilon_1 - 2)\tau_3 \cdot \tau_2 = 0, \tag{4.17}$$

where $\epsilon_1, \epsilon_2 \in \mathbb{C}$ are constants. Then, necessarily any two relations corresponding to connected edges entail the other two relations.

Proof. The lemma asserts two claims. First that (4.14) (4.15) together, are logically equivalent to (4.16) (4.17) together. The other assertion is that (4.14) (4.16) together are logically equivalent to (4.15) (4.17) together. We will demonstrate how (4.14) (4.15) together imply (4.16). All the other demonstrations can be argued analogously, and so we omit them.

Begin by setting

$$w_{ij}(z) = -z - \frac{\tau_i'(z)}{\tau_i(z)} + \frac{\tau_j'(z)}{\tau_j(z)}, \quad i \in \{0,1,2,3\},\ i \neq j, \tag{4.18}$$

$$U_i(z) = z^2 - 2(\log \tau_i)''(z), \quad i \in \{0,1,2,3\}. \tag{4.19}$$

By (4.6) of Proposition 6,

$$U_0 = w_{01}^2 + w_{01}' + \epsilon_1 + 1 = w_{03}^2 + w_{03}' + \epsilon_2 + 1,$$

which we rewrite as

$$\frac{w_{03}' - w_{01}'}{w_{03} - w_{01}} = -w_{03} - w_{01} + \frac{\alpha}{w_{03} - w_{01}}, \quad \alpha = \epsilon_1 - \epsilon_2. \tag{4.20}$$

Write

$$\mathrm{Wr}[\tau_1, \tau_3] = \tau_1 \tau_3 \left(\frac{\tau_3'}{\tau_3} - \frac{\tau_1'}{\tau_1} \right) = \tau_1 \tau_3 (w_{03} - w_{01}).$$

Hence, by (4.13),

$$\tau_0 \tau_2 = \tau_1 \tau_3 (w_{03} - w_{01})$$

$$\frac{\tau_2'}{\tau_2} = -\frac{\tau_0'}{\tau_0} + \frac{\tau_1'}{\tau_1} + \frac{\tau_3'}{\tau_3} + \frac{w_{03}' - w_{01}'}{w_{03} - w_{01}}$$

which by (4.18) (4.20) maybe rewritten as

$$w_{12} = w_{03} + \frac{w_{03}' - w_{01}'}{w_{03} - w_{01}} = -w_{01} + \frac{\alpha}{w_{03} - w_{01}},$$

$$(w_{01} + w_{12})(w_{03} - w_{01}) = \alpha. \tag{4.21}$$

It follows that

$$\frac{w_{12}' + w_{01}'}{w_{12} + w_{01}} + \frac{w_{03}' - w_{01}'}{w_{03} - w_{01}} = 0.$$

Hence,

$$w_{12} - w_{01} + \frac{w_{12}' + w_{01}'}{w_{12} + w_{01}} - \frac{\alpha}{w_{12} + w_{01}} =$$

$$= w_{03} + \frac{w_{03}' - w_{01}'}{w_{03} - w_{01}} - w_{01} + \frac{w_{12}' + w_{01}'}{w_{12} + w_{01}} - \frac{\alpha}{w_{12} + w_{01}} = 0.$$

Equivalently,

$$w_{12}^2 - w_{01}^2 + w_{12}' + w_{01}' - \alpha = 0. \tag{4.22}$$

Hence, by (4.6)

$$U_1 = w_{01}^2 - w_{01}' + \epsilon_1 - 1 = w_{12}^2 + w_{12}' + \epsilon_2 - 1.$$

Relation (4.16) now follows by Proposition 6. ■

Proof of Theorem 2. Suppose that $M_2 = M_1 \cup \{m\}$, $m \notin M_1$. We claim that (4.11) holds. By Proposition 3 no generality is lost if we assume that M_1 is in standard form, and hence that

$$\tau_{M_1} = \tau(t_q, \dots t_1), \quad t_q < \dots < t_1$$

is a pure Wronskian. We proceed by induction on q, the number of positive elements of M_1. If $q = 0$, then $\tau_{M_1} = 1$ and $\tau_{M_2} = H_m$. The bilinear relation (4.11) is then nothing but the classical Hermite differential equation

$$H_m''(z) - 2zH_m'(z) + 2mH_m(z) = 0.$$

Suppose that $q > 0$ and that the claim has been shown to be true for all M_1 with fewer positive elements. Set

$$M_0 = M_1 \setminus \{t_q\}, \qquad M_2 = M_0 \cup \{m\}.$$

By an elementary calculation,

$$\begin{aligned}
\deg \tau_{M_1} &= \sum_{i=1}^{q} m_i - \frac{1}{2}q(q-1) \\
\deg \tau_{M_2} &= \sum_{i=1}^{q} m_i + m - \frac{1}{2}q(q+1) \\
\deg \tau_{M_0} &= \sum_{i=1}^{q-1} m_i - \frac{1}{2}(q-1)(q-2) \\
\deg \tau_{M_3} &= \sum_{i=1}^{q-1} m_i + m - \frac{1}{2}q(q-1).
\end{aligned}$$

Hence, by the inductive hypothesis, (4.14) (4.15) hold with

$$\begin{aligned}
\epsilon_1 &= 2(\deg \tau_{M_1} - \deg \tau_{M_0}) = 2m_q - 2q + 2, \\
\epsilon_2 &= 2(\deg \tau_{M_3} - \deg \tau_{M_0}) = 2m - 2q + 2.
\end{aligned}$$

Observe that

$$2(\deg \tau_{M_2} - \deg \tau_{M_1}) = 2m - 2q.$$

Hence, by Lemma 2 (4.11) holds also.

Conversely, suppose that (4.11) holds for some $\lambda \in \mathbb{C}$. We claim that $M_2 = M_1 \cup \{m\}$, $m \notin M_1$. Without loss of generality, suppose that

$$M_1 = (\,|\, t_q, \dots, t_1), \quad t_1 > \cdots > t_q > 0$$

is in standard form and hence that

$$\tau_{M_1} = \tau(t_q, \dots, t_1).$$

Using Proposition 3, assume without loss of generality that

$$\tau_{M_2} = \tau(\hat{t}_{\hat{q}}, \dots, \hat{t}_1),$$

is also a pure Wronskian with $\hat{q} \geq q$. Using the reduction argument above it then suffices to demonstrate this claim for the case where M_1 is the trivial Maya diagram; $q = 0$. In this case, (4.11) reduces to the Hermite differential equation. The Hermite polynomials are the unique polynomial solutions and so the claim follows. ■

We see thus that Hermite-type τ-functions are indexed by Maya diagrams, and that flip operations on Maya diagrams correspond to bilinear relations between these τ-functions. Since p-cyclic chains of bilinear relations (4.7) correspond to rational solutions of the A_{p-1} Painlevé system, it now becomes feasible to construct rational solutions in terms of cycles of Maya diagrams.

Definition 7. For $p = 1, 2, \dots$ and $\boldsymbol{\mu} = (\mu_0, \dots, \mu_{p-1}) \in \mathbb{Z}^p$, let

$$\phi_{\boldsymbol{\mu}} = \phi_{\mu_0} \circ \cdots \circ \phi_{\mu_{p-1}}, \tag{4.23}$$

denote the indicated multi-flip. We will call $\boldsymbol{\mu} \in \mathbb{Z}^p$ non-degenerate if the set $\{\mu_0, \dots, \mu_{p-1}\}$ has cardinality p, that is if $\mu_i \neq \mu_j$ for $i \neq j$. If $\hat{\boldsymbol{\mu}} \subset \mathbb{Z}$ is a set of cardinality q, we let $\phi_{\hat{\boldsymbol{\mu}}} = \phi_{\boldsymbol{\mu}}$, where $\boldsymbol{\mu} \in \mathbb{Z}^q$ is any non-degenerate enumeration of $\hat{\boldsymbol{\mu}}$. Finally, given a finite set $\hat{\boldsymbol{\mu}} \subset \mathbb{Z}$ we will call a sequence $\boldsymbol{\mu} = (\mu_0, \dots, \mu_{p-1})$ an *odd enumeration* of $\hat{\boldsymbol{\mu}}$ if

$$\hat{\boldsymbol{\mu}} = \{a \in \mathbb{Z} \colon m_{\boldsymbol{\mu}}(a) \equiv 1 \mod 2\},$$

where $m_{\boldsymbol{\mu}}(a) \geq 0$ is the number of times that $a \in \mathbb{Z}$ occurs in $\boldsymbol{\mu}$.

Proposition 8. *For $\boldsymbol{\mu} \in \mathbb{Z}^p$ and a finite $\hat{\boldsymbol{\mu}} \subset \mathbb{Z}$ we have $\phi_{\boldsymbol{\mu}} = \phi_{\hat{\boldsymbol{\mu}}}$ if and only if $\boldsymbol{\mu}$ is an odd enumeration of $\hat{\boldsymbol{\mu}}$.*

We are now ready to introduce the basic concept of this section.

Definition 8. We say that M is p-cyclic with shift k, or (p, k) cyclic, if there exists a $\boldsymbol{\mu} \in \mathbb{Z}^p$ such that

$$\phi_{\boldsymbol{\mu}}(M) = M + k. \tag{4.24}$$

We will say that M is p-cyclic if it is (p, k) cyclic for some $k \in \mathbb{Z}$.

Proposition 9. *For Maya diagrams $M, M' \in \mathcal{M}$, define the set*

$$\Upsilon(M, M') = (M \setminus M') \cup (M' \setminus M). \tag{4.25}$$

Then $\phi_{\hat{\boldsymbol{\mu}}}$ where $\hat{\boldsymbol{\mu}} = \Upsilon(M, M')$ is the unique multi-flip such that $M' = \phi_{\hat{\boldsymbol{\mu}}}(M)$ and $M = \phi_{\hat{\boldsymbol{\mu}}}(M')$.

As an immediate corollary, we have the following.

Proposition 10. *Let k be an integer, $M \in \mathcal{M}$ a Maya diagram, and let p be the cardinality of $\Upsilon(M, M+k)$. Then M is $(p+2j, k)$-cyclic for every $j = 0, 1, 2, \ldots$.*

Proof. Let $\hat{\boldsymbol{\mu}} = \Upsilon(M, M+k)$. Then $\phi_{\hat{\boldsymbol{\mu}}}(M) = M + k$, by the preceding proposition, and hence M is (p, k) cyclic. Let $\boldsymbol{\mu} \in \mathbb{Z}^p$ be an odd enumeration of $\hat{\boldsymbol{\mu}}$; i.e., $\boldsymbol{\mu}$ is obtained by adjoining j pairs of repeated indices to the elements of $\hat{\boldsymbol{\mu}}$. By Proposition 8,

$$\phi_{\boldsymbol{\mu}}(M) = \phi_{\hat{\boldsymbol{\mu}}}(M) = M + k,$$

and therefore M is also $(p+2j, k)$-cyclic. ■

The following result should be regarded as a refinement of Proposition 7.

Proposition 11. *Let $M \in \mathcal{M}$ be a Maya diagram, k a non-zero integer, and $(\mu_0, \ldots, \mu_{p-1}) \in \mathbb{Z}^p$ an odd enumeration of $\Upsilon(M, M+k)$. Extend $\boldsymbol{\mu}$ to an infinite p-quasiperiodic sequence by letting*

$$\mu_{i+p} = \mu_i + k, \quad , k = 0, 1, 2, \ldots$$

and recursively define

$$\begin{aligned} M_0 &= M, \\ M_{i+1} &= \phi_{\mu_i}(M_i), \quad i = 0, 1, 2, \ldots \end{aligned} \tag{4.26}$$

so that $M_{i+p} = M_i + k$ by construction. Next, for $i = 0, 1, \ldots$, let

$$\begin{aligned} \tau_i &= \tau_{M_i}, \\ a_i &= 2(\mu_i - \mu_{i+1}), \\ \sigma_i &= \begin{cases} +1 & \text{if } \mu_i \in M_i \\ -1 & \text{if } \mu_i \in M_{i+1} \end{cases}, \\ w_i(z) &= \sigma_i\, z + \frac{\tau'_{i+1}(z)}{\tau_{i+1}(z)} - \frac{\tau'_i(z)}{\tau_i(z)}, \end{aligned} \tag{4.27}$$

which are all p-periodic by construction. The just-defined $w_i(z), a_i,\ i = 0, 1, \ldots, p-1$ mod p constitute a rational solution to the p-cyclic dressing chain (2.12) with shift $\Delta = 2k$.

Proof. Set

$$\epsilon_i = 2\sigma_i(\deg \tau_i - \tau_{i+1}), \quad i = 0, 1, \dots p-1 \mod p.$$

Theorem 2 then implies that

$$(D^2 + 2\sigma_i z D + \epsilon_i)\tau_i \cdot \tau_{i+1} = 0.$$

Proposition 3 allows us to assume, without loss of generality, that

$$M_i = M(|\, t_{i,q_i}, \dots, t_{i,1})$$

and that $\mu_i \geq 0$ for all i. Such an outcome can be imposed by applying a sufficiently positive translation to the Maya diagrams in question, without altering the log-derivatives of the τ-functions. Since each τ_i is a Wronskian of polynomials of degrees $t_{i,1}, \dots, t_{i,q_i}$, we have

$$\deg \tau_i = \sum_{j=1}^{q_i} t_{i,j} - \frac{1}{2} q_i(q_i - 1).$$

If $\sigma_i = -1$, then

$$\{t_{i+1,1}, \dots, t_{i+1,q_{i+1}}\} = \{t_{i,1}, \dots, t_{i,q_i}\} \cup \{\mu_i\}, \qquad q_{i+1} = q_i + 1.$$

If $\sigma_i = +1$, then

$$\{t_{i,1}, \dots, t_{i+1,q_i}\} = \{t_{i+1,1}, \dots, t_{i+,q_{i+1}}\} \cup \{\mu_i\}, \quad q_{i+1} = q_i - 1.$$

It follows that

$$\begin{aligned} \epsilon_i &= 2\sigma_i(\deg \tau_i - \deg \tau_{i+1}) = 2\mu_i - 2q_i + 1 + \sigma_i, \\ \epsilon_i - \epsilon_{i+1} + \sigma_i + \sigma_{i+1} &= 2\mu_i - 2\mu_{i+1} + 2(q_{i+1} - q_i) + \sigma_i - \sigma_{i+1} + \sigma_i + \sigma_{i+1} \\ &= 2(\mu_i - \mu_{i+1}). \end{aligned}$$

Hence, the definition of a_i in (4.27) agrees with the definition in (4.8). Therefore, $w_i(z), a_i, i = 0, 1, \dots, p-1$ satisfy (2.12) by Proposition 7. Finally, by (2.11),

$$\Delta = -\sum_{i=0}^{p-1} a_i = 2\sum_{i=0}^{p-1}(\mu_{i+1} - \mu_i) = 2(\mu_p - \mu_0) = 2k.$$

■

The remaining part of the construction is to classify cyclic Maya diagrams for a given period, which we tackle next. Under the correspondence described by Proposition 11, the reversal symmetry (2.13) manifests as the transformation

$$(M_0, \dots, M_p) \mapsto (M_p, \dots, M_0), \quad (\mu_1, \dots, \mu_p) \mapsto (\mu_p, \dots, \mu_1), \quad k \mapsto -k.$$

In light of the above remark, there is no loss of generality if we restrict our attention to cyclic Maya diagrams with a positive shift $k > 0$.

5 Cyclic Maya diagrams

In this section we introduce the key concepts of *genus* and *interlacing* to achieve a full classification of cyclic Maya diagrams.

Definition 9. For $p = 1, 2, \ldots$ let $\mathcal{Z}_p \subset \mathbb{Z}^p$ denote the set of non-decreasing integer sequences[1] $\beta_0 \leq \beta_1 \leq \cdots \leq \beta_{p-1}$ of integers. For $\boldsymbol{\beta} \in \mathcal{Z}_{2g+1}$ define the Maya diagram

$$\Xi(\boldsymbol{\beta}) = (-\infty, \beta_0) \cup [\beta_1, \beta_2) \cup \cdots \cup [\beta_{2g-1}, \beta_{2g}) \tag{5.1}$$

where

$$[m, n) = \{j \in \mathbb{Z} \colon m \leq j < n\}.$$

Let $\hat{\mathcal{Z}}_p \subset \mathbb{Z}^p$ denote the set of strictly increasing integer sequences $\beta_0 < \beta_1 < \cdots < \beta_{p-1}$. Equivalently, we may regard a $\boldsymbol{\beta} \in \hat{\mathcal{Z}}_p$ as a p-element subset of $\mathbb{Z}$.

Proposition 12. *Every Maya diagram $M \in \mathcal{M}$ has a unique representation of the form $M = \Xi(\boldsymbol{\beta})$ where $\boldsymbol{\beta} \in \hat{\mathcal{Z}}_{2g+1}$. Moreover, M is in standard form if and only if* $\min \boldsymbol{\beta} = 0$.

Proof. After removal of the initial infinite ▣ segment and the trailing infinite □ segment, a given Maya diagram M consists of $2g$ alternating empty □ and filled ▣ segments of variable length. The genus g counts the number of such pairs. The even block coordinates β_{2i} indicate the starting positions of the empty segments, and the odd block coordinates β_{2i+1} indicate the starting positions of the filled segments. See Figure 1 for a visual illustration of this construction. ■

Definition 10. We call the integer $g \geq 0$ the genus of $M = \Xi(\boldsymbol{\beta})$, $\boldsymbol{\beta} \in \hat{\mathcal{Z}}_p$ and $(\beta_0, \beta_1, \ldots, \beta_{2g})$ the block coordinates of M.

Proposition 13. *Let $M = \Xi(\boldsymbol{\beta})$, $\boldsymbol{\beta} \in \hat{\mathcal{Z}}_p$ be a Maya diagram specified by its block coordinates. We then have*

$$\boldsymbol{\beta} = \Upsilon(M, M+1).$$

Proof. Observe that

$$M + 1 = (-\infty, \beta_0] \cup (\beta_1, \beta_2] \cup \cdots \cup (\beta_{2g-1}, \beta_{2g}],$$

where

$$(m, n] = \{j \in \mathbb{Z} \colon m < j \leq n\}.$$

It follows that

$$(M+1) \setminus M = \{\beta_0, \ldots, \beta_{2g}\}$$
$$M \setminus (M+1) = \{\beta_1, \ldots, \beta_{2g-1}\}.$$

The desired conclusion follows immediately. ■

[1] Equivalently, we may regard $\mathcal{Z}_p$ as the set of p-element integer multi-sets. A multi-set is generalization of the concept of a set that allows for multiple instances for each of its elements.

Let $\mathcal{M}_g$ denote the set of Maya diagrams of genus g. The above discussion may be summarized by saying that the mapping (5.1) defines a bijection $\Xi : \hat{\mathcal{Z}}_{2g+1} \to \mathcal{M}_g$, and that the block coordinates are precisely the flip sites required for a translation $M \mapsto M+1$.

	−3	−2	−1	0	1	2	3	4	5	6	7	8	9	10	11	
…	●	●	●	●	●		●	●			●	●	●			…
						β_0	β_1		β_2		β_3			β_4		

Figure 1. Block coordinates $(\beta_0, \ldots, \beta_4) = (2, 3, 5, 7, 10)$ of a genus 2 Maya diagram $M = (-\infty, \beta_0) \cup [\beta_1, \beta_2) \cup [\beta_3, \beta_4)$. Note that the genus is both the number of finite-size empty blocks and the number of finite-size filled blocks.

The next concept we need to introduce is the interlacing and modular decomposition.

Definition 11. Fix a $k \in \mathbb{N}$ and let $M^{(0)}, M^{(1)}, \ldots M^{(k-1)} \subset \mathbb{Z}$ be sets of integers. We define the interlacing of these to be the set

$$\Theta\left(M^{(0)}, M^{(1)}, \ldots M^{(k-1)}\right) = \bigcup_{i=0}^{k-1} (kM^{(i)} + i), \tag{5.2}$$

where

$$kM + j = \{km + j \colon m \in M\}, \quad M \subset \mathbb{Z}.$$

Dually, given a set of integers $M \subset \mathbb{Z}$ and a $k \in \mathbb{N}$ define the sets

$$M^{(i)} = \{m \in \mathbb{Z} \colon km + i \in M\}, \quad i = 0, 1, \ldots, k-1.$$

We will call the k-tuple of sets $\left(M^{(0)}, M^{(1)}, \ldots M^{(k-1)}\right)$ the k-modular decomposition of M.

The following result follows directly from the above definitions.

Proposition 14. *We have* $M = \Theta\left(M^{(0)}, M^{(1)}, \ldots M^{(k-1)}\right)$ *if and only if* $\left(M^{(0)}, M^{(1)}, \ldots M^{(k-1)}\right)$ *is the k-modular decomposition of M.*

Even though the above operations of interlacing and modular decomposition apply to general sets, they have a well-defined restriction to Maya diagrams. Indeed, it is not hard to check that if $M = \Theta\left(M^{(0)}, M^{(1)}, \ldots M^{(k-1)}\right)$ and M is a Maya diagram, then $M^{(0)}, M^{(1)}, \ldots M^{(k-1)}$ are also Maya diagrams. Conversely, if the latter are all Maya diagrams, then so is M. Another important case concerns the interlacing of finite sets. The definition (5.2) implies directly that if $\boldsymbol{\mu}^{(i)} \in \mathcal{Z}_{p_i}$, $i = 0, 1, \ldots, k-1$ then

$$\boldsymbol{\mu} = \Theta\left(\boldsymbol{\mu}^{(0)}, \ldots, \boldsymbol{\mu}^{(k-1)}\right)$$

is a finite set of cardinality $p = p_0 + \cdots + p_{k-1}$.

Visually, each of the k Maya diagrams is dilated by a factor of k, shifted by one unit with respect to the previous one and superimposed, so the interlaced Maya diagram incorporates the information from $M^{(0)}, \ldots M^{(k-1)}$ in k different modular classes. An example can be seen in Figure 2. In other words, the interlaced Maya diagram is built by copying sequentially a filled or empty box as determined by each of the k Maya diagrams.

−4 −3 −2 −1 0 1 2 3 4 5 6

$M_0 = \Xi(0, 1, 4), \qquad g_0 = 1$

$M_1 = \Xi(-1, 1, 3, 5, 6), \qquad g_1 = 2$

$M_2 = \Xi(4), \qquad g_2 = 0$

−5 −4 −3 −2 −1 0 1 2 3 4 5 6 7 8 9 10 11 12 13 14 15 16 17

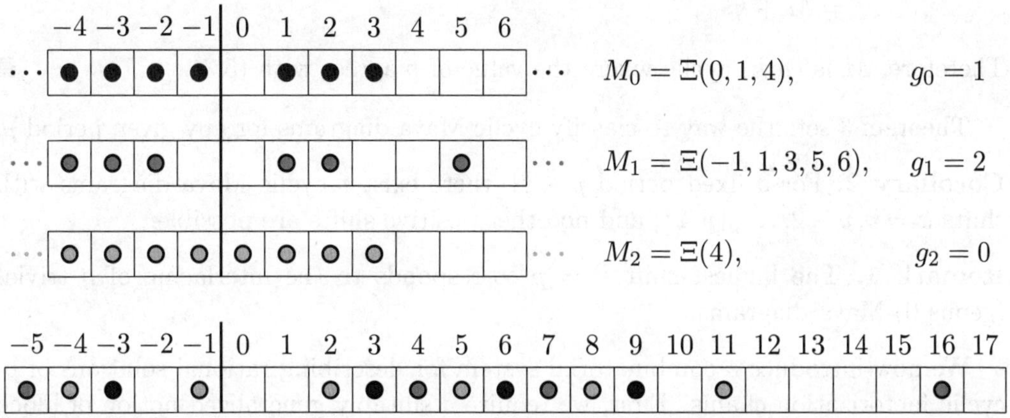

$M = \Xi_3(0, 1, 4 | -1, 1, 3, 5, 6 | 4) = \Xi(-2, -1, 0, 2, 10, 11, 12, 16, 17)$

Figure 2. Interlacing of three Maya diagrams with genus 1, 2 and 0 with block coordinates and 3-block coordinates for the interlaced Maya diagram.

Equipped with these notions of genus and interlacing, we are now ready to state the main result for the classification of cyclic Maya diagrams.

Theorem 3. *Let $M = \Theta\left(M^{(0)}, M^{(1)}, \ldots M^{(k-1)}\right)$ be the k-modular decomposition of a given Maya diagram M. Let g_i be the genus of $M^{(i)}$, $i = 0, 1, \ldots, k-1$. Then, M is (p, k)-cyclic where*

$$p = p_0 + p_1 + \cdots + p_{k-1}, \qquad p_i = 2g_i + 1. \tag{5.3}$$

Proof. Let $\boldsymbol{\beta}^{(i)} = \Upsilon\left(M^{(i)}, M^{(i+1)}\right) \in \mathcal{Z}_{p_i}$ be the block coordinates of $M^{(i)}$, $i = 0, 1, \ldots, k-1$. Consider the interlacing $\boldsymbol{\mu} = \Theta\left(\boldsymbol{\beta}^{(0)}, \ldots, \boldsymbol{\beta}^{(k-1)}\right)$. From Proposition 13 we have that,

$$\phi_{\boldsymbol{\beta}^{(i)}}\left(M^{(i)}\right) = M^{(i)} + 1.$$

so it follows that

$$\begin{aligned}\phi_{\boldsymbol{\mu}}(M) &= \phi_{\Theta(\boldsymbol{\beta}^{(0)},\ldots,\boldsymbol{\beta}^{(k-1)})}\Theta\left(M^{(0)},\ldots,M^{(k-1)}\right)\\ &= \Theta\left(\phi_{\boldsymbol{\beta}^{(0)}}(M^{(0)}),\ldots,\phi_{\boldsymbol{\beta}^{(k-1)}}(M^{(k-1)})\right)\\ &= \Theta\left(M^{(0)}+1,\ldots,M^{(k-1)}+1\right)\\ &= \Theta\left(M^{(0)},\ldots,M^{(k-1)}\right)+k\\ &= M+k.\end{aligned}$$

Therefore, M is (p,k) cyclic where the value of p agrees with (5.3). ■

Theorem 3 sets the way to classify cyclic Maya diagrams for any given period p.

Corollary 2. For a fixed period $p \in \mathbb{N}$, there exist p-cyclic Maya diagrams with shifts $k = p, p-2, \ldots, \lfloor p/2 \rfloor$, and no other positive shifts are possible.

Remark 1. The highest shift $k = p$ corresponds to the interlacing of p trivial (genus 0) Maya diagrams.

We now introduce a combinatorial system for describing rational solutions of p-cyclic factorization chains. First, we require a suitably generalized notion of block coordinates suitable for describing p-cyclic Maya diagrams.

Definition 12. For $p_0, \ldots, p_{k-1} \in \mathbb{N}$ set

$$\mathcal{Z}_{p_0,\ldots,p_{k-1}} := \mathcal{Z}_{p_0} \times \cdots \times \mathcal{Z}_{p_{k-1}} \subset \mathbb{Z}^{p_0+\cdots+p_{k-1}}.$$

Thus, an element of $\mathcal{Z}_{p_0,\ldots,p_{k-1}}$ is a concatenation $(\boldsymbol{\beta}^{(0)}|\boldsymbol{\beta}^{(1)}|\ldots|\boldsymbol{\beta}^{(k-1)})$ of k non-decreasing subsequences, $\boldsymbol{\beta}^{(i)} \in \mathcal{Z}_{p_i}$, $i = 0, 1, \ldots, k-1$. Let $\Xi \colon \mathcal{Z}_{p_0,\ldots,p_{k-1}} \to \mathcal{M}$ be the mapping with action

$$\Xi \colon (\boldsymbol{\beta}^{(0)}|\boldsymbol{\beta}^{(1)}|\ldots|\boldsymbol{\beta}^{(k-1)}) \mapsto \Theta\left(\Xi(\boldsymbol{\beta}^{(0)}),\ldots,\Xi(\boldsymbol{\beta}^{(k-1)})\right).$$

Let $M \in \mathcal{M}$ be a (p,k) cyclic Maya diagram, and let

$$M = \Theta\left(M^{(0)},\ldots M^{(k-1)}\right)$$

be the corresponding k-modular decomposition. Let $p_i = 2g_i + 1$, $i = 0, 1, \ldots, k-1$ where g_i is the genus of $M^{(i)}$ and let $\boldsymbol{\beta}^{(i)} \in \hat{\mathcal{Z}}_{2g_i+1}$ be the block coordinates of $M^{(i)} \in \mathcal{M}_{g_i}$. In light of the fact that

$$M = \Theta\left(\Xi(\boldsymbol{\beta}^{(0)}),\ldots,\Xi(\boldsymbol{\beta}^{(k-1)})\right),$$

we will refer to the concatenated sequence

$$(\boldsymbol{\beta}^{(0)}|\boldsymbol{\beta}^{(1)}|\ldots|\boldsymbol{\beta}^{(k-1)}) = \left(\beta_0^{(0)},\ldots,\beta_{p_0-1}^{(0)}|\beta_0^{(1)},\ldots,\beta_{p_1-1}^{(1)}|\cdots|\beta_0^{(k-1)},\ldots,\beta_{p_{k-1}-1}^{(k-1)}\right)$$

as the k-block coordinates of M.

Definition 13. Fix a $k \in \mathbb{N}$. For $m \in \mathbb{Z}$ let $[m]_k \in \{0, 1, \dots, k-1\}$ denote the residue class of m modulo division by k. For $m, n \in \mathbb{Z}$ say that $m \preccurlyeq_k n$ if and only if

$$[m]_k < [n]_k, \quad \text{or} \quad [m]_k = [n]_k \text{ and } m \leq n.$$

In this way, the transitive, reflexive relation $\preccurlyeq_k$ forms a total order on $\mathbb{Z}$.

Proposition 15. *Let M be a (p,k) cyclic Maya diagram. There exists a unique p-tuple $\boldsymbol{\mu} \in \mathbb{Z}^p$ ordered relative to $\preccurlyeq_k$ such that*

$$\phi_{\boldsymbol{\mu}}(M) = M + k \tag{5.4}$$

Proof. Let $(\beta_0, \dots, \beta_{p-1}) = (\boldsymbol{\beta}^{(0)} | \boldsymbol{\beta}^{(1)} | \dots | \boldsymbol{\beta}^{(k-1)})$ be the k-block coordinates of M. Set

$$\boldsymbol{\mu} = \Theta\left(\boldsymbol{\beta}^{(0)}, \dots, \boldsymbol{\beta}^{(k-1)}\right)$$

so that (5.4) holds by the proof to Theorem 3. The desired enumeration of $\boldsymbol{\mu}$ is given by

$$(k\beta_0, \dots, k\beta_{p-1}) + (0^{p_0}, 1^{p_1}, \dots, (k-1)^{p_{k-1}})$$

where the exponents indicate repetition. Explicitly, $(\mu_0, \dots, \mu_{p-1})$ is given by

$$\left(k\beta_0^{(0)}, \dots, k\beta_{p_0-1}^{(0)}, k\beta_0^{(1)} + 1, \dots, k\beta_{p_1-1}^{(1)} + 1, \dots, k\beta_0^{(k-1)} + k - 1, \dots, k\beta_{p_{k-1}-1}^{(k-1)} + k - 1\right).$$

■

Definition 14. In light of (5.4) we will refer to the tuple $(\mu_0, \mu_1, \dots, \mu_{p-1})$ as the k-canonical flip sequence of M and refer to the tuple $(p_0, p_1, \dots, p_{k-1})$ as the k-signature of M.

By Proposition 11 a rational solution of the p-cyclic dressing chain requires a (p,k) cyclic Maya diagram, and an additional item data, namely a fixed ordering of the canonical flip sequence. We will specify such ordering as

$$\boldsymbol{\mu}_{\boldsymbol{\pi}} = (\mu_{\pi_0}, \dots, \mu_{\pi_{p-1}})$$

where $\boldsymbol{\pi} = (\pi_0, \dots, \pi_{p-1})$ is a permutation of $(0, 1, \dots, p-1)$. With this notation, the chain of Maya diagrams described in Proposition 11 is generated as

$$M_0 = M, \qquad M_{i+1} = \phi_{\mu_{\pi_i}}(M_i), \qquad i = 0, 1, \dots, p-1. \tag{5.5}$$

Remark 2. Using a translation it is possible to normalize M so that $\mu_0 = 0$. Using a cyclic permutation it is possible to normalize $\boldsymbol{\pi}$ so that $\pi_p = 0$. The net effect of these two normalizations is to ensure that $M_0, M_1, \dots, M_{p-1}$ have standard form.

Remark 3. In order to obtain a full classification of rational solutions, it will be necessary to account for degenerate chains which include multiple flips at the same site. For this reason, we must allow $\boldsymbol{\beta}^{(i)} \in \mathcal{Z}_{p_i}$ to be merely non-decreasing sequences.

6 Examples of Hermite-type rational solutions

In this section we will put together all the results derived above in order to describe an effective way of labelling and constructing Hermite-type rational solutions to the Noumi-Yamada-Painlevé system using cyclic Maya diagrams. As an illustrative example, we describe rational solutions of the P_{IV} and P_V systems, because these are known to be reductions of the A_2 and A_3 systems, respectively. We then give examples of rational solutions to the A_4 system.

In order to specify a Hermite-type rational solution of a p-cyclic dressing chain, we require three items of data.

1. We begin by specifying a signature sequence $(p_0, \dots, p_{k-1})$ consisting of odd positive integers that sum to p. This sequence determines the genus $g_i = 2p_i + 1$, $i = 0, 1, \dots, k-1$ of the k interlaced Maya diagrams that give rise to a (p, k)-cyclic Maya diagram M. The possible values of k are given by Corollary 2.

2. Once the signature is fixed, we specify an element of $\mathcal{Z}_{p_0,\dots,p_{k-1}}$; i.e., k-block coordinates

$$(\beta_0, \dots, \beta_{p-1}) = (\boldsymbol{\beta}^{(0)} | \dots | \boldsymbol{\beta}^{(k-1)})$$

which determine a (p, k)-cyclic Maya diagram $M = \Xi(\boldsymbol{\beta}^{(0)} | \dots | \boldsymbol{\beta}^{(k-1)})$, and a canonical flip sequence $\boldsymbol{\mu} = (\mu_0, \dots, \mu_{p-1})$ as per Proposition 15.

3. Once the k-block coordinates and canonical flip sequence $\boldsymbol{\mu}$ are fixed, we specify a permutation $(\pi_0, \dots, \pi_{p-1})$ of $(0, 1, \dots, p-1)$ that determines the actual flip sequence $\boldsymbol{\mu}_\pi$, i.e. the order in which the flips in the canonical flip sequence are applied to build a p-cycle of Maya diagrams.

4. With the above data, we apply Proposition 11 with M and $\boldsymbol{\mu}_\pi$ to construct the rational solution.

For any signature of a Maya p-cycle, we need to specify the p integers in the canonical flip sequence, but following Remark 2, we can get rid of translation invariance by imposing $\mu_0 = \beta_0^{(0)} = 0$, leaving only $p-1$ free integers. The remaining number of degrees of freedom is $p-1$, which coincides with the number of generators of the symmetry group $A_{p-1}^{(1)}$. This is a strong indication that the class described above captures a generic orbit of a seed solution under the action of the symmetry group. Moreover, it is sometimes advantageous to consider only permutations such that $\pi_p = 0$ in order to remove the invariance under cyclic permutations.

6.1 Painlevé IV

As was mentioned in the introduction, the A_2 Noumi-Yamada system is equivalent to the P_{IV} equation [1, 38]. The reduction of (1.2) to the P_{IV} equation (1.1) is accomplished via the following substitutions

$$\begin{aligned}
&\sqrt{2}f_0(x) = y(t), \quad -\sqrt{8}f_{1,2}(x) = y(t) + 2t \pm \frac{y'(t) - \sqrt{-2b}}{y(t)} \\
&x = -\sqrt{2}t, \qquad a = \alpha_2 - \alpha_1, \; b = -2\alpha_0^2.
\end{aligned} \tag{6.1}$$

It is known [6] that every rational solution of P_{IV} can be described in terms of either generalized Hermite (GH) or Okamoto (O) polynomials, both of which may be given as a Wronskian of classical Hermite polynomials. We now exhibit these solutions using the framework described in the preceding section.

The 3-cyclic Maya diagrams fall into exactly one of two classes:

k	$(p_0, \ldots, p_{k-1})$	$(\beta_0, \beta_1, \beta_2)$	(μ_0, μ_1, μ_2)
1	(3)	$(0, n_1, n_1 + n_2)$	$(0, n_1, n_1 + n_2)$
3	(1, 1, 1)	$(0 \vert n_1 \vert n_2)$	$(0, 3n_1 + 1, 3n_2 + 2)$

The corresponding Maya diagrams are

$$\begin{aligned}
M_{\mathrm{GH}}(n_1, n_2) &= \Xi(0, n_1, n_1 + n_2) = (-\infty, 0) \cup [n_1, n_1 + n_2) \\
M_{\mathrm{O}}(n_1, n_2) &= \Xi_3(0|n_1|n_2) \\
&= \Theta((-\infty, 0), (-\infty, n_1), (-\infty, n_2)) \\
&= (-\infty, 0) \cup \{3j + 1 \colon 0 \le j < n_1\} \cup \{3j + 2 \colon 0 \le j \le n_2\}
\end{aligned}$$

The generalized Hermite and Okamoto polynomials, denoted below by $\tau_{\mathrm{GH}(n_1,n_2)}$, $n_1, n_2 \geq 0$ and $\tau_{\mathrm{O}(n_1,n_2)}$, $n_1, n_2 \geq 0$, respectively, are two-parameter families of Hermite Wronskians that correspond to the above diagrams. For example (see (3.8) for definition):

$$\begin{aligned}
\tau_{\mathrm{GH}(3,5)} &= \tau(3, 4, 5, 6, 7), \\
\tau_{\mathrm{O}(3,2)} &= \tau(1, 2, 4, 5, 7).
\end{aligned}$$

Having chosen one of the above polynomials as τ_0 there are $6 = 3!$ distinct rational solutions of the A_2 system (1.2) corresponding the possible permutations of the canonical flip sites (μ_0, μ_1, μ_2). The translations of the block coordinates $(\beta_0, \beta_1, \beta_2)$ engendered by these permutations is enumerated in the table below.

	(012)	(102)	(021)	(210)	(120)	(201)
τ_0	000	000	000	000	000	000
τ_1	100	010	100	001	010	001
τ_2	110	110	101	011	011	101
τ_3	111	111	111	111	111	111

Example. The Maya diagram, $M_{\mathrm{GH}}(3, 5)$ has the canonical flip sequence $\mu = (0, 3, 8)$. Applying the permutation $\pi = (2, 1, 0)$ yields the flip sequence $\mu_\pi =$

(8, 3, 0) and the following sequence of polynomials and block coordinates:

$$\begin{aligned}
\tau_0 &= \tau(3,4,5,6,7), && (0,3,8)\\
\tau_1 &= \tau(3,4,5,6,7,8), && (0,3,9)\\
\tau_2 &= \tau(4,5,6,7,8), && (0,4,9)\\
\tau_3 &= \tau(0,4,5,6,7,8) \propto \tau(3,4,5,6,7), && (1,4,9)
\end{aligned}$$

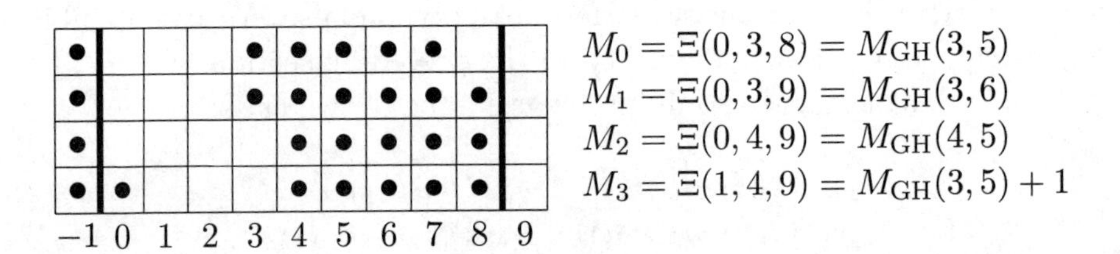

Figure 3. A cycle generated by the (3, 1)-cyclic Maya diagram $M_{\mathrm{GH}}(3,5)$ and permutation $\pi = (210)$.

Hence, by (4.27)

$$(a_0, a_1, a_2) = (10, 6, -18), \qquad (\sigma_0, \sigma_1, \sigma_2) = (-1, 1, -1).$$

Applying (4.8) gives the following rational solution of the 3-cyclic dressing chain (2.12):

$$w_0 = z + \frac{\tau_1'}{\tau_1} - \frac{\tau_0'}{\tau_0}, \quad w_1 = -z + \frac{\tau_2'}{\tau_2} - \frac{\tau_1'}{\tau_1} \quad w_2 = z + \frac{\tau_0'}{\tau_0} - \frac{\tau_2'}{\tau_2}, \qquad w_i = w_i(z),\ \tau_i = \tau_i(z).$$

Applying (2.18) and (6.1) gives the following rational solution of P_{IV}:

$$y(t) = \frac{d}{dt}\log\frac{\tau_0(t)}{\tau_2(t)}, \quad a = \frac{1}{2}(a_1 - a_2) = 12,\ b = -\frac{a_0^2}{2} = -50.$$

Example. The Maya diagram $M_{\mathrm{O}}(3,2)$ has the canonical flip sequence

$$\mu = \Theta(0|3|2) = (0, 10, 8).$$

Using the permutation (2, 0, 1) by way of example, we generate the following sequence of polynomials and block coordinates:

$$\begin{aligned}
\tau_0 &= \tau(1,2,4,5,7), && (0|3|2)\\
\tau_1 &= \tau(1,2,4,5,7,8), && (0|3|3)\\
\tau_2 &= \tau(0,1,2,4,5,7,8), && (1|3|3)\\
\tau_3 &= \tau(0,1,2,4,5,7,8,10) \propto \tau(1,2,4,5,7) && (1|4|3)
\end{aligned}$$

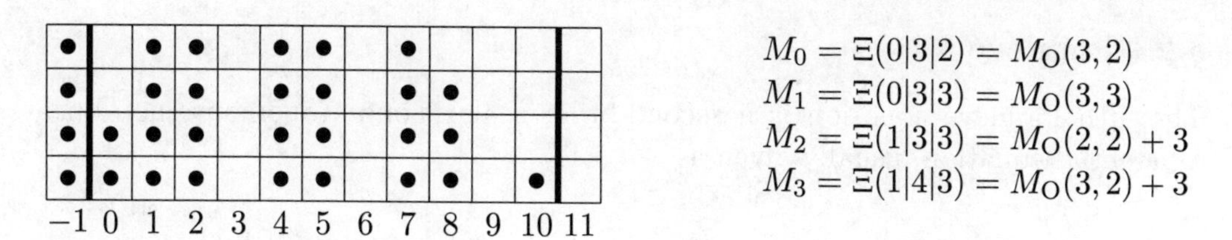

Figure 4. A cycle generated by the $(3,3)$-cyclic Maya diagram $M_{\mathrm{O}}(3,2)$ and permutation $\pi = (201)$.

Hence, by (4.27)

$$(a_0, a_1, a_2) = (16, -20, -2), \qquad (\sigma_0, \sigma_1, \sigma_2) = (-1, -1, -1).$$

Applying (4.8) gives the following rational solution of the 3-cyclic dressing chain (2.12):

$$w_0 = \frac{\tau_1'}{\tau_1} - \frac{\tau_0'}{\tau_0}, \quad w_1 = \frac{\tau_2'}{\tau_2} - \frac{\tau_1'}{\tau_1} \quad w_2 = \frac{\tau_0'}{\tau_0} - \frac{\tau_2'}{\tau_2}.$$

Applying (2.18) and (6.1) gives the following rational solution of $\mathrm{P_{IV}}$:

$$y(t) = -\frac{2}{3}t + \frac{d}{dt}\log\frac{\tau_2(Kt)}{\tau_0(Kt)}, \; K = \frac{1}{\sqrt{3}}, \quad a = \frac{1}{6}(a_1 - a_2) = -3, \; b = -\frac{a_0^2}{18} = -\frac{128}{9}.$$

For each of the above classes of solutions, observe that the permutations (π_0, π_1, π_2) and (π_1, π_0, π_2), while producing distinct solutions of the A_2 system, give the same solution of $\mathrm{P_{IV}}$. This means that there are 3 distinct GH classes and 3 distinct O classes of rational solution of $\mathrm{P_{IV}}$. These are enumerated below.

$$y = \frac{d}{dt}\log\frac{\tau_{\mathrm{GH}(n_1,n_2)}(t)}{\tau_{\mathrm{GH}(n_1,n_2+1)}(t)}, \qquad a = -(1 + n_1 + 2n_2), \; b = -2n_1^2;$$

$$y = \frac{d}{dt}\log\frac{\tau_{\mathrm{GH}(n_1,n_2)}(t)}{\tau_{\mathrm{GH}(n_1-1,n_2)}(t)}, \qquad a = 2n_1 + n_2 - 1, \; b = -2n_2^2;$$

$$y = -2t + \frac{d}{dt}\log\frac{\tau_{\mathrm{GH}(n_1,n_2)}(t)}{\tau_{\mathrm{GH}(n_1+1,n_2-1)}(t)}, \quad a = n_2 - n_1 - 1, \; b = -2(n_1 + n_2)^2;$$

$$y = -\frac{2}{3}t + \frac{d}{dt}\log\frac{\tau_{\mathrm{O}(n_1,n_2)}(z)}{\tau_{\mathrm{O}(n_1-1,n_2-1)}(z)}, \quad a = n_1 + n_2, \; b = -\frac{2}{9}(1 - 3n_1 + 3n_2)^2;$$

$$y = -\frac{2}{3}t + \frac{d}{dt}\log\frac{\tau_{\mathrm{O}(n_1,n_2)}(z)}{\tau_{\mathrm{O}(n_1+1,n_2)}(z)}, \quad a = -1 - 2n_1 + n_2, \; b = -\frac{2}{9}(2 + 3n_2)^2;$$

$$y = -\frac{2}{3}t + \frac{d}{dt}\frac{\tau_{\mathrm{O}(n_1,n_2)}(z)}{\tau_{\mathrm{O}(n_1,n_2+1)}(z)}, \qquad a = -2 - 2n_2 + n_1, \; b = -\frac{2}{9}(1 + 3n_1)^2,$$

where $z = \frac{t}{\sqrt{3}}$.

6.2 Painlevé V

The fifth Painlevé equation is a second-order scalar non-autonomous, non-linear differential equation, usually given as

$$y'' = (y')^2\left(\frac{1}{2y}+\frac{1}{y-1}\right)-\frac{y'}{t}+\frac{(y-1)^2}{t^2}\left(ay+\frac{b}{y}\right)+\frac{cy}{t}+\frac{dy(y+1)}{y-1}, \quad y=y(t), \tag{6.2}$$

where a,b,c,d are complex-valued parameters. An equivalent form is

$$\begin{aligned}\phi &= \alpha_0(y-1)+\alpha_2\left(1-\frac{1}{y}\right)-t\,\frac{y'}{y}, \quad \phi=\phi(t)\\ \phi' &= \frac{1}{2t}\left(\frac{\phi(y(2-\phi)-\phi-2)}{y-1}+2\phi(\alpha_0+\alpha_2-1)-2ct\right)-dt\,\frac{y+1}{y-1}\end{aligned}$$

where

$$a=\frac{\alpha_0^2}{2},\ b=-\frac{\alpha_2^2}{2}.$$

The A_3 Noumi-Yamada system, the specialization of (2.17) with $n=2$, has the form

$$\begin{aligned}xf_0' &= 2f_0f_2(f_1-f_3)+(1-2\alpha_2)f_0+2\alpha_0f_2\\ xf_1' &= 2f_1f_3(f_2-f_0)+(1-2\alpha_3)f_1+2\alpha_1f_3\\ xf_2' &= 2f_0f_2(f_3-f_1)+(1-2\alpha_0)f_2+2\alpha_2f_0\\ xf_3' &= 2f_1f_3(f_0-f_2)+(1-2\alpha_1)f_3+2\alpha_3f_1\end{aligned} \tag{6.3}$$

where $f_i=f_i(x)$, $i=0,1,2,3$, and which is subject to normalizations

$$f_0+f_2=f_1+f_3=\frac{x}{2}, \qquad \alpha_0+\alpha_1+\alpha_2+\alpha_3=1. \tag{6.4}$$

The transformation of (6.3) to (6.2) is given by the following relations:

$$\begin{aligned}&y=-\frac{f_2}{f_0}, \qquad \phi=\frac{1}{2}x(f_1-f_3),\\ &y=y(t),\ \phi=\phi(t),\quad f_i=f_i(x),\quad t=\frac{x^2}{\Delta}\\ &a=\frac{\alpha_0^2}{2},\ b=-\frac{\alpha_2^2}{2},\ c=\frac{\Delta}{4}(\alpha_3-\alpha_1),\ d=a-\frac{\Delta^2}{32}.\end{aligned} \tag{6.5}$$

The 4-cyclic Maya diagrams fall into exactly one of three classes:

k	$(p_0,\dots,p_{k-1})$	$(\beta_0,\beta_1,\beta_2,\beta_3)$	$(\mu_0,\mu_1,\mu_2,\mu_3)$
2	$(3,1)$	$(0,n_1,n_1+n_2\vert n_3)$	$(0,2n_1,2n_1+2n_2,2n_3+1)$
2	$(1,3)$	$(0\vert n_1,n_1+n_2,n_1+n_2+n_3)$	$(0,2n_1+1,2(n_1+n_2)+1,2(n_1+n_2+n_3)+1)$
4	$(1,1,1,1)$	$(0\vert n_1\vert n_2\vert n_3)$	$(0,4n_1+1,4n_2+2,4n_2+3)$

We will denote the corresponding Maya diagrams as

$$\begin{aligned} M(n_1, n_2|n_3) &= \Xi(0, n_1, n_1 + n_2|n_3), \\ M(|n_1, n_2, n_3) &= \Xi(0|n_1, n_1 + n_2, n_1 + n_2 + n_3), \\ M(|n_1|n_2|n_3) &= \Xi(0|n_1|n_2|n_3). \end{aligned}$$

For example (see (3.8) for definition):

$$\begin{aligned} \tau_{M(3,1|2)} &= \tau(1, 3, 6), \\ \tau_{M(|3,1,2)} &= \tau(1, 3, 5, 9, 11), \\ \tau_{M(|3|1|2)} &= \tau(1, 2, 3, 5, 7, 9). \end{aligned}$$

Having chosen one of the above polynomials as τ_0, there are $24 = 4!$ distinct rational solutions of the A_3 system (1.2) corresponding the possible permutations of the canonical flip sites $(\mu_0, \mu_1, \mu_2, \mu_4)$. However, the projection from the set of A_3 solutions to the set of $\mathrm{P_V}$ solutions is not one-to-one. The action of the permutation group $\mathfrak{S}_4$ on the set of $\mathrm{P_V}$ solutions has non-trivial isotropy corresponding to the Klein 4-group: $(0, 1, 2, 3), (1, 0, 2, 3), (0, 1, 3, 2), (1, 0, 3, 2)$. Therefore each of the above τ-functions generates $6 = 24/4$ distinct rational solutions of $\mathrm{P_V}$.

Example. We exhibit a $(4, 2)$-cyclic Maya diagram in the signature class $(3, 1, 1)$ by taking

$$M_0 = M(3, 1|2) = \Xi(0, 3, 4|2),$$

depicted in the first row of Figure 5. The canonical flip sequence is $\boldsymbol{\mu} = (0, 6, 8, 5)$. By way of example, we choose the permutation (0132), which gives the chain of Maya diagrams shown in Figure 5 and the following τ functions:

$$\begin{aligned} \tau_0 &= \tau(1, 3, 6) \propto z(8z^6 - 12z^4 - 6z^2 - 3) \\ \tau_1 &= \tau(0, 1, 3, 6) \propto 4z^4 - 4z^2 - 1 \\ \tau_2 &= \tau(0, 1, 3) \propto z \\ \tau_3 &= \tau(0, 1, 3, 5) \propto z^3 \\ \tau_4 &= \tau(0, 1, 3, 5, 8) \propto \tau(1, 3, 6). \end{aligned}$$

By (4.27),

$$(a_0, \ldots, a_3) = (-12, 2, -6, 12), \qquad (\sigma_0, \ldots, \sigma_3) = (-1, 1, -1, -1).$$

Hence, by (4.8) and (2.18),

$$\begin{aligned} f_0 &= \frac{6x^5 - 24x^3 - 24x}{x^6 - 6x^4 - 12x^2 - 24} \\ f_1 &= -\frac{3}{x} + \frac{4x^3 - 8x}{x^4 - 4x^2 - 4} \end{aligned}$$

with f_2, f_3 given by (6.4). By (6.5) the corresponding rational solution of P_V (6.2) is given by

$$y = \frac{7}{6} - \frac{t}{3} + \frac{4t+2}{12t^2 - 12t - 3}$$
$$a = \frac{9}{2},\ b = -\frac{9}{8},\ c = -\frac{5}{2},\ d = -\frac{1}{2}.$$

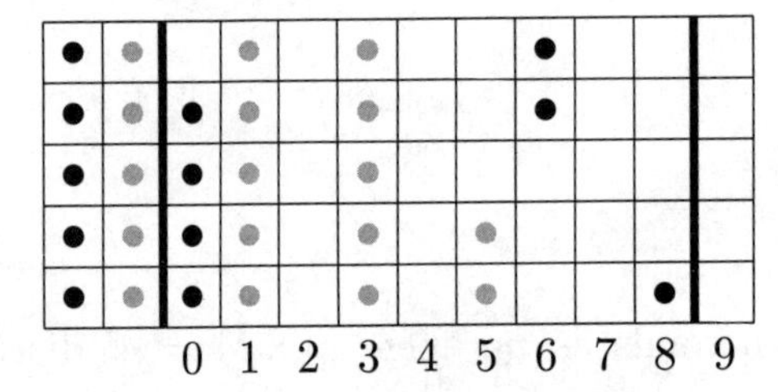

$$M_0 = \Xi\,(0,3,4|2) = M(3,1|2)$$
$$M_1 = \Xi\,(1,3,4|2) = M(2,1|1) + 2$$
$$M_2 = \Xi\,(1,4,4|2) = M(2,0|1) + 2$$
$$M_2 = \Xi\,(1,4,4|3) = M(2,0|2) + 2$$
$$M_2 = \Xi\,(1,4,5|3) = M(2,1|2) + 2$$

Figure 5. A Maya 4-cycle with shift $k = 2$ for the choice $(n_1, n_2|n_3) = (3,1|2)$ and permutation $\pi = (0,1,3,2)$.

Example. We exhibit a (4,2)-cyclic Maya diagram in the signature class (1,1,3) by taking

$$M_0 = M(|3,1,2) = \Xi(0|3,4,6),$$

depicted in the first row of Figure 6. The canonical flip sequence is $\boldsymbol{\mu} = (0,7,9,13)$. By way of example, we choose the permutation (0132), which gives the chain of Maya diagrams shown in Figure 6 and the following τ functions:

$$\tau_0 = \tau(1,3,5,9,11) \propto z^{15}(4z^4 - 36z^2 + 99)$$
$$\tau_1 = \tau(0,1,3,5,9,11) \propto z^{10}(4z^4 - 28z^2 + 63)$$
$$\tau_2 = \tau(0,1,3,5,7,9,11) \propto z^{15}$$
$$\tau_3 = \tau(0,1,3,5,7,9,11,13) \propto z^{21}$$
$$\tau_4 = \tau(0,1,3,5,7,11,13) \propto \tau(1,3,5,9,11).$$

By (4.27),

$$(a_0, \ldots, a_3) = (-12, 2, -6, 12), \qquad (\sigma_0, \ldots, \sigma_3) = (-1,-1,-1,1).$$

Hence, by (4.8) and (2.18),

$$f_0 = \frac{x}{2} + \frac{4x^3 - 72x^2}{x^4 - 36x^2 + 396}$$
$$f_1 = -\frac{11}{x} + \frac{x}{2} + \frac{4x^3 - 56x}{x^4 - 28x^2 + 252}$$

with f_2, f_3 given by (6.4). By (6.5) the corresponding rational solution of P_V (6.2) is given by

$$y = \frac{8t - 36}{4t^2 - 28t + 63}, \qquad a = \frac{49}{8},\ b = -2,\ c = -\frac{13}{2},\ d = -\frac{1}{2}.$$

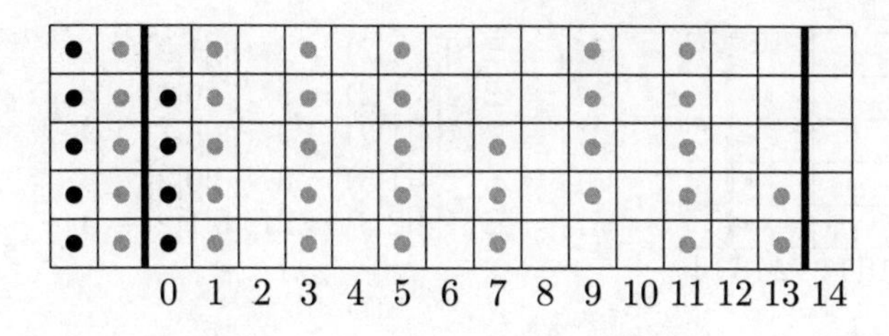

$$
\begin{aligned}
M_0 &= \Xi\,(0|3,4,6) = M(|3,1,2)\\
M_1 &= \Xi\,(1|3,4,6) = M(|2,1,2)+2\\
M_2 &= \Xi\,(1|4,4,6) = M(|3,0,2)+2\\
M_3 &= \Xi\,(1|4,4,7) = M(|3,0,3)+2\\
M_4 &= \Xi\,(1|4,5,7) = M(|3,1,2)+2
\end{aligned}
$$

Figure 6. A Maya 4-cycle with shift $k = 2$ for the choice $(|n_1, n_2, n_3) = (|3,1,2)$ and permutation $\pi = (0,1,3,2)$.

Example. We exhibit a $(4,4)$-cyclic Maya diagram in the signature class $(1,1,1)$ by taking

$$M_0 = M(|3|1|2) = \Xi(0|3|1|2),$$

depicted in the first row of Figure 7. The canonical flip sequence is $\boldsymbol{\mu} = (0,13,6,11)$. By way of example, we choose the permutation (0132), which gives the chain of Maya diagrams shown in Figure 7 and the following τ functions:

$$
\begin{aligned}
\tau_0 &= \tau(1,2,3,5,7,9) \propto z^{10}(2z^2+9)\\
\tau_1 &= \tau(0,1,2,3,5,7,9) \propto z^{6}\\
\tau_2 &= \tau(0,1,2,3,5,7,9,13) \propto z^{10}(2z^2-9)\\
\tau_3 &= \tau(0,1,2,3,5,7,9,11,13) \propto z^{15}\\
\tau_4 &= \tau(0,1,2,3,5,6,7,9,11,13) \propto \tau(1,2,3,5,7,9)
\end{aligned}
$$

By (4.27),

$$(a_0,\ldots,a_3) = (-26,4,10,4), \qquad (\sigma_0,\ldots,\sigma_3) = (-1,-1,-1,-1)$$

Hence, by (4.8) and (2.18),

$$
\begin{aligned}
f_0 &= -\frac{1}{x-6} - \frac{1}{x+6} + \frac{x}{4} + \frac{2x}{x^2+36}\\
f_1 &= -\frac{9}{x} + \frac{x}{4},
\end{aligned}
$$

with f_2, f_3 given by (6.4). By (6.5) the corresponding rational solution of $\mathrm{P_V}$ (6.2) is given by

$$y = -1 - \frac{72}{4t^2-117}, \qquad a = \frac{169}{32},\ b = -\frac{25}{32},\ c = 0,\ d = -2$$

6.3 Rational solutions of the A_4 Noumi-Yamada system

In this section we describe the rational Hermite-type solutions to the A_4-Painlevé system, and give examples in each signature class.

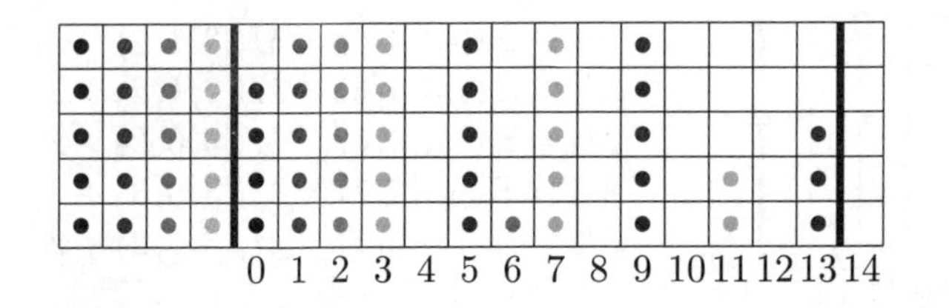

$$
\begin{aligned}
M_0 &= \Xi\,(0|3|1|2) = M(|3|1|2)\\
M_1 &= \Xi\,(1|3|1|2) = M(|2|0|1) + 4\\
M_2 &= \Xi\,(1|4|1|2) = M(|3|0|1) + 4\\
M_3 &= \Xi\,(1|4|1|3) = M(|3|0|2) + 4\\
M_4 &= \Xi\,(1|4|2|3) = M(|3|1|2) + 4
\end{aligned}
$$

Figure 7. A Maya 4-cycle with shift $k = 4$ for the choice $(|n_1|n_2|n_3) = (|3|1|2)$ and permutation $\boldsymbol{\pi} = (0, 1, 3, 2)$.

The A_4 Painlevé system consists of 5 equations in 5 unknowns $f_i = f_i(x)$, $i = 0, \dots, 4$ and complex parameters α_i, $i = 0, \dots, 5$:

$$
\begin{aligned}
f_0' &= f_0(f_1 - f_2 + f_3 - f_4) + \alpha_0\\
f_1' &= f_1(f_2 - f_3 + f_4 - f_0) + \alpha_1\\
f_2' &= f_2(f_3 - f_4 + f_0 - f_1) + \alpha_2,\\
f_3' &= f_3(f_4 - f_0 + f_1 - f_2) + \alpha_3\\
f_4' &= f_4(f_0 - f_1 + f_2 - f_3) + \alpha_4
\end{aligned}
\tag{6.6}
$$

with normalization

$$f_0 + f_1 + f_2 + f_3 + f_4 = x.$$

Rational Hermite-type solutions of the A_4 system (6.6) correspond to chains of 5-cyclic Maya diagrams belonging to one of the following signature classes:

$$(5), (3, 1, 1), (1, 1, 3), (1, 1, 3), (1, 1, 1, 1, 1).$$

With the normalizations $\mu_0 = 0$, each 5-cyclic Maya diagram may be uniquely labelled by one of the above signatures, and a 4-tuple of non-negative integers (n_1, n_2, n_3, n_4). For each of the above signatures, the corresponding k-block coordinates are then given by

$$
\begin{array}{lll}
k = 1 & (5) & M(n_1, n_2, n_3, n_4) := \Xi(0, n_1, n_1 + n_2, n_1 + n_2 + n_3, n_1 + \dots + n_4)\\
k = 3 & (3, 1, 1) & M(n_1, n_2|n_3|n_4) := \Xi(0, n_1, n_1 + n_2|n_3|n_4)\\
k = 3 & (1, 3, 1) & M(|n_1, n_2, n_3|n_4) := \Xi(0|n_1, n_1 + n_2, n_1 + n_2 + n_3|n_4)\\
k = 3 & (1, 1, 3) & M(|n_1|n_2, n_3, n_4) := \Xi(0|n_1|n_2, n_2 + n_3, n_2 + n_3 + n_4)\\
k = 5 & (1, 1, 1, 1, 1) & M(|n_1|n_2|n_3|n_4) := \Xi(0|n_1|n_2|n_3|n_4).
\end{array}
$$

Below, we exhibit examples belonging to each of these classes.

Example. We exhibit a $(5, 1)$-cyclic Maya diagram in the signature class (5) by taking

$$M_0 = M(2, 3, 1, 1) = \Xi(0, 2, 5, 6, 7),$$

depicted in the first row of Figure 8. The canonical flip sequence is $\boldsymbol{\mu} = (0, 2, 5, 6, 7)$. By way of example, we choose the permutation (34210), which gives the chain of

Maya diagrams shown in Figure 8. Note that the permutation specifies the sequence of block coordinates that get shifted by one at each step of the cycle. Solutions with signature (5) were already studied in [10], and they are based on a genus 2 generalization of the generalized Hermite polynomials that appear in the solution of $P_{IV}(A_2\text{-Painlevé})$.

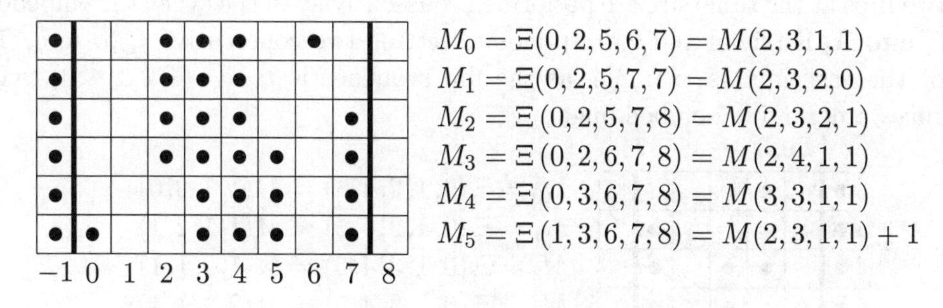

Figure 8. A Maya 5-cycle with shift $k = 1$ for the choice $(n_1, n_2, n_3, n_4) = (2, 3, 1, 1)$ and permutation $\boldsymbol{\pi} = (34210)$.

We shall now provide the explicit construction of the rational solution to the A_4-Painlevé system (6.6), by using Proposition 1 and Proposition 11. The permutation $\boldsymbol{\pi} = (34210)$ on the canonical sequence $\boldsymbol{\mu} = (0, 2, 5, 6, 7)$ produces the flip sequence $\boldsymbol{\mu}_{\boldsymbol{\pi}} = (6, 7, 5, 2, 0)$. The pseudo-Wronskians corresponding to each Maya diagram in the cycle are ordinary Wronskians, which will always be the case with the normalization imposed in Remark 2. They read (see Figure 8):

$$\begin{aligned}
\tau_0 &= \tau(2,3,4,6)\\
\tau_1 &= \tau(2,3,4)\\
\tau_2 &= \tau(2,3,4,7)\\
\tau_3 &= \tau(2,3,4,5,7)\\
\tau_4 &= \tau(3,4,5,7)\\
\tau_5 &= \tau(0,3,4,5,7) \propto \tau(2,3,4,6).
\end{aligned}$$

The rational solution to the 5-cyclic dressing chain (2.12) is given by (4.27), with

$$(\sigma_0, \ldots, \sigma_4) = (1, -1, -1, 1, -1), \quad (a_0, \ldots, a_4) = (-2, 4, 6, 4, -14).$$

Finally, the corresponding rational solution $f_i(x), \alpha_i$ $i = 0, 1, \ldots, 4 \mod 5$ to the A_4-Painlevé system (2.16) is given by

$$f_i(x) = -(\sigma_i + \sigma_{i+1})\frac{x}{2} + \frac{1}{\sqrt{2}}\frac{d}{dx}\log\frac{\tau_i(z)}{\tau_{i+2}(z)}, \quad z = \frac{x}{\sqrt{2}}, \quad \alpha_i = -\frac{a_i}{2}.$$

Example. We construct a degenerate example belonging to the (5) signature class, by choosing $(n_1, n_2, n_3, n_4) = (1, 1, 2, 0)$. The presence of $n_4 = 0$ means M_0 and M_4 have genus 1 instead of the generic genus 2. The degeneracy occurs because the canonical flip sequence $\boldsymbol{\mu} = (0, 1, 2, 4, 4)$ contains two flips at the same site.

Choosing the permutation (42130), by way of example, produces the chain of Maya diagrams shown in Figure 9. The explicit construction of the rational solutions follows the same steps as in the previous example, and we shall omit it here. It is worth noting, however, that due to the degenerate character of the chain, three linear combinations of $f_0, \ldots, f_4$ will provide a solution to the lower rank A_2-Painlevé. If the two flips at the same site are performed consecutively in the cycle, the embedding of A_2 into A_4 is trivial and corresponds to setting two consecutive f_i to zero. This is not the case in this example, as the flip sequence is $\boldsymbol{\mu}_\pi = (4, 2, 1, 4, 0)$, which produces a non-trivial embedding.

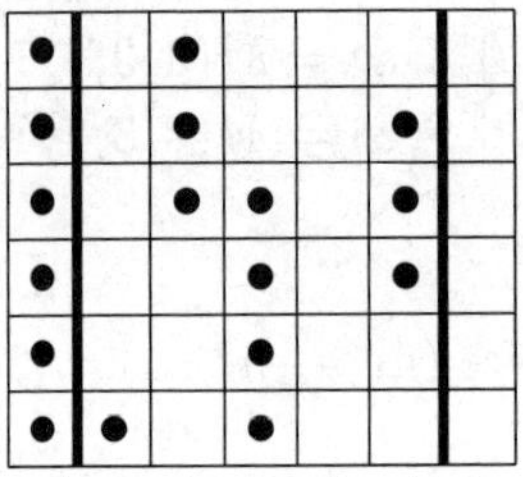

$$
\begin{aligned}
M_0 &= \Xi\,(0,1,2,4,4) = M(1,1,2,0)\\
M_1 &= \Xi\,(0,1,2,4,5) = M(1,1,2,1)\\
M_2 &= \Xi\,(0,1,3,4,5) = M(1,2,1,1)\\
M_3 &= \Xi\,(0,2,3,4,5) = M(2,1,1,1)\\
M_4 &= \Xi\,(0,2,3,5,5) = M(2,1,2,0)\\
M_5 &= \Xi\,(1,2,3,5,5) = M(1,1,2,0)+1
\end{aligned}
$$

Figure 9. A degenerate Maya 5-cycle with $k = 1$ for the choice $(n_1, n_2, n_3, n_4) = (1, 1, 2, 0)$ and permutation $\boldsymbol{\pi} = (42130)$.

Example. We construct a $(5, 3)$-cyclic Maya diagram in the signature class $(1, 1, 3)$ by choosing $(n_1, n_2, n_3, n_4) = (3, 1, 1, 2)$, which means that the first Maya diagram has 3-block coordinates $(0|3|1, 2, 4)$. The canonical flip sequence is given by $\boldsymbol{\mu} = \Theta\,(0|3|1, 2, 4) = (0, 10, 5, 8, 14)$. The permutation (41230) gives the chain of Maya diagrams shown in Figure 10. Note that, as in Example 6.3, the permutation specifies the order in which the 3-block coordinates are shifted by $+1$ in the subsequent steps of the cycle. Solutions in the signature class $(1, 1, 3)$ were not given in [10], and they are new to the best of our knowledge.

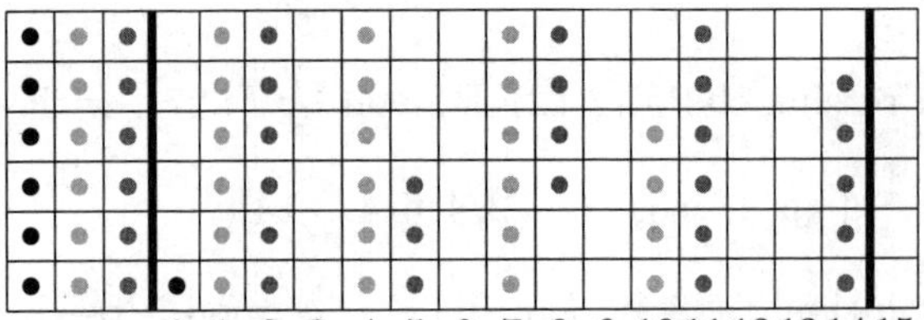

$$
\begin{aligned}
M_0 &= \Xi\,(0|3|1,2,4) = M(|3|1,1,2)\\
M_1 &= \Xi\,(0|3|1,2,5) = M(|3|1,1,3)\\
M_2 &= \Xi\,(0|4|1,2,5) = M(|4|1,1,3)\\
M_3 &= \Xi\,(0|4|2,2,5) = M(|4|2,0,3)\\
M_4 &= \Xi\,(0|4|2,3,5) = M(|4|2,1,2)\\
M_5 &= \Xi\,(1|4|2,3,5) = M(|3|1,1,2)+3
\end{aligned}
$$

Figure 10. A Maya 5-cycle with shift $k = 3$ for the choice $(|n_1|n_2, n_3, n_4) = (|3|1, 1, 2)$ and permutation $\boldsymbol{\pi} = (41230)$.

We proceed to build the explicit rational solution to the A_4-Painlevé system (6.6). In this case, the permutation $\boldsymbol{\pi} = (41230)$ on the canonical sequence $\boldsymbol{\mu} = (0, 10, 5, 8, 14)$ produces the flip sequence $\boldsymbol{\mu}_\pi = (14, 10, 5, 8, 0)$, so that the values of the a_i parameters given by (4.27) become $(a_0, a_1, a_2, a_3, a_4) = (8, 10, -6, 16, -34)$.

The pseudo-Wronskians corresponding to each Maya diagram in the cycle are ordinary Wronskians, which will always be the case with the normalization imposed in Remark 2. They read (see Figure 10):

$$\begin{aligned}
\tau_0 &= \tau(1,2,4,7,8,11)\\
\tau_1 &= \tau(1,2,4,7,8,11,14)\\
\tau_2 &= \tau(1,2,4,7,8,10,11,14)\\
\tau_3 &= \tau(1,2,4,5,7,8,10,11,14)\\
\tau_4 &= \tau(1,2,4,5,7,10,11,14)
\end{aligned}$$

The rational solution to the 5-cyclic dressing chain (2.12) is given by (4.27), with

$$(\sigma_0,\ldots,\sigma_4) = (-1,1,-1,-1,-1), \quad (a_0,\ldots,a_4) = (-6,-12,8,20,-16).$$

The corresponding rational solution $f_i(x), \alpha_i\ i = 0,1,\ldots,4 \mod 5$ to the A_4-Painlevé system (2.16) is given by

$$f_i(x) = -(\sigma_i+\sigma_{i+1})\frac{x}{6} + \frac{1}{\sqrt{6}}\frac{d}{dx}\log\frac{\tau_i(z)}{\tau_{i+2}(z)}, \quad z = \frac{x}{\sqrt{6}}, \quad \alpha_i = -\frac{a_i}{6}.$$

Example. We construct a $(5,5)$-cyclic Maya diagram in the signature class $(1,1,1,1,1)$ by choosing $(n_1 n_2 n_3 n_4) = (2,3,0,1)$, which means that the first Maya diagram has 5-block coordinates $(0|2|3|0|1)$. The canonical flip sequence is given by

$$\boldsymbol{\mu} = \Theta\,(0|2|3|0|1) = (0,11,17,3,9).$$

The permutation (32410) gives the chain of Maya diagrams shown in Figure 11. Note that, as it happens in the previous examples, the permutation specifies the order in which the 5-block coordinates are shifted by $+1$ in the subsequent steps of the cycle. These types of solutions with signature $(1,1,1,1,1)$ were already studied in [10], and they are based on a generalization of the Okamoto polynomials that appear in the solution of $\mathrm{P_{IV}}$ (A_2-Painlevé).

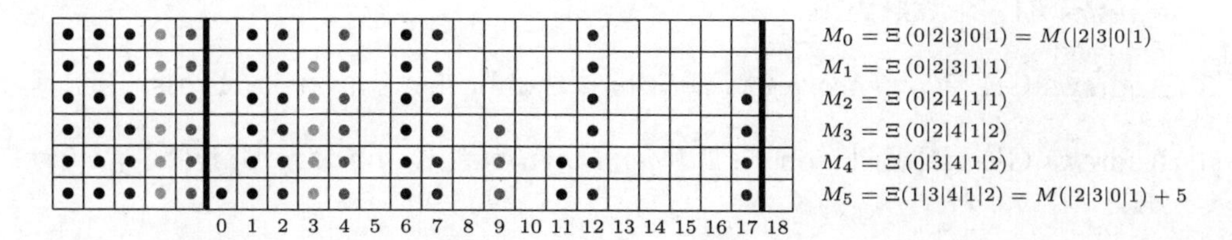

Figure 11. A Maya 5-cycle with shift $k = 5$ for the choice $(n_1, n_2, n_3, n_4) = (2,3,0,1)$ and permutation $\boldsymbol{\pi} = (32410)$.

We proceed to build the explicit rational solution to the A_4-Painlevé system (6.6). In this case, the flip sequence is given by $\boldsymbol{\mu}_{\boldsymbol{\pi}} = (3,17,9,11,0)$. The corresponding

Hermite Wronskians are shown below (see Figure 11):

$$
\begin{aligned}
\tau &= \tau(1,2,4,6,7,12)\\
\tau &= \tau(1,2,3,4,6,7,12)\\
\tau &= \tau(1,2,3,4,6,7,12,17)\\
\tau &= \tau(1,2,3,4,6,7,9,12,17)\\
\tau &= \tau(1,2,3,4,6,7,9,11,12,17)
\end{aligned}
$$

The rational solution to the 5-cyclic dressing chain (2.12) is given by (4.27), with

$$(\sigma_0,\ldots,\sigma_4) = (-1,-1,-1,-1,-1), \quad (a_0,\ldots,a_4) = (-28,16,-4,22,-16).$$

The corresponding rational solution to the A_4-Painlevé system (2.16) is given by

$$f_i(x) = \frac{x}{5} + \frac{1}{\sqrt{10}}\frac{d}{dx}\log\frac{\tau_i(z)}{\tau_{i+2}(z)}, \quad z = \frac{x}{\sqrt{10}}, \quad \alpha_i = -\frac{a_i}{10},$$

where $i = 0,1,\ldots,4 \mod 5$.

Acknowledgments

The research of DGU has been supported in part by Spanish MINECO-FEDER Grant MTM2015-65888-C4-3 and by the ICMAT-Severo Ochoa project SEV-2015-0554. The research of RM was supported in part by NSERC grant RGPIN-228057-2009.

References

[1] Adler VE, Nonlinear chains and Painlevé equations, *Physica D: Nonlinear Phenomena* **73**, 335–351, 1994.

[2] Airault H, Rational solutions of Painlevé equations, *Studies in Applied Mathematics* **61**, 31–53, 1979.

[3] Andrews GE, *The Theory of Partitions*, Cambridge University Press, 1998.

[4] Andrews GE and Eriksson K, *Integer Partitions*, Cambridge University Press, 2004.

[5] Bermúdez D, Complex SUSY transformations and the Painlevé IV equation, *Symmetry, Integrability and Geometry: Methods and Applications* **8**, 69–10, 2012.

[6] Clarkson PA, The fourth Painlevé equation and associated special polynomials, *Journal of Mathematical Physics* **44**, 5350–5374, 2003.

[7] ———, Painlevé equations—nonlinear special functions, *Journal of Computational and Applied Mathematics* **153**, 127–140, 2003.

[8] ———, Special polynomials associated with rational solutions of the Painlevé equations and applications to soliton equations, *Computational Methods and Function Theory* **6**, 329–401, 2006.

[9] ———, Special polynomials associated with rational solutions of the defocusing nonlinear Schrödinger equation and the fourth Painlevé equation, *European Journal of Applied Mathematics* **17**, 293–322, 2006.

[10] Filipuk G and Clarkson PA, The symmetric fourth Painlevé hierarchy and associated special polynomials, *Studies in Applied Mathematics* **121**, 157–188, 2008.

[11] Fokas AS, Its AR, Kapaev AA, Kapaev AI, Novokshenov VY, and Novokshenov VIU, *Painlevé Transcendents: the Riemann-Hilbert Approach*, American Mathematical Soc., Rhode Island, 2006.

[12] Fukutani S, Okamoto K, and Umemura H, Special polynomials and the Hirota bilinear relations of the second and the fourth Painlevé equations, *Nagoya Mathematical Journal* **159**, 179–200, 2000.

[13] Garcia-Ferrero M, Gómez-Ullate D, and Milson R, A Bochner type classification theorem for exceptional orthogonal polynomials, *J. Math. Anal. Appl.* **472**, 584626, 2019.

[14] Gómez-Ullate D, Kamran N, and Milson R, An extended class of orthogonal polynomials defined by a Sturm–Liouville problem, *J. Math. Anal. Appl.* **359**, 352–367, 2009.

[15] ———, An extension of Bochner's problem: exceptional invariant subspaces, *Journal of Approximation Theory* **162**, 987–1006, 2010.

[16] ———, A conjecture on exceptional orthogonal polynomials, *Foundations of Computational Mathematics* **13**, 615–666, 2013.

[17] Gomez-Ullate D, Grandati Y and Milson R, Rational extensions of the quantum harmonic oscillator and exceptional Hermite polynomials, *Journal of Physics A* **47**, 015203, 2013.

[18] ———, Durfee rectangles and pseudo-Wronskian equivalences for Hermite polynomials, *Studies in Applied Mathematics* **141**, 596–625, 2018.

[19] Grandati Y, Exceptional orthogonal polynomials and generalized Schur polynomials, *Journal of Mathematical Physics* **55**, 083509, 2014.

[20] Kajiwara K and Ohta Y, Determinant structure of the rational solutions for the Painlevé II equation, *Journal of Mathematical Physics* **37**, 4693–4704, 1996.

[21] ———, Determinant structure of the rational solutions for the Painlevé IV equation, *Journal of Physics A* **31**, 2431, 1998.

[22] Kitaev AV, Law CK, McLeod JB, et al., Rational solutions of the fifth Painlevé equation, *Differential Integral Equations* **7**, 967–1000, 1994.

[23] Masuda T, Ohta Y, and Kajiwara K, A determinant formula for a class of rational solutions of Painlevé V equation, *Nagoya Mathematical Journal*, **168** 1–25, 2002.

[24] Matsuda K, Rational solutions of the Noumi and Yamada system of type A_4, *Journal of Mathematical Physics* **53**, 023504, 2012.

[25] Miwa T, Jinbo M, Jimbo M, and Date E, *Solitons: Differential Equations, Symmetries and Infinite Dimensional Algebras*, vol. 135, Cambridge University Press, 2000.

[26] Murata Y, Classical solutions of the third Painlevé equation, *Nagoya Mathematical Journal* **139**, 37–65, 1995.

[27] Noumi M and Yamada Y, Higher order Painlevé equations of type a, *Funkcialaj Ekvacioj* **41**, 483–503, 1998.

[28] ———, Umemura polynomials for the Painlevé V equation, *Physics Letters A* **247**, 65–69, 1998.

[29] ———, Symmetries in the fourth Painlevé equation and Okamoto polynomials, *Nagoya Mathematical Journal* **153**, 53–86, 1999.

[30] Noumi M, *Painlevé Equations Through Symmetry*, vol. 223, Springer Science & Business, 2004.

[31] Oblomkov AA, Monodromy-free Schrödinger operators with quadratically increasing potentials, *Theoretical and Mathematical Physics* **121**, 1574–1584, 1999.

[32] Okamoto K, Studies on the Painlevé equations IV, Third Painlevé equation PIII, *Funkcial. Ekvac* **30**, 305–32, 1987.

[33] Olsson JB, *Combinatorics and Representations of Finite Groups*, FB Mathematik, Universität Essen, 1994.

[34] Takasaki K, Spectral curve, Darboux coordinates and Hamiltonian structure of periodic dressing chains, *Communications in mathematical physics* **241**, 111–142, 2003.

[35] Tsuda T, Universal characters, integrable chains and the Painlevé equations, *Advances in Mathematics* **197**, 587–606, 2005.

[36] Umemura H, Special polynomials associated with the Painlevé equation I, *Proceedings of Workshop on the Painlevé equations*, 1996.

[37] Umemura H and Watanabe H, Solutions of the second and fourth Painlevé equations, *Nagoya Mathematical Journal* **148**, 151–198, 1997.

[38] Veselov AP and Shabat AB, Dressing chains and the spectral theory of the Schrödinger operator, *Functional Analysis and Its Applications* **27**, 81–96, 1993.

[39] Vorobev AP, On the rational solutions of the second Painlevé equation, *Differentsial'nye Uravneniya* **1**, 79–81, 1965.

[40] Wilson G, Bispectral commutative ordinary differential operators, *J. Reine Angew. Math* **442**, 177–204, 1993.

[41] Yablonskii AI, On rational solutions of the second Painlevé equation, *Vesti Akad. Navuk. BSSR Ser. Fiz. Tkh. Nauk.* **3**, 30–35, 1959.

B3. Cluster algebras and discrete integrability

Andrew N.W. Hone[1], *Philipp Lampe and Theodoros E. Kouloukas*

School of Mathematics, Statistics & Actuarial Science
University of Kent, Canterbury CT2 7FS, UK.

Abstract

Cluster algebras are a class of commutative algebras whose generators are defined by a recursive process called mutation. We give a brief introduction to cluster algebras, and explain how discrete integrable systems can appear in the context of cluster mutation. In particular, we give examples of birational maps that are integrable in the Liouville sense and arise from cluster algebras with periodicity, as well as examples of discrete Painlevé equations that are derived from Y-systems.

1 Introduction

Cluster algebras are a special class of commutative algebras that were introduced by Fomin and Zelevinsky almost twenty years ago [21], and rapidly became the hottest topic in modern algebra. Rather than being defined a priori by a given set of generators and relations, the generators of a cluster algebra are produced recursively by iteration of a process called mutation. In certain cases, a sequence of mutations in a cluster algebra can correspond to iteration of a birational map, so that a discrete dynamical system is generated. The reason why cluster algebras have attracted so much attention is that cluster mutations and associated discrete dynamical systems or difference equations arise in such a wide variety of contexts, including Teichmuller theory [19, 20], Poisson geometry [30], representation theory [11], and integrable models in statistical mechanics and quantum field theory [14, 34, 68], to name but a few.

The purpose of this review is to give a brief introduction to cluster algebras, and describe certain situations where the associated dynamics is completely integrable, in the sense that a discrete version of Liouville's theorem in classical mechanics is valid. Furthermore, within the context of cluster algebras, we will describe a way to detect whether a given discrete system is integrable, based on an associated tropical dynamical system and its connection to the notion of algebraic entropy. Finally, we describe how discrete Painlevé equations can arise in the context of cluster algebras.

2 Cluster algebras: definition and examples

A cluster algebra with coefficients, of rank N, is generated by starting from a seed $(B, \mathbf{x}, \mathbf{y})$ consisting of an *exchange matrix* $B = (b_{ij}) \in \mathrm{Mat}_N(\mathbb{Z})$, an N-tuple of *cluster variables* $\mathbf{x} = (x_1, x_2, \ldots, x_N)$, and another N-tuple of *coefficients*

[1]Currently on leave at UNSW, Sydney, Australia.

$\mathbf{y} = (y_1, y_2, \ldots, y_N)$. The exchange matrix is assumed to be skew-symmetrizable, meaning that there is a diagonal matrix D, consisting of positive integers, such that DB is skew-symmetric. For each integer $k \in [1, N]$, there is a mutation μ_k which produces a new seed $(B', \mathbf{x}', \mathbf{y}') = \mu_k(B, \mathbf{x}, \mathbf{y})$. The mutation μ_k consists of three parts: *matrix mutation*, which is applied to B to produce $B' = (b'_{ij}) = \mu_k(B)$, where

$$b'_{ij} = \begin{cases} -b_{ij} & \text{if } i = k \text{ or } j = k, \\ b_{ij} + \mathrm{sgn}(b_{ik})[b_{ik}b_{kj}]_+ & \text{otherwise,} \end{cases} \tag{2.1}$$

with $\mathrm{sgn}(a)$ being ± 1 for positive/negative $a \in \mathbb{R}$ and 0 for $a = 0$, and

$$[a]_+ = \max(a, 0);$$

coefficient mutation, defined by $\mathbf{y}' = (y'_j) = \mu_k(\mathbf{y})$ where

$$y'_j = \begin{cases} y_k^{-1} & \text{if } j = k, \\ y_j \left(1 + y_k^{-\mathrm{sgn}(b_{jk})}\right)^{-b_{jk}} & \text{otherwise;} \end{cases} \tag{2.2}$$

and *cluster mutation*, given by $\mathbf{x}' = (x'_j) = \mu_k(\mathbf{x})$ with the *exchange relation*

$$x'_k = \frac{y_k \prod_{i=1}^N x_i^{[b_{ki}]_+} + \prod_{i=1}^N x_i^{[-b_{ki}]_+}}{(1 + y_k)x_k}, \tag{2.3}$$

and $x'_j = x_j$ for $j \neq k$.

Given an initial seed, one can apply an arbitrary sequence of mutations, which produces a sequence of seeds. This can be visualized by attaching the initial seed to the root of an N-regular tree $\mathbb{T}_N$ (with N branches attached to each vertex), and then labelling the seeds as $(B_\mathbf{t}, \mathbf{x}_\mathbf{t}, \mathbf{y}_\mathbf{t})$ with "time" $\mathbf{t} \in \mathbb{T}_N$. Note that mutation is an involution, $\mu_k \cdot \mu_k = \mathrm{id}$, but in general two successive mutations do not commute, i.e., typically $\mu_j \cdot \mu_k \neq \mu_k \cdot \mu_j$ for $j \neq k$. Moreover, in general the exponents and coefficients appearing in the exchange relation (2.3) change at each stage, because the matrix B and the $\mathbf{y}$ variables are altered by each of the previous mutations.

Definition 1. The cluster algebra $\mathcal{A}(B, \mathbf{x}, \mathbf{y})$ is the algebra over $\mathbb{C}(\mathbf{y})$ generated by the cluster variables produced by all possible sequences of mutations applied to the seed $(B, \mathbf{x}, \mathbf{y})$.

We will also consider the case of coefficient-free cluster algebras, for which the $\mathbf{y}$ variables are absent, the seeds are just $(B, \mathbf{x})$, and the cluster mutation is defined by the simpler exchange relation

$$x'_k = \frac{\prod_{i=1}^N x_i^{[b_{ki}]_+} + \prod_{i=1}^N x_i^{[-b_{ki}]_+}}{x_k}. \tag{2.4}$$

Remark 1. The original definition of a cluster algebra in [21] involves a more general setting in which the coefficients $\mathbf{y}$ are elements of a semifield $\mathbb{P}$, that is, an abelian multiplicative group together with a binary operation $\oplus$ that is commutative, associative and distributive with respect to multiplication. In that setting, with the N-tuple $\mathbf{y} \in \mathbb{P}^N$, the algebra $\mathcal{A}(B, \mathbf{x}, \mathbf{y})$ is defined over $\mathbb{Z}[\mathbb{P}]$, and the addition in the denominator of (2.3) is given by $\oplus$. The case we consider here corresponds to $\mathbb{P} = \mathbb{P}_{univ}$, the universal semifield, consisting of subtraction-free rational functions in the variables y_j, in which case $\oplus$ becomes ordinary addition in the field of rational functions $\mathbb{C}(\mathbf{y})$. However, starting with the more general setting, we can also consider the case of the trivial semifield with one element, $\mathbb{P} = \{1\}$, which yields the coefficient-free case (2.4).

In order to illustrate the above definitions, we now present a number of concrete examples. For the sake of simplicity, we concentrate on the coefficient-free case in the rest of this section, and return to the equations with coefficients $\mathbf{y}$ at a later stage.

Example 1. The cluster algebra of type B_2: A particular cluster algebra of rank $N = 2$ is given by taking the exchange matrix

$$B = \begin{pmatrix} 0 & 2 \\ -1 & 0 \end{pmatrix}, \tag{2.5}$$

and the initial cluster $\mathbf{x} = (x_1, x_2)$, to define a seed $(B, \mathbf{x})$. The matrix B is skew-symmetrizable: the diagonal matrix $D = \mathrm{diag}(1, 2)$ is such that

$$DB = \begin{pmatrix} 0 & 2 \\ -2 & 0 \end{pmatrix}$$

is skew-symmetric. Applying the mutation μ_1 and using the rule (2.1) gives a new exchange matrix

$$B' = \mu_1(B) = \begin{pmatrix} 0 & -2 \\ 1 & 0 \end{pmatrix} = -B,$$

while the coefficient-free exchange relation (2.4) gives a new cluster $\mathbf{x}' = (x_1', x_2)$ with

$$x_1' = \frac{x_2^2 + 1}{x_1}.$$

Since mutation acts as an involution, we have $\mu_1(B', \mathbf{x}') = (B, \mathbf{x})$, so nothing new is obtained by applying μ_1 to this new seed. Thus we consider $\mu_2(B', \mathbf{x}') = \mu_2 \cdot \mu_1(B, \mathbf{x})$ instead, which produces $\mu_2(B') = B$ and $\mu_2(\mathbf{x}') = (x_1', x_2')$, where

$$x_2' = \frac{x_1' + 1}{x_2} = \frac{x_1 + x_2^2 + 1}{x_1 x_2}.$$

Once again, a repeat application of the same mutation μ_2 returns to the previous seed, so instead we consider applying μ_1 to obtain $\mu_1 \cdot \mu_2 \cdot \mu_1(B) = -B$ and $\mu_1 \cdot \mu_2 \cdot \mu_1(\mathbf{x}) = (x_1'', x_2')$, with

$$x_1'' = \frac{x_1^2 + 2x_1 + x_2^2 + 1}{x_1 x_2^2}.$$

Repeating this sequence of mutations, it is clear that the exchange matrix just changes by an overall sign at each step. Perhaps more surprising is the fact that after obtaining $(\mu_2 \cdot \mu_1)^2(\mathbf{x}) = (x_1'', x_2'')$, with

$$x_2'' = \frac{x_1 + 1}{x_2},$$

the variable x_1 reappears in the cluster after a further step, i.e., $\mu_1 \cdot (\mu_2 \cdot \mu_1)^2(\mathbf{x}) = (x_1, x_2'')$, and finally $(\mu_2 \cdot \mu_1)^3(\mathbf{x}) = (x_1, x_2) = \mathbf{x}$, so that the initial seed $(B, \mathbf{x})$ is restored after a total of six mutations. Thus the cluster algebra has a finite number of generators in this case, since there are only the six cluster variables $x_1, x_2, x_1', x_2', x_1'', x_2''$. This example is called the cluster algebra of type B_2, since the initial matrix B is derived from the Cartan matrix of the B_2 root system, that is

$$C = \begin{pmatrix} 2 & -2 \\ -1 & 2 \end{pmatrix},$$

by replacing the diagonal entries in C with 0, and changing signs of the off-diagonal entries so that b_{ij} and b_{ji} have opposite signs for $i \neq j$.

There are two significant features of the preceding example, namely the fact that there are only finitely many clusters, and the fact that the cluster variables are all Laurent polynomials (polynomials in x_1, x_2 and their reciprocals) with integer coefficients. The first feature is rare: a cluster algebra is said to be of finite type if there are only finitely many clusters, and it was shown in [23] that all such cluster algebras are generated from seeds corresponding to the finite root systems that appear in the Cartan-Killing classification of finite-dimensional semisimple Lie algebras. The second feature (the Laurent phenomenon) is ubiquitous [22], and follows from the following result, proved in [22].

Proposition 1. *All cluster variables in a coefficient-free cluster algebra $\mathcal{A}(B, \mathbf{x})$ are Laurent polynomials in the variables from the initial cluster, with integer coefficients, i.e., they are elements of the ring $\mathbb{Z}[\mathbf{x}^{\pm 1}] := \mathbb{Z}[x_1^{\pm 1}, x_2^{\pm 1}, \ldots, x_N^{\pm 1}]$.*

There is an analogous statement in the case that coefficients are included, and in fact it is possible to prove the stronger result that all of the coefficients of the cluster variables have positive integer coefficients, so they belong to $\mathbb{Z}_{>0}[\mathbf{x}^{\pm 1}]$ (see [36, 54], for instance).

Example 2. The cluster algebra of type $\tilde{A}_{1,3}$: As an example of rank $N = 4$, we take the skew-symmetric matrix

$$B = \begin{pmatrix} 0 & 1 & 0 & 1 \\ -1 & 0 & 1 & 0 \\ 0 & -1 & 0 & 1 \\ -1 & 0 & -1 & 0 \end{pmatrix}, \tag{2.6}$$

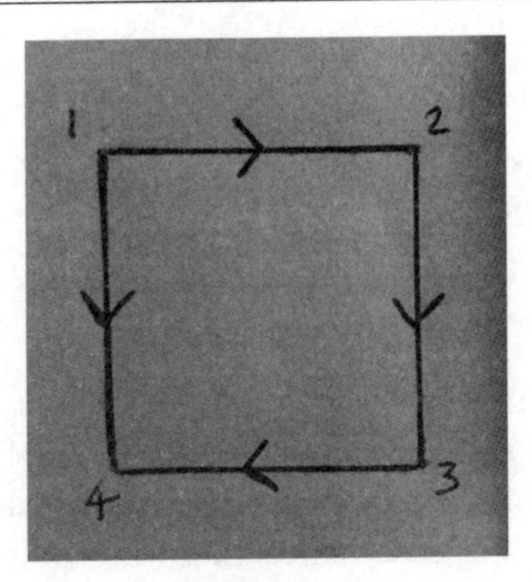

Figure 1. The quiver Q corresponding to the exchange matrix (2.6).

which is obtained from the Cartan matrix of the affine root system $A_3^{(1)}$ [49], namely

$$C = \begin{pmatrix} 2 & -1 & 0 & -1 \\ -1 & 2 & -1 & 0 \\ 0 & -1 & 2 & -1 \\ -1 & 0 & -1 & 2 \end{pmatrix},$$

by replacing each of the diagonal entries of C with 0, and making a suitable adjustment of signs for the off-diagonal entries, such that $b_{ij} = -b_{ji}$. Since B is a skew-symmetric integer matrix, it can be associated with a quiver Q without 1- or 2-cycles, that is, a directed graph specified by the rule that b_{ij} is equal to the number of arrows $i \to j$ if it is non-negative, and minus the number of arrows $j \to i$ otherwise (see Fig. 1). If the mutation μ_1 is applied, then the new exchange matrix is

$$B' = \mu_1(B) = \begin{pmatrix} 0 & -1 & 0 & -1 \\ 1 & 0 & 1 & 0 \\ 0 & -1 & 0 & 1 \\ 1 & 0 & -1 & 0 \end{pmatrix},$$

which corresponds to a new quiver Q' obtained by a cyclic permutation of the vertices of the original Q (see Fig. 2), while the initial cluster $\mathbf{x} = (x_1, x_2, x_3, x_4)$ is mutated to $\mu_1(\mathbf{x}) = (x_1', x_2, x_3, x_4)$, where x_1' is defined by the relation

$$x_1 x_1' = x_2 x_4 + 1.$$

Rather than trying to describe the effect of every possible choice of mutation, we consider what happens when μ_1 is followed by μ_2, and once more observe that, at the level of the associated quiver, this just corresponds to applying the same cyclic permutation as before to the vertex labels 1,2,3,4. The new cluster obtained from this is $\mu_2 \cdot \mu_1(\mathbf{x}) = (x_1', x_2', x_3, x_4)$, with x_2' defined by

$$x_2 x_2' = x_3 x_1' + 1,$$

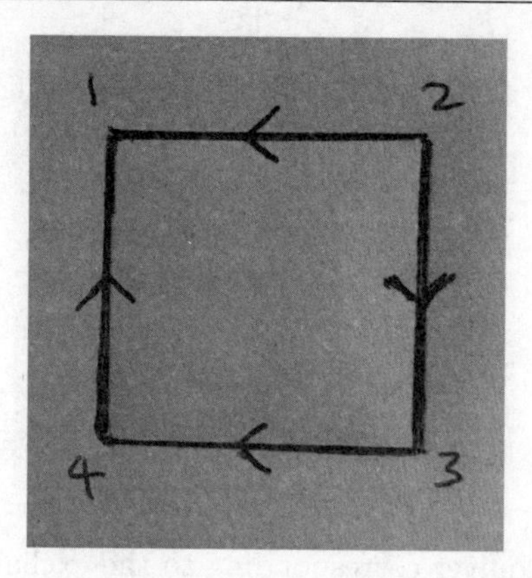

Figure 2. The mutated quiver $Q' = \mu_1(Q)$ obtained by applying μ_1 to (2.6).

and if μ_3 is applied next, then $\mu_3 \cdot \mu_2 \cdot \mu_1(\mathbf{x}) = (x_1', x_2', x_3', x_4)$, with

$$x_3 x_3' = x_4 x_2' + 1.$$

Continuing in this way, it is not hard to see that the composition $\mu_4 \cdot \mu_3 \cdot \mu_2 \cdot \mu_1$ takes the original B to itself, and applying this sequence of mutations repeatedly in the same order generates a new cluster variable at each step, with the sequence of cluster variables satisfying the nonlinear recurrence relation

$$x_n x_{n+4} = x_{n+1} x_{n+3} + 1 \tag{2.7}$$

(where we have made the identification $x_1' = x_5$, $x_2' = x_6$, and so on). Regardless of other possible choices of mutations, this particular sequence of mutations alone generates an infinite set of distinct cluster variables, as can be seen by fixing some numerical values for the initial cluster. In fact, as was noted in [43], for any orbit of (2.7) there is a constant K such that the iterates satisfy the linear recurrence

$$x_{n+6} + x_n = K x_{n+3}. \tag{2.8}$$

Upon fixing $(x_1, x_2, x_3, x_4) = (1, 1, 1, 1)$, the nonlinear recurrence generates the integer sequence

$$1, 1, 1, 1, 2, 3, 4, 9, 14, 19, 43, 76, \ldots,$$

which also satisfies the linear recurrence (2.8) with $K = 5$; so the terms grow exponentially with n, and the integers x_n are distinct for $n \geq 4$. This is called the $\tilde{A}_{1,3}$ cluster algebra, because the corresponding quiver is an orientation of the edges of an affine Dynkin diagram of type A with one anticlockwise arrow and three clockwise arrows.

The skew-symmetry of B is preserved under matrix mutation, and for any skew-symmetric integer matrix there is an equivalent operation of *quiver mutation* which acts on the associated quiver Q: to obtain the mutated quiver $\mu_k(Q)$ one should (i) add pq arrows $i \xrightarrow{pq} j$ whenever Q has a path of length two passing through vertex

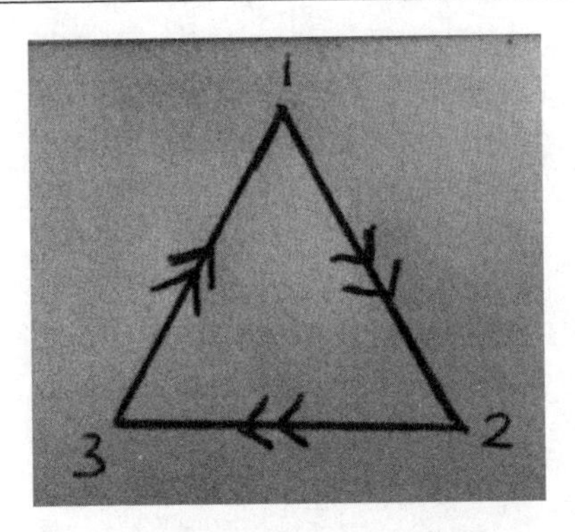

Figure 3. The quiver corresponding to the exchange matrix (2.9).

k with p arrows $i \xrightarrow{p} k$ and q arrows $k \xrightarrow{q} j$; (ii) reverse all arrows in Q that go in/out of vertex k; (iii) delete any 2-cycles created in the first step.

Unlike the B_2 cluster algebra, the above example is not of finite type, because there are infinitely many clusters. However, it turns out that it is of finite mutation type, in the sense that there are only a finite number of exchange matrices produced under mutation from the initial B. Cluster algebras of finite mutation type have also been classified [16, 17]: as well as those of finite type, they include cluster algebras associated with triangulated surfaces [20, 19], cluster algebras of rank 2, plus a finite number of exceptional cases.

Example 3. Cluster algebra related to Markoff's equation: For $N = 3$, consider the exchange matrix

$$B = \begin{pmatrix} 0 & 2 & -2 \\ -2 & 0 & 2 \\ 2 & -2 & 0 \end{pmatrix}, \tag{2.9}$$

which is associated with the quiver in Fig. 3. After any sequence of matrix mutations, one can obtain only B or $-B$, so this is another example of finite mutation type: it is connected to the moduli space of once-punctured tori, and the Markoff equation

$$x^2 + y^2 + z^2 = 3xyz \tag{2.10}$$

which arises in that context as well as in Diophantine approximation theory [7, 9]. Upon applying μ_1 to the initial cluster (x_1, x_2, x_3), the result is (x_1', x_2, x_3) with

$$x_1 x_1' = x_2^2 + x_3^2,$$

and a subsequent application of μ_2 yields (x_1', x_2', x_3), where

$$x_2 x_2' = x_3^2 + x_1'^2.$$

Repeated application of the mutations $\mu_3 \cdot \mu_2 \cdot \mu_1$ in that order produces a new cluster variable at each step, and upon identifying $x_4 = x_1'$, $x_5 = x_2'$, and so on,

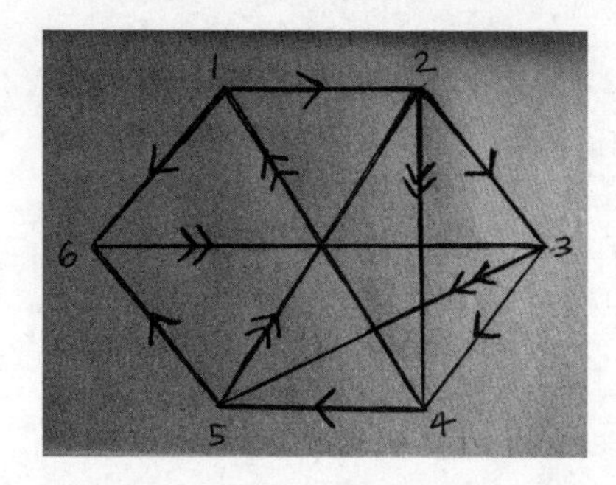

Figure 4. The Somos-6 quiver corresponding to the exchange matrix (2.14).

the sequence of cluster variables (x_n) is generated by a recurrence of third order, namely

$$x_n x_{n+3} = x_{n+1}^2 + x_{n+2}^2. \tag{2.11}$$

It can also be shown that on each orbit of (2.11) there is a constant K such that the nonlinear relation

$$x_{n+3} + x_n = K x_{n+1} x_{n+2}$$

holds for all n, and by using the latter to eliminate x_{n+3} it follows that

$$K = \frac{x_n^2 + x_{n+1}^2 + x_{n+2}^2}{x_n x_{n+1} x_{n+2}} \tag{2.12}$$

is an invariant for (2.11), independent of n. In particular, taking the initial values to be $(1, 1, 1)$ gives $K = 3$, and each adjacent triple $(x, y, z) = (x_n, x_{n+1}, x_{n+2})$ in the resulting sequence

$$1, 1, 1, 2, 5, 29, 433, 37666, 48928105, \ldots \tag{2.13}$$

is an integer solution of Markoff's equation (2.10). The terms of this sequence have double exponential growth: $\log x_n$ grows exponentially with n.

The next example is generic, in the sense that there are both infinitely many clusters and infinitely many exchange matrices.

Example 4. A Somos-6 recurrence: A sequence that is generated by a quadratic recurrence relation of the form

$$x_n x_{n+k} = \sum_{j=1}^{\lfloor k/2 \rfloor} \alpha_j \, x_{n+j} x_{n+k-j},$$

where α_j are coefficients, is called a Somos-k sequence (see [22, 29, 41, 64, 65]). A certain class of Somos-6 sequences can be generated by starting from the exchange matrix

$$B = \begin{pmatrix} 0 & 1 & 0 & -2 & 0 & 1 \\ -1 & 0 & 1 & 2 & -2 & 0 \\ 0 & -1 & 0 & 1 & 2 & -2 \\ 2 & -2 & -1 & 0 & 1 & 0 \\ 0 & 2 & -2 & -1 & 0 & 1 \\ -1 & 0 & 2 & 0 & -1 & 0 \end{pmatrix}, \tag{2.14}$$

which corresponds to the quiver in Fig. 4. Upon applying cyclic sequences of mutations ordered as $\mu_6 \cdot \mu_5 \cdot \mu_4 \cdot \mu_3 \cdot \mu_2 \cdot \mu_1$, a sequence of cluster variables (x_n) is produced which satisfies the particular Somos-6 recurrence

$$x_n x_{n+6} = x_{n+1} x_{n+5} + x_{n+3}^2. \tag{2.15}$$

If six 1s are chosen as initial values, then an integer Somos-6 sequence beginning with

$$1, 1, 1, 1, 1, 1, 2, 3, 4, 8, 17, 50, 107, 239, \ldots$$

is produced. For this sequence, $\log x_n$ grows like n^2. However, applying successive mutations other than these cyclic ones generally causes the magnitude of the entries of the exchange matrices to grow, for instance,

$$\mu_5 \cdot \mu_4 \cdot \mu_2(B) = \begin{pmatrix} 0 & -1 & 1 & 0 & 0 & 1 \\ 1 & 0 & -3 & -10 & 4 & 0 \\ -1 & 3 & 0 & -1 & 0 & -2 \\ 0 & 10 & 1 & 0 & -3 & 3 \\ 0 & -4 & 0 & 3 & 0 & -1 \\ -1 & 0 & 2 & -3 & 1 & 0 \end{pmatrix};$$

and (e.g., starting with the initial seed evaluated as $(1, 1, 1, 1, 1, 1)$ as before) typically this results in the values of cluster variables showing double exponential growth with the number of steps.

3 Cluster algebras with periodicity

The exchange relation (2.3) can be regarded as a birational map in $\mathbb{C}^N$. Alternatively, $\mathbf{x} \in (\mathbb{C}^*)^N$ can be viewed as coordinates in a toric chart for some algebraic variety, and a mutation $\mathbf{x} \mapsto \mu_k(\mathbf{x}) = \mathbf{x}'$ as a change of coordinates to another chart. The latter point of view is passive, in the sense that there is some fixed variety and mutation just selects different choices of coordinate charts. Instead of this, we would like to take an active view, regarding each mutation as an iteration in a discrete dynamical system. However, there is a problem with this, because a general sequence of mutations is specified by a "time" $\mathbf{t}$ belonging to the tree $\mathbb{T}_N$, and (except for the case of rank $N = 2$) this cannot naturally be identified with a discrete time belonging to the set of integers $\mathbb{Z}$. Furthermore, there is the additional problem that matrix mutation, as in (2.1), typically changes the exponents appearing in the exchange relation, so that in general it is not possible to interpret successive mutations as iterations of the same map.

Despite the above comments, it turns out that the most interesting cluster algebras appearing "in nature" have special symmetries, in the sense that they display periodic behaviour with respect to at least some subset of the possible mutations. In fact, all of the examples in the previous section are of this kind. Here we consider a notion of periodicity that was introduced by Fordy and Marsh [26] in the context of skew-symmetric exchange matrices B, which correspond to quivers.

Definition 2. An exchange matrix B is said to be *cluster mutation-periodic* with period m if (for a suitable labelling of indices) $\mu_m \cdot \mu_{m-1} \cdot \ldots \cdot \mu_1(B) = \rho^m(B)$, where ρ is the cyclic permutation $(1, 2, 3, \ldots, N) \mapsto (N, 1, 2, \ldots, N-1)$.

In the context of quiver mutation, the case of cluster mutation-periodicity with period $m = 1$ means that the action of mutation μ_1 on Q is the same as the action of ρ, which is such that the number of arrows $i \to j$ in Q is the same as the number of arrows $\rho^{-1}(i) \to \rho^{-1}(j)$ in $\rho(Q)$. This means that the cluster map $\varphi = \rho^{-1} \cdot \mu_1$ acts as the identity on Q (or equivalently, on B), but in general $\mathbf{x} \mapsto \varphi(\mathbf{x})$ has a non-trivial action on the cluster. Mutation-periodicity with period 1 implies that iterating this map is equivalent to iterating a single recurrence relation.

Example 5. Although it is not skew-symmetric, the exchange matrix (2.5) in Example 1 is cluster mutation-periodic with period 2, since $\mu_2 \cdot \mu_1(B) = \rho^2(B) = B$ where ρ is the switch $1 \leftrightarrow 2$. Defining the cluster map to be $\varphi = \rho^{-2} \cdot \mu_2 \cdot \mu_1$, the action of φ is periodic with period 3 for any choice of initial cluster, i.e., $\varphi^3 =$id.

Example 6. The exchange matrix (2.6) in Example 2 is cluster mutation-periodic with period 1. The cluster map is given by

$$\varphi: \ (x_1, x_2, x_3, x_4) \mapsto \left(x_2, x_3, x_4, \frac{x_2 x_4 + 1}{x_1} \right), \tag{3.1}$$

whose iterates are equivalent to those of the nonlinear recurrence (2.7).

Example 7. The exchange matrix (2.9) in Example 3 is cluster mutation-periodic with period 2. The cluster map $\varphi = \rho^{-2} \cdot \mu_2 \cdot \mu_1$ is given by

$$\varphi: \ (x_1, x_2, x_3) \mapsto \left(x_3, x_1', \frac{x_3^2 + x_1'^2}{x_2} \right), \quad \text{with} \quad x_1' = \frac{x_2^2 + x_3^2}{x_1}. \tag{3.2}$$

Each iteration of (3.2) is equivalent to two iterations of the nonlinear recurrence (2.11). This period 2 example is exceptional because, in general, cluster mutation-periodicity with period $m > 2$ does not give rise to a single recurrence relation (see [26] for more examples).

Example 8. The exchange matrix (2.14) in Example 4 is cluster mutation-periodic with period 1. The cluster map is given by

$$(x_1, x_2, x_3, x_4, x_5, x_6) \mapsto \left(x_2, x_3, x_4, x_5, x_6, \frac{x_2 x_6 + x_4^2}{x_1} \right), \tag{3.3}$$

whose iterates are equivalent to those of the nonlinear recurrence (2.15).

Remark 2. There is a more general notion of periodicity, due to Nakanishi [59], which extends Definition 2. This yields broad generalizations of Zamolodchikov's Y-systems [68], a set of functional relations, arising from the thermodynamic Bethe ansatz for certain integrable quantum field theories, that were the prototype for the coefficient mutation (2.2) in a cluster algebra. The original Y-systems are associated with Dynkin diagrams, and Zamolodchikov's conjecture that these yield purely periodic dynamics was proved in [24], while the corresponding result for pairs of Dynkin diagrams was shown in [51]. We shall introduce examples of generalized Y-systems in the sequel.

Fordy and Marsh gave a complete classification of period 1 quivers. Their result can be paraphrased as follows.

Theorem 1. *Let $(a_1, \ldots, a_{N-1})$ be an $(N-1)$-tuple of integers that is palindromic, i.e., $a_j = a_{N-j}$ for all $j \in [1, N-1]$. Then the skew-symmetric exchange matrix $B = (b_{ij})$ with entries specified by*

$$b_{1,j+1} = a_j \quad \textit{and} \quad b_{i+1,j+1} = b_{ij} + a_i[-a_j]_+ - a_j[-a_i]_+,$$

for all $i, j \in [1, N-1]$, is cluster mutation-periodic with period 1, and every period 1 skew-symmetric B arises in this way.

The above result says that a period 1 skew-symmetric B matrix is completely determined by the entries in its first row (or equivalently, its first column), and these form a palindrome after removing b_{11}. The entries a_j in the palindrome are precisely the exponents that appear in the exchange relation defining the cluster map φ, whose iterates are equivalent to those of the nonlinear recurrence relation

$$x_n x_{n+N} = \prod_{j:\, a_j>0} x_{n+j}^{a_j} + \prod_{j:\, a_j<0} x_{n+j}^{-a_j}. \tag{3.4}$$

Thus (3.4) corresponds to a special sequence of mutations in a particular subclass of cluster algebras. Such a nonlinear recurrence is an example of a generalized T-system, in the terminology of [59].

Next we would like to turn to the question of which recurrences of this special type correspond to discrete integrable systems. We begin our approach to this question in the next section, by considering the notion of algebraic entropy, which gives a measure of the growth of iterates in a discrete dynamical system defined by iteration of rational functions.

4 Algebraic entropy and tropical dynamics

There a various different ways of quantifying the growth, or complexity, of a discrete dynamical system (see [1], for instance). In the context of discrete integrability of birational maps, Bellon and Viallet introduced the concept of algebraic entropy, and proposed that zero algebraic entropy should be a criterion for integrability [4]. For a birational map φ, one can calculate the degree $d_n = \deg \varphi^n$, given by the maximum of the degrees of the components of φ^n, and then the algebraic entropy is defined to be

$$\mathcal{E} := \lim_{n\to\infty} \frac{\log d_n}{n}.$$

Typically, the degree d_n grows exponentially with n, so $\mathcal{E} > 0$, but in rare cases there can be subexponential growth, leading to vanishing entropy. In the case of birational maps in two dimensions, the types of degree growth have been fully classified [12], and there are only four possibilities: bounded degrees, linear growth, quadratic growth, or exponential growth; the first three cases, with zero entropy, coincide with the existence of invariant foliations. Thus, at least for maps of the plane, the requirement of zero entropy identifies symplectic maps that are integrable in the sense that they satisfy the conditions needed for a discrete analogue of the Liouville-Arnold theorem to hold [6, 55, 67].

Measuring the degree growth and seeking maps with zero algebraic entropy is a useful tool for identifying discrete integrable systems. (For another approach, based on the growth of heights in orbits defined over $\mathbb{Q}$ or a number field, see [37].) Once such a map has been identified, it leaves open the question of Liouville integrability; this is discussed in the next section. For now, we concentrate on the case of maps arising from cluster algebras, and consider algebraic entropy in that setting.

The advantage of working with cluster maps is that, due to the Laurent property, it is sufficient to consider the growth of degrees of the denominators of the cluster variables in order to determine the algebraic entropy. In particular, in the period 1 case, by Proposition 1 every iterate of (3.4) can be written in the form

$$x_n = \frac{\mathrm{P}_n(\mathbf{x})}{\mathbf{x}^{\mathbf{d}_n}}, \tag{4.1}$$

where the polynomial P_n is not divisible by any of the x_j from the initial cluster, and the monomial $\mathbf{x}^{\mathbf{d}_n} = \prod_{j=1}^N x_j^{d_n^{(j)}}$ is specified by the integer vector

$$\mathbf{d}_n = (d_n^{(1)}, \ldots, d_n^{(N)})^T,$$

known as a d-vector. From the fact that cluster variables are subtraction-free rational expressions in $\mathbf{x}$ (or, a fortiori, from the fact that these Laurent polynomials are now known to have positive integer coefficients [36]), it follows that the d-vectors in a cluster algebra satisfy the max-plus tropical analogue of the exchange relations for the corresponding cluster variables [19, 25], where the latter is obtained from (2.4) by replacing each addition with max, and each multiplication with addition. In the case of (3.4), this implies the following result.

Proposition 2. *If x_n given by (4.1) satisfies (3.4), then the sequence of vectors $\mathbf{d}_n$ satisfies the tropical recurrence relation*

$$\mathbf{d}_n + \mathbf{d}_{n+N} = \max\left(\sum_{j:\, a_j>0} a_j \mathbf{d}_{n+j}, - \sum_{j:\, a_j<0} a_j \mathbf{d}_{n+j} \right). \tag{4.2}$$

Note that the equality in (4.2) holds componentwise. For a detailed proof of this result, see [27].

In the period 1 situation, the problem of determining the evolution of d-vectors can be simplified further, upon noting that the first component of $\mathbf{d}_n$ has the initial values

$$d_1^{(1)} = -1, \qquad d_j^{(1)} = 0 \quad \text{for} \quad 2 \le j \le N, \tag{4.3}$$

while each of the other components $d_n^{(k)}$ for $k \in [2, N]$ has the same set of initial values but shifted by $k - 1$ steps (so, for instance, $d_2^{(2)} = -1$ and $d_j^{(2)} = 0$ for $3 \le j \le N + 1$, etc.). The total degree of the monomial $\mathbf{x}^{\mathbf{d}_n}$ is $\sum_{k=1}^{N} d_n^{(k)}$, the sum of the components of the d-vector, and if the components are all non-negative then this coincides with the degree of the denominator of the rational function (4.1). Unless there is periodicity of d-vectors, corresponding to degrees remaining bounded (which can only happen in finite type cases like Example 1), then all these components are positive for large enough n. Moreover, it is not hard to see that the growth of the degree of the numerators P_n appearing in the Laurent polynomials (4.1) is controlled by that of the denominators. Thus, to determine the growth of degrees of Laurent polynomials generated by (3.4), it is sufficient to consider the solution of the scalar version of (4.2), with initial data given by (4.3), and the growth of this determines the algebraic entropy.

Example 9. For the recurrence (2.7) in Example 2, the tropical equation for determining the degrees of d-vectors is given in scalar form by

$$d_n + d_{n+4} = \max(d_{n+1} + d_{n+3}, 0). \tag{4.4}$$

If we take initial values $d_1 = -1$, $d_2 = d_3 = d_4 = 0$, corresponding to (4.3), then by induction it follows that $d_n \ge 0$ for all $n \ge 2$, so that the max on the right-hand side of (4.4) can be replaced by its first entry, to yield the linear recurrence

$$d_n + d_{n+4} = d_{n+1} + d_{n+3} \quad \text{for} \quad n \ge 1.$$

The characteristic polynomial of the latter factorizes as $(\lambda - 1)^2(\lambda^2 + \lambda + 1) = 0$, leading to the solution

$$d_n = n/3 - 1 + \frac{\mathrm{i}}{3\sqrt{3}}(\varepsilon^n - \varepsilon^{-n}), \quad \varepsilon = (-1 + \mathrm{i}\sqrt{3})/2.$$

Thus we have a sequence that grows linearly with n, beginning with

$$-1, 0, 0, 0, 1, 1, 1, 2, 2, 2, 3, 3, 3, 4, 4, 4, \ldots,$$

where each positive integer appears three times in succession, which corresponds to the degree of the denominator of x_n in each of the variables x_1, x_2, x_3, x_4 separately. Clearly the total degree of the denominator also grows linearly, and the algebraic entropy is equal to the limit $\lim_{n\to\infty}(\log d_n)/n = 0$ in this case.

Example 10. The exchange matrix (2.9) in Example 3 is period 2 rather than period 1, but we can still calculate the growth of d-vectors in the recurrence (2.11) by taking its tropical version, namely

$$d_n + d_{n+3} = 2\max(d_{n+1}, d_{n+2}),$$

and choosing the initial values $d_1 = -1$, $d_2 = d_3 = 0$, which produces a sequence beginning

$$-1, 0, 0, 1, 2, 4, 7, 12, 20, 33, 54, 88, \ldots.$$

By induction one can show that $d_{n+2} \geq d_{n+1}$ for $n \geq 0$, so in fact the linear recurrence

$$d_{n+3} + d_n = 2d_{n+2}$$

holds for this sequence, with characteristic equation $(\lambda - 1)(\lambda^2 - \lambda - 1) = 0$, and it turns out that the differences

$$F_n = d_{n+3} - d_{n+2}$$

are just the Fibonacci numbers. Hence there is a constant $\mathrm{C} > 0$ such that

$$d_n \sim \mathrm{C}\left(\frac{1+\sqrt{5}}{2}\right)^n,$$

and the algebraic entropy $\mathcal{E} = \log((1+\sqrt{5})/2)$, which is the same as the limit $\lim_{n\to\infty}(\log\log x_n)/n$ for the sequence (2.13); see [40].

Example 11. For the period 1 exchange matrix (2.14) in Example 4, we consider the recurrence

$$d_n + d_{n+6} = \max(d_{n+1} + d_{n+5}, 2d_{n+3}), \tag{4.5}$$

which is the max-plus analogue of (2.15), and take initial data

$$d_1 = -1, \quad d_2 = d_3 = \cdots = d_6 = 0, \tag{4.6}$$

which generates a degree sequence beginning

$$-1, 0, 0, 0, 0, 0, 1, 1, 1, 2, 2, 3, 3, 3, 5, 5, 6, 7, 7, 9, 9, 10, 12, 12, 14, 15, 16, \ldots. \tag{4.7}$$

In order to simplify the analysis of (4.5), we observe that the combination

$$U_n = d_{n+2} - 2d_{n+1} + d_n \tag{4.8}$$

satisfies a recurrence of fourth order, namely

$$U_{n+4} + 2U_{n+3} + 3U_{n+2} + 2U_{n+1} + U_n = \max(U_{n+3} + 2U_{n+2} + U_{n+1}, 0). \tag{4.9}$$

(The origin of the substitution (4.8) will be explained in the next section.) The values in (4.6) correspond to the initial conditions

$$U_1 = -1, \quad U_2 = U_3 = U_4 = 0$$

for (4.9), which generate a sequence (U_n) beginning with

$$-1,0,0,0,1,-1,0,1,-1,1,-1,0,2,-2,1,0,-1,2,-2,1,1,-2,2,-1,0,1,-2,\ldots,$$

and further calculation with a computer shows that this sequence does not repeat for the first 40 steps, but then $U_{42} = -1$ and $U_{43} = U_{44} = U_{45} = 0$, so it is periodic with period 41. Thus in terms of the shift operator $\mathcal{S}$, which sends $n \to n+1$,

$$(\mathcal{S}^{41} - 1)U_n = (\mathcal{S}^{41} - 1)(d_{n+2} - 2d_{n+1} + d_n) = (\mathcal{S}^{41} - 1)(\mathcal{S} - 1)^2 d_n = 0,$$

which is a linear recurrence of order 43 satisfied by the degree sequence (4.7). Clearly the characteristic polynomial of the latter has $\lambda = 1$ as a triple root, and all other characteristic roots have modulus 1. Therefore, for some constant $\mathrm{C}' > 0$,

$$d_n \sim \mathrm{C}' n^2$$

as $n \to \infty$, which implies that (2.15) has algebraic entropy $\mathcal{E} = 0$.

The preceding examples indicate that we should regard (2.7) and (2.15) as being integrable in some sense, and (2.11) as non-integrable. According to the relation (2.8), we know that (2.7) has at least one conserved quantity K; and it turns out to have three independent conserved quantities [43]. The recurrence (2.11) also has a conserved quantity, given by (2.12), but it is possible to show that it can have no other algebraic conserved quantities, independent of this one. In the next section we will derive two independent conserved quantities for (2.15), and we will discuss the interpretation of all these examples from the viewpoint of Liouville integrability.

In [27], a detailed analysis of the behaviour of the tropical recurrences (4.2) led to the conjecture that the algebraic entropy of (3.4) should be positive if and only if the following condition holds:

$$\max\left(\sum_{j=1}^{N-1}[a_j]_+, -\sum_{j=1}^{N-1}[-a_j]_+\right) \geq 3. \tag{4.10}$$

In other words, in order for the cluster map defined by (3.4) to have a zero entropy, the degree of nonlinearity cannot be too large. The analysis of algebraic entropy for other types of cluster maps has been carried out more recently using methods based on Newton polytopes [28], and using the same methods it is also possible to prove the above conjecture[2]. By enumerating the possible choices of exponents that lie below the bound (4.10), this leads to a complete proof of a classification result for nonlinear recurrences of the form (3.4), as stated in [27].

Theorem 2. *A cluster map φ given by a recurrence (3.4) has algebraic entropy $\mathcal{E} = 0$ if and only if it belongs to one of the following four families:*
(i) For even $N = 2m$, recurrences of the form

$$x_n x_{n+2m} = x_{n+m} + 1. \tag{4.11}$$

[2]P. Galashin, private communication, 2017.

(ii) For $N \geq 2$ and $1 \leq q \leq \lfloor N/2 \rfloor$, recurrences of the form

$$x_n x_{n+N} = x_{n+q} x_{n+N-q} + 1. \tag{4.12}$$

(iii) For even $N = 2m$ and $1 \leq q \leq m-1$, recurrences of the form

$$x_n x_{n+2m} = x_{n+q} x_{n+2m-q} + x_{n+m}. \tag{4.13}$$

(iv) For $N \geq 2$ and $1 \leq p < q \leq \lfloor N/2 \rfloor$, recurrences of the form

$$x_n x_{n+N} = x_{n+p} x_{n+N-p} + x_{n+q} x_{n+N-q}. \tag{4.14}$$

Case (i) is somewhat trivial: the recurrence (4.11) is equivalent to taking m copies of the Lyness 5-cycle

$$x_n x_{n+2} = x_{n+1} + 1,$$

for which every orbit has period 5, corresponding to the cluster algebra of finite type associated with the root system A_2; so in this case the dynamics is purely periodic and there is no degree growth. Both case (ii), which corresponds to affine quivers of type $\tilde{A}_{q,N-q}$, and case (iii) display linear degree growth, similar to Example 9. Case (iv) consists of Somos-N recurrences, which display quadratic degree growth [58], as in Example 11. Hence only zero, linear, quadratic or exponential growth is displayed by the cluster recurrences (3.4). Interestingly, these are the only types of growth found in the other families of cluster maps considered in [28]. We do not know if other types of growth are possible; are there cluster maps with cubic degree growth, for instance?

5 Poisson and symplectic structures

So far we have alluded to the concept of integrability, but have skirted around the issue of giving a precise definition of what it means for a map to be integrable. An expected feature of integrability is the ability to find explicit solutions of the equations being considered; the recurrence (2.7) displays this feature, because all of its iterates satisfy a linear recurrence of the form (2.8), which can be solved exactly. There are many other criteria that can be imposed: existence of sufficiently many conserved quantities or symmetries, or compatibility of an associated linear system (Lax pair), for instance; and not all of these requirements may be appropriate in different circumstances. It is an unfortunate fact that the definition of an integrable system varies depending on the context, i.e., whether it be autonomous or non-autonomous ordinary differential equations, partial differential equations, difference equations, maps or something else that is being considered. Thus we need to address this problem and clarify the context, in order to specify what integrability means for maps associated with cluster algebras.

There is a precise definition of Liouville integrability in the context of finite-dimensional Hamiltonian mechanics, on a real symplectic manifold M of dimension $2m$, with associated Poisson bracket $\{\,,\,\}$: given a particular function H, the

Hamiltonian flow generated by H is completely integrable, in the sense of Liouville, if there exist m independent functions on M (including the Hamiltonian), say $H_1 = H$, $H_2, \dots, H_m$, which are in involution with respect to the Poisson bracket, i.e., $\{H_j, H_k\} = 0$ for all j, k. In the context of classical mechanics, this notion of integrability provides everything one could hope for. To begin with, systems satisfying these requirements have (at least) m independent conserved quantities: all of the first integrals $H_1, \dots, H_m$ are preserved by the time evolution, so each of the trajectories lies on an m-dimensional intersection of level sets for these functions. Furthermore, Liouville proved that the solution of the equations of motion for such systems can be reduced to a finite number of quadratures, so they really are "able to be integrated" as one would expect; and Arnold showed in addition that the flow reduces to quasiperiodic motion on compact m-dimensional level sets, which are diffeomorphic to tori T^m [2], so nowadays the combined result is referred to as the Liouville-Arnold theorem. Another approach to integrability is to require a sufficient number of symmetries, and this is a consequence of the Liouville definition: the Hamilton's equations arising from H have the maximum number of commuting symmetries, namely the flows generated by each of the first integrals H_j.

The notion of Liouville integrability can be extended to symplectic maps in a natural way [6, 55, 67]. However, the requirement of working in even dimensions is too restrictive for our purposes, so instead of a symplectic form we start with a (possibly degenerate) Poisson structure and consider Poisson maps φ, defined in terms of the pullback of functions, written as $\varphi^* F = F \cdot \varphi$.

Definition 3. Given a Poisson bracket $\{\,,\}$ on a manifold M, a map $\varphi : M \to M$ is called a Poisson map if

$$\{\varphi^* F, \varphi^* G\} = \varphi^* \{F, G\}$$

holds for all functions F, G on M.

(We are being deliberately vague about what sort of Poisson manifold $(M, \{,\})$ is being considered, e.g., a real smooth manifold, or a complex algebraic variety, and what sort of functions, e.g., smooth/analytic/rational, because this may vary according to the context.)

In order to have a suitable notion of integrability for cluster maps, we require a Poisson structure of some kind. In general, given a difference equation or map, there is no canonical way to find a compatible Poisson bracket. Fortunately, it turns out that for cluster algebras there is often a natural Poisson bracket, of log-canonical type, that is compatible with cluster mutations, and there is always a log-canonical presymplectic form [18, 30, 31, 48].

Example 12. Somos-5 Poisson bracket: The skew-symmetric exchange matrix

$$\begin{pmatrix} 0 & 1 & -1 & -1 & 1 \\ -1 & 0 & 2 & 0 & -1 \\ 1 & -2 & 0 & 2 & -1 \\ 1 & 0 & -2 & 0 & 1 \\ -1 & 1 & 1 & -1 & 0 \end{pmatrix} \tag{5.1}$$

is cluster mutation-periodic with period 1. Its associated cluster map is a Somos-5 recurrence, which belongs to family (iv) above, given by (4.14) with $N = 5$, $p = 1$, $q = 2$. The skew-symmetric matrix $P = (p_{ij})$ given by

$$P = \begin{pmatrix} 0 & 1 & 2 & 3 & 4 \\ -1 & 0 & 1 & 2 & 3 \\ -2 & -1 & 0 & 1 & 2 \\ -3 & -2 & -1 & 0 & 1 \\ -4 & -3 & -2 & -1 & 0 \end{pmatrix}$$

defines a Poisson bracket, given in terms of the original cluster variables $\mathbf{x} = (x_1, x_2, x_3, x_4, x_5)$ by

$$\{x_i, x_j\} = p_{ij}x_ix_j. \tag{5.2}$$

This bracket is called log-canonical because it is just given by the constant matrix P in terms of the logarithmic coordinates $\log x_i$. It is also compatible with the cluster algebra structure, in the sense that it remains log-canonical under the action of any mutation, i.e. writing $\mathbf{x} \mapsto \mu_k(\mathbf{x}) = \mathbf{x}' = (x_i')$, in the new cluster variables it takes the form

$$\{x_i', x_j'\} = p_{ij}'x_i'x_j'$$

for some constant skew-symmetric matrix $P' = (p_{ij}')$. Moreover, under the cluster map $\varphi = \rho^{-1} \cdot \mu_1$ defined by

$$\varphi : (x_1, x_2, x_3, x_4, x_5) \mapsto \left(x_2, x_3, x_4, x_5, \frac{x_2x_5 + x_3x_4}{x_1} \right), \tag{5.3}$$

the bracket (5.2) is preserved, in the sense that for all $i, j \in [1, 5]$ the pullback of the map on the coordinates satisfies

$$\{\varphi^* x_i, \varphi^* x_j\} = \varphi^*\{x_i, x_j\}.$$

Hence φ is a Poisson map with respect to this bracket.

Given a Poisson map, we can give a definition of discrete integrability, by adapting a definition from [66], that applies in the continuous case of Hamiltonian flows on Poisson manifolds.

Definition 4. Suppose that the Poisson tensor is of constant rank $2m$ on a dense open subset of a Poisson manifold M of dimension N, and that the algebra of Casimir functions is maximal, i.e. it contains $N - 2m$ independent functions. A Poisson map $\varphi : M \to M$ is said to be completely integrable if it preserves $N - m$ independent functions $F_1, \ldots, F_{N-m}$ which are in involution, including the Casimirs.

Example 13. Complete integrability of the $\tilde{A}_{1,3}$ cluster map: Setting $P = B$ with the exchange matrix (2.6) in Example 2, the bracket

$$\{x_i, x_j\} = b_{ij}x_ix_j$$

is compatible with the cluster algebra structure, and is preserved by the cluster map φ corresponding to (2.7). At points where all coordinates x_j are non-zero, the Poisson tensor has full rank 4, since B is invertible; so there are no Casimirs. Note that $B^{-1} = -\frac{1}{2}B$, so B is proportional to its own inverse, and the map φ is symplectic, i.e. $\varphi^*\omega = \omega$, where up to overall rescaling the symplectic form is

$$\omega = \sum_{i<j} \frac{b_{ij}}{x_i x_j} \, \mathrm{d}x_i \wedge \mathrm{d}x_j. \tag{5.4}$$

Now, observe that the recurrence can be rewritten with a 2×2 determinant, as

$$|D_n| = 1, \quad \text{where} \quad D_n = \begin{pmatrix} x_n & x_{n+1} \\ x_{n+3} & x_{n+4} \end{pmatrix},$$

and construct the 3×3 matrix sequence

$$\tilde{D}_n = \begin{pmatrix} x_n & x_{n+1} & x_{n+2} \\ x_{n+3} & x_{n+4} & x_{n+5} \\ x_{n+6} & x_{n+7} & x_{n+8} \end{pmatrix}.$$

Then, by the method of Dodgson condensation [13], the determinant can be expanded as

$$|\tilde{D}_n| = \frac{1}{x_{n+4}} \begin{vmatrix} |D_n| & |D_{n+1}| \\ |D_{n+3}| & |D_{n+4}| \end{vmatrix} = 0.$$

Further calculation shows that the kernel of $\tilde{D}_n$ is spanned by the vector $(1, -J_n, 1)^T$, where J_n is periodic with period 3, so there is a linear relation

$$x_{n+2} - J_n x_{n+1} + x_n = 0, \quad \text{with} \quad J_{n+3} = J_n. \tag{5.5}$$

Similarly, the linear relation (2.8), with invariant K (independent of n), corresponds to the fact that the kernel of $\tilde{D}_n^T$ is spanned by $(1, -K, 1)^T$. The J_i can be considered as functions of the phase space coordinates x_1, x_2, x_3, x_4, by writing

$$J_1 = \frac{x_1 + x_3}{x_2},$$

and similarly for J_2, J_3. Computing the Poisson bracket between these functions yields

$$\{J_i, J_{i+1}\} = 2J_i J_{i+1} - 2,$$

where the indices are read mod 3, so that J_1, J_2, J_3 form a Poisson subalgebra of dimension 3; and

$$K = J_1 J_2 J_3 - J_1 - J_2 - J_3$$

is a Casimir for this subalgebra, in the sense that $\{J_i, K\} = 0$ for $i = 1, 2, 3$. The map φ preserves any symmetric function of the J_i, so picking the three independent functions

$$F_1 = K, \quad F_2 = J_1 J_2 + J_2 J_3 + J_3 J_1, \quad F_3 = J_1 + J_2 + J_3$$

we have $\varphi^* F_j = F_j$ for all j, but at most two of these can be in involution: $\{F_1, F_2\} = 0 = \{F_1, F_3\}$, but $\{F_2, F_3\} \neq 0$. Thus, choosing just F_1 and F_2, say, the conditions of Definition 4 are satisfied, and the map φ given by (3.1) is completely integrable.

Remark 3. The fact that cluster variables obtained from affine quivers satisfy linear relations with constant coefficients, such as (2.8), has been shown in various different ways: for type A in [26, 27], using Dodgson condensation (equivalently, the Desnanot-Jacobi formula); for types A and D in the context of frieze relations [3]; and for all simply-laced types A, D, E in [50], using cluster categories (but see also [11, 63] for another family of quivers made from products of finite and affine Dynkin types A). The fact that there are additional linear relations with periodic coefficients, like (5.5), was shown for all $\tilde{A}_{p,q}$ quivers in [27], where it was shown that the quantities J_i are coordinates in the dressing chain for Schrödinger operators, and this has recently been extended to affine types D and E [62].

Example 14. Non-existence of a log-canonical bracket for $\tilde{A}_{1,2}$: For the cluster algebra of type $\tilde{A}_{1,2}$, defined by the skew-symmetric exchange matrix

$$B = \begin{pmatrix} 0 & 1 & 1 \\ -1 & 0 & 1 \\ -1 & -1 & 0 \end{pmatrix}$$

it is easy to verify that there is no bracket of log-canonical form, like (5.2), that is compatible with cluster mutations. However, iterates of the cluster map, defined by the recurrence

$$x_n x_{n+3} = x_{n+1} x_{n+2} + 1,$$

satisfy the linear relation $x_{n+4} - K x_{n+2} + x_n = 0$, for a first integral K. In fact, setting $u_n = x_n x_{n+1}$ yields a recurrence of second order,

$$u_n u_{n+2} = u_{n+1}(u_{n+1} + 1), \tag{5.6}$$

and, rewriting K in terms of u_1, u_2, this corresponds to a symplectic map $\hat{\varphi}$ in the (u_1, u_2) plane with symplectic form $\hat{\omega} = \mathrm{d}\log u_1 \wedge \mathrm{d}\log u_2$ and one first integral; so the map $\hat{\varphi}$ is completely integrable.

Example 15. Casimirs for Somos-5: The Poisson tensor for the Somos-5 map (5.3), defined by (5.2), has rank 2 on $\mathbb{C}^5 \setminus \{x_i = 0\}$ (away from the coordinate hyperplanes). The kernel of the matrix P is spanned by the vectors

$$\tilde{\mathbf{v}}_1 = (1, -2, 1, 0, 0)^T, \ \tilde{\mathbf{v}}_2 = (0, 1, -2, 1, 0)^T, \ \tilde{\mathbf{v}}_3 = (0, 0, 1, -2, 1)^T, \tag{5.7}$$

which correspond to three independent Casimir functions

$$F_1 = \mathbf{x}^{\tilde{\mathbf{v}}_1} = \frac{x_1 x_3}{x_2^2}, \quad F_2 = \mathbf{x}^{\tilde{\mathbf{v}}_2} = \frac{x_2 x_4}{x_3^2}, \quad F_3 = \mathbf{x}^{\tilde{\mathbf{v}}_3} = \frac{x_3 x_5}{x_4^2},$$

whose Poisson bracket with any other function G vanishes: $\{F_j, G\} = 0$ for $j = 1, 2, 3$. There are two independent first integrals H_1, H_2, i.e. functions that are preserved by the action of φ, so that $\varphi^* H_i = H_i \cdot \varphi = H_i$ for $i = 1, 2$; and these are themselves Casimirs because they can be written in terms of the F_j [41]:

$$H_1 = F_1 F_2 F_3 + \frac{1}{F_1} + \frac{1}{F_2} + \frac{1}{F_3} + \frac{1}{F_1 F_2 F_3}, \tag{5.8}$$

$$H_2 = F_1F_2 + F_2F_3 + \frac{1}{F_1F_2} + \frac{1}{F_2F_3} + \frac{1}{F_1F_2^2F_3}. \tag{5.9}$$

However, the full algebra of Casimirs is not preserved by the map φ, because F_1, F_2, F_3 transform as

$$\varphi^*F_1 = F_2, \quad \varphi^*F_2 = F_3, \quad \varphi^*F_3 = \frac{F_2F_3 + 1}{F_1F_2^2F_3^2}.$$

Hence the Somos-5 map is not completely integrable with respect to this bracket.

The previous two examples show that if the exchange matrix B is degenerate, then the cluster coordinates may not be the correct ones to use, as either there is no invariant log-canonical bracket in these coordinates, as in the case of $\tilde{A}_{1,2}$, or even if there is such a bracket, a full set of Casimirs is not preserved by the cluster map. (A Poisson map sends Casimirs to other Casimirs, but need not preserve each Casimir individually.) The way out of this quandary, which was already hinted at in Example 14, is to work on a reduced space where the map φ reduces to a symplectic map $\hat{\varphi}$. It turns out that there is a canonical way to do this, based on the presymplectic form ω associated with the cluster algebra, which in general, for any skew-symmetric exchange matrix $B = (b_{ij})$, is given by the formula (5.4) above.

In the case that B is nondegenerate (which is possible for even N only, as in Example 2), ω is a closed, nondegenerate 2-form, so the cluster map is symplectic, but otherwise ω has a null distribution, generated by vector fields of the form

$$\sum_{j=1}^{N} w_j x_j \frac{\partial}{\partial x_j}, \qquad \text{for} \quad \mathbf{w} = (w_j) \in \ker B.$$

These vector fields all commute with other, and can be integrated to yield a commuting set of scaling symmetries: each $\mathbf{w} \in \ker B$ generates a one-parameter scaling group

$$\mathbf{x} \mapsto \tilde{\mathbf{x}} = \lambda^{\mathbf{w}} \cdot \mathbf{x}, \qquad \lambda \in \mathbb{C}^*, \tag{5.10}$$

where the notation means that each component is scaled so that $\tilde{x}_j = \lambda^{w_j} x_j$. Regarding B as a linear transformation on $\mathbb{Q}^N$, skew-symmetry means that there is an orthogonal direct sum decomposition $\mathbb{Q}^N = \operatorname{im} B \oplus \ker B$. If B has rank $2m$, then an integer basis $\mathbf{v}_1, \mathbf{v}_2, \ldots, \mathbf{v}_{2m}$ for $\operatorname{im} B$ yields a complete set of rational functions invariant under the symmetries (5.10), given by the monomials

$$u_j = \mathbf{x}^{\mathbf{v}_j}, \quad j = 1, \ldots, 2m. \tag{5.11}$$

In the case that B has period 1, it was shown in [27] that by choosing the basis suitably, the rational map $\pi : \mathbf{x} \mapsto \mathbf{u} = (u_j)$ reduces φ to a birational symplectic map $\hat{\varphi}$, with symplectic form $\hat{\omega}$, in the sense that $\hat{\varphi} \cdot \pi = \pi \cdot \varphi$, and $\pi^*\hat{\omega} = \omega$, where

$$\hat{\omega} = \sum_{i<j} \frac{\hat{b}_{ij}}{u_i u_j} \, \mathrm{d}u_i \wedge \mathrm{d}u_j \tag{5.12}$$

is also log-canonical. In [45] it was further shown that (up to an overall sign) there is a canonical choice of basis for $\mathrm{im}\, B \cap \mathbb{Z}^N$ with the property that

$$\varphi^* \mathbf{x}^{\mathbf{v}_j} = \mathbf{x}^{\mathbf{v}_{j+1}}, \quad \text{for} \quad j \in [1, 2m-1].$$

This is called a *palindromic basis*, because the first $N - 2m + 1$ entries of $\mathbf{v}_1$ form a palindrome, with the remaining $2m - 1$ entries being zero, and this palindrome is just shifted along to get the other basis elements; the basis is fixed uniquely if the first entry of $\mathbf{v}_1$ is chosen to be positive. The advantage of a palindromic basis is that the birational map $\hat{\varphi}$ is equivalent to an iteration of a single recurrence relation.

Definition 5. Given a cluster mutation-periodic skew-symmetric exchange matrix B with period 1, of rank $2m$, and the symplectic coordinates $(u_j) \in \mathbb{C}^{2m}$ defined by (5.11) with a palindromic basis, the *U-system* is the recurrence corresponding to the reduced cluster map $\hat{\varphi}$, which, for some rational function $\mathcal{F}$, has the form

$$u_n u_{n+2m} = \mathcal{F}(u_{n+1}, \ldots, u_{n+2m-1}). \tag{5.13}$$

We have already seen an example of a U-system, namely the reduced recurrence (5.6) for $\tilde{A}_{1,2}$. An integrable U-system corresponds to the canonical version of integrability for maps: the U-system is equivalent to a symplectic map in dimension $2m$, so m independent first integrals in involution are needed for complete integrability.

Example 16. Complete integrability of the Somos-5 U-system: With B given by (5.1), a palindromic basis for $\mathrm{im}\, B$ is written using (5.7) as

$$\mathbf{v}_1 = \tilde{\mathbf{v}}_1 + \tilde{\mathbf{v}}_2 = (1, -1, -1, 1, 0)^T, \quad \mathbf{v}_2 = \tilde{\mathbf{v}}_2 + \tilde{\mathbf{v}}_3 = (0, 1, -1, -1, 1)^T,$$

so the reduced coordinates are

$$u_1 = \frac{x_1 x_4}{x_2 x_3} = F_1 F_2, \quad u_2 = \frac{x_2 x_5}{x_3 x_4} = F_2 F_3,$$

and $\omega = \sum_{i<j} b_{ij} \mathrm{d}\log x_i \wedge \mathrm{d}\log x_j$ reduces to the symplectic form

$$\hat{\omega} = \frac{\mathrm{d}u_1 \wedge \mathrm{d}u_2}{u_1 u_2}$$

in these coordinates. The cluster map (5.3) reduces to an iteration of the U-system

$$u_n u_{n+2} = \frac{u_{n+1} + 1}{u_{n+1}}, \tag{5.14}$$

and although H_1 does not survive this reduction, the first integral H_2 can be rewritten in terms of u_1, u_2, to yield the function

$$H = u_1 + u_2 + \frac{1}{u_1} + \frac{1}{u_2} + \frac{1}{u_1 u_2},$$

so the U-system corresponds to a completely integrable symplectic map in two dimensions. The generic level sets of H are cubic curves of genus 1, and this is an example of a symmetric QRT map (see [39] and references).

The general Somos-6 recurrence, with constant coefficients α, β, γ, has the form

$$x_n x_{n+6} = \alpha x_{n+1} x_{n+5} + \beta x_{n+2} x_{n+4} + \gamma x_{n+3}^2, \tag{5.15}$$

which has the Laurent property [22], but cannot come from a cluster algebra when $\alpha\beta\gamma \neq 0$, due to there being too many terms on the right-hand side. In fact, it appears in the more general setting of mutations in LP algebras, which allow exchange relations with more terms [53]. Being quadratic relations, Somos recurrences are reminiscent of Hirota bilinear equations for tau functions in soliton theory, and indeed, the general Somos-6 recurrence is a reduction of Miwa's equation [10], which is the bilinear discrete BKP equation, also known as the cube recurrence in algebraic combinatorics. Here we conclude our discussion of Example 4, by setting $\beta = 0$, to obtain a bilinear equation with a total of three terms, which can be obtained as a reduction of the discrete Hirota equation (bilinear discrete KP, or octahedron recurrence), that is

$$T_1 T_{-1} = T_2 T_{-2} + T_3 T_{-3}, \tag{5.16}$$

where the tau function $T = T(m_1, m_2, m_3)$ and the subscript $\pm j$ denotes a shift in the jth independent variable, so e.g., $T_{\pm 1} = T(m_1 \pm 1, m_2, m_3)$, and so on. The advantage of making a reduction from this equation with more independent variables is that it has a Lax pair, which reduces to a Lax pair for the Somos recurrence, and there is an associated spectral curve, whose coefficients provide first integrals.

Example 17. A Somos-6 U-system: Setting $\beta = 0$ in (5.15) produces

$$x_n x_{n+6} = \alpha x_{n+1} x_{n+5} + \gamma x_{n+3}^2. \tag{5.17}$$

This differs from (2.15) and (3.3) by the inclusion of coefficients α, γ, which can be achieved by augmenting the cluster algebra with frozen variables that appear in the exchange relations but do not themselves mutate (see [26] and references, for instance), and does not change other features such as Poisson brackets or the (pre)symplectic forms. Upon applying the method in [46], we can obtain (5.15) as a plane wave reduction of (5.16), by setting

$$T(m_1, m_2, m_3) = a_1^{m_1^2} a_2^{m_2^2} a_3^{m_3^2} x_n, \qquad n = m_0 + 3m_1 + 2m_3,$$

with m_0 arbitrary, and taking $\alpha = a_3^2/a_1^2$, $\gamma = a_2^2/a_1^2$. Under this reduction, the linear system whose compatibility gives the discrete KP equation becomes

$$\begin{aligned} Y_n \psi_{n+3} + \alpha\zeta\psi_{n+2} &= \xi\psi_n, \\ \psi_{n+3} - X_n \psi_{n+1} &= \zeta\psi_n, \end{aligned} \tag{5.18}$$

where ψ_n is a wave function, ζ, ξ are spectral parameters, and

$$X_n = \frac{x_{n+2} x_{n+3}}{x_{n+4} x_{n+1}}, \quad Y_n = \frac{x_{n+4} x_n}{x_{n+3} x_{n+1}}.$$

The equation (5.17) is the compatibility condition for these two linear equations for ψ_n (to be precise, the parameter γ arises as an integration constant). This is more conveniently seen by writing the second linear equation in matrix form, with a vector $\Psi_n = (\psi_n, \psi_{n+1}, \psi_{n+2})^T$, as

$$\Psi_{n+1} = \mathbf{M}_n \psi_n, \quad \mathbf{M}_n = \begin{pmatrix} 0 & 1 & 0 \\ 0 & 0 & 1 \\ \zeta & X_n & 0 \end{pmatrix}, \tag{5.19}$$

and then using the second linear equation in (5.18) to reformulate the first one as an eigenvalue problem with Ψ_n as the eigenvector, that is

$$\mathbf{L}_n \Psi_n = \xi \Psi_n, \quad \mathbf{L}_n = \begin{pmatrix} \zeta Y_n & u_n & \zeta\alpha \\ \zeta^2\alpha & \zeta(Y_{n+1} + \alpha X_n) & u_{n+1} \\ \zeta u_{n+2} & \zeta^2\alpha + u_{n+1}^{-1} & \zeta(Y_{n+2} + \alpha X_{n+1}) \end{pmatrix}. \tag{5.20}$$

In the above expression for the Lax matrix $\mathbf{L}_n$, we have introduced the quantities

$$u_n = X_n Y_n = \frac{x_n x_{n+2}}{x_{n+1}^2},$$

which for $n = 1, 2, 3, 4$ give a set of symplectic coordinates obtained from the palindromic basis $\mathbf{v}_1 = (1, -2, 1, 0, 0, 0)^T$, $\mathbf{v}_2 = (0, 1, -2, 1, 0, 0)^T$, $\mathbf{v}_3 = (0, 0, 1, -2, 1, 0)^T$, $\mathbf{v}_4 = (0, 0, 0, 1, -2, 1)^T$ for $\operatorname{im} B$, with B as in (2.14), and satisfy the U-system

$$u_n u_{n+4} = \frac{\alpha u_{n+1} u_{n+2}^2 u_{n+3} + \gamma}{u_{n+1}^2 u_{n+2}^3 u_{n+3}^2} \tag{5.21}$$

(which should be compared with the tropical formulae (4.8) and (4.9) above), corresponding to the reduced cluster map $\hat{\varphi}$. The symplectic form $\hat{\omega}$, such that $\hat{\varphi}^* \hat{\omega} = \hat{\omega}$, is

$$\hat{\omega} = \sum_{i<j} \hat{b}_{ij} \mathrm{d} \log u_i \wedge \mathrm{d} \log u_j, \quad \hat{B} = (\hat{b}_{ij}) = \begin{pmatrix} 0 & 1 & 2 & 1 \\ -1 & 0 & 2 & 2 \\ -2 & -2 & 0 & 1 \\ -1 & -2 & -1 & 0 \end{pmatrix},$$

so the associated nondegenerate Poisson bracket for these coordinates is given by $\{u_i, u_j\} = \hat{p}_{ij} u_i u_j$ with $(\hat{p}_{ij}) = \hat{B}^{-1}$. The compatibility condition of the matrix system given by (5.19) and (5.20) is the discrete Lax equation

$$\mathbf{L}_{n+1} \mathbf{M}_n = \mathbf{M}_n \mathbf{L}_n,$$

which is equivalent to the U-system (5.21). So this is an isospectral evolution, and the spectral curve

$$\det(\mathbf{L}_n - \xi \mathbf{1}) = -\xi^3 + H_1 \xi^2 + (1 - H_2 \zeta^2)\xi + \alpha^3 \zeta^5 + \gamma \zeta^3 = 0 \tag{5.22}$$

is independent of n, with the non-trivial coefficients being H_1, given by

$$(u_n + u_{n+3}) u_{n+1} u_{n+2} + \alpha \left(\frac{1}{u_n u_{n+1}} + \frac{1}{u_{n+1} u_{n+2}} + \frac{1}{u_{n+2} u_{n+3}} \right) + \frac{\gamma}{u_n u_{n+1}^2 u_{n+2}^2 u_{n+3}},$$

and

$$\begin{aligned} H_2 &= \alpha\left(\frac{u_n u_{n+1}}{u_{n+3}} + \frac{u_{n+2}u_{n+3}}{u_n}\right) + \gamma\left(\frac{1}{u_n u_{n+1} u_{n+2}} + \frac{1}{u_{n+1}u_{n+2}u_{n+3}}\right) \\ &\quad +\alpha^2\left(\frac{1}{u_n u_{n+1}^2 u_{n+2}} + \frac{1}{u_{n+1}u_{n+2}^2 u_{n+3}}\right) + \frac{\alpha\gamma}{u_n u_{n+1}^3 u_{n+2}^3 u_{n+3}}, \end{aligned}$$

which provide two independent first integrals. It can be verified directly that $\{H_1, H_2\} = 0$, which shows that each iteration of (5.21) corresponds to a completely integrable symplectic map $\hat{\varphi}$ (in different coordinates, the involutivity of these quantities was also shown in [44]). The trigonal spectral curve (5.22) has genus 4, and admits the involution $(\zeta, \xi) \mapsto (-\zeta, -\xi)$, giving a quotient curve of genus 2, with a Prym variety that is isomorphic to the Jacobian of a second genus 2 curve, analogous to the situation for the general Somos-6 map in [15]. However, in this case there is a more direct way to find the second genus 2 curve, as the hyperelliptic spectral curve of a 2×2 Lax pair obtained by deriving (5.17) as a reduction of a discrete time Toda equation on a 5-point lattice [46, 47]. For explicit analytic formulae for the solutions in terms of genus 2 sigma functions, see [44, 15].

6 Discrete Painlevé equations from coefficient mutation

The continuous Painlevé equations are a special set of non-autonomous ordinary differential equations of second order that are characterized by the absence of movable critical points in their solutions, which is known as the Painlevé property. Discrete Painlevé equations are a particular class of ordinary difference equations which, like their continuous counterparts, are non-autonomous (meaning that the independent variable appears explicitly); in many cases, they appeared from the search for an appropriate discrete analogue of the Painlevé property [35]. The resulting notion of singularity confinement turned out to be much weaker than the Painlevé property for differential equations, and is not sufficient for integrability, although it is a very useful tool when used judiciously in tandem with other techniques for identifying integrable maps or discrete Painlevé equations [57]. In fact, singularity confinement seems to be very closely related to the Laurent property [42], and it is interesting to speculate whether all discrete integrable systems are related to a system with the Laurent property by introducing a tau function or some other lift of the coordinates [8, 38, 56].

Recently there have been various studies that show how certain discrete Painlevé equations and their higher order analogues can arise from mutation of coefficients in cluster algebras. Here we concentrate on the methods used in [45], but for other related approaches see the work of Okubo [60, 61] and that of Bershtein et al. [5].

A *Y-system* is a set of difference equations arising as relations between coefficients appearing from a sequence of mutations in a cluster algebra with periodicity. The original Y-systems were obtained by Zamolodchikov as a set of functional equations in certain quantum field theories associated with simply-laced affine Lie algebras [68], yet they arise from cluster algebras of finite type obtained from the corresponding finite-dimensional root systems, and display purely periodic dynamics.

Generalized Y-systems were defined by Nakanishi [59] starting from a general notion of periodicity in a cluster algebra, and typically display complicated dynamical behaviour.

Here we concentrate on the case of cluster mutation-periodic quivers with period 1, for which the Y-system can be written as a single scalar difference equation, given by

$$y_n y_{n+N} = \frac{\prod_{j=1}^{N-1}(1+y_{n+j})^{[a_j]_+}}{\prod_{j=1}^{N-1}(1+y_{n+j}^{-1})^{[-a_j]_+}}, \tag{6.1}$$

where, as in Theorem 1, $a_j = b_{1,j+1}$ are the components of the palindromic $(N-1)$-tuple that determines the exchange matrix. (Here we assume that the first non-zero component a_j is positive; there is no loss of generality in doing so, due to the freedom to replace $B \to -B$, but some signs are reversed compared with [45] and [59].) In this context, the coefficient-free recurrence (3.4) that defines the cluster map is referred to as the *T-system*. It was first observed in [25] that there is a relation between the evolution of coefficients $\mathbf{y}$ under mutations (2.2) in a cluster algebra, and the evolution of cluster variables $\mathbf{x}$ due to the associated coefficient-free cluster mutations given by (2.4), which can be summarized by the slogan that "the T-system provides a solution of the Y-system." In the case at hand, the precise statement is that making the substitution

$$y_n = \prod_{j=1}^{N-1} x_{n+j}^{a_j} \tag{6.2}$$

in (6.1) provides a solution of the Y-system whenever x_n satisfies the coefficient-free T-system (3.4).

Although the equations (3.4) and (6.1) are both of order N, there can be a discrepancy between the solutions of the T-system and the Y-system, in the sense that the general solution of the former does not yield the general solution of the latter. This discrepancy is determined by the following result.

Proposition 3. *Let x_n satisfy the modified T-system*

$$x_n x_{n+N} = \mathcal{Z}_n \left(\prod_{j=1}^{N-1} x_{n+j}^{[a_j]_+} + \prod_{j=1}^{N-1} x_{n+j}^{[-a_j]_+} \right). \tag{6.3}$$

Then the substitution (6.2) yields a solution of the Y-system (6.1) if and only if $\mathcal{Z}_n$ satisfies the Z-system

$$\prod_{j=1}^{N-1} \mathcal{Z}_{n+j}^{a_j} = 1. \tag{6.4}$$

Each iteration of the modified T-system (6.3) with non-autonomous coefficients evolving according to (6.4) preserves the presymplectic form given by (5.4) in terms of the entries of the exchange matrix B, and if B is degenerate we can use a palindromic basis for $\operatorname{im} B$ to reduce this to a non-autonomous recurrence in lower dimension that preserves the symplectic form (5.12).

Definition 6. The pair of equations (6.3) and (6.4) is called the *T_z-system.* The *U_z-system* associated with (6.3) is given by (6.4) together with

$$u_n u_{n+2m} = \mathcal{Z}_n \mathcal{F}(u_{n+1}, \ldots, u_{n+2m-1}),$$

where the rational function $\mathcal{F}$ is the same as in (5.13).

We conclude this section with a couple of examples.

Example 18. Somos-5 Y-system and q-Painlevé II: The Y-system associated with the exchange matrix (5.1) is

$$y_n y_{n+5} = \frac{(1+y_{n+1})(1+y_{n+4})}{(1+y_{n+2}^{-1})(1+y_{n+3}^{-1})},$$

and (noting that on the right-hand side of the substitution (6.2) there is the freedom to shift $n \to n+1$) the general solution of this can be written as

$$y_n = \frac{x_n x_{n+3}}{x_{n+1} x_{n+2}},$$

where x_n satisfies the non-autonomous Somos-5 relation

$$x_n x_{n+5} = \mathcal{Z}_n (x_{n+1} x_{n+4} + x_{n+2} x_{n+3}), \quad \text{with} \quad \frac{\mathcal{Z}_n \mathcal{Z}_{n+3}}{\mathcal{Z}_{n+1} \mathcal{Z}_{n+2}} = 1.$$

Equivalently, we can identify $y_n = u_n$ and solve the third order Z-system for $\mathcal{Z}_n$ to write the U_z-system as a non-autonomous version of the QRT map (5.14), that is

$$u_n u_{n+2} = \mathcal{Z}_n (1 + u_n^{-1}), \quad \text{with} \quad \mathcal{Z}_n = \beta_n \mathfrak{q}^n, \ \beta_{n+2} = \beta_n.$$

The latter is equivalent to a q-Painlevé II equation identified in [52], having a continuum limit to the Painlevé II differential equation

$$\frac{d^2 u}{dz^2} = 2u^3 + zu + \alpha.$$

Example 19. A q-Somos-6 relation: The Y-system corresponding to (2.14) is

$$y_n y_{n+6} = \frac{(1+y_{n+1})(1+y_{n+5})}{(1+y_{n+3}^{-1})^2}.$$

Its general solution can be written as

$$y_n = \frac{x_n x_{n+4}}{y_{n+2}^2},$$

where x_n satisfies a q-Somos-6 relation given by

$$x_n x_{n+6} = \mathcal{Z}_n (x_{n+1} x_{n+5} + x_{n+3}^2), \quad \text{with} \quad \mathcal{Z}_n = \alpha_\pm \mathfrak{q}_\pm^n, \tag{6.5}$$

with the solution of the fourth order Z-system

$$\frac{\mathcal{Z}_n \mathcal{Z}_{n+4}}{\mathcal{Z}_{n+2}^2} = 1$$

being given in terms of quantities $\alpha_\pm$ and $\mathfrak{q}_\pm$ that alternate with the parity of n. Alternatively, one can write

$$y_n = u_n u_{n+1}^2 u_{n+2}$$

with u_n satisfying a non-autonomous version of (5.21), that is

$$u_n u_{n+4} = \frac{\mathcal{Z}_n (u_{n+1} u_{n+2}^2 u_{n+3} + 1)}{u_{n+1}^2 u_{n+2}^3 u_{n+3}^2},$$

with $\mathcal{Z}_n$ as in (6.5). The latter should be regarded as a fourth order analogue of a discrete Painlevé equation.

7 Conclusions

We have just scratched the surface in this brief introduction to cluster algebras and discrete integrability. Among other important examples that we have not described here, we would like to mention pentagram maps [32] and cluster integrable systems related to dimer models [14, 34]. A slightly different viewpoint, with some different choices of topics, can be found in the review [33].

Acknowledgments

The work of all three authors is supported by EPSRC Fellowship EP/M004333/1. ANWH is grateful to the School of Mathematics & Statistics, UNSW for hospitality and additional support under the Distinguished Researcher Visitor Scheme.

References

[1] Abarenkova N, Anglès d'Auriac J-C, Boukraa S, Hassani S and J.-M. Maillard J M, Real Arnold complexity versus real topological entropy for birational transformations, *J. Phys. A: Math. Gen.* **33**, 1465–1501, 2000.

[2] Arnold V I, *Mathematical Methods of Classical Mechanics*, Graduate Texts in Mathematics **60**, Springer-Verlag, 1978.

[3] Assem I, Reutenauer C and Smith D, Friezes, *Adv. Math.* **225**, 3134–3165, 2010.

[4] Bellon M P and Viallet C M, Algebraic entropy, *Commun. Math. Phys.* **204**, 425–437, 1999.

[5] Bershtein M, Gavrylenko P and Marshakov A, Cluster integrable systems, q-Painlevé equations and their quantization, *J. High Energ. Phys.* **2018**, 77, 2018.

[6] Bruschi M, Ragnisco O, Santini P M, Gui-Zhang T, Integrable symplectic maps, *Physica D* **49**, 273–294, 1991.

[7] Cassels J W S, *An Introduction to Diophantine Approximation*, Cambridge University Press, New York, 1957.

[8] Chang X K, Hu X B and Xin G, Hankel determinant solutions to several discrete integrable systems and the Laurent property, *SIAM J. Discrete Math.* **29**, 667–682, 2015.

[9] Cohn H, An approach to Markoff's minimal forms through modular functions, *Ann. Math.* **61**, 1–12, 1955.

[10] Date E, Jimbo M and Miwa T, Method for generating discrete soliton equations III, *J. Phys. Soc. Japan* **52**, 388–393, 1983.

[11] Di Francesco P and Kedem R, Q-systems, heaps, paths and cluster positivity, *Commun. Math. Phys.* **293**, 727–802, 2010.

[12] Diller J and Favre C, Dynamics of bimeromorphic maps of surfaces, *Amer. J. Math.* **123**, 1135–1169, 2001.

[13] Dodgson C L, Condensation of determinants, *Proc. R. Soc. Lond.* **15**, 150–55, 1866.

[14] Eager R, Franco S and Schaeffer K, Dimer models and integrable systems, *J. High Energy Phys.* **2012**, 106, 2012.

[15] Fedorov Yu N and Hone A N W, Sigma-function solution to the general Somos-6 recurrence via hyperelliptic Prym varieties, *J. Integrable Systems* **1**, xyw012, 1–34, 2016.

[16] Felikson A, Shapiro M and Tumarkin P, Skew-symmetric cluster algebras of finite mutation type, *J. Eur. Math. Soc.* **14**, 1135–1180, 2012.

[17] Felikson A, Shapiro M and Tumarkin P, Cluster algebras of finite mutation type via unfoldings, *Int. Math. Res. Not.* **8**, 1768–1804, 2012.

[18] Fock V V and Goncharov A B, Cluster ensembles, quantization and the dilogarithm, *Ann. Sci. Éc. Norm. Supér.* **42**, 865–930, 2009.

[19] Fomin S, Shapiro M and Thurston D, Cluster algebras and triangulated surfaces. Part I: Cluster complexes, *Acta Math.* **201**, 83–146, 2008.

[20] Fomin S and Thurston D, Cluster algebras and triangulated surfaces. Part II: Lambda lengths, *Mem. Am. Math. Soc.* **255**, no. 1223, 2018.

[21] Fomin S and Zelevinsky A, Cluster algebras I: foundations, *J. Amer. Math. Soc.* **15**. 497–529, 2002.

[22] Fomin S and Zelevinsky A, The Laurent phenomenon, *Adv. Appl. Math.* **28**, 119–144, 2002.

[23] Fomin S and Zelevinsky A, Cluster algebras II: finite type classification, *Invent. Math.* **154**, 63–121, 2003.

[24] Fomin S and Zelevinsky A, Y-systems and generalized associahedra, *Ann. of Math.* **158**, 977–1018, 2003.

[25] Fomin S and Zelevinsky A, Cluster algebras IV: coefficients, *Comp. Math.* **143**, 112–164, 2007.

[26] Fordy A P and Marsh R J, Cluster mutation-periodic quivers and associated Laurent sequences, *Journal of Algebraic Combinatorics* **34**, 19–66, 2011.

[27] Fordy A P and Hone A N W, Discrete integrable systems and Poisson algebras from cluster maps, *Commun. Math. Phys.* **325**, 527–584, 2014.

[28] Galashin P and Pylyavskyy P, Quivers with additive labelings: classification and algebraic entropy, `arXiv:1704.05024`, 2017.

[29] Gale D, The strange and surprising saga of the Somos sequences, *Math. Intelligencer* **13**, issue 1, 40–42, 1991.

[30] Gekhtman M, Shapiro M and Vainshtein A, Cluster algebras and Poisson geometry, *Mosc. Math. J.* **3**, 899–934, 2003.

[31] Gekhtman M, Shapiro M and Vainshtein A, Cluster algebras and Weil-Petersson forms, *Duke Math. J.* **127**, 291–311, 2005.

[32] Gekhtman M, Shapiro M, Tabachnikov S and Vainshtein A, Integrable cluster dynamics of directed networks and pentagram maps, *Adv. Math.* **300**, 390–450, 2016.

[33] Glick M and Rupel D, Introduction to Cluster Algebras, in Levi D, Rebelo R and Winternitz P (Eds.), *Symmetries and Integrability of Difference Equations*, CRM Series in Mathematical Physics, Springer, Cham, 2017.

[34] Goncharov A B and Kenyon R, Dimers and cluster integrable systems, *Ann. Sci. Éc. Norm. Supér.* **46**, 747–813, 2013.

[35] Grammaticos B, Ramani A and Papageorgiou V, Do integrable mappings have the Painlevé property?, *Phys. Rev. Lett.* **67**, 1825–1828, 1991.

[36] Gross M, Hacking P, Keel S and Kontsevich M, Canonical bases for cluster algebras *J. Amer. Math. Soc.* **31**, 497–608, 2018.

[37] Halburd R G, Diophantine integrability, *J. Phys. A: Math. Gen.* **38**, L263–L269, 2005.

[38] Hamad K and van der Kamp P H, From discrete integrable equations to Laurent recurrences, *J. Differ. Equ. Appl.* **22**, 789–816, 2016.

[39] Hamad K, Hone A N W, van der Kamp P H and Quispel G R W, QRT maps and related Laurent systems, *Adv. Appl. Math.* **96**, 216–248, 2018.

[40] Hone A N W, Diophantine non-integrability of a third-order recurrence with the Laurent property, *J. Phys. A: Math. Gen.* **39**, L171–L177, 2006.

[41] Hone A N W, Sigma function solution of the initial value problem for Somos 5 sequences, *Trans. Amer. Math. Soc.* **359**, 5019–5034, 2007.

[42] Hone A N W, Singularity confinement for maps with the Laurent property, *Phys. Lett. A* **361**, 341–345, 2007.

[43] Hone A N W, Laurent polynomials and superintegrable maps, *SIGMA* **3**, 022, 2007.

[44] Hone A N W, Analytic solutions and integrability for bilinear recurrences of order six, *Appl. Anal.* **89**, 473–492, 2010.

[45] Hone A N W and Inoue R, Discrete Painlevé equations from Y-systems, *J. Phys. A: Math. Theor.* **47**, 474007, 2014.

[46] Hone A N W, Kouloukas T E and Ward C, On reductions of the Hirota-Miwa equation, *SIGMA* **13**, 057, 2017.

[47] Hone A N W, Kouloukas T E and Quispel G R W, Some integrable maps and their Hirota bilinear forms, *J. Phys. A: Math. Theor.* **51**, 044004, 2018.

[48] Inoue R and Nakanishi T, Difference equations and cluster algebras I: Poisson bracket for integrable difference equations, *RIMS Kôkyûroku Bessatsu* **B28**, 63–88, 2011.

[49] Kac V G, *Infinite Dimensional Lie Algebras*, Cambridge University Press, 1990.

[50] Keller B and Scherotzke S, Linear recurrence relations for cluster variables of affine quivers, *Adv. Math.* **228**, 1842–1862, 2011.

[51] Keller B, The periodicity conjecture for pairs of Dynkin diagrams, *Ann. of Math.* **177**, 111–170, 2013.

[52] Kruskal M D, Tamizhmani K M, Grammaticos B and Ramani A, Asymmetric discrete Painlevé equations, *Regul. Chaotic Dyn.* **5**, 274–280, 2000.

[53] Lam T and Pylyavskyy P, Laurent phenomenon algebras, *Cambridge J. Math.* **4**, 121–162, 2016.

[54] Lee K and Schiffler R, Positivity for cluster algebras, *Ann. Math.* **182**, 73–125, 2015.

[55] Maeda S, Completely integrable symplectic mapping, *Proc. Japan Acad. Ser. A Math. Sci.* **63**, 198–200, 1987.

[56] Mase T, The Laurent phenomenon and discrete integrable systems, *RIMS Kôkyûroku Bessatsu* **B41**, 043–064, 2013.

[57] Mase T, Willox R, Grammaticos B and Ramani A, Deautonomization by singularity confinement: an algebro-geometric justification, *Proc. R. Soc. A* **471**, 20140956, 2015.

[58] Mase T, Investigation into the role of the Laurent property in integrability, *J. Math. Phys.* **57**, 022703, 2016.

[59] Nakanishi T, Periodicities in cluster algebras and dilogarithm identities, in Skowronski A and Yamagata K (Eds.), *Representations of Algebras and Related Topics*, EMS Series of Congress Reports, Eur. Math. Soc., 407–444, 2011.

[60] Okubo N, Discrete integrable systems and cluster algebras, *RIMS Kôkyûroku Bessatsu* **B41**, 025–041, 2013.

[61] Okubo N, Bilinear equations and q-discrete Painlevé equations satisfied by variables and coefficients in cluster algebras, *J. Phys. A: Math. Theor.* **48**, 355201, 2015.

[62] Pallister J, Linearisability and integrability of discrete dynamical systems from cluster algebras, presentation at *SIDE 13*, Fukuoka, Japan, 11-17 November 2018, `http://side13conference.net/presentations/JoePallister.pdf`.

[63] Pylyavkskyy P, Zamolodchikov integrability via rings of invariants, *J. Integrable Systems* **1**, xyw010, 1–23, 2016.

[64] van der Poorten A J and Swart C S, Recurrence relations for elliptic sequences: Every Somos 4 is a Somos k, *Bull. Lond. Math. Soc.* **38**, 546–554, 2006.

[65] Robinson R, Periodicity of Somos sequences, *Proc. Amer. Math. Soc.* **116**, 613–619, 1992.

[66] Vanhaecke P, *Integrable Systems in the Realm of Algebraic Geometry*, Springer, 1996.

[67] Veselov A P, Integrable maps, *Russ. Math. Surv.* **46**, 1–51, 1991.

[68] Zamolodchikov Al B, On the thermodynamic Bethe ansatz equations for reflectionless ADE scattering theories, *Phys. Lett. B* **253**, 391–394, 1991.

B4. A review of elliptic difference Painlevé equations

Nalini Joshi [a] *and Nobutaka Nakazono* [b]

[a] *School of Mathematics and Statistics, The University of Sydney, New South Wales 2006, Australia.*
nalini.joshi@sydney.edu.au

[b] *Institute of Engineering, Tokyo University of Agriculture and Technology, 2-24-16 Nakacho Koganei, Tokyo 184-8588, Japan.*
nakazono@go.tuat.ac.jp

Abstract

Discrete Painlevé equations are nonlinear, nonautonomous difference equations of second-order. They have coefficients that are explicit functions of the independent variable n and there are three different types of equations according to whether the coefficient functions are linear, exponential or elliptic functions of n. In this chapter, we focus on the elliptic type and give a review of the construction of such equations on the E_8 lattice. The first such construction was given by Sakai [38]. We focus on recent developments giving rise to more examples of elliptic discrete Painlevé equations.

1 Introduction

Discrete Painlevé equations are nonlinear integrable ordinary difference equations of second order. They have a long history (see §1.1), but it is only in the past two decades that striking developments have led to their recognition as one of the most important classes of equations in the theory of integrable systems [15, 18, 25].

Almost all of the currently known collection of discrete Painlevé equations were derived by Grammaticos, Ramani and collaborators [15], and recognized as having fundamental properties that parallel those of the Painlevé equations, such as Lax pairs, Bäcklund transformations and special solutions. Sakai [38] unified these discrete Painlevé equations and also discovered a new equation whose coefficients are iterated on elliptic curves. Sakai's equation is an elliptic difference Painlevé equation.

Sakai's unification is based on a deep geometric theory shared by all the discrete Painlevé equations, first described by Okamoto [33] for the classical Painlevé equations (see §1.1). The fundamental unifying property is based on the fact that the initial-value (or phase) space of the Painlevé equations can be compactified and regularized by a minimum of eight blow-ups on a Hirzebruch surface. This beautiful observation also leads to symmetries of the equation, arising from an isomorphism between the intersection diagram of the resulting space and affine root systems.

This led Sakai to describe discrete Painlevé equations as the result of translations on lattices defined by affine Weyl groups [19]. In particular, Sakai's elliptic difference equation [38] is iterated on the lattice generated by the affine exceptional Lie group $E_8^{(1)}$ (see also [30,32]). More recently, other elliptic difference equations of Painlevé type have been discovered [4,6,21,36] through other approaches. This review concentrates on describing the construction of such elliptic difference Painlevé equations by using Sakai's geometric way.

To describe the construction that underlies and explains all of these examples, we rely on the following mathematical description. Fix a point in the $E_8^{(1)}$ lattice [7]. Then there are 240 nearest neighbors of this point in the lattice, lying at a distance whose squared length is equal to 2. We refer to the 120 vectors between the initial fixed point and its possible nearest neighbors as nearest-neighbor-connecting vectors (NVs). Similarly, there are 2160 next-nearest neighbors, lying at a distance whose squared length is 4. The 1080 vectors between the fixed point and such next-nearest neighbors will be referred to as next-nearest-neighbor-connecting vectors (NNVs). Sakai's elliptic difference equation is constructed in terms of translations expressed in terms of NVs. However, more recently deduced examples are obtained from NNVs.

The example that led us to this key observation is the following elliptic difference Painlevé equation, originally found by Ramani, Carstea and Grammaticos [36]:

$$\begin{cases} y_{n+1}\Big(k^2(\mathrm{cg}_\mathrm{e}^2-\mathrm{cz}_n^2)\mathrm{cz}_n\mathrm{dz}_n x_n^2 y_n-(1-k^2\mathrm{sz}_n^4)\mathrm{cg}_\mathrm{e}\mathrm{dg}_\mathrm{e}x_n \\ \qquad +(1-k^2\mathrm{sg}_\mathrm{e}^2\mathrm{sz}_n^2)\mathrm{cz}_n\mathrm{dz}_n y_n\Big) \\ \quad =(1-k^2\mathrm{sz}_n^4)\mathrm{cg}_\mathrm{e}\mathrm{dg}_\mathrm{e}x_n y_n-(\mathrm{cg}_\mathrm{e}^2-\mathrm{cz}_n^2)\mathrm{cz}_n\mathrm{dz}_n \\ \qquad -(1-k^2\mathrm{sg}_\mathrm{e}^2\mathrm{sz}_n^2)\mathrm{cz}_n\mathrm{dz}_n x_n^2, \\ x_{n+1}\Big(k^2(\mathrm{cg}_\mathrm{o}^2-\widehat{\mathrm{cz}}_n^2)\widehat{\mathrm{cz}}_n\widehat{\mathrm{dz}}_n y_{n+1}^2 x_n-(1-k^2\widehat{\mathrm{sz}}_n^4)\mathrm{cg}_\mathrm{o}\mathrm{dg}_\mathrm{o}y_{n+1} \\ \qquad +(1-k^2\mathrm{sg}_\mathrm{o}^2\widehat{\mathrm{sz}}_n^2)\widehat{\mathrm{cz}}_n\widehat{\mathrm{dz}}_n x_n\Big) \\ \quad =(1-k^2\widehat{\mathrm{sz}}_n^4)\mathrm{cg}_\mathrm{o}\mathrm{dg}_\mathrm{o}y_{n+1}x_n-(\mathrm{cg}_\mathrm{o}^2-\widehat{\mathrm{cz}}_n^2)\widehat{\mathrm{cz}}_n\widehat{\mathrm{dz}}_n \\ \qquad -(1-k^2\mathrm{sg}_\mathrm{o}^2\widehat{\mathrm{sz}}_n^2)\widehat{\mathrm{cz}}_n\widehat{\mathrm{dz}}_n y_{n+1}^2, \end{cases} \tag{1.1}$$

where

$$\mathrm{sz}_n=\mathrm{sn}\,(z_n),\quad \widehat{\mathrm{sz}}_n=\mathrm{sn}\,(z_n+\gamma_\mathrm{e}+\gamma_\mathrm{o}),\quad \mathrm{sg}_\mathrm{e}=\mathrm{sn}\,(\gamma_\mathrm{e}), \tag{1.2a}$$

$$\mathrm{sg}_\mathrm{o}=\mathrm{sn}\,(\gamma_\mathrm{o}),\quad \mathrm{cz}_n=\mathrm{cn}\,(z_n),\quad \widehat{\mathrm{cz}}_n=\mathrm{cn}\,(z_n+\gamma_\mathrm{e}+\gamma_\mathrm{o}), \tag{1.2b}$$

$$\mathrm{cg}_\mathrm{e}=\mathrm{cn}\,(\gamma_\mathrm{e}),\quad \mathrm{cg}_\mathrm{o}=\mathrm{cn}\,(\gamma_\mathrm{o}),\quad \mathrm{dz}_n=\mathrm{dn}\,(z_n), \tag{1.2c}$$

$$\widehat{\mathrm{dz}}_n=\mathrm{dn}\,(z_n+\gamma_\mathrm{e}+\gamma_\mathrm{o}),\quad \mathrm{dg}_\mathrm{e}=\mathrm{dn}\,(\gamma_\mathrm{e}),\quad \mathrm{dg}_\mathrm{o}=\mathrm{dn}\,(\gamma_\mathrm{o}), \tag{1.2d}$$

and $z_n=z_0+2(\gamma_\mathrm{e}+\gamma_\mathrm{o})n$. We will refer to this equation as the RCG equation. Here, sn, cn and dn are Jacobi elliptic functions and k is the modulus of the elliptic sine. For more information about Jacobian elliptic functions and notations, see [9, Chapter 22] and [1,40].

It turns out that the RCG equation (1.1) is a *projective reduction* of an NNV, i.e., the iterative step is not a translation on the $E_8^{(1)}$ lattice, but its square is a translation corresponding to a NNV. In general, we can derive various discrete Painlevé equations from elements of infinite order in the affine Weyl group that are not necessarily translations by taking a projection on a certain subspace of the parameters. Kajiwara *et al.* studied such procedures [23, 24] to obtain non-translation type discrete Painlevé equations and named these "projective reductions". The RCG equation (1.1) is the first known case of an elliptic difference Painlevé equation obtained from such a reduction.

We constructed a discrete Painlevé equation, which has the RCG equation as a projective reduction, by using NNVs on the $E_8^{(1)}$ lattice in [21]. Most discrete Painlevé equations admit the special solutions expressible in terms of solutions of linear equations when some of the parameters take special values (see, for example, [25] and references therein). It is known that projectively-reduced equations have different types of such solutions from translation-type discrete Painlevé equations on the same lattice [23, 24]. In this chapter, we will also show the special solutions of Equation (1.1), which is quite different from those of a translation-type elliptic difference Painlevé equation reported in [22, 25].

1.1 Background

Shohat studied polynomials $\Phi_n(x)$ indexed by degree $n \in \mathbb{N}$, defined in an interval $(-\infty, \infty)$, with a weight function $p(x) = \exp(-x^4/4)$ such that

$$\int_{-\infty}^{\infty} p(x)\Phi_m(x)\,\Phi_n(x)dx = 0, \quad (m \neq n,\ m, n \in \mathbb{N}). \tag{1.3}$$

Shohat obtained the 3-term recurrence relation (see Equation (39) of [39], obtained through methods attributed to Laguerre [26])

$$\Phi_n(x) - (x - c_n)\,\Phi_{n-1}(x) + \lambda_n\,\Phi_{n-2}(x) = 0, \quad n \geq 2, \tag{1.4}$$

where $\Phi_0(x) \equiv 1$, $\Phi_1(x) = x - c_1$, where c_1 is independent of x, and deduced the following difference equation for λ_n:

$$\lambda_{n+2}\big(\lambda_{n+1} + \lambda_{n+2} + \lambda_{n+3}\big) = n + 1. \tag{1.5}$$

We now know that this equation is intimately related to one of the six classical Painlevé equations, universal classes of second-order ordinary differential equations (ODEs) studied by Painlevé [34], Fuchs [13] and Gambier [14]. Fokas *et al.* [12] showed that the solutions of Equation (1.5) are solutions of the fourth Painlevé equation:

$$\mathrm{P_{IV}}: \quad w'' = \frac{w'^2}{2w} + \frac{3w^3}{2} + 4tw^2 + 2(t^2 - \alpha)w + \frac{\beta}{w}.$$

Actually, solutions of $\mathrm{P_{IV}}$: $w = w_n$, $n \in \mathbb{Z}$, satisfy a more general version of Shohat's equation given by

$$w_n\left(w_{n+1} + w_n + w_{n-1}\right) = a\,n + b + c\,(-1)^n + d\,w_n, \tag{1.6}$$

where a, b, c, d are constants (see [11] and [27]). This equation is an integrable equation in its own right, with fundamental properties such as a Lax pair [8].

Equation (1.6) is one of many equations now known as discrete Painlevé equations. In general, there exist three types of discrete Painlevé equations. They are distinguished by the types of function t_n appearing in the coefficient of each equation.

(i) If there exists $k \in \mathbb{Z}_{>0}$ such that $t_{n+k} - t_n$ is a constant, then the equation is said to be of *additive-type*.

(ii) If there exists $k \in \mathbb{Z}_{>0}$ such that t_{n+k}/t_n is a constant, denoted q ($\neq 0, 1$), then the equation is said to be of *multiplicative-type* or *q-difference-type*.

(iii) If t_n can be expressed by the elliptic function of n, then the equation is said to be of *elliptic-type*.

We list a few discrete Painlevé equations here.

$$\text{d-P}_{\text{II}}\ [35]: \ X_{n+1} + X_{n-1} = \frac{t_n X_n + a}{1 - {X_n}^2}, \tag{1.7}$$

$$q\text{-P}_{\text{III}}\ [37]: \ X_{n+1} X_{n-1} = \frac{(X - a t_n)(X - a^{-1} t_n)}{(X - b)(X - b^{-1})}, \tag{1.8}$$

$$\text{d-P}_{\text{IV}}\ [37]: \ (X_{n+1} + X_n)(X_n + X_{n-1}) = \frac{({X_n}^2 - a^2)({X_n}^2 - b^2)}{(X_n - t_n)^2 - c^2}, \tag{1.9}$$

$$\begin{aligned} q\text{-P}_{\text{V}}\ [37]: \ &(X_{n+1} X_n - 1)(X_n X_{n-1} - 1) \\ &= {t_n}^2 \frac{(X_n - a)(X_n - a^{-1})(X_n - b)(X_n - b^{-1})}{(X_n - c t_n)(X_n - c^{-1} t_n)}, \end{aligned} \tag{1.10}$$

where a, b and c are constants. Here, $t_{n+1} - t_n$ is a constant for Equations (1.7) and (1.9) and t_{n+1}/t_n is a constant for Equations (1.8) and (1.10), that is, Equations (1.7) and (1.9) are additive-type, while Equations (1.8) and (1.10) are multiplicative-type. Note that the notation for each equation originates from their discrete types and continuum limits. Moreover, an example of elliptic-type is given by Equation (1.1).

Okamoto [33] described a geometric framework for studying the Painlevé equations. He showed that the initial-value (or phase) space of the Painlevé equations, which is a foliated vector bundle [28], can be compactified and regularised by a minimum of eight blow-ups on a Hirzebruch surface (or nine in $\mathbb{P}^2$).

This geometric theory also leads to a description of their symmetry groups, described in terms of affine Weyl groups [31]. Such symmetries lead to transformations of the Painlevé equations called *Bäcklund transformations.*

Sakai's geometric description of discrete Painlevé equations, based on types of space of initial values, is well known [38]. This picture relies on compactifying and regularizing the space of initial values. The spaces of initial values are constructed by a blow up of $\mathbb{P}^1 \times \mathbb{P}^1$ at base points (see §3) and are classified into 22 types according to the configuration of the base points as follows:

Discrete type	Type of surface
Elliptic	$A_0^{(1)}$
Multiplicative	$A_0^{(1)*}$, $A_1^{(1)}$, $A_2^{(1)}$, $A_3^{(1)}$, ..., $A_8^{(1)}$, $A_7^{(1)'}$
Additive	$A_0^{(1)**}$, $A_1^{(1)*}$, $A_2^{(1)*}$, $D_4^{(1)}$, ..., $D_8^{(1)}$, $E_6^{(1)}$, $E_7^{(1)}$, $E_8^{(1)}$

In each case, the root system characterizing the surface forms a subgroup of the 10-dimensional Picard lattice. The symmetry group of each equation, formed by Cremona isometries, arises from the orthogonal complement of this root system inside the Picard lattice.

1.2 Periodic reduction of the Q4-equation

In this section, we recall how to obtain Equation (1.1) from the lattice Krichever-Novikov system [2, 17] (or, Q4 in the terminology of Adler-Bobenko-Suris [3]):

$$\begin{aligned}&\operatorname{sn}(\alpha_l)\left(u_{l,m}u_{l+1,m}+u_{l,m+1}u_{l+1,m+1}\right)-\operatorname{sn}(\beta_m)\left(u_{l,m}u_{l,m+1}+u_{l+1,m}u_{l+1,m+1}\right)\\&\quad-\operatorname{sn}(\alpha_l-\beta_m)\left(u_{l+1,m}u_{l,m+1}+u_{l,m}u_{l+1,m+1}\right)\\&\quad+\operatorname{sn}(\alpha_l)\operatorname{sn}(\beta_m)\operatorname{sn}(\alpha_l-\beta_m)\left(1+k^2u_{l,m}u_{l+1,m}u_{l,m+1}u_{l+1,m+1}\right)=0,\qquad(1.11)\end{aligned}$$

where α_l and β_m are parameters, $u_{l,m}$ is the dependent variable, l and m are independent variables (often taken to be integer) and k is the modulus of the elliptic function sn. Taking a periodic reduction $u_{l+1,m-1}=u_{l,m}$ of Equation (1.11), and identifying $n=l+m$, leads to an autonomous second order ordinary difference equation [20]:

$$\begin{aligned}&\left(\operatorname{sn}(\alpha_0)-\operatorname{sn}(\beta_0)\right)u_n(u_{n+1}+u_{n-1})-\operatorname{sn}(\alpha_0-\beta_0)\left(u_{n+1}u_{n-1}+{u_n}^2\right)\\&\quad+\operatorname{sn}(\alpha_0)\operatorname{sn}(\beta_0)\operatorname{sn}(\alpha_0-\beta_0)\left(1+k^2{u_n}^2u_{n+1}u_{n-1}\right)=0.\qquad(1.12)\end{aligned}$$

Ramani *et al.* [36] deautonomised Equation (1.12) by the method of singularity confinement after a change of variables $\alpha_0\to\gamma+z$, $\beta_0\to\gamma-z$. The resulting equation then becomes Equation (1.1) with $x_n=u_{2n}$ and $y_n=u_{2n-1}$.

1.3 Outline of the chapter

In §2, we recall the basic definitions of reflection group theory and define translations, for the interested reader. The initial-value space of the RCG equation (1.1) is constructed in §3, where we also introduce related algebro-geometric concepts. In §4, we construct Cremona isometries on this initial value space, which roughly speaking, are mappings that preserve its geometric structure. In §5, we give the resulting birational actions on the coordinates and parameters of The initial-value space. We show that by using these birational actions, we arrive at the RCG equation. Some explicit special solutions of the RCG equation are described in §6. In Appendix A we illustrate the geometric ideas for the case of $A_4^{(1)}$, and in Appendix B we describe the generic known examples of elliptic difference Painlevé equations.

2 $E_8^{(1)}$-lattice

To understand how to construct discrete Painlevé equations from symmetry groups, we need the theory of finite and affine reflection groups. In this section, we recall the basic definitions of reflection group theory before defining the transformation group $W(E_8^{(1)})$ and describing its translations operators.

Consider two n-dimensional real vector spaces V and V^*, spanned by the basis sets $\Delta = \{\alpha_1, \ldots, \alpha_n\}$ and $\Delta^\vee = \{\alpha_1^\vee, \ldots, \alpha_n^\vee\}$ respectively. The elements of Δ are called *simple roots*, while those of $\Delta^\vee$ are *simple coroots.* To define reflections, we use a bilinear pairing given by the entries of an $n \times n$ Cartan matrix $A = (A_{ij})$ (see the definition of Cartan matrices in [5]):

$$\langle \alpha_i^\vee, \alpha_j \rangle = A_{ij}, \tag{2.1}$$

for all $i, j \in \{1, \ldots, n\}$. If V and V^* are Euclidean spaces, this bilinear pairing is the usual inner product.

It is also important to define the fundamental *weights* h_i, $i = 1, \ldots, n$, which are given by

$$\langle \alpha_i^\vee, h_j \rangle = \delta_{ij}, \quad (1 \leq i, j \leq n). \tag{2.2}$$

Correspondingly, the integer linear combinations (or $\mathbb{Z}$-modules)

$$Q = \sum_{k=1}^{n} \mathbb{Z}\alpha_k, \quad Q^\vee = \sum_{k=1}^{n} \mathbb{Z}\alpha_k^\vee, \quad P = \sum_{k=1}^{n} \mathbb{Z}h_k, \tag{2.3}$$

are called the *root lattice*, *coroot lattice* and *weight lattice* respectively. We are now in a position to define reflections. For each $i \in \{1, \ldots, n\}$, the linear operator defined by

$$s_{\alpha_i}(\alpha_j) = \alpha_j - A_{ji}\alpha_i, \quad s_{\alpha_i}(\alpha_j^\vee) = \alpha_j^\vee - A_{ji}\alpha_i^\vee \tag{2.4}$$

for every $j \in \{1, \ldots, n\}$ is a *reflection operator.* That is, it has the following properties:

(i) $s_{\alpha_i}(\alpha_i) = -\alpha_i, \quad s_{\alpha_i}(\alpha_i^\vee) = -\alpha_i^\vee.$

(ii) $s_{\alpha_i}^2 = 1.$

(iii) $s_{\alpha_i}.\langle \alpha_j^\vee, \alpha_k \rangle = \langle s_{\alpha_i}(\alpha_j^\vee), s_{\alpha_i}(\alpha_k) \rangle = \langle \alpha_j^\vee, \alpha_k \rangle.$

The group W generated by s_{α_1}, …, s_{α_n} is called a Weyl group. The root system of W is defined to be the subset Φ of Q given by $\Phi = W(\Delta)$. A root system is said to be irreducible if it is not a combination of mutually orthogonal root systems. Each irreducible root system Φ contains a unique root given by

$$\widetilde{\alpha} = \sum_i C_i \alpha_i, \tag{2.5}$$

whose height (i.e., the sum of coefficients in the expansion) is maximal, which is called the *highest root*.

For our case, $n = 8$, and the starting point is the Cartan matrix of type E_8

$$A = \begin{pmatrix} 2 & -1 & 0 & 0 & 0 & 0 & 0 & 0 \\ -1 & 2 & -1 & 0 & 0 & 0 & 0 & 0 \\ 0 & -1 & 2 & -1 & 0 & 0 & 0 & -1 \\ 0 & 0 & -1 & 2 & -1 & 0 & 0 & 0 \\ 0 & 0 & 0 & -1 & 2 & -1 & 0 & 0 \\ 0 & 0 & 0 & 0 & -1 & 2 & -1 & 0 \\ 0 & 0 & 0 & 0 & 0 & -1 & 2 & 0 \\ 0 & 0 & -1 & 0 & 0 & 0 & 0 & 2 \end{pmatrix}. \tag{2.6}$$

The astute reader might notice that this is not the standard one given in Bourbaki [5], but it is equivalent to it (under conjugate transforms of the bases). We make this choice because it corresponds to the identification of simple roots made in Section 4. Moreover, the highest root and coroot are given respectively by

$$\widetilde{\alpha} = 2\alpha_1 + 4\alpha_2 + 6\alpha_3 + 5\alpha_4 + 4\alpha_5 + 3\alpha_6 + 2\alpha_7 + 3\alpha_8, \tag{2.7a}$$

$$\widetilde{\alpha}^\vee = 2\alpha_1^\vee + 4\alpha_2^\vee + 6\alpha_3^\vee + 5\alpha_4^\vee + 4\alpha_5^\vee + 3\alpha_6^\vee + 2\alpha_7^\vee + 3\alpha_8^\vee. \tag{2.7b}$$

We now expand the root system by defining α_0 by $\alpha_0 + \widetilde{\alpha} = 0$. The corresponding extension of the coroot system is defined by $\alpha_0^\vee$ and $\delta^\vee$ (called the *null coroot*): $\alpha_0^\vee + \widetilde{\alpha}^\vee = \delta^\vee$. To construct the corresponding affine Weyl group, we now extend the Cartan matrix A by adding a row and column given respectively by

$$A_{j0} = \langle -\widetilde{\alpha}^\vee, \alpha_j \rangle, \quad A_{0j} = \langle \alpha_j^\vee, -\widetilde{\alpha} \rangle, \tag{2.8}$$

along with $A_{00} = 2$. The extended Cartan matrix and corresponding root systems and groups are now denoted with a superscript containing (1) to denote this extension.

The extended Cartan matrix of type $E_8^{(1)}$ is given by

$$A^{(1)} = (A_{ij})_{i,j=0}^{8} = \begin{pmatrix} 2 & 0 & 0 & 0 & 0 & 0 & 0 & -1 & 0 \\ 0 & 2 & -1 & 0 & 0 & 0 & 0 & 0 & 0 \\ 0 & -1 & 2 & -1 & 0 & 0 & 0 & 0 & 0 \\ 0 & 0 & -1 & 2 & -1 & 0 & 0 & 0 & -1 \\ 0 & 0 & 0 & -1 & 2 & -1 & 0 & 0 & 0 \\ 0 & 0 & 0 & 0 & -1 & 2 & -1 & 0 & 0 \\ 0 & 0 & 0 & 0 & 0 & -1 & 2 & -1 & 0 \\ -1 & 0 & 0 & 0 & 0 & 0 & -1 & 2 & 0 \\ 0 & 0 & 0 & -1 & 0 & 0 & 0 & 0 & 2 \end{pmatrix}. \tag{2.9}$$

The reflections have the form

$$s_i(\lambda) = \lambda - \langle \alpha_i^\vee, \lambda \rangle \alpha_i, \quad i = 0, \ldots, 8, \ \lambda \in P. \tag{2.10}$$

In particular, this gives $s_i(h_i) = h_i - \alpha_i, i = 0, \ldots, 8$. On the other hand, each root α_i can be expressed in terms of weights h_i by using the relationship

$$\alpha_i = \sum_{j=0}^{8} A_{ij} h_j.$$

Therefore, we obtain the reflections on the fundamental weights as follows:

$$\begin{aligned}
&s_0(h_0) = -h_0 + h_7, && s_1(h_1) = -h_1 + h_2,\\
&s_2(h_2) = h_1 - h_2 + h_3, && s_3(h_3) = h_2 - h_3 + h_4 + h_8,\\
&s_4(h_4) = h_3 - h_4 + h_5, && s_5(h_5) = h_4 - h_5 + h_6,\\
&s_6(h_6) = h_5 - h_6 + h_7, && s_7(h_7) = h_0 + h_6 - h_7,\\
&s_8(h_8) = h_3 - h_8.
\end{aligned} \tag{2.11}$$

It is useful to express the actions of the reflections on the roots, and we obtain

$$\begin{aligned}
&s_0 : (\alpha_0, \alpha_7) \mapsto (-\alpha_0, \alpha_7 + \alpha_0), \qquad s_1 : (\alpha_1, \alpha_2) \mapsto (-\alpha_1, \alpha_2 + \alpha_1),\\
&s_2 : (\alpha_1, \alpha_2, \alpha_3) \mapsto (\alpha_1 + \alpha_2, -\alpha_2, \alpha_3 + \alpha_2),\\
&s_3 : (\alpha_2, \alpha_3, \alpha_4, \alpha_8) \mapsto (\alpha_2 + \alpha_3, -\alpha_3, \alpha_4 + \alpha_3, \alpha_8 + \alpha_3),\\
&s_4 : (\alpha_3, \alpha_4, \alpha_5) \mapsto (\alpha_3 + \alpha_4, -\alpha_4, \alpha_5 + \alpha_4),\\
&s_5 : (\alpha_4, \alpha_5, \alpha_6) \mapsto (\alpha_4 + \alpha_5, -\alpha_5, \alpha_6 + \alpha_5),\\
&s_6 : (\alpha_5, \alpha_6, \alpha_7) \mapsto (\alpha_5 + \alpha_6, -\alpha_6, \alpha_7 + \alpha_6),\\
&s_7 : (\alpha_6, \alpha_7) \mapsto (\alpha_6 + \alpha_7, -\alpha_7), \qquad s_8 : (\alpha_3, \alpha_8) \mapsto (\alpha_3 + \alpha_8, -\alpha_8).
\end{aligned} \tag{2.12}$$

Note that fundamental weights and simple roots which are not explicitly shown in Equation (2.11) and (2.12) remain unchanged under the action of the corresponding reflections.

Under the linear actions on the weight lattice (2.10), $W(E_8^{(1)}) = \langle s_0, \ldots, s_8 \rangle$ forms an affine Weyl group of type $E_8^{(1)}$. Indeed, the following fundamental relations hold:

$$(s_i s_j)^{l_{ij}} = 1, \qquad l_{ij} = \begin{cases} 1, & i = j \\ 3, & (i,j) = (1,2), (2,3), \ldots, (6,7), (3,8), (7,0), \\ 2, & \text{otherwise.} \end{cases} \tag{2.13}$$

Note that representing the simple reflections s_i by nodes and connecting i-th and j-th nodes by $(l_{ij} - 2)$ lines, we obtain the Dynkin diagram of type $E_8^{(1)}$ shown in Figure 2.1.

Remark 2.1. We can also define the action of $W(E_8^{(1)})$ on the coroot lattice $Q^\vee$ by replacing α_i with $\alpha_i^\vee$ in the actions (2.12). Note that $\delta^\vee$ is invariant under the action of $W(E_8^{(1)})$. Then, the transformations in $W(E_8^{(1)})$ preserve the form $\langle \cdot, \cdot \rangle$, that is, the following holds:

$$w.\langle \gamma, \lambda \rangle = \langle w(\gamma), w(\lambda) \rangle = \langle \gamma, \lambda \rangle, \tag{2.14}$$

for arbitrary $w \in W(E_8^{(1)})$, $\gamma \in Q^\vee$ and $\lambda \in P$.

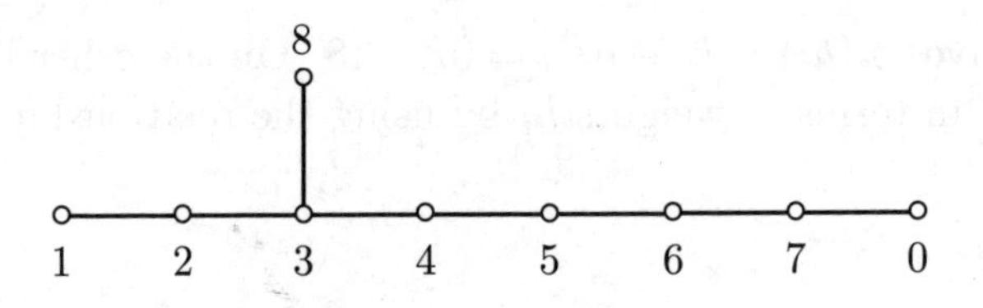

Figure 2.1. Dynkin diagram of type $E_8^{(1)}$.

For each root $\alpha \in Q$, we define the Kac translation T_α on the weight lattice P by

$$T_\alpha(\lambda) = \lambda - \langle \delta^\vee, \lambda \rangle \alpha, \quad \lambda \in P, \tag{2.15}$$

and on the coroot lattice Q by

$$T_\alpha(\lambda^\vee) = \lambda^\vee + \langle \lambda^\vee, \alpha \rangle \delta^\vee, \quad \lambda^\vee \in Q. \tag{2.16}$$

We can easily verify that under the linear actions above, the translation T_α has the following properties.

(i) For any $\alpha, \beta \in Q$, $T_\alpha \circ T_\beta = T_{\alpha+\beta}$.

(ii) For any $w \in W(E_8^{(1)})$ and $\alpha \in Q$,

$$w \circ T_\alpha = T_{w(\alpha)} \circ w. \tag{2.17}$$

(iii) For any $\alpha, \beta \in Q$, $T_\alpha(\beta) = \beta$.

For any $\alpha = \sum_{i=0}^{8} c_i \alpha_i \in Q$, we define the squared length of T_α by

$$|T_\alpha|^2 := \langle \alpha^\vee, \alpha \rangle, \tag{2.18}$$

where $\alpha^\vee = \sum_{i=0}^{8} c_i \alpha_i^\vee$. Then, from (2.14) and (2.17), we obtain the following lemma.

Lemma 2.2. *For any $\alpha, \beta \in Q$, if T_α and T_β are conjugate to each other in $W(E_8^{(1)})$, then the squared lengths of T_α and T_β are equal.*

Proof. Assume that T_α and T_β, where $\alpha, \beta \in Q$, are conjugate to each other in $W(E_8^{(1)})$. Let $w \circ T_\alpha \circ w^{-1} = T_\beta$, where $w \in W(E_8^{(1)})$. Since of (2.17), we obtain $w \circ T_\alpha \circ w^{-1} = T_{w(\alpha)}$, which gives

$$T_{w(\alpha)} = T_\beta. \tag{2.19}$$

Since of (2.14) and (2.18), the statement follows from

$$|T_\alpha|^2 = |T_{w(\alpha)}|^2 = |T_\beta|^2. \tag{2.20}$$

■

In [30] Murata *et al.* consider the following translation as the time evolution of the Sakai's elliptic difference equation:

$$T^{(M)} = s_{12384325438654327654380765432123456708345672345683452348 32}, \tag{2.21}$$

whose action on the coroot lattice $Q^\vee$ is given by

$$T^{(M)}(\alpha_1^\vee) = \alpha_1^\vee + 2\delta^\vee, \quad T^{(M)}(\alpha_2^\vee) = \alpha_2^\vee - \delta^\vee, \tag{2.22}$$

while in [4, 21] Joshi *et al.* showed that the squared time evolution of the RCG equation corresponds to the following translation:

$$T^{(JN)} = s^2_{56453483706756452348321564534837067564523483217067348304 68}, \tag{2.23}$$

whose action on the coroot lattice $Q^\vee$ is given by

$$T^{(JN)}(\alpha_1^\vee) = \alpha_1^\vee - 2\delta^\vee, \quad T^{(JN)}(\alpha_5^\vee) = \alpha_5^\vee + \delta^\vee. \tag{2.24}$$

Note that for convenience we use the following notations for the composition of the reflections s_i:

$$s_{i_1\cdots i_m} = s_{i_1} \circ \cdots \circ s_{i_m}, \quad s^2_{i_1\cdots i_m} = s_{i_1\cdots i_m} \circ s_{i_1\cdots i_m}, \tag{2.25}$$

where $i_1 \cdots i_m \in \{0, \ldots, 8\}$. Comparing (2.16) and (2.22) and comparing (2.16) and (2.24), we can respectively express the translations $T^{(M)}$ and $T^{(JN)}$ by the Kac translations as the following:

$$T^{(M)} = T_{\alpha_1}, \quad T^{(JN)} = T_{\alpha_0+2\alpha_2+4\alpha_3+4\alpha_4+4\alpha_5+3\alpha_6+2\alpha_7+2\alpha_8}, \tag{2.26}$$

where

$$|T_{\alpha_1}|^2 = 2, \quad |T_{\alpha_0+2\alpha_2+4\alpha_3+4\alpha_4+4\alpha_5+3\alpha_6+2\alpha_7+2\alpha_8}|^2 = 4. \tag{2.27}$$

Therefore, the translations $T^{(M)}$ and $T^{(JN)}$ are not conjugate to each other in $W(E_8^{(1)})$ and respectively correspond to an NV and an NNV.

Remark 2.3. As another example, we will show the lattice and transformation group of type $A_4^{(1)}$ in §A.

3 The initial-value space of the RCG equation

Consider the RCG equation (1.1) as a discrete dynamical system. It is reversible and so the system maps (x_n, y_n) to (x_{n+1}, y_{n+1}) at each forward time step or to (x_{n-1}, y_{n-1}) at each backward step. Because the coefficients of y_{n+1} or x_{n+1} may vanish, the iterates may become unbounded and we compactify this system by embedding it in the projective space $\mathbb{P}^1 \times \mathbb{P}^1$.

Compactification is not enough to avoid all problems. Let ϕ and ϕ^{-1} be respectively the forward and backward time evolution of Equation (1.1). We denote the action of these mappings ϕ by

$$\phi : (x, y; \gamma_e, \gamma_o, z_0) \mapsto \big(\tilde{x}, \tilde{y}; \gamma_e, \gamma_o, z_0 + 2(\gamma_e + \gamma_o)\big), \tag{3.1a}$$

$$\phi^{-1} : (x, y; \gamma_e, \gamma_o, z_0) \mapsto \big(\underset{\sim}{x}, \underset{\sim}{y}; \gamma_e, \gamma_o, z_0 - 2(\gamma_e + \gamma_o)\big). \tag{3.1b}$$

(We obtain Equation (1.1) by writing $x_n = \phi^n(x)$, $y_n = \phi^n(y)$, $z_n = \phi^n(z_0)$.) Then there exist points where the image values of ϕ or ϕ^{-1} approach $0/0$. We refer to such points as *base points* because they are equivalent to the usual definition used in the case of algebraic curves [16]. Below, we describe the base points of the RCG equation explicitly.

In general, the process of blowing up a finite sequence of points p_i, possibly infinitely near each other, in $\mathbb{P}^1 \times \mathbb{P}^1$, leads to a variety X, called The initial-value space. Assuming there are 8 blow-ups, then let the sequence of blow-ups be $\pi_i : X_i \mapsto X_{i-1}$ of p_i in X_{i-1}, with $X = X_8 \to \ldots \to X_0 = \mathbb{P}^1 \times \mathbb{P}^1$. Each blow-up replaces p_i by an exceptional line $\mathcal{E}_i$. We refer to the total sequence of blow-ups by $\epsilon : X \to \mathbb{P}^1 \times \mathbb{P}^1$, and moreover, denote the linear equivalence classes of the total transform of vertical and horizontal lines in $\mathbb{P}^1 \times \mathbb{P}^1$ respectively by H_0 and H_1.

To find base points of Equation (1.1), we need to find simultaneous zeroes of pairs of polynomials. For example, base points arising from the component $\tilde{y}$ in the action of ϕ lie at the simultaneous solutions of

$$\begin{cases} (1 - k^2\mathrm{sz}^4)\mathrm{cg_e dg_e}\, xy - (\mathrm{cg_e^2} - \mathrm{cz}^2)\mathrm{cz\, dz} - (1 - k^2\mathrm{sg_e^2 sz^2})\mathrm{cz\, dz}\, x^2 = 0, \\ k^2(\mathrm{cg_e^2} - \mathrm{cz}^2)\mathrm{cz\, dz}\, x^2 y - (1 - k^2\mathrm{sz}^4)\mathrm{cg_e dg_e}\, x + (1 - k^2\mathrm{sg_e^2 sz^2})\mathrm{cz\, dz}\, y = 0, \end{cases}$$

where the dependence on n has been suppressed. These polynomial equations can be solved explicitly. For example, the first equation in the above pair can be solved for y in terms of x, and substituting into the second equation leads to a quartic equation for x. The four solutions of this equation can be expressed explicitly in terms of the coefficient elliptic functions by using elliptic function identities. A similar argument gives us four more base points arising from the remaining equations; for $\tilde{x}$ in the mapping ϕ and for $\underset{\sim}{x}$ and $\underset{\sim}{y}$ in the mapping ϕ^{-1}.

These lead us to the eight base points listed below:

$$p_1 : (x, y) = \big(\mathrm{cd}\,(\gamma_o + \kappa)\,, \mathrm{cd}\,(z_0 - \gamma_e - \gamma_o + \kappa)\,\big), \tag{3.2a}$$

$$p_2 : (x, y) = \big(\mathrm{cd}\,(\gamma_o + \mathrm{i}K')\,, \mathrm{cd}\,(z_0 - \gamma_e - \gamma_o + \mathrm{i}K')\,\big), \tag{3.2b}$$

$$p_3 : (x, y) = \big(\mathrm{cd}\,(\gamma_o + 2K)\,, \mathrm{cd}\,(z_0 - \gamma_e - \gamma_o + 2K)\,\big), \tag{3.2c}$$

$$p_4 : (x, y) = \big(\mathrm{cd}\,(\gamma_o)\,, \mathrm{cd}\,(z_0 - \gamma_e - \gamma_o)\,\big), \tag{3.2d}$$

$$p_5 : (x, y) = \big(\mathrm{cd}\,(z_0 + \kappa)\,, \mathrm{cd}\,(\gamma_e + \kappa)\,\big), \tag{3.2e}$$

$$p_6 : (x, y) = \big(\mathrm{cd}\,(z_0 + \mathrm{i}K')\,, \mathrm{cd}\,(\gamma_e + \mathrm{i}K')\,\big), \tag{3.2f}$$

$$p_7 : (x, y) = \big(\mathrm{cd}\,(z_0 + 2K)\,, \mathrm{cd}\,(\gamma_e + 2K)\,\big), \tag{3.2g}$$

$$p_8 : (x, y) = \big(\mathrm{cd}\,(z_0)\,, \mathrm{cd}\,(\gamma_e)\,\big), \tag{3.2h}$$

where $K = K(k)$ and $K' = K'(k)$ are complete elliptic integrals and

$$\kappa = 2K + \mathrm{i}K', \tag{3.3}$$

which lie on the elliptic curve

$$\mathrm{sn}\,(z_0 - \gamma_\mathrm{e})^2\,(1 + k^2x^2y^2) + 2\mathrm{cn}\,(z_0 - \gamma_\mathrm{e})\,\mathrm{dn}\,(z_0 - \gamma_\mathrm{e})\,xy - (x^2 + y^2) = 0. \tag{3.4}$$

The base points (3.2) can be generalized to

$$p_i : (x, y) = \big(\mathrm{cd}\,(c_i + \eta)\,, \mathrm{cd}\,(\eta - c_i)\big), \quad i = 1, \ldots, 8, \tag{3.5}$$

where c_i, $i = 1, \ldots, 8$, and η are non-zero complex parameters. These points lie on the elliptic curve

$$\mathrm{sn}\,(2\eta)^2\,(1 + k^2x^2y^2) + 2\mathrm{cn}\,(2\eta)\,\mathrm{dn}\,(2\eta)\,xy - (x^2 + y^2) = 0. \tag{3.6}$$

The generalized base points (3.5) and elliptic curve (3.6) can be respectively reduced to the points (3.2) and curve (3.4) by taking

$$c_2 = c_1 + 2K, \quad c_3 = c_1 + \mathrm{i}K', \quad c_4 = c_1 + \kappa, \quad c_6 = c_5 + 2K, \tag{3.7a}$$

$$c_7 = c_5 + \mathrm{i}K', \quad c_8 = c_5 + \kappa, \tag{3.7b}$$

and letting

$$z_0 = \eta + c_5 + \kappa, \quad \gamma_\mathrm{e} = c_5 - \eta + \kappa, \quad \gamma_\mathrm{o} = \eta + c_1 + \kappa. \tag{3.7c}$$

Note that the two biquadratic curves (3.4) and (3.6) are non-singular, for $k \neq 0, \pm 1$. It follows that each curve is parametrized by elliptic functions. The resulting initial value space obtained after resolution is of type $A_0^{(1)}$ as shown in the following section.

4 Cremona isometries

As explained in §3, we now investigate a variety X, obtained after a sequence of blow-ups. We focus on surfaces X defined by blowing up base points on biquadratic curves, such as Equations (3.5) and (3.6). The resulting structure contains equivalence classes of lines and information about their intersections. In this section, we construct Cremona isometries, which are roughly speaking, mappings of X that preserve this structure. As a result, they provide symmetries of the dynamical system iterated on X.

An important object in this framework is given by the Picard lattice of X, or $\mathrm{Pic}(X)$, which is defined by

$$\mathrm{Pic}(X) = \mathbb{Z}H_0 + \mathbb{Z}H_1 + \mathbb{Z}E_1 + \cdots + \mathbb{Z}E_8, \tag{4.1}$$

where $E_i = \epsilon^{-1}(p_i)$, $i = 1, \ldots, 8$, are exceptional divisors obtained from blow-up of the base points (3.5). We define a symmetric bilinear form, called the intersection form, on $\mathrm{Pic}(X)$ by

$$(H_i|H_j) = 1 - \delta_{ij}, \quad (H_i|E_j) = 0, \quad (E_i|E_j) = -\delta_{ij}, \tag{4.2}$$

where $\delta_{ij} = 0$, if $i \neq j$, or 1, if $i = j$. The anti-canonical divisor of X is given by

$$-K_X = 2H_0 + 2H_1 - \sum_{i=1}^{8} E_i. \tag{4.3}$$

For later convenience, let $\delta = -K_X$. The anti-canonical divisor δ corresponds to the curve of bi-degree $(2,2)$ passing through the base points p_i, $i = 1, \ldots, 8$, with multiplicity 1, that is, the curve (3.6). Since this curve is non-singular, for $k \neq 0, \pm 1$, the anti-canonical divisor cannot be decomposed. Therefore, we can identify the surface X as being of type $A_0^{(1)}$ in Sakai's classification [38].

We define the root lattice

$$Q(A_0^{(1)\perp}) = \sum_{k=0}^{8} \mathbb{Z}\beta_k \tag{4.4}$$

in Pic(X) that are orthogonal to the anti-canonical divisor δ. The simple roots β_i, $i = 0, \ldots, 8$, are given by

$$\begin{aligned} &\beta_1 = H_1 - H_0, \quad \beta_2 = H_0 - E_1 - E_2, \quad \beta_i = E_{i-1} - E_i, \quad i = 3, \ldots, 7, \\ &\beta_8 = E_1 - E_2, \quad \beta_0 = E_7 - E_8, \end{aligned} \tag{4.5}$$

where

$$\delta = 2\beta_1 + 4\beta_2 + 6\beta_3 + 5\beta_4 + 4\beta_5 + 3\beta_6 + 2\beta_7 + 3\beta_8 + \beta_0. \tag{4.6}$$

We can easily verify that

$$(\beta_i|\beta_j) = \begin{cases} -2, & i = j \\ 1, & i = j-1 \quad (j = 2, \ldots, 7), \quad \text{or if} \quad (i,j) = (3,8), (7,0) \\ 0, & \text{otherwise.} \end{cases} \tag{4.7}$$

Representing intersecting β_i and β_j by a line between nodes i and j, we obtain the Dynkin diagram of $E_8^{(1)}$ shown in Figure 2.1.

Remark 4.1. From Pic(X) we can obtain the coroot lattice, weight lattice and root lattice in §2. Indeed, the coroot lattice $Q^\vee$ is given from Pic(X) by

$$\alpha_i^\vee = \beta_i, \quad i = 0, \ldots, 8, \tag{4.8}$$

the weight lattice P is given from Pic$(X)/(\mathbb{Z}\delta)$ by

$$h_0 \equiv -\mathcal{E}_8, \quad h_1 \equiv -H_0, \quad h_2 \equiv -H_0 - H_1, \tag{4.9a}$$

$$h_3 \equiv -\mathcal{E}_3 - \mathcal{E}_4 - \mathcal{E}_5 - \mathcal{E}_6 - \mathcal{E}_7 - \mathcal{E}_8, \tag{4.9b}$$

$$h_4 \equiv -\mathcal{E}_4 - \mathcal{E}_5 - \mathcal{E}_6 - \mathcal{E}_7 - \mathcal{E}_8, \quad h_5 \equiv -\mathcal{E}_5 - \mathcal{E}_6 - \mathcal{E}_7 - \mathcal{E}_8, \tag{4.9c}$$

$$h_6 \equiv -\mathcal{E}_6 - \mathcal{E}_7 - \mathcal{E}_8, \quad h_7 \equiv -\mathcal{E}_7 - \mathcal{E}_8, \tag{4.9d}$$

$$h_8 \equiv -H_0 - H_1 + \mathcal{E}_1, \tag{4.9e}$$

and the root lattice Q is given from $\mathrm{Pic}(X)/(\mathbb{Z}\delta)$ by

$$\alpha_i \equiv \beta_i, \quad i = 0, \ldots, 8. \tag{4.10}$$

Note that in this case we define the bilinear pairing $\langle \cdot, \cdot \rangle : Q^\vee \times P \to \mathbb{Z}$ for arbitrary $\alpha^\vee \in Q^\vee$ and $h \in P$ by

$$\langle \alpha^\vee, h \rangle = -(\alpha^\vee | h). \tag{4.11}$$

Therefore, in a similar manner as in §2, we can define the transformation group $W(E_8^{(1)})$ as follows.

Definition 4.2 ([10]). An automorphism of Pic(X) is called a Cremona isometry if it preserves

(i) the intersection form $(\,|\,)$ on Pic(X);

(ii) the canonical divisor K_X;

(iii) the effectiveness of each effective divisor of Pic(X).

It is well known that the reflections are Cremona isometries. In this case we define the reflections s_i, $i = 0, \ldots, 8$, by the following linear actions:

$$s_i(v) = v + (v|\, \beta_i)\, \beta_i, \tag{4.12}$$

for all $v \in \mathrm{Pic}(X)$. They collectively form an affine Weyl group of type $E_8^{(1)}$, denoted by $W\big(E_8^{(1)}\big)$. Namely, we can easily verify that under the actions (4.12) the fundamental relations (2.13) hold. Moreover, for each root $\beta \in Q(A_0^{(1)\perp})$, we can define the Kac translation T_β on the Picard lattice by

$$T_\beta(\lambda) = \lambda + (\delta|\lambda)\beta - \left(\frac{(\beta|\beta)(\delta|\lambda)}{2} + (\beta|\lambda)\right)\delta, \quad \lambda \in \mathrm{Pic}(X). \tag{4.13}$$

5 Birational actions of the Cremona isometries for the Jacobi setting

In this section, we give the birational actions of the Cremona isometries on the coordinates and parameters of the base points (3.5). By using these birational actions, we reconstruct Equation (1.1).

We focus on a particular example first to explain how to deduce such birational actions. Recall that H_0 and H_1 are given by the linear equivalence classes of vertical lines $x =$ constant and horizontal lines $y =$ constant, respectively. Applying the reflection operator s_2 given by (4.12) to H_1, we find that $s_2(H_1) = H_0 + H_1 - \mathcal{E}_1 - \mathcal{E}_2$, which means that $s_2(y)$ can be described by the curve of bi-degree (1, 1) passing through base points p_1 and p_2 with multiplicity 1. (See [25] for for more detail.) This result leads us to the birational action given below in Equation (5.1b). Similarly, from the linear actions of s_i, $i = 0, \ldots, 8$, we obtain their birational actions on the coordinates and parameters of the base points (3.5) as follows. The actions of the generators of $W\big(E_8^{(1)}\big)$ on the coordinates (x, y) are given by

$$s_1(x) = y, \quad s_1(y) = x, \tag{5.1a}$$

$$\left(\frac{s_2(y) - \mathrm{cd}\left(2\eta - \frac{c_1-c_2}{2}\right)}{s_2(y) - \mathrm{cd}\left(2\eta + \frac{c_1-c_2}{2}\right)}\right)\left(\frac{x - \mathrm{cd}\,(\eta + c_1)}{x - \mathrm{cd}\,(\eta + c_2)}\right)\left(\frac{y - \mathrm{cd}\,(\eta - c_2)}{y - \mathrm{cd}\,(\eta - c_1)}\right)$$
$$= \left(\frac{1 - \dfrac{\mathrm{cd}\,(\eta - c_2)}{\mathrm{cd}\,(\eta)}}{1 - \dfrac{\mathrm{cd}\,(\eta - c_1)}{\mathrm{cd}\,(\eta)}}\right)\left(\frac{1 - \dfrac{\mathrm{cd}\,(\eta + c_1)}{\mathrm{cd}\,(\eta)}}{1 - \dfrac{\mathrm{cd}\,(\eta + c_2)}{\mathrm{cd}\,(\eta)}}\right)\left(\frac{1 - \dfrac{\mathrm{cd}\left(2\eta - \frac{c_1-c_2}{2}\right)}{\mathrm{cd}\left(\frac{c_1+c_2}{2}\right)}}{1 - \dfrac{\mathrm{cd}\left(2\eta + \frac{c_1-c_2}{2}\right)}{\mathrm{cd}\left(\frac{c_1+c_2}{2}\right)}}\right), \tag{5.1b}$$

while those on the parameters c_i, $i = 1, \ldots, 8$, and η are given by

$$s_0(c_7) = c_8, \quad s_0(c_8) = c_7, \quad s_1(\eta) = -\eta, \quad s_2(\eta) = \eta - \frac{2\eta + c_1 + c_2}{4}, \tag{5.1c}$$

$$s_2(c_i) = \begin{cases} c_i - \dfrac{3(2\eta + c_1 + c_2)}{4}, & i = 1, 2, \\ c_i + \dfrac{2\eta + c_1 + c_2}{4}, & i = 3, \ldots, 8, \end{cases} \tag{5.1d}$$

$$s_k(c_{k-1}) = c_k, \quad s_k(c_k) = c_{k-1}, \quad k = 3, \ldots, 7, \tag{5.1e}$$

$$s_8(c_1) = c_2, \quad s_8(c_2) = c_1. \tag{5.1f}$$

Note that $\lambda = \sum_{i=1}^{8} c_i$ is invariant under the action of $W\big(E_8^{(1)}\big)$.

For Jacobi's elliptic function $\mathrm{cd}\,(u)$ it is well known that shifts by half periods give the following relations:

$$\mathrm{cd}\,(u + 2K) = -\mathrm{cd}\,(u), \quad \mathrm{cd}\,(u + \mathrm{i}K') = \frac{1}{k\,\mathrm{cd}\,(u)}. \tag{5.2}$$

These identities motivate our search for the transformations that are identity mappings on the $\mathrm{Pic}(X)$. Indeed, we define such transformations ι_i, $i = 1, \ldots, 4$, by the following actions:

$$\iota_1 : (c_1, \ldots, c_8, \eta, x, y) \mapsto \left(c_1 - \frac{\mathrm{i}K'}{2}, \ldots, c_8 - \frac{\mathrm{i}K'}{2}, \eta - \frac{\mathrm{i}K'}{2}, \frac{1}{kx}, y\right), \tag{5.3a}$$

$$\iota_2 : (c_1, \ldots, c_8, \eta, x, y) \mapsto \left(c_1 - \frac{\mathrm{i}K'}{2}, \ldots, c_8 - \frac{\mathrm{i}K'}{2}, \eta + \frac{\mathrm{i}K'}{2}, x, \frac{1}{ky}\right), \tag{5.3b}$$

$$\iota_3 : (c_1, \ldots, c_8, \eta, x, y) \mapsto (c_1 - K, \ldots, c_8 - K, \eta - K, -x, y), \tag{5.3c}$$

$$\iota_4 : (c_1, \ldots, c_8, \eta, x, y) \mapsto (c_1 - K, \ldots, c_8 - K, \eta + K, x, -y). \tag{5.3d}$$

Adding the transformations ι_i, we extend $W\big(E_8^{(1)}\big)$ to

$$\widetilde{W}\big(E_8^{(1)}\big) = \langle \iota_1, \iota_2, \iota_3, \iota_4 \rangle \rtimes W\big(E_8^{(1)}\big). \tag{5.4}$$

In general, for a function $F = F(c_i, \eta, x, y)$, we let an element $w \in \widetilde{W}(E_8^{(1)})$ act as $w.F = F(w.c_i, w.\eta, w.x, w.y)$, that is, w acts on the arguments from the left. Under the birational actions (5.1) and (5.3), the generators of $\widetilde{W}(E_8^{(1)})$ satisfy the fundamental relations of type $E_8^{(1)}$ (2.13) and the following relations:

$$(\iota_i\iota_j)^{m_{ij}} = 1, \quad i,j = 1,2,3,4, \quad \iota_k s_l = s_l \iota_k, \quad k = 1,2,3,4,\ l \neq 1,2, \tag{5.5a}$$

$$\iota_{\{1,2,3,4\}} s_1 = s_1 \iota_{\{2,1,4,3\}}, \quad \iota_1 s_2 = s_2 \iota_1 \iota_2, \quad \iota_2 s_2 = s_2 \iota_2, \tag{5.5b}$$

$$\iota_3 s_2 = s_2 \iota_3 \iota_4, \quad \iota_4 s_2 = s_2 \iota_4, \tag{5.5c}$$

where

$$m_{ij} = \begin{cases} 1, & i = j \\ 2, & \text{otherwise.} \end{cases} \tag{5.6}$$

Now we are in a position to derive Equation (1.1) from the Cremona transformations. Note that for convenience we use the notation (2.25) for the composition of the reflections s_i and the notation

$$c_{j_1 \cdots j_n} = c_{j_1} + \cdots + c_{j_n}, \quad j_1 \cdots j_n \in \{1, \ldots, 8\}, \tag{5.7}$$

for the summation of the parameters c_i. Let

$$R_{J,1} = s_{56453483706756452348321564534837067564523483217067348304 68}\iota_4\iota_3\iota_2\iota_1. \tag{5.8}$$

The action of $R_{J,1}$ on the root lattice $Q(A_0^{(1)\perp})$:

$$R_{J,1}: \begin{pmatrix} \alpha_0 \\ \alpha_1 \\ \alpha_2 \\ \alpha_3 \\ \alpha_4 \\ \alpha_5 \\ \alpha_6 \\ \alpha_7 \\ \alpha_8 \end{pmatrix} \mapsto \begin{pmatrix} -1 & 0 & 0 & 0 & 0 & 0 & 0 & 0 & 0 \\ -1 & -1 & -4 & -6 & -5 & -4 & -3 & -2 & -3 \\ 0 & 0 & 1 & 2 & 1 & 0 & 0 & 0 & 1 \\ 0 & 0 & 0 & -1 & 0 & 0 & 0 & 0 & 0 \\ 0 & 0 & 0 & 0 & -1 & 0 & 0 & 0 & 0 \\ 1 & 1 & 2 & 4 & 4 & 3 & 3 & 2 & 2 \\ 0 & 0 & 0 & 0 & 0 & 0 & -1 & 0 & 0 \\ 0 & 0 & 0 & 0 & 0 & 0 & 0 & -1 & 0 \\ 0 & 0 & 0 & 0 & 0 & 0 & 0 & 0 & -1 \end{pmatrix} \begin{pmatrix} \alpha_0 \\ \alpha_1 \\ \alpha_2 \\ \alpha_3 \\ \alpha_4 \\ \alpha_5 \\ \alpha_6 \\ \alpha_7 \\ \alpha_8 \end{pmatrix}, \tag{5.9}$$

is not translational, and that on the parameter space:

$$R_{J,1}(c_i) = -c_i + \frac{c_{1234} - c_{5678}}{4} - \kappa, \quad i = 1, \ldots, 4, \tag{5.10}$$

$$R_{J,1}(c_j) = -c_j + \frac{c_{1234} + 3c_{5678}}{4} - \kappa, \quad j = 5, \ldots, 8, \quad R_{J,1}(\eta) = \eta + \frac{\lambda}{2}, \tag{5.11}$$

where κ is defined by (3.3), is also not translational. However, when the parameters take special values (3.7), the action of $R_{J,1}$ becomes the translational motion in the parameter subspace:

$$R_{J,1}: (\gamma_{\rm e}, \gamma_{\rm o}, z_0) \mapsto (\gamma_{\rm e}, \gamma_{\rm o}, z_0 + 2(\gamma_{\rm e} + \gamma_{\rm o}) - 2\kappa), \tag{5.12}$$

and then the action on the coordinates (x, y) with $x_n = R_{J,1}^n(x)$, $y_n = R_{J,1}^n(y)$, $z_n = R_{J,1}^n(z_0)$, gives the time evolution of Equation (1.1), that is, $R_{J,1} = \phi$. Note that we can without loss of generality ignore "2κ" in

$$R_{J,1}(z_0) = 2(\gamma_{\rm e} + \gamma_{\rm o}) - 2\kappa, \tag{5.13}$$

because of the form of Equation (1.1) and the following relations:

$$\mathrm{sn}\,(u + 2\kappa) = \mathrm{sn}\,(u)\,, \quad \mathrm{cn}\,(u + 2\kappa) = -\mathrm{cn}\,(u)\,, \quad \mathrm{dn}\,(u + 2\kappa) = -\mathrm{dn}\,(u)\,. \tag{5.14}$$

6 Special solutions of the RCG equation

In this section, we show the special solutions of the RCG equation.

Let us consider Equation (1.1) under the following condition:

$$\gamma_{\rm o} = \frac{\mathrm{i}K'}{2}, \tag{6.1}$$

which gives

$$\mathrm{sg_o} = \frac{\mathrm{i}}{k^{1/2}}, \quad \mathrm{cg_o} = \frac{(1+k)^{1/2}}{k^{1/2}}, \quad \mathrm{dg_o} = (1+k)^{1/2}. \tag{6.2}$$

Then, the base points p_i, $i = 1, 2, 3, 4$, in (3.2) can be expressed by

$$p_1 : (x, y) = \left(-\frac{1}{k^{1/2}}, -\mathrm{cd}\left(z_0 - \gamma_{\rm e} + \frac{\mathrm{i}K'}{2}\right)\right), \tag{6.3a}$$

$$p_2 : (x, y) = \left(\frac{1}{k^{1/2}}, \mathrm{cd}\left(z_0 - \gamma_{\rm e} + \frac{\mathrm{i}K'}{2}\right)\right), \tag{6.3b}$$

$$p_3 : (x, y) = \left(-\frac{1}{k^{1/2}}, -\mathrm{cd}\left(z_0 - \gamma_{\rm e} - \frac{\mathrm{i}K'}{2}\right)\right), \tag{6.3c}$$

$$p_4 : (x, y) = \left(\frac{1}{k^{1/2}}, \mathrm{cd}\left(z_0 - \gamma_{\rm e} - \frac{\mathrm{i}K'}{2}\right)\right). \tag{6.3d}$$

This means that there exist the bi-degree $(1, 0)$ curve $x = -k^{-1/2}$, passing through p_1 and p_3, and the bi-degree $(1, 0)$ curve $x = k^{-1/2}$, passing through p_2 and p_4, which correspond to $H_0 - \mathcal{E}_1 - \mathcal{E}_3$ and $H_0 - \mathcal{E}_2 - \mathcal{E}_4$, respectively. Moreover, the action

$$\phi : \ H_0 - \mathcal{E}_1 - \mathcal{E}_3 \leftrightarrow H_0 - \mathcal{E}_2 - \mathcal{E}_4, \tag{6.4}$$

implies that there exist the special solutions when

$$x_n = \pm\frac{(-1)^n}{k^{1/2}}. \tag{6.5}$$

Therefore, we obtain the following lemma.

Lemma 6.1. *The following are special solutions of Equation* (1.1):

$$(x_n, y_n) = \left(\frac{(-1)^n}{k^{1/2}}, \frac{(-1)^n}{k^{1/2}}\right), \qquad (x_n, y_n) = \left(\frac{(-1)^n}{k^{1/2}}, \frac{(-1)^{n+1}}{k^{1/2}}\right), \tag{6.6a}$$

$$(x_n, y_n) = \left(\frac{(-1)^{n+1}}{k^{1/2}}, \frac{(-1)^n}{k^{1/2}}\right), \qquad (x_n, y_n) = \left(\frac{(-1)^{n+1}}{k^{1/2}}, \frac{(-1)^{n+1}}{k^{1/2}}\right), \tag{6.6b}$$

and

$$(x_n, y_n) = \left(\frac{(-1)^n}{k^{1/2}}, \frac{\mathrm{i}\tan(u_n)}{k^{1/2}}\right), \quad (x_n, y_n) = \left(\frac{(-1)^{n+1}}{k^{1/2}}, -\frac{\mathrm{i}\tan(u_n)}{k^{1/2}}\right), \tag{6.7}$$

where $\mathrm{i} = \sqrt{-1}$ *and* u_n *is the solution of the following linear equation:*

$$u_{n+1} + u_n = \tan^{-1}\left(-\mathrm{i}\,\frac{(1 - k\,\mathrm{sg}_\mathrm{e}^2)\mathrm{cz}_n\mathrm{dz}_n}{\mathrm{cg}_\mathrm{e}\mathrm{dg}_\mathrm{e}(1 - k\,\mathrm{sz}_n^2)}\right). \tag{6.8}$$

Proof. Under the condition

$$x_n = \frac{(-1)^n}{k^{1/2}}, \tag{6.9}$$

Equations (1.1) are reduced to the following discrete Riccati equation:

$$y_{n+1} + y_n = \frac{\mathrm{i}\,A_n}{k^{1/2}}(1 + k\,y_{n+1}y_n), \tag{6.10}$$

where A_n is given by

$$A_n = -\mathrm{i}\,\frac{(1 - k\,\mathrm{sg}_\mathrm{e}^2)\mathrm{cz}_n\mathrm{dz}_n}{\mathrm{cg}_\mathrm{e}\mathrm{dg}_\mathrm{e}(1 - k\,\mathrm{sz}_n^2)}. \tag{6.11}$$

Note that under the condition

$$x_n = \frac{(-1)^{n+1}}{k^{1/2}}, \tag{6.12}$$

Equations (1.1) can be reduced to

$$y_{n+1} + y_n = -\frac{\mathrm{i}\,A_n}{k^{1/2}}(1 + k\,y_{n+1}y_n), \tag{6.13}$$

which can be rewritten as the discrete Riccati equation (6.10) by the transformation $y_n \mapsto -y_n$. Therefore, it is enough for us to just consider the case (6.9).

Let us consider the solutions of the discrete Riccati equation (6.10). If

$$1 + k\,y_{n+1}y_n = 0, \tag{6.14}$$

then we obtain

$$y_{n+1} + y_n = 0. \tag{6.15}$$

Therefore, we obtain

$$y_n = \frac{(-1)^n}{k^{1/2}}, \ \frac{(-1)^{n+1}}{k^{1/2}}, \tag{6.16}$$

which gives the special solutions (6.6). In the following we assume

$$1 + k\, y_{n+1} y_n \neq 0. \tag{6.17}$$

Then, the discrete Riccati equation (6.10) can be rewritten as the following:

$$\frac{y_{n+1} + y_n}{1 + k\, y_{n+1} y_n} = \frac{\mathrm{i}\, A_n}{k^{1/2}}. \tag{6.18}$$

By letting

$$y_n = \frac{\mathrm{i}\, \tan(u_n)}{k^{1/2}}, \tag{6.19}$$

and using the tangent addition formula, the discrete Riccati equation (6.18) can be rewritten as

$$\tan(u_{n+1} + u_n) = A_n, \tag{6.20}$$

which gives the linear equation (6.8). Therefore, we have completed the proof. ■

Acknowledgments

This research was supported by an Australian Laureate Fellowship # FL120100094 and grant # DP160101728 from the Australian Research Council and JSPS KAKENHI Grant Number JP17J00092.

Appendix A $A_4^{(1)}$-lattice

In this section, we give a more detailed description of the weight lattice and affine Weyl group by using the lattice of type $A_4^{(1)}$ as an example.

We consider the following $\mathbb{Z}$-modules:

$$Q^\vee = \sum_{k=0}^{4} \mathbb{Z}\alpha_k^\vee, \quad P = \sum_{k=0}^{4} \mathbb{Z}h_k, \tag{A.1}$$

with the bilinear pairing $\langle \cdot, \cdot \rangle : Q^\vee \times P \to \mathbb{Z}$ defined by

$$\langle \alpha_i^\vee, h_j \rangle = \delta_{ij}, \quad 0 \le i, j \le 4. \tag{A.2}$$

We also define the submodule of P by $Q = \sum_{k=0}^{4} \mathbb{Z}\alpha_k$, where α_i, $i = 0, \ldots, 4$, are defined by

$$\begin{pmatrix} \alpha_0 \\ \alpha_1 \\ \alpha_2 \\ \alpha_3 \\ \alpha_4 \end{pmatrix} = (A_{ij})_{i,j=0}^{4} \begin{pmatrix} h_0 \\ h_1 \\ h_2 \\ h_3 \\ h_4 \end{pmatrix}, \tag{A.3}$$

and satisfy

$$\langle \alpha_i^\vee, \alpha_j \rangle = A_{ij}. \tag{A.4}$$

Here, $(A_{ij})_{i,j=0}^4$ is the Generalized Cartan matrix of type $A_4^{(1)}$:

$$(A_{ij})_{i,j=0}^4 = \begin{pmatrix} 2 & -1 & 0 & 0 & -1 \\ -1 & 2 & -1 & 0 & 0 \\ 0 & -1 & 2 & -1 & 0 \\ 0 & 0 & -1 & 2 & -1 \\ -1 & 0 & 0 & -1 & 2 \end{pmatrix}. \tag{A.5}$$

Then, the generators $\{\alpha_0^\vee, \ldots, \alpha_4^\vee\}$, $\{h_0, \ldots, h_4\}$ and $\{\alpha_0, \ldots, \alpha_4\}$ are identified with simple coroots, fundamental weights and simple roots of type $A_4^{(1)}$, respectively. We note that the following relation holds:

$$\alpha_0 + \alpha_1 + \alpha_2 + \alpha_3 + \alpha_4 = 0, \tag{A.6}$$

and we call the corresponding coroot the null-coroot, denoted by $\delta^\vee$:

$$\delta^\vee = \alpha_0^\vee + \alpha_1^\vee + \alpha_2^\vee + \alpha_3^\vee + \alpha_4^\vee. \tag{A.7}$$

In the following subsections, we consider the transformation group acting on these lattices.

A.1 Affine Weyl group of type $A_4^{(1)}$

In this section, we consider the transformations which collectively form an affine Weyl group of type $A_4^{(1)}$.

We define the transformations s_i, $i = 0, \ldots, 4$, by the reflections for the roots $\{\alpha_0, \ldots, \alpha_4\}$:

$$s_i(\lambda) = \lambda - \langle \alpha_i^\vee, \lambda \rangle \alpha_i, \quad i = 0, \ldots, 4, \quad \lambda \in P, \tag{A.8}$$

which give

$$\begin{aligned} &s_0(h_0) = -h_0 + h_1 + h_4, \quad s_1(h_1) = h_0 - h_1 + h_2, \quad s_2(h_2) = h_1 - h_2 + h_3, \\ &s_3(h_3) = h_2 - h_3 + h_4, \quad s_4(h_4) = h_0 + h_3 - h_4. \end{aligned} \tag{A.9}$$

From definitions (A.2), (A.3), (A.4) and (A.8), we can compute actions on the simple roots α_i, $i = 0, \ldots, 4$, as the following:

$$s_i(\alpha_j) = \begin{cases} -\alpha_j, & i = j \\ \alpha_j + \alpha_i, & i = j \pm 1 \pmod 5 \\ \alpha_j, & i = j \pm 2 \pmod 5). \end{cases} \tag{A.10}$$

Under the linear actions on the weight lattice (A.9), $W(A_4^{(1)}) = \langle s_0, s_1, s_2, s_3, s_4\rangle$ forms an affine Weyl group of type $A_4^{(1)}$, that is, the following fundamental relations hold:

$$(s_i s_j)^{l_{ij}} = 1, \quad \text{where} \quad l_{ij} = \begin{cases} 1, & i = j \\ 3, & i = j \pm 1 \pmod 5 \\ 2, & i = j \pm 2 \pmod 5. \end{cases} \tag{A.11}$$

Note that representing the simple reflections s_i by nodes and connecting i-th and j-th nodes by $(l_{ij} - 2)$ lines, we obtain the Dynkin diagram of type $A_4^{(1)}$ shown in Figure A.1.

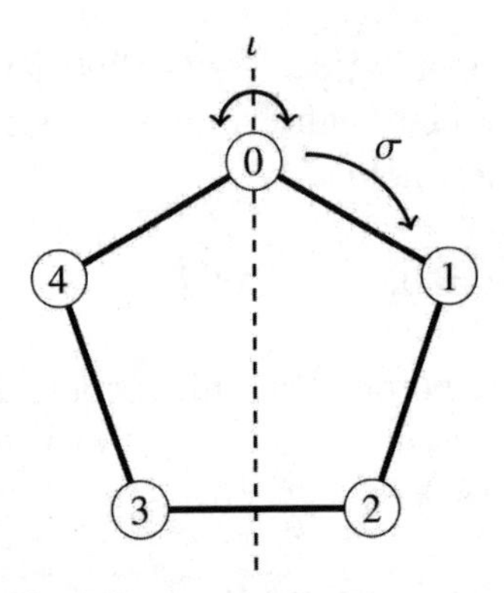

Figure A.1. Dynkin diagram of type $A_4^{(1)}$. The transformation ι is the reflection with respect to the dotted line, and the transformation σ is the rotation symmetry with respect to an angle of $2\pi/5$ in a clockwise manner.

Remark A.1. We can also define the action of $W(A_4^{(1)})$ on the coroot lattice $Q^\vee$ by replacing α_i with $\alpha_i^\vee$ in the actions (A.10). Note that $\delta^\vee$ is invariant under the action of $W(A_4^{(1)})$. Then, the transformations in $W(A_4^{(1)})$ preserve the form $\langle \cdot, \cdot \rangle$.

Let M_i, $i = 0, \ldots, 4$, be the orbits of h_i, $i = 0, \ldots, 4$, defined by

$$M_i = \left\{ w(h_i) \,\middle|\, w \in W(A_4^{(1)}) \right\}, \tag{A.12}$$

and T_i, $i = 0, \ldots, 4$, be the transformations defined by

$$T_0 = s_{04321234}, \quad T_1 = s_{10432340}, \quad T_2 = s_{21043401}, \tag{A.13a}$$

$$T_3 = s_{32104012}, \quad T_4 = s_{43210123}, \tag{A.13b}$$

whose actions on the fundamental weights and simple roots are given by

$$T_i(h_j) = h_j - \alpha_i, \quad T_i(\alpha_j) = \alpha_j, \qquad i, j = 0, \ldots, 4. \tag{A.14}$$

Note that $T_0 \circ T_1 \circ T_2 \circ T_3 \circ T_4 = 1$. Then, the following lemma holds.

Lemma A.2. *The following hold:*

$$M_i = \{h_i + \alpha \mid \alpha \in Q\}\,, \quad i = 0, \dots, 4. \tag{A.15}$$

Proof. The relation $M_0 \subset \{h_0 + \alpha \mid \alpha \in Q\}$ is obvious from the definition (A.8), and the relation $M_0 \supset \{h_0 + \alpha \mid \alpha \in Q\}$ follows from

$$h_0 + \sum_{i=0}^{4} k_i \alpha_i = {T_0}^{k_0} \circ {T_1}^{k_1} \circ {T_2}^{k_2} \circ {T_3}^{k_3} \circ {T_4}^{k_4}(h_0), \tag{A.16}$$

where $k_i \in \mathbb{Z}$. Therefore, the case $i = 0$ holds. In a similar manner, we can also prove the statements for M_i $i = 1, \dots, 4$. Therefore, we have completed the proof. ■

Since of Lemma A.2, we can express the lattices M_i by

$$M_i = \{T(h_i) \mid T \in \langle T_1, \dots, T_4 \rangle\}\,, \quad i = 0, \dots, 4. \tag{A.17}$$

In general, the translations on the orbits M_i, $i = 0, \dots, 4$, are given by the Kac translations defined on the weight lattice P:

$$T_\alpha(\lambda) = \lambda - \langle \delta^\vee, \lambda \rangle \alpha, \quad \lambda \in P, \quad \alpha \in Q. \tag{A.18}$$

The translations T_i, $i = 0, \dots, 4$, can be expressed by the Kac translations as the following:

$$T_i = T_{\alpha_i}, \quad i = 0, \dots, 4. \tag{A.19}$$

Note that the translations T_i, $i = 0, \dots, 4$, are called the fundamental translations on the weight lattice P or in affine Weyl group $W(A_4^{(1)})$. Indeed, because of the following property of the Kac translations:

$$T_\alpha \circ T_\beta = T_{\alpha+\beta}, \tag{A.20}$$

where $\alpha, \beta \in Q$, all Kac translations in $W(A_4^{(1)})$ can be expressed by the compositions of T_i, $i = 0, \dots, 4$.

In this case we do not have translations moving a fundamental weight h_i to another fundamental weight h_j. However, by extending $W(A_4^{(1)})$ with the automorphisms of the Dynkin diagram, we can define such translations as shown in the following subsection.

A.2 Extended affine Weyl group of type $A_4^{(1)}$

In this section, we consider the extended affine Weyl group of type $A_4^{(1)}$.

We define the transformations σ and ι by

$$\sigma : h_0 \to h_1, \quad h_1 \to h_2, \quad h_2 \to h_3, \quad h_3 \to h_4, \quad h_4 \to h_0, \tag{A.21a}$$

$$\iota : h_1 \leftrightarrow h_4, \quad h_2 \leftrightarrow h_3, \tag{A.21b}$$

whose actions on the root lattice are given by

$$\sigma : \alpha_0 \to \alpha_1, \quad \alpha_1 \to \alpha_2, \quad \alpha_2 \to \alpha_3, \quad \alpha_3 \to \alpha_4, \quad \alpha_4 \to \alpha_0, \tag{A.22a}$$

$$\iota : \alpha_1 \leftrightarrow \alpha_4, \quad \alpha_2 \leftrightarrow \alpha_3. \tag{A.22b}$$

Moreover, we also define their actions on the coroot lattice $Q^\vee$ by replacing α_i with $\alpha_i^\vee$ in the actions (A.22). Then, the transformations σ and ι satisfy

$$\sigma^5 = \iota^2 = 1, \quad \sigma \circ \iota = \iota \circ \sigma^{-1}, \tag{A.23}$$

and the relations with $W(A_4^{(1)}) = \langle s_0, s_1, s_2, s_3, s_4 \rangle$ are given by

$$\sigma \circ s_i = s_{i+1} \circ \sigma, \quad \iota \circ s_i = s_{-i} \circ \iota, \tag{A.24}$$

that is, the transformations σ and ι are automorphisms of the Dynkin diagram of type $A_4^{(1)}$ (see Figure A.1). Therefore, we call

$$\widetilde{W}(A_4^{(1)}) = W(A_4^{(1)}) \rtimes \langle \sigma, \iota \rangle \tag{A.25}$$

as the extended affine Weyl group of type $A_4^{(1)}$.

Remark A.3. We can easily verify that $\delta^\vee$ is also invariant under the action of $\widetilde{W}(A_4^{(1)})$, and the transformations in $\widetilde{W}(A_4^{(1)})$ preserve the form $\langle \cdot, \cdot \rangle$.

Let M be the orbits of h_0 defined by

$$M = \left\{ w(h_0) \,\middle|\, w \in \widetilde{W}(A_4^{(1)}) \right\}. \tag{A.26}$$

Moreover, we also define the following transformations:

$$T_{01} = \sigma^4 s_{2340}, \quad T_{12} = \sigma^4 s_{3401}, \quad T_{23} = \sigma^4 s_{4012}, \tag{A.27a}$$

$$T_{34} = \sigma^4 s_{0123}, \quad T_{40} = \sigma^4 s_{1234}, \tag{A.27b}$$

whose actions on the fundamental weights are translational as the following:

$$T_{i\,i+1}(h_j) = h_j + v_{i\,i+1}, \quad T_{i\,i+1}(v_{j\,j+1}) = v_{j\,j+1}, \quad i, j \in \mathbb{Z}/(5\mathbb{Z}), \tag{A.28}$$

where

$$v_{01} = h_0 - h_1,\ v_{12} = h_1 - h_2,\ v_{23} = h_2 - h_3,\ v_{34} = h_3 - h_4,\ v_{40} = h_4 - h_0. \tag{A.29}$$

Note that

$$\begin{pmatrix}\alpha_0\\ \alpha_1\\ \alpha_2\\ \alpha_3\\ \alpha_4\end{pmatrix} = \begin{pmatrix}1 & 0 & 0 & 0 & -1\\ -1 & 1 & 0 & 0 & 0\\ 0 & -1 & 1 & 0 & 0\\ 0 & 0 & -1 & 1 & 0\\ 0 & 0 & 0 & -1 & 1\end{pmatrix} \cdot \begin{pmatrix}v_{01}\\ v_{12}\\ v_{23}\\ v_{34}\\ v_{40}\end{pmatrix}, \tag{A.30}$$

$$\begin{pmatrix}v_{01}\\ v_{12}\\ v_{23}\\ v_{34}\\ v_{40}\end{pmatrix} = \frac{1}{5}\begin{pmatrix}4 & 0 & 1 & 2 & 3\\ 3 & 4 & 0 & 1 & 2\\ 2 & 3 & 4 & 0 & 1\\ 1 & 2 & 3 & 4 & 0\\ 0 & 1 & 2 & 3 & 4\end{pmatrix} \cdot \begin{pmatrix}\alpha_0\\ \alpha_1\\ \alpha_2\\ \alpha_3\\ \alpha_4\end{pmatrix}, \tag{A.31}$$

$$v_{01} + v_{12} + v_{23} + v_{34} + v_{40} = 0, \tag{A.32}$$

$$T_{01} \circ T_{12} \circ T_{23} \circ T_{34} \circ T_{40} = 1. \tag{A.33}$$

In a similar manner as the proof of Lemma A.2, by using the translations T_{ij} we can prove the following:

$$M = \{h_0 + v \mid v \in V\} = \{T(h_0) \mid T \in \langle T_{01}, T_{12}, T_{23}, T_{34}, T_{40}\rangle\}, \tag{A.34}$$

where $V = \mathbb{Z}v_{01} + \mathbb{Z}v_{12} + \mathbb{Z}v_{23} + \mathbb{Z}v_{34} + \mathbb{Z}v_{40}$. Note that since $T_{v_{i\,i+1}} : h_{i+1} \mapsto h_i$, the following hold:

$$h_i \in M, \quad i = 0, \ldots, 4. \tag{A.35}$$

The translations on the weight lattice P spanned by V is given by

$$T_v(\lambda) = \lambda - \langle \delta^\vee, \lambda\rangle v, \quad \lambda \in P, \quad v \in V. \tag{A.36}$$

In this case, the translations $T_{i\,i+1}$, $i \in \mathbb{Z}/(5\mathbb{Z})$, can be expressed by

$$T_{01} = T_{v_{01}}, \quad T_{12} = T_{v_{12}}, \quad T_{23} = T_{v_{23}}, \quad T_{34} = T_{v_{34}}, \quad T_{40} = T_{v_{40}}, \tag{A.37}$$

and are called the fundamental translations in $\widetilde{W}(A_4^{(1)})$, that is, all translations in $\widetilde{W}(A_4^{(1)})$ can be expressed as the compositions of these translations. Note that the fundamental translations in $W(A_4^{(1)})$ can be expressed as the compositions of the fundamental translations in $\widetilde{W}(A_4^{(1)})$ as the following:

$$T_i = T_{i\,i+1} \circ T_{i-1,i}{}^{-1}, \quad i \in \mathbb{Z}/(5\mathbb{Z}). \tag{A.38}$$

For any $v = \sum_{i=0}^{4} c_i\alpha_i \in V$, where $c_i \in \mathbb{R}$, we define the squared length of T_v by

$$|T_v|^2 := \langle v^\vee, v\rangle = (c_0 - c_1)^2 + (c_1 - c_2)^2 + (c_2 - c_3)^2 + (c_3 - c_4)^2 + (c_4 - c_0)^2, \tag{A.39}$$

where $v^\vee = \sum_{i=0}^{4} c_i\alpha_i^\vee$. Note that we here extended the domain of the bilinear pairing $\langle\cdot,\cdot\rangle$ from $Q^\vee \times P$ to $\overline{Q}^\vee \times P$, where $\overline{Q}^\vee = \sum_{k=0}^{4} \mathbb{R}\alpha_k^\vee$. We can easily verify that the squared length of fundamental translations in $W(A_4^{(1)})$: T_i is 2, while the squared length of fundamental translations in $\widetilde{W}(A_4^{(1)})$: $T_{i\,i+1}$ is 4/5.

Appendix B General elliptic difference equations

In this section, we provide two generic elliptic difference equations. The first is Sakai's $A_0^{(1)}$-surface equation. This was re-expressed by Murata [29] as follows:

$$T^{(M)}:(f,g;t,b_1,\ldots,b_8)\mapsto\left(\overline{f},\overline{g};t+\frac{\delta}{2},b_1,\ldots,b_8\right), \tag{B.1}$$

where $\overline{f}$ and $\overline{g}$ are given by

$$\begin{aligned}&\det\left(v(f,g),v_1,\ldots,v_8,v_c\right)\det\left(v(\overline{f},g),\check{v}_1,\ldots,\check{v}_8,\check{v}_c\right)\\&\quad=P_+\,(f-f_c)(\overline{f}-\overline{f_c})\prod_{i=1}^{8}(g-g_i),\end{aligned} \tag{B.2a}$$

$$\begin{aligned}&\det\left(v(g,\overline{f}),\widehat{u}_1,\ldots,\widehat{u}_8,\widehat{u}_c\right)\det\left(v(\overline{g},\overline{f}),\overline{u}_1,\ldots,\overline{u}_8,\overline{u}_c\right)\\&\quad=\overline{P}_-\,(g-g_c)(\overline{g}-\overline{g}_c)\prod_{i=1}^{8}(\overline{f}-\overline{f}_i).\end{aligned} \tag{B.2b}$$

Here,

$$\delta=\sum_{k=1}^{8}b_k,\quad v_i=v(f_i,g_i),\quad \check{v}_i=v(\overline{f}_i,g_i),\quad \widehat{u}_i=v(g_i,\overline{f}_i),\quad \overline{u}_i=v(\overline{g}_i,\overline{f}_i), \tag{B.3}$$

for i=1,...,8,c, and

$$\begin{aligned}&f_i=\wp(t-b_i),\quad g_i=\wp(t+b_i),\quad i=1,\ldots,8,\\&f_c=\wp\left(t+\frac{t^2}{\delta}\right),\quad g_c=\wp\left(t-\frac{t^2}{\delta}\right).\end{aligned} \tag{B.4}$$

Moreover, $v(a,b)$ and $P_\pm$ are given by

$$v(a,b)=(ab^4,ab^3,ab^2,ab,a,b^4,b^3,b^2,b,1)^T,$$

$$P_\pm=\frac{\sigma(4t)^4\sigma(4t\pm\delta)^4}{\sigma\left(t\mp\frac{t^2}{\delta}\right)^{16}}\prod_{1\le i<j\le 8}\sigma(b_i-b_j)^2\prod_{i=1}^{8}\frac{\sigma\left(\frac{t^2}{\delta}-b_i\right)\sigma\left(2t\pm\frac{t^2}{\delta}\pm b_i\right)}{\sigma(t\pm b_i)^{14}\sigma(t\mp b_i)^2\sigma\left(t\mp b_i\pm\frac{\delta}{2}\right)^2},$$

where σ is the Weierstrass sigma function; see [9, Chapter 23]. Note that $\overline{P}_-=P_-|_{t\to\overline{t}}$. The above system was obtained by deducing translations on the lattice of type $E_8^{(1)}$. We note that this translation corresponds to NVs in the lattice.

The second case of an elliptic difference equation was found by Joshi and Nakazono [21]. It has a projective reduction to the RCG equation (1.1). The generic equation is given by

$$\begin{aligned}T^{(JN)}:&(x,y;c_1,\ldots,c_4,c_5,\ldots,c_8,\eta)\\&\mapsto(\overline{x},\overline{y};c_1-\lambda,\ldots,c_4-\lambda,c_5+\lambda,\ldots,c_8+\lambda,\eta+\lambda),\end{aligned} \tag{B.5}$$

where $\overline{x}$ and $\overline{y}$ are given by

$$\left(\frac{k\,\mathrm{cd}\,(\eta-c_8+\kappa)\,\overline{y}+1}{k\,\mathrm{cd}\,(\eta-c_7+\kappa)\,\overline{y}+1}\right)\left(\frac{\tilde{x}-\mathrm{cd}\,(\eta-c_7+\frac{c_{5678}}{2}+\lambda+\kappa)}{\tilde{x}-\mathrm{cd}\,(\eta-c_8+\frac{c_{5678}}{2}+\lambda+\kappa)}\right)$$
$$=G_{\frac{c_{5678}-2c_5+\lambda}{2},\frac{c_{5678}-2c_6+\lambda}{2},\frac{c_{5678}-2c_7+\lambda}{2},\frac{c_{5678}-2c_8+\lambda}{2},\eta+\frac{\lambda}{2}+\kappa}\frac{P_{\frac{c_{5678}-2c_5+\lambda}{2},\frac{c_{5678}-2c_6+\lambda}{2},\frac{c_{5678}-2c_7+\lambda}{2},\eta+\frac{\lambda}{2}+\kappa}(\tilde{x},\tilde{y})}{P_{\frac{c_{5678}-2c_5+\lambda}{2},\frac{c_{5678}-2c_6+\lambda}{2},\frac{c_{5678}-2c_8+\lambda}{2},\eta+\frac{\lambda}{2}+\kappa}(\tilde{x},\tilde{y})},\tag{B.6a}$$

$$\left(\frac{k\,\mathrm{cd}\,(\eta+c_4+\kappa)\,\overline{x}+1}{k\,\mathrm{cd}\,(\eta+c_3+\kappa)\,\overline{x}+1}\right)\left(\frac{k\,\mathrm{cd}\,(\eta-c_3+2\lambda+\kappa)\,\overline{y}+1}{k\,\mathrm{cd}\,(\eta-c_4+2\lambda+\kappa)\,\overline{y}+1}\right)$$
$$=G_{\eta-c_1+\frac{c_{1234}}{4}+\lambda,\eta-c_2+\frac{c_{1234}}{4}+\lambda,\eta-c_3+\frac{c_{1234}}{4}+\lambda,\eta-c_4+\frac{c_{1234}}{4}+\lambda,\frac{c_{5678}+2\lambda}{4}+\kappa}$$
$$\frac{P_{\eta-c_1+\frac{c_{1234}}{4}+\lambda,\eta-c_2+\frac{c_{1234}}{4}+\lambda,\eta-c_3+\frac{c_{1234}}{4}+\lambda,\frac{c_{5678}+2\lambda}{4}+\kappa}\left(\frac{-1}{k\overline{y}},\tilde{x}\right)}{P_{\eta-c_1+\frac{c_{1234}}{4}+\lambda,\eta-c_2+\frac{c_{1234}}{4}+\lambda,\eta-c_4+\frac{c_{1234}}{4}+\lambda,\frac{c_{5678}+2\lambda}{4}+\kappa}\left(\frac{-1}{k\overline{y}},\tilde{x}\right)},\tag{B.6b}$$

and $\tilde{x}$ and $\tilde{y}$ are given by

$$\left(\frac{k\,\mathrm{cd}\,(\eta+c_8-\frac{c_{5678}}{2})\,\tilde{y}+1}{k\,\mathrm{cd}\,(\eta+c_7-\frac{c_{5678}}{2})\,\tilde{y}+1}\right)\left(\frac{x-\mathrm{cd}\,(\eta+c_7)}{x-\mathrm{cd}\,(\eta+c_8)}\right)$$
$$=G_{c_5,c_6,c_7,c_8,\eta}\frac{P_{c_5,c_6,c_7,\eta}(x,y)}{P_{c_5,c_6,c_8,\eta}(x,y)},\tag{B.6c}$$

$$\left(\frac{k\,\mathrm{cd}\,(\eta-c_4+\frac{c_{1234}}{2})\,\tilde{x}+1}{k\,\mathrm{cd}\,(\eta-c_3+\frac{c_{1234}}{2})\,\tilde{x}+1}\right)\left(\frac{k\,\mathrm{cd}\,(\eta+c_3+\frac{c_{5678}}{2})\,\tilde{y}+1}{k\,\mathrm{cd}\,(\eta+c_4+\frac{c_{5678}}{2})\,\tilde{y}+1}\right)$$
$$=G_{\eta+c_1+\frac{c_{5678}}{4},\eta+c_2+\frac{c_{5678}}{4},\eta+c_3+\frac{c_{5678}}{4},\eta+c_4+\frac{c_{5678}}{4},\frac{c_{5678}}{4}}$$
$$\frac{P_{\eta+c_1+\frac{c_{5678}}{4},\eta+c_2+\frac{c_{5678}}{4},\eta+c_3+\frac{c_{5678}}{4},\frac{c_{5678}}{4}}\left(\frac{-1}{k\tilde{y}},x\right)}{P_{\eta+c_1+\frac{c_{5678}}{4},\eta+c_2+\frac{c_{5678}}{4},\eta+c_4+\frac{c_{5678}}{4},\frac{c_{5678}}{4}}\left(\frac{-1}{k\tilde{y}},x\right)}.\tag{B.6d}$$

Here, $\lambda=\sum_{k=1}^{8}c_k$, κ is defined by (3.3), $c_{j_1\cdots j_n}$ is the summation of the parameters c_i (see (5.7)), and the functions $G_{a_1,a_2,a_3,a_4,b}$, $Q_{a_1,a_2,a_3,a_4,a_5,b}(X)$ and $P_{a_1,a_2,a_3,b}(X,Y)$ are given by

$$G_{a_1,a_2,a_3,a_4,b}=\left(\frac{1-\frac{\mathrm{cd}\left(a_4+\frac{a_1+a_2}{2}\right)}{\mathrm{cd}\left(a_2+\frac{a_1+a_2}{2}\right)}}{1-\frac{\mathrm{cd}\left(a_3+\frac{a_1+a_2}{2}\right)}{\mathrm{cd}\left(a_2+\frac{a_1+a_2}{2}\right)}}\right)\left(\frac{1-\frac{\mathrm{cd}(b-a_4)}{\mathrm{cd}(b-a_1)}}{1-\frac{\mathrm{cd}(b-a_3)}{\mathrm{cd}(b-a_1)}}\right)$$
$$\left(\frac{1-\frac{\mathrm{cd}\left(b+a_4-\frac{a_1+a_2+a_3+a_4}{2}\right)}{\mathrm{cd}\left(b+a_2+\frac{a_1+a_2+a_3+a_4}{2}\right)}}{1-\frac{\mathrm{cd}\left(b+a_3-\frac{a_1+a_2+a_3+a_4}{2}\right)}{\mathrm{cd}\left(b+a_2+\frac{a_1+a_2+a_3+a_4}{2}\right)}}\right)\left(\frac{1-\frac{\mathrm{cd}\left(a_3+\frac{a_1+a_2}{2}\right)}{\mathrm{cd}\left(2b+a_2-\frac{a_1+a_2}{2}\right)}}{1-\frac{\mathrm{cd}\left(a_4+\frac{a_1+a_2}{2}\right)}{\mathrm{cd}\left(2b+a_2-\frac{a_1+a_2}{2}\right)}}\right),\tag{B.7}$$

$$
\begin{aligned}
&Q_{a_1,a_2,a_3,a_4,a_5,b}(X) \\
&= \left(\mathrm{cd}\left(b+a_3-\frac{a_5}{2}\right)-\mathrm{cd}\left(b+a_2+\frac{a_5}{2}\right)\right)\left(\mathrm{cd}\left(b+a_1+\frac{a_5}{2}\right)-\mathrm{cd}\left(b+a_4+\frac{a_5}{2}\right)\right) \\
&\left(\mathrm{cd}\,(b+a_4)\,\mathrm{cd}\,(b+a_1)+\mathrm{cd}\,(b+a_2)\,X\right)+\left(\mathrm{cd}\left(b+a_3-\frac{a_5}{2}\right)-\mathrm{cd}\left(b+a_1+\frac{a_5}{2}\right)\right) \\
&\quad \left(\mathrm{cd}\left(b+a_4+\frac{a_5}{2}\right)-\mathrm{cd}\left(b+a_2+\frac{a_5}{2}\right)\right)\left(\mathrm{cd}\,(b+a_4)\,\mathrm{cd}\,(b+a_2)+\mathrm{cd}\,(b+a_1)\,X\right) \\
&-\left(\mathrm{cd}\left(b+a_3-\frac{a_5}{2}\right)-\mathrm{cd}\left(b+a_4+\frac{a_5}{2}\right)\right)\left(\mathrm{cd}\left(b+a_1+\frac{a_5}{2}\right)-\mathrm{cd}\left(b+a_2+\frac{a_5}{2}\right)\right) \\
&\quad \left(\mathrm{cd}\,(b+a_1)\,\mathrm{cd}\,(b+a_2)+\mathrm{cd}\,(b+a_4)\,X\right), \qquad \text{(B.8)}
\end{aligned}
$$

$$
P_{a_1,a_2,a_3,b}(X,Y)=C_1XY+C_2X+C_3Y+C_4, \qquad \text{(B.9)}
$$

where

$$
\begin{aligned}
C_1 &= \Big(\mathrm{cd}\,(b-a_3)-\mathrm{cd}\,(b-a_2)\Big)\mathrm{cd}\,(b+a_1)+\Big(\mathrm{cd}\,(b-a_1)-\mathrm{cd}\,(b-a_3)\Big)\mathrm{cd}\,(b+a_2) \\
&\quad +\Big(\mathrm{cd}\,(b-a_2)-\mathrm{cd}\,(b-a_1)\Big)\mathrm{cd}\,(b+a_3), \\
C_2 &= \Big(\mathrm{cd}\,(b-a_2)-\mathrm{cd}\,(b-a_3)\Big)\mathrm{cd}\,(b-a_1)\,\mathrm{cd}\,(b+a_1) \\
&\quad +\Big(\mathrm{cd}\,(b-a_3)-\mathrm{cd}\,(b-a_1)\Big)\mathrm{cd}\,(b-a_2)\,\mathrm{cd}\,(b+a_2) \\
&\quad +\Big(\mathrm{cd}\,(b-a_1)-\mathrm{cd}\,(b-a_2)\Big)\mathrm{cd}\,(b-a_3)\,\mathrm{cd}\,(b+a_3), \\
C_3 &= \Big(\mathrm{cd}\,(b+a_3)-\mathrm{cd}\,(b+a_2)\Big)\mathrm{cd}\,(b-a_1)\,\mathrm{cd}\,(b+a_1) \\
&\quad +\Big(\mathrm{cd}\,(b+a_1)-\mathrm{cd}\,(b+a_3)\Big)\mathrm{cd}\,(b-a_2)\,\mathrm{cd}\,(b+a_2) \\
&\quad +\Big(\mathrm{cd}\,(b+a_2)-\mathrm{cd}\,(b+a_1)\Big)\mathrm{cd}\,(b-a_3)\,\mathrm{cd}\,(b+a_3), \\
C_4 &= \Big(\mathrm{cd}\,(b+a_2)\,\mathrm{cd}\,(b-a_3)-\mathrm{cd}\,(b-a_2)\,\mathrm{cd}\,(b+a_3)\Big)\mathrm{cd}\,(b-a_1)\,\mathrm{cd}\,(b+a_1) \\
&\quad +\Big(\mathrm{cd}\,(b+a_3)\,\mathrm{cd}\,(b-a_1)-\mathrm{cd}\,(b-a_3)\,\mathrm{cd}\,(b+a_1)\Big)\mathrm{cd}\,(b-a_2)\,\mathrm{cd}\,(b+a_2) \\
&\quad +\Big(\mathrm{cd}\,(b+a_1)\,\mathrm{cd}\,(b-a_2)-\mathrm{cd}\,(b-a_1)\,\mathrm{cd}\,(b+a_2)\Big)\mathrm{cd}\,(b-a_3)\,\mathrm{cd}\,(b+a_3).
\end{aligned}
$$

Similar to the earlier case of Sakai, this equation also arises from a translation on the lattice of type $E_8^{(1)}$. We note that this translation corresponds to NNVs in the lattice.

References

[1] Abramowitz M and Stegun I. *Handbook of Mathematical Functions.* Dover Books on Mathematics. Dover Publications, New York, N.Y., 2012.

[2] Adler V E. Bäcklund transformation for the Krichever-Novikov equation. *Internat. Math. Res. Notices,* (1):1–4, 1998.

[3] Adler V E, Bobenko A I, and Suris Y B. Classification of integrable equations on quad-graphs. The consistency approach. *Comm. Math. Phys.*, 233(3):513–543, 2003.

[4] Atkinson J, Howes P, Joshi N, and Nakazono N. Geometry of an elliptic difference equation related to Q4. *J. Lond. Math. Soc. (2)*, 93(3):763–784, 2016.

[5] Bourbaki, N. *Groupes et algèbres de Lie: Chapitres 4, 5 et 6*. Bourbaki, Nicolas. Springer, Berlin, Heidelberg, 2007.

[6] Carstea A S, Dzhamay A, and Takenawa T. Fiber-dependent deautonomization of integrable 2D mappings and discrete Painlevé equations. *J. Phys. A*, 50(40):405202, 41, 2017.

[7] Conway J, Sloane N, Bannai E, Borcherds R, Leech J, Norton S, Odlyzko A, Parker R, Queen L, and Venkov B. *Sphere Packings, Lattices and Groups*. Grundlehren der mathematischen Wissenschaften. Springer, New York, N.Y., 2013.

[8] Cresswell C, and Joshi N. The discrete first, second and thirty-fourth Painlevé hierarchies. *J. Phys. A: Mathematical and General*, 32(4):655, 1999.

[9] *NIST Digital Library of Mathematical Functions*. http://dlmf.nist.gov/, Release 1.0.14 of 2016-12-21. Olver F W J, Olde Daalhuis A B, Lozier D W, Schneider B I, Boisvert R F, Clark C W, Miller B R and Saunders B V, eds.

[10] Dolgachev I V. *Classical Algebraic Geometry*. Cambridge University Press, Cambridge, 2012; Dolgachev I and Ortland D. Point sets in projective spaces and theta functions. *Astérisque*, (165):210 pp. (1989), 1988; Looijenga E. Rational surfaces with an anticanonical cycle. *Ann. of Math. (2)*, 114(2):267–322, 1981.

[11] Fokas A S., Grammaticos B, and Ramani A. From continuous to discrete Painlevé equations, *J. Math. Anal. and Appl.*, 180(2):342–360, 1993.

[12] Fokas A S., Its A R., and Kitaev A V. Discrete Painlevé equations and their appearance in quantum gravity. *Communications in Mathematical Physics*, 142(2):313–344, 1991.

[13] Fuchs R. Sur quelques équations différentielles linéaires du second ordre. *Comptes Rendus de l'Academie des Sciences Paris*, 141:555–558, 1905.

[14] Gambier B. Sur les équations différentielles du second ordre et du premier degré dont l'intégrale générale est à points critiques fixes. *Acta Mathematica*, 33(1):1–55, 1910.

[15] Grammaticos B and Ramani A. Discrete Painlevé equations: a review. In *Discrete integrable systems*, volume 644 of *Lecture Notes in Phys.*, pages 245–321. Springer, Berlin, 2004.

[16] P. Griffiths. *Introduction to Algebraic Curves*, volume 76. American Mathematical Soc., Providence, RI, 1989.

[17] Hietarinta J. Searching for CAC-maps. *J. Nonlinear Math. Phys.*, 12(suppl. 2):223–230, 2005.

[18] Hietarinta J, Joshi N, and Nijhoff F W. *Discrete Systems and Integrability*. Cambridge Texts in Applied Mathematics. Cambridge University Press, Cambridge, 2016.

[19] Humphreys J. *Reflection Groups and Coxeter Groups*. Cambridge Studies in Advanced Mathematics. Cambridge University Press, Cambridge, 1992.

[20] Joshi N, Grammaticos B, Tamizhmani T, and Ramani A. From integrable lattices to non-QRT mappings. *Lett. Math. Phys.*, 78(1):27–37, 2006.

[21] Joshi N and Nakazono N. Elliptic Painlevé equations from next-nearest-neighbor translations on the $E_8^{(1)}$ lattice. *J. Phys. A*, 50(30):305205, 17, 2017.

[22] Kajiwara K, Masuda T, Noumi M, Ohta Y, and Yamada Y. ${}_{10}E_9$ solution to the elliptic Painlevé equation. *J. Phys. A*, 36(17):L263–L272, 2003.

[23] Kajiwara K and Nakazono N. Hypergeometric solutions to the symmetric q-Painlevé equations. *Int. Math. Res. Not.*, (4):1101–1140, 2015.

[24] Kajiwara K, Nakazono N, and Tsuda T. Projective reduction of the discrete Painlevé system of type $(A_2 + A_1)^{(1)}$. *Int. Math. Res. Not.*, (4):930–966, 2011.

[25] Kajiwara K, Noumi M, and Yamada Y. Geometric aspects of Painlevé equations. *J. Phys. A*, 50(7):073001, 164, 2017.

[26] Laguerre E. Sur la réduction en fractions continues d'une fraction qui satisfait à une équation différentielle linéaire du premier ordre dont les coefficients sont rationnels. *Journal de Mathématiques Pures et Appliquées*, 1:135–166, 1885.

[27] Magnus, A. Painlevé-type differential equations for the recurrence coefficients of semi-classical orthogonal polynomials. *J. of computational and applied mathematics.* 57(1):215–237, 1995.

[28] Milnor, J. *Foliations and Foliated Vector Bundles.* Institute for Advanced Study, Princeton, N.J., 1970.

[29] Murata, M. New expressions for discrete Painlevé equations. *Funkcial. Ekvac.*, 47(2):291–305, 2004.

[30] Murata M, Sakai H, and Yoneda J. Riccati solutions of discrete Painlevé equations with Weyl group symmetry of type $E_8^{(1)}$. *J. Math. Phys.*, 44(3):1396–1414, 2003.

[31] M. Noumi. *Painlevé Equations through Symmetry.* Translations of mathematical monographs. American Mathematical Society, Providence, R.I., 2004.

[32] Ohta Y, Ramani A, and Grammaticos B. An affine Weyl group approach to the eight-parameter discrete Painlevé equation. *J. Phys. A*, 34(48):10523–10532, 2001. Symmetries and integrability of difference equations (Tokyo, 2000).

[33] Okamoto K. Sur les feuilletages associés aux équations du second ordre à points critiques fixes de P. Painlevé. *Japan. J. Math. (N.S.)*, 5(1):1–79, 1979.

[34] Painlevé P. Sur les équations différentielles du second ordre et d'ordre supérieur dont l'intégrale générale est uniforme. *Acta Mathematica*, 25(1):1–85, 1902.

[35] Periwal V and Shevitz D. Unitary-matrix models as exactly solvable string theories. *Phys. Rev. Lett.*, 64(12):1326–1329, 1990.

[36] Ramani A, Carstea A S, and Grammaticos B. On the non-autonomous form of the Q_4 mapping and its relation to elliptic Painlevé equations. *J. Phys. A*, 42(32):322003, 8, 2009.

[37] Ramani A, Grammaticos B, and Hietarinta J. Discrete versions of the Painlevé equations. *Phys. Rev. Lett.*, 67(14):1829–1832, 1991.

[38] Sakai H. Rational surfaces associated with affine root systems and geometry of the Painlevé equations. *Comm. Math. Phys.*, 220(1):165–229, 2001.

[39] Shohat J. A differential equation for orthogonal polynomials. *Duke Mathematical Journal*, 5(2):401–417, 1939.

[40] Whittaker E and Watson G. *A Course of Modern Analysis.* Cambridge University Press, Cambridge, 1996.

B5. Linkage mechanisms governed by integrable deformations of discrete space curves

Shizuo Kaji [a], *Kenji Kajiwara* [a], *and Hyeongki Park* [b]

[a] *Institute of Mathematics for Industry, Kyushu University*
744 Motooka, Nishi-ku, Fukuoka, Japan, 819-0395
[b] *Graduate School of Mathematics, Kyushu University*
744 Motooka, Nishi-ku, Fukuoka, Japan, 819-0395

Abstract

A *linkage mechanism* consists of rigid bodies assembled by joints which can be used to translate and transfer motion from one form in one place to another. In this chapter, we are particularly interested in a family of spatial linkage mechanisms which consist of n-copies of a rigid body joined together by hinges to form a ring. Each hinge joint has its own axis of revolution and rigid bodies joined to it can be freely rotated around the axis. The family includes the famous threefold symmetric Bricard 6R linkage, also known as the Kaleidocycle, which exhibits a characteristic "turning-over" motion. We can model such a linkage as a discrete closed curve in $\mathbb{R}^3$ of constant torsion up to sign. Then, its motion is described as the deformation of the curve preserving torsion and arc length. We describe certain motions of this object that are governed by the semi-discrete mKdV and sine-Gordon equations, where infinitesimally the motion of each vertex is confined in the osculating plane.

1 Introduction

A linkage is a mechanical system consisting of rigid bodies (called *links*) joined together by *joints*. They are used to transform one motion to another as in the famous Watt parallel motion and a lot of examples are found in engineering as well as in natural creatures (see, for example, [7]).

Mathematical study of linkage dates back to Euler, Chebyshev, Sylvester, Kempe, and Cayley and since then the topology and the geometry of the configuration space have attracted many researchers (see [12, 24, 31] for a survey). Most of the research focuses on *pin joint linkages*, which consist of only one type of joint called pin joints. A pin joint constrains the positions of ends of adjacent links to stay together. To a pin joint linkage we can associate a graph whose vertices are joints and edges are links, where edges are assigned its length. The state of a pin joint linkage is effectively specified by the coordinates of the joint positions, where the distance of two joints connected by a link is constrained to its length. Thus, its configuration space can be modelled by the space of isometric imbeddings of the corresponding graph to some Euclidean space. Note that in practice, joints and links have sizes and they collide to have limited mobility, but here we consider ideal linkages with which joints and links can pass through each other.

While the configuration spaces of (especially planar) pin joint linkages are well studied, there are other types of linkages which are not so popular. In this chapter, we are mainly interested in linkages consisting of hinges (revolute joints). To set up a framework to study linkages with various types of joints, we first introduce a mathematical model of general linkages as graphs decorated with groups (§2.1), extending previous approaches (see [34] and references therein). This formulation can be viewed as a special type of constraint network (e.g., [14]). Then in §2.2, we focus on linkages consisting of hinges. Unlike a pin joint which constrains only the relative positions of connected links, a hinge has an axis so that it also constrains the relative orientation of connected links.

We are particularly interested in a simple case when n links in $\mathbb{R}^3$ are joined by hinges to form a circle (§3). Such a linkage can be roughly thought of as a discrete closed space curve, where hinge axes are identified with the lines spanned by the binormal vectors. Properties of such linkages can thus be translated and stated in the language of discrete curves. An example of such linkage is the threefold symmetric Bricard 6R linkage consisting of six hinges (Fig. 1), which exhibits a turning-over motion and has the configuration space homeomorphic to a circle. As a generalisation to the threefold symmetric Bricard 6R linkage, we consider a family of linkages consisting of copies of an identical links connected by hinges, which we call *Kaleidocycles*, and they are characterised as discrete curves of constant speed and constant torsion.

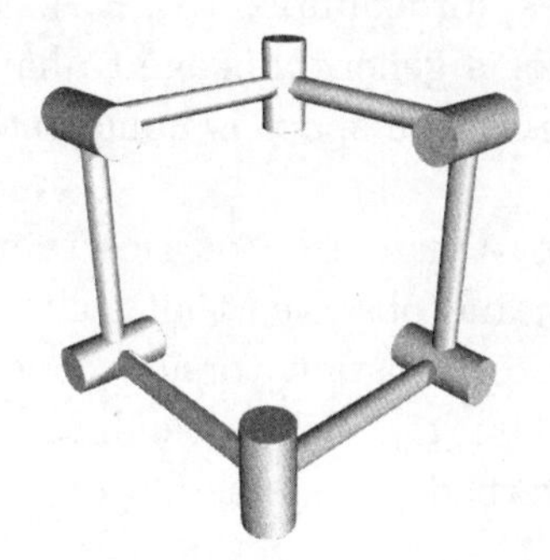

Figure 1. Threefold symmetric Bricard 6R linkage.

The theory of discrete space curves has been studied by many authors. The simplest way to discretise a space curve is by a polygon, that is, an ordered sequence of points $\gamma = (\gamma_0, \gamma_1, \ldots), \gamma_i \in \mathbb{R}^3$. Deformation of a curve is a time-parametrised sequence of curves $\gamma(t)$, where t runs through (an interval of) the real numbers. Deformations of a given smooth/discrete space curve can be described by introducing an appropriate frame such as the Frenet frame, which satisfies the system of linear partial differential/differential-difference equations. The compatibility condition gives rise to nonlinear partial differential/differential-difference equation(s), which are often integrable. It is sometimes possible to construct deformations for the space curves using integrable systems which preserve some geometric properties of the space curve such as length, curvature, and torsion. For example, a deformation is said to be *isoperimetric* if the deformation preserves the arc length. In this case, the modified Korteweg-de Vries (mKdV) or the nonlinear Schrödinger (NLS) equation and their hierarchies naturally arise as the compatibility condition [6, 10, 15, 18, 27, 28, 37, 41]. Various continuous deformations for the discrete space curves have been studied in [11, 16, 18, 36, 38], where the deformations are described by the differential-difference analogue of the mKdV and the NLS equations.

The motion of Kaleidocycles corresponds to isoperimetric and torsion-preserving deformation of discrete closed space curves of constant torsion. In §4, we define a flow on the configuration space of a Kaleidocycle by the differential-difference analogue of the mKdV and the sine-Gordon equations (*semi-discrete mKdV and sine-Gordon equations*). This flow generates the characteristic turning-over motion of the Kaleidocycle.

Kaleidocycles exhibit interesting properties and pose some topological and geometrical questions. In §5 we indicate some directions of further study to close this exposition.

We list some more preceding works in different fields which are relevant to our topic in some ways.

Mobility analysis of a linkage mechanism studies how many degrees of freedom a particular state of the linkage has, which corresponds to determination of the local dimension at a point in the configuration space (see, for example, [35]). On the other hand, rigidity of linkages consisting of hinges are studied in the context of the body-hinge framework (see, for example, [20, 25]). The main focus of the study is to give a characterisation for a generic linkage to have no mobility. That is, the question is to see if the configuration space is homeomorphic to a point or isolated points.

Sato and Tanaka [42] study the motion of a certain linkage mechanism with a constrained degree of freedom and observed that soliton solutions appear.

Closed (continuous) curves of constant torsion have attracted sporadic interest of geometers, e.g., [1, 5, 19, 45, 46]. In particular, [6] discusses an evolution of a constant torsion curve governed by a sine-Gordon equation in the continuous setting.

2 A mathematical model of linkage

The purpose of this section is to set up a general mathematical model of linkages. This section is almost independent of later sections, and can be skipped if the reader is concerned only with our main results on the motion of Kaleidocycles.

2.1 A group theoretic model of linkage

We define an abstract linkage as a decorated graph, and its realisation as a certain imbedding of the graph in a Euclidean space. Our definition generalises the usual graphical model of a pin joint linkage to allow different types of joints.

Denote by $SO(n)$ the group of orientation preserving linear isometries of the n-dimensional Euclidean space $\mathbb{R}^n$. An element of $SO(n)$ is identified with a sequence of n-dimensional column vectors $[f_1, f_2, \ldots, f_n]$ which are mutually orthogonal and have unit length with respect to the standard inner product $\langle x, y\rangle$ of $x, y \in \mathbb{R}^n$. Denote by $SE(n)$ the group of n-dimensional orientation preserving Euclidean transformations. That is, it consists of the affine transformations $\mathbb{R}^n \to \mathbb{R}^n$ which preserves the orientation and the standard metric. We represent the elements of $SE(n)$ by $(n+1) \times (n+1)-$homogeneous matrices acting on

$$\mathbb{R}^n \simeq \{{}^t(x_1, x_2, \ldots, x_n, 1) \in \mathbb{R}^{n+1}\}$$

by multiplication from the left. For example, an element of $SE(3)$ is represented by a matrix

$$\begin{pmatrix} a_{11} & a_{12} & a_{13} & l_1 \\ a_{21} & a_{22} & a_{23} & l_2 \\ a_{31} & a_{32} & a_{33} & l_3 \\ 0 & 0 & 0 & 1 \end{pmatrix}.$$

The vector $l = {}^t(l_1, l_2, l_3)$ is called the translation part. The upper-left 3×3-block of A is called the linear part and denoted by $\bar{A} \in SO(3)$. Thus, the action on $v \in \mathbb{R}^3$ is also written by $v \mapsto \bar{A}v + l$.

Definition 1. An n-dimensional *abstract linkage* L consists of the following data:

- a connected oriented finite graph $G = (V, E)$
- a subgroup $J_v \subset SE(n)$ assigned to each $v \in V$, which defines the joint symmetry
- an element $C_e \in SE(n)$ assigned to each $e \in E$, which defines the link constraint.

In practical applications, we are interested in the case when $n = 2$ or 3. When $n = 2$ linkages are said to be *planar*, and when $n = 3$ linkages are said to be *spatial.*

We say a linkage L is *homogeneous* if for any pair $v_1, v_2 \in V$, the following conditions are satisfied:

- there exists a graph automorphism which maps v_1 to v_2 (i.e., $Aut(G)$ acts transitively on V),
- $J_{v_1} = J_{v_2}$,
- and $C_{e_1} = C_{e_2}$ for any $e_1, e_2 \in E$.

A *state* or *realisation* ϕ of an abstract linkage L is an assignment of a coset to each vertex

$$\phi : v \mapsto SE(n)/J_v$$

such that for each edge $e = (v_1, v_2) \in E$, the following condition is satisfied:

$$\phi(v_2)J_{v_2} \cap \phi(v_1)J_{v_1}C_e \neq \emptyset, \tag{2.1}$$

where cosets are identified with subsets of $SE(n)$.

Let us give an intuitive description of (2.1). Imagine a reference joint sitting at the origin in a reference orientation. The subset $\phi(v_1)J_{v_1}$ consists of all the rigid transformations which maps the reference joint to the joint at v_1 with a specified position and an orientation $\phi(v_1)$ up to the joint symmetry J_{v_1}. The two subsets $\phi(v_2)J_{v_2}$ and $\phi(v_1)J_{v_1}C_e$ intersects if and only if the joint at v_1 can be aligned to that at v_2 by the transformation C_e.

Example 1. The usual pin joints v_1, v_2 connected by a bar-shaped link e of length l are represented by $J_{v_1} = J_{v_2} = SO(n)$ and C_e being any translation by l. Note that $SE(3)/J_{v_1} \simeq \mathbb{R}^3$. It is easy to see that (2.1) amounts to saying the difference in the translation part of $\phi(v_2)$ and $\phi(v_1)$ should have the norm equal to l.

Two revolute joints (hinges) v_1, v_2 in $\mathbb{R}^3$ connected by a link e of length l making an angle α are represented by $J_{v_1} = J_{v_2}$ being the group generated by rotations around the z-axis and the π-rotation around the x-axis, and C_e being the rotation by α around x-axis followed by the translation along x-axis by l; that is

$$J_{v_1} = J_{v_2} = \left\{ \left. \begin{pmatrix} \cos\theta & \mp\sin\theta & 0 & 0 \\ \sin\theta & \pm\cos\theta & 0 & 0 \\ 0 & 0 & \pm 1 & 0 \\ 0 & 0 & 0 & 1 \end{pmatrix} \right| \theta \in \mathbb{R} \right\}, \quad C_e = \begin{pmatrix} 1 & 0 & 0 & l \\ 0 & \cos\alpha & -\sin\alpha & 0 \\ 0 & \sin\alpha & \cos\alpha & 0 \\ 0 & 0 & 0 & 1 \end{pmatrix}.$$

Note that $SE(3)/J_{v_1}$ is the space of based lines (i.e., lines with specified origins) in $\mathbb{R}^3$, and the line is identified with the axis of the hinge.

The space $\overline{\mathcal{C}}(L)$ of all realisations of a given linkage L admits an action of $SE(n)$ defined by $\phi \mapsto g\phi(v)$ for $g \in SE(n)$. The quotient of $\overline{\mathcal{C}}(L)$ by $SE(n)$ is denoted by $\mathcal{C}(L)$ and called the *configuration space* of L. Each connected component of $\mathcal{C}(L)$ corresponds to the mobility of the linkage L in a certain state. When a connected component is a manifold, its dimension is what mechanists call *the (internal) degrees of freedom* (DOF, for short). Given a pair of points on $\mathcal{C}(L)$, the problem of finding an explicit path connecting the points is called *motion planning* and has been one of the main topics in mechanics [30]. In a similar manner, many questions about a linkage can be phrased in terms of the topology and the geometry of its configuration space.

Example 2. Consider the following spatial linkages consisting of pin joints depicted in Figure 2. In the latter, we assume the two joints a and b are fixed to the wall. Up to the action of the global rigid transformation $SE(3)$, these two linkages are equivalent and share the same configuration space $\mathcal{C}(L)$; in the left linkage, the global action is killed by fixing the positions of three joints except for p. The topology of $\mathcal{C}(L)$ changes with respect to the parameter l which is the length of the bars. Namely, we have

$$\mathcal{C}(L) = \{x_p \in \mathbb{R}^3 \mid |x_p - x_a|^2 = |x_p - x_b|^2 = l^2\} = \begin{cases} S^1 & (l > 2h) \\ pt & (l = 2h) \\ \emptyset & (l < 2h) \end{cases}.$$

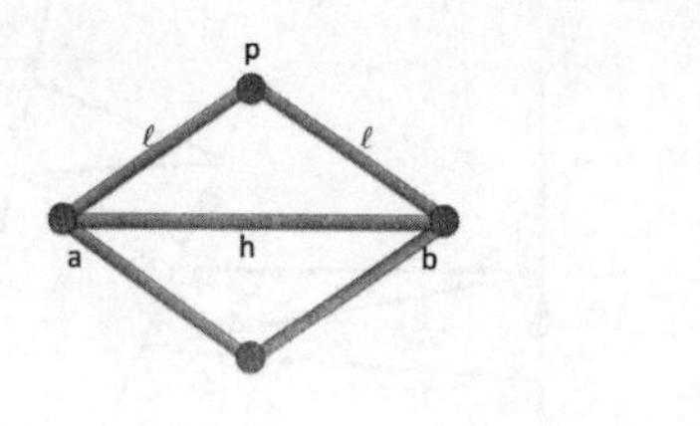

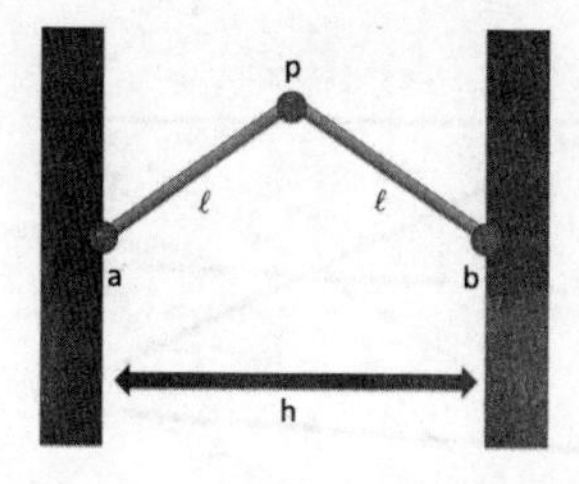

Figure 2. Example of equivalent pin joint linkages.

This seemingly trivial example is indeed related to a deeper and subtle question on the topology of the configuration space; the space is identified with the *real* solutions to a system of algebraic equations.

2.2 Hinged linkage in three space

Now, we focus on a class of spatial linkages consisting of hinges, known also as three dimensional body-hinge frameworks [20]. In this case, the definition in the previous section can be reduced to a simpler form.

Notice that in $\mathbb{R}^3$ a pair of hinges connected by a link can be modelled by a tetrahedron. A hinge is an isometrically embedded real line in $\mathbb{R}^3$. Given a pair of hinges, unit-length segments on the hinges containing the base points in the centre span a tetrahedron, or a quadrilateral when the two hinges are parallel (see Fig. 3 Left). It is sometimes convenient to decompose the link constraint $C_{(v_1,v_2)} \in SE(3)$ into three parts; a translation along the hinge direction at v_1, a screw motion along an axis perpendicular to both hinges, and a translation along the hinge direction at v_2. This corresponds to a common presentation among mechanists called the Denavit–Hartenberg parameters [9]. We can find the decomposition geometrically as follows: Find a line segment which is perpendicular to both hinges connected by the link e, which we call the *core segment*. It is unique unless the hinges are parallel. The intersection points of the core segment and the hinges are called the *marked* points. Form a tetrahedron from the line segments on hinges containing the marked points in the centre. By construction, this tetrahedron has a special shape that the line connecting the centre of two hinge edges (the core segment) is perpendicular to the hinge edges. Such a tetrahedron is called a *disphenoid*. The shape of the disphenoid defines a screw motion along the core segment up to a π-rotation. The translations along the hinge directions are to match the marked points to the base points (see Fig. 3). To sum up, a spatial hinged linkage can be considered as a collection of lines connected by disphenoids at marked points. Thus, we arrive in the following definition.

Definition 2. A *hinged network* consists of

- a connected oriented finite graph $G = (V, E)$,
- two edge labels $\nu : E \to [0, \pi)$ called the *torsion angle* and $\varepsilon : E \to \mathbb{R}_{\geq 0}$ called the *segment length*,

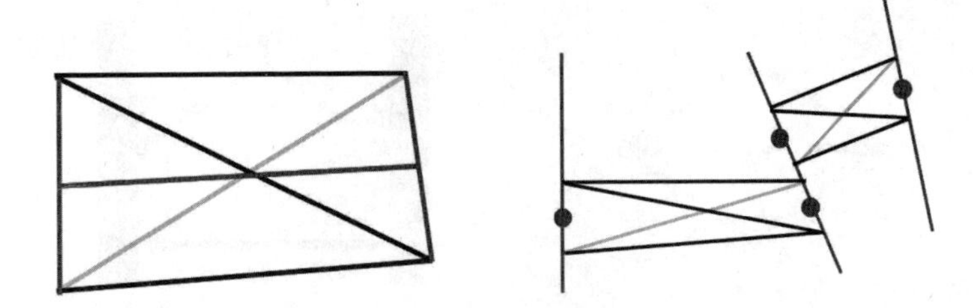

Figure 3. Left: a disphenoid formed by two hinge edges, Right: three hinges connected by disphenoids. The dots indicate the marked points.

- and a vertex label $\iota_v : E(v) \to \mathbb{R}$ called the *marking*, where $E(v) \subset E$ is the set of edges adjacent to $v \in V$.

A state of a hinged network is an assignment to each vertex $v \in V$ of an isometric embedding $h_v : \mathbb{R} \to \mathbb{R}^3$ such that for any $(v_1, v_2) \in E$

1. $|l| = \varepsilon(v_1, v_2)$, where $l = h_{v_1} \circ \iota_{v_1}(v_1, v_2) - h_{v_2} \circ \iota_{v_2}(v_1, v_2)$
2. $l \perp h_{v_1}(\mathbb{R})$ and $l \perp h_{v_2}(\mathbb{R})$
3. $\angle h_{v_1}(\mathbb{R}) h_{v_2}(\mathbb{R}) = \nu$, where the angle is measured in the left-hand screw manner with respect to l.

Intuitively, $h_v(\mathbb{R})$ is the line spanned by the hinges, and the first two conditions demand that the marked points are connected by the core segments l, whereas the last condition dictates the torsion angle of adjacent hinges $h_{v_1}(\mathbb{R})$ and $h_{v_2}(\mathbb{R})$.

A hinged network is said to be *serial* when the graph G is a line graph; i.e., a connected graph of the shape $\bullet \to \bullet \to \bullet \to \cdots \to \bullet$. It is said to be *closed* when the graph G is a circle graph; i.e., a connected finite graph with every vertex having outgoing degree one and incoming degree one. A hinged network is homogeneous if

- $Aut(G)$ acts on G transitively,
- $\nu(e), \varepsilon(e)$, and ι_v do not depend on $e \in E$ and $v \in V$. That is, it is made of congruent tetrahedral links.

Example 3. A planar pin joint linkage is a special type of hinged network with $\nu(e) = 0$ for all $e \in E$ and $\iota_v = 0$ for all $v \in V$. That is, all hinges are parallel and marked points are all at the origin. On the other hand, any hinged network can be thought of as a spatial pin joint linkage by replacing every tetrahedral link with four bar links connected by four pin joints forming the tetrahedron. Therefore, hinged networks form an intermediate class of linkages which sits between planar pin joint linkages and spatial pin joint linkages.

Example 4. The hinged network depicted in Fig. 5 is over the wedge sum of two circle graphs. It exhibits a jump roping motion. A similar but more complex network is found in [7, §6].

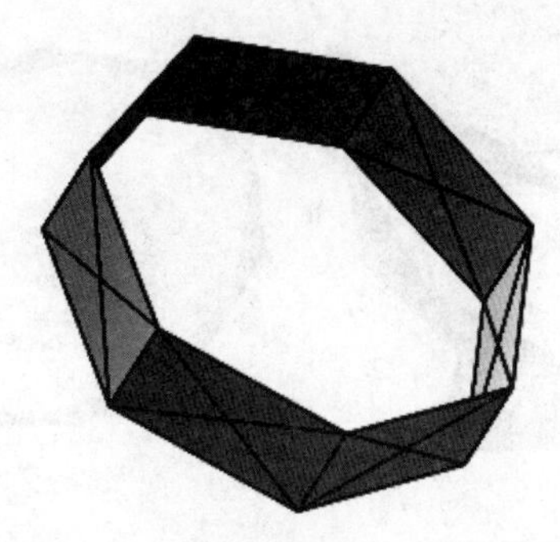

Figure 4. A degenerate hinged network over a circle corresponding to a planar six-bar pin joint linkage.

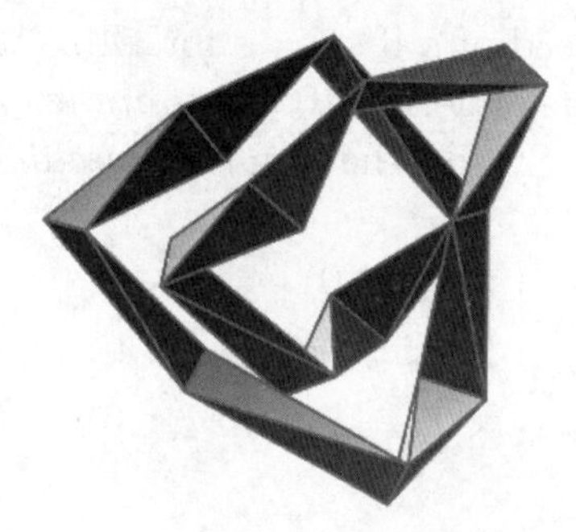

Figure 5. A hinged network over the wedge of two circles.

Example 5. Closed hinged networks with $\varepsilon(e) = 0$ (that is, adjacent hinge lines intersect) for all $e \in E$ provide a linkage model for discrete *developable strips* studied recently by K. Naokawa and C. Müller (see Fig. 6). They are made of (planar) quadrilaterals joined together by the pair of non-adjacent edges as hinges.

3 Hinged network and discrete space curve

In this section, we describe a connection between spatial closed hinged networks and discrete closed space curves. This connection is the key idea of this chapter which provides a way to study certain linkages using tools in discrete differential geometry.

First, we briefly review the basic formulation of discrete space curves (see, for example, [18]). A *discrete space curve* is a map

$$\gamma : \mathbb{Z} \to \mathbb{R}^3, \quad (i \mapsto \gamma_n).$$

For simplicity, in this chapter we always assume that $\gamma_n \neq \gamma_{n+1}$ for any n and that three points γ_{-1}, γ_0 and γ_1 are not colinear. The *tangent vector* $T : \mathbb{Z} \to S^2$ is defined by

$$T_n = \frac{\gamma_{n+1} - \gamma_n}{\varepsilon_n}, \quad \varepsilon_n = |\gamma_{n+1} - \gamma_n|. \tag{3.1}$$

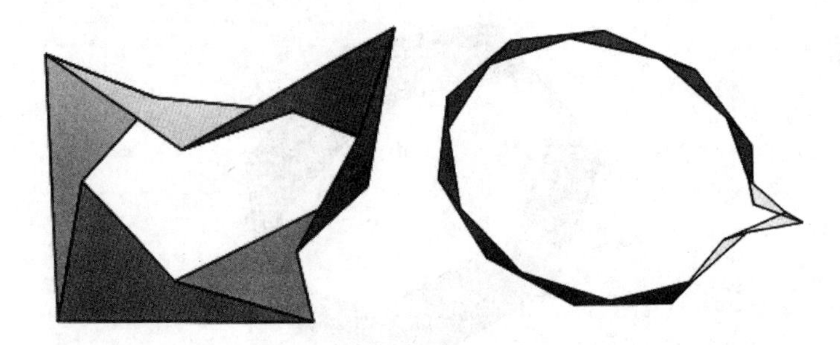

Figure 6. Developable discrete Möbius strip consisting of 6 (respectively 12) congruent quadrilateral links.

We say γ has a constant speed of ε if $\varepsilon_n = \varepsilon$ for all n. A discrete space curve with a constant speed is sometimes referred to as an *arc length parametrised curve* [17]. The *normal vector* $N : \mathbb{Z} \to S^2$ and the *binormal vector* $B : \mathbb{Z} \to S^2$ are defined by

$$B_n = \begin{cases} \frac{T_{n-1}\times T_n}{|T_{n-1}\times T_n|} & (T_{n-1} \times T_n \neq 0) \\ B_{n-1} & (T_{n-1} \times T_n = 0 \text{ and } n > 0) \\ B_{n+1} & (T_{n-1} \times T_n = 0 \text{ and } n < 0), \end{cases} \tag{3.2}$$

$$N_n = B_n \times T_n, \tag{3.3}$$

respectively. Then, $[T_n, N_n, B_n] \in SO(3)$ is called the *Frenet frame* of γ. For our purpose, it is more convenient to use a modified version of the ordinary Frenet frame, which we define as follows. Set $b_0 = B_0$ and define $b_n = \pm B_n$ recursively so that $\langle b_n \times b_{n-1}, T_{n-1}\rangle \geq 0$ and $\langle b_{n-1}, b_n\rangle \neq -1$. Then, $\Phi_n = [T_n, \widetilde{N}_n, b_n] \in SO(3)$, where $\widetilde{N}_n = b_n \times T_n$ (see Fig. 7).

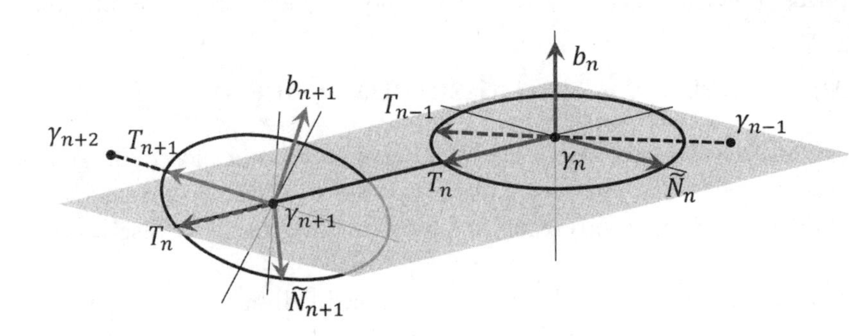

Figure 7. A discrete space curve with the frame Φ_n.

For $\theta \in \mathbb{R}$, we define $R_1(\theta), R_3(\theta) \in SO(3)$ by

$$R_1(\theta) = \begin{bmatrix} 1 & 0 & 0 \\ 0 & \cos\theta & -\sin\theta \\ 0 & \sin\theta & \cos\theta \end{bmatrix}, \quad R_3(x) = \begin{bmatrix} \cos\theta & -\sin\theta & 0 \\ \sin\theta & \cos\theta & 0 \\ 0 & 0 & 1 \end{bmatrix}. \tag{3.4}$$

There exist $\kappa : \mathbb{Z} \to [-\pi, \pi)$ and $\nu : \mathbb{Z} \to [0, \pi)$ such that

$$\Phi_{n+1} = \Phi_n L_n, \quad L_n = R_1(-\nu_{n+1}) R_3(\kappa_{n+1}). \tag{3.5}$$

We call κ the *signed curvature angle* and ν the *torsion angle*. Fig. 8 illustrates how to obtain Φ_{n-1} from Φ_n by (3.5). Note that we have

$$\begin{gathered} \langle T_n, T_{n-1}\rangle = \cos\kappa_n, \quad \langle b_n, b_{n-1}\rangle = \cos\nu_n, \quad \langle b_n, \widetilde{N}_{n-1}\rangle = \sin\nu_n, \\ \langle b_n, T_n\rangle = \langle b_{n+1}, T_n\rangle = 0. \end{gathered} \tag{3.6}$$

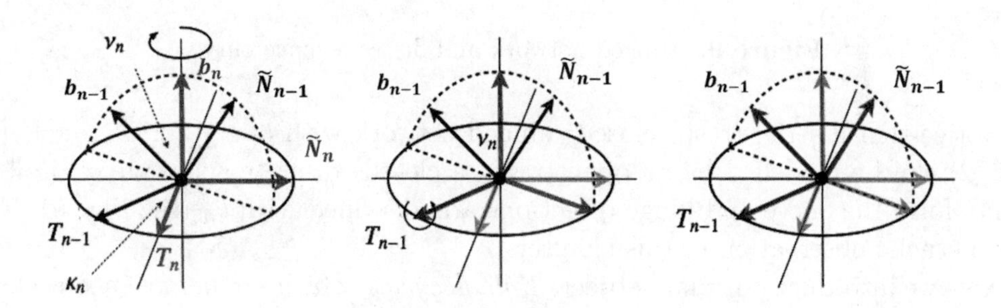

Figure 8. The curvature angle κ and the torsion angle ν.

The reason why we introduce the modified frame is that the ordinary Frenet frame behaves discontinuously under deformation when the ordinary curvature angle vanishes at a point. During the turning-over motion of a Kaleidocycle, it goes through such a state at some points, and the above modified frame behaves consistently even under the situation.

Fix a natural number N. A discrete space curve γ is said to be *closed* of length N if $\gamma_{n+kN} = \gamma_n$ for any $k \in \mathbb{Z}$. Unlike the ordinary Frenet frame, closedness does not imply $\Phi_{n+kN} = \Phi_n$ but they may differ by rotation by π around T_n. We say b is oriented (resp. anti-oriented) if $b_n = b_{n+N}$ (resp. $b_n = -b_{n+N}$) for all n.

We can consider a discrete version of the Darboux form [8, 46], which gives a correspondence between spherical curves and space curves. Given $b : \mathbb{Z} \to S^2$ with $b_n \times b_{n-1} \neq 0$ for all n and $\varepsilon : \mathbb{Z} \to \mathbb{R}_{\geq 0}$, we can associate a discrete space curve satisfying

$$\gamma_0 = 0, \quad \gamma_n = \gamma_{n-1} + \varepsilon_{n-1} \frac{b_n \times b_{n-1}}{|b_n \times b_{n-1}|}, \tag{3.7}$$

which we denote by $\gamma^{b,\varepsilon}$. The curve $\gamma^{b,\varepsilon}$ is closed of length N if

$$\sum_{n=0}^{N-1} \left(\varepsilon_{k+n} \frac{b_{k+n+1} \times b_{k+n}}{|b_{k+n+1} \times b_{k+n}|} \right) = 0 \tag{3.8}$$

for all k.

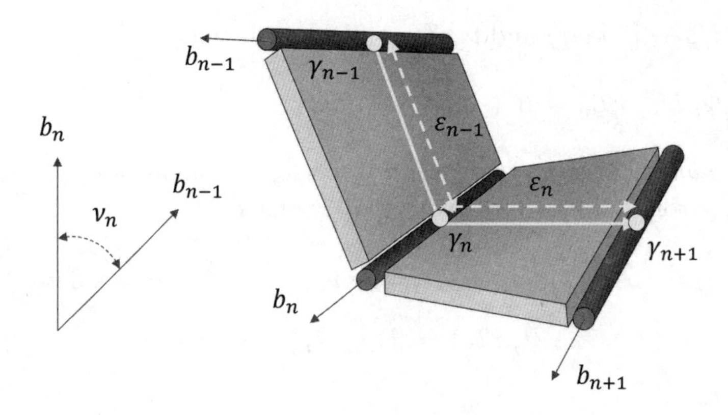

Figure 9. Hinged network and discrete space curve.

Notice that a serial (resp. closed) hinged network with $\iota_v = 0$ for all $v \in V$ (see Def. 2) can be modelled by an open (resp. a closed) discrete space curve; its base points form the curve and hinge directions are identified with b_n (see Fig. 9). This is the crucial observation of this chapter.

Now we introduce our main object, *Kaleidocycles*, which are homogeneous closed hinged networks. We model them as constant speed discrete space curves of constant torsion. They are a generalisation to a popular paper toy called the Kaleidocycle (see, e.g., [4, 43]). A serial hinged network similar to our Kaleidocycle is proposed in [33].

Definition 3. Fix $\nu \in [0, \pi]$ and $\epsilon > 0$. An N-Kaleidocycle with a speed ε and a torsion angle ν is a closed discrete space curve γ of length N which has a constant speed $\varepsilon_n = \varepsilon$ and a constant torsion angle $\nu_n = \nu$. It is said to be *oriented* (resp. *anti-oriented*) when associated b is oriented (resp. anti-oriented).

When ν is either 0 or π, the corresponding Kaleidocycles are planar, and we call them *degenerate*. For fixed N and ε, an oriented (resp. anti-oriented) non-

Figure 10. Left: anti-oriented Kaleidocycle with $N = 9$. Right: a Kaleidocycle with a knotted topology.

degenerate Kaleidocycle with a torsion angle ν is determined by the Darboux form $\gamma^{b,\varepsilon}$ by a map $b : \mathbb{Z} \to S^2$ satisfying p

- $b_{n+N} = b_n$ (resp. $b_{n+N} = -b_n$),
- $\langle b_n, b_{n+1} \rangle = \cos\nu$,
- $\sum_{n=0}^{N-1} b_{n+1} \times b_n = 0$.

We use b and γ interchangeably to represent a Kaleidocycle.

Consider the real algebraic variety $\overline{\mathcal{M}}_N$ defined by the following system of quadratic equations ([44, Ex. 5.2, 8.13]):

$$\langle b_n, b_{n+1} \rangle = c \quad (0 \le n < N), \qquad \sum_{n=0}^{N-1} b_{n+1} \times b_n = 0, \qquad b_N = \pm b_0, \tag{3.9}$$

where c is considered as an indeterminate. The orthogonal group $O(3)$ acts on b_i's in the standard way, and hence, on $\overline{\mathcal{M}}_N$. Denote by $\mathcal{M}_N$ the quotient of $\overline{\mathcal{M}}_N$ by the action of $O(3)$. The variety $\mathcal{M}_N$ serves as the configuration space of all N-Kaleidocycles with varying $c = \cos\nu$. It decomposes into two disjoint sub-spaces $\mathcal{M}_N^+$ consisting of all oriented Kaleidocycles ($b_N = b_0$) and $\mathcal{M}_N^-$ consisting of anti-oriented ones ($b_N = -b_0$).

As $\mathcal{M}_N^-$ (resp. $\mathcal{M}_N^+$) is a closed variety, its image under the projection π_c onto the c-axis is a union of closed intervals. Notice that the image $\pi_c(\mathcal{M}_N^-)$ does not coincide with the whole interval $[-1, 1]$; $c = 1$ means b_i are all equal so we cannot have $b_N = -b_0$. The fibre $\pi_c^{-1}(c)$ consists of N-Kaleidocycles with a fixed c. With a generic value of c, a simple dimension counting in (3.9) shows that $\dim(\pi_c^{-1}(c)) = N - 6$. Hence, the degree of freedom (DOF) of the Kaleidocycle with a torsion angle $\nu = \arccos(c)$ is generally $N - 6$. For $N > 6$, a generic Kaleidocycle is *reconfigurable* meaning that it can continuously change its shape. We will investigate a particular series of reconfiguration in the next section.

Remark 1. The most popular Kaleidocycle with $N = 6$ has $c = 0$, which is equivalent to the threefold symmetric Bricard 6R linkage (Fig. 1). This Kaleidocycle is highly symmetric and not generic, resulting in 1 DOF [13].

4 Deformation of discrete curves

4.1 Continuous isoperimetric deformations on discrete curves

Kaleidocycles exhibit a characteristic turning-over motion (see Fig. 11 and see [22] for some animations). In general, an N-Kaleidocycle has $N - 6$ degrees of freedom so that it wobbles in addition to turning-over. With special values of torsion angle, however, the DOF of the Kaleidocycle seems to degenerate to exactly one, leaving only the turning-over motion as we will discuss in §5. In this case, the motion of the core segment looks to be orthogonal to the hinge directions. In the following, we would like to model the motion explicitly. It turns out that we can construct the motion of Kaleidocycles using semi-discrete mKdV and sine-Gordon equations.

Figure 11. Turning-over motion of a Kaleidocycle with $N = 7$.

In this section, we consider certain continuous deformations of discrete space curves which correspond to motion of homogeneous serial and closed hinged networks. Our approach is to construct a flow on the configuration space by differential-difference equations. We use the same notations as in Section 3. Observe that a hinged network moves in such a way that its tetrahedral links are not distorted. In the language of discrete space curves, the motion corresponds to a deformation which preserves the speed ϵ_n and the torsion angle ν_n for all n.

Let $\gamma(0) : \mathbb{Z} \to \mathbb{R}^3$ be an (open) discrete space curve which has a constant speed $\varepsilon_n(0) = \varepsilon_*(0)$ and a constant torsion angle $\nu_n(0) = \nu_*(0)$. Given a family of functions $w(t) : \mathbb{Z} \to \mathbb{R}$ with the deformation parameter $t \in \mathbb{R}$ and a constant $\rho > 0$, we consider a family of discrete space curves $\gamma(t)$ defined by

$$\frac{d\gamma_n}{dt} = \frac{\varepsilon_n}{\rho}\left(\cos w_n T_n + \sin w_n \widetilde{N}_n\right) \qquad (n \in \mathbb{Z}). \tag{4.1}$$

That is, the motion of each point γ_n is confined in the osculating plane and its speed depends only on the length of the segment $\varepsilon_n = |\gamma_{n+1} - \gamma_n|$. We say a deformation is *isoperimetric* if the segment length ε_n does not depend on t for all n, We would like to find conditions on w under which the above deformation is isoperimetric. From (3.1), (3.5) and (4.1), we have

$$\begin{aligned}
\frac{d\varepsilon_n}{dt} &= \frac{\varepsilon_n}{\rho}\left\langle \Phi_{n+1}\begin{bmatrix}\cos w_{n+1}\\ \sin w_{n+1}\\ 0\end{bmatrix} - \Phi_n\begin{bmatrix}\cos w_n\\ \sin w_n\\ 0\end{bmatrix}, \Phi_n\begin{bmatrix}1\\0\\0\end{bmatrix}\right\rangle \\
&= \frac{\varepsilon_n}{\rho}\left\langle \Phi_n\begin{bmatrix}\cos(\kappa_{n+1} + w_{n+1}) - \cos w_n\\ \cos\nu_n \sin(\kappa_{n+1} + w_{n+1}) - \sin w_n\\ -\sin\nu_n \sin(\kappa_{n+1} + w_{n+1})\end{bmatrix}, \Phi_n\begin{bmatrix}1\\0\\0\end{bmatrix}\right\rangle \\
&= \frac{\varepsilon_n}{\rho}\big(\cos(\kappa_{n+1} + w_{n+1}) - \cos w_n\big).
\end{aligned}$$

Therefore, for each n, $d\varepsilon_n/dt$ vanishes if and only if

$$\cos(\kappa_{n+1} + w_{n+1}) - \cos w_n = 0, \tag{4.2}$$

which yields

$$w_n = -w_{n-1} - \kappa_n, \tag{4.3}$$

or

$$w_n = w_{n-1} - \kappa_n. \tag{4.4}$$

We consider a deformation when (4.3) (resp. (4.4)) simultaneously holds for all n. Note that in this case $w_n(t)$ for all n is determined once $w_0(t)$ is given.

Those deformations are characterised by the following propositions:

Proposition 1. *Let $\gamma(0) : \mathbb{Z} \to \mathbb{R}^3$ be a discrete space curve with a constant speed $\varepsilon_n(0) = \varepsilon_*(0)$ and a constant torsion angle $\nu_n(0) = \nu_*(0)$. Let $\gamma(t)$ be its deformation according to (4.1) with $w : \mathbb{Z} \to \mathbb{R}$ satisfying the condition (4.3). Then we have:*

1. *The speed $\varepsilon_n(t)$ and the torsion angle $\nu_n(t)$ do not depend on t nor n. That is, $\varepsilon_n(t) = \varepsilon_*(0)$ and $\nu_n(t) = \nu_*(0)$ for all t and n.*

2. *The signed curvature angle $\kappa_n = \kappa_n(t)$ and $w_n = w_n(t)$ satisfy*

$$\frac{d\kappa_n}{dt} = \alpha \left(\sin w_{n-1} - \sin w_n\right), \tag{4.5}$$

where $\alpha = \frac{1+\cos\nu_*(0)}{\rho}$.

3. *The deformation of the frame $\Phi_n(t) = [T_n(t), \widetilde{N}_n(t), b_n(t)]$ is given by*

$$\frac{d\Phi_n}{dt} = \Phi_n M_n,$$

$$M_n = \frac{1}{\rho}\begin{bmatrix} 0 & (1+\cos\nu_*(0))\sin w_n & -\sin\nu_*(0)\sin w_n \\ -(1+\cos\nu_*(0))\sin w_n & 0 & \sin\nu_*(0)\cos w_n \\ \sin\nu_*(0)\sin w_n & -\sin\nu_*(0)\cos w_n & 0 \end{bmatrix}. \tag{4.6}$$

Proposition 2. *Let $\gamma(0) : \mathbb{Z} \to \mathbb{R}^3$ be a discrete space curve with a constant speed $\varepsilon_n(0) = \varepsilon_*(0)$ and a constant torsion angle $\nu_n(0) = \nu_*(0)$. Let $\gamma(t)$ be its deformation according to (4.1) with $w : \mathbb{Z} \to \mathbb{R}$ satisfying the condition (4.4). Then we have:*

1. *The speed $\varepsilon_n(t)$ and the torsion angle $\nu_n(t)$ do not depend on t nor n. That is, $\varepsilon_n(t) = \varepsilon_*(0)$ and $\nu_n(t) = \nu_*(0)$ for all t and n.*

2. *The signed curvature angle $\kappa_n = \kappa_n(t)$ and $w_n = w_n(t)$ satisfy*

$$\frac{d\kappa_n}{dt} = -\hat{\alpha} \left(\sin w_n + \sin w_{n-1}\right), \tag{4.7}$$

where $\hat{\alpha} = \frac{1-\cos\nu_*(0)}{\rho}$.

3. The deformation of the frame $\Phi_n(t) = [T_n(t), \widetilde{N}_n(t), b_n(t)]$ *is given by*

$$\frac{d\Phi_n}{dt} = \Phi_n M_n,$$
$$M_n = \frac{1}{\rho}\begin{bmatrix} 0 & (1-\cos\nu_*(0))\sin w_n & \sin\nu_*(0)\sin w_n \\ -(1-\cos\nu_*(0))\sin w_n & 0 & -\sin\nu_*(0)\cos w_n \\ -\sin\nu_*(0)\sin w_n & \sin\nu_*(0)\cos w_n & 0 \end{bmatrix}. \tag{4.8}$$

Proof. We only prove Proposition 1 since Proposition 2 can be proved in the same manner. We first show the second and the third statements. We denote $\dot{f} = \frac{df}{dt}$, $\nu = \nu_*(0)$ and $\varepsilon = \varepsilon_*(0)$ for simplicity. Since ε is a constant by the preceding argument, the deformation of T_n can be computed from (4.1) and (4.4) as

$$\begin{aligned}
\dot{T}_n &= \frac{1}{\rho}\Phi_n\left(L_n\begin{bmatrix}\cos w_{n+1}\\ \sin w_{n+1}\\ 0\end{bmatrix} - \begin{bmatrix}\cos w_n\\ \sin w_n\\ 0\end{bmatrix}\right)\\
&= \frac{1}{\rho}\Phi_n\begin{bmatrix}\cos(\kappa_{n+1}+w_{n+1}) - \cos w_n\\ \cos\nu\sin(\kappa_{n+1}+w_{n+1}) - \sin w_n\\ -\sin\nu\sin(\kappa_{n+1}+w_{n+1})\end{bmatrix}\\
&= \frac{1}{\rho}\Phi_n\begin{bmatrix}0\\ -(1+\cos\nu)\sin w_n\\ \sin\nu\sin w_n\end{bmatrix}.
\end{aligned} \tag{4.9}$$

Differentiating $\cos\kappa_n = \langle T_n, T_{n-1}\rangle$ with respect to t, we have

$$-\dot{\kappa_n}\sin\kappa_n = \langle \dot{T}_n, T_{n-1}\rangle + \langle T_n, \dot{T}_{n-1}\rangle. \tag{4.10}$$

Noting

$$T_{n-1} = \Phi_n L_{n-1}^{-1}\begin{bmatrix}1\\0\\0\end{bmatrix} = \Phi_n\begin{bmatrix}\cos\kappa_n\\ -\sin\kappa_n\\ 0\end{bmatrix}, \tag{4.11}$$

and

$$\begin{aligned}
\dot{T}_{n-1} &= \frac{1}{\rho}\Phi_n L_n^{-1}\begin{bmatrix}0\\ -(1+\cos\nu)\sin w_{n-1}\\ \sin\nu\sin w_{n-1}\end{bmatrix}\\
&= \frac{1}{\rho}\Phi_n\begin{bmatrix}-(1+\cos\nu)\sin\kappa_n\sin w_{n-1}\\ -(1+\cos\nu)\cos\kappa_n\sin w_{n-1}\\ -\sin\nu\sin w_{n-1}\end{bmatrix},
\end{aligned} \tag{4.12}$$

we get from (4.9) and (4.10)

$$\dot{\kappa_n} = \frac{1+\cos\nu}{\rho}\left(\sin w_{n-1} - \sin w_n\right), \tag{4.13}$$

which is equivalent to (4.7). This proves the second statement. Next, we see from the definition of b_n

$$\dot{b}_n = \frac{d}{dt}\left(\frac{1}{|T_{n-1}\times T_n|}\right) T_{n-1}\times T_n + \frac{1}{|T_{n-1}\times T_n|}\left(\dot{T}_{n-1}\times T_n + T_{n-1}\times \dot{T}_n\right). \quad (4.14)$$

Noting

$$T_{n-1}\times T_n = \Phi_n \begin{bmatrix} \cos\kappa_n \\ -\sin\kappa_n \\ 0 \end{bmatrix} \times \Phi_n \begin{bmatrix} 1 \\ 0 \\ 0 \end{bmatrix} = \Phi_n \begin{bmatrix} 0 \\ 0 \\ \sin\kappa_n \end{bmatrix}, \quad (4.15)$$

$$\begin{aligned} \dot{T}_{n-1}\times T_n &= \frac{1}{\rho}\Phi_n \begin{bmatrix} -(1+\cos\nu)\sin\kappa_n \sin w_{n-1} \\ -(1+\cos\nu)\cos\kappa_n \sin w_{n-1} \\ -\sin\nu \sin w_{n-1} \end{bmatrix} \times \Phi_n \begin{bmatrix} 1 \\ 0 \\ 0 \end{bmatrix} \\ &= \frac{1}{\rho}\Phi_n \begin{bmatrix} 0 \\ -\sin\nu\sin w_{n-1} \\ (1+\cos\nu)\cos\kappa_n \sin w_{n-1} \end{bmatrix}, \end{aligned} \quad (4.16)$$

and

$$\begin{aligned} T_{n-1}\times \dot{T}_n &= \Phi_n \begin{bmatrix} \cos\kappa_n \\ -\sin\kappa_n \\ 0 \end{bmatrix} \times \frac{1}{\rho}\Phi_n \begin{bmatrix} 0 \\ -(1+\cos\nu)\sin w_n \\ \sin\nu\sin w_n \end{bmatrix} \\ &= \frac{1}{\rho}\Phi_n \begin{bmatrix} -\sin\nu\sin\kappa_n\sin w_n \\ -\sin\nu\cos\kappa_n\sin w_n \\ -(1+\cos\nu)\cos\kappa_n\sin w_n \end{bmatrix}, \end{aligned} \quad (4.17)$$

we get from (4.13) and (4.14)

$$\dot{b}_n = \frac{1}{\rho}\Phi_n \begin{bmatrix} -\sin\nu\sin w_n \\ \sin\nu\cos w_n \\ 0 \end{bmatrix}. \quad (4.18)$$

We immediately obtain $\dot{\widetilde{N}}_n$ from (4.9) and (4.14) as

$$\dot{\widetilde{N}} = \dot{b}_n \times T_n + b_n \times \dot{T}_n = \frac{1}{\rho}\Phi_n \begin{bmatrix} (1+\cos\nu)\sin w_n \\ 0 \\ -\sin\nu\cos w_n \end{bmatrix}. \quad (4.19)$$

Then we have (4.8) from (4.9), (4.14) and (4.19), which proves the third statement. Finally, differentiating $\cos\nu = \langle b_n, b_{n-1}\rangle$ with respect to t, it follows from (4.18) and (4.2) that

$$-\dot{\nu}\sin\nu = \langle \dot{b}_n, b_{n-1}\rangle + \langle b_n, \dot{b}_{n-1}\rangle = -\frac{\sin^2\nu}{\rho}\left(\cos(\kappa_n + w_n) - \cos w_{n-1}\right) = 0,$$

which implies $\dot{\nu} = 0$. This completes the proof of the first statement. ■

Remark 2. The condition (4.3) suggests the *potential function* θ_n in Proposition 1 such that we have

$$\kappa_n = \frac{\theta_{n+1} - \theta_{n-1}}{2}, \quad w_n = \frac{\theta_n - \theta_{n+1}}{2}. \tag{4.20}$$

Then, (4.5) is rewritten as

$$\frac{d}{dt}\left(\theta_{n+1} + \theta_n\right) = 2\alpha \sin\left(\frac{\theta_{n+1} - \theta_n}{2}\right). \tag{4.21}$$

To the best of the authors' knowledge, this is a novel form of the *semi-discrete potential mKdV equation.* In fact, the continuum limit $\alpha = \frac{2}{\epsilon}$, $X = \epsilon n + t$, $T = \frac{\epsilon^2}{12}t$, $\epsilon \to 0$ yields the potential mKdV equation

$$\theta_T + \frac{1}{2}(\theta_X)^3 + \theta_{XXX} = 0. \tag{4.22}$$

Similarly, introducing the potential function θ_n in Proposition 2 such that

$$\kappa_n = \frac{\theta_{n+1} - \theta_{n-1}}{2}, \quad w_n = -\frac{\theta_{n+1} + \theta_n}{2}, \tag{4.23}$$

suggested by (4.4), we can rewrite (4.7) as

$$\frac{d}{dt}\left(\theta_{n+1} - \theta_n\right) = 2\alpha \sin\left(\frac{\theta_{n+1} + \theta_n}{2}\right), \tag{4.24}$$

which is nothing but the *semi-discrete sine-Gordon equation* [3, 39, 40].

Remark 3. In the above argument, we assume that the speed of the deformation ρ in (4.1) is a constant and does not depend on n. Then, by demanding that the deformation preserve arc length ((4.3) or (4.4)), it followed that the torsion angle is also preserved. Conversely, it seems to be the case that for the deformation to preserve both the arc length and the torsion angle, the speed ρ is required not to depend on n.

Remark 4 (Continuum limit)**.** The isoperimetric torsion-preserving discrete deformations for the discrete space curves of constant torsion have been considered in [18], where the deformations are governed by the discrete sine-Gordon and the discerte mKdV equations. It is possible to obtain the continuous deformations discussed in this section by suitable continuum limits from those discrete deformations. More precisely, let γ_n^m ($m \in \mathbb{Z}$) be a family of discrete curves obtained by applying the discrete deformations m times to $\gamma_n^0 = \gamma_n$, where γ_n is the discrete curve with a constant speed ε and a constant torsion angle ν. Then the above discrete deformation is given by

$$\gamma_n^{m+1} = \gamma_n^m + \delta_m \left(\cos w_n^m T_n^m + \sin w_n^m N_n^m\right). \tag{4.25}$$

Then if we choose δ_m and w_0^m so that the sign of $\sigma_n^m = \sin(w_{n+1}^m + \kappa_{n+1}^m - w_{n-1}^m)$ does not depend on n, the isoperimetric condition and the compatibility condition of the Frenet frame yield the discrete mKdV equation

$$\frac{w_{n+1}^{m+1} - w_n^m}{2} = \arctan\left(\frac{b+a}{b-a}\tan\frac{w_n^{m+1}}{2}\right) - \arctan\left(\frac{b+a}{b-a}w_{n+1}^m\right), \tag{4.26}$$

when $\sigma_n^m > 0$, and the discrete sine-Gordon equation

$$\frac{w_{n+1}^{m+1} + w_n^m}{2} = \arctan\left(\frac{b+a}{b-a}\tan\frac{w_n^{m+1}}{2}\right) + \arctan\left(\frac{b+a}{b-a}w_{n+1}^m\right), \tag{4.27}$$

when $\sigma_n^m < 0$ with

$$a = \left(1 + \tan^2\frac{\nu}{2}\right)\varepsilon, \quad b = \left(1 + \tan^2\frac{\nu}{2}\right)\delta. \tag{4.28}$$

For the discrete mKdV equation (4.26), in the limit of

$$a = \frac{2\varepsilon}{\rho\alpha}, \quad m = \frac{\rho}{\varepsilon\delta}t, \quad b \to 0\ (\delta \to 0), \tag{4.29}$$

(4.26) is reduced to the semi-discrete mKdV equation (4.5). Similarly, the discrete sine-Gordon equation (4.27) is reduced to the semi-discrete sine-Gordon equation (4.7) in the limit

$$a = \frac{\alpha\rho}{\varepsilon}, \quad m = \frac{\rho}{\varepsilon\delta}t, \quad b \to \infty\ (\delta \to 0). \tag{4.30}$$

Obviously, the discrete deformation equation of the discrete curve (4.25) is reduced to the continuous deformation equation (4.1). Moreover, it is easily verified that the discrete deformation equations of the Frenet frame in [18] are reduced to (4.6) and (4.8).

4.2 Turning-over motion of Kaleidocycles

An N-Kaleidocycle corresponds to a closed discrete curve γ of length N having a constant speed ε and a constant torsion angle ν whose b is oriented. Since γ is closed, for (4.1) to define a deformation of γ, we need a periodicity condition $w_{n+N} = w_n$ (when oriented) or $w_{n+N} = -w_n$ (when anti-oriented) for any $n \in \mathbb{Z}$.

When N is odd and the Kaleidocycle is oriented, the equation (4.3) together with $w_0 = w_N$ forms a linear system for w_n $(0 \le n \le N)$ which is regular. Therefore, we can find w_n $(0 \le n \le N)$ uniquely as the solution to the system. Then, the equation (4.1) generates a deformation of γ which preserves the segment length and the torsion angle, while γ remains closed. That is, the turning-over motion of the Kaleidocycle is governed by the semi-discrete mKdV equation (4.5) (see Fig. 12). Note that by (4.5), the total curvature angle $\sum_{i=0}^{N-1}\kappa_n(t)$ is also preserved.

When the Kaleidocycle is anti-oriented, the equation (4.4) together with $w_0 = -w_N$ forms a linear system for w_n $(0 \le n \le N)$ which is regular for any N. Similarly

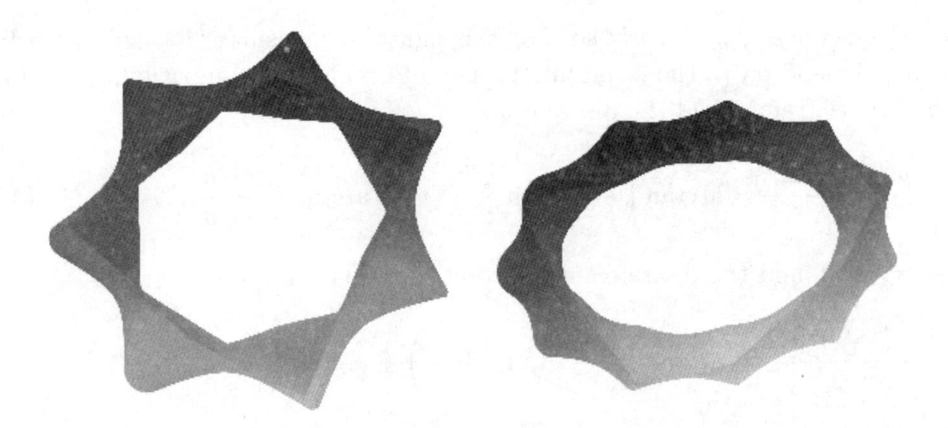

Figure 12. Surface drawn by the evolution of the center curves of Kaleidocycles with $N = 7$ and $N = 25$ respectively.

to the above, in this case the turning-over motion of the Kaleidocycle is governed by the semi-discrete sine-Gordon equation (4.7).

Note that if an N-Kaleidocycle with an odd N is anti-oriented $b_0 = -b_N$, we can define an oriented Kaleidocycle by taking its "mirrored image" $b_i \mapsto (-1)^i b_i$ which conforms to the definition 3. Thus, for an odd Kaleidocycle, both the semi-discrete mKdV equation and the semi-discrete sine-Gordon equation generate the turning-over motion.

5 Extreme Kaleidocycles

We defined Kaleidocycles in Def. 3 and saw the torsion angle cannot be chosen arbitrarily. A natural question is for what torsion angle ν there exists an N-Kaleidocycle for each N. It seems there are no Kaleidocycles with $\nu \in (0, \pi)$ for $N \leq 5$. For $6 \leq N \leq 50$, we conducted numerical experiments with [22] and found that there exists $c_N^* \in [0, 1]$ which satisfy the following. Recall that $\pi_c : \mathcal{M}_N \to \mathbb{R}$ is the projection of the configuration space $\mathcal{M}_N$ onto the c-axis, where $c = \cos \nu$.

1. When N is odd, $\pi_c(\mathcal{M}_N^+) = [-c_N^*, 1]$ and $\pi_c(\mathcal{M}_N^-) = [-1, c_N^*]$.

2. When N is even, $\pi_c(\mathcal{M}_N^+) = [-1, 1]$ and $\pi_c(\mathcal{M}_N^-) = [-c_N^*, c_N^*]$.

Moreover, $N \arccos(c_N^*)$ converges monotonously to a constant, where arccos takes the principal value in $[0, \pi]$. Interestingly, at the boundary values $c = \pm c_N^*$, the fibre of π_c seems to be exactly one-dimensional for any $N \geq 6$. This means, they are exactly the one-dimensional orbits defined in §4.2.

We summarise our numerical findings.

Conjecture 1. Let $N \geq 6$. We have the following:

1. The space $\pi_c^{-1}(c_N^*) \cap \mathcal{M}_N^-$ is a circle. Moreover, the involution defined by $b_n \mapsto (-1)^n b_n$ induces isomorphisms $\pi_c^{-1}(-c_N^*) \cap \mathcal{M}_N^+ \simeq \pi_c^{-1}(c_N^*) \cap \mathcal{M}_N^-$ when N is odd and $\pi_c^{-1}(-c_N^*) \cap \mathcal{M}_N^- \simeq \pi_c^{-1}(c_N^*) \cap \mathcal{M}_N^-$ when N is even.

2. The orbit of any element $\gamma \in \pi_c^{-1}(c_N^*) \cap \mathcal{M}_N^-$ of the flow generated by the semi-discrete sine-Gordon equation described in §4.2 coincides with $\pi_c^{-1}(c_N^*) \cap \mathcal{M}_N^- \simeq S^1$.

3. When N is odd, the orbit of any element $\gamma \in \pi_c^{-1}(-c_N^*) \cap \mathcal{M}_N^+$ generated by the semi-discrete mKdV equation described in §4.2 coincides with $\pi_c^{-1}(-c_N^*) \cap \mathcal{M}_N^+ \simeq S^1$. Moreover, on $\pi_c^{-1}(-c_N^*) \cap \mathcal{M}_N^+$ we have $\sum_{n=0}^{N-1} \kappa_n = 0$ and we can also define its deformation by the semi-discrete sine-Gordon equation if we define w by (4.4) and $\sum_{n=0}^{N-1} \dot{\kappa}_n = 2\alpha \sum_{n=0}^{N-1} \sin(w_n) = 0$. The orbit coincides with $\pi_c^{-1}(-c_N^*) \cap \mathcal{M}_N^+$ as well. That is, for an oriented Kaleidocycle with $\nu = \arccos(-c_N^*)$, we can define two motions one by the semi-discrete sine-Gordon equation (4.4), the other by the semi-discrete mKdV equation (4.3), and they coincide up to rigid transformations.

4. Any strip $(\gamma^{b,\varepsilon}, b)$ corresponding to $b \in \pi_c^{-1}(c_N^*) \cap \mathcal{M}_N^-$ is a 3-half twisted Möbius strip (see §5.2). There are no Kaleidocycles with one or two half twisting.

5. When N tends to infinity, $N \arccos c_N^*$ converges to a constant. There exists a unique limit curve up to congruence for any sequence $\gamma_N \in \pi_c^{-1}(c_N^*) \cap \mathcal{M}_N^-$, and it has a constant torsion up to sign.

We call those Kaleidocycles having the extremal torsion angle *extreme Kaleidocycles.*

Remark 5. The extreme Kaleidocycles were discovered by the first named author and his collaborators [21, 23]. In particular, when it is anti-oriented, it is called the *Möbius Kaleidocycle* because they are a discrete version of the Möbius strip with a 3π-twist. Coincidentally, Möbius is the first one to give the dimension counting formula for generic linkages [32] (although it is often attributed to Maxwell), and our Möbius Kaleidocycles are exceptions to his formula.

We end this chapter with a list of interesting properties, questions and some supplementary materials of Kaleidocycles for future research.

5.1 Kinematic energy

Curves with adapted frames serve as a model of elastic rods and are studied, for example, in Langer and Singer [29] in a continuous setting, and in [2] in a discrete setting. Serial and closed hinged networks are discrete curves with specific frames as we saw in §3. From this viewpoint, we consider some energy functionals defined for discrete curves with frames and investigate how they behave on the configuration space $\mathcal{M}_N$ of Kaleidocycles.

Let γ be a constant speed discrete closed curve of length N. The *elastic energy* $\mathcal{E}_e$ and the *twisting energy* $\mathcal{E}_t$ are defined respectively by

$$\mathcal{E}_e(\gamma) = \sum_{n=0}^{N-1} \kappa_n^2, \qquad \mathcal{E}_t(\gamma) = \sum_{n=0}^{N-1} \nu_n^2.$$

By the definition of Kaleidocycle, $\mathcal{E}_t$ takes a constant value when a Kaleidocycle undergoes any motion.

Interestingly, a numerical simulation by [22] suggests that on $\pi_c^{-1}(c_N^*) \cap \mathcal{M}_N^-$ (and also on $\pi_c^{-1}(-c_N^*) \cap \mathcal{M}_N^+$ for an odd N and on $\pi_c^{-1}(-c_N^*) \cap \mathcal{M}_N^-$ for an even N) for a fixed N, $\mathcal{E}_e$ takes an almost constant value. The summands of $\mathcal{E}_e$ are locally determined and vary depending on the states, however, the total is almost stable so that only small force should be applied to rotate the Kaleidocycle. It is also noted that the sum $\mathcal{E}_e + \mathcal{E}_t$ is a discrete version of the elastic energy of the Kirchoff rod defined by the strip, and it also takes almost constant values.

Similarly, we introduce the following three more energy functionals, which are observed to take almost constant values on $\pi_c^{-1}(-c_N^*) \cap \mathcal{M}_N^+$. The *dipole energy* is defined to be

$$\mathcal{E}_d(\gamma) := 2\left(\sum_{i<j} \frac{\langle b_i, b_j \rangle}{|\gamma_i - \gamma_j|^3} - 3\frac{\langle b_i, \gamma_i - \gamma_j \rangle \langle b_j, \gamma_i - \gamma_j \rangle}{|\gamma_i - \gamma_j|^5}\right).$$

The *Coulomb energy* with an exponent $\alpha > 0$ is defined to be

$$\mathcal{E}_c(\gamma) := 2\sum_{i<j} \frac{1}{|\gamma_i - \gamma_j|^\alpha}.$$

The *averaged hinge magnitude* is defined to be

$$\mathcal{E}_a(\gamma) := \frac{1}{N}\left|\sum_{n=0}^{N-1} b_n\right|.$$

However, we have no rigorous statements about them. It may be the case that one needs some other discretisation of the continuous counterparts of these energies to show their behaviour theoretically. It is also interesting to characterise or generalise extreme Kaleidocycles in terms of variational calculus on the space of discrete closed curves.

5.2 Topological invariants

As noted in [29], for a curve to be closed, topological constraints come into the story. This quantises some continuous quantity and makes it an isotopy invariant.

Let γ be a constant speed discrete closed curve of length N. First, interpolate γ_n and b_n for $(0 \le n < 2N)$ linearly to obtain a continuous vector field $\bar{b}$ defined on the polygonal curve $\bar{\gamma}$, which goes around the polygon twice. We define the *twisting number* $\mathcal{T}$ of γ as the linking number between twice the centre curve $\bar{\gamma}$ and the boundary curve $\bar{\gamma} + \epsilon\bar{b}$, where $\epsilon > 0$ is small enough. Intuitively, it is the number of half-twists of the strip defined by γ and b. The Călugăreanu-White formula relates this topological invariant to the sum of two conformal invariants and provides a direct discretisation without the need of interpolation (cf. [26]):

$$\mathcal{T} = 2(Tw + Wr), \tag{5.1}$$

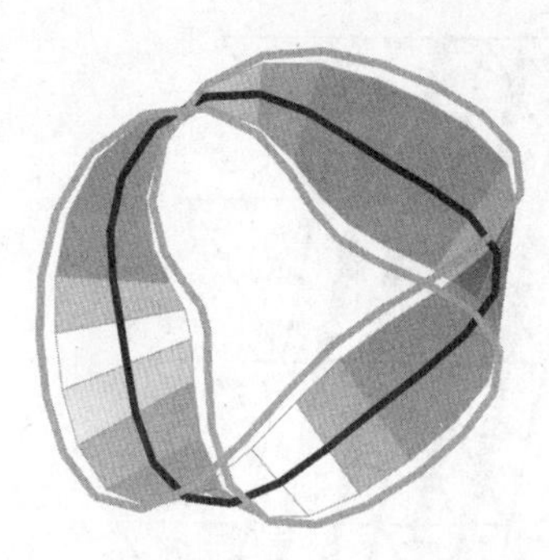

Figure 13. Twisting number as the linking number between centre and boundary curves.

where Wr is the *writhe* of the polygonal curve γ which can be computed as a double summation [26, Eq. (13)] and

$$Tw = \frac{1}{2\pi} \sum_{n=0}^{N-1} \nu$$

is the *total twist.* The twisting number $\mathcal{T}$ takes values in the integers, enforcing topological constraints to the curve.

Recall by definition that anti-oriented extreme Kaleidocycles are discrete closed space curves of constant speed and constant torsion which have the minimum odd twisting number. Our numerical experiments suggest that the minimum is not one but three.

Let γ be a discrete closed space curve of constant speed and constant torsion corresponding to a Kaleidocycle. Under any motion of the Kaleidocycle, Tw stays constant by definition. By (5.1) the corresponding deformation of the curve preserves the writhe as well. This can equivalently be phrased in terms of the *Gauss map* $G(\gamma) : n \mapsto T_n$ $(0 \leq n \leq N-1)$. The Gauss-Bonnet theorem tells us that $A + 2\pi Tw = 0 \mod \pi$, where A is the area on the sphere enclosed by $G(\gamma)$. By (5.1) we have $Wr = A/2\pi \mod 1/2$. Thus, the deformation of the closed discrete space curve considered in §4.2 induces one of the closed discrete spherical curves which preserves the enclosed area A.

Kaleidocycles can be folded from a piece of paper. We include a development plan for the extreme Kaleidocycle with $N = 8$ so that the readers can personally make and investigate its motion.

Acknowledgments. *The first named author is partially supported by JST, PRESTO Grant Number JPMJPR16E3, Japan. The second named author is partially supported by JSPS KAKENHI Grant Numbers JP16H03941, JP16K13763. The last named author acknowledges the support from the "Leading Program in Mathematics for Key Technologies" of Kyushu University.*

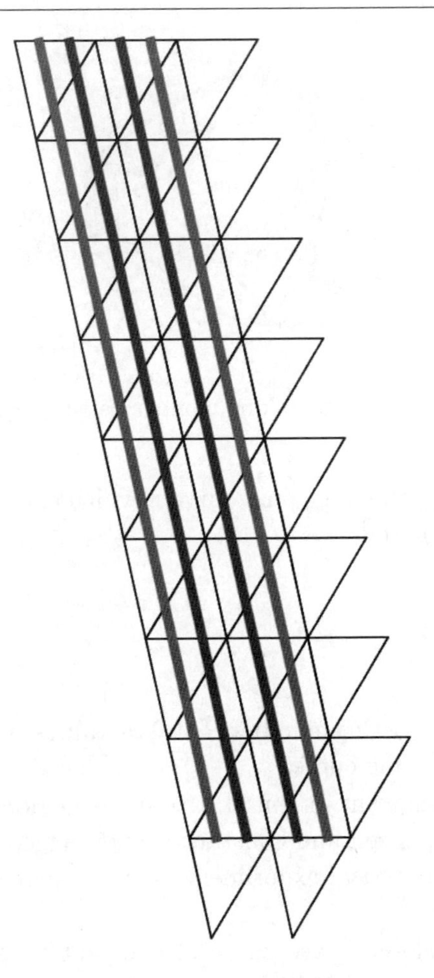

Figure 14. Development plan of an extreme Kaleidocycle with eight hinges. Black horizontal lines indicate valley folds and black slanted lines indicate mountain folds.

References

[1] Bates L M and Melko O M, On curves of constant torsion I, *Journal of Geometry*, **104**(2), 213–227, 2013.

[2] Bergou M, Wardetzky M, Robinson S, Audoly B and Grinspun E, Discrete elastic rods, *ACM Trans. Graph.*, **27**(3), Article 63, 2008.

[3] Boiti M, Pempinelli F and Prinari B, Integrable discretization of the sine-Gordon equation, *Inverse Problems* **18**(5), 1309–1324, 2002.

[4] Byrnes R, Metamorphs: Transforming Mathematical Surprises, Tarquin Pubns, 1999.

[5] Calini A M, Ivey T A, Bäcklund transformations and knots of constant torsion, *J. Knot Theory Ram.*, **7**, 719–746, 1998.

[6] Calini A M an Ivey T A, Topology and sine-Gordon evolution of constant torsion curves, *Phys. Lett. A*, **254**(3–4), 170–178, 1999.

[7] You Z and Chen Y, Motion Structures: Deployable Structural Assemblies of Mechanisms, Taylor & Francis, 2011.

[8] Darboux G, Leçcons sur la Théorie Générale des Surfaces, Gauthier-Villars, 1917.

[9] Denavit J and Hartenberg R S, A kinematic notation for lower-pair mechanisms based on matrices, *Trans ASME J. Appl. Mech.* **23**, 215–221, 1955.

[10] Doliwa A and Santini P M, An elementary geometric characterization of the integrable motions of a curve, *Phys. Lett. A*, **185**, 373–384, 1994.

[11] Doliwa A and Santini P M, Integrable dynamics of a discrete curve and the Ablowitz-Ladik hierarchy, *J. Math. Phys.* **36**, 1259–1273, 1995.

[12] Farber M, Invitation to Topological Robotics, European Mathematical Society, Zurich, 2008.

[13] Fowler P W and Guest S D, A symmetry analysis of mechanisms in rotating rings of tetrahedra, *Proc. R. Soc. A* **461**, 1829–1846, 2005.

[14] Freuder E C, Synthesizing constraint expressions, *Commun. ACM* **21**(11), 958–966, 1978.

[15] Hasimoto H, A soliton on a vortex filament, *J. Fluid. Mech.* **51**, 477–485, 1972.

[16] Hisakado M and Wadati M, Moving discrete curve and geometric phase, *Phys. Lett. A*, **214**, 252–258, 1996.

[17] Hoffmann T, Discrete Differential Geometry of Curves and Surfaces, *MI Lecture Notes* vol. 18, Kyushu University, 2009.

[18] Inoguchi J, Kajiwara K, Matsuura N and Ohta Y, *Discrete mKdV and discrete sine-Gordon flows on discrete space curves*, *J. Phys. A: Math. Theor.* **47**, 2014.

[19] Ivey T A, Minimal Curves of Constant Torsion, *Proc. AMS*, **128**(7), 2095–2103, 2000.

[20] Jordán T, Király C and Tanigawa S, Generic global rigidity of body-hinge frameworks, *J. Comb. Theory B*, **117**, 59–76, 2016.

[21] Kaji S, A closed linkage mechanism having the shape of a discrete Möbius strip, *the Japan Society for Precision Engineering Spring Meeting Symposium Extended Abstracts*, 62–65, 2018. The original is in Japanese but an English translation is available at arXiv:1909.02885.

[22] Kaji S, Geometry of the moduli space of a closed linkage: a Maple code, available at `https://github.com/shizuo-kaji/Kaleidocycle`

[23] Kaji S, Schönke S, Grunwald M and Fried E, Möbius Kaleidocycle, patent filed, JP2018-033395, 2018.

[24] Kapovich M and Millson J, Universality theorems for configuration spaces of planar linkages, *Topology* **41**(6), 1051–1107, 2002.

[25] Katoh N and Tanigawa S, A proof of the molecular conjecture, *Discrete Comput Geom.*, **45**, 647–700, 2011.

[26] Klenin K and Langowski J, Computation of writhe in modeling of supercoiled DNA, *Biopolymers*, **54**, 307–317, 2000.

[27] Lamb G L Jr., Solitons and the motion of helical curves, *Phys. Rev. Lett.* **37**, 235–237, 1976.

[28] Langer J and Perline R, Curve motion inducing modified Korteweg-de Vries systems, *Phys. Lett. A*, **239**, 36–40, 1998.

[29] Langer J and Singer D A, Lagrangian aspects of the Kirchhoff elastic rod, *SIAM Review* **38**(4), 605–618, 1996.

[30] LaValle S M, Planning algorithms, Cambridge University Press, 2006.

[31] Magalhães M L S and Pollicott M, Geometry and dynamics of planar linkages, *Comm. Math. Phys.*, **317**(3), 615–634, 2013.

[32] Möbius A F, Lehrbuch der Statik, **2**, Leipzig, 1837.

[33] Moses M S and Ackerman M K and Chirikjian G S, ORIGAMI ROTORS: Imparting continuous rotation to a moving platform using compliant flexure hinges, *Proc. IDETC/CIE* 2013.

[34] Müller A, Representation of the kinematic topology of mechanisms for kinematic analysis, *Mech. Sci.*, **6**, 1–10, 2015.

[35] Müller A, Local kinematic analysis of closed-loop linkages – mobility, singularities, and shakiness, *J. Mechanisms Robotics* **8**(4), 041013, 2016.

[36] Nakayama K, Elementary vortex filament model of the discrete nonlinear Schrödinger equation, *J. Phys. Soc. Jpn.* **76**, 074003, 2007.

[37] Nakayama K, Segur H and Wadati M, Integrability and the motions of curves, *Phys. Rev. Lett.* **69**, 2603–2606, 1992.

[38] Nishinari K, A discrete model of an extensible string in three-dimensional space, *J. Appl. Mech.* **66**, 695–701, 1999.

[39] Orfanidis S J, Discrete sine-Gordon equations, *Phys. Rev. D*, **18**(10), 3822–3827, 1978.

[40] Orfanidis S J, Sine-Gordon equation and nonlinear σ model on a lattice, *Phys. Rev. D*, **18**(10), 3828–3832, 1978.

[41] Rogers C and Schief W K, Bäcklund and Darboux Transformations: Geometry and Modern Applications in Soliton Theory, Cambridge University Press, Cambridge, 2002.

[42] Sato K and Tanaka R, Solitons in one-dimensional mechanical linkage *Phys. Rev. E*, **98**, 2018.

[43] Schattschneider D and Walker W M, M. C. Escher Kaleidocycles, Pomegranate Communications: Rohnert Park, CA, 1987. (TASCHEN; Reprint edition, 2015).

[44] Sommese A J, Hauenstein J D, Bates D J and Wampler C W, Numerically Solving Polynomial Systems with Bertini, Software, Environments, and Tools, Vol. 25, SIAM, Philadelphia, PA, 2013.

[45] Weiner L J, Closed curves of constant torsion, *Arch. Math.* (Basel) **25**, 313–317, 1974.

[46] Weiner L J, Closed curves of constant torsion II, *Proc. AMS*, **67**(2), 1977.

B6. The Cauchy problem of the Kadomtsev-Petviashvili hierarchy and infinite-dimensional groups

Jean-Pierre Magnot [a] *and Enrique G. Reyes* [b]

[a] *LAREMA, Université d'Angers, 2 Bd Lavoisier, 49045 Angers cedex 1, France; Lycée Jeanne d'Arc, Avenue de Grande Bretagne, 63000 Clermont-Ferrand, France.*
jean-pierr.magnot@ac-clermont.fr

[b] *Departamento de Matemática y Ciencia de la Computación, Universidad de Santiago de Chile, Casilla 307 Correo 2, Santiago, Chile.*
e_g_reyes@yahoo.ca ; enrique.reyes@usach.cl

Abstract

We introduce basic concepts of generalized Differential Geometry of Frölicher and diffeological spaces; we consider formal and non-formal pseudodifferential operators in one independent variable, and we use them to build regular Frölicher Lie groups and Lie algebras on which we set up the Kadomtsev-Petviashvili hierarchy. The geometry of our groups allows us to prove smooth versions of the algebraic Mulase factorization of infinite dimensional groups based on formal pseudodifferential operators, and also an Ambrose-Singer theorem for infinite dimensional bundles. Using these tools we sketch proofs of the well-posedness of the Cauchy problem for the Kadomtsev-Petviashvili (KP) hierarchy in a smooth category. We also introduce a version of the KP hierarchy on infinite dimensional groups of series of non-formal pseudodifferential operators and we solve its initial value problem.

1 Introduction

Let $L = \partial + a_1\partial^{-1} + a_2\partial^{-2} + \cdots$ be a (formal) pseudodifferential operator. The Kadomtsev-Petviashvili (KP) hierarchy is the system

$$\frac{\partial}{\partial t_m} L = [(L^m)_+, L] \,, \qquad m \geq 1 \,, \tag{1.1}$$

in which $(L^m)_+$ denotes the projection of L^m into the space of differential operators. Equation (1.1) encodes, for each m, an infinite number of equations for the coefficients of L and, very importantly, it implies that if L solves the "t_m" KP system, so does $L + \epsilon \frac{\partial}{\partial t_n} L$ for any n to first order in ϵ. Also, the KP hierarchy generates symmetries of any equation of the form

$$\frac{\partial}{\partial t_m}(L^n)_+ - \frac{\partial}{\partial t_n}(L^m)_+ = [(L^m)_+, (L^n)_+] \,,$$

a finite system of partial differential equations in three independent variables which becomes (for $n = 3$, $m = 2$, $t_2 = y$, $t_3 = t$, $2a_1 = u$) the standard Kadomtsev-Petviashvili equation

$$3u_{yy} = \partial\left(4u_t - \partial^3(u) - 6u\partial(u)\right) . \tag{1.2}$$

These remarks appear in the classical book [13, Chp. 4] by L. Dickey. It follows that integrating the KP hierarchy corresponds to finding the flows of an infinite number of infinitesimal symmetries of equations such as (1.2). In this chapter we explain our approach to finding these flows in a smooth category.

Our basic idea is to exploit factorization theorems of Lie groups as in the work by Reyman and Semenov-Tian-Shansky [49, 52] and Mulase [40, 41]. Now, the papers [49, 52] are written in an ODEs context, while Mulase's factorization theorem (a far-reaching generalization of the Birkhoff decomposition of loop groups, see [48]) is essentially an algebraic result. We need to adapt the arguments appearing therein if we are to obtain non-formal results on well-posedness: we need to work with *bona fide* Lie groups and Lie algebras, to count with exponential mappings having properties similar to "finite-dimensional" exponential mappings, *and* to prove smooth versions of Mulase's theorem.

We have considered and solved the above problems in the papers [20, 38, 39] building upon our earlier works [19, 35, 36]. In addition, we have realized that the work in [35] on the Ambrose-Singer theorem also yields a well-posedness proof for KP, see [38]. We have found that an appropriate rigorous framework for the analysis of KP is the category of diffeology spaces and its subcategory of Frölicher spaces. This general approach to smoothness was introduced in mathematical physics by J.-M. Souriau in the 1980s, see [53] and the recent treatise [25]; we also refer to [35, 38, 55] and [1] for several applications. The subcategory of Frölicher spaces was first considered in [21] and then in [26]. These spaces are close to standard spaces appearing in functional analysis in the sense that "Frölicher smooth" is the same as "Gâteaux smooth", if we restrict ourselves to a Fréchet space context. But, there are many important spaces which have natural Frölicher structures and not Fréchet structures, see *e.g.* [38, 37]. Even more importantly, there are groups which are "regular Frölicher Lie groups" with regular Lie algebras and well-defined and useful exponential maps, see [36], but which are very difficult to work with (if possible at all!) using only tools available in the Fréchet category.

We think that the field of "non-algebraic" integrable systems is an important one. In this chapter we summarize our work on these issues and we point out some possible directions of research.

2 Diffeologies, Frölicher spaces and the Ambrose-Singer theorem

2.1 Diffeologies and Frölicher Lie groups

We present the general definition of a diffeological space after Souriau's [53], see also [25, 42], and [9] for a related notion appearing in algebraic topology.

Definition 1. Let X be a set.
• A **p-parametrization** of dimension p (or p-plot) on X is a map from an open subset O of $\mathbb{R}^p$ to X.
• A **diffeology** on X is a set $\mathcal{P}$ of parametrizations on X, called plots of the diffeology, such that

- For all $p \in \mathbb{N}$, any constant map $\mathbb{R}^p \to X$ is in $\mathcal{P}$;
- For each arbitrary set of indexes I and family $\{f_i : O_i \to X\}_{i \in I}$ of compatible maps that extend to a map $f : \bigcup_{i \in I} O_i \to X$, if $\{f_i : O_i \to X\}_{i \in I} \subset \mathcal{P}$, then $f \in \mathcal{P}$.
- For each $f \in \mathcal{P}$ defined on $O \subset \mathbb{R}^p$ and each smooth map (in the usual sense) g from $O' \subset \mathbb{R}^q$ to O we have $f \circ g \in \mathcal{P}$.

• If $\mathcal{P}$ is a diffeology on X, we say that $(X, \mathcal{P})$ is a **diffeological space**, and if $(X, \mathcal{P})$ and $(X', \mathcal{P}')$ are diffeological spaces, a map $f : X \to X'$ is **differentiable** (or, smooth) if and only if $f \circ \mathcal{P} \subset \mathcal{P}'$.

Recent applications of diffeologies to mathematical physics (including general relativity) appear in [25] and [1]. Any diffeological space $(X, \mathcal{P})$ can be endowed with the weakest topology such that all the maps that belong to $\mathcal{P}$ are continuous. This topology is called the D-topology, see [25, p. 54].

We now introduce Frölicher spaces, see [21], following terminology used in Kriegl and Michor's treatise [26].

Definition 2. • A **Frölicher** space is a triple $(X, \mathcal{F}, \mathcal{C})$ such that:

- $\mathcal{C}$ is a set of paths $\mathbb{R} \to X$;
- $\mathcal{F}$ is the set of functions from X to $\mathbb{R}$ such that the function $f : X \to \mathbb{R}$ is in $\mathcal{F}$ if and only if for any $c \in \mathcal{C}$, $f \circ c \in C^\infty(\mathbb{R}, \mathbb{R})$;
- A path $c : \mathbb{R} \to X$ is in $\mathcal{C}$ (i.e. is a **contour**) if and only if for any $f \in \mathcal{F}$, $f \circ c \in C^\infty(\mathbb{R}, \mathbb{R})$.

• Let $(X, \mathcal{F}, \mathcal{C})$ and $(X', \mathcal{F}', \mathcal{C}')$ be two Frölicher spaces; a map $f : X \to X'$ is **differentiable** (or, smooth) if and only if $\mathcal{F}' \circ f \circ \mathcal{C} \subset C^\infty(\mathbb{R}, \mathbb{R})$.

Any family of maps $\mathcal{F}_g$ from X to $\mathbb{R}$ generates a Frölicher structure $(X, \mathcal{F}, \mathcal{C})$ by setting, after [26]:

- $\mathcal{C} = \{c : \mathbb{R} \to X$ such that $\mathcal{F}_g \circ c \subset C^\infty(\mathbb{R}, \mathbb{R}); \}$
- $\mathcal{F} = \{f : X \to \mathbb{R}$ such that $f \circ \mathcal{C} \subset C^\infty(\mathbb{R}, \mathbb{R})\}$.

Clearly, $\mathcal{F}_g \subset \mathcal{F}$. We call $\mathcal{F}_g$ a **generating set of functions** for the Frölicher structure $(X, \mathcal{F}, \mathcal{C})$. This notion is useful for us since it allows us to describe in a simple way a Frölicher structure. A Frölicher space $(X, \mathcal{F}, \mathcal{C})$ carries a natural topology, the pull-back topology of $\mathbb{R}$ via $\mathcal{F}$. We also note that in the case of a finite dimensional differentiable manifold X it is natural to take $\mathcal{F}$ as the set of all smooth maps from X to $\mathbb{R}$, and $\mathcal{C}$ as the set of all smooth paths from $\mathbb{R}$ to X. In this case the underlying topology of the Frölicher structure is the same as the manifold topology, see [26]. In the infinite dimensional case there is, to our knowledge, no complete study of the relation between the Frölicher topology and the manifold topology; our intuition is that these two topologies are, in general, different.

Now, we note that if $(X, \mathcal{F}, \mathcal{C})$ is a Frölicher space, we can define a natural diffeology on X by using the following family of maps f defined on open domains $D(f)$ of Euclidean spaces (see [33]):

$$\mathcal{P}_\infty(\mathcal{F}) = \coprod_{p \in \mathbb{N}} \{ f : D(f) \to X;\ \mathcal{F} \circ f \in C^\infty(D(f), \mathbb{R}) \quad \text{(in the usual sense)}\}.$$

If X is a differentiable manifold, this diffeology has been called the **nébuleuse diffeology** by J.-M. Souriau in [53], and the nebulae diffeology in [25]. We can easily show the following, see for instance [33]:

Proposition 3. *Let $(X, \mathcal{F}, \mathcal{C})$ and $(X', \mathcal{F}', \mathcal{C}')$ be two Frölicher spaces. A map $f : X \to X'$ is smooth in the sense of Frölicher if and only if it is smooth for the underlying diffeologies $\mathcal{P}_\infty(\mathcal{F})$ and $\mathcal{P}_\infty(\mathcal{F}')$.*

In fact, it is known that the following implications hold, see for example [55].

Smooth manifold $\Rightarrow$ Frölicher space $\Rightarrow$ Diffeological space

Remark 4. We notice that the set of contours $\mathcal{C}$ of the Frölicher space $(X, \mathcal{F}, \mathcal{C})$ **does not** give us a diffeology, because a diffeology needs to be stable under restriction of domains. In the case of paths in $\mathcal{C}$ the domain is always $\mathbb{R}$, whereas the domain of 1-plots can (and has to) be any interval of $\mathbb{R}$. However, $\mathcal{C}$ defines a "minimal diffeology" $\mathcal{P}_1(\mathcal{F})$ whose plots are smooth parametrizations which are locally of the form $c \circ g$, where $g \in \mathcal{P}_\infty(\mathbb{R})$ and $c \in \mathcal{C}$. Using this setting, we can replace $\mathcal{P}_\infty$ by $\mathcal{P}_1$ in Proposition 3.

This technical remark is inspired by [55] and [33]; it is based on [26, p. 26, Boman's theorem].

Remark 5. Frölicher and Gâteaux smoothness are the same notion if we restrict to a Fréchet context. Indeed, for a smooth map $f : (F, \mathcal{P}_1(F)) \to \mathbb{R}$ defined on a Fréchet space with its 1-dimensional diffeology, we have that $\forall (x, h) \in F^2$, the map $t \mapsto f(x + th)$ is smooth as a classical map in $C^\infty(\mathbb{R}, \mathbb{R})$, and so f is Gâteaux smooth. The converse is obvious.

Diffeologies on cartesian products, projective limits, quotients, subsets, as well as pull-back and push-forward diffeologies are described in [25, Chapter 1]. We will use them, but we must refer the reader to the literature for details. Short expositions appear in [53] and in [35, 38, 37].

2.1.1 Functional diffeology

Let $(X, \mathcal{P})$ and $(X', \mathcal{P}')$ be two diffeological spaces. Let $S \subset C^\infty(X, X')$ be a set of smooth maps. The **functional diffeology** on S is the diffeology $\mathcal{P}_S$ made of all plots

$$\rho : D(\rho) \subset \mathbb{R}^k \to S$$

such that, for each $p \in \mathcal{P}$, the maps $\Phi_{\rho,p} : (x, y) \in D(p) \times D(\rho) \mapsto \rho(y)(x) \in X'$ are plots of $\mathcal{P}'$. With this definition, we can prove the following classical property:

Proposition 6. [25, p. 35] *Let X, Y, Z be diffeological spaces,*

$$C^\infty(X \times Y, Z) = C^\infty(X, C^\infty(Y, Z)) = C^\infty(Y, C^\infty(X, Z))$$

as diffeological spaces equipped with functional diffeologies.

As explained in [25, Chp. 5], functional diffeologies are crucial for the development of a homotopy theory for diffeological spaces. They are also of interest for the calculus of variations, see [25, p. 38], and for our study of diffeomorphism groups.

2.1.2 Tangent space

Let X be a diffeological space. There exist two main approaches to the notion of the tangent space at a point $x \in X$. The **internal tangent space** at $x \in X$ described in [10], and the **external tangent space** ${}^eTX$, defined simply as the set of derivations on $C^\infty(X, \mathbb{R})$, see [26, 25]. It is known that these two constructions coincide in the case of finite dimensional manifolds and in other important cases, see [26, Section 28] and [10]. For us, it is more convenient to begin with the definition of a **tangent cone** after [15]:

For each $x \in X$, we consider

$$C_x = \{c \in C^\infty(\mathbb{R}, X) | c(0) = x\}$$

and take the equivalence relation $\mathcal{R}$ given by

$$c\mathcal{R}c' \Leftrightarrow \forall f \in C^\infty(X, \mathbb{R}), \partial_t(f \circ c)|_{t=0} = \partial_t(f \circ c')|_{t=0} \, .$$

The tangent cone at x is the quotient

$${}^iT_xX = C_x/\mathcal{R} \, .$$

Equivalence classes of $\mathcal{R}$ are called **germs** and denoted by $V = \partial_t c(t)|_{t=0} = \partial_t c(0) \in {}^iT_xX$. We also use the notation

$$Df(V) = \partial_t(f \circ c)|_{t=0} \, .$$

It is shown in [15] that there exist examples of diffeological spaces for which the tangent cone at a point x is not a vector space, hence the need for more sophisticated definitions as in [10, 25, 26]. Fortunately, tangent cones are all we need in the case of diffeological groups, as we now see.

2.1.3 Regular Lie groups

If we start with an algebraic structure, we can define a corresponding compatible diffeological structure. For example, following [25, pp. 66-68], an $\mathbb{R}-$vector space equipped with a diffeology is called a diffeological vector space if addition and scalar multiplication are smooth, and an analogous definition holds for Frölicher vector spaces. In a similar vein we define:

Definition 7. Let G be a group equipped with a diffeology $\mathcal{P}$. We say that G is a **diffeological group** if both multiplication and inversion are smooth, and an analogous definition holds for Frölicher groups.

We have, after [29, Proposition 1.6.], see also [38].

Proposition 8. *Let G be a diffeological group with unit $e \in G$. Then the tangent cone at the unit element, ${}^iT_eG$, is a diffeological vector space called the tangent space of G at e.*

For example, we define $X + Y = \partial_t(c.d)(0)$, where $(c,d) \in \mathcal{C}^2$, $c(0) = d(0) = e_G$, $X = \partial_t c(0)$ and $Y = \partial_t d(0)$. Now we wish to transform ${}^iT_eG$ into a Lie algebra. This is not easy and a general criterion appears in [29, Definition 1.13 and Theorem 1.14]. For us, the most direct way to proceed, since we work with a well-specified class of examples, is via the following definition.

Definition 9. Let G be a diffeological group and $\mathfrak{g} = ^i T_eG$ be its corresponding tangent space at the unit e. We have:

- Let $(X, g) \in \mathfrak{g} \times G$; then, $Ad_g(X) = \partial_t(gcg^{-1})(0)$ where $c \in \mathcal{C}$, $c(0) = e_G$, and $X = \partial_t c(0)$.
- Let $(X, Y) \in \mathfrak{g}^2$; then, $[X, Y] = \partial_t(Ad_{c(t)}Y)$ where $c \in \mathcal{C}$, $c(0) = e_G$, $X = \partial_t c(0)$.

The diffeological group G is a **diffeological Lie group** if and only if the derivative of the adjoint action of G on ${}^iT_eG$ defines a smooth Lie bracket. In this case, we call $\mathfrak{g} = ^i T_eG$ the **diffeological Lie algebra** of G.

This definition implies that the basic properties of adjoint and coadjoint actions, and of Lie brackets, remain globally the same as in the case of finite-dimensional Lie groups, and the proofs are similar: we only need to replace manifold charts by plots of the underlying diffeologies (see e.g. [29] for further details, and [2] for the case of Frölicher Lie groups). The following two definitions allow us to work with smooth exponential mappings.

Definition 10. [29] A diffeological Lie group G with diffeological Lie algebra $\mathfrak{g}$ is said to be **regular** if and only if there is a smooth map

$$Exp : C^\infty([0;1], \mathfrak{g}) \to C^\infty([0,1], G)$$

such that $g(t) = Exp(v(t))$ is the unique solution to the differential equation

$$\begin{cases} g(0) &= e\,, \\ \dfrac{dg(t)}{dt} g(t)^{-1} &= v(t)\,. \end{cases} \tag{2.1}$$

The exponential function is the map

$$\begin{aligned} exp : \mathfrak{g} &\to G \\ v &\mapsto exp(v) = g(1)\,, \end{aligned}$$

in which v is considered as a constant path in $C^\infty([0;1], \mathfrak{g})$, and g is the image by Exp of v.

Definition 11. [29] Let V be a Frölicher vector space. The space V is **integral** (or, **regular**) if there is a smooth map

$$\int_0^{(.)} : C^\infty([0;1];V) \to C^\infty([0;1],V)$$

such that $\int_0^{(.)} v = u$ if and only if u is the unique solution of the differential equation

$$\begin{cases} u(0) &= 0 \\ u'(t) &= v(t) \, . \end{cases}$$

The word "integral" is used in [29, Definition 1.3], while "regular" is used in [35, Definition 1.16]. Definition 11 applies, for instance, if V is a complete locally convex topological vector space equipped with its natural Frölicher structure given by the Frölicher completion of its nébuleuse diffeology, see [25, 33, 35]. The best behaved class of groups (to which the groups considered herein belong) is the following:

Definition 12. Let G be a Frölicher Lie group with Lie algebra $\mathfrak{g}$. We say that G is **fully regular** if $\mathfrak{g}$ is integral and G is regular.

The properties of the specific groups which we use in the following sections are consequences of two structural results which we quote for completeness. Proofs appear in [35].

Theorem 13. [35] *Let $(A_n)_{n\in\mathbb{N}^*}$ be a sequence of integral (Frölicher) vector spaces equipped with a graded smooth multiplication operation on $\bigoplus_{n\in\mathbb{N}^*} A_n$, i.e. a multiplication such that for each $n, m \in \mathbb{N}^*$, $A_n.A_m \subset A_{n+m}$ is smooth with respect to the corresponding Frölicher structures. Let us define the (non-unital) algebra of formal series:*

$$\mathcal{A} = \left\{ \sum_{n\in\mathbb{N}^*} a_n \,|\, \forall n \in \mathbb{N}^*, a_n \in A_n \right\} ,$$

equipped with the Frölicher structure of the infinite product. Then, the set

$$1 + \mathcal{A} = \left\{ 1 + \sum_{n\in\mathbb{N}^*} a_n \,|\, \forall n \in \mathbb{N}^*, a_n \in A_n \right\}$$

is a fully regular Frölicher Lie group with $\mathfrak{g} = \mathcal{A}$. Moreover, the exponential map defines a smooth bijection $\mathcal{A} \to 1 + \mathcal{A}$.

The definition of a product Frölicher structure is in [26], see also [35, 38]. Hereafter we write $[u]_n$ for the A_n-component of u, $u \in \mathcal{A}$.

Theorem 14. [35] *Let*

$$1 \longrightarrow K \stackrel{i}{\longrightarrow} G \stackrel{p}{\longrightarrow} H \longrightarrow 1$$

be an exact sequence of Frölicher Lie groups, such that there is a smooth section $s : H \to G$, and such that the subset diffeology from G on $i(K)$ coincides with *the push-forward diffeology from K to $i(K)$. We consider also the corresponding sequence of Lie algebras*

$$0 \longrightarrow \mathfrak{k} \xrightarrow{i'} \mathfrak{g} \xrightarrow{p} \mathfrak{h} \longrightarrow 0.$$

Then,

- *The Lie algebras $\mathfrak{k}$ and $\mathfrak{h}$ are integral if and only if the Lie algebra $\mathfrak{g}$ is integral*
- *The Frölicher Lie groups K and H are regular if and only if the Frölicher Lie group G is regular.*

Similar results, as in Theorem 14, are valid for Fréchet Lie groups, see [26].

2.2 Principal bundles, connections and the Ambrose-Singer theorem

Let P be a diffeological space and let G be a Frölicher Lie group. We assume that there exists a smooth right-action $P \times G \to P$, such that $\forall (p, p', g) \in P \times P \times G$ we have $p.g = p'.g \Rightarrow p = p'$. We set $M = P/G$ and we equip it with the quotient diffeology. P is said to be a **principal G−bundle** with base M.

We define connections on a principal G-bundle in terms of paths on the total space P following Iglesias-Zemmour's approach, see [25, Article 8.32].

Definition 15. Let G be a diffeological group, and let $\pi\colon P \to X$ be a principal G-bundle. Denote by $\mathrm{Paths}_{\mathrm{loc}}(P)$ the diffeological space of **local paths** (see [25, Article 1.63]), and by $tpath(P)$ the **tautological bundle of local paths**

$$tpath(P) := \{(\gamma, t) \in \mathrm{Paths}_{\mathrm{loc}}(P) \times \mathbb{R} \mid t \in domain(\gamma)\}\,.$$

A **diffeological connection** is a smooth map $H_\theta\colon tpath(P) \to \mathrm{Paths}_{\mathrm{loc}}(P)$ satisfying the following properties for any $(\gamma, t_0) \in tpath(P)$:

1. the domain of γ equals the domain of $H_\theta(\gamma, t_0)$,
2. $\pi \circ \gamma = \pi \circ H_\theta(\gamma, t_0)$,
3. $H_\theta(\gamma, t_0)(t_0) = \gamma(t_0)$,
4. $H_\theta(\gamma \cdot g, t_0) = H_\theta(\gamma, t_0) \cdot g$ for all $g \in G$,
5. $H_\theta(\gamma \circ f, s) = H_\theta(\gamma, f(s)) \circ f$ for any smooth map f from an open subset of $\mathbb{R}$ into $D(\gamma)$,
6. $H_\theta(\theta(\gamma, t_0), t_0) = H_\theta(\gamma, t_0)$.

Another formulation of this definition can be found in [35] under the terminology of **path-lifting**. Diffeological connections allow us to construct unique horizontal lifts of paths in $\mathrm{Paths}_{\mathrm{loc}}(M)$, see [25, Article 8.32], and they pull back by smooth maps [25, Article 8.33]. We now introduce connection forms. We need the following preliminary result:

Proposition 16. *Let V be a diffeological vector space. The diffeological group G acts smoothly on the right on the space of V-valued differential forms $\Omega(P,V)$ by setting*

$$\forall (g,\alpha) \in G \times \Omega^n(P,V) \times G, \forall p \in \mathcal{P}(P), \quad (g_*\alpha)_{g.p} = \alpha_p \circ (dg^{-1})^n .$$

This proposition allows us to make the following definition:

Definition 17. Let $\alpha \in \Omega(P;\mathfrak{g})$. The differential form α is **right-invariant** if and only if for each $p \in \mathcal{P}(P)$, and for each $g \in G$,

$$\alpha_{g.p} = Ad_{g^{-1}} \circ g_*\alpha_p .$$

A connection form is now defined as a particular right-invariant one-form:

Definition 18. A **connection** on P is a right-invariant $\mathfrak{g}-$valued $1-$form θ such that, for each $v \in \mathfrak{g}$, for any path $c : \mathbb{R} \to G$ such that $\left\{ \begin{array}{rcl} c(0) & = & e_G \\ \partial_t c(t)|_{t=0} & = & v \end{array} \right.$, and for each $p \in P$,

$$\theta(\partial_t(p.c(t))_{t=0}) = v .$$

Let $p \in P$ and let γ be a smooth path in P starting at p and defined on $[0;1]$. We set $H_\theta\gamma(t) = \gamma(t)g(t)$, in which $g(t) \in C^\infty([0;1];\mathfrak{g})$ is a path satisfying the differential equation

$$\left\{ \begin{array}{rcl} \theta\left(\partial_t H_\theta\gamma(t)\right) & = & 0 \\ H_\theta\gamma(0) & = & \gamma(0) . \end{array} \right.$$

The first line of this equation is equivalent to the differential equation $g^{-1}(t)\partial_t g(t) = -\theta(\partial_t\gamma(t))$ which is integrable, and the second to the initial condition $g(0) = e_G$. This shows that horizontal lifts are well-defined, as in the case of manifolds; the map $H_\theta(.)$ defines a diffeological connection, and we can consider the holonomy group of the connection as in classical differential geometry. As in this case, it can be shown that the holonomy group is invariant (up to conjugation) under the choice of the basepoint p.

We remark that in full generality, see [35], we now need to make a technical assumption on the "dimension" of P (a notion discussed in [25, p. 47]) but we omit it since it is always satisfied by the bundles appearing in this work.

Definition 19. Let $\alpha \in \Omega(P;\mathfrak{g})$ be any $G-$invariant form; the horizontal derivative of α is $\nabla\alpha = d\alpha - \frac{1}{2}[\theta,\alpha]$. The curvature 2-form induced by a connection θ on P is

$$\Omega = \nabla\theta .$$

We fix a principal G-bundle P with base M. We have, after [35]:

Theorem 20. *We assume that G_1 and G are fully regular Frölicher Lie groups with Lie algebras $\mathfrak{g}_1$ and $\mathfrak{g}$, and we let $\rho : G_1 \mapsto G$ be an injective morphism of Lie groups. If there is a connection θ on P with curvature Ω so that, for any smooth* *one-parameter family $H_\theta c_t$ of horizontal paths starting at $p \in P$, and for any smooth vector fields X, Y in M, the map*

$$(s,t) \in [0,1]^2 \quad \rightarrow \quad \Omega_{Hc_t(s)}(X,Y) \tag{2.2}$$

is a smooth $\mathfrak{g}_1$-valued map, and if M is simply connected, then the structure group G of P reduces to G_1, and the connection θ also reduces.

We can now state our version of the Ambrose-Singer theorem. Part (3) below is technical and written down here only for completeness: it uses the terminology of [50] for the classification of groups via properties of the exponential maps:

Theorem 21. [35] *Let P be a principal bundle with basis M and a fully regular Frölicher structure group G. Let θ be a connection on P and let H_θ be the associated diffeological connection.*

1. *For each $p \in P$, the holonomy group at p is a diffeological subgroup of G which does not depend on the choice of p, up to conjugation.*
2. *There exists a second holonomy group H^{red} which is the smallest structure group for which there is a sub-bundle P' to which θ reduces. Its Lie algebra is spanned by the curvature elements, i.e. it is the smallest integrable Lie algebra which contains the curvature elements.*
3. *If G is a Lie group (in the classical sense) of type I or II, there is a (minimal) closed Lie subgroup $\bar{H}^{red}$ (in the classical sense) such that $H^{red} \subset \bar{H}^{red}$, and whose Lie algebra is the closure in $\mathfrak{g}$ of the Lie algebra of H^{red}. The group $\bar{H}^{red}$ is the smallest closed Lie subgroup of G among the structure groups of closed sub-bundles $\bar{P}'$ of P to which θ reduces.*

The following result is the consequence of Theorem 21 which we use in order to integrate the KP hierarchy.

Proposition 22. *If the connection θ is flat and if M is connected and simply connected then, for any path γ starting at $p \in P$, the map*

$$\gamma \mapsto H\gamma(1)$$

depends only on $\pi(\gamma(1)) \in M$ and defines a global smooth section $M \rightarrow P$. Therefore, $P = M \times G$.

This proposition is in [35], see also [25, Article 8.35 Note 1]. We also state:

Theorem 23. *Let $(G, \mathfrak{g})$ be a fully regular Lie group and let X be a simply connected Frölicher space. Let $\alpha \in \Omega^1(X, \mathfrak{g})$ such that*

$$d\alpha + [\alpha, \alpha] = 0\ . \tag{$*$}$$

Then, there exists a smooth map

$$f : X \rightarrow G$$

such that

$$df.f^{-1} = \alpha\,.$$

Moreover, one can obtain one solution f from another one by applying (pointwise for $x \in X$) the adjoint action of G on $\mathfrak{g}$.

An analogous result appears in [26, section 40.2] in the "convenient setting" considered by Kriegl and Michor. This theorem can be also written down assuming that α satisfies the equation

$$d\alpha - [\alpha, \alpha] = 0$$

instead of $(*)$. The correspondence between solutions to these two equations is given by the smooth map $f \mapsto f^{-1}$ on the group $C^\infty(X, G)$.

2.3 Groups of diffeomorphisms

Let us make some comments on the structure of diffeomorphism groups, as an introduction to our work to be reported in the following sections.

Let M be a locally compact Riemannian manifold equipped with its nébuleuse diffeology. We equip the group of diffeomorphisms $Diff(M)$ with the topology of uniform convergence on compact sets of derivatives at all orders, what is usually called the $C^\infty-$compact-open topology or weak topology, see [23]. Traditionally, $Vect(M)$ is considered to be the Lie algebra of $Diff(M)$, but [26, Section 43.1] shows that this identification is not always correct. It is true when M is compact. In this case, $Vect(M)$ *is* the Lie algebra of $Diff(M)$, a result which can be obtained using Omori's regularity theorems [44, 45], see also [10]. In general, what is well known is that infinitesimal actions (i.e. elements of the internal tangent space at identity) of $Diff(M)$ on $C^\infty(M, \mathbb{R})$ generate vector fields on M. The bracket on vector fields is given by

$$(X, Y) \in Vect(M) \mapsto [X, Y] = \nabla_X Y - \nabla_Y X\ ,$$

in which ∇ is the Levi-Civita connection on TM. This is a Lie bracket, stable under the adjoint action of $Diff(M)$. Moreover, the compact-open topology on $Diff(M)$ generates a corresponding $C^\infty-$compact-open topology on $Vect(M)$ which is precisely the $D-$topology for the functional diffeology on $Diff(M)$. It follows that we can apply Leslie's results appearing in [29]: $Vect(M)$ equipped with the C^∞ compact-open topology is a Fréchet vector space, the Lie bracket is smooth, and $Diff(M)$ is a diffeological Lie group in the sense of [29, Definition 1.13 and Theorem 1.14] with Lie algebra $\mathfrak{g} \subset Vect(M)$.

Now we ask about regularity results for the group $Diff(M)$ equipped with the smooth compact-open topology. They are the following:

Theorem 24. *Let M be a finite dimensional manifold without boundary.*

1. *If M is not compact, $Diff(M)$ is a Frölicher Lie group which is not regular.*
2. *If M is compact, $Diff(M)$ is a fully regular Frölicher Lie group, and in fact, it is an ILH Lie group.*

Part (1) is proven in [37] while an extensive exposition about (2) is given in [45].

3 Infinite-dimensional Lie groups and pseudodifferential operators

We introduce the groups and algebras of pseudodifferential operators needed to set up a KP hierarchy and to investigate its Cauchy problem. We have considered several different instances of "pseudodifferential operators" in our papers [35, 19, 38, 20, 39] and in this section we summarize our work. We begin with non-formal pseudodifferential operators and then we consider the algebraic setting of formal pseudodifferential operators. We also present our smooth versions of the Mulase factorization theorem.

3.1 Frölicher Lie groups of pseudodifferential and Fourier integral operators

In this section E is a real or complex finite-dimensional vector bundle over S^1; below we will specialize our considerations to the case $E = S^1 \times V$ in which V is a finite-dimensional vector space. We follow [3, Section 2.1]:

Definition 25. The graded algebra of differential operators acting on the space of smooth sections $C^\infty(S^1, E)$ is the algebra $DO(E)$ generated by:

• Elements of $End(E)$, the group of smooth maps $E \to E$ leaving each fibre globally invariant and which restrict to linear maps on each fibre. This group acts on sections of E via (matrix) multiplication;

• The differentiation operators

$$\nabla_X : g \in C^\infty(S^1, E) \mapsto \nabla_X g$$

where ∇ is a connection on E and X is a vector field on S^1.

Multiplication operators are operators of order 0; differentiation operators and vector fields are operators of order 1. In local coordinates, a differential operator of order k has the form $P(u)(x) = \sum p_{i_1 \cdots i_r} \nabla_{x_{i_1}} \cdots \nabla_{x_{i_r}} u(x)$, $\quad r \leq k$, in which the coefficients $p_{i_1 \cdots i_r}$ can be matrix-valued functions on S^1.

The algebra $DO(E)$ is graded by order. We note by $DO^k(S^1)$,$k \geq 0$, the differential operators of order less than or equal to k. $DO(E)$ is a subalgebra of the algebra of classical pseudodifferential operators $Cl(S^1, E)$. We need to assume knowledge of the basic theory of pseudodifferential operators as it appears for example in [22]. A global symbolic calculus for pseudodifferential operators has been defined independently by J. Bokobza-Haggiag, see [4] and H. Widom, see [56]. In these papers is shown, among other things, how the geometry of the base manifold M furnishes an obstruction to generalizing local formulas of composition and inversion of symbols; we do not recall these formulas here since they are not involved in our computations.

After [36], we assume henceforth that S^1 is equipped with charts such that the changes of coordinates are translations.

Notations. We follow [39]. We note by $PDO(S^1, E)$ (resp. $PDO^o(S^1, E)$, resp. $Cl(S^1, E)$) the space of pseudodifferential operators (resp. pseudodifferential operators of order o, resp. classical pseudodifferential operators) acting on smooth sections of E, and by $Cl^o(S^1, E) = PDO^o(S^1, E) \cap Cl(S^1, E)$ the space of classical pseudodifferential operators of order o. We also denote by $Cl^{o,*}(S^1, E)$ the group of units of $Cl^o(S^1, E)$. If the vector bundle E is trivial, i.e. $E = S^1 \times V$ or $E = S^1 \times \mathbb{K}^p$ with $\mathbb{K} = \mathbb{R}$ or $\mathbb{C}$, we use the notation $Cl(S^1, V)$ or $Cl(S^1, \mathbb{K}^p)$ instead of $Cl(S^1, E)$.

A topology on spaces of classical pseudodifferential operators has been described by Kontsevich and Vishik in [27]; other descriptions appear in [6, 46, 51]. We shall use the Kontsevich-Vishik topology. This is a Fréchet topology such that each space $Cl^o(S^1, E)$ is closed in $Cl(S^1, E)$. We set

$$PDO^{-\infty}(S^1, E) = \bigcap_{o \in \mathbb{Z}} PDO^o(S^1, E) .$$

It is well-known that $PDO^{-\infty}(S^1, E)$ is a two-sided ideal of $PDO(S^1, E)$, see e.g. [22, 51]. Therefore, we can define the quotients

$$\mathcal{F}PDO(S^1, E) = PDO(S^1, E)/PDO^{-\infty}(S^1, E),$$

$$\mathcal{F}Cl(S^1, E) = Cl(S^1, E)/PDO^{-\infty}(S^1, E),$$

$$\mathcal{F}Cl^o(S^1, E) = Cl^o(S^1, E)/PDO^{-\infty}(S^1, E) .$$

The script font $\mathcal{F}$ stands for *formal pseudodifferential operators.* The quotient $\mathcal{F}PDO(S^1, E)$ is an algebra isomorphic to the set of formal symbols, see [4], and the identification is a morphism of $\mathbb{C}$-algebras for the usual multiplication on formal symbols (see e.g. [22]). We denote by $\mathcal{F}Cl^{0,*}(S^1, E)$ the group of units of the algebra $\mathcal{F}Cl^0(S^1, E)$.

Theorem 26. *The groups $Cl^{0,*}(S^1, E)$, $\mathcal{F}Cl^{0,*}(S^1, E)$, and $Diff_+(S^1)$, in which $Diff_+(S^1)$ is the group of all the orientation-preserving diffeomorphisms of S^1, are regular Frölicher Lie groups.*

In actual fact, it has been noticed in [34] that the group $Cl^{0,*}(S^1, V)$ (resp. $\mathcal{F}Cl^{0,*}(S^1, V)$) is open in $Cl^0(S^1, V)$ (resp. $\mathcal{F}Cl^0(S^1, V)$) and also that it is a regular *Fréchet* Lie group. It follows from [16, 45] that $Diff_+(S^1)$ is open in the Fréchet manifold $C^\infty(S^1, S^1)$. This fact makes it a Fréchet manifold and, following [45], a regular Fréchet Lie group.

Definition 27. A classical pseudodifferential operator A on S^1 is said to be an odd class operator if and only if for all $n \in \mathbb{Z}$ and all $(x, \xi) \in T^*S^1$ we have:

$$\sigma_n(A)(x, -\xi) = (-1)^n \sigma_n(A)(x, \xi) ,$$

in which σ_n is the symbol of A.

This particular class of pseudodifferential operators has been introduced in [27, 28]. Hereafter, the notation Cl_{odd} will refer to odd class classical pseudodifferential operators.

Proposition 28. *The algebra $Cl^0_{odd}(S^1, V)$ is a closed subalgebra of $Cl^0(S^1, V)$. Moreover, the group $Cl^{0,*}_{odd}(S^1, V)$ of all invertible odd class bounded classical pseudodifferential operators is*

- *an open subset of $Cl^0(S^1, V)$ and,*
- *a regular Fréchet Lie group.*

Proof. Let us sketch a proof of this important result. We note by $\sigma(A)(x, \xi)$ the total formal symbol of $A \in Cl^0(S^1, V)$, and we define the map

$$\phi : Cl^0(S^1, V) \to \mathcal{F}Cl^0(S^1, V)$$

by

$$\phi(A) = \sum_{n \in \mathbb{N}} \sigma_{-n}(x, \xi) - (-1)^n \sigma_{-n}(x, -\xi).$$

This map is smooth, and

$$Cl^0_{odd}(S^1, V) = Ker(\phi),$$

which shows that $Cl^0_{odd}(S^1, V)$ is a closed subalgebra of $Cl^0(S^1, V)$. Moreover, if $H = L^2(S^1, V)$,

$$Cl^{0,*}_{odd}(S^1, V) = Cl^0_{odd}(S^1, V) \cap GL(H),$$

which proves that $Cl^{0,*}_{odd}(S^1, V)$ is open in the Fréchet algebra $Cl^0(S^1, V)$, and it follows that it is a regular Fréchet Lie group by arguing along the lines of [42]. ■

In addition to groups of pseudodifferential operators, we also need a restricted class of groups of Fourier integral operators which we call $Diff_+(S^1)$–pseudodifferential operators following [36]. These groups appear as central extensions of $Diff_+(S^1)$ by groups of (often bounded) pseudodifferential operators. We do not state the basic facts on Fourier integral operators here (they can be found in the classical paper [24]), but we recall the following theorem.

Theorem 29. [36] *Let H be a regular Lie group of pseudodifferential operators acting on smooth sections of a trivial bundle $E \sim V \times S^1 \to S^1$. The group $Diff(S^1)$ acts smoothly on $C^\infty(S^1, V)$, and it is assumed to act smoothly on H by adjoint action. If H is stable under the $Diff(S^1)$–adjoint action, then there exists a regular Lie group G of Fourier integral operators defined through the exact sequence:*

$$0 \to H \to G \to Diff(S^1) \to 0\,.$$

If H is a Frölicher Lie group, then G is a Frölicher Lie group.

This result is an application of Theorem 14, see [36, Theorem 4]. The pseudodifferential operators considered in this theorem can be classical, odd class, or anything else. Applying the formulas of "changes of coordinates" of e.g. [22], we obtain that odd class pseudodifferential operators are stable under adjoint action of $Diff(S^1)$. Thus, we can define the following group:

Definition 30. The group $FCl^{0,*}_{Diff(S^1),odd}(S^1,V)$ is the regular Frölicher Lie group G obtained in Theorem 29 with $H = Cl^{0,*}_{odd}(S^1,V)$.

Operators A in this group can be understood as operators in $Cl^{0,*}_{odd}(S^1,V)$ twisted by diffeomorphisms, this is,

$$A = B \circ g\,, \tag{3.1}$$

where $g \in Diff(S^1)$ and $B \in Cl^{0,*}_{odd}(S^1,V)$, see [36]. The diffeomorphism g is the phase of the operators, but here the phase (and hence the decomposition (3.1)) is unique, which is not the case for general Fourier integral operators, see e.g. [24].

We can also consider the group $Diff_+(S^1)$ introduced in the first theorem of this section. Theorem 29 allows us to make the following definition:

Definition 31. The group $FCl^{0,*}_{Diff_+(S^1),odd}(S^1,V)$ is the regular Frölicher Lie group of all operators in $FCl^{0,*}_{Diff(S^1),odd}(S^1,V)$ whose phase diffeomorphisms lie in $Diff_+(S^1)$.

We observe that the central extension $FCl^{0,*}_{Diff(S^1),odd}(S^1,V)$ contains, as a subgroup, the group

$$DO^{0,*}(S^1,V) \rtimes Diff_+(S^1)\,, \tag{3.2}$$

where $DO^{0,*}(S^1,V) = C^\infty(S^1, GL(V))$ is the loop group of $GL(V)$.

Let us now assume that V is a complex vector space. By the symmetry property stated in Definition 27, an odd class pseudodifferential operator A has a partial symbol of non-negative order n that reads

$$\sigma_n(A)(x,\xi) = \gamma_n(x)(i\xi)^n\,, \tag{3.3}$$

where $\gamma_n \in C^\infty(S^1, L(V))$. This fact allows us to check the following direct sum decomposition:

Proposition 32.

$$Cl_{odd}(S^1,V) = Cl^{-1}_{odd}(S^1,V) \oplus DO(S^1,V)\,.$$

Notation. Let $A \in Cl_{odd}(S^1,V)$. We note by A_D and A_S the operators such that $A = A_S + A_D$ in the previous decomposition.

The second summand of this expression will be specialized in the sequel to differential operators having symbols of order 1. Because of (3.3), we can understand these symbols as elements of $Vect(S^1) \otimes Id_V$.

We have finished introducing the spaces of non-formal pseudodifferential and Fourier operators that we use in our work. We end this subsection by considering series of pseudodifferential operators. This step is needed because we wish to control the growth of the objects appearing in our KP hierarchy.

We fix a formal parameter h and we consider the following algebras and groups, setting val_h as the obvious valuation of $h-$series, following [35, 38, 39]:

Definition 33. The set of odd class $h-$pseudodifferential operators is the set of formal series

$$Cl_{h,odd}(S^1,V) = \left\{ \sum_{n\in\mathbb{N}} a_n h^n \,|\, a_n \in Cl^n_{odd}(S^1,V) \right\} . \tag{3.4}$$

Theorem 34. *The set $Cl_{h,odd}(S^1,V)$ is a Fréchet algebra, and its group of units given by*

$$Cl^*_{h,odd}(S^1,V) = \left\{ \sum_{n\in\mathbb{N}} a_n h^n \,|\, a_n \in Cl^n_{odd}(S^1,V), a_0 \in Cl^{0,*}_{odd}(S^1,V) \right\} , \tag{3.5}$$

is a regular Fréchet Lie group.

We will use in the final section of this work that the decomposition $A = A_S + A_D$ explained after Proposition 32 extends trivially to $Cl_{h,odd}(S^1,V)$.

3.2 Frölicher Lie groups and formal pseudodifferential operators

Let A be a commutative $\mathbb{K}-$algebra with unit 1, in which for simplicity we assume that $\mathbb{K}$ is either $\mathbb{R}$ or $\mathcal{C}$. Let A^* be the group of invertible elements (or, *units*) of A. We assume that A is equipped with a derivation, that is, with a $\mathbb{K}$-linear map $\partial : A \to A$ satisfying the Leibnitz rule $\partial(f\cdot g) = (\partial f)\cdot g + f\cdot(\partial g)$ for all $f, g \in A$.

Let ξ be a formal variable not in A. The *algebra of symbols* over A is the vector space

$$\Psi_\xi(A) = \left\{ P_\xi = \sum_{\nu\in\mathbf{Z}} a_\nu\, \xi^\nu : a_\nu \in A\,,\ a_\nu = 0 \text{ for } \nu \gg 0 \right\}$$

equipped with the associative multiplication $\circ$ given by

$$P_\xi \circ Q_\xi = \sum_{k\geq 0} \frac{1}{k!}\, \frac{\partial^k P_\xi}{\partial \xi^k}\, \partial^k Q_\xi \,, \tag{3.6}$$

with the prescription that multiplication on the right-hand side of (3.6) is standard multiplication of Laurent series in ξ with coefficients in A, see [13, 43]. The algebra A is included in $\Psi_\xi(A)$. The *algebra of formal pseudodifferential operators* over A is the vector space

$$\Psi(A) = \left\{ P = \sum_{\nu\in\mathbb{Z}} a_\nu\, \partial^\nu : a_\nu \in A\,,\ a_\nu = 0 \text{ for } \nu \gg 0 \right\}$$

equipped with the unique multiplication which makes the map $\sum_{\nu\in\mathbb{Z}} a_\nu\, \xi^\nu \mapsto \sum_{\nu\in\mathbb{Z}} a_\nu\, \partial^\nu$ an algebra homomorphism. The algebra $\Psi(A)$ is associative but not commutative. It becomes a Lie algebra over K if we define, as usual,

$$[P, Q] = P\,Q - Q\,P\,. \tag{3.7}$$

The *order* of $P \neq 0 \in \Psi(A)$, $P = \sum_{\nu \in \mathbb{Z}} a_\nu \partial^\nu$, is N if $a_N \neq 0$ and $a_\nu = 0$ for all $\nu > N$. If P is of order N, the coefficient a_N is called the *leading term* or *principal symbol* of P. We note by $\Psi^N(A)$ the vector space of pseudodifferential operators P as above satisfying $m > N \Rightarrow a_m = 0$, that is,

$$\Psi^N(A) = \{a \in \Psi(A) | \forall m > N, a_m = 0\}.$$

The special case of $\Psi^0(A)$ is of particular interest, since it is an algebra. Adapting the notations used in the previous subsection, we write $\Psi^{0,*}(A)$ for its group of units, i.e. the group of invertible elements of $\Psi^0(A)$.

We assume now that the algebra A is a **Frölicher algebra**, and that therefore addition, scalar multiplication and multiplication are smooth, and that inversion is a smooth operation on A^*, in which A^* is equipped with the subset Frölicher structure. We also assume that the derivation ∂ is smooth. Then, identifying a formal operator $a \in \Psi(A)$ with its sequence of partial symbols, $\Psi(A)$, as a linear subspace of $A^{\mathbb{Z}}$, carries a natural Frölicher structure, and we obtain:

Proposition 35. *$\Psi(A)$ is a Frölicher algebra.*

Now let us assume that A^* is a Frölicher Lie group with Lie algebra $\mathfrak{g}_A$, and that A is an integral Frölicher vector space. Our Theorems 13 and 14 imply the following two results:

Lemma 36. *$1 + \Psi^{-1}(A)$ is a regular Frölicher Lie group with regular Lie algebra $\Psi^{-1}(A)$.*

Theorem 37. *There exists a short exact sequence of groups:*

$$1 \longrightarrow 1 + \Psi^{-1}(A) \longrightarrow \Psi^{0,*}(A) \longrightarrow A^* \longrightarrow 1$$

such that:

1. *the injection $1 + \Psi^{-1}(A) \to \Psi^{0,*}(A)$ is smooth*
2. *the principal symbol map $\Psi^{0,*}(A) \to A^*$ is smooth and has a global section which is the restriction to A^* of the canonical inclusion $A \to \Psi^0(A)$.*

As a consequence, A^* is a fully regular Frölicher Lie group if and only if $\Psi^{0,*}(A)$ is a fully regular Frölicher Lie group with Lie algebra $\mathfrak{g}_A \oplus \Psi^{-1}(A)$.

It is natural to wonder if we can extend these results to the full algebra $\Psi(A)$. As we show in [38], this is not possible and motivates us, after [40, 41], to "regularize" $\Psi(A)$ into some "deformed" algebra. In order to do this, we need to restrict the algebra A. We follow Mulase's works [40, 41] and choose a specific A which takes into account "dependence in time", something essential since we wish to write down evolution equations.

After Bourbaki [5, Algebra, Chapters 1-3, p. 454-457], we let T be the additive monoid of all sequences $(n_i)_{i \in \mathbb{N}^*}$, $n_i \in \mathbb{N}$, such that $n_i = 0$ except for a finite number of indices, and R as a commutative ring equipped with a derivation ∂. A *formal* *power series* is a function u from T to R, $u = (u_t)_{t\in T}$. Consider an infinite number of formal variables τ_i, $i \in I$, and set $\tau = (\tau_1, \tau_2, \cdots)$. Then, the formal power series u can be written as $u = \sum_{t\in T} u_t \tau^t$ in which $\tau^t = \tau_1^{n_1}\tau_2^{n_2}\cdots$. We say that $u_t \in R$ is a coefficient and that $u_t\tau^t$ is a term.

Operations are defined in a usual manner: If $u = (u_t)_{t\in T}$, $v = (v_t)_{t\in T}$, then

$$u + v = (u_t + v_t)_{t\in T} \quad \text{and} \quad u\,v = w\,,$$

in which $w = (w_t)_{t\in T}$ and

$$w_t = \sum_{\substack{r,s\in T\\ r+s=t}} u_r v_s\,.$$

It is shown in [5, p. 455] that this multiplication is well defined, and that A_t, the set of all these formal power series equipped with these two operations, is a commutative algebra with unit. The derivation ∂ on R extends to a derivation on A via

$$\partial u = \sum_{t\in T} (\partial u_t)\tau^t.$$

We define the *valuation* of a power series following [40, p. 59]: Let $u \in A_t$, $u \neq 0$. We write $u = \sum_{t\in T} u_t\tau^t$, and if $t = (n_i)_{i\in I}$, we set $\mid t \mid = \sum i n_i$. The terms $u_t\tau^t$ such that $\mid t \mid = p$ are called *monomials of valuation* p. The formal power series u_p whose terms of valuation p are those of u, and whose other terms are zero, is called *the homogeneous part of* u *of valuation* p. The series u_0, the homogeneous part of u of valuation 0, is identified with an element of R called the *constant term* of u. For a formal series $u \neq 0$, the least integer $p \geqslant 0$ such that $u_p \neq 0$ is called the **valuation** of u, and it is denoted by $val_t(u)$. We extend this definition to the case $u = 0$ by setting $val_t(0) = +\infty$ (see [5, p. 457]). The following properties hold: if u, v are formal power series then

$$val_t(u + v) \geqslant inf(val_t(u), val_t(v))\,, \quad \text{if } u + v \neq 0\,, \tag{3.8}$$

$$val_t(u + v) = inf(val_t(u), val_t(v))\,, \quad \text{if } val_t(u) \neq val_t(v)\,, \tag{3.9}$$

$$val_t(uv) \geqslant val_t(u) + val_t(v)\,, \quad \text{if } uv \neq 0\,. \tag{3.10}$$

Definition 38 (Mulase, [41]). The spaces of formal pseudodifferential and differential operators of infinite order, $\widehat{\Psi}(A_t)$ and $\widehat{\mathcal{D}}_{A_t}$ respectively, are

$$\widehat{\Psi}(A_t) = \left\{ P = \sum_{\alpha\in\mathbb{Z}} a_\alpha\,\partial^\alpha \;\middle|\; a_\alpha \in A_t \text{ and } \exists (C, N) \in \mathbb{R}\times\mathbb{N} \text{ such that } val_t(a_\alpha) > C\alpha - N\ \forall\,\alpha \gg 0 \right\} \tag{3.11}$$

and

$$\widehat{\mathcal{D}}_{A_t} = \left\{ P = \sum_{\alpha\in\mathbb{Z}} a_\alpha\,\partial^\alpha \;\middle|\; P \in \widehat{\Psi}(A_t) \text{ and } a_\alpha = 0 \text{ for } \alpha < 0 \right\}. \tag{3.12}$$

In addition, we define

$$\mathcal{I}_{A_t} = \Psi^{-1}(A_t) = \left\{ P \in \widehat{\Psi}(A_t) \mid \forall \alpha \geq 0, a_\alpha = 0 \right\}$$

and

$$G_{A_t} = 1 + \mathcal{I}(A_t) \ .$$

We notice that $A_t \subset \widehat{\Psi}(A_t)$, that

$$G_{A_t} \subset \Psi(A_t) \subset \widehat{\Psi}(A_t) \ ,$$

and that $\widehat{\mathcal{D}}_{A_t} \not\subset \Psi(A_t)$. The operations on $\widehat{\Psi}(A_t)$ are extensions of the operations on $\Psi(A_t)$. More precisely we have ([41], see also [19] for explicit computations):

Lemma 39. *The space $\widehat{\Psi}(A_t)$ has an algebra structure, and $\widehat{\mathcal{D}}_{A_t}$ is a subalgebra of $\widehat{\Psi}(A_t)$.*

Definition 40. Let $\mathcal{K}$ be the ideal of A_t generated by $\tau_1, \tau_2, \cdots$. If $P \in \widehat{\Psi}(A_t)$, we denote by $P|_{\tau=0}$ the equivalence class $P \, mod \, \mathcal{K}$ (i.e. the projection on the constant coefficient of the τ−series), and we identify it with an element of $\Psi(A)$. We also set $G_A = 1 + \mathcal{I}_A$, and we define the spaces

$$G(\widehat{\Psi}(A_t)) = \{ P \in \widehat{\Psi}(A_t) : P|_{\tau=0} \in G_{A_t} \} \tag{3.13}$$

and

$$\widehat{\mathcal{D}}^{\times}_{A_t} = \{ P \in \widehat{\mathcal{D}}_{A_t} : P|_{\tau=0} = 1 \} \ . \tag{3.14}$$

Hereafter we assume that R is, in actual fact, a Frölicher $\mathbb{K}$−algebra and that $\partial : R \to R$ is smooth. These assumptions imply the following two results, see [38]:

Theorem 41. *The following algebras are Frölicher algebras:*

(1) A_t ; (2) $\Psi(A_t)$; (3) $\widehat{\Psi}(A_t)$; (4) $\widehat{\mathcal{D}}_{A_t}$.

Theorem 42. *1. $\widehat{\mathcal{D}}^{\times}_{A_t}$ is a Frölicher group.*

2. $G_R = 1 + \Psi^{-1}(R)$ and $G_{A_t} = 1 + \Psi^{-1}(A_t)$ are Frölicher Lie groups.

3. $G(\widehat{\Psi}(A_t))$ is a Frölicher group.

We remark that we have not stated the existence of an exponential map. In fact, it seems difficult to show the existence of the exponential map on $\widehat{\Psi}(A_t)$, and very difficult to determine the tangent space $T_1 G(\widehat{\Psi}(A_t))$; it also looks hard to differentiate the "adjoint action" of $G(\widehat{\Psi}(A_t))$ on it. Most of the difficulties come from the very general definition of $\widehat{\Psi}(A_t)$. This is why we construct a Frölicher subalgebra $\overline{\Psi}(A_t) \subset \widehat{\Psi}(A_t)$ as follows:

Definition 43. The regularized space of formal pseudodifferential and differential operators of infinite order are, respectively, $\overline{\Psi}(A_t)$ and $\overline{\mathcal{D}}_{A_t}$, in which

$$\overline{\Psi}(A_t) = \left\{ \sum_{\alpha \in \mathbb{Z}} a_\alpha \, \partial^\alpha \in \widehat{\Psi}(A_t) \mid val_t(a_\alpha) \geq \alpha \right\} \tag{3.15}$$

and

$$\overline{\mathcal{D}}_{A_t} = \left\{ P = \sum_{\alpha \in \mathbb{Z}} a_\alpha \, \partial^\alpha \mid P \in \overline{\Psi}(A_t) \text{ and } a_\alpha = 0 \text{ for } \alpha < 0 \right\} . \tag{3.16}$$

In addition, we define

$$G(\overline{\Psi}(A_t)) = \{ P \in \overline{\Psi}(A_t) \mid P|_{\mathcal{T}=0} \in G_{A_t} \} \tag{3.17}$$

and

$$\overline{\mathcal{D}}^{\times}_{A_t} = \{ P \in \overline{\mathcal{D}}_A \mid P|_{\mathcal{T}=0} = 1 \} , \tag{3.18}$$

and we can prove, see [38], that $G(\overline{\Psi}(A_t))$ and $\overline{\mathcal{D}}^{\times}_{A_t}$ are fully regular Frölicher Lie groups.

3.3 Formal pseudodifferential operators with arbitrary coefficients

We can generalize our foregoing discussion. We hope this generalization will be of use, for instance, in the development of p-adic KP theory. Our basic idea is to replace algebras of power series by general algebras equipped with non-archimidean valuations. We only present the main points of this generalization and we refer to [20] for details.

Let A be an associative (but not necessarily commutative!) K–algebra with unit 1, in which K is an arbitrary field of characteristic zero. We assume that A is a *diffeological algebra* and that A is equipped with a smooth derivation. We consider $\Psi(A)$ as in the previous subsection, but instead of assuming that A is an algebra of series, we equip it with a valuation, adapting an idea from [11, 12].

Definition 44. A valuation on an algebra A as above is a map $\sigma : A \to \mathbb{Z} \cup \{\infty\}$ which satisfies the following properties for all a, b in A and $k \in K$:
$\sigma(a) = \infty$ if and only if $a = 0$, and $\sigma(1) = 0$
$\sigma(ab) = \sigma(a) + \sigma(b)$, and $\sigma(ka) = \sigma(a)$
$\sigma(a + b) \geq min(\sigma(a), \sigma(b))$.

The following conventions are implicit in the definition above: for all $n \in \mathbb{Z}$ we set $n + \infty = \infty$; $\infty + \infty = \infty$; $n < \infty$. We also note the following three obvious properties of a valuation σ: $\sigma(-1) = 0$; $\sigma(-x) = \sigma(x)$ for all $x \in A$; $\sigma(a + b) = min(\sigma(a), \sigma(b))$ whenever $\sigma(a) \neq \sigma(b)$.

We also assume that the derivation D is compatible with the valuation σ in the sense that $\sigma(D(x)) \geq \sigma(x)$ for all $x \in A$. The reader can check that these assumptions do hold in the case of the algebra A_t appearing in the previous subsection.

A valuation allows us to equip A with a topology. For $\alpha \in \mathbb{Z}$ and $x_0 \in A$ we set

$$V_\alpha(x_0) = \{x \in A : \sigma(x - x_0) > \alpha\} ;$$

we easily see that for each $x_0 \in A$, the collection $\{V_\alpha(x_0)\}_{\alpha \in \mathbb{Z}}$ is a basis of neighbourhoods for a first countable Hausdorff topology on A. It is a classical fact that this topology is metrizable: we define the absolute value $|x| = c^{\sigma(x)}$ for $x \in A$ and a fixed real number $0 < c < 1$. Then, we easily check that the following holds:

$|x| \geq 0$, $|x| = |kx|$ for $k \in K^*$, and $|x| = 0$ if and only if $x = 0$;
$|xy| = |x|\,|y|$;
$|x + y| \leq max\{|x|, |y|\}$, and $|x + y| = max\{|x|, |y|\}$ if $|x| \neq |y|$.

If we set $d(x, y) = |x - y|$ we obtain a metric d on A and the metric topology coincides with the topology introduced above. In this topology the sets $V_\alpha(x_0)$ coincide with the balls $\{x \in A : |x| < c^\alpha\}$, and are both open and closed. A becomes a topological algebra and D a continuous derivation. We also remark that compatibility of derivation and valuation implies that $|D(a)| \leq |a|$ for all $a \in A$.

Now we let $\hat{A}$ be the completion of the metric space (A, d), and we extend D to a continuous derivation on $\hat{A}$. We continue denoting the extension of the absolute value on A by $|\cdot|$, and the extension of the continuous derivation on A by D.

In the previous subsection we used Frölicher algebras and obtained Frölicher structures on spaces of (regularized) formal pseudodifferential operators. Due to the presence of quotients, diffeologies are more natural than Frölicher spaces, see [55] and references therein. We proceed as follows, after [20]:

We extend the quotient vector space projection $\pi_p : A \to A/A_p$, $p \in \mathbb{Z}$, in which $A_p = \{a \in A : \sigma(a) \geq p\}$, in the following way: for $\hat{a} \in \hat{A}$ we set $\pi_p(\hat{a}) = a + A_p$ if and only if $\sigma(a) = \hat{\sigma}(\hat{a})$ and $\hat{\sigma}(a - \hat{a}) \geq p$. It is possible to find such an $a \in A$ because of standard properties of valuations. Now we define:

Definition 45. We equip the quotients A/A_p with their quotient diffeology. The completion $\hat{A}$ is equipped with the pull-back diffeology with respect to the family of maps $\{\pi_p : p \in \mathbb{Z}\}$, see [25, p. 32]. The valuation σ is a **diffeological valuation** if the diffeology of A is the pull-back of the diffeology of $\hat{A}$ and all plots are continuous in the valuation topologies of A and $\hat{A}$.

We assume that A and $\hat{A}$ are equipped with diffeological valuations, and that D is smooth on $\hat{A}$ (and so, its restriction to A is also smooth).

Definition 46. [20] The space of regularized formal pseudodifferential and differential operators are, respectively, $\widehat{\Psi}(\hat{A})$ and $\widehat{\mathcal{D}}_{\hat{A}}$, in which

$$\widehat{\Psi}(\hat{A}) = \Bigg\{ P = \sum_{\alpha \in \mathbb{Z}} a_\alpha D^\alpha : a_\alpha \in \hat{A} \text{ and } \exists A_P, B_P \in \mathbb{R}^+, M_P, N_P, L_P \in \mathbb{Z}^+ \text{ so that}$$
$$M_P \geq N_P\,,\ |a_\alpha| < \frac{A_P}{\alpha - N_P}\ \forall\, \alpha > M_P\,,\ \text{and } |a_\alpha| < B_P\ \forall\, \alpha < -L_P \Bigg\} \tag{3.19}$$

and

$$\widehat{\mathcal{D}}_{\hat{A}} = \left\{ P = \sum_{\alpha\in\mathbb{Z}} a_\alpha D^\alpha : P \in \widehat{\Psi}(\hat{A}) \text{ and } a_\alpha = 0 \text{ for } \alpha < 0 \right\} . \tag{3.20}$$

The definition of the absolute value $|\cdot|$ implies that $\hat{A}$ is contained in $\widehat{\Psi}(\hat{A})$ and, as a by-product, we note that our assumptions on A and $\hat{A}$ imply that $\widehat{\Psi}(\hat{A})$ and $\widehat{\mathcal{D}}(\hat{A})$ are diffeological spaces.

Lemma 47. [20] *The space $\widehat{\Psi}(\hat{A})$ has an algebra structure and $\widehat{\mathcal{D}}_{\hat{A}}$ is a subalgebra of $\widehat{\Psi}(\hat{A})$. Moreover, if $\hat{A}$ is a diffeological K-algebra, $\widehat{\Psi}(\hat{A})$ and $\widehat{\mathcal{D}}_{\hat{A}}$ are diffeological K-algebras.*

Now we construct groups. There exist two standard structures associated to the valuation $\hat{\sigma}$ on $\hat{A}$. The subring $\mathcal{O}_{\hat{A}} = \{a \in \hat{A} : \hat{\sigma}(a) \geq 0\}$, and the two-sided ideal $\mathcal{P}_{\hat{A}} = \{a \in \mathcal{O}_{\hat{A}} : \hat{\sigma}(a) > 0\}$. If A is a diffeological algebra, these algebraic constructions carry natural underlying diffeologies. Since we can check that the derivation D on $\hat{A}$ is compatible with $\hat{\sigma}$, we have $D(\mathcal{P}_{\hat{A}}) \subset \mathcal{P}_{\hat{A}}$; it follows that the derivation D is well defined on the quotient ring $\mathcal{O}_{\hat{A}}/\mathcal{P}_{\hat{A}}$. We let $\pi : \mathcal{O}_{\hat{A}} \to \mathcal{O}_{\hat{A}}/\mathcal{P}_{\hat{A}}$ be the canonical projection. Since A is a diffeological algebra, the map π is smooth.

The set $G_{\hat{A}} = 1 + \mathcal{I}_{\mathcal{O}_{\hat{A}}/\mathcal{P}_{\hat{A}}}$ is a multiplicative group, see [12], and a diffeological group according to Theorem 13. For $P = \sum_{\nu\in\mathbb{Z}} a_\nu D^\nu \in \widehat{\Psi}(\mathcal{O}_{\hat{A}})$ we set $\pi(P) = \sum_{\nu\in\mathbb{Z}} \pi(a_\nu) D^\nu$.

Definition 48. We define the spaces

$$\widehat{\Psi}(\hat{A})^\times = \{P \in \widehat{\Psi}(\mathcal{O}_{\hat{A}}) : \pi(P) \in G_{\hat{A}}\} \tag{3.21}$$

and

$$\widehat{\mathcal{D}}^\times_{\hat{A}} = \{P \in \widehat{\mathcal{D}}_{\mathcal{O}_{\hat{A}}} : \pi(P) = 1\} . \tag{3.22}$$

Proposition 49. *The space $\widehat{\Psi}(\hat{A})^\times$ is a group: each element P in $\widehat{\Psi}(\hat{A})^\times$ has an inverse of the form*

$$P^{-1} = \sum_{n\geq 0} (1 - P)^n .$$

In addition, the space $\widehat{\mathcal{D}}^\times_{\hat{A}}$ is a subgroup of $\widehat{\Psi}(\hat{A})^\times$.

We can prove the following result on smoothness, see [20]. It is to be compared with the results presented at the end of the previous subsection.

Proposition 50. *The group $G(\Psi(\hat{A})) := \widehat{\Psi}(\hat{A})^\times$ is a diffeological Lie group with Lie algebra $\Psi(\mathcal{O}_{\hat{A}})$; the group $G_+(\mathcal{D}_{\hat{A}}) := \widehat{\mathcal{D}}^\times_{\hat{A}}$ is a diffeological Lie group with Lie* *algebra $\mathcal{D}_{\mathcal{O}_{\hat{A}}}$; the group $G_-(\mathcal{I}_{\hat{A}}) := 1 + \mathcal{I}_{\mathcal{O}_{\hat{A}}}$ is a diffeological Lie group with Lie algebra $\mathcal{I}_{\mathcal{O}_{\hat{A}}}$. Moreover, the exponential map*

$$\exp : P \in \mathcal{I}_{\mathcal{O}_{\hat{A}}} \mapsto \sum_{n \in \mathbb{N}} \frac{(sP)^n}{n!} \in G_-(\mathcal{I}_{\hat{A}})$$

is one-to-one and onto; its inverse is the classical logarithmic series log, *and both* exp *and* log *are smooth. As a consequence, the inversion is smooth in $G_-(\mathcal{I}_{\hat{A}})$.*

3.4 On matched pairs of diffeological Lie groups

The notion of a matched pair of groups arose in physics from observations on R-matrices, see [14], and was studied in [17, 18, 31, 54]. They are also mentioned in the literature as bicross product groups, see [7, 8, 30]. Two groups G, H form a matched pair if they act on each other and these actions α, β obey the following conditions for all $(x, y, u, v) \in G^2 \times H^2$:

$$\alpha_{x^{-1}}(e_H) = e_H\,, \tag{3.23}$$
$$\beta_{u^{-1}}(e_G) = e_G\,, \tag{3.24}$$
$$\alpha_{x^{-1}}(uv) = \alpha_{x^{-1}}(u)\alpha_{\beta_{u^{-1}}(x)^{-1}}(v)\,, \tag{3.25}$$
$$\beta_{u^{-1}}(xy) = \beta_{u^{-1}}(x)\beta_{\alpha_{x^{-1}}(u)^{-1}}(y)\,. \tag{3.26}$$

These conditions allow us to define a group multiplication on the set $G \times H$ thus:

$$(x, u).(y, v) = \left(\beta_{u^{-1}}(x^{-1})^{-1}y, u\alpha_x(u)\right) \tag{3.27}$$
$$(x, u)^{-1} = \left(\beta_u(x)^{-1}, \alpha_{x^{-1}}(u^{-1})\right)\,. \tag{3.28}$$

The set $G \times H$ equipped with this multiplication is called the bicross product group $G \bowtie H$. Analogous definitions hold for Lie groups, see [17, 18] and, following [32, Proposition 6.2.15], we note that if a Lie group K decomposes into two Lie subgroups G, H such that $G.H = K$ and $G \cap H = \{e_K\}$, there exists a matched pair construction with α, β satisfying conditions (3.23-3.26). Related discussions on Lie algebras can be found in [14, 31]. We adapt these algebraic constructions to diffeological groups:

Definition 51. Let G, H be diffeological groups (resp. diffeological Lie groups). Then $G \bowtie H$ is a diffeological group (resp. diffeological Lie group) if and only if α and β are smooth actions.

The following open question is quite natural:

If G and H are regular, is $G \bowtie H$ regular?

3.5 The Mulase-Birkhoff decomposition and its smoothness

The Mulase-Birkhoff decomposition consists of "lifting" the decomposition $A = A_S + A_D$ of Subsection 3.1 (and the analogous decomposition for the case of formal pseudodifferential operators) to a decomposition of the corresponding groups.

Mulase's algebraic theorem is the following, see [40, 41] and the later exposition [19]:

Theorem 52. *For any $W \in G(\widehat{\Psi}(A_t))$ there exist unique $S \in G_{A_t}$ and $Y \in \widehat{\mathcal{D}}^{\times}_A$ such that*

$$W = S^{-1} Y . \tag{3.29}$$

In other words, there exists a unique global factorization of the formal Lie group $G(\widehat{\Psi}(A))$ as a product group,

$$G(\widehat{\Psi}(A_t)) = G_{A_t} \widehat{\mathcal{D}}^{\times}_A .$$

This decomposition is a decomposition of matched pairs of (formal) Lie groups and, as explained in [41], it generalizes the classical Birkhoff decomposition for loop groups appearing in [48]. We summarize our results on smooth versions of Theorem 52.

Theorem 53. *Decomposition* (3.29) *holds in the following cases:*

1. *following* [20], $W \in G(\Psi(\hat{A}))$, $S \in G_-(\mathcal{I}_{\hat{A}})$ *and* $Y \in G_+(\mathcal{D}_{\hat{A}})$;
2. *following* [35, 38], $W \in G(\overline{\Psi}(A_t))$, $S \in G_{A_t}$ *and* $Y \in \overline{\mathcal{D}}^{\times}_{A_t}$.

In each of these settings, the map $W \mapsto (S, Y)$ is smooth, see [35, 38, 20].

A further instance of this theorem appears in our second proof of the well-posedness of the Cauchy problem for KP. We use notations introduced in Subsection 3.2:

Let $K = \mathbb{R}$ or $\mathcal{C}$, and let R be a diffeological commutative ring over K equipped with a continuous and smooth derivation ∂.

Definition 54. In what follows, val_h denotes standard valuation of $h-$series. We define:

$$\Psi_h(R) = \left\{ \sum_{\alpha \in \mathbb{Z}} a_\alpha \partial^\alpha \in \Psi(R)[[h]] \mid val_h(a_\alpha) \geq \alpha \right\} ,$$

$$G\Psi_h(R) = \left\{ \sum_{\alpha \in \mathbb{Z}} a_\alpha \partial^\alpha \in \Psi_h(R) \mid a_0 = 1 + b_0, \quad val_h(b_0) \geq 1 \right\},$$

$$G_{R,h} = \left\{ A \in G\Psi_h(R) \mid A = 1 + B, \quad B \in \Psi^{-1}(R)[[h]] \right\} ,$$

$$\mathcal{D}_h(R) = \left\{ \sum_{\alpha \in \mathbb{Z}} a_\alpha \partial^\alpha \in \Psi_h(R) \mid a_\alpha = 0 \text{ if } \alpha < 0 \text{ and } a_0 = 1 + b_0, \quad val_h(b_0) \geq 1 \right\} .$$

The next Lemma is proved in [35]; a shorter proof appears in [38].

Lemma 55. *$\Psi_h(R)$ is a Frölicher algebra and an integral Lie algebra, and the groups $G\Psi_h(R)$, $G_{R,h}$ and $\mathcal{D}_h(R)$ are regular Frölicher Lie groups with Lie algebras given respectively by:*

$$\mathfrak{g}\Psi_h(R) = \left\{ \sum_{\alpha \in \mathbb{Z}} a_\alpha \partial^\alpha \in \Psi_h(R) \mid val_h(a_0) \geq 1 \right\} ,$$

$$\mathfrak{g}_{R,h} = \Psi^{-1}(R)[[h]] ,$$

and

$$\mathfrak{d}_h(R) = \left\{ \sum_{\alpha \in \mathbb{Z}} a_\alpha \partial^\alpha \in D_h(R) \mid a_0 = 0 \text{ if } \alpha < 0 \quad val_h(a_0) \geq 1 \right\} .$$

Theorem 56. [38] *Decomposition* (3.29) *holds for $W \in G\Psi_h(R)$, $S \in G_{R,h}$ and $Y \in \mathcal{D}_h(R)$ and, in particular, with $T = \cup_{n \in \mathbb{N}} \mathbb{R}^n$, $R = C^\infty(T, C^\infty(S^1, \mathbb{R})) = C^\infty(T \times S^1, \mathbb{R})$ and $\partial = \frac{d}{dx}$ on S^1, thereby recovering* [35]. *The map $W \mapsto (S, Y)$ is smooth.*

4 The Cauchy problem for the KP hierarchy

We start with the setting described in Section 3.3. The derivations d/dt_k can be introduced as fixed derivations on $\mathcal{O}_{\hat{A}}$ which commute with D.

Theorem 57. [20] *Consider the KP system of equations*

$$\frac{dL}{dt_k} = \left[(L^k)_+, L \right] \tag{4.1}$$

with initial condition $L(0) = L_0 = \sum_{\nu \in \mathbb{Z}} a_\nu D^b \nu \in \Psi(\mathcal{O}_{\hat{A}})$, such that $\hat{\sigma}(a_\nu) \geq 1$ for all $\nu \geq 0$, and let $Y \in G_+ = G_+(\mathcal{D}_A)$ and $S \in G_- = G_-(\mathcal{I}_A)$ be the unique solution to the factorization problem

$$\exp(t_k L_0^k) = S^{-1}(t_k) Y(t_k) .$$

The unique solution to Equation (4.1) *with $L(0) = L_0$ is*

$$L(t_k) = Y L_0 Y^{-1} .$$

Moreover, the map $L_0 \in \Psi(\mathcal{O}_{\hat{A}}) \mapsto L \in \hat{\Psi}(\hat{A})$ is smooth.

This theorem is proven in [20] using Reyman and Semenov-tian-Shansky's approach appearing in [49], see also [47]. We provide some details below in the more restricted context of algebras of series.

We note that in this abstract setting we cannot discuss smoothness with respect to "time" in a non-trivial way, see [20]. We now turn to the setting of Section 3.2 after [38]. We set

$$T = \{(t_1, t_2, ...) \mid \forall \mathbb{N} \in \mathbb{N}^*, t_i \in \mathbb{R}\} \sim \mathbb{R}^{\mathbb{N}^*} .$$

The KP system reads as

$$\frac{dL}{dt_k} = \left[(L^k)_+, L\right] \tag{4.2}$$

with initial condition $L(0) = L_0$. We assume that L_0 is a pseudodifferential operator of order 1, $L_0 \in \partial + \Psi^{-1}(R)$. We state the following result, after [38]:

Theorem 58. *Consider the KP system (4.2) with initial condition $L(0) = L_0$.*

1. *There exists a pair $(S, Y) \in G_{A_t} \times \overline{\mathcal{D}}(A_t)^\times$ such that the unique solution to Equation (4.2) with $L(0) = L_0$ is*

$$L(t) = Y\, L_0\, Y^{-1} = S L_0 S^{-1} .$$

2. *The pair (S, Y) is given by the unique solution to the decomposition problem*

$$exp\left(\sum_{k\in\mathbb{N}} t_k L_0^k\right) = S^{-1} Y .$$

3. *The solution operator L depends smoothly on the "time" variables t_k and on the initial value L_0, that is, the map (we use notation introduced after Theorem 13)*

$$(L_0, s) \in (\partial + \Psi^{-1}(R)) \times T \mapsto \sum_{n\in\mathbb{N}}\left(\sum_{|t|=n} [L(s)]_t\right) \in (\partial + \Psi^{-1}(R))^{\mathbb{N}}$$

is smooth, for $s \in T = \cup_{n\in\mathbb{N}} \mathbb{K}^n$ equipped with the structure of locally convex topological space given by the inductive limit.

4.1 Integration and well-posedness of the KP hierarchy via R-matrices

As explained above, our first proof of Theorem 58 is modelled after Reyman and Semenov-tian-Shansky's [49], see also [47]. We set

$$U = \exp\left(\sum_{k\in\mathbb{N}} t_k L_0^k\right) ,$$

and we can prove that $U \in G(\overline{\Psi}(A_t)$, see [38]. Then we consider the smooth Mulase-Birkhoff decomposition

$$U \mapsto (S, Y) \in G_{A_t} \times \overline{\mathcal{D}}(A_t)^\times$$

and we set $L = Y\, L_0\, Y^{-1}$. Calculations appearing in [19, 38, 20] allow us to conclude that indeed L satisfies (4.2) and that $L(0) = L_0$.

Smoothness with respect to t is proved by construction and smoothness of the map $L_0 \mapsto Y$ follows from our smooth Mulase-Birkhoff decomposition. Thus, the map

$$L_0 \mapsto L(t) = Y(t) L_0 Y^{-1}(t)$$

is smooth in $\overline{\Psi}(A_t)$.

We also need to check that the solution L indeed belongs to $\Psi(A_t)$. This is carried out using the relation $L^k = S L_0^k S^{-1}$ which is obtained from the definition of L, see [19, 38, 20]. It follows that

$$L = L_0 + S[L_0, S^{-1}] \, .$$

Now, $[L_0, S^{-1}] \in \Psi^{-1}(A_t)$, $S \in \Psi^0(A_t)$ and $L_0 \in \partial + \Psi^{-1}(R)$, so that

$$L \in \partial + \Psi^{-1}(A_t) \, .$$

Remark 59. The above process allows us to construct a solution to any fixed equation of the KP hierarchy. Setting $t_k = 0$, $k \neq k_0$, we get an operator $U_1(t_{k_0}) = exp(t_{k_0} L_0) = S_{k_0}^{-1} Y_{k_0}$ such that $L_{k_0} = Y_{k_0} L_0 Y_{k_0}^{-1} = S_{k_0} L_0 S_{k_0}^{-1}$ is the unique solution of the $t_{k_0}-$equation of the KP-hierarchy

$$\frac{dL}{dt_{k_0}} = \left[L_+^{k_0}, L \right] \, .$$

4.2 Integration and well-posedness of the KP hierarchy via the Ambrose-Singer theorem

This proof uses the material appearing in subsection 2.2, after [35], and it is partially motivated by [40, 41]. We equip $T = \cup_{n \in \mathbb{N}} \mathbb{K}^n$ with the structure of a locally convex topological space given by the inductive limit. Since we have $[L^n, L] = 0$, we conclude

$$\frac{dL}{dt_n} = [L_+^n, L] = -[L_-^n, L].$$

Then, setting $dt = \sum_{k \in \mathbb{N}^*} dt_k$ we can define

$$Z dt = \sum_{n \in \mathbb{N}^*} L_+^n dt_n \, , \qquad Z^c dt = - \sum_{n \in \mathbb{N}^*} L_-^n dt_n \, ,$$

and we can check that $Z^c dt \in \Omega^1(T, \Psi^{-1}(A_t))$. We consider the equation

$$dZ^c - [Z^c, Z^c] = 0 \, . \tag{4.3}$$

Our set-up implies that (4.3) is a rigorous zero-curvature equation whenever L solves KP. Our plan is to use zero curvature equations such as (4.3), and to build a solution to the KP hierarchy with the help of our Ambrose-Singer theorem. We use the scaling $t_n \mapsto h^n t_n$, see [35], to define the operator

$$\tilde{L}(t_1, t_2, ...) = h L(h t_1, h^2 t_2, ...)$$

in $\Psi(A_t)[[h]]$, and the 1-forms

$$\tilde{Z}_{\tilde{L},+}(t_1,...) = \sum_{n=1}^{+\infty} h^n (L^n)_+ dt_n \ , \qquad \tilde{Z}_{\tilde{L},-}(t_1,...) = -\sum_{n=1}^{+\infty} h^n (L^n)_+ dt_n \ .$$

The KP hierarchy becomes the "deformed KP hierarchy"

$$d(\tilde{L}) = [\tilde{Z}_+, \tilde{L}] = [\tilde{Z}_-, \tilde{L}] \ ,$$

and we also have the equations

$$d\tilde{Z}_{\tilde{L},+} + [\tilde{Z}_{\tilde{L},+}, \tilde{Z}_{\tilde{L},+}] = 0 \ , \qquad d\tilde{Z}_{\tilde{L},-} - [\tilde{Z}_{\tilde{L},-}, \tilde{Z}_{\tilde{L},-}] = 0 \ ,$$

whenever $\tilde{L}$ solves deformed KP.

Now we fix an initial condition $L_0 \in \partial + \Psi^{-1}(R)$ and we observe that $h\partial$ *is* a true solution to our deformed KP hierarchy. We consider the one-forms $Z_{h\partial,+}$ and $Z_{h\partial,-}$ defined as above with $h\partial$ instead of $\tilde{L}$. Then, the equations

$$d\tilde{Z}_{h\partial,+} + [\tilde{Z}_{h\partial,+}, \tilde{Z}_{h\partial,+}] = 0 \ , \qquad d\tilde{Z}_{h\partial,-} - [\tilde{Z}_{h\partial,-}, \tilde{Z}_{h\partial,-}] = 0$$

obviously hold. These equations allow us to apply our Ambrose-Singer theorem (Theorem 23) on the trivial principal bundle $T \times G\Psi_h(R)$. We obtain:

Theorem 60. *Let $S_0 \in G_{R,h}$ such that $L_0 = S_0 h\partial S_0^{-1}$. There exists a unique function $\tilde{U} \in C^\infty(T, G\Psi_h(R))$ satisfying*

$$d\tilde{U} \cdot \tilde{U}^{-1} = \tilde{Z}_{h\partial,+}$$

with initial condition $\tilde{U}(0) = S_0^{-1} \cdot Y_0$, in which $Y_0 = 1$.

Our factorization theorems yield the unique factorization $\tilde{U} = \tilde{S}^{-1} \cdot \tilde{Y}$ with $\tilde{S} \in G_{R,h}$, $\tilde{Y} \in \mathcal{D}_h(R)$, $\tilde{S}(0) = S_0$, and $\tilde{Y}(0) = Y_0 = 1$; we set

$$\tilde{L} = \tilde{S}(h\partial)\tilde{S}^{-1} \in C^\infty(T, \Psi_h(R)) \ .$$

We can check that $\tilde{L}$ indeed solves the deformed KP hierarchy. For this, we need to "dress" the connection one-forms $Z_{h\partial,+}$ and $Z_{h\partial,-}$ using $\tilde{S}$, and to prove that these dressed one-forms are in fact the one-forms $\tilde{Z}_{\tilde{L},+}$ and $\tilde{Z}_{\tilde{L},-}$ corresponding to our choice of $\tilde{L}$. Details are in [38]. It follows that $\tilde{L}$ satisfies the equations

$$d\tilde{L} = [\tilde{Z}_-, \tilde{L}] \ , \qquad d\tilde{L} = [\tilde{Z}_+, \tilde{L}] \ . \tag{4.4}$$

Now we check that our solution to (4.4) is unique. This fact follows from the equation

$$\tilde{L} = \tilde{Y} L_0 \tilde{Y}^{-1} \ ,$$

which is not difficult to derive following Mulase's hints appearing in [41].

It remains to prove that, in a sense, "we can take the limit $h \to 1$" in the above relations, so as to recover Theorem 58. We omit this step, since we need to use infinite jets in order to proceed rigorously. We refer the reader to [38].

5 A non-formal KP hierarchy

As we have already pointed out in Subsection 3.1, the decomposition $A = A_S + A_D$, $a_S \in Cl^{-1}_{odd}(S^1, V)$, $A_D \in DO(S^1, V)$, which is valid on $Cl_{odd}(S^1, V)$, see Proposition 32, extends straightforwardly to the algebra $Cl_{h,odd}(S^1, V)$. In the non-formal context of Subsection 3.1, our Theorem 58 becomes the following result:

Theorem 61. [39] *Let* $S_0 \in Cl^{-1,*}_{odd}(S^1, V)$, $L_0 = S_0 h(d/dt) S_0^{-1}$, *and* $U_h(t_1, ...) = \exp\left(\sum_{n \in N^*} h^n t_n (L_0)^n\right) \in Cl_{h,odd}(S^1, V)$. *Then:*

- *There exists a unique pair* (S, Y) *such that*

  1. $U_h = S^{-1} Y$,
  2. $Y \in Cl^*_{h,odd}(S^1, V)_D$
  3. $S \in Cl^*_{h,odd}(S^1, V)$ *and* $S - 1 \in Cl_{h,odd}(S^1, V)_S$.

  Moreover, the map

  $$(S_0, t_1, ..., t_n, ...) \in Cl^{0,*}_{odd}(S^1, V) \times T \mapsto (U_h, Y) \in (Cl^*_{h,odd}(S^1, V))^2$$

  is smooth.

- *The operator* $L \in Cl_{h,odd}(S^1, V)[[ht_1, ..., h^n t_n ...]]$ *given by* $L = S L_0 S^{-1} = Y L_0 Y^{-1}$ *is the unique solution to the hierarchy of equations*

  $$\begin{cases} \dfrac{d}{dt_n} L &= [(L^n)_D(t), L(t)] = -[(L^n)_S(t), L(t)] \\ L(0) &= L_0 \end{cases}, \tag{5.1}$$

  in which the operators in this infinite system are understood as formal operators.

- *The operator* $L \in Cl_{h,odd}(S^1, V)[[ht_1, ..., h^n t_n ...]]$ *given by* $L = S L_0 S^{-1} = Y L_0 Y^{-1}$ *is the unique solution of the hierarchy of equations*

  $$\begin{cases} \dfrac{d}{dt_n} L &= [(L^n)_D(t), L(t)] = -[(L^n)_S(t), L(t)] \\ L(0) &= L_0 \end{cases} \tag{5.2}$$

  in which the operators in this infinite system are understood as odd class, non-formal operators.

We refer the reader to [39] for details of the proof. In particular, checking that the announced solution is the unique solution to the non-formal hierarchy (5.2) is not completely obvious, since in principle there may exist two solutions which differ by a smoothing operator.

Finally, we mention that in [39] we also relate the operator

$$U_h(t_1, ..., t_n, ...) = \exp\left(\sum_{n \in N^*} h^n t_n (L_0)^n\right),$$

which generates the solutions of the $h-$deformed KP hierarchy described in Theorem 61, to the Taylor expansion of functions in the image of the twisted operator

$$A : f \in C^{\infty}(S^1, V) \mapsto S_0^{-1}(f) \circ g \,,$$

in which $g \in Diff_+(S^1)$. A non-trivial technical step needed in order to do so is to relate the times t_i to the derivatives of g at x_0. Once this is done (see [39, Section 6]) the following theorem results:

Theorem 62. [39] *Let $f \in C^{\infty}(S^1, V)$ and set $c = S_0^{-1}(f) \circ g \in C^{\infty}(S^1, V)$. The Taylor series at x_0 of the function c is given by*

$$c(x_0 + h) \sim_{x_0} S_0^{-1}\left(U_h(t_1/h, t_2/h^2, ...)(f)\right)(g(x_0)) \,.$$

Acknowledgments. *Both authors have been partially supported by CONICYT (Chile) via the Fondo Nacional de Desarrollo Científico y Tecnológico grant # 1161691.*

References

[1] Blohmann, C; Barbosa Fernandes, M.C.; Weinstein, A; Groupoid symmetry and constraints in general relativity. *Commun. Contemp. Math.* **15** 1250061 (2013) [25 pages].

[2] Batubenge, A.; Ntumba, P.; On the way to Frölicher Lie groups. *Quaestionnes mathematicae* **28** (2005), 73–93.

[3] Berline, N.; Getzleer, E.; Vergne, M.; *Heat Kernels and Dirac Operators*, Springer (2004).

[4] Bokobza-Haggiag, J.; Opérateurs pseudo-différentiels sur une variété différentiable. *Ann. Inst. Fourier, Grenoble* **19** (1969), 125–177.

[5] Bourbaki, N.; *Elements of Mathematics*, Springer-Verlag, Berlin (1998).

[6] Cardona, A.; Ducourtioux, C.; Magnot, J-P.; Paycha, S.; Weighted traces on pseudo-differential operators and geometry on loop groups. *Infin. Dimens. Anal. Quantum Probab. Relat. Top.* **5** (2002), 503–541.

[7] Chapovskii, Yu.A.; Kalyuzhnyi, A.A.; Podkolzin, G.B.; Construction of cocycles for bicross product of Lie groups. *Functional Analysis and Its Applications* **40** (2006), 139–142.

[8] Chapovskii, Yu.A.; Kalyuzhnyi, A.A.; Podkolzin, G.B.; Finding cocycles in the bicrossed product construction for Lie groups. *Ukrainian Mathematical Journal* **59** (2007), 1693–1707.

[9] Chen, K.T.; Iterated Path Integrals. *Bull. Am. Math. Soc.* **83** (1977), 831–879.

[10] Christensen, J.D. and Wu, E.; Tangent spaces and tangent bundles for diffeological spaces. *Cahiers de Topologie et Géométrie Différentielle* Volume **LVII** (2016), 3–50.

[11] Demidov, E.E.; On the Kadomtsev-Petviashvili hierarchy with a noncommutative timespace. *Funct. Anal. Appl.* **29** no. 2, (1995), 131–133.

[12] Demidov, E.E.; Noncommutative deformation of the Kadomtsev-Petviashvili hierarchy. In Algebra. 5, Vseross. Inst. Nauchn. i Tekhn. Inform. (VINITI), Moscow, 1995. (Russian). *J. Math. Sci. (New York)* **88** no. 4, (1998), 520–536 (English).

[13] Dickey, L.A.; *Soliton Equations and Hamiltonian Systems. Second Edition*, Advanced Series in Mathematical Physics 12, World Scientific Publ. Co., Singapore (2003).

[14] Drinfel'd, V.G.; Hamiltonian structures on Lie groups, Lie bialgebras and the geometric meaning of the classical Yang-Baxter equations. *Sov. Math. DokL*, **27** (1983), 68.

[15] Dugmore, D.; Ntumba, P.; On tangent cones of Frölicher spaces. *Quaestiones mathematicae* **30** no.1 (2007), 67–83.

[16] Eells, J.; A setting for global analysis. *Bull. Amer. Math. Soc.* **72** (1966), 751–807.

[17] Esen, O; Sütlü; Lagrangian dynamics on matched pairs. *J. Geom. Phys.* **111** (2017), 142–157.

[18] Esen, O; Sütlü; Hamiltonian dynamics on matched pairs. *International Journal of Geometric Methods in Modern Physics* **13** No. 10, 1650128 (2016).

[19] Eslami Rad, A.; Reyes, E. G.; The Kadomtsev-Petviashvili hierarchy and the Mulase factorization of formal Lie groups. *J. Geom. Mech.* **5** (2013) 345–363.

[20] Eslami Rad, A.; Magnot, J.-P.; Reyes, E. G.; The Cauchy problem of the Kadomtsev-Petviashvili hierarchy with arbitrary coefficient algebra. *J. Nonlinear Math. Phys.* **24**:sup1 (2017), 103–120.

[21] Frölicher, A; Kriegl, A; *Linear Spaces and Differentiation Theory*, Wiley series in Pure and Applied Mathematics, Wiley Interscience (1988).

[22] Gilkey, P; *Invariance Theory, the Heat Equation and the Atiyah-Singer Index Theorem*, Publish or Perish (1984).

[23] Hirsch M., *Differential Topology*, Springer (1997).

[24] Hörmander,L.; Fourier integral operators. I. *Acta Mathematica* **127** (1971), 79–189.

[25] Iglesias-Zemmour, P.; *Diffeology*, AMS Mathematical Monographs **185** (2013).

[26] Kriegl, A.; Michor, P.W.; *The Convenient Setting for Global Analysis*, Math. Surveys and Monographs **53**, American Mathematical Society, Providence, USA (2000).

[27] Kontsevich, M.; Vishik, S.; Determinants of elliptic pseudo-differential operators. Max Plank Institut fur Mathematik, Bonn, Germany, preprint n. 94-30 (1994).

[28] Kontsevich, M.; Vishik, S.; Geometry of determinants of elliptic operators. Functional Analysis on the Eve of the 21st Century, Vol. 1 (New Brunswick, NJ, 1993), *Progr. Math.* **131**,173–197 (1995).

[29] Leslie, J.; On a diffeological group realization of certain generalized symmetrizable Kac-Moody Lie algebras. *J. Lie Theory* **13** (2003), 427–442.

[30] Majid, S.; Physics for algebraists: non-commutative and non-cocommutative Hopf algebras by a bicross product construction. *J. Algebra*, **130** (1990), 17–64.

[31] Majid, S.; Matched pairs of Lie groups associated to solutions of the Yang-Baxter equations. *Pacific Journal of Mathematics* **141**(2) (1990), 311–332.

[32] Majid, S.; *Foundations of Quantum Group Theory*, Cambridge University Press, Cambridge (1995).

[33] Magnot, J-P.; Difféologie du fibré d'holonomie d'une connexion en dimension infinie. *C. R. Math. Rep. Acad. Sci. Canada* **28** no 4 (2006), 121–128.

[34] Magnot, J-P.; Chern forms on mapping spaces. *Acta Appl. Math.* **91** (2006), 67–95.

[35] Magnot, J-P.; Ambrose-Singer theorem on diffeological bundles and complete integrability of KP equations. *Int. J. Geom. Meth. Mod. Phys.* **10**, no 9 (2013) Article ID 1350043.

[36] Magnot, J-P.; On $Diff(M)-$pseudodifferential operators and the geometry of non-linear grassmannians. *Mathematics* **4**, 1; doi:10.3390/math4010001 (2016).

[37] Magnot, J-P.; The group of diffeomorphisms of a non-compact manifold is not regular. *Demonstr. Math.* **51** (2018), 8–16.

[38] Magnot, J-P.; Reyes, E. G.; Well-posedness of the Kadomtsev-Petviashvili hierarchy, Mulase factorization, and Frölicher Lie groups `arXiv:1608.03994`.

[39] Magnot, J-P.; Reyes, E. G.; $Diff_+(S^1)-$pseudo-differential operators and the Kadomtsev-Petviashvili hierarchy `arXiv:1808.03791`.

[40] Mulase, M.; Complete integrability of the Kadomtsev-Petvishvili equation. *Advances in Math.* **54** (1984), 57–66.

[41] Mulase, M.; Solvability of the super KP equation and a generalization of the Birkhoff decomposition. *Invent. Math.* **92** (1988), 1–46.

[42] Neeb, K-H.; Towards a Lie theory of locally convex groups. *Japanese J. Math.* **1** (2006), 291–468.

[43] Olver, P.J.; *Applications of Lie Groups to Differential Equations. Second Edition*, Springer-Verlag, New York (1993).

[44] Omori H., Groups of diffeomorphisms and their subgroups. *Trans. Amer. Math. Soc.* **179** (1973) 85–122.

[45] Omori, H.; *Infinite dimensional Lie groups*, AMS Translations of Mathematical Monographs **158** (1997).

[46] Paycha, S; *Regularised Integrals, Sums and Traces. An Analytic Point of View.* University Lecture Series **59**, AMS (2012).

[47] Perelomov, A.M.; *Integrable Systems of Classical Mechanics and Lie Algebras*, Birkhäuser Verlag, Berlin (1990).

[48] Pressley, A.; Segal, G.B.; *Loop Groups*, Oxford University Press (1986).

[49] Reyman, A.G.; Semenov-Tian-Shansky, M.A.; Reduction of Hamiltonian Systems, Affine Lie Algebras and Lax Equations II. *Invent. Math.* **63** (1981), 423–432.

[50] Robart, T.; Sur l'intégrabilité des sous-algèbres de Lie en dimension infinie; *Can. J. Math.* **49** (1997), 820–839.

[51] Scott, S.; *Traces and Determinants of Pseudodifferential Operators*, OUP (2010).

[52] Semenov-Tian-Shansky, M.A.; What is a classical r-matrix? *Funct. Anal. Appl.* **17** (1983), 259–272.

[53] Souriau, J.M.; Un algorithme générateur de structures quantiques. *Astérisque*, Hors Série, 341–399 (1985).

[54] Takeuchi, M.; Matched pairs of groups and bismash products of Hopf algebras. *Comm. Algebra*, **9** (1981), 841–882.

[55] Watts, J.; *Diffeologies, Differentiable Spaces and Symplectic Geometry.* PhD thesis, University of Toronto. arXiv:1208.3634v1

[56] Widom, H.; A complete symbolic calculus for pseudo-differential operators;. *Bull. Sc. Math. 2e serie* **104** (1980) 19–63.

B7. Wronskian solutions of integrable systems

Da-jun Zhang

Department of Mathematics, Shanghai University
Shanghai 200444, P.R. China

Abstract

The Wronski determinant (Wronskian) provides a compact form for τ-functions that play roles in a large range of mathematical physics. In 1979 Matveev and Satsuma, independently, obtained solutions in Wronskian form for the Kadomtsev-Petviashvili equation. Later, in 1981 these solutions were constructed from Sato's approach. Then in 1983, Freeman and Nimmo invented the so-called Wronskian technique, which allows directly verifying bilinear equations when their solutions are given in terms of Wronskians. In this technique the considered bilinear equation is usually reduced to the Plücker relation on Grassmannians; and finding solutions of the bilinear equation is transferred to find a Wronskian vector that is defined by a linear differential equation system. General solutions of such differential equation systems can be constructed by means of triangular Toeplitz matrices. In this monograph we review the Wronskian technique and solutions in Wronskian form, with supporting instructive examples, including the Korteweg-de Vries (KdV) equation, modified KdV equation, the Ablowitz-Kaup-Newell-Segur hierarchy and reductions, and lattice potential KdV equation. (Dedicated to Jonathan J C Nimmo).

1 Introduction

One of the kernel figures in the realm of integrable theory is the τ-function, in terms of which multi-solitons of integrable systems are expressed. There are several remarkable ways to solve integrable systems and provide τ-functions with explicit forms. In the Inverse Scattering Transform (IST), τ-functions are written by means of the Cauchy matrices (cf. [3]). Same expressions are also employed in some direct approaches (eg. Cauchy matrix approach [29, 45, 53] and operator approach [38]). Hirota's exponential polynomials provide a second form for τ-functions which can be derived in the bilinear method [17] or constructed using vertex operators [27]. A third form for τ-functions is the Wronskian form which was constructed using Darboux transformations [26] or the Wronskian technique [7].

In 1979 Matveev and Satsuma, independently, derived solutions in Wronskian form for the Kadomtsev-Petviashvili (KP) equation [23, 37]. Two years later these solutions were reconstructed from the celebrated Sato approach [36]. Then in 1983 Freeman and Nimmo invented the Wronskian technique, which provides a procedure to verify bilinear KP and Korteweg-de Vries (KdV) equations when their solutions are given in terms of Wronskians [7]. Soon after, it proved popular in integrable systems [9, 8, 31, 32, 33, 34]. In this technique the considered bilinear equation will be reduced to a known identity, say, usually, the Plücker relation on Grassmannians; and seeking solutions of the bilinear equation is conveyed to find a Wronskian vector that is defined by a linear differential equation system (LDES for short). Take the KdV equation as an example, which has the following LDES

$$\phi_{xx} = A\phi, \quad \phi_t = -4\phi_{xxx}, \tag{1.1}$$

where $\phi = (\phi_1, \phi_2, \cdots, \phi_N)^T$ and A is arbitrary in $\mathbb{C}_{N\times N}$. In Freeman-Nimmo's consideration the coefficient matrix A is diagonal with distinct nonzero eigenvalues [7]. This was generalised to the case of A being a Jordan block [40] in 1988. Explicit general solutions of such a LDES with arbitrary A can be written out by means of either the variation of constants method of ordinary differential equations [22] or triangular Toeplitz matrices [51], and solutions can then be classified according to canonical forms of A.

In this chapter we review the Wronskian technique and solutions in Wronskian form for integrable equations. Instructive examples include the KdV equation, modified KdV (mKdV) equation, the Ablowitz-Kaup-Newell-Segur (AKNS) hierarchy and reductions, and lattice potential KdV (lpKdV) equation, which cover Wronskian, double Wronskian and Casoratian forms of solutions.

The review is organized as follows. Sec. 2 serves as a preliminary in which we introduce bilinear equation and Bäcklund transformation (BT), notations of Wronskians, some determinantal identities, and triangular Toeplitz matrices. In Sec. 3, for the KdV equation we show how the Wronskian technique works in verifying bilinear equation and BT, and how to present explicit general solutions of its LDES. Limit relations between multiple pole and simple pole solutions are also explained. Sec. 4 is for solutions of the mKdV equation, which exhibits many aspects different from the KdV case. Sec. 5 serves as a part for double Wronskians and reduction technique, and Sec. 6 introduces Casoratian technique with the lpKdV equation as a fully discrete example. Finally, conclusions are given in Sec. 7.

2 Preliminary

2.1 The KdV stuff

Let us go through the stuff of the KdV equation

$$u_t + 6uu_x + u_{xxx} = 0, \tag{2.1}$$

which will serve as a demonstration in Wronskian technique. It has a Lax pair

$$\phi_{xx} + u\phi = -\lambda\phi, \tag{2.2a}$$

$$\phi_t = -4\phi_{xxx} - 6u\phi_x - 3u_x\phi, \tag{2.2b}$$

where λ is a spectral parameter. Employing the transformation

$$u = 2(\ln f)_{xx}, \tag{2.3}$$

the KdV equation (2.1) is written as its bilinear form

$$(D_tD_x + D_x^4)f \cdot f = 0, \tag{2.4}$$

where D is the well-known Hirota's bilinear operator defined by [14, 16]

$$D_x^m D_y^n f(x,y)\cdot g(x,y) = (\partial_x - \partial_{x'})^m(\partial_y - \partial_{y'})^n f(x,y)g(x',y')|_{x'=x,y'=y}. \quad (2.5)$$

Hirota gave the following compact form for the N-soliton solution of (2.4) [14]:

$$f = \sum_{\mu=0,1} \exp\left(\sum_{j=1}^{N} \mu_j\eta_j + \sum_{1\le i<j}^{N} \mu_i\mu_j a_{ij}\right), \quad (2.6)$$

where $\eta_j = k_i x - k_i^3 t + \eta_i^{(0)}$ with $k_i, \eta_i^{(0)} \in \mathbb{R}$, $e^{a_{ij}} = \left(\frac{k_i-k_j}{k_i+k_j}\right)^2$, and the summation of μ means to take all possible $\mu_j = 0,1$ ($j = 1,2,\cdots,N$). A proof of (2.6) satisfying (2.4) can be found in [14] and [3].

The bilinear KdV equation (2.4) admits a bilinear BT [16]

$$D_x^2 f\cdot g = \lambda f g, \quad (2.7a)$$

$$(D_x^3 + D_t + 3\lambda D_x) f\cdot g = 0, \quad (2.7b)$$

which indicates that if f is a solution of (2.4) and we solve the BT (2.7) to get g, then g will be a solution of (2.4) as well and $u = 2(\ln g)_{xx}$ provides a second solution to the KdV equation. Note that taking $\phi = g/f$ together with (2.3), the BT (2.7) will recover the Lax pair (2.2), and vice versa, from (2.2) to (2.7).

2.2 Wronskians

Wronskian is the determinant of a square matrix where its columns are arranged with consecutively increasing order derivatives of the first column. Consider

$$\phi = (\phi_1, \phi_2, \cdots, \phi_N)^T \quad (2.8)$$

where $\phi_i = \phi_i(x)$ are C^∞ functions. Then a Wronskian with ϕ as the first (elementary) column is $W = |\phi, \phi^{(1)}, \phi^{(2)}, \cdots, \phi^{(N-1)}|$, where $\phi^{(i)} = \partial_x^i \phi$. It can be more compactly expressed as (cf. [7])

$$W = |0,1,2,\cdots,N-1| = |\widehat{N-1}|.$$

Due to its special structure, derivatives of a Wronskian have quite simple expressions. For example,

$$W_x = |\widehat{N-2}, N|, \quad W_{xx} = |\widehat{N-3}, N-1, N| + |\widehat{N-2}, N+1|,$$

and if $\phi = \phi(x,y,t)$ with dispersion relation $\phi_y = \phi_{xx}$ and $\phi_t = \phi_{xxx}$, then one has

$$W_y = |\widehat{N-2}, N+1| - |\widehat{N-3}, N-1, N|,$$

$$W_t = |\widehat{N-2}, N+2| - |\widehat{N-3}, N-1, N+1| + |\widehat{N-4}, N-2, N-1, N|.$$

A double Wronskian is generated by two elementary column vectors

$$\varphi = (\varphi_1, \varphi_2, \cdots, \varphi_{N+M})^T, \quad \psi = (\psi_1, \psi_2, \cdots, \psi_{N+M})^T, \quad (2.9)$$

with the form $W = |\varphi, \varphi^{(1)}, \varphi^{(2)}, \cdots, \varphi^{(N-1)}; \psi, \psi^{(1)}, \psi^{(2)}, \cdots, \psi^{(M-1)}|$, and can be simply written as (cf. [30])

$$W = |0, 1, 2, \cdots, N-1; 0, 1, 2, \cdots, M-1| = |\widehat{N-1}; \widehat{M-1}|.$$

Taking the advantage of its structure, derivatives of a double Wronskian is simple as well, e.g. $W_x = |\widehat{N-2}, N; \widehat{M-1}| + |\widehat{N-1}; \widehat{M-2}, M|$.

2.3 Determinantal identities

The Wronskian technique allows directly verifying a solution in Wronskian form of a bilinear equation. Although a Wronskian provides simple expressions for its derivatives, during the verification one needs some determinantal identities to simplify high order derivatives. Finally, the bilinear equation to be verified is reduced to the Plücker relation (Laplace expansion of a zero-valued determinant). Let us go through these determinantal identities.

Theorem 1. *[51] Let $\Xi \in \mathbb{C}_{N\times N}$ and denote its column vectors as $\{\Xi_j\}$; let $\Omega = (\Omega_{i,j})_{N\times N}$ be an operator matrix (i.e. $\Omega_{i,j}$ are operators), and denote its column vectors as $\{\Omega_j\}$. The following relation holds,*

$$\sum_{j=1}^{N} |\Omega_j * \Xi| = \sum_{j=1}^{N} |(\Omega^T)_j * \Xi^T|, \tag{2.10}$$

where

$$|A_j * \Xi| = |\Xi_1, \cdots, \Xi_{j-1},\ A_j \circ \Xi_j,\ \Xi_{j+1}, \cdots, \Xi_N|,$$

and $A_j \circ \Xi_j$ stands for

$$A_j \circ B_j = (A_{1,j}B_{1,j},\ A_{2,j}B_{2,j}, \cdots, A_{N,j}B_{N,j})^T,$$

in which $A_j = (A_{1,j},\ A_{2,j}, \cdots, A_{N,j})^T$, $B_j = (B_{1,j},\ B_{2,j}, \cdots, B_{N,j})^T$ are Nth-order vectors.

Theorem 2. *Let $\mathbf{a}_j = (a_{1,j}, a_{2,j}, \cdots, a_{N,j})^T$, $j = 1, \cdots, 2N$ be Nth-order column vectors over $\mathbb{C}$. The Plücker relation is described as*

$$\sum_{j=1}^{N+1} (-1)^{N+1-j} |\mathbf{a}_1, \mathbf{a}_2, \cdots, \mathbf{a}_{j-1}, \mathbf{a}_{j+1}, \cdots, \mathbf{a}_{N+1}| \cdot |\mathbf{a}_j, \mathbf{a}_{N+2}, \cdots, \mathbf{a}_{2N}| = 0. \tag{2.11}$$

In fact, (2.11) is a Laplace expansion w.r.t. the first N rows of the following zero-valued determinant

$$\begin{vmatrix} a_{1,1} & \cdots & a_{1,N+1} & 0 & \cdots & 0 \\ \vdots & \vdots & \vdots & \vdots & \vdots & \vdots \\ a_{N,1} & \cdots & a_{N,N+1} & 0 & \cdots & 0 \\ a_{1,1} & \cdots & a_{1,N+1} & a_{1,N+2} & \cdots & a_{1,2N} \\ \vdots & \vdots & \vdots & \vdots & \vdots & \vdots \\ a_{N,1} & \cdots & a_{N,N+1} & a_{N,N+2} & \cdots & a_{N,2N} \end{vmatrix}.$$

Special cases of (2.11) are the following.

Corollary 1. Let $P \in \mathbb{C}_{N\times(N-1)}$, $Q \in \mathbb{C}_{N\times(N-k+1)}$ be the remainder of P after removing its arbitrary $k-2$ columns where $3 \leq k < N$, and $\mathbf{a}_i$, $i = 1, 2, \cdots, k$, are Nth-order column vectors. Then one has

$$\sum_{i=1}^{k}(-1)^{i-1}|P, \mathbf{a}_i| \cdot |Q, \mathbf{a}_1, \cdots, \mathbf{a}_{i-1}, \mathbf{a}_{i+1}, \cdots, \mathbf{a}_k| = 0, \qquad k \geq 3. \tag{2.12}$$

This is a practical formula to generate identities used in Wronskian verification. For example, when $k = 4$, (2.12) yields

$$\begin{aligned}&|P, \mathbf{a}_1| \cdot |Q, \mathbf{a}_2, \mathbf{a}_3, \mathbf{a}_4| - |P, \mathbf{a}_2| \cdot |Q, \mathbf{a}_1, \mathbf{a}_3, \mathbf{a}_4|\\ &+ |P, \mathbf{a}_3| \cdot |Q, \mathbf{a}_1, \mathbf{a}_2, \mathbf{a}_4| - |P, \mathbf{a}_4| \cdot |Q, \mathbf{a}_1, \mathbf{a}_2, \mathbf{a}_3| = 0,\end{aligned}$$

and when $k = 3$,

$$|M, \mathbf{a}, \mathbf{b}||M, \mathbf{c}, \mathbf{d}| - |M, \mathbf{a}, \mathbf{c}||M, \mathbf{b}, \mathbf{d}| + |M, \mathbf{a}, \mathbf{d}||M, \mathbf{b}, \mathbf{c}| = 0, \tag{2.13}$$

where we have taken $P = (Q, \mathbf{p}_{N-1})$, $M = Q$, $\mathbf{a} = \mathbf{p}_{N-1}, \mathbf{b} = \mathbf{a}_1, \mathbf{c} = \mathbf{a}_2, \mathbf{d} = \mathbf{a}_3$, where $\mathbf{p}_{N-1}$ is the last column of P.

2.4 Triangular Toeplitz matrices

Triangular Toeplitz matrices are used to express general solutions of the LDES like (1.1). A lower triangular Toeplitz matrix (LTTM) of order N is defined as

$$\mathcal{A} = \begin{pmatrix} a_0 & 0 & 0 & \cdots & 0 & 0 \\ a_1 & a_0 & 0 & \cdots & 0 & 0 \\ \cdots & \cdots & \cdots & \cdots & \cdots & \cdots \\ a_{N-1} & a_{N-2} & a_{N-3} & \cdots & a_1 & a_0 \end{pmatrix} \in \mathbb{C}_{N\times N}. \tag{2.14}$$

All such matrices of order N compose a commutative set, denoted by $\widetilde{G}_N$, w.r.t. matrix multiplication. It is easy to find that $G_N = \left\{\mathcal{A} \middle| \mathcal{A} \in \widetilde{G}_N, |\mathcal{A}| \neq 0\right\}$ is an Abelian group.

Lemma 1. *[51] 1. A C^∞ function $\alpha(k)$ can generate an LTTM* (2.14) *via*

$$a_j = \frac{1}{j!}\partial_k^j \alpha(k), \quad j = 0, 1, \cdots, N-1. \tag{2.15}$$

2. On the other hand, for any $\mathcal{A}$ defined as (2.14)*, there exists a complex polynomial $\alpha(z) = \sum_{j=0}^{N-1} \alpha_j z^{N-1-j}$, such that*

$$\partial_z^j \alpha(z)|_{z=k} = j! a_j, \quad j = 0, 1, \cdots, N-1.$$

3. For any given $\mathcal{A} \in G_N$, there exist $\pm\mathcal{B} \in G_N$ such that $\mathcal{B}^2 = \mathcal{A}$.

We may also consider a block LTTM defined as

$$\mathcal{A}^B_{[*]} = \begin{pmatrix} A_0 & 0 & 0 & \cdots & 0 & 0 \\ A_1 & A_0 & 0 & \cdots & 0 & 0 \\ \cdots & \cdots & \cdots & \cdots & \cdots & \cdots \\ A_{N-1} & A_{N-2} & A_{N-3} & \cdots & A_1 & A_0 \end{pmatrix}_{2N\times 2N}, \tag{2.16}$$

where for $\mathcal{A}^B_{[D]}$ we take $A_j = \begin{pmatrix} a_{j1} & 0 \\ 0 & a_{j2} \end{pmatrix}$, for $\mathcal{A}^B_{[T]}$ we take $A_j = \begin{pmatrix} a_j & 0 \\ b_j & a_j \end{pmatrix}$, and for $\mathcal{A}^B_{[\epsilon]}$ we take $A_j = \begin{pmatrix} a_j & \epsilon b_j \\ b_j & a_j \end{pmatrix}$ with $\epsilon \in \mathbb{R}$ and usually taking $\epsilon = \pm 1$. Note that with regard to the block LTTM $\mathcal{A}^B_{[*]}$, for each case of "*" with different A_j, there are similar properties as $\mathcal{A}$ holds [51].

3 The KdV equation

3.1 Wronskian solution of the bilinear KdV

Let us repeat Freeman-Nimmo's Wronskian technique [7] with a slight extension.

Theorem 3. *The bilinear KdV equation* (2.4) *admits a Wronskian solution*

$$f = |\widehat{N-1}| \tag{3.1}$$

which is composed of the first column vector $\phi(x,t) = (\phi_1, \phi_2, \cdots, \phi_N)^T$ *satisfying the LDES*

$$\phi_{xx} = -A\phi, \tag{3.2a}$$
$$\phi_t = -4\phi_{xxx}, \tag{3.2b}$$

where $A = (a_{ij})_{N\times N} \in \mathbb{C}_{N\times N}$ *is arbitrary.*

Proof. We calculate derivatives

$$\begin{aligned} f_x &= |\widehat{N-2}, N|, \quad f_{xx} = |\widehat{N-3}, N-1, N| + |\widehat{N-2}, N+1|, \\ f_{xxx} &= |\widehat{N-4}, N-2, N-1, N| + 2|\widehat{N-3}, N-1, N+1| + |\widehat{N-2}, N+2|, \\ f_{xxxx} &= |\widehat{N-5}, N-3, N-2, N-1, N| + 3|\widehat{N-4}, N-2, N-1, N+1| \\ &\quad + 2|\widehat{N-3}, N, N+1| + 3|\widehat{N-3}, N-1, N+2| + |\widehat{N-2}, N+3|, \\ f_t &= -4(|\widehat{N-4}, N-2, N-1, N| - |\widehat{N-3}, N-1, N+1| + |\widehat{N-2}, N+2|), \\ f_{tx} &= -4(|\widehat{N-5}, N-3, N-2, N-1, N| - |\widehat{N-3}, N, N+1| \\ &\quad + |\widehat{N-2}, N+3|). \end{aligned}$$

By substitution the bilinear KdV equation (2.4) yields

$$
\begin{aligned}
&f_{xt}f - f_xf_t + f_{xxxx}f - 4f_{xxx}f_x + 3(f_{xx})^2 \\
= &-3|\widehat{N-1}|\Big(|\widehat{N-5}, N-3, N-2, N-1, N| - |\widehat{N-4}, N-2, N-1, N+1| \\
&-2|\widehat{N-3}, N, N+1| - |\widehat{N-3}, N-1, N+2| + |\widehat{N-2}, N+3|\Big) \\
&-12|\widehat{N-2}, N||\widehat{N-3}, N-1, N+1| \\
&+3\Big(|\widehat{N-3}, N-1, N| + |\widehat{N-2}, N+1|\Big)^2.
\end{aligned} \tag{3.3}
$$

Next, we make use of Theorem 1 in which we take $\Omega_{ij} \equiv \partial_x^2$ and $\Xi = (\widehat{N-1})$. It follows from (2.10) that

$$
\mathrm{Tr}(A)f = |\widehat{N-3}, N-1, N| - |\widehat{N-2}, N+1|. \tag{3.4}
$$

Similarly, we can calculate $\mathrm{Tr}(A)(\mathrm{Tr}(A)f)$. Then, using equality $f\mathrm{Tr}(A)(\mathrm{Tr}(A)f) = (\mathrm{Tr}(A)f)^2$ we find

$$
\begin{aligned}
&|\widehat{N-1}|\Big(|\widehat{N-5}, N-3, N-2, N-1, N| + 2|\widehat{N-3}, N, N+1| \\
&\quad - |\widehat{N-3}, N-1, N+2| - |\widehat{N-4}, N-2, N-1, N+1| + |\widehat{N-2}, N+3|\Big) \\
=&\Big(|\widehat{N-3}, N-1, N| - |\widehat{N-2}, N+1|\Big)^2,
\end{aligned}
$$

which can be used to eliminate some terms in (3.3) generated from f_{xxxx} and f_{xt}. As a result, we have

$$
\begin{aligned}
&f_{xt}f - f_xf_t + f_{xxxx}f - 4f_{xxx}f_x + 3(f_{xx})^2 \\
=&12\Big(|\widehat{N-1}||\widehat{N-3}, N, N+1| - |\widehat{N-2}, N||\widehat{N-3}, N-1, N+1| \\
&+ |\widehat{N-3}, N-1, N||\widehat{N-2}, N+1|\Big),
\end{aligned}
$$

which is zero if we make use of (2.13) and take $M = (\widehat{N-3})$, $\mathbf{a} = N-2$, $\mathbf{b} = N-1$, $\mathbf{c} = N$, $\mathbf{d} = N+1$. Thus, a direct verification of the Wronskian solution of the bilinear KdV equation is completed. ■

Remark 1. Note that within the complex field $\mathbb{C}$ any square matrix A is similar to a lower triangular canonical form A' by a transform matrix P, i.e. $A = P^{-1}A'P$. Define $\psi = P\phi$. Then (3.2) yields $\psi_{xx} = -A'\psi$, $\psi_t = -4\psi_{xxx}$ and the Wronskians composed of ϕ and ψ are related to each other as $f(\psi) = |P|f(\phi)$. Obviously, if $f(\phi)$ solves the bilinear KdV equation (2.4), so does $f(\psi)$. In this sense, we only need to consider the LDES (3.2) with a canonically formed A.

3.2 Wronskian solution of the bilinear BT

The bilinear BT (2.7) can transform solutions of the KdV equation from $N-1$ solitons to N solitons by taking $g=|\widehat{N-2}|$ and $f=|\widehat{N-1}|$. This was proved by Nimmo and Freeman in [34]. In the following we give a proof for a more general case.

Theorem 4. *The Wronskians*

$$g=|\widehat{N-2},\sigma_N|,\quad f=|\widehat{N-1}| \tag{3.5}$$

satisfy the bilinear BT (2.7) *with* $\lambda=-a_{NN}$, *where* ϕ *is governed by the LDES* (3.2) *in which* A *is lower triangular, and* σ_j *is defined as*

$$\sigma_j=(\delta_{j,1},\delta_{j,2},\cdots,\delta_{j,N})^T,\quad \delta_{j,i}=\begin{cases}1 & j=i,\\ 0 & j\neq i.\end{cases} \tag{3.6}$$

Proof. We only prove (2.7a). Note that

$$g_x=|\widehat{N-3},N-1,\sigma_N|,\quad g_{xx}=|\widehat{N-4},N-2,N-1,\sigma_N|+|\widehat{N-3},N,\sigma_N|.$$

From (2.7a) we have

$$\begin{aligned}&f_{xx}g-2f_xg_x+fg_{xx}-\lambda fg\\ =&\Big(|\widehat{N-3},N-1,N|+|\widehat{N-2},N+1|\Big)|\widehat{N-2},\sigma_N|\\ &-2|\widehat{N-2},N||\widehat{N-3},N-1,\sigma_N|\\ &+|\widehat{N-1}|\Big(|\widehat{N-4},N-2,N-1,\sigma_N|+|\widehat{N-3},N,\sigma_N|-\lambda|\widehat{N-2},\sigma_N|\Big).\end{aligned} \tag{3.7}$$

To simplify the above form, similar to (3.4), for g we have

$$\mathrm{Tr}(A)g=|\widehat{N-4},N-2,N-1,\sigma_N|-|\widehat{N-3},N,\sigma_N|+\sum_{j=1}^{N}a_{jN}|\widehat{N-2},\sigma_j|.$$

Then, making use of the equality generated from $g(\mathrm{Tr}(A)f)=f(\mathrm{Tr}(A)g)$, (3.7) is reduced to

$$\begin{aligned}&2\Big(|\widehat{N-3},N-1,N||\widehat{N-2},\sigma_N|-|\widehat{N-2},N||\widehat{N-3},N-1,\sigma_N|\\ &+|\widehat{N-1}||\widehat{N-3},N,\sigma_N|\Big)-|\widehat{N-1}||\widehat{N-2},\widetilde{\sigma}_N|,\end{aligned}$$

where $\widetilde{\sigma}_N=(a_{1N},a_{2N},\cdots,a_{NN}+\lambda)^T$. The last term is zero if we require A to be lower triangular and $\lambda=-a_{NN}$; the first three terms together contribute a zero value due to (2.13) with $M=(\widehat{N-3})$, $\mathbf{a}=N-2$, $\mathbf{b}=N-1$, $\mathbf{c}=N$, $\mathbf{d}=\sigma_N$.

Bilinear BT (2.7b) can be proved in a similar way. One can refer to [46] for more details. ∎

Note that bilinear BT (2.7) provides a transformation between two solutions (3.5). As a starting point, g must be a solution of the bilinear KdV equation. The simplest case is $g=1$, i.e. $N=1$, $\sigma_1=1$.

3.3 Classification of solutions

Based on Remark 1, it is possible to classify solutions of the KdV equation according to canonical forms of A. In general, from the viewpoint of the IST, under strict complex analysis in direct scattering, transmission coefficient $T(k)$ admits only distinct simple poles, which are defined on the imaginary axis of the k-plane. Each soliton is identified by one of these simple poles. However, in the bilinear approach or Darboux transformation, solutions are derived through more direct ways without restriction on those poles. For example, multiple-pole solutions can be obtained when A in (3.2) is a Jordan block.

In the following, we list out elementary Wronskian column vectors in terms of the canonical forms of A. Note that mathematically there is no need to differentiate whether the eigenvalues of A are real or complex.

Case 1

$$A = \mathrm{Diag}(-k_1^2, -k_2^2, \cdots, -k_N^2), \tag{3.8}$$

where $\{\lambda_j = -k_j^2\}$ are distinct nonzero numbers. In this case, ϕ satisfying (3.2) is given as (2.8) with

$$\phi_j = b_j^+ e^{\xi_j} + b_j^- e^{-\xi_j}, \tag{3.9}$$

or equivalently

$$\phi_j = a_j^+ \cosh \xi_j + a_j^- \sinh \xi_j, \tag{3.10}$$

where

$$\xi_j = k_j x - 4k_j^3 t + \xi_j^{(0)}, \tag{3.11}$$

and $a_j^\pm, b_j^\pm, \xi_j^{(0)} \in \mathbb{C}$.

Note that if all $k_j, a_j^\pm, b_j^\pm, \xi_j^{(0)}$ are real, together with taking $a_j^\pm = [1 \mp (-1)^j]/2$ and $0 < k_1 < k_2 < \cdots < k_N$ in (3.10), the corresponding Wronskian f generates an N-soliton solution. In fact, such a Wronskian can be written as

$$f = K \cdot \left(\prod_{j=1}^{N} e^{-\xi_j}\right) \sum_{\mu=0,1} \exp\left\{\sum_{j=1}^{N} 2\mu_j \xi_j' + \sum_{1\le j<l\le N} \mu_j \mu_l a_{jl}\right\}, \tag{3.12}$$

where the sum over $\mu = 0, 1$ refers to each of $\mu_j = 0, 1$ for $j = 0, 1, \cdots, N$, and

$$\xi_j' = \xi_j - \frac{1}{4} \sum_{l=1, l\neq j}^{N} a_{jl}, \quad e^{a_{jl}} = \left(\frac{k_l - k_j}{k_l + k_j}\right)^2, \quad K = \prod_{1\le j<l\le N} (k_l - k_j).$$

This is nothing but Hirota's expression (2.6) for the N-soliton solution. A similar proof for (3.12) can be found in Ref. [52].

If all k_j are pure imaginary, we replace k_j with $i\kappa_j$ where $\kappa_j \in \mathbb{R}$ and i is the imaginary unit, ϕ_j (3.10) can be rewritten as

$$\phi_j^+ = a_j^+ \cos\theta_j + a_j^- \sin\theta_j, \quad a_j^\pm \in \mathbb{C}, \tag{3.13}$$

where

$$\theta_j = \kappa_j x + 4\kappa_j^3 t + \theta_j^{(0)}, \quad \theta_j^{(0)} \in \mathbb{R}. \tag{3.14}$$

Case 2

$$A = -\widetilde{J}_N[k_1], \quad \widetilde{J}_N[k_1] \doteq \begin{pmatrix} k_1^2 & 0 & 0 & \cdots & 0 & 0 & 0 \\ 2k_1 & k_1^2 & 0 & \cdots & 0 & 0 & 0 \\ 1 & 2k_1 & k_1^2 & \cdots & 0 & 0 & 0 \\ \cdots & \cdots & \cdots & \cdots & \cdots & \cdots & \cdots \\ 0 & 0 & 0 & \cdots & 1 & 2k_1 & k_1^2 \end{pmatrix}_{N\times N}. \tag{3.15}$$

A general solution of the LDES (3.2) of this case is

$$\phi = \mathcal{A}\mathcal{Q}^+ + \mathcal{B}\mathcal{Q}^-, \quad \mathcal{A}, \mathcal{B} \in \widetilde{G}_N, \tag{3.16}$$

where

$$\mathcal{Q}^\pm = (\mathcal{Q}_0^\pm, \mathcal{Q}_1^\pm, \cdots, \mathcal{Q}_{N-1}^\pm)^T, \quad \mathcal{Q}_j^\pm = \frac{1}{j!}\partial_{k_1}^j \phi_1^\pm, \tag{3.17}$$

with $\phi_1^+ = a_1^+ \cosh\xi_1$, $\phi_1^- = a_1^- \sinh\xi_1$, or $\phi_1^\pm = b_1^\pm e^{\pm\xi_1}$.

One can verify that $\mathcal{Q}^\pm$ are two independent solutions of the LDES (3.2). Since (3.2) is linear and LTTMs $\mathcal{A}$ and $\mathcal{B}$ contain enough $2N$ arbitrary constants, (3.16) provides a general solution to (3.2) when A takes Jordan block (3.15). Due to the gauge property mentioned in Remark 1, a significant form of (3.16) is

$$\phi = \mathcal{A}\mathcal{Q}^+ + \mathcal{Q}^-, \text{ or } \phi = \mathcal{Q}^+ + \mathcal{B}\mathcal{Q}^-, \quad \mathcal{A}, \mathcal{B} \in G_N. \tag{3.18}$$

Note also that one can take $a_1^\pm$ or $b_1^\pm$ to be functions of k_1, but their contribution to $\mathcal{Q}_j^\pm$ through $\partial_{k_1}^j \phi_1^\pm$ can be balanced by LTTMs $\mathcal{A}$ and $\mathcal{B}$ in light of Lemma 1.

Case 3

$$A = J_N[0] = \begin{pmatrix} 0 & 0 & 0 & \cdots & 0 & 0 \\ 1 & 0 & 0 & \cdots & 0 & 0 \\ \cdots & \cdots & \cdots & \cdots & \cdots & \cdots \\ 0 & 0 & 0 & \cdots & 1 & 0 \end{pmatrix}_{N\times N}. \tag{3.19}$$

In this case, we get rational solutions to the KdV equation. General solution of this case is given by

$$\phi = \mathcal{A}\mathcal{R}^+ + \mathcal{B}\mathcal{R}^-, \quad \mathcal{A}, \mathcal{B} \in \widetilde{G}_N \tag{3.20}$$

where $\mathcal{R}^\pm = (\mathcal{R}_0^\pm, \mathcal{R}_1^\pm, \cdots, \mathcal{R}_{N-1}^\pm)^T$ with

$$R_j^+ = \frac{1}{(2j)!}\Big[\frac{\partial^{2j}}{\partial {k_1}^{2j}} \cosh\xi_1\Big]_{k_1=0}, \quad R_j^- = \frac{1}{(2j+1)!}\Big[\frac{\partial^{2j+1}}{\partial {k_1}^{2j+1}} \sinh\xi_1\Big]_{k_1=0}.$$

Significant form of (3.20) is

$$\phi = \mathcal{A}\mathcal{R}^+ + \mathcal{R}^-, \text{ or } \phi = \mathcal{R}^+ + \mathcal{B}\mathcal{R}^-, \quad \mathcal{A}, \mathcal{B} \in G_N. \tag{3.21}$$

3.4 Notes

In Freeman-Nimmo's proof [7] the coefficient matrix A in the LDES (3.2) takes a diagonal form. This was generalised to a case of A being Jordan form by a trick through taking derivatives for ϕ_1 w.r.t. k_1 [40]. With the help of Theorem 1, one can implement Wronskian verification starting from the LDES (3.2) with arbitrary A, as we have done in Sec. 3.1 for the bilinear KdV equation and in Sec. 3.2 for the bilinear BT. Nimmo and Freeman also developed a procedure to get rational solutions in Wronskian form [33]. Although, as mentioned in [33], they failed to find more examples than the KdV equation, their technique is indeed general and valid for a large group of integrable equations (cf. [44]). With regard to finding general solutions for the LDES (3.2) with a general A, taking advantage of the linearity of (3.2) and the gauge property of its solutions (see Remark 1), one only needs to consider A being a diagonal or a Jordan block. To find general solutions of the Jordan block case, [22] employed a variation of constants method of ordinary differential equations, while here we make use of LTTMs, which is more explicit and convenient.

The name "positons" was introduced by Matveev [24, 25] for those solutions generated from (3.13). Since when taking $u = 0$, (3.13) is a solution of the Schrödinger equation (2.2a) with positive eigenvalue $\lambda = \kappa_j^2$, the corresponding solutions of the KdV equation bear the name of "positons". Along this line, solitons belong to "negatons" [35], and when $\lambda \in \mathbb{C}$ the solutions are called "complexitons" [21] (see also [18]). However, such a classification depends on nonlinear equations (see next section for the mKdV equation), and it seems that all these solutions of the KdV equation, except solitons, are singular, because by direct scattering analysis all discrete eigenvalues $\{\lambda_j\}$ have to be simple and negative.

Multiple-pole solutions can be explained as a special limit of simple-pole solutions [26, 51]. To understand this, consider the following scaled Wronskian

$$\frac{W(\phi_1, \phi_2, \cdots, \phi_N)}{\prod_{j=2}^{N}(k_1 - k_j)^{j-1}} \tag{3.22}$$

where $\phi_1 = \phi_1(k_1)$ is given as (3.10) and $\phi_j = \phi_1(k_j)$ for $j = 2, 3, \cdots, N$. Implementing Taylor expansion successively for ϕ_j, $j > 1$, at k_1, (3.22) turns out to be $W(\phi_1, \partial_{k_1}\phi_1, \frac{1}{2!}\partial_{k_1}^2\phi_1, \cdots, \frac{1}{(N-1)!}\partial_{k_1}^{N-1}\phi_1)$, which is a Wronskian for multiple-pole solutions. Rational solutions can be obtained with a more elaborate procedure by taking $k_1 \to 0$ after expansion.

4 The mKdV equation

4.1 Wronskian solutions

The mKdV$^+$ (in the following, the mKdV for short) equation

$$v_t + 6v^2 v_x + v_{xxx} = 0 \tag{4.1}$$

is another typical $(1+1)$-dimensional integrable equation, with a Lax pair

$$\psi_{xx} + 2iv\psi_x = \lambda\psi, \tag{4.2a}$$
$$\psi_t = -4\psi_{xxx} + 12iv\psi_{xx} + (6iv_x - v^2)\psi_x. \tag{4.2b}$$

It also has a Lax pair with the AKNS spectral problem (see (5.2) with $q = -r = v$). This equation has solutions in Wronskian form as well, but its LDES contains a complex operation (see (4.6)), which creates problems in obtaining general solutions.

Employing the transformation

$$v = i\left(\ln \frac{f^*}{f}\right)_x, \tag{4.3}$$

the mKdV equation (4.1) is bilinearized as (cf. [15])

$$(D_t + D_x^3) f^* \cdot f = 0, \tag{4.4a}$$
$$D_x^2 f^* \cdot f = 0, \tag{4.4b}$$

where $*$ stands for the complex conjugate.

Theorem 5. *The bilinear mKdV equation* (4.4) *has a Wronskian solution*

$$f = |\widehat{N-1}|, \tag{4.5}$$

where the elementary column vector ϕ obeys the LDES

$$\phi_x = B\phi^*, \tag{4.6a}$$
$$\phi_t = -4\phi_{xxx}, \tag{4.6b}$$

and $|B| \neq 0$ is required.

A proof can be found in the Appendix of [55].

Remark 2. In the proof of Theorem 5, f^* is written as

$$f^* = |B^*||-1, \widehat{N-2}|, \tag{4.7}$$

which requires $|B| \neq 0$. Therefore rational solutions of the mKdV equation cannot be derived from Theorem 5. Rational solutions will be considered separately in Sec. 4.3.

Remark 3. There is no gauge property for the LDES (4.6). Noting that both ϕ and ϕ^* are involved in (4.6a), for any $\widetilde{B} = PBP^{-1}$ which is similar to B, the new defined vector $\psi = P\phi$ does not satisfy $\psi_x = \widetilde{B}\psi^*$. Without a gauge property we cannot construct general solutions of the LDES (4.6) and conduct a completed classification of solutions according to the canonical forms of B. However, we do solve this problem if we make use of the LDES (3.2) of the KdV equation and its gauge property. See Theorem 6.

4.2 Classification of solutions

4.2.1 Gauge property: revisit

Consider (3.2a) in the LDES of the KdV equation and introduce $\mathbb{A} = P^{-1}AP$ which is similar to A with a transform matrix P. Defining $\varphi = P^{-1}\phi$ and $\mathbb{B} = P^{-1}BP^*$, from (4.6a) we have $\varphi_x = \mathbb{B}\varphi^*$. Obviously Wronskian $f(\phi) = |P|f(\varphi)$, i.e. ϕ and φ yield the same solution for the mKdV equation through (4.3). We conclude the above analysis by the following theorem.

Theorem 6. *The Wronskian* (4.5) *provides a solution to the bilinear mKdV equation* (4.4) *where the elementary column vector* φ *satisfies*

$$\varphi_{xx} = \mathbb{A}\varphi, \tag{4.8a}$$

$$\varphi_x = \mathbb{B}\varphi^*, \tag{4.8b}$$

$$\varphi_t = -4\varphi_{xxx}, \tag{4.8c}$$

$|\mathbb{B}| \neq 0$ *and*

$$\mathbb{A} = \mathbb{B}\mathbb{B}^*. \tag{4.9}$$

With this theorem, and noting the fact that the eigenvalues of $\mathbb{A}$ defined by (4.9) are either real or appear as conjugate pairs if there are any complex ones [55], one can construct general solutions to the LDES (4.8) and implement a full classification of solutions according to the canonical forms of $\mathbb{A}$ instead of $\mathbb{B}$.

4.2.2 Solitons

It can be proved that the case of $\mathbb{A}$ containing N negative eigenvalues yields only trivial solutions of the mKdV equation (4.1) [55]. In the following we consider the case

$$\mathbb{A} = \mathrm{Diag}(\lambda_1^2, \lambda_2^2, \cdots, \lambda_N^2), \tag{4.10}$$

where, without loss of generality, we let $\lambda_j = \varepsilon_j||k_j|| \neq 0$, in which $\varepsilon_j = \pm 1$ and k_j can be either real or complex numbers with distinct absolute values (modulus).

When $k_j \in \mathbb{R}$, i.e. $\lambda_j^2 = k_j^2$ in (4.10), $\mathbb{B}$ takes a form

$$\mathbb{B} = \mathrm{Diag}(k_1, k_2, \cdots, k_N), \tag{4.11}$$

where $k_j \in \mathbb{R}$. A solution to the LDES (4.8) is $\varphi = (\varphi_1, \varphi_2, \cdots, \varphi_N)^T$ where

$$\varphi_j = a_j^+ e^{\xi_j} + ia_j^- e^{-\xi_j}, \quad \xi_j = k_j x - 4k_j^3 t + \xi_j^{(0)}, \quad a_j^+, a_j^-, k_j, \xi_j^{(0)} \in \mathbb{R}. \tag{4.12}$$

Particularly, when $a_j^+ = (-1)^{j-1}$, $a_j^- \equiv 1$, the Wronskian solution $f(\varphi)$ can be written as [50]

$$f = K \cdot \left(\prod_{j=1}^{N} e^{\xi_j}\right) \sum_{\mu=0,1} \exp\Big\{\sum_{j=1}^{N} \mu_j(2\eta_j + \frac{\pi}{2}i) + \sum_{1\leq j<l\leq N} \mu_j\mu_l a_{jl}\Big\},$$

where the sum over $\mu = 0, 1$ refers to each of $\mu_j = 0, 1$ for $j = 1, 2, \cdots, N$, and

$$\eta_j = -\xi_j - \frac{1}{4}\sum_{l=2, l\neq j}^{N} a_{jl}, \quad e^{a_{jl}} = \left(\frac{k_l - k_j}{k_l + k_j}\right)^2, \quad K = \prod_{1\leq j<l\leq N}(k_j - k_l).$$

This coincides with the N-soliton solution in Hirota's form [3].

When $k_j = k_{j1} + ik_{j2} \in \mathbb{C}$, i.e. $\lambda_j^2 = k_{j1}^2 + k_{j2}^2$ in (4.10), $\mathbb{B}$ is taken as (4.11) but with $k_j \in \mathbb{C}$, and the solution φ of (4.8) is composed of

$$\varphi_j = \gamma_j(a_j^+ e^{\xi_j} + ia_j^- e^{-\xi_j}), \ \xi_j = \lambda_j x - 4\lambda_j^3 t + \xi_j^{(0)}, \ \ a_j^+, a_j^-, \ \xi_j^{(0)} \in \mathbb{R}, \quad (4.13a)$$

where

$$\gamma_j = 1 + \frac{i(\lambda_j - k_{j1})}{k_{j2}}, \quad \lambda_j = \varepsilon_j ||k_j||. \quad (4.13b)$$

However, this yields the same solution to the mKdV equation as (4.12) does. Note that the solution obtained in [34] by Nimmo and Freeman is the case $k_j = ik_{j2}$.

With the above discussions we come to the following remark.

Remark 4. No matter whether the diagonal matrix $\mathbb{B}$ (4.11) is real or complex, the related Wronskian $f(\varphi)$ ALWAYS generates N-soliton solutions to the mKdV equation when all $\{k_j\}$ have distinct absolute values, and each soliton is identified by $\lambda_j = ||k_j||$. In this sense, unlike the KdV equation, there is no positon-negaton-complexiton classification for the mKdV equation (4.1). Define an equivalent relation $\sim$ on the complex plane $\mathbb{C}$ by

$$k_i \sim k_j, \ \text{iff} \ ||k_i|| = ||k_j||. \quad (4.14)$$

Then the quotient space $\mathbb{C}/\sim$ denotes the positive half real axis (or positive half imaginary axis), on which we choose distinct $\{k_j\}$ for solitons.

4.2.3 Limit solutions of solitons

This is the case in which both $\mathbb{A}$ and $\mathbb{B}$ are LTTMs. To find a general solution of this case, we consider $\mathbb{A} = \widetilde{J}_N[k_1]$ defined as (3.15) with $k_1 \in \mathbb{R}$. Thus we can make use of (3.16) and write

$$\varphi = \mathcal{A}\mathcal{Q}^+ + \mathcal{B}\mathcal{Q}^-, \quad \mathcal{A}, \mathcal{B} \in \widetilde{G}_N, \quad (4.15)$$

which solves (4.8a) and (4.8c), where

$$\mathcal{Q}^\pm = (\mathcal{Q}_0^\pm, \mathcal{Q}_1^\pm, \cdots, \mathcal{Q}_{N-1}^\pm)^T, \quad \mathcal{Q}_s^\pm = \frac{1}{s!}\partial_{k_1}^s e^{\pm\xi_1}, \quad (4.16)$$

and ξ_1 is defined in (4.12). The matrix $\mathbb{B}$ that satisfies (4.9) is a standard Jordan block

$$\mathbb{B} = J_N[k_1] \doteq \begin{pmatrix} k_1 & 0 & 0 & \cdots & 0 & 0 \\ 1 & k_1 & 0 & \cdots & 0 & 0 \\ \cdots & \cdots & \cdots & \cdots & \cdots & \cdots \\ 0 & 0 & 0 & \cdots & 1 & k_1 \end{pmatrix}_N. \quad (4.17)$$

In the following we impose extra conditions on $\mathcal{A}$ and $\mathcal{B}$ so that (4.15) solves (4.8b) as well. To do that, substituting (4.15) into (4.8b), one has

$$\mathcal{A}\mathcal{Q}_x^+ + \mathcal{B}\mathcal{Q}_x^- = \mathbb{B}(\mathcal{A}^*\mathcal{Q}^+ + \mathcal{B}^*\mathcal{Q}^-). \tag{4.18}$$

Meanwhile, it can be verified that $\mathcal{Q}_{0,x}^\pm = \pm\mathbb{B}\mathcal{Q}_0^\pm$, Then, noting that $\mathbb{B}$ is real and $\mathbb{B}$, $\mathcal{A}$ and $\mathcal{B}$ are commutative, it follows that

$$\mathbb{B}(\mathcal{A}\mathcal{Q}^+ - \mathcal{B}\mathcal{Q}^-) = \mathbb{B}(\mathcal{A}^*\mathcal{Q}^+ + \mathcal{B}^*\mathcal{Q}^-). \tag{4.19}$$

Since $\mathcal{Q}^+$ and $\mathcal{Q}^-$ are linearly independent, this indicates that $\mathcal{A}$ is real and $\mathcal{B}$ is pure imaginary. Thus, a general solution to the LDES (4.8) of the mKdV equation is

$$\varphi = \mathcal{A}^+\mathcal{Q}^+ + i\mathcal{A}^-\mathcal{Q}^-, \tag{4.20}$$

where $\mathcal{A}^\pm$ are real LTTMs.

The case of $k_1 \in \mathbb{C}$ contributes the same solutions. For more details of analysis of this case, one can refer to [55]. Note that we use the subtitle "limit solutions of solitons" for this part because the Wronskian f composed of (4.20) can also be obtained by taking limits from solitons. Of course, this is a case in which multiple-pole solutions are related to solitons.

4.2.4 Breathers and limit breathers

Breathers are obtained when $\mathbb{A}$ has N distinct complex conjugate-pairs of eigenvalues, with a canonical form

$$\mathbb{A} = \mathrm{Diag}(k_1^2, k_1^{*2}, \ \cdots \ , \ k_N^2, k_N^{*2}), \ \ k_j = k_{1j} + ik_{2j}, \ \ k_{1j}k_{2j} \neq 0, \tag{4.21}$$

and $\mathbb{B}$ takes a special block diagonal form

$$\mathbb{B} = \mathrm{Diag}(\Theta_1, \Theta_2, \cdots, \Theta_N), \quad \Theta_j = \begin{pmatrix} 0 & k_j \\ k_j^* & 0 \end{pmatrix}. \tag{4.22}$$

Solution to the LDES (4.8) is

$$\varphi = (\varphi_{11}, \varphi_{12}, \varphi_{21}, \varphi_{22}, \cdots, \varphi_{N1}, \varphi_{N2})^T, \tag{4.23a}$$

where

$$\varphi_{j1} = a_j e^{\xi_j} + b_j e^{-\xi_j}, \quad \varphi_{j2} = a_j^* e^{\xi_j^*} - b_j^* e^{-\xi_j^*}, \tag{4.23b}$$

$$\xi_j = k_j x - 4k_j^3 t + \xi_j^{(0)}, \quad a_j, b_j, \xi_j^{(0)} \in \mathbb{C}. \tag{4.23c}$$

Limit solutions of breathers are obtained when

$$\mathbb{A} = \begin{pmatrix} \mathcal{K} & 0 & 0 & \dots & 0 & 0 & 0 \\ \widetilde{\mathcal{K}} & \mathcal{K} & 0 & \dots & 0 & 0 & 0 \\ I_2 & \widetilde{\mathcal{K}} & \mathcal{K} & \dots & 0 & 0 & 0 \\ \dots & \dots & \dots & \dots & \dots & \dots & \dots \\ 0 & 0 & 0 & \dots & I_2 & \widetilde{\mathcal{K}} & \mathcal{K} \end{pmatrix}_{2N\times 2N}, \tag{4.24}$$

and

$$\mathbb{B}=\begin{pmatrix}\Theta_1 & 0 & \dots & 0 & 0\\ \widetilde{I}_2 & \Theta_1 & \dots & 0 & 0\\ \dots & \dots & \dots & \dots & \dots\\ 0 & 0 & \dots & \widetilde{I}_2 & \Theta_1\end{pmatrix}_{2N\times 2N}, \tag{4.25}$$

where

$$\mathcal{K}=\begin{pmatrix}k_1^2 & 0\\ 0 & k_1^{*2}\end{pmatrix},\ \widetilde{\mathcal{K}}=\begin{pmatrix}2k_1 & 0\\ 0 & 2k_1^*\end{pmatrix},\ I_2=\begin{pmatrix}1 & 0\\ 0 & 1\end{pmatrix},\ \widetilde{I}_2=\begin{pmatrix}0 & 1\\ 1 & 0\end{pmatrix}.$$

In this case, a general solution of (4.8) is given through rewriting $\varphi=(\varphi^{+T},\varphi^{-T})^T$ where $\varphi^{\pm}=(\varphi_1^{\pm},\varphi_2^{\pm},\cdots,\varphi_N^{\pm})^T$ satisfy

$$\varphi_{xx}^{+}=\mathbb{A}'\varphi^{+},\quad \varphi_{xx}^{-}=\mathbb{A}'^{*}\varphi^{-}, \tag{4.26a}$$
$$\varphi_{x}^{+}=\mathbb{B}'\varphi^{-*},\quad \varphi_{x}^{-}=\mathbb{B}'^{*}\varphi^{+*}, \tag{4.26b}$$
$$\varphi_{t}^{\pm}=-4\varphi_{xxx}^{\pm}, \tag{4.26c}$$

in which $\mathbb{A}'=\widetilde{J}_N[k_1]$ and $\mathbb{B}=J_N[k_1]$ with $k_1\in\mathbb{C}$. Explicit forms of $\varphi^{\pm}$ are

$$\varphi^{+}=\mathcal{A}\mathcal{Q}^{+}+\mathcal{B}\mathcal{Q}^{-},\quad \varphi^{-}=\mathcal{A}^{*}\mathcal{Q}^{+*}-\mathcal{B}^{*}\mathcal{Q}^{-*},\quad \mathcal{A},\mathcal{B}\in\widetilde{G}_N. \tag{4.27}$$

4.3 Rational solutions

Recall Remark 2 that indicates the LDES (4.6) fails in generating rational solutions for the mKdV equation (4.1) due to $|B|\neq 0$. To derive rational solutions, we make use of the Galilean transformation (GT)

$$v(x,t)=v_0+V(X,t),\quad x=X+6{v_0}^2t,\quad v_0\in\mathbb{R},\ v_0\neq 0, \tag{4.28}$$

and consider the transformed equation (known also as the KdV-mKdV equation)

$$V_t+12v_0VV_X+6V^2V_X+V_{XXX}=0, \tag{4.29}$$

which admits rational solutions in Wronskian form. Once its rational solutions is obtained, one can reverse the GT (4.28) and get rational solutions of the mKdV equation (4.1).

By the transformation

$$V=i\left(\ln\frac{f^*}{f}\right)_X, \tag{4.30}$$

(4.29) is bilinearised as [42]

$$(D_t+D_X^3)f^*\cdot f=0, \tag{4.31a}$$
$$(D_X^2-2iv_0D_X)f^*\cdot f=0. \tag{4.31b}$$

Theorem 7. *[55] The bilinear equation* (4.31) *admits Wronskian solution* $f = |\widehat{N-1}|$, *where the elementary column vector* φ *is determined by*

$$i\varphi_X = v_0\varphi + \mathbb{B}\varphi^*, \tag{4.32a}$$

$$\varphi_t = -4\varphi_{XXX}. \tag{4.32b}$$

Note that solutions of the above system can be classified as in Sec. 4.2 by introducing

$$\varphi_{XX} = \mathbb{A}\varphi, \tag{4.33}$$

where $\mathbb{A} = \mathbb{B}\mathbb{B}^* - v_0^2 I_N$ and I_N is the Nth-order unit matrix.

The N-soliton solution is obtained when taking

$$\mathbb{B} = \mathrm{diag}(-\sqrt{v_0^2 + k_1^2}, -\sqrt{v_0^2 + k_2^2}, \cdots, -\sqrt{v_0^2 + k_N^2}), \tag{4.34}$$

and φ satisfying (4.32) is composed of

$$\varphi_j = \sqrt{2v_0 + 2ik_j}\, e^{\eta_j} + \sqrt{2v_0 - 2ik_j}\, e^{-\eta_j}, \quad \eta_j = k_j X - 4k_j^3 t, \tag{4.35}$$

where $\{k_j\}$ are N distinct real positive numbers.

To obtain rational solutions, we take $\mathbb{B}$ to be defined as (2.14) with

$$a_j = \frac{-1}{(2j)!} \frac{\partial^{2j}}{\partial {k_1}^{2j}} \sqrt{v_0^2 + k_1^2}\,\Big|_{k_1=0}.$$

In this case φ_j is composed of

$$\varphi_{j+1} = \frac{1}{(2j)!} \frac{\partial^{2j}}{\partial {k_1}^{2j}} \varphi_1 \,\Big|_{k_1=0}, \quad (j = 0, 1, \cdots, N-1), \tag{4.36}$$

where φ_1 is defined as in (4.35).

Write the corresponding Wronskian as $f = F_1 + iF_2$, where $F_1 = \mathrm{Re}[f]$, $F_2 = \mathrm{Im}[f]$. From (4.30), solutions of the KdV-mKdV equation (4.29) are expressed as

$$V(X,t) = -2\,\frac{F_{1,X}F_2 - F_1 F_{2,X}}{F_1^2 + F_2^2}, \tag{4.37}$$

and for the mKdV equation (4.1),

$$v(x,t) = v_0 - \frac{2(F_{1,x}F_2 - F_1 F_{2,x})}{F_1^2 + F_2^2}, \tag{4.38}$$

where one needs to replace X in F_j with $X = x - 6v_0^2 t$. The simplest rational solution of the mKdV equation (4.1) is

$$v = v_0 - \frac{4v_0}{4v_0^2(x - 6v_0^2 t)^2 + 1}. \tag{4.39}$$

4.4 Notes

Unlike the KdV equation, there is no positon-negaton-complexiton classification for the solutions of the mKdV equation according to the eigenvalues of $\mathbb{B}$ in the LDES (4.8). In the simple-pole case, one can choose distinct $\{k_j\}$ from $\mathbb{C}/\sim$ to get solitons where $\sim$ is defined by (4.14).

Analysis of direct scattering of the mKdV equation (4.1) shows that the transmission coefficient $T(k)$ can have multiple poles [43]. Therefore multiple-pole solutions (limit solutions of solitons in this chapter) are not singular. Note that a typical characteristic of double pole solutions is that at large time the two solitons asymptotically travel along logarithmic trajectories with a linear background. This is also true for limit breathers [55]. Such a typical behavior was probably first found by Zakharov and Shabat for the double pole solution of the nonlinear Schrödinger (NLS) equation [48]. For more details of strict asymptotic analysis of this type of solutions one can refer to [55, 57].

Note specially that the mKdV equation (4.1) serves as an integrable model that describes, by its breathers, the ultra-short pulse propagation in a medium described by a two-level Hamiltonian [19]. Mixed solutions of the mKdV equation is potentially used to generate rogue waves [41]. Such solutions correspond to the case that $\mathbb{B}$ in (4.8) or (4.32) is a block diagonal with diagonal and different Jordan cells, and the elementary column vector φ is composed accordingly.

5 The AKNS and reductions

5.1 The AKNS hierarchy and double Wronskian solutions

A typical equation to admit double Wronskian solutions is the NLS equation [30], which belongs to the well-known AKNS hierarchy. Let us recall some results of this hierarchy.

The AKNS hierarchy

$$u_{t_n} = K_n = \begin{pmatrix} K_{1,n} \\ K_{2,n} \end{pmatrix} = L^n \begin{pmatrix} -q \\ r \end{pmatrix}, \quad u = \begin{pmatrix} q \\ r \end{pmatrix}, \quad n = 1, 2, \cdots, \tag{5.1}$$

is derived from the AKNS spectral problem [1]

$$\Phi_x = \begin{pmatrix} \lambda & q \\ r & -\lambda \end{pmatrix} \Phi, \quad \Phi = \begin{pmatrix} \phi_1 \\ \phi_2 \end{pmatrix}, \tag{5.2}$$

where q and r are functions of (x,t), λ is the spectral parameter, and the recursion operator is

$$L = \begin{pmatrix} -\partial_x + 2q\partial_x^{-1}r & 2q\partial_x^{-1}q \\ -2r\partial_x^{-1}r & \partial_x - 2r\partial_x^{-1}q \end{pmatrix}.$$

By imposing

$$r(x,t) = \delta q^*(\sigma x, t), \quad \delta, \sigma = \pm 1 \tag{5.3}$$

on the even-indexed members of the AKNS hierarchy

$$iu_{t_{2l}} = -K_{2l}, \quad l = 1, 2, \cdots, \tag{5.4}$$

where we have replaced t_{2l} by it_{2l}, one gets the NLS hierarchy

$$iq_{t_{2l}} = -K_{1,2l}|_{(5.3)}, \quad l = 1, 2, \cdots. \tag{5.5}$$

Note that $(\sigma, \delta) = (1, \mp 1)$ yields the classical focusing and defocusing NLS hierarchy, while $\sigma = -1$ yields the so-called nonlocal NLS hierarchy (cf. [2]). There are more reductions (cf. [4]), while here let us only focus on the NLS hierarchy as an instructive example on double Wronskians and reductions.

After rewriting the AKNS hierarchy (5.1) as its recursive form

$$\begin{pmatrix} q_{t_{n+1}} \\ r_{t_{n+1}} \end{pmatrix} = L \begin{pmatrix} q_{t_n} \\ r_{t_n} \end{pmatrix}, \quad n = 1, 2, \cdots \tag{5.6}$$

and introducing transformation

$$q = \frac{h}{f}, \quad r = -\frac{g}{f}, \tag{5.7}$$

(5.6) is bilinearised as (with $t_1 = x$) [28]

$$(D_{t_{n+1}} - D_x D_{t_n}) g \cdot f = 0, \tag{5.8a}$$

$$(D_{t_{n+1}} - D_x D_{t_n}) f \cdot h = 0, \tag{5.8b}$$

$$D_x^2 f \cdot f = 2gh. \tag{5.8c}$$

Theorem 8. *[20, 47] The bilinear AKNS hierarchy* (5.8) *allows us the following double Wronskian solutions,*

$$f = |\widehat{N-1}; \widehat{M-1}|, \tag{5.9a}$$

$$g = 2^{N-M+1} |\widehat{N}; \widehat{M-2}|, \tag{5.9b}$$

$$h = 2^{M-N+1} |\widehat{N-2}; \widehat{M}|, \tag{5.9c}$$

where the elementary column vectors (2.9) *are defined as*

$$\varphi = \exp\left(\frac{1}{2}\sum_{j=1}^{\infty} A^j t_j\right) C^+, \quad \psi = \exp\left(-\frac{1}{2}\sum_{j=1}^{\infty} A^j t_j\right) C^-, \tag{5.10}$$

in which $A \in \mathbb{C}_{(N+M)\times(N+M)}$ is an arbitrary constant matrix and

$$C^{\pm} = (c_1^{\pm}, c_2^{\pm}, \cdots, c_{N+M}^{\pm})^T, \quad c_i^{\pm} \in \mathbb{C}.$$

The proof is similar to the single Wronskian case as of Theorem 3. One can also refer to [47] for more details.

Note that (5.10) provides solutions for the whole AKNS hierarchy (5.6) as well as any special equation $u_{t_n} = K_n$ in that hierarchy due to the theory of symmetries. In the latter case, the terms $e^{\pm\frac{1}{2}\sum_{j\neq n} A^j t_j}$ in (5.10) are absorbed into $C^{\pm}$. For the hierarchy (5.4), φ and ψ are taken as (with $t_1 = x$)

$$\varphi = \exp\left(\frac{1}{2}Ax + \frac{i}{2}\sum_{j=1}^{\infty} A^{2j} t_{2j}\right) C^+, \quad \psi = \exp\left(-\frac{1}{2}Ax - \frac{i}{2}\sum_{j=1}^{\infty} A^{2j} t_{2j}\right) C^-. \tag{5.11}$$

5.2 Reductions of double Wronskians

As we have seen, (5.10) provides solutions through double Wronskians f, g, h for the unreduced hierarchy (5.1). In the following we present a simple reduction procedure that enables us to obtain double Wronskian solutions for the reduced hierarchies. Let us take (5.4) and (5.5) as an example. Note that $\sigma, \delta = \pm 1$.

Theorem 9. *Consider double Wronskians (5.9) with φ and ψ defined by (5.11), which provides solutions to the unreduced hierarchy (5.4). Impose the following constraints on φ and ψ: taking $M = N$, and*

$$C^- = TC^{+*}, \tag{5.12}$$

where the $2N \times 2N$ matrix T obeys

$$AT + \sigma TA^* = 0, \tag{5.13a}$$

$$TT^* = \delta\sigma I_{2N}. \tag{5.13b}$$

Then the NLS hierarchy (5.5) has solution

$$q(x,t) = 2\frac{|\widehat{N-2}; \widehat{N}|}{|\widehat{N-1}; \widehat{N-1}|}, \tag{5.14}$$

where

$$\psi(x,t) = T\varphi^*(\sigma x, t). \tag{5.15}$$

Proof. For convenience we introduce notation

$$\widehat{\varphi}^{(N)}(ax)_{[bx]} = \big(\varphi(ax), \partial_{bx}\varphi(ax), \partial^2_{bx}\varphi(ax), \cdots, \partial^N_{bx}\varphi(ax)\big), \quad a, b = \pm 1. \tag{5.16}$$

First, we show that under (5.13) we have

$$\begin{aligned}
\psi(\sigma x, t) =& \exp\Big(-\frac{1}{2}\sigma Ax - \frac{i}{2}\sum_{j=1}^{\infty} A^{2j}t_{2j}\Big)C^- \\
=& \exp\Big(\frac{1}{2}TA^*T^{-1}x - \frac{i}{2}\sum_{j=1}^{\infty}(TA^*T^{-1})^{2j}t_{2j}\Big)TC^{+*} \\
=& T\exp\Big(\frac{1}{2}A^*x - \frac{i}{2}\sum_{j=1}^{\infty} A^{*2j}t_{2j}\Big)C^{+*} \\
=& T\varphi^*(x,t),
\end{aligned}$$

which gives (5.15). Next, with the notation (5.16) and assumption (5.13b), we have

$$\begin{aligned}
f^*(\sigma x, t) &= |\widehat{\varphi}^{*(N-1)}(\sigma x)_{[\sigma x]}; T^*\widehat{\varphi}^{(N-1)}(\sigma^2 x)_{[\sigma x]}| \\
&= |T^*|(\sigma\delta)^N|T\widehat{\varphi}^{*(N-1)}(\sigma x)_{[\sigma x]}; \widehat{\varphi}^{(N-1)}(x)_{[\sigma x]}| \\
&= |T^*|(\sigma\delta)^N(-1)^N|\widehat{\varphi}^{(N-1)}(x)_{[x]}; T\widehat{\varphi}^{*(N-1)}(\sigma x)_{[x]}| \\
&= |T^*|(\sigma\delta)^N(-1)^N f(x,t).
\end{aligned}$$

In a similar manner,

$$g^*(\sigma x,t)=|T^*|\delta^{N+1}\sigma^N(-1)^{N-1}h(x,t).$$

Thus, we immediately reach

$$r^*(\sigma x,t)=-\frac{g^*(\sigma x,t)}{f^*(\sigma x,t)}=\frac{\delta h(x,t)}{f(x,t)}=\delta q(x,t),$$

i.e. the reduction (5.3) for the NLS hierarchy. ■

As for solutions T and A of (5.13), if we assume they are block matrices of the form

$$T=\begin{pmatrix} T_1 & T_2\\ T_3 & T_4\end{pmatrix},\quad A=\begin{pmatrix} K_1 & \mathbf{0}\\ \mathbf{0} & K_4\end{pmatrix}, \tag{5.17}$$

where T_i and K_i are $N\times N$ matrices, then, solutions to (5.13) are given in Table 1 where $\mathbf{K}_N\in\mathbb{C}_{N\times N}$.

(σ,δ)	T	A
$(1,-1)$	$T_1=T_4=\mathbf{0}$, $T_3=-T_2=\mathbf{I}_N$	$K_1=-K_4^*=\mathbf{K}_N$
$(1,1)$	$T_1=T_4=\mathbf{0}$, $T_3=T_2=\mathbf{I}_N$	$K_1=-K_4^*=\mathbf{K}_N$
$(-1,-1)$	$T_1=T_4=\mathbf{0}$, $T_3=T_2=\mathbf{I}_N$	$K_1=K_4^*=\mathbf{K}_N$
$(-1,1)$	$T_1=T_4=\mathbf{0}$, $T_3=-T_2=\mathbf{I}_N$	$K_1=K_4^*=\mathbf{K}_N$

Table 1. T and A for the NLS hierarchy

When $\mathbf{K}_N=\mathrm{Diag}(k_1,k_2,\cdots,k_N)$, we have

$$\varphi=\big(c_1e^{\theta(k_1)},c_2e^{\theta(k_2)},\cdots,c_Ne^{\theta(k_N)},d_1e^{\theta(-\sigma k_1^*)},d_2e^{\theta(-\sigma k_2^*)},\cdots,d_Ne^{\theta(-\sigma k_N^*)}\big)^T,$$

where

$$\theta(k_l)=\frac{1}{2}k_lx+\frac{i}{2}\sum_{j=1}^{\infty}k_l^{2j}t_{2j}. \tag{5.18}$$

When $\mathbf{K}_N$ is the Jordan matrix $J_N(k)$ defined as in (4.17), we have

$$\varphi=\Big(ce^{\theta(k)},\frac{\partial_k}{1!}(ce^{\theta(k)}),\cdots,\frac{\partial_k^{N-1}}{(N-1)!}(ce^{\theta(k)}),$$
$$de^{\theta(-\sigma k^*)},\frac{\partial_{k^*}}{1!}(de^{\theta(-\sigma k^*)}),\cdots,\frac{\partial_{k^*}^{N-1}}{(N-1)!}(de^{\theta(-\sigma k^*)})\Big)^T.$$

Note that one more solution for the case $(\sigma,\delta)=(-1,\pm1)$ is

$$T_1=-T_4=\sqrt{-\delta}\mathbf{I}_N,\ T_2=T_3=\mathbf{0}_N,\quad K_1=\mathbf{K}_N,\quad K_4=-\mathbf{H}_N, \tag{5.19}$$

where $\mathbf{K}_N, \mathbf{H}_N \in \mathbb{R}_{N\times N}$. When

$$\mathbf{K}_N = \mathrm{Diag}(k_1, k_2, \cdots, k_N), \quad \mathbf{H}_N = \mathrm{Diag}(h_1, h_2, \cdots, h_N),$$

we have

$$\varphi = \left(c_1 e^{\theta(k_1)}, c_2 e^{\theta(k_2)}, \cdots, c_N e^{\theta(k_N)}, d_1 e^{\theta(-h_1)}, d_2 e^{\theta(-h_2)}, \cdots, d_N e^{\theta(-h_N)}\right)^T,$$

and when $\mathbf{K}_N = J_N[k]$, $\mathbf{H}_N = J_N[h]$ as defined in (4.17), we have

$$\begin{aligned}\varphi = \Bigg(& ce^{\theta(k)}, \frac{\partial_k}{1!}(ce^{\theta(k)}), \cdots, \frac{\partial_k^{N-1}}{(N-1)!}(ce^{\theta(k)}), \\ & de^{\theta(-h)}, \frac{\partial_h}{1!}(de^{\theta(-h)}), \cdots, \frac{\partial_h^{N-1}}{(N-1)!}(de^{\theta(-h)})\Bigg)^T,\end{aligned}$$

where $\theta(k)$ is defined in (5.18).

5.3 Notes

The reduction technique we presented in this section is first introduced in [5]. The technique is also valid for all one-field reductions of the AKNS hierarchy (cf. [4]) as well as for one-field reductions of other systems that allows us double Wronskian solutions.

It is known that the matrix A in (5.10) and its similar forms lead to the same solutions for the AKNS hierarchy (5.1), and the eigenvalues of A correspond to discrete spectrum of the AKNS spectral problem (5.2). Therefore the structures of A in Table 1 indicate how the distribution of the discrete spectrum changes with different reductions.

6 Discrete case: the lpKdV equation

6.1 The lpKdV equation and discrete stuff

There has been a surge of interest in discrete integrable systems in the last two decades (cf. [10] and the references therein). Let us get familiar with some notations in discrete. Suppose $u(n, m)$ is a function defined on $\mathbb{Z} \times \mathbb{Z}$ where n and m are two discrete independent variables. The basic operation on $u(n, m)$ is a shift instead of differentiation. In this section we employ notations

$$u \doteq u(n, m), \quad \widetilde{u} \doteq u(n+1, m), \quad \widehat{u} \doteq u(n, m+1), \quad \widehat{\widetilde{u}} \doteq u(n+1, m+1). \quad (6.1)$$

A discrete Hirota's bilinear equation is written as (cf. [11])

$$\sum_j c_j\, f_j(n + \nu_j^+, m + \mu_j^+)\, g_j(n + \nu_j^-, m + \mu_j^-) = 0, \quad (6.2)$$

where $\nu_i^+ + \nu_i^- = \nu_k^+ + \nu_k^-, \mu_i^+ + \mu_i^- = \mu_k^+ + \mu_k^-, \forall i, k$. Casoratian is a discrete version of Wronskian. Let

$$\varphi(n,m,l) = (\varphi_1(n,m,l), \varphi_2(n,m,l), \cdots, \varphi_N(n,m,l))^T, \tag{6.3}$$

where φ_i are functions of (n,m,l) defined on $\mathbb{Z}\times\mathbb{Z}\times\mathbb{Z}$. A Casoratian w.r.t. l together with its compact expression is

$$\begin{aligned} C(\varphi) =& |\varphi(n,m,0), \varphi(n,m,1), \cdots, \varphi(n,m,N-1)| \\ =& |\varphi(0), \varphi(1), \cdots, \varphi(N-1)| = |0, 1, \cdots, N-1| = |\widehat{N-1}|. \end{aligned} \tag{6.4}$$

Similarly, we have $|\widehat{N-2}, N| = |0, 1, \cdots, N-2, N|$.

The lpKdV equation with the notations in (6.1) is written as

$$(u - \widehat{\widetilde{u}})(\widetilde{u} - \widehat{u}) = b^2 - a^2, \tag{6.5}$$

which is a discrete version of the potential KdV equation, where a and b are respectively the spacing parameters of n and m directions. By the transformation

$$u = an + bm + c_0 - \frac{g}{f}, \tag{6.6}$$

where c_0 is a constant, (6.5) is bilinearised as [11]

$$\mathcal{H}_1 = \widehat{g}\widetilde{f} - \widetilde{g}\widehat{f} + (a-b)(\widehat{f}\widetilde{f} - f\widehat{\widetilde{f}}) = 0, \tag{6.7a}$$

$$\mathcal{H}_2 = g\widehat{\widetilde{f}} - \widehat{\widetilde{g}}f + (a+b)(f\widehat{\widetilde{f}} - \widehat{f}\widetilde{f}) = 0. \tag{6.7b}$$

6.2 Casoratian solutions

Theorem 10. *The bilinear lpKdV equation* (6.7) *admits Casoratian solutions*

$$f(\varphi) = |\widehat{N-1}|,\ g(\varphi) = |\widehat{N-2}, N|, \tag{6.8}$$

where the elementary column vector $\varphi(n,m,l)$ satisfies

$$\widetilde{\varphi} - \overline{\varphi} = (a-c)\varphi, \tag{6.9a}$$

$$\widehat{\varphi} - \overline{\varphi} = (b-c)\varphi, \tag{6.9b}$$

and there exists an auxiliary vector ψ and an invertible matrix $\Gamma = \Gamma(m)$ such that

$$\varphi = \Gamma\psi, \tag{6.9c}$$

$$\psi + \overline{\psi} = (b+c)\widehat{\psi}. \tag{6.9d}$$

Here c is a constant and $\overline{f}(n,m,l) \doteq f(n,m,l+1)$. Note that Γ is independent of n, l, and it then follows from (6.9b) *and* (6.9c) *that*

$$\widetilde{\psi} - \overline{\psi} = (a-c)\psi. \tag{6.10}$$

Proof. Making use of evolution relations (6.9a) and (6.9b), one can derive shift relations of f and g [11],

$$-(a-c)^{N-2}\underset{\sim}{f} = |\widehat{N-2}, \underset{\sim}{\varphi}(N-2)|, \tag{6.11a}$$
$$-(b-c)^{N-2}\underset{\wedge}{f} = |\widehat{N-2}, \underset{\wedge}{\varphi}(N-2)|, \tag{6.11b}$$
$$-(a-c)^{N-2}[\underset{\sim}{g} + (a-c)\underset{\sim}{f}] = |\widehat{N-3}, N-1, \underset{\sim}{\varphi}(N-2)|, \tag{6.11c}$$
$$-(b-c)^{N-2}[\underset{\wedge}{g} + (b-c)\underset{\wedge}{f}] = |\widehat{N-3}, N-1, \underset{\wedge}{\varphi}(N-2)|. \tag{6.11d}$$

Then, for down-tilde-hat-shifted (6.7a), we find

$$\begin{aligned}
&[(a-c)(b-c)]^{N-2}[-(a-b)f\underset{\sim}{\underset{\wedge}{f}} + \underset{\wedge}{f}(\underset{\sim}{g} + (a-c)\underset{\sim}{f}) - \underset{\sim}{f}(\underset{\wedge}{g} + (b-c)\underset{\wedge}{f})] \\
=& -|\widehat{N-1}||\widehat{N-3}, \underset{\wedge}{\varphi}(N-2), \underset{\sim}{\varphi}(N-2)| \\
&+ |\widehat{N-2}, \underset{\wedge}{\varphi}(N-2)||\widehat{N-3}, N-1, \underset{\sim}{\varphi}(N-2)| \\
&- |\widehat{N-2}, \underset{\sim}{\varphi}(N-2)||\widehat{N-3}, N-1, \underset{\wedge}{\varphi}(N-2)|
\end{aligned}$$

which vanishes in light of (2.13) where $M = (\widehat{N-3})$, and **a**, **b**, **c** and **d** take $\varphi(N-2)$, $\varphi(N-1)$, $\underset{\wedge}{\varphi}(N-2)$ and $\underset{\sim}{\varphi}(N-2)$, respectively.

To prove (6.7b), we make use of the auxiliary vector ψ and consider Casoratian $f(\psi)$, which evolutes in m direction as

$$(b+c)^{N-2}\widehat{f}(\psi) = |\psi(0), \psi(1), \cdots, \psi(N-2), \widehat{\psi}(N-2)|.$$

Then, noting that $\varphi = \Gamma\psi$ and $f(\varphi) = |\Gamma| f(\psi)$, we recover $\widehat{f}(\varphi)$ as

$$\begin{aligned}
(b+c)^{N-2}\widehat{f}(\varphi) =& (b+c)^{N-2}|\widehat{\Gamma}|\widehat{f}(\psi) \\
=& |\widehat{\Gamma}||\Gamma^{-1}||\varphi(0), \cdots, \varphi(N-2), \mathring{E}\varphi(N-2)|,
\end{aligned}$$

i.e.

$$(b+c)^{N-2}\widehat{f}(\varphi) = |\widehat{\Gamma}||\Gamma^{-1}||\widehat{N-2}, \mathring{E}\varphi(N-2)|, \tag{6.12}$$

where $\mathring{E}\varphi(l) = \Gamma\widehat{\Gamma}^{-1}\widehat{\varphi}(l)$. With the help of $f(\psi)$ and $g(\psi)$, we also get

$$(a+b)[(a-c)(b+c)]^{N-2}\underset{\sim}{\widehat{f}}(\varphi) = |\widehat{\Gamma}||\Gamma^{-1}||\widehat{N-3}, \underset{\sim}{\varphi}(N-2), \mathring{E}\varphi(N-2)| \tag{6.13}$$

and

$$(b+c)^{N-2}[\widehat{g}(\varphi) - (b+c)\widehat{f}(\varphi)] = |\widehat{\Gamma}||\Gamma^{-1}||\widehat{N-3}, N-1, \mathring{E}\varphi(N-2)|. \tag{6.14}$$

Then, with formulas (6.11a), (6.11c), (6.12), (6.13) and (6.14), one can verify (6.7b) in its down-tilde-shifted version.

Thus we complete the proof. ∎

Explicit φ and ψ that satisfy (6.9) can be given according to the canonical forms of Γ. When

$$\Gamma=\mathrm{Diag}(\gamma_1,\gamma_2,\cdots,\gamma_N),\quad \gamma_j=(b^2-k_j^2)^m \tag{6.15}$$

with distinct k_j, one can take

$$\varphi_i(n,m,l)=\varphi_i^{+}+\varphi_i^{-},\quad \varphi_i^{\pm}=\varrho_i^{\pm}(c\pm k_i)^l(a\pm k_i)^n(b\pm k_i)^m, \tag{6.16}$$

$$\psi_i(n,m,l)=\psi_i^{+}+\psi_i^{-},\quad \psi_i^{\pm}=\varrho_i^{\pm}(c\pm k_i)^l(a\pm k_i)^n(b\mp k_i)^{-m}, \tag{6.17}$$

where $\varrho_i^{\pm}\in\mathbb{C}$. When Γ is a LTTM defined as (2.14) with $a_j=\frac{1}{j!}\partial_{k_1}^j\gamma_1$ where γ_1 is defined in (6.15), one can take

$$\varphi(m,n,l)=\mathcal{A}^{+}\mathcal{Q}^{+}+\mathcal{A}^{-}\mathcal{Q}^{-},\quad \mathcal{A}^{\pm}\in\widetilde{G}_N, \tag{6.18a}$$

$$\psi(m,n,l)=\mathcal{B}^{+}\mathcal{P}^{+}+\mathcal{B}^{-}\mathcal{P}^{-},\quad \mathcal{B}^{\pm}\in\widetilde{G}_N, \tag{6.18b}$$

where

$$\mathcal{Q}^{\pm}=(Q_0^{\pm},Q_1^{\pm},\cdots,Q_{N-1}^{\pm})^T,\quad Q_s^{\pm}=\frac{1}{s!}\partial_{k_1}^s\varphi_1^{\pm}, \tag{6.18c}$$

$$\mathcal{P}^{\pm}=(P_0^{\pm},P_1^{\pm},\cdots,P_{N-1}^{\pm})^T,\quad P_s^{\pm}=\frac{1}{s!}\partial_{k_1}^s\psi_1^{\pm}. \tag{6.18d}$$

and $\varphi_1^{\pm}$ and $\psi_1^{\pm}$ are defined in (6.16) and (6.17) respectively. When Γ is an LTTM defined as (2.14) with $a_j=\frac{1}{(2j)!}\partial_{k_1}^{2j}\gamma_1|_{k_1=0}$ where γ_1 defined in (6.15), one can take

$$\varphi(m,n,l)=\mathcal{A}^{+}\mathcal{R}^{+}+\mathcal{A}^{-}\mathcal{R}^{-},\quad \mathcal{A}^{\pm}\in\widetilde{G}_N, \tag{6.19a}$$

where $\mathcal{R}^{\pm}=(\mathcal{R}_0^{\pm},\mathcal{R}_1^{\pm},\cdots,\mathcal{R}_{N-1}^{\pm})^T$ and

$$\mathcal{R}_s^{+}=\frac{1}{(2s)!}\partial_{k_1}^{2s}\varphi_1|_{k_1=0},\ \text{with}\ \varrho_1^{-}=\varrho_1^{+}, \tag{6.19b}$$

$$\mathcal{R}_s^{-}=\frac{1}{(2s+1)!}\partial_{k_1}^{2s+1}\varphi_1|_{k_1=0},\ \text{with}\ \varrho_1^{-}=-\varrho_1^{+}, \tag{6.19c}$$

with φ_1 defined in (6.16). This case yields rational solutions.

6.3 Notes

The system (6.9) is the "LDES" of the bilinear lpKdV equation, in which ψ and Γ have been used as auxiliaries in order to implement Casoration verifications. More formulas of shifted Casoratians can be found in [11] with the help of auxiliary vectors. Examples that Theorem 1 plays its role in the discrete case can be found in [12, 13]. It can be proved that Γ and its similar forms lead to the same solutions to the lpKdV equation (6.5), which means one can make use of canonical forms of Γ to derive and classify solutions, including rational solutions. For more results on rational solutions in Casoration form for fully discrete 2D integrable systems one can refer to [39, 49, 56].

Equation (6.16) is a discrete counterpart of the continuous exponential function (e.g. (3.9)), where $\pm k_i$ satisfy $x^2 - k_i^2 = 0$. In principle, for fully discrete 2D integrable systems, their dispersion relations are defined by the curve

$$P_M(x,k) = \sum_{i=1}^{M} a_i(x^i - k^i) = 0, \tag{6.20}$$

solutions of which are used to define discrete exponential functions. For more details one can refer to [13, 54].

Besides, the techniques and treatments used in Sec. 3 and 4 for Wronskians are also valid for Casoratians. For example, the technique to verify BT with Wronskian solutions in Sec. 3.2 has applied to Casoratians [57], and the reduction procedure of double Wronskians described in Sec. 5.2 has been also generalised to double Casoratians [6], although [6] and [57] describe semidiscrete models.

7 Conclusions

We have reviewed the Wronskian technique and solutions in Wronskian/Casoratian forms for continuous and discrete integrable systems. In this context four instructive examples were employed. By the KdV equation we showed standard verifying procedures of Wronskian solutions of the bilinear KdV equation and the bilinear BT. It also served as an example that displays the construction of general solutions of the LDES and furthermore the classification of solutions for the KdV equation according to the canonical forms of the coefficient matrix A in its LDES. The second example is the mKdV equation which is special in many aspects. Note that there is a complex conjugate operation in its LDES; there is no solution classification as the KdV equation; breathers in Wronskian form result from block diagonal $\mathbb{B}$ (4.22); solitons or breathers of double pole case are not singular and travel asymptotically with logarithmic curves instead of straight lines; and rational solutions are obtained by means of GT (4.28). The third example is the AKNS hierarchy, together with its double Wronskian solutions and reductions. A reduction technique was shown to get double Wronskian solutions for the reduced hierarchy. Finally, the lpKdV equation served as an example of fully discrete integrable systems which have received significant progress in the recent two decades. We described how to obtain shift formulas of Casoratians by introducing auxiliary vectors, so that Casoratians solutions of discrete bilinear equations can be verified as in the continuous case.

More than thirty years have passed since the Wronskian technique was proposed in 1983 [7]. Almost no secret is left behind in this technique. As a matter of fact, solutions in Wronskian form have their own advantage in presenting explicit multiple-pole solutions (including rational solutions) and understanding relations between simple-pole and multiple-pole solutions by taking limits w.r.t. poles $\{k_j\}$. This benefits from the regular structure of a Wronskian that each row is governed by a single k_j. Note that the IST [3], Cauchy matrix approach [29] and operator approach (see [38] and the references therein) yield solutions in terms of the Cauchy matrix, but the expression for multiple-pole solutions are not as simple as those in

Wronskian case (cf. [53]), and so far no rational solutions are presented in terms of the Cauchy matrix.

There are other compact expressions for solutions of integrable systems, such as Grammian and Pfaffian, mainly derived from bilinear methods. For more details one may refer to [17].

Acknowledgments

The project is supported by the NSF of China (Nos.11631007 and 11875040).

References

[1] Ablowitz M J, Kaup D J, Newell A C and Segur H, Nonlinear-evolution equations of physical significance, *Phys. Rev. Lett.* **31**, 125–7, 1973.

[2] Ablowitz M J and Musslimani Z H, Integrable nonlocal nonlinear Schrödinger equation, *Phys. Rev. Lett.* **110**, No.064105 (5pp), 2013.

[3] Ablowitz M J and Segur H, *Solitons and the Inverse Scattering Transform*, SIAM, Philadelphia, 1981.

[4] Chen K, Deng X, Lou S Y and Zhang D J, Solutions of nonlocal equations reduced from the AKNS hierarchy, *Stud. Appl. Math.* **141**, 113–41, 2018.

[5] Chen K and Zhang D J, Solutions of the nonlocal nonlinear Schrödinger hierarchy via reduction, *Appl. Math. Lett.* **75**, 82–8, 2018.

[6] Deng X, Lou S Y and Zhang D J, Bilinearisation-reduction approach to the nonlocal discrete nonlinear Schrödinger equations, *Appl. Math. Comput.* **332**, 477–83, 2018.

[7] Freeman N C and Nimmo J J C, Soliton solutions of the KdV and KP equations: the Wronskian technique, *Phys. Lett. A* **95**, 1–3, 1983.

[8] Freeman N C and Nimmo J J C, Soliton solitons of the KdV and KP equations: the Wronskian technique, *Proc. R. Soc. Lond.* **A389**, 319–29, 1983.

[9] Freeman N C, Soliton solutions of nonlinear evolution equations, *IMA J. Appl. Math.* **32**, 125–45, 1984.

[10] Hietarinta J, Joshi N and Nijhoff F W, *Discrete Systems and Integrablity*, Cambridge Univ. Press, Cambridge, 2016.

[11] Hietarinta J and Zhang D J, Soliton solutions for ABS lattice equations: II. Casoratians and bilinearization, *J. Phys. A: Math. Theor.* **42**, No.404006 (30pp), 2009.

[12] Hietarinta J and Zhang D J, Multisoliton solutions to the lattice Boussinesq equation, *J. Math. Phys.* **51**, No.033505 (12pp), 2010.

[13] Hietarinta J and Zhang D J, Soliton taxonomy for a modification of the lattice Boussinesq equation, *Symmetry Integrability Geom. Methods Appl.* **7**, No.061 (14pp), 2011.

[14] Hirota R, Exact solution of the Korteweg-de Vries equation for multiple collisions of solitons, *Phys. Rev. Lett.* **27**, 1192–4, 1971.

[15] Hirota R, Exact solution of the modified Korteweg-de Vries equation for multiple collisions of solitons, *J. Phys. Soc. Jpn.* **33**, 1456–8, 1972.

[16] Hirota R, A new form of Bäcklund transformations and its relation to the inverse scattering problem, *Prog. Theore. Phys.* **52**, 1498–512, 1974.

[17] Hirota R, *The Direct Method in Soliton Theory* (in English). Cambridge Univ. Press, Cambridge, 2004.

[18] Jaworski M, Breather-like solution to the Korteweg-de Vries equation, *Phys. Lett. A* **104**, 245–7, 1984.

[19] Leblond H and Mihalache D, Models of few optical cycle solitons beyond the slowly varying envelope approximation, *Phys. Reports* **523**, 61–126, 2013.

[20] Liu Q M, Double Wronskian solutions of the AKNS and the classical Boussinesq hierarchies, *J. Phys. Soc. Jpn.* **59**, 3520–7, 1990.

[21] Ma W X, Complexiton solution to the KdV equation, *Phys. Lett. A* **301**, 35–44, 2002.

[22] Ma W X and You Y C, Solving the Korteweg-de Vries equation by its bilinear form: Wronskian solutions, *Trans. Americ. Math. Soc.* **357**, 1753–78, 2005.

[23] Matveev V B, Darboux transformation and explicit solutions of the Kadomtsev-Petviashvili equation, depending on functional parameters, *Lett. Math. Phys.* **3**, 213–6, 1979.

[24] Matveev V B, Generalized Wronskian formula for solutions of the KdV equations: First applications, *Phys. Lett. A* **166**, 205–8, 1992.

[25] Matveev V B, Positon-positon and soliton-positon collisions: KdV case, *Phys. Lett. A* **166**, 209–12, 1992.

[26] Matveev V B and Salle M A, *Darboux Transformations and Solitons*, Springer-Verlag, Berlin, 1991.

[27] Miwa T, Jimbo M and Date E, *Solitons: Differential Equations, Symmetries and Infinite Dimensional Algebras*, Cambridge Univ. Press, Cambridge, 2000.

[28] Newell A C, *Solitons in Mathematics and Physics*, SIAM, Philadelphin, 1985.

[29] Nijhoff F W, Atkinson J and Hietarinta J, Soliton solutions for ABS lattice equations. I. Cauchy matrix approach, *J. Phys. A: Math. Theor.* **42**, No.404005 (34pp), 2009.

[30] Nimmo J J C, A bilinear Bäcklund transformation for the nonlinear Schrödinger equation, *Phys. Lett. A* **99**, 279–80, 1983.

[31] Nimmo J J C, Soliton solutions of three differential-difference equations in Wronskian form, *Phys. Lett. A* **99**, 281–6, 1983.

[32] Nimmo J J C and Freeman N C, A method of obtaining the N-soliton solution of the Boussinesq equation in terms of a Wronskian, *Phys. Lett. A* **95**, 4–6, 1983.

[33] Nimmo J J C and Freeman N C, Rational solutions of the KdV equation in Wronskian form, *Phys. Lett. A* **96**, 443–6, 1983.

[34] Nimmo J J C and Freeman N C, The use of Bäcklund transformations in obtaining N-soliton solutions in Wronskian form, *J. Phys. A: Math. Gen.* **17**, 1415–24, 1984.

[35] Rasinariu C, Sukhatme U and Khare A, Negaton and positon solutions of the KdV and mKdV hierarchy, *J. Phys. A: Math. Gen.* **29**, 1803–23, 1996.

[36] Sato M, Soliton equations as dynamical systems on an infinite dimensional Grassmann manifolds, *RIMS Kokyuroku Kyoto Univ.* **439**, 30–46, 1981.

[37] Satsuma J, A Wronskian representation of N-soliton solutions of nonlinear evolution equations, *J. Phys. Soc. Jpn.* **46**, 359–60, 1979.

[38] Schiebold C, Cauchy-type determinants and integrable systems, *Linear Algebra Appl.* **433** 447–75, 2010.

[39] Shi Y and Zhang D J, Rational solutions of the H3 and Q1 models in the ABS lattice list, *Symmetry Integrability Geom. Methods Appl.* **7**, No.046 (11pp), 2011.

[40] Sirianunpiboon S, Howard S D and Roy S K, A note on the Wronskian form of solutions of the KdV equation, *Phys. Lett. A* **134**, 31–3, 1988.

[41] Slunyaev A V and Pelinovsky E N, Role of multiple soliton interactions in the generation of rogue waves: the modified Korteweg-de Vries framework, *Phys. Rev. Lett.* **117**, No.214501(5pp), 2016.

[42] Wadati M, Wave propagation in nonlinear lattice I, *J. Phys. Soc. Jpn.* **38**, 673–80, 1975.

[43] Wadati M and Ohkuma K, Multiple-pole solutions of the modified Korteweg de Vries equation, *J. Phys. Soc. Jpn.* **51**, 2029–35, 1982.

[44] Wu H and Zhang D J, Mixed rational-soliton solutions of two differential-difference equations in Casorati determinant form, *J. Phys. A: Gen. Math.* **36**, 4867–73, 2003.

[45] Xu D D, Zhang D J and Zhao S L, The Sylvester equation and integrable equations: I. The Korteweg-de Vries system and sine-Gordon equation, *J. Nonlin. Math. Phys.* **21**, 382–406, 2014.

[46] Xuan Q F, Ou M Y and Zhang D J, Wronskian solutions to the KdV equation via Bäcklund transformation, arXiv:nlin.SI/0706.3487, 2007.

[47] Yin F M, Sun Y P, Cai F Q and Chen D Y, Solving the AKNS hierarchy by its bilinear form: Generalized double Wronskian solutions, *Comm. Theore. Phys.* **49**, 401–8, 2008.

[48] Zakharov V E and Shabat A B, Exact theory of two-dimensional self-focusing and one-dimensional self-modulation of waves in nonlinear media, *Sov. Phys. JETP* **34**, 62–9, 1972. (*Zh. Eksp. Teor. Fiz.* **61**, 118–34, 1971.)

[49] Zhang D D and Zhang D J, Rational solutions to the ABS list: transformation approach, *Symmetry Integrability Geom. Methods Appl.* **13**, No.078 (24pp), 2017.

[50] Zhang D J, The N-soliton solutions for the modified KdV equation with self-consistent sources, *J. Phys. Soc. Jpn.* **71**, 2649–56, 2002.

[51] Zhang D J, Notes on solutions in Wronskian form to soliton equations: KdV-type, arXiv:nlin.SI/0603008, 2006.

[52] Zhang D J and Chen D Y, The N-soliton solutions of the sine-Gordon equation with self-consistent sources, *Physica A* **321**, 467–81, 2003.

[53] Zhang D J and Zhao S L, Solutions to ABS lattice equations via generalized Cauchy matrix approach, *Stud. Appl. Math.* **131**, 72–103, 2013.

[54] Zhang D J, Zhao S L and Nijhoff F W, Direct linearization of an extended lattice BSQ system, *Stud. Appl. Math.* **129**, 220–48, 2012.

[55] Zhang D J, Zhao S L, Sun Y Y and Zhou J, Solutions to the modified Korteweg-de Vries equation, *Rev. Math. Phys.* **26**, No.1430006 (42pp), 2014.

[56] Zhao S L and Zhang D J, Rational solutions to Q3$_\delta$ in the Adler-Bobenko-Suris list and degenerations, *J. Nonl. Math. Phys.* **26**, 107–32, 2019.

[57] Zhou J, Zhang D J and Zhao S L, Breathers and limit solutions of the nonlinear lumped self-dual network equation, *Phys. Lett. A* **373**, 3248–58, 2009.

C1. Global gradient catastrophe in a shallow water model: evolution unfolding by stretched coordinates

Roberto Camassa

University of North Carolina at Chapel Hill, Carolina Center for Interdisciplinary Applied Mathematics, Department of Mathematics, Chapel Hill, NC 27599, USA

Abstract

Continuous, piecewise differentiable exact solutions with discontinuous derivatives of the shallow water (Airy's) hyperbolic system are constructed by splicing together self-similar parabolae with the evolution from a constant background state. These new solutions are used to illustrate the mechanism by which "vacuum states," mathematically corresponding to the hyperbolic-parabolic transition points of the governing equations, can be filled by the evolution of the hyperbolic system, and in particular how dry spots persist until a non-generic gradient catastrophe develops at the dry point(s). The continuation of solutions asymptotically for short times beyond the catastrophe is then investigated analytically, in its weak form, with an approach inspired by the stretched coordinates used in singular perturbation theory.

1 Introduction

It is well known that under certain circumstances (see, e.g., [17, 18]), model equations such as the Airy's shallow water system,

$$\eta_t + (u\eta)_x = 0\,, \qquad u_t + uu_x + \eta_x = 0\,, \qquad x \in \mathbb{R}, \quad t \in \mathbb{R}^+\,, \tag{1.1}$$

here written in suitable nondimensional space-time (x,t) coordinates, with η representing the water layer thickness and u the layer-averaged horizontal velocity, can capture some of the fundamental dynamics of the parent Euler equations for a free-surface, inviscid fluid under gravity, extending laterally to infinity and confined below by a flat bottom. Much of the model effectiveness at the qualitative and even quantitative level ultimately resides in the fact that the conservation laws of mass and horizontal momentum are captured either exactly, as in the first equation, or asymptotically in the limit of long waves, as in the second equation. With this in mind, we have recently found and studied exact solutions that can be used to shed some light on the peculiar features of the dynamics when the layer thickness vanishes (as set by initial data). In fact, it can be shown, even for the parent Euler equations, that "dry" points where $\eta(x,t) = 0$, so that the free surface touches the bottom of the fluid layer, tend to persist as long as the dependent variables $\eta(x,\cdot)$ and $u(x,\cdot)$ remain sufficiently regular at these contact points. (This persistence can be viewed in the broader context of continuum mechanics assumptions for boundaries, see, e.g., [4].) As a consequence, detachment of the free surface from the bottom can only happen through a loss of regularity of the solution. Viewed from within the mathematical structure of system (1.1), such points where $\eta = 0$ correspond to location where the hyperbolic property of the system is lost, that is, the characteristic speeds

$$\lambda_\pm \equiv u \pm \sqrt{\eta} \equiv \frac{3R_\pm + R_\mp}{4}$$

corresponding to the Riemann invariant form of system (1.1)

$$\partial_t R_\pm + \lambda_\pm \partial_x R_\pm = 0\,, \qquad \text{where} \quad R_\pm \equiv u \pm \sqrt{\eta}$$

coincide, and the system becomes parabolic. Setting aside the question of the physical implications of the dynamics supported by the shallow water model when loss of regularity of its solutions occurs, in a way that the long-wave assumption can no longer be attained, it is nonetheless of interest to study exact solutions that can be followed analytically even past the occurrence of a singularity. Thus, the nature of a hyperbolic-parabolic transition can be examined in great detail analytically, and the results used to shed some light on general properties of the dynamics. Furthermore, as demonstrated in [1], besides their inherent mathematical interest the closed form solutions can actually illustrate, qualitatively and even quantitatively, the evolution of the parent Euler equations for certain classes of initial data.

2 Exact solutions

We consider the class of initial data (see, e.g., [1, 15])

$$\eta = \gamma(t)x^2 + \mu(t)\,, \qquad u = \nu(t)x\,. \tag{2.1}$$

where the time evolution of the coefficients $\gamma(t)$, $\mu(t)$ and $\nu(t)$ is governed by the ODEs

$$\dot{\nu} + \nu^2 + 2\gamma = 0\,, \qquad \dot{\gamma} + 3\nu\gamma = 0\,, \qquad \dot{\mu} + \nu\mu = 0\,. \tag{2.2}$$

This system admits closed form solutions for generic initial data $\gamma(0) = \gamma_0$, $\mu(0) = \mu_0$ and $\nu(0) = \nu_0$. Of particular interest for our purposes is the case of contact $\eta(x, 0) = 0$ which occurs for $\mu_0 = 0$; then, the third equation in system (2.2) implies $\mu(t) = 0$ for all times $t > 0$ as long as the system's solution exists. This means that before a singularity, for as long as the parabolic form of $\eta(x, t)$ is maintained, the parabola's vertex at $x = 0$ stays at the bottom, or $\eta(0, t) = 0$ for this time interval.

The solutions $\nu(t)$ and $\gamma(t)$ corresponding to the initial conditions $\nu(0) = 0$ and $\gamma(0) = \gamma_0 > 0$ are best defined in terms of the auxiliary time-like variable $\sigma \in [1, \infty)$ defined implicitly by

$$t(\sigma) = \frac{\sqrt{\sigma - 1} + \sigma \arctan\left(\sqrt{\sigma - 1}\right)}{2\sqrt{\gamma_0}\,\sigma}\,, \tag{2.3}$$

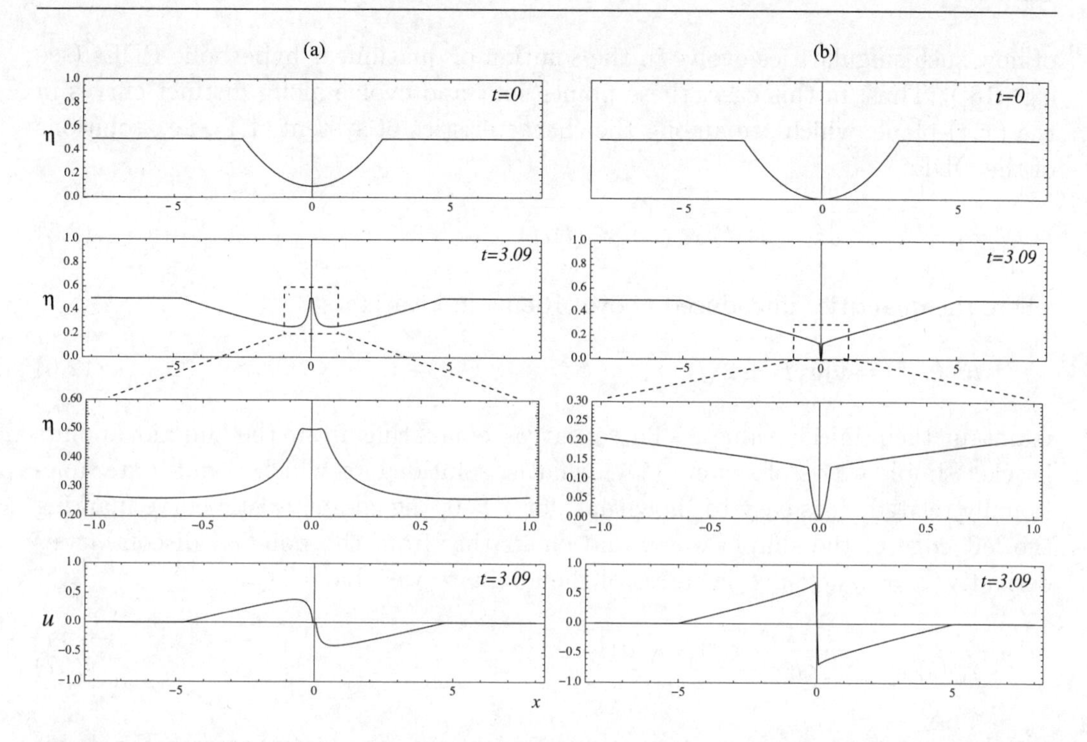

Figure 1. Snapshots of the evolution governed by system (1.1) for the class of initial data constructed by splicing together a parabolic core (2.2) for η with a constant background state $\eta = Q$. Here $\gamma_0 = 1/16$ and $\nu_0 = 0$: (a) $\mu_0 = 0.1$; (b) $\mu_0 = 0$. At the time $t = 3.09$, the flat region around $x = 0$ immediately following the parabola's disappearance can be noticed in the η-magnification (bottom panel) for case (a), while the central region connected to $\eta = 0$ shrinks in the limit $t \to t_c$ ($t_c = \pi$ for this choice of the parameter γ_0) to a finite height segment $\eta(t_c, 0) = Q/4$ but maintains its parabolic shape, as seen from the enlargements (bottom panels) for case (b).

so that

$$\gamma(t) = \gamma_0 \big(\sigma(t)\big)^3, \qquad \nu(t) = 2\sqrt{\gamma_0}\,\big(\sigma(t)\big)^2 \sqrt{\sigma(t) - 1}. \tag{2.4}$$

Therefore $\gamma(t)$ and $|\nu(t)|$ are strictly increasing functions of time, $\gamma(t) \geq \gamma_0$ with $\gamma \to \infty$ and $\nu \to -\infty$ as $\sigma \to \infty$, which corresponds to $t \to \pi/(4\sqrt{\gamma_0})$. The solutions of system (2.2) thus develop a singularity in finite time $t = t_c = \pi/(4\sqrt{\gamma_0})$, which corresponds to the η-parabola collapsing onto a vertical half line at $x = 0$.

Of course, the thickness of the water layer represented by (2.1) is unbounded for $|x| \to \infty$, making this solution (2.1) *per se* devoid of physical meaning. However, the parabola can be chopped at some elevation, $\eta = Q$ say, and spliced continuously with a constant background state $\eta(x, 0) = Q$ for $|x| > \sqrt{Q/\gamma_0}$, to form physically relevant initial data (see top ($t = 0$) panels of figure 1 for a plot of initial data constructed this way).

The evolution of the points $|x| = \sqrt{Q/\gamma_0}$, where the parabola joins the constant background, which are locations of a discontinuity in the derivative of η, must follow the characteristics originating at those locations, based on the general theory of how such singularities evolve in the solution of quasilinear hyperbolic PDEs (see e.g. [18]). Thus, in this case, these points split and evolve along distinct curves in the (x,t)-plane, which are among the characteristics of system (1.1), i.e., solutions of the ODEs

$$\dot{x}_\pm = \lambda_\pm \equiv u(x_\pm(t),t) \pm \sqrt{\eta(x_\pm(t),t)} \tag{2.5}$$

where the quantities introduced above (Riemann invariants)

$$R_\pm(x,t) = u(x,t) \pm 2\sqrt{\eta(x,t)} \tag{2.6}$$

maintain their initial values. These curves emanating from the junction points bracket simple waves of system (1.1), that is, solutions for which η and u are functionally related. It is easy to show that, for $x>0$, the characteristic curve marking the left edge of the simple wave, and emanating from the point of discontinuous derivative $x=\sqrt{Q/\gamma_0}$ is, in terms of the auxiliary variable $\sigma(t)$,

$$x_Q(t) = \sqrt{Q/\gamma_0}\,\frac{\sqrt{\sigma(t)}-\sqrt{\sigma(t)-1}}{\sigma(t)}, \tag{2.7}$$

and hence, as $\sigma\to\infty \Rightarrow t\to t_c$ defined above, the curve crosses the t-axis $x=0$, which of course corresponds to the parabola's collapse we have seen before. Moreover, in the half domain $x\geq 0$ such simple wave solution, $\eta\equiv N(x,t)$ and $u\equiv V(x,t)$ say, can be expressed in closed form, albeit implicitly, through an auxiliary variable $\sigma_0(x,t)\in[1,\infty)$ (emulating its counterpart for the solutions of the ODEs (2.2) governing the evolution of the coefficients pair (γ,ν)),

$$N(x,t) = \sigma_0 Q\left(\sqrt{\sigma_0}-\sqrt{\sigma_0-1}\right)^2, \qquad V(x,t) = 2\sqrt{N(x,t)}-2\sqrt{Q}\,. \tag{2.8}$$

Here σ_0 as a function of x and t is defined by the solution of

$$x = \Lambda(\sigma_0)\left(t-\frac{\sqrt{\sigma_0-1}+\sigma_0\arctan\left(\sqrt{\sigma_0-1}\right)}{2\,\sigma_0\sqrt{\gamma_0}}\right)+\frac{\sqrt{Q}\,\sigma_0-\sqrt{Q(\sigma_0-1)}}{\sigma_0\sqrt{\gamma_0}} \tag{2.9}$$

for any given point (x,t) in the spatio-temporal half-plane $x>0$ and $t>0$ where the simple wave exists. Here we have used the shorthand notation Λ for the characteristic velocity λ_-,

$$\Lambda(\sigma_0) = 3\sqrt{Q}\,\sigma_0\left(1-\sqrt{1-\frac{1}{\sigma_0}}-\frac{2}{3\sigma_0}\right). \tag{2.10}$$

Symmetric and antisymmetric extensions, respectively for η and u, when $x<0$ complete the exact solution form. Note that according to (2.7), as $t\to t_c$, the left edge of the simple wave N reaches $x=0$ at the elevation

$$\eta = \lim_{t\to t_c}\gamma(t)\left(x_Q(t)\right)^2 = Q\lim_{\sigma\to\infty}\sigma^2\left(1-\sqrt{1-\frac{1}{\sigma}}\right)^2 = Q/4\,.$$

A typical evolution to the initial data according to the piecewise exact solution constructed with the above results is depicted in figure 1, where we contrast the dry case with its "wet" counterpart, where the initial condition of the coefficient μ in (2.2)) is taken to be different from zero. (Closed form solutions for the wet case $\mu(0) > 0$ can be obtained by analogous calculations as those for the dry case above. It can be shown [1] that the gradient catastrophe is avoided by recovering a local rest state $u = 0$ and $\eta = Q$ in a neighbourhood of the origin, with two simple waves propagating symmetrically outward from $x = 0$ and extending the rest-interval where $u = 0$ until a standard shock point develops, at times $t_s > t_c$ along the simple waves' sides closer to the origin). The last panel in figure 1 is in agreement with the solution developing a singularity in finite time corresponding to $\gamma(t) \to \infty$, as the parabolic core of the piecewise solution collapses onto a vertical segment. As shown above, this segment has finite length $Q/4$, the elevation at which the characteristics (λ_- and λ_+ respectively for $x > 0$ and $x < 0$) emanating from the splicing points $|x| = \sqrt{Q/\gamma_0}$ intersect at $x = 0$.

In summary, from the solution of the ODEs (2.2), and/or from the implicit expressions of the bracketing simple waves (2.9), (2.8) by finding the first time at which the partial derivative N_x becomes infinite, the verticality at the origin is a "global gradient catastrophe" (as the derivative of η and u become infinite not just at single points but for all points in an interval of the range of $\eta(x, \cdot)$), and occurs at the collapse time

$$t = t_c \equiv \frac{\pi}{4\sqrt{\gamma_0}} \,. \tag{2.11}$$

A remark is now in order: even though the choice of initial data (2.1) may seem rather restrictive, for more general data $u(x, 0) = f(x)$ and $\eta(x, 0) = g(x)$ analytic in x, a solution in the form of a Taylor series can always be sought,

$$u(x,t) = \sum_{n=0}^{\infty} u_n(t) x^{2n+1} \,, \qquad \eta(x,t) = \sum_{n=0}^{\infty} \eta_n(t) x^{2n+2} \,,$$

where for simplicity we have assumed the functions f and g to be antisymmetric and symmetric, respectively, and the lack of a constant term in the series reflects the property $u(0, t) = 0$ and $\eta(0, t)$. Hence, as long as the series converge locally, η maintains a tangent contact point with the bottom at $x = 0$, with curvature $2\eta_0(t)$ starting from $\eta_0(0) = g''(0)/2$. A simple calculation shows that putting these expressions in system (1.1) yields, at the leading order, the same ODEs for $\nu(t)$ and $\gamma(t)$ as (2.2) (with $\mu \equiv 0$), namely

$$\dot{u}_0 + u_0^2 + 2\eta_0 = 0 \,, \qquad \dot{\eta}_0 + 3u_0\eta_0 = 0 \,. \tag{2.12}$$

This system for (η_0, u_0) is uncoupled from all higher order coefficients $n > 0$. For $n \geq 1$, the pairs (η_n, u_n) belong to an infinite hierarchy of linear ODEs

$$\dot{u}_n + (2n+2)\eta_n + (n+1)\sum_{k=0}^{n} u_k u_{n-k} = 0 \,, \quad \dot{\eta}_n + (2n+3)\sum_{k=0}^{n} \eta_k u_{n-k} = 0 \,, \tag{2.13}$$

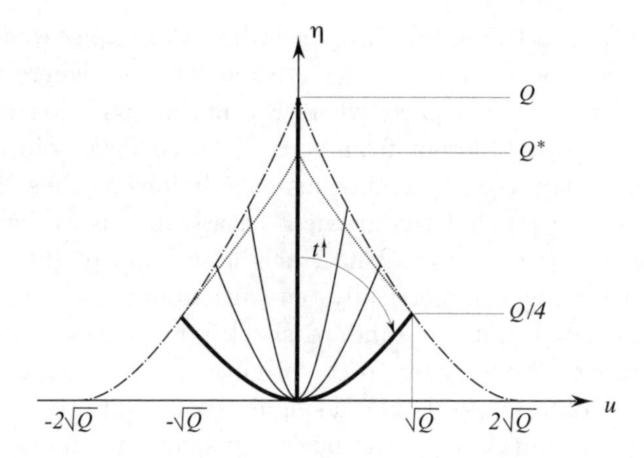

Figure 2. Schematic of the evolution in the hodograph plane (u, η) from the initial condition $\eta = \gamma_0 x^2$ for $x \in \sqrt{Q/\gamma_0}$, $\eta = Q$ otherwise. The initial data are mapped into the segment vertical segment $u = 0$, $\eta \in [0, Q]$. This segment opens into a one-parameter family of parabolic arcs (thin curves) of increasing curvature as time increases, limiting to the arc (thick curve) $\eta = 1/4u^2$ as $t \to t_c^-$. The end points of the arcs in the family slide along the simple waves $u \pm 2\sqrt{\eta} = \pm 2\sqrt{Q}$ (dash dotted curves), respectively for the right $x > 0$, $u < 0$, and left $x < 0$, $u > 0$ wings of the solution. Dotted lines mark the shock curves from $Q/4$ to Q^* immediately after collision at time $t = t_c^+$.)

whose solution can be determined recursively from all the lower order pairs, starting from the first one in (2.12). Thus, the same exact formula (2.11) for time of gradient catastrophe applies to system (2.12) and hence, by linearity, to the whole hierarchy, thereby showing that the collapse time formula is generic for all smooth (symmetric) initial data with $u = 0$ and *finite* second derivative of η at the dry contact point. Moreover, an immediate application of the recursion formulae (2.13) shows that, for the class of data with zero initial velocity, if the curvature at the contact point is initially zero then $\eta_0(t) = 0$ at all times, and the evolution of the next coefficient pair $\eta_1(t)$, $u_1(t)$, is governed by

$$\dot{u}_1 = -4\eta_1 \,, \qquad \dot{\eta}_1 = 0 \,,$$

that is, the local form of $\eta \sim \eta_1(0)x^4$ in a neighbourhood of the origin is time independent, while the cubic velocity profile $u = u_1(t)x^3 = -4\eta_1(0)\, t\, x^3$ "steepens" at a linear rate in time. Higher order zeros for the contact point $\eta = 0$ can be dealt with similarly.

The gradient catastrophe for the case $\eta_0(0) > 0$, $u_0(0) = 0$ admits an interesting representation in the hodograph plane (u, η) (see figure 2 for a sketch of the evolution in these variables). The initial parabola corresponds to the vertical segment $\eta \in [0, Q]$ at $u = 0$. As time increases, this segment opens up as a one-parameter (t) family of parabolic arcs of time-dependent decreasing curvature, just the opposite of what happens in the physical space x, t, limiting onto the final arc of parabola $\eta = u^2/4$, $u \in (-\sqrt{Q}, \sqrt{Q})$ at time $t = t_c$, which corresponds to the vertical segment $\eta \in [0, Q/4]$ in physical space. As it happens, this limiting form of the core part of the exact solution is also itself a simple wave, though its mapping to the physical plane is singular and corresponds to the vertical segment $\eta \in [0, Q/4]$. For $0 < t < t_c$ the two end-points of each member of the family of parabolic arcs "slide" along the simple waves (themselves arcs of parabolae) $u - 2\sqrt{\eta} = -2\sqrt{Q}$ and $u + 2\sqrt{\eta} = 2\sqrt{Q}$, to the right and to left of the origin $u = 0$, respectively for $x > 0$ and $x < 0$ in physical space. Thus, for an instant at time $t = t_c$ the whole exact solution is comprised of simple waves, piecewise joint continuously together to bound a region of the hodograph plane between the arcs (dash-dot lines) of parabolae corresponding to the "wing" simple waves and the arc of simple wave to which the core limits to as $t \to t_c$.

3 Evolution beyond the gradient catastrophe

As illustrated in the last time snapshots of figure 1(b), the gradient catastrophe singularity of the dependent variables (η, u) at the time $t = t_c$ consists of a jump discontinuity for both fields. The physical interpretation of the segment connecting the free surface to the bottom at $x = 0$ is that of colliding water masses at that point and time; the spatial symmetry of the fields u and η and the assumption of no viscosity in the derivation of the Airy model (1.1) offer the alternative interpretation of a water mass (from the right say) moving left and colliding with a vertical wall at $x = 0$. The latter interpretation clearly suggests that a sudden rise of the water level along the wall has to be expected as time progresses. This section will study properties of this part of the evolution that can be analyzed in closed, albeit approximate, form.

First, in view of the considerations above, it is natural to remove the vertical segment at $t = t_c$ to obtain a connected domain for the water layer at time $t = t_c^+$. However, the discontinuity from the values $\sqrt{Q}$ (left) to $-\sqrt{Q}$ (right) in the velocity u would remain at $x = 0$, and as discussed above it is to be expected that this velocity jump would result in the instantaneous rising of the fluid in a neighbourhood of this location. This can be determined precisely by studying the initial value problem for system (1.1) corresponding to initial data obtained by the wing solutions $N(x, t_c)$ and $V(x, t_c)$ in (2.8), appropriately reflected across the origin.

Looking at the limit $x \to 0^+$ of $N(x, t_c)$, it can be seen that this function is not differentiable at the origin; more precisely

$$N(x, t_c) \sim \frac{Q}{4} + \frac{3^{2/3} {\gamma_0}^{1/3} Q^{2/3}}{8} x^{2/3} + o(x^{2/3}) \qquad \text{as} \quad x \to 0\,. \tag{3.1}$$

The consequences of this branch point singularity on the evolution of the "new" initial data after gradient catastrophe are mostly of technical nature, and it is worth studying a class of data that removes the fractional power obstacle, yet captures the main features of the time advancement past $t = t_c$.

Consider the initial data obtained by splicing together two "Stoker" simple waves (see [17]) crossing at $x = 0$ and $\eta(0,0) = \eta_0 = Q/4$, respectively for η,

$$N_S(x,t) = \begin{cases} Q & x \geq x_Q, \\ \dfrac{1}{9}\left(\dfrac{x - x_d}{t + t_c} + 2\sqrt{Q}\right)^2, & 0 < x < x_Q, \\ \dfrac{1}{9}\left(-\dfrac{x + x_d}{t + t_c} + 2\sqrt{Q}\right)^2, & -x_Q < x < 0 \\ Q, & x \leq -x_Q \end{cases} \tag{3.2}$$

and u,

$$V_S(x,t) = \begin{cases} 0, & x \geq x_Q \\ \dfrac{2}{3}\dfrac{x - x_d}{t + t_c} - \dfrac{2}{3}\sqrt{Q}, & 0 < x < x_Q \\ \dfrac{2}{3}\dfrac{x + x_d}{t + t_c} + \dfrac{2}{3}\sqrt{Q}, & -x_Q < x < 0 \\ 0, & x \leq -x_Q \end{cases} . \tag{3.3}$$

With the following choice of parameters

$$x_d = \frac{1}{4}\sqrt{\frac{3Q}{g_0}}, \quad t_c = \frac{1}{2}\sqrt{\frac{3}{g_0}}, \quad x_Q = \sqrt{Q}\,t + \frac{3}{4}\sqrt{\frac{3Q}{g_0}}, \tag{3.4}$$

one can obtain a configuration that closely resembles that of the pair $N(x, t_c)$, $V(x, t_c)$ in (2.8) sketched in figure 3, dashed curves. As depicted in this figure, the evolution out of these initial data for times $t > 0$ (corresponding to times $t > t_c$ for the original case of (2.1)) involves the generation of two symmetrically placed shocks in both the η and u fields, moving away from the origin, with an intermediate region between the shock corresponding to fluid slowly filling the initial time "hole" at $x = 0$, $\eta = Q/4$. These shocks move over the background given by the evolution of the Stoker waves (3.2), (3.3) for $t > 0$.

3.1 Unfolding variables

Let $x_s(t)$ be the position of the right-going shock. We introduce the "unfolding" coordinates (ξ, τ) to set the shock positions at fixed locations in time, e.g., $\xi = \pm 1$, which can be achieved by

$$\xi \equiv \frac{x}{x_s(t)}, \qquad \tau \equiv \log(x_s(t)), \tag{3.5}$$

so that the Jacobian of this transformation reads

$$\partial_x = \frac{1}{x_s}\partial_\xi, \qquad \partial_t = \frac{\dot{x}_s}{x_s}(\partial_\tau - \xi\,\partial_\xi). \tag{3.6}$$

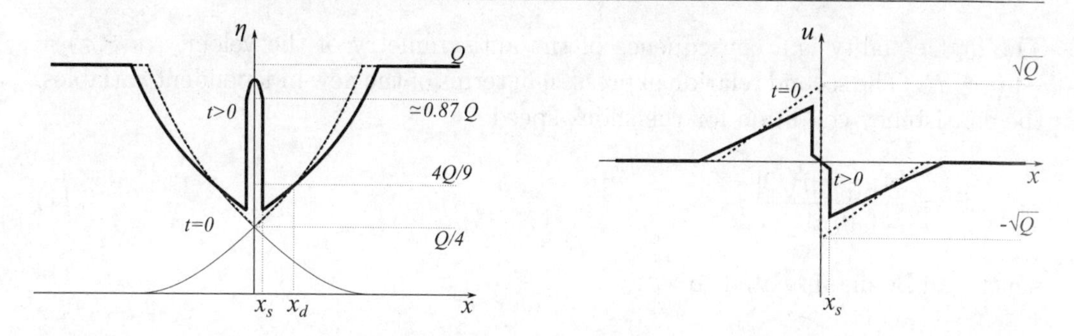

Figure 3. Schematics of the initial condition (dash) and evolution (solid) with shock development at short times $t > t_c$ for the Airy solutions (3.2), (3.3). The thin segments of parabolae are removed from the initial data and the shocks develop from the initial discontinuity in the velocity (right panel).

With this mapping, the Airy system (1.1) assumes the form (with a little abuse of notation by maintaining the same symbols for dependent variables)

$$\frac{\partial \eta}{\partial \tau} - \xi \frac{\partial \eta}{\partial \xi} + \frac{1}{\dot{x}_s} \frac{\partial}{\partial \xi} (\eta u) = 0 \,, \qquad \frac{\partial u}{\partial \tau} - \xi \frac{\partial u}{\partial \xi} + \frac{1}{\dot{x}_s} \frac{\partial}{\partial \xi} \left(\frac{u^2}{2} + \eta \right) = 0 \,. \tag{3.7}$$

In this system, the role of $\dot{x}_s$ is that of a placeholder for the expression that couples the evolution equation for the physical time t to the new evolution variable τ. The shock position evolves according to the equation that defines $\dot{x}_s$ in terms of the jump amplitudes $[\eta]$ and $[\eta u]$:

$$\begin{aligned} &\frac{d(e^\tau)}{dt} = \dot{x}_s = \frac{[\eta u]}{[\eta]} = \frac{N(x_s(t),t)\, V(x_s(t),t) - \eta(1,\tau(t))\, u(1,\tau(t))}{N(x_s(t),t) - \eta(1,\tau(t))} \,, \\ &x_s(0) = 0, \end{aligned} \tag{3.8}$$

so that, in the new variables,

$$\begin{aligned} &\frac{dt}{d\tau} = \frac{e^\tau \left(N(e^\tau, t(\tau)) - \eta(1,\tau) \right)}{N(e^\tau, t(\tau))\, V(e^\tau, t(\tau)) - \eta(1,\tau)\, u(1,\tau)} \,, \\ &t(\tau) \to 0 \quad \text{as } \tau \to -\infty \,. \end{aligned} \tag{3.9}$$

Here and in the following, we have suppressed the subscript $\cdot_S$ for the Stoker simple-waves, and adopted the usual square-bracket notation for jumps defined as the difference between values $\cdot_\pm$ at the right and left of the shock location, respectively (see e.g., [18]). From these expressions, the system governing the evolution of the "inner" solution between shocks consists of this ordinary differential equation together with the partial differential equations (3.7), to be solved within the strip $\xi \in [0,1]$ (by symmetry only half the ξ domain $[-1,1]$ may be used) subject to the boundary conditions

$$\begin{aligned} &u(0,\tau) = 0 \,, \\ &u(1,\tau) = V(e^\tau, t(\tau)) + \sqrt{\frac{\left(N(e^\tau, t(\tau)) - \eta(1,\tau)\right)^2 \left(N(e^\tau, t(\tau)) + \eta(1,\tau)\right)}{2N(e^\tau, t(\tau))\, \eta(1,\tau)}} \,. \end{aligned} \tag{3.10}$$

The first equality is a consequence of the antisymmetry of the velocity, $u(\xi,\tau) = -u(-\xi,\tau)$. The second relation expresses, in terms of the new independent variables, the consistency condition for the shock speed

$$\dot{x}_s = \frac{[\eta\, u^2 + \eta^2/2]}{[\eta\, u]}\,, \tag{3.11}$$

which can be manipulated to

$$[u]^2 = \frac{[\eta]^2(\eta_+ + \eta_-)}{2\eta_+\,\eta_-}\,, \tag{3.12}$$

whence the second relation in (3.11) follows.

The boundary-value problem (3.11) for the evolution equations (3.7) and (3.9), by depending on the unknown η at the boundary, $\eta(1,\tau)$, is reminiscent in its structure of the classical (irrotational) water-wave problem, wherein the unknowns, the free surface location and the velocity potential along it, determine and in turn are determined by the solution of a PDE (for the water wave problem, the Laplace equation for the velocity potential in the fluid domain). Just as in that case, an additional equation must be provided at the boundary, which has its analog in (3.9). Similarly to the water-wave problem, the resulting structure is highly nonlinear and hence hardly amenable to closed form solutions. To make progress, observe that the "initial" data as $\tau \to -\infty$, i.e., $t = 0$, are given by $u(\xi,-\infty) = 0$ and $\eta(\xi,-\infty) = Q^*$, where Q^* is the constant solution of the cubic equation

$$-2\sqrt{Q} + 2\sqrt{N(0,0)} + \sqrt{\frac{\left(N(0,0) - Q^*\right)^2\left(N(0,0) + Q^*\right)}{2N(0,0)\,Q^*}} = 0\,, \tag{3.13}$$

subject to the condition $Q^* > N(0,0)$. This value of η is reached instantaneously at $t = 0^+$ (or $t = t_c^+$ for the original problem with simple waves $N(x,t_c)$ and $V(x,t_c)$ of (2.8)), and it is sketched in the hodograph plane (u,η) in figure 2 by the dotted arcs of the cubic relation joined at Q^* on the η-axis $u = 0$. Thus, the initial time evolution can be followed by the linearization of system (3.7)-(3.9) around $\eta = Q^*$ and $u = 0$. This approach is mostly straightforward, though the details are bit involved, and will be reported elsewhere. It suffices to say that the initial evolution as $\tau \to -\infty$ (and so $t \to 0^+$) is asymptotic to

$$\begin{aligned}\eta(\xi,\tau) &= Q^* + \frac{1}{2}\left(F(e^\tau\xi - \Phi(\tau)) + F(-e^\tau\xi - \Phi(\tau))\right),\\ u(\xi,\tau) &= \frac{1}{2\sqrt{Q^*}}\left(F(e^\tau\xi - \Phi(\tau)) - F(-e^\tau\xi - \Phi(\tau))\right),\end{aligned} \tag{3.14}$$

for any function $F(\cdot)$ of sufficient regularity, with the function Φ defined by

$$\Phi(\tau) \equiv \sqrt{Q^*}\int_{-\infty}^{\tau} e^{\tau'}\phi_0(\tau')d\tau'\,, \tag{3.15}$$

where the integrand $\phi_0(\tau)$ is

$$\phi_0(\tau) = \frac{N(e^\tau, t(\tau)) - Q^*}{N(e^\tau, t(\tau))\, V(e^\tau, t(\tau))} \,. \tag{3.16}$$

Here s_0 is the initial shock speed $\dot{x}_s$ at $t = 0^+$, and the initial conditions on the system's solution (η, u) as $\tau \to -\infty$ require $F(0) = 0$. Substitution of these expressions evaluated at $\xi = 1$ into the linearized version of the boundary condition (3.11) leads to a functional equation for F, coupled to the (linearized) evolution equation (3.9) for $t(\tau)$. By assuming sufficient regularity for F, the functional equation can be solved approximately by Taylor series. The result to second order is

$$\eta(\xi,\tau) \sim Q^* - F'(0)\Phi_0\, e^\tau + \left(-F'(0)\Phi_1 + \frac{1}{2}\, F''(0)(\xi^2 + \Phi_0^2)\right) e^{2\tau} \,, \tag{3.17}$$

$$u(\xi,\tau) \sim \frac{F'(0)}{\sqrt{Q^*}}\, \xi\, e^\tau - \frac{F''(0)}{\sqrt{Q^*}}\, \Phi_0\, \xi\, e^{2\tau} \,, \tag{3.18}$$

where the Taylor coefficients $F'(0)$, $F''(0)$ and those for the asymptotic expansion of $\Phi(\tau) \sim \Phi_0\, e^\tau + \Phi_1\, e^{2\tau}$ are, respectively,

$$F'(0) \simeq -0.22215\sqrt{g_0\, Q} \,, \qquad F''(0) \simeq -0.58487\, g_0 \,, \tag{3.19}$$

and

$$\Phi_0 = \frac{\sqrt{Q^*}(4Q^* - Q)}{Q^{3/2}} \simeq 2.33087 \,,$$

$$\Phi_1 = \frac{\sqrt{g_0\, Q^*}}{3\sqrt{3}} \frac{(Q + 4Q^*)^2}{Q^3} \simeq -3.6325 \sqrt{\frac{g_0}{Q}} \,, \tag{3.20}$$

with the parametrization (3.4) for the initial data. The asymptotics for the shock location x_s is similarly computed as $\tau \to -\infty$.

Returning to the original physical variables $\eta(x,t)$, $u(x,t)$, the asymptotic expressions (3.17),(3.18) lead to

$$\eta(x,t) \sim Q^* - F'(0)\Phi_0\, s_0\, t - \left(F'(0)\Phi_0\, s_1 + \left(F'(0)\Phi_1 - \frac{1}{2}\, F''(0)\Phi_0^2\right) s_0^2\right) t^2 + \\ + \frac{1}{2}\, F''(0) x^2 \,, \tag{3.21}$$

$$u(x,t) \sim \frac{x}{\sqrt{Q^*}} \left(F'(0) - F''(0)\, \Phi_0\, (s_0\, t + s_1\, t^2)\right) \,, \tag{3.22}$$

$$x_s(t) \sim s_0 t + s_1 t^2 \,, \quad \text{with} \quad s_0 \simeq 0.400969\sqrt{Q} \,, \quad s_1 \simeq 0.11703\sqrt{g_0 Q} \,, \tag{3.23}$$

as $t \to 0^+$. Note that the asymptotic validity of these expressions is established within the unfolding variable formulation, and hence it is not necessarily maintained after the mapping (3.5), unless this is also consistently expanded with the asymptotics. Nonetheless, the above expressions for the η and u fields reveal that the layer thickness η jumps, at time $t = 0^+$, to $\eta = Q^*$, i.e., about 87% of the background rest thickness Q, and grows linearly in time while maintaining a local parabolic shape. Similarly, the local velocity u jumps from a discontinuity of amplitude $2\sqrt{Q}$ at $x = 0$ to a continuous linear profile between shocks, with negative slope which evolves linearly in time.

It is worth mentioning at this point that a numerical approach to solving for the time evolution of system (1.1) with the initial data we have studied might run into subtle issues in resolving the gradient catastrophe. Of course, a shock capturing algorithm must be employed, as the jet emanating from the gradient collapse at the origin at $t = t_c$ (from the new initial data $\eta = N(x, t_c)$ and $u = V(x, t_c)$, appropriately symmetrized across the origin) evolves into propagating shocks. However, we remark that standard WENO algorithms (see, e.g.,[16]) we have implemented, when compared with the analytical results we have obtained, fare rather poorly, mainly owing to the fact that the singularity formation of our case differs substantially from that of a generic shock at $t = t_c^+$. Alternatively, a numerical approach that would solve the boundary value problem for the evolution system in unfolding variables can be used with the stretched domain between shocks. This eliminates resolution issues around $x = 0$, and allows the use of spectrally accurate codes. The main details of this and other cases are reported elsewhere [2].

4 Discussion and conclusions

The dynamics of water layers under gravity when the free surface contacts a horizontal boundary poses interesting questions from both physical and mathematical perspectives. While highly idealized, the hyperbolic approximation of the Airy shallow water model may offer insight into fundamental mechanisms whose influence can persist even for the more realistic parent Euler and Navier-Stokes equations [1]. In the spirit of classical results [12, 13] for Euler fluids, we have studied special solutions that illustrate some of these points and are amenable to analytical approaches. In particular, the persistence of the contact dry point is shown to be terminated by a formation of a singularity, in both the horizontal velocity and the fluid layer thickness η. Continuation of the solution beyond the singularity time can only be achieved in the weak form, as shocks develop. This presents some challenges, which we are able to handle by introducing stretched coordinates, inspired by techniques of singular perturbation theory [9]. The new coordinates magnify the region around the singularity and, in particular, show how shocks develop and bracket a locally rising "jet" of fluid. Interestingly, the jet height is measurably smaller, by more than 10%, than the analogous jet that would form when an arbitrarily small wetting layer is initially maintained over the contact point. The alternative interpretation of the jet as a splash of fluid along a vertical wall may even be investigated experimentally with the proper set-up, where other physical effects such a viscosity and surface tension play a secondary role.

From a mathematical perspective, the shallow-water Airy system is an illustrative example of hyperbolic systems of conservation laws [10], and in particular of equations governing gas dynamics. Contact points, where characteristic speeds coincide, can then be interpreted as vacuum states and mark transitions between hyperbolic and parabolic behaviour of the system. The dynamics in neighbourhoods of such states are notoriously difficult to characterise [11]. The new stretched variables we have introduced bridge regular and singular regimes, when shocks form. As mentioned above, the dynamics of the system between shocks is reminiscent of the classical water wave problem where the Laplace equation in the fluid domain must be solved with boundary conditions that depend on the solution itself. Thus, in §3.1, the shock location equation (3.9) plays the role of the kinematic boundary condition while (3.11) is analogous to the dynamics boundary condition at the free surface of the water-wave problem; these simultaneously determine and are determined by the solution of a partial differential equation (in this case, the hyperbolic system (3.7)) in the bulk of the domain. Also worth mentioning in this context, is that the interpretation of the gradient catastrophe we have studied as colliding water masses, as well as the contact point of the water surface with the bottom itself, is reminiscent of the questions arising in the context of the so-called "splash" contact singularities which can form in self-intersection of the free-surface of Euler fluids and related models [3]. Finally, the exact solutions we have analyzed can be used as starting points to inspire further studies of more general initial data. In fact, the highly non-generic type of singularity formation we have studied reveals aspects that numerical and alternative analytical techniques may miss [14] for similar, smoother classes of initial data. Other directions, some of which are currently underway, will extend this investigation to higher dimensions, and study the influence and relevance of additional physical effects, such as dispersion (see e.g., [5, 6, 7, 8]), dissipation and surface tension, on the results in this work.

Acknowledgments

This is joint work with Gregorio Falqui & Giovanni Ortenzi (Università of Milano-Bicocca), Marco Pedroni (Università di Bergamo), and Giuseppe Pitton (Imperial College London), and is largely based on [2]. The author thanks the hospitality of the ICERM's program "Singularities and Waves in Incompressible Fluids" in the Spring of 2017, the support by the National Science Foundation under grants RTG DMS-0943851, CMG ARC-1025523, DMS-1009750, DMS-1517879, and by the Office of Naval Research under grants N00014-18-1-2490 and DURIP N00014-12-1-0749. All investigators gratefully acknowledge the auspices of the GNFM Section of INdAM under which part of this work was carried out, and the Dipartimento di Matematica e Applicazioni of Università Milano-Bicocca for its hospitality; travel support by grant H2020-MSCA-RISE-2017 Project No. 778010 IPaDEGAN is also acknowledged.

References

[1] Camassa R, Falqui G, Ortenzi G, Pedroni M & Thompson C, Hydrodynamic models and confinement effects by horizontal boundaries, *J. Nonlinear Sci.* **29**,

1445–1498, 2019.

[2] Camassa R, Falqui G, Ortenzi G, Pedroni M & Pitton G., Singularity formation as a wetting mechanism in a dispersionless water wave mode, *Nonlinearity* **32**, 4079–4116, 2019.

[3] Castro A, Cordoba D, Fefferman C & Gancedo F, Breakdown of smoothness for the Muskat problem. *Arch. Rat. Mech. Anal.* **208**, 805–909, 2013.

[4] Childress S, *An Introduction to Theoretical Fluid Mechanics*, American Mathematical Society, Providence RI, 2009.

[5] El G A, Grimshaw R H J & Smyth N F, Unsteady undular bores in fully nonlinear shallow-water theory, *Phys. Fluids* **18**, 027104, 2006.

[6] Esler J G & Pearce J D, Dispersive dam-break and lock-exchange flows in a two-layer fluid, *J. Fluid Mech.* **667**, 555–585, 2011.

[7] Forest M G & McLaughlin K T-R, Onset of oscillations in nonsoliton pulses in nonlinear dispersive fibers, *J. Nonlinear Sci.* **7**, 43–62, 1998.

[8] Gurevich A V & Krylov A L, Dissipationless shock waves in media with positive dispersion, *Zh. Eksp. Teor. Fiz.* **92**, 1684–1699, 1987.

[9] Kevorkian J & Cole J D, *Perturbation Methods in Applied Mathematics*, Springer, New York NY, 1981.

[10] Lax P D, Development of singularities of solutions of nonlinear hyperbolic partial differential equations, *J. Math. Phys.* **5**, 611–613, 1964.

[11] Liu T-P & Smoller J A, On the vacuum state for the isentropic gas dynamics equations, *Adv. Appl. Math.* **1**, 345–359, 1980.

[12] Longuet-Higgins M S, Self-similar, time-dependent flow with a free surface, *J. Fluid Mech.* **73**, 603–620, 1976.

[13] Longuet-Higgins M S, Parametric solutions for breaking waves, *J. Fluid Mech.* **121**, 403–424, 1982.

[14] Moro A & Trillo S, Mechanism of wave breaking from a vacuum point in the defocusing nonlinear Schrödinger equation, *Phys. Rev. E* **89**, 023202, 2014.

[15] Ovsyannikov L V, Two-layer "shallow water" model, *J. Appl. Mech. Tech. Phys.* **20**, 127–135, 1979.

[16] Shu C-W, High order weighted essentially non-oscillatory schemes for convection dominated problems, *SIAM Review* **51**, 82–126, 2009.

[17] Stoker J J, *Water Waves: The Mathematical Theory with Applications*, Wiley-Interscience, New York NY, 1957.

[18] Whitham G B, *Linear and Nonlinear Waves*, Wiley, New York NY, 1999.

C2. Vibrations of an elastic bar, isospectral deformations, and modified Camassa-Holm equations

Xiangke Chang [a] *and Jacek Szmigielski* [b]

[a] *LSEC, ICMSEC, Academy of Mathematics and Systems Science, Chinese Academy of Sciences, P.O. Box 2719, Beijing 100190, PR China; and School of Mathematical Sciences, University of Chinese Academy of Sciences, Beijing 100049, PR China;*
changxk@lsec.cc.ac.cn

[b] *Department of Mathematics and Statistics, University of Saskatchewan, 106 Wiggins Road, Saskatoon, Saskatchewan, S7N 5E6, Canada;*
szmigiel@math.usask.ca

Abstract

Vibrations of an elastic rod are described by a Sturm-Liouville system. We present a general discussion of isospectral (spectrum preserving) deformations of such a system. We interpret one family of such deformations in terms of a two-component modified Camassa-Holm equation (2-mCH) and solve completely its dynamics for the case of discrete measures (multipeakons). We show that the underlying system is Hamiltonian and prove its Liouville integrability. The present paper generalizes our previous work on interlacing multipeakons of the 2-mCH and multipeakons of the 1-mCH. We give a unified approach to both equations, emphasizing certain natural family of interpolation problems germane to the solution of the inverse problem for 2-mCH as well as to this type of a Sturm-Liouville system with singular coefficients.

1 Introduction

One of the most important applications of the Sturm-Liouville systems is provided by the longitudinal vibrations of an elastic bar of stiffness p and density ρ [2, Chapter 10.3]. The longitudinal displacement v satisfies the wave equation

$$\rho(x)\frac{\partial^2 v}{\partial t^2} = \frac{\partial}{\partial x}[p(x)\frac{\partial v}{\partial x}], \tag{1.1}$$

which after the separation of variables $v = u(x)\cos\omega t$ leads to

$$D_x[p(x)D_x u] + \omega^2\rho(x)u = 0, \tag{1.2}$$

where $D_x = \frac{d}{dx}$. In a different area of applications, in geophysics, the *Love waves* were proposed by an early 20 century British geophysicist Augustus Edward Hugh Love who predicted the existence of *horizontal surface waves* causing Earth shifting during an earthquake. The wave amplitudes of these waves satisfy [1]

$$D_x(\mu D_x)u + (\omega^2\rho - k^2\mu)u = 0, \quad 0 < x < \infty, \tag{1.3}$$

where μ is the *sheer modulus*, x is the depth below the Earth surface and the boundary conditions are $D_x u(0) = u(\infty) = 0$ which can be interpreted as the Neumann condition on one end and the Dirichlet condition on the other. In applications to geophysics the frequency ω is fixed and the phase velocity is ω/k. In particular, in the infinity speed limit ($k = 0$), we obtain the same Sturm-Liouville system as in (1.2). In either case the problem can conveniently be written as the first order system:

$$D_x \Phi = \begin{bmatrix} 0 & n \\ -\omega^2 \rho & 0 \end{bmatrix} \Phi, \tag{1.4}$$

where $n = \frac{1}{\mu}$, $\Phi = \begin{bmatrix} \phi_1 \\ \phi_2 \end{bmatrix}$, $\phi_1 = \phi$, $\phi_2 = \mu D_x \phi$. In the present paper we study (1.4) on the whole real axis and impose the boundary conditions $\phi_1(-\infty) = \phi_2(\infty) = 0$ which can be interpreted as the Dirichlet condition at $-\infty$ and the Neumann condition at $+\infty$. Our motivation to study this problem comes from yet another area of applied mathematics dealing with integrable nonlinear partial differential equations. To focus our discussion we begin by considering the nonlinear partial differential equation

$$m_t + \left((u^2 - u_x^2)m\right)_x = 0, \qquad m = u - u_{xx}, \tag{1.5}$$

which is one of many variants of the famous Camassa-Holm equation (CH) [3]:

$$m_t + u m_x + 2u_x m = 0, \qquad m = u - u_{xx}, \tag{1.6}$$

for the shallow water waves. We will call (1.5) the mCH equation for short. The history of the mCH equation is long and convoluted: (1.5) appeared in the papers of Fokas [7], Fuchssteiner [8], Olver and Rosenau[15] and was, later, rediscovered by Qiao [16, 17].

Subsequently, in [19], Song, Qu and Qiao proposed a natural two-component generalization of (1.5)

$$\begin{aligned} &m_t + [(u - u_x)(v + v_x)m]_x = 0, \\ &n_t + [(u - u_x)(v + v_x)n]_x = 0, \\ &m = u - u_{xx}, \qquad n = v - v_{xx}, \end{aligned} \tag{1.7}$$

which, for simplicity, we shall call the 2-mCH. Formally, the 2-mCH reduces to the mCH when $v = u$.

We are interested in the class of non-smooth solutions of (1.7) given by the *peakon ansatz* [3, 4, 11]:

$$u = \sum_{j=1}^{N} m_j(t) e^{-|x - x_j(t)|}, \qquad v = \sum_{j=1}^{N} n_j(t) e^{-|x - x_j(t)|}, \tag{1.8}$$

where all smooth coefficients $m_j(t), n_j(t)$ are taken to be positive, and hence

$$m = u - u_{xx} = 2\sum_{j=1}^{N} m_j \delta_{x_j}, \qquad n = v - v_{xx} = 2\sum_{j=1}^{N} n_j \delta_{x_j}$$

are positive discrete measures.

For the above ansatz, (1.7) can be viewed as a distribution equation, requiring in particular that we define the products Qm and Qn, where

$$Q = (u - u_x)(v + v_x). \tag{1.9}$$

It is shown in Appendix A that the choice consistent with the Lax integrability discussed in Section 2 is to take Qm, Qn to mean $\langle Q\rangle m$, $\langle Q\rangle n$ respectively, where $\langle f\rangle$ denotes the average function (the arithmetic average of the right and left limits). Substituting (1.8) into (1.7) and using the multiplication rule mentioned above leads to the system of ODEs:

$$\dot{m}_j = 0, \qquad \dot{n}_j = 0, \tag{1.10a}$$

$$\dot{x}_j = \langle Q\rangle(x_j). \tag{1.10b}$$

In the present paper, we shall develop an inverse spectral approach to solve the peakon ODEs (1.10) and hence (1.7) under the following assumptions:

1. all m_k, n_k are positive,

2. the initial positions are assumed to be ordered as $x_1(0) < x_2(0) < \cdots < x_n(0)$.

We emphasize that the second condition is not restrictive since it can be realized by relabeling positions as long as positions $x_j(0)$ are distinct.

The present paper generalizes our previous work on interlacing multipeakons of the 2-mCH in [4] and multipeakons of the 1-mCH in [6]. It is worth mentioning, however, that the technique of the present paper is a modification of the one employed in [6] and is distinct from the inhomogeneous string approach adapted in [4]. As a result we give a unified approach to both equations; this is accomplished by putting common interpolation problems front and center of the solution to the inverse problem for (1.4). Moreover, by solving (1.10), we furnish a family of isospectral flows for the Sturm-Liouville system (1.4). The full explanation of the connection between (1.4) and (1.7) is reviewed in the following sections.

2 The Lax formalism: the boundary value problem

The Lax pair for (1.7) can be written:

$$\frac{\partial}{\partial x}\Psi_x = \frac{1}{2}U\Psi, \quad \Psi_t = \frac{1}{2}V\Psi, \quad \Psi = \begin{bmatrix}\Psi_1\\ \Psi_2\end{bmatrix}, \tag{2.1}$$

where

$$U = \begin{pmatrix} -1 & \lambda m\\ -\lambda n & 1\end{pmatrix}, \quad V = \begin{pmatrix} 4\lambda^{-2} + Q & -2\lambda^{-1}(u - u_x) - \lambda m Q\\ 2\lambda^{-1}(v + v_x) + \lambda n Q & -Q\end{pmatrix}, \tag{2.2}$$

with $Q = (u - u_x)(v + v_x)$. This form of the Lax pair is a slight modification (in particular, V have slightly different diagonal terms) of the original Lax pair in [19].

The modification is needed for consistency with the boundary value problem to be discussed below in Remark 1.

We recall that for smooth solutions the role of the Lax pair is to provide the Zero Curvature representation $\frac{\partial U}{\partial t} - \frac{\partial V}{\partial x} + \frac{1}{2}[U, V] = 0$ of the original non-linear partial differential equation, in our case (1.7). In the non-smooth case the situation is more subtle as explained in Appendix A.

Following [4] we perform the gauge transformation $\Phi = \mathrm{diag}(\frac{e^{\frac{x}{2}}}{\lambda}, e^{-\frac{x}{2}})\Psi$ which leads to a simpler x-equation

$$\Phi_x = \begin{bmatrix} 0 & h \\ -zg & 0 \end{bmatrix} \Phi, \qquad g = \sum_{j=1}^{N} g_j \delta_{x_j}, \qquad h = \sum_{j=1}^{N} h_j \delta_{x_j}, \tag{2.3}$$

where $g_j = n_j e^{-x_j}$, $h_j = m_j e^{x_j}$, $z = \lambda^2$, and thus $g_j h_j = m_j n_j$. We note that (2.3), that is the x member of the Lax pair, has the form of the Sturm-Liouville problem given by (1.4), provided we specify the boundary conditions. Our initial goal is to solve (2.3) subject to boundary conditions $\Phi_1(-\infty) = 0$, $\Phi_2(+\infty) = 0$ which are chosen in such a way as to remain invariant under the flow in t whose infinitesimal change is generated by the matrix V in the Lax equation.

Since the coefficients in the boundary value problem are distributions (measures), to make

$$\Phi_x = \begin{bmatrix} 0 & h \\ -zg & 0 \end{bmatrix} \Phi, \qquad \Phi_1(-\infty) = \Phi_2(+\infty) = 0, \tag{2.4}$$

well posed, we need to define the multiplication of the measures h and g by Φ. As suggested by the results in Appendix A we require that Φ be left continuous and we subsequently define the terms $\Phi_a \delta_{x_j} = \Phi_a(x_j)\delta_{x_j}, a = 1, 2$. This choice makes the Lax pair well defined as a distributional Lax pair and, as it is shown in the Appendix A, the compatibility condition of the x and t members of the Lax pair indeed implies (1.10). The latter result is more subtle than a routine check of compatibility for smooth Lax pairs.

Since the right-hand side of (2.4) is zero on the complement of the support of g and h, which in our case consists of points $\{x_1, \ldots, x_N\}$, the solution Φ is a piecewise constant function, which solves a finite difference equation.

Lemma 1. *Let $q_k = \Phi_1(x_k+)$, $p_k = \Phi_2(x_k+)$, then the difference form of the boundary value problem (2.4) reads:*

$$\begin{bmatrix} q_k \\ p_k \end{bmatrix} = T_k \begin{bmatrix} q_{k-1} \\ p_{k-1} \end{bmatrix}, \qquad T_k = \begin{bmatrix} 1 & h_k \\ -zg_k & 1 \end{bmatrix}, \qquad 1 \le k \le N, \tag{2.5}$$

where $q_0 = 0$, $p_0 = 1$, and the boundary condition on the right end (see (2.4)) is satisfied whenever $p_N(z) = 0$.

By inspection we obtain the following corollary.

Corollary 1. $q_k(z)$ is a polynomial of degree $\lfloor \frac{k-1}{2} \rfloor$ in z, and $p_k(z)$ is a polynomial of degree $\lfloor \frac{k}{2} \rfloor$, respectively.

Remark 1. Note that $\det(T_k) = 1 + zh_kg_k = 1 + zm_kn_k \neq 1$. In other words the setup we are developing goes beyond an SL_2 theory. In order to understand the origin of this difference we go back to the original Lax pair (2.2). If we assumed the matrix V to be traceless, with some coefficient $\alpha(\lambda)$ on the diagonal, then in the asymptotic region $x >> 0$ the second equation in (2.1) would read

$$\begin{bmatrix} \dot{a}e^{-x/2} \\ \dot{b}e^{x/2} \end{bmatrix} = \frac{1}{2} \begin{bmatrix} \alpha(\lambda) & -4\lambda^{-1}u_+e^{-x} \\ 0 & -\alpha(\lambda) \end{bmatrix} \begin{bmatrix} ae^{-x/2} \\ be^{x/2} \end{bmatrix},$$

where $u(x) = u_+e^{-x}$ in the asymptotic region. The simplest way to implement the isospectrality is to require that $\dot{b} = 0$, which requires gauging away $-\alpha(\lambda)$. This is justified on general grounds by observing that for any Lax equation $\dot{L} = [B, L]$, B is not uniquely defined. In particular, any term commuting with L can be added to B without changing Lax equations. In our case, we are adding a multiple of the identity to the original formulation in [19]. Furthermore, the gauge transformation leading up to (2.3) is not unimodular which takes us outside of the SL_2 theory.

The polynomials p_k, q_k can be constructed integrating directly the initial value problem

$$\Phi_x = \begin{bmatrix} 0 & h \\ -zg & 0 \end{bmatrix} \Phi, \qquad \Phi_1(-\infty) = 0, \quad \Phi_2(-\infty) = 1, \tag{2.6}$$

with the same rule regarding the multiplication of discrete measures g, h by piecewise smooth left-continuous functions as specified for (2.4).

With this convention in place we obtain the following characterization of $\Phi_1(x)$ and $\Phi_2(x)$, proven in its entirety in [6, Lemma 2.4].

Lemma 2. *Let us set*

$$\Phi_1(x) = \sum_{0 \le k} \Phi_1^{(k)}(x) z^k, \qquad \Phi_2(x) = \sum_{0 \le k} \Phi_2^{(k)}(x) z^k.$$

Then

$$\Phi_1^{(0)}(x) = \int_{\eta_0 < x} h(\eta_0) d\eta_0, \qquad \Phi_2^{(0)}(x) = 1$$

for $k = 0$, otherwise

$$\Phi_1^{(k)}(x) = (-1)^k \int_{\eta_0 < \xi_1 < \eta_1 < \cdots < \xi_k < \eta_k < x} \Big[\prod_{p=1}^{k} h(\eta_p) g(\xi_p)\Big] h(\eta_0)\, d\eta_0 d\xi_1 \ldots d\eta_k, \tag{2.7a}$$

$$\Phi_2^{(k)}(x) = (-1)^k \int_{\xi_1 < \eta_1 < \cdots < \xi_k < \eta_k < x} \Big[\prod_{p=1}^{k} g(\eta_p) h(\xi_p)\Big]\, d\xi_1 \ldots d\eta_k. \tag{2.7b}$$

If the points of the support of the discrete measure g (and h) are ordered $x_1 < x_2 < \cdots < x_N$ then

$$\Phi_1^{(k)}(x) = (-1)^k \sum_{\substack{j_0<i_1<j_1<\cdots<i_k<j_k \\ x_{j_k}<x}} \Big[\prod_{p=1}^{k} h_{j_p} g_{i_p}\Big]\, h_{j_0}, \tag{2.8a}$$

$$\Phi_2^{(k)}(x) = (-1)^k \sum_{\substack{i_1<j_1<\cdots<i_k<j_k \\ x_{j_k}<x}} \Big[\prod_{p=1}^{k} g_{j_p} h_{i_p}\Big]\,. \tag{2.8b}$$

To simplify the formulas in Lemma 2 we introduce the following notation. Our basic set of indices is $\{1,2,\dots,N\}$ which we denote by $[N]$ and if $k \le N$ we set $[k] = \{1,2,\dots,k\}$. We will denote by capital letters I and J any subsets of these sets and use the notation $\binom{[k]}{j}$ for the set of all j-element subsets of $[k]$, listed in increasing order; for example $I \in \binom{[k]}{j}$ means that $I = \{i_1, i_2, \dots, i_j\}$ for some increasing sequence $i_1 < i_2 < \cdots < i_j \le k$. Furthermore, given a multi-index $I = \{i_1, i_2, \dots, i_j\}$ and a set of numbers $a_{i_1}, \dots, a_{i_j}$ indexed by I, we will abbreviate $a_I = a_{i_1} a_{i_2} \dots a_{i_j}$ etc.

Definition 1. Let $I, J \in \binom{[k]}{l}$, or $I \in \binom{[k]}{l+1}, J \in \binom{[k]}{l}$.
Then I, J are said to be *interlacing*, denoted $I < J$, if

$$i_1 < j_1 < i_2 < j_2 < \cdots < i_l < j_l$$

or,

$$i_1 < j_1 < i_2 < j_2 < \cdots < i_l < j_l < i_{l+1},$$

in the latter case. The same notation is used in the degenerate case $I \in \binom{[k]}{1}, J \in \binom{[k]}{0}$.

Using this notation we can now express the results of Lemma 2 in a compact form.

Corollary 2. The unique solutions q_k and p_k to the recurrence equations (2.5) with initial conditions $q_0 = 0, p_0 = 1$ are given by

$$q_k(z) = \sum_{l=0}^{\lfloor \frac{k-1}{2} \rfloor} \Big(\sum_{\substack{I\in\binom{[k]}{l+1}, J\in\binom{[k]}{l} \\ I<J}} h_I g_J \Big)(-z)^l, \tag{2.9a}$$

$$p_k(z) = 1 + \sum_{l=1}^{\lfloor \frac{k}{2} \rfloor} \Big(\sum_{\substack{I,J\in\binom{[k]}{l} \\ I<J}} h_I g_J \Big)(-z)^l. \tag{2.9b}$$

We can now make a brief comment about the spectrum of the boundary value problem (2.4). We observe that a complex number z is an *eigenvalue* of the boundary value problem (2.4) if there exists a solution $\{q_k(z), p_k(z)\}$ to (2.5) for which $p_N(z) = 0$. The set of all eigenvalues comprises the *spectrum* of the boundary value problem (2.4). Our choice of boundary conditions was picked to ensure the invariance of the spectrum under the time evolution. To verify that the flow is isospectral (spectrum preserving) we examine the t part of the Lax pair (2.1) in the region $x > x_N$, as indicated in Remark 1 and perform the gauge transformation to determine the flow of Φ.

Lemma 3. *Let* $\{q_k, p_k\}$ *satisfy the system of difference equations* (2.5). *Then the Lax equations* (2.1) *imply*

$$\dot{q}_N = \frac{2}{z} q_N - \frac{2u_+}{z} p_N, \qquad \dot{p}_N = 0, \tag{2.10}$$

where $u_+ = \sum_{j=1}^N h_j$.

This lemma implies that the polynomial $p_N(z)$ is independent of time and, in particular, its zeros, i.e. the spectrum, are time invariant. Furthermore, Corollary 2 allows one to write the coefficients of $p_N(z)$ in terms of the variables g_j, h_j (or equivalently m_j, n_j, x_j) and thus identify $\lfloor \frac{N}{2} \rfloor$ constants of motion of the system (1.10):

$$M_j = \sum_{\substack{I,J \in \binom{[N]}{j} \\ I<J}} h_I g_J, \qquad 1 \le j \le \lfloor \frac{N}{2} \rfloor. \tag{2.11}$$

In the next section we will investigate the role of these constants in the integrability of (1.10).

3 Liouville integrability

3.1 Bi-Hamiltonian structure

The results of the previous section, especially the existence of $\lfloor \frac{N}{2} \rfloor$ constants, suggests that the system (1.10) might be integrable in a classical Liouville sense which is proven below. For smooth solutions $u(x,t), v(x,t)$ of (1.7) the Hamiltonian structure, in fact a bi-Hamiltionian one, of the 2-mCH equation (1.7) was given by Tian and Liu in [21]. By employing two compatible Hamiltonian operators

$$\mathscr{L}_1 = \begin{pmatrix} D_x m D_x^{-1} m D_x & D_x m D_x^{-1} n D_x \\ D_x n D_x^{-1} m D_x & D_x n D_x^{-1} n D_x \end{pmatrix}, \quad \mathscr{L}_2 = \begin{pmatrix} 0 & -D_x^2 - D_x \\ D_x^2 - D_x & 0 \end{pmatrix}$$

and the Hamiltonians

$$H_1 = \int n(u_x - u)dx, \qquad H_2 = \frac{1}{2}\int n(v + v_x)(u - u_x)^2 dx, \tag{3.1}$$

the 2-mCH equation (1.7) can be written as

$$\begin{pmatrix} m_t \\ \\ n_t \end{pmatrix} = \mathscr{L}_1 \begin{pmatrix} \frac{\delta H_1}{\delta m} \\ \\ \frac{\delta H_1}{\delta n} \end{pmatrix} = \mathscr{L}_2 \begin{pmatrix} \frac{\delta H_2}{\delta m} \\ \\ \frac{\delta H_2}{\delta n} \end{pmatrix}. \tag{3.2}$$

We note that the word *compatible* mentioned above means that an arbitrary linear combination of the two Hamiltonian operators is also Hamiltonian. Since we work in the non-smooth context the results obtained for smooth functions will not hold in the non-smooth region, and one either has to formulate a limiting procedure leading to the non-smooth sector or study the non-smooth sector independently. At present, we prefer the second approach mainly because it is technically simpler, and also because it is not clear at this point which Hamiltonian structures have meaningful limits.

3.2 Hamiltonian vector field

We focus on the peakon sector of (1.7) described by the system of equations (1.10).

Theorem 1. *The equations* (1.10) *for the motion of N peakons of the original PDE* (1.7) *are given by Hamilton's equations of motion:*

$$\dot{x}_j = \{x_j, H\}, \qquad \dot{m}_j = \{m_j, H\}, \qquad \dot{n}_j = \{n_j, H\}, \tag{3.3}$$

for the Hamiltonian

$$H = -\frac{1}{2}\int n(\xi)(u_\xi(\xi) - u(\xi))d\xi = 2M_1 + \sum_{k=1}^{N} m_k n_k.$$

Here M_1 is a constant of motion appearing in (2.11), *the Poisson bracket* $\{,\}$ *is given by*

$$\{x_i, x_k\} = \operatorname{sgn}(x_i - x_k), \tag{3.4a}$$
$$\{m_i, m_k\} = \{m_i, x_k\} = \{n_i, n_k\} = \{n_i, x_k\} = \{n_i, m_k\} = 0, \tag{3.4b}$$

and the ordering condition $x_1 < x_2 < \cdots < x_N$ is in place.

Proof. Clearly,

$$\{m_j, h\} = \{n_j, h\} = 0$$

under the above Poisson bracket, hence

$$\dot{m}_j = \dot{n}_j = 0.$$

We proceed with the computation of $\{x_j, H\}$:

$$\begin{aligned}
\{x_j, H\} &= \left\{ x_j,\ 2 \sum_{1 \le i < k \le N} m_i n_k e^{x_i - x_k} + \sum_{k=1}^{N} m_k n_k \right\} \\
&= 2 \sum_{1 \le i < k \le N} m_i n_k \left\{ x_j, e^{x_i - x_k} \right\} \\
&= 2 \sum_{1 \le i < k \le N} m_i n_k e^{x_i - x_k} \left(\operatorname{sgn}(x_j - x_i) - \operatorname{sgn}(x_j - x_k) \right) \\
&= 2 \sum_{k=1}^{j-1} m_k n_j e^{x_k - x_j} + 2 \sum_{k=j+1}^{N} m_j n_k e^{x_j - x_k} + 4 \sum_{1 \le i < j < k \le N} m_i n_k e^{x_i - x_k} \\
&\overset{(1.9)}{=} \langle Q \rangle (x_j),
\end{aligned}$$

thus proving the results. ■

3.3 Liouville integrability

We will introduce a natural Poisson manifold (M, π) defined by the Poisson bracket (3.4). Since m_j, n_j are constant we can restrict our considerations to the non-trivial part of the Poisson structure involving only x_j. Let us denote

$$M = \left\{ x_1 < x_2 < \cdots < x_N \right\} \tag{3.5}$$

and define

$$\pi(f, g) = \{f, g\} = \sum_{1 \le i < j \le N} \{x_i, x_j\} \frac{\partial f}{\partial x_i} \frac{\partial g}{\partial x_j} \tag{3.6}$$

for all differentiable functions f, g on M. Then M has a structure of a Poisson manifold M to be denoted (M, π).

One can check directly from (3.4) that regardless whether $N = 2K$ or $N = 2K + 1$.

Lemma 4. *$rank(\pi) = 2K$.*

Our objective now is to identify an appropriate number of Poisson commuting quantities. We will break down our analysis according to whether N is even or odd.

1. **Case** $N = 2K$. It follows from Lemma 3 and (2.11) that the quantities

$$M_j = \sum_{\substack{I, J \in \binom{[2K]}{j} \\ I < J}} h_I g_J, \qquad 1 \le j \le K, \tag{3.7}$$

with $h_i = m_i e^{x_i}$, $g_i = n_i e^{-x_i}$, form a set of K constants of motion for the system (1.10). We claim that these constants of motion Poisson commute.

Short of giving a detailed proof, we would like to outline the argument which goes back to J. Moser in [14]. Since M_j commute with the Hamiltonian H (see Theorem 1) their Poisson bracket $\{M_j, M_k\}$ commutes with H and thus $\{M_j, M_k\}$ is a constant of motion for every pair of indices j,k. For cases for which the inverse spectral methods allow one to express M_j in terms of leading asymptotic positions, in particular exploiting the asymptotic result that particles corresponding to to adjacent positions x_j, x_{j+1} pair up, while distinct pairs do not interact, leads to a suppression of the majority of terms in M_j. The precise argument is presented in [5, Theorem 3.8] while needed asymptotic results can be found in Theorem 14.

Theorem 2. *The Hamiltonians* $M_1, \cdots, M_K$ *Poisson commute.*

2. **Case $N = 2K+1$.** Again, following Lemma 3 and (2.11) we see

$$M_j = \sum_{\substack{I,J \in \binom{[2K+1]}{j} \\ I<J}} h_I g_J, \qquad 1 \le j \le K, \tag{3.8}$$

with $h_i = m_i e^{x_i}, \quad g_i = n_i e^{-x_i}$, are constants of motion for the system (1.10) in the odd case.

In the odd case, there is an extra constant of motion, which can be computed from the value of the Weyl function $W(z)$ at $z = \infty$ (see Section 2 and Section 4 for details regarding the Weyl function). We point out that this constant is 0 in the even case. The computation is routine and produces

$$c = \frac{\sum\limits_{\substack{I \in \binom{[2K+1]}{K+1}, J \in \binom{[2K+1]}{K} \\ I<J}} h_I g_J}{\sum\limits_{\substack{I,J \in \binom{[2K+1]}{K} \\ I<J}} h_I g_J} = \frac{\sum\limits_{\substack{I \in \binom{[2K+1]}{K+1}, J \in \binom{[2K+1]}{K} \\ I<J}} h_I g_J}{M_K}, \tag{3.9}$$

which, in turn, gives an extra constant of motion

$$M_c = \sum_{\substack{I \in \binom{[2K+1]}{K+1}, J \in \binom{[2K+1]}{K} \\ I<J}} h_I g_J = \prod_{j=1}^{K+1} m_{2j-1} e^{x_j} \prod_{j=1}^{K} n_{2j} e^{-x_j},$$

so that $\{M_1, M_2, \cdots, M_K, M_c\}$ form a set of $K+1$ constants of motion for the system (1.10) in this case.

It is not hard to see by using the same argument as in [5, Theorem 3.9] and the asymptotic results in Theorem 18, that the following theorem holds.

Theorem 3. *The Hamiltonians* $M_1, \cdots, M_K, M_c$ *Poisson commute.*

Combining now both theorems above we conclude (the proof is similar to the argument in [5, Theorem 3.10]).

Theorem 4. *The conservative peakon system given by* (1.10) *is Liouville integrable.*

4 Forward map: spectrum and spectral data

The spectrum of the boundary value problem (2.4) (or equivalently, (2.5)) is given by the zeros of the polynomial $p_N(z)$. However, one cannot recover the measures g and h from the spectrum alone. One needs extra data and the right object to turn to is the *Weyl function*

$$W(z) = \frac{q_N(z)}{p_N(z)}, \tag{4.1}$$

which in our case is a rational function with poles located at the spectrum of the boundary value problem. Another compelling reason for using the Weyl function is that, as we will show below, the residues of W evolve linearly in time, while the value of W at $z = \infty$ is a constant of motion. The investigation of the analytic properties of W can be greatly simplified by observing that W is built out of solutions to the recurrence (2.5). This suggests forming a recurrence of Weyl functions whose solution at step N is $W(z)$. This leads to the following result which is an immediate consequence of (2.5).

Lemma 5. *Let* $\{q_k, p_k\}$ *be the solution to* (2.5) *and let* $w_{2k} = \frac{q_k}{p_k}, w_{2k-1} = \frac{q_{k-1}}{p_k}$. *Then*

$$w_1 = 0, \qquad w_{2k} = (1 + z m_k n_k) w_{2k-1} + h_k, \qquad 1 \le k \le N, \tag{4.2a}$$

$$\frac{1}{w_{2k}} = \frac{1}{w_{2k+1}} + z g_{k+1}, \qquad 1 \le k \le N-1. \tag{4.2b}$$

We will now show that all these Weyl functions, including the original $W(z)$, have the following properties in common:

1. they all have simple poles located on $\mathbf{R}_+$;

2. all the residues are positive;

3. the values at $z = \infty$ are non-negative.

The rational functions of this type have been studied, as a special case, in the famous memoir by T. Stieltjes [20]. The most relevant for our studies is the following theorem which is a special case of a more general theorem proved by Stieltjes.

Theorem 5 (T. Stieltjes). *Any rational function* $F(z)$ *admitting the integral representation*

$$F(z) = c + \int \frac{d\nu(x)}{x - z}, \tag{4.3}$$

where $d\nu(x)$ *is the (Stieltjes) measure corresponding to the piecewise constant non-decreasing function* $\nu(x)$ *with finitely many jumps in* $\mathbf{R}_+$ *has a finite (terminating) continued fraction expansion*

$$F(z) = c + \cfrac{1}{a_1(-z) + \cfrac{1}{a_2 + \cfrac{1}{a_3(-z) + \cfrac{1}{\ddots}}}}, \tag{4.4}$$

where all $a_j > 0$ and, conversely, any rational function with this type of a continued fraction expansion has the integral representation (4.3).

We now apply Stieltjes' result to our case.

Lemma 6. *Given $h_j, g_j > 0, h_j g_j = m_j n_j > 0, 1 \leq j \leq N$, let $w_j s$ satisfy the recurrence relations of Lemma 5. Then $w_j s$ are shifted Stieltjes transforms of finite, discrete Stieltjes measures supported on $\mathbf{R}_+$, with nonnegative shifts. More precisely:*

$$w_{2k-1}(z) = \int \frac{d\mu^{(2k-1)}(x)}{x - z},$$
$$w_{2k}(z) = c_{2k} + \int \frac{d\mu^{(2k)}(x)}{x - z},$$

where $c_{2k} > 0$ when k is odd, otherwise, $c_{2k} = 0$. Furthermore, the number of points in the support $d\mu^{(2k)}(x)$ and $d\mu^{(2k-1)}$ is $\lfloor \frac{k}{2} \rfloor$.

Proof. We only sketch the proof, for further details we refer to [6, Lemma 3.6]. The proof goes by induction on k. The base case $k = 1$ is elementary. Assuming the induction hypothesis to hold up to $2k$ we invert (4.2b) to get:

$$w_{2k+1}(z) = \frac{1}{-z g_{k+1} + \frac{1}{w_{2k}}}$$

which, by induction hypothesis, implies that w_{2k+1} has the required continued fraction expansion covered by Stieltjes' theorem, and thus has the required integral representation. We subsequently feed this integral representation into (4.2a) to obtain the Stietljes integral representation for w_{2k+2}. The analysis of the signs of the values of the Weyl functions at $z = \infty$ is carried out in [6, Lemma 3.6].

■

Remark 2. The recurrence in Lemma 5 can be viewed as the recurrence on the Weyl functions corresponding to shorter bars (keeping in mind the interpretation in terms of the longitudinal vibrations of an elastic bar) obtained by truncating at the index k. Then W_{2k} is precisely the Weyl function corresponding to the measures $\sum_{j=1}^{k} h_j \delta_{x_j}$ and $\sum_{j=1}^{k} g_j \delta_{x_j}$, while W_{2k-1} corresponds to the measures $\sum_{j=1}^{k-1} h_j \delta_{x_j}$ and $\sum_{j=1}^{k} g_j \delta_{x_j}$ respectively.

Now, in particular, we note that by Lemma 6

$$W(z) = \frac{q_N(z)}{p_N(z)} = c_{2N} + \int \frac{d\mu^{(2N)}(x)}{x-z}, \qquad d\mu^{(2N)} = \sum_{j=1}^{\lfloor \frac{N}{2} \rfloor} b_j^{(2N)} \delta_{\zeta_j} \tag{4.5}$$

and thus the following theorem holds.

Theorem 6. *$W(z)$ is a (shifted) Stieltjes transform of a positive, discrete measure $d\mu$ with support inside $\mathbf{R}_+$. More precisely:*

$$W(z) = c + \int \frac{d\mu(x)}{x-z}, \quad d\mu = \sum_{i=1}^{\lfloor \frac{N}{2} \rfloor} b_j \delta_{\zeta_j},\ 0 < \zeta_1 < \cdots < \zeta_{\lfloor \frac{N}{2} \rfloor},\ 0 < b_j,\ 1 \le j \le \left\lfloor \frac{N}{2} \right\rfloor,$$

where $c > 0$ when N is odd and $c = 0$ when N is even.

The next corollary summarizes the properties of the spectrum of the boundary value problem (2.4), or equivalently (2.5).

Corollary 3.

1. The spectrum of the boundary value problem (2.3) is positive and simple.

2. $W(z) = c + \sum_{j=1}^{\lfloor \frac{N}{2} \rfloor} \frac{b_j}{\zeta_j - z}$, where all residues satisfy $b_j > 0$ and $c \ge 0$.

5 Inverse problem

5.1 The first inverse problem and the interpolation problem

The initial inverse problem we are interested in solving can be stated as follows:

Definition 2. Given a rational function (see Theorem 6)

$$W(z) = c + \int \frac{d\mu(x)}{x-z}, \quad d\mu = \sum_{i=1}^{\lfloor \frac{N}{2} \rfloor} b_j \delta_{\zeta_j},\ 0 < \zeta_1 < \cdots < \zeta_{\lfloor \frac{N}{2} \rfloor},\ 0 < b_j,\ 1 \le j \le \left\lfloor \frac{N}{2} \right\rfloor, \tag{5.1}$$

where $c > 0$ when N is odd and $c = 0$ when N even, as well as positive constants $m_1, m_2, \ldots, m_N, n_1, n_2, \cdots, n_N$, such that the products $m_j n_j$ are distinct, find positive constants g_j, h_j, $1 \le j \le N$, for which

1. $g_j h_j = m_j n_j$,

2. the unique solution of the initial value problem:

$$\begin{bmatrix} q_k \\ p_k \end{bmatrix} = \begin{bmatrix} 1 & h_k \\ -z g_k & 1 \end{bmatrix} \begin{bmatrix} q_{k-1} \\ p_{k-1} \end{bmatrix}, \qquad 1 \le k \le N, \tag{5.2}$$

$$\begin{bmatrix} q_0 \\ p_0 \end{bmatrix} = \begin{bmatrix} 0 \\ 1 \end{bmatrix},$$

satisfies

$$\frac{q_N(z)}{p_N(z)} = W(z).$$

Remark 3. The restriction that the products $m_j n_j$ be distinct has been made to facilitate the argument and will be eventually relaxed by taking appropriate limits of the generic case (see the comments below Theorem 10).

The basic idea of the solution of the above inverse problem is to associate to it a certain interpolation problem. We now proceed to explain how such an interpolation problem appears already in the solution of the forward problem, that is, in the solution of the difference boundary value problem (2.5), or equivalently (5.2) above. First, let us denote by

$$T_k(z) = \begin{bmatrix} 1 & h_k \\ -zg_k & 1 \end{bmatrix}, \tag{5.3}$$

the transition matrix appearing in (5.2). Clearly,

$$\begin{bmatrix} q_N(z) \\ p_N(z) \end{bmatrix} = T_N(z)T_{N-1}(z)\cdots T_{N-k+1}(z) \begin{bmatrix} q_{N-k}(z) \\ p_{N-k}(z) \end{bmatrix}. \tag{5.4}$$

Let us introduce a different indexing $i' = N - i + 1$ which is a bijection of the set $[1, N]$ and represents counting points of the beam from right to left rather than left to right. Moreover, let us denote by $\hat{T}_j(z)$ the classical adjoint of $T_{j'}$. Thus

$$\hat{T}_j(z) = \begin{bmatrix} 1 & -h_{N-j+1} \\ zg_{N-j+1} & 1 \end{bmatrix} = \begin{bmatrix} 1 & -h_{j'} \\ zg_{j'} & 1 \end{bmatrix}. \tag{5.5}$$

Then (5.4) implies

$$\hat{T}_k(z)\cdots \hat{T}_2\hat{T}_1(z) \begin{bmatrix} W(z) \\ 1 \end{bmatrix} = \frac{\prod_{j=1}^{k} (1 + zm_{j'}n_{j'})}{p_N(z)} \begin{bmatrix} q_{(k+1)'}(z) \\ p_{(k+1)'}(z) \end{bmatrix}, \tag{5.6}$$

where we used that $\det(\hat{T}_{j'}) = 1 + zg_{j'}h_{j'}$ and, subsequently, $g_{j'}h_{j'} = m_{j'}n_{j'}$. We clearly have

Lemma 7. *Let us fix $1 \le k \le N$ and denote $\hat{S}_k(z) = \hat{T}_k(z)\cdots \hat{T}_2\hat{T}_1(z)$. Then for every $1 \le j \le k$ the vector $\begin{bmatrix} W(z_j) \\ 1 \end{bmatrix}$, where $z_j = -\frac{1}{m_{j'}n_{j'}}$, is in the $\ker(\hat{S}_k(z_j))$.*

We proceed by explicitly writing the conditions that the vectors $\begin{bmatrix} W(z_i) \\ 1 \end{bmatrix}$ be null vectors of $\hat{S}_k(z_i)$. To this end we write

$$\hat{S}_k(z) = \begin{bmatrix} \hat{q}_k(z) & \hat{Q}_k(z) \\ \hat{p}_k(z) & \hat{P}_k(z) \end{bmatrix}, \tag{5.7}$$

from which two sets of interpolation conditions

$$\hat{q}_k(z_j)W(z_j) + \hat{Q}_k(z_j) = 0, \quad 1 \le j \le k, \tag{5.8a}$$

$$\hat{p}_k(z_j)W(z_j) + \hat{P}_k(z_j) = 0, \quad 1 \le j \le k, \tag{5.8b}$$

emerge. Whether these conditions, given $W(z)$, determine the polynomials $\hat{q}_k, \hat{Q}_k, \hat{p}_k, \hat{P}_k$ will depend on their degrees and this is the subject of the next result whose proof is an easy exercise in induction.

Lemma 8. *For any k, $1 \le k \le N$,*

1. $\deg \hat{q}_k(z) = \lfloor \frac{k}{2} \rfloor$, $\det \hat{Q}_k(z) = \lfloor \frac{k-1}{2} \rfloor$, $\deg \hat{p}_k(z) = \lfloor \frac{k+1}{2} \rfloor$, $\det \hat{P}_k(z) = \lfloor \frac{k}{2} \rfloor$,
2. $\hat{q}_k(0) = 1$, $\hat{p}_k(0) = 0$, $\hat{P}_k(0) = 1$.

Now it is elementary to check that the number of interpolation conditions in equations (5.8) is the same as the number of unknown coefficients in $\hat{q}_k, \hat{Q}_k, \hat{p}_k, \hat{Q}_k$, so, in principle, the solution exists. Before we state our next lemma we revisit the notation introduced in Definition 1. For any multi-index $I \in \binom{[k]}{j}$, where we recall $I = \{i_1, i_2, \cdots, i_j\}$ is an ordered set associated to an increasing sequence $1 \le i_1 < i_2 \cdots < i_j \le k$, we assign its ordered image I' obtained by applying the bijection $i \to N+1-i$ to I and reordering. The following result can be demonstrated by using induction on k and the definition of $\hat{S}_k(z)$ (5.7).

Lemma 9. *For any k, $1 \le k \le N$,*

$$\hat{q}_k(z) = 1 + \sum_{j=1}^{\lfloor \frac{k}{2} \rfloor} \Big(\sum_{\substack{I,J \in \binom{[k]}{j} \\ I<J}} g_{I'} h_{J'} \Big)(-z)^j, \tag{5.9a}$$

$$\hat{Q}_k(z) = -\sum_{j=0}^{\lfloor \frac{k-1}{2} \rfloor} \Big(\sum_{\substack{I \in \binom{[k]}{j+1}, J \in \binom{[k]}{j} \\ I<J}} h_{I'} g_{J'} \Big)(-z)^j, \tag{5.9b}$$

$$\hat{p}_k(z) = -\sum_{j=1}^{\lfloor \frac{k+1}{2} \rfloor} \Big(\sum_{\substack{I \in \binom{[k]}{j}, J \in \binom{[k]}{j-1} \\ I<J}} g_{I'} h_{J'} \Big)(-z)^j, \tag{5.9c}$$

$$\hat{P}_k(z) = 1 + \sum_{j=1}^{\lfloor \frac{k}{2} \rfloor} \Big(\sum_{\substack{I,J \in \binom{[k]}{j} \\ I<J}} h_{I'} g_{J'} \Big)(-z)^j, \tag{5.9d}$$

with the convention that $\hat{q}_1(z) = 1$, $\hat{Q}_1(z) = -h_N$, $\hat{p}_1(z) = zg_N$, $\hat{P}_1(z) = 1$.

Example 1. It is instructive to display $\hat{S}_k(z)$ for small k. The notation is that of (5.7).

a)

$$\hat{S}_1(z) = \begin{bmatrix} \hat{q}_1(z) & \hat{Q}_1(z) \\ \hat{p}_1(z) & \hat{P}_1(z) \end{bmatrix} = \begin{bmatrix} 1 & -h_{1'} \\ g_{1'}z & 1 \end{bmatrix}.$$

b)

$$\hat{S}_2(z) = \begin{bmatrix} \hat{q}_2(z) & \hat{Q}_2(z) \\ \hat{p}_2(z) & \hat{P}_2(z) \end{bmatrix} = \begin{bmatrix} 1-(g_{1'}h_{2'})z & -(h_{1'}+h_{2'}) \\ (g_{1'}+g_{2'})z & 1-(h_{1'}g_{2'})z \end{bmatrix}.$$

c)

$$\hat{S}_3(z) = \begin{bmatrix} \hat{q}_3(z) & \hat{Q}_3(z) \\ \hat{p}_3(z) & \hat{P}_3(z) \end{bmatrix} =$$
$$\begin{bmatrix} 1-(g_{1'}h_{2'}+g_{1'}h_{3'}+g_{2'}h_{3'})z & -(h_{1'}+h_{2'}+h_{3'})+(h_{1'}g_{2'}h_{3'})z \\ (g_{1'}+g_{2'}+g_{3'})z-(g_{1'}h_{2'}g_{3'})z^2 & 1-(h_{1'}g_{2'}+h_{1'}g_{3'}+h_{2'}g_{3'})z \end{bmatrix}.$$

For a polynomial $f(z)$ let us denote by f^+ the coefficient at the highest power in z and use the convention $\hat{q}_0 = 1$. Then, by inspection, we obtain

$$g_{1'} = \frac{\hat{p}_1^+}{\hat{q}_0^+}, \quad g_{2'} = \frac{\hat{P}_2^+}{\hat{Q}_1^+}, \quad g_{3'} = \frac{\hat{p}_3^+}{\hat{q}_2^+},$$

and continuing with the help of induction we are led to

Theorem 7. *For any* $1 \le k \le N$,

$$g_{k'} = \frac{\hat{p}_k^+}{\hat{q}_{k-1}^+}, \qquad \textit{if } k \textit{ is odd,} \tag{5.10a}$$

$$g_{k'} = \frac{\hat{P}_k^+}{\hat{Q}_{k-1}^+}, \qquad \textit{if } k \textit{ is even.} \tag{5.10b}$$

Remark 4. Since, $g_{k'}h_{k'} = m_{k'}n_{k'}$, knowing $g_{k'}$ determines uniquely $h_{k'}$.

Now we can state our strategy for solving the original inverse problem stated in Definition 2:

1. given $W(z)$ we solve the interpolation problems (5.8) for all $1 \le k \le N$;
2. from the solution to the interpolation problem at stage $1 \le k \le N$, we use Theorem 7 to recover $g_{k'}, h_{k'}$ and thus $\hat{T}_k = \begin{bmatrix} 1 & -h_{k'} \\ zg_{k'} & 1 \end{bmatrix}$, finally all transitions matrices T_k (see (5.3));
3. we then define

$$\begin{bmatrix} q_k(z) \\ p_k(z) \end{bmatrix} = T_k(z)T_{k-1}(z)\cdots T_1(z)\begin{bmatrix} 0 \\ 1 \end{bmatrix}; \tag{5.11}$$

4. we define $\tilde{W}(z) = \frac{q_N(z)}{p_N(z)}$. The fact that $W(z) = \tilde{W}(z)$ follows from (5.4) and (5.6) with $k = N$.

The interpolation problem is linear so the solution will be expressed in terms of determinants. Before, however, we present the final formulae it is helpful to introduce a family of generalized Cauchy-Vandermonde matrices [9, 12, 13] attached to a Stieltjes transform of a positive measure. Matrices of this type arise in some interpolation problems, including the current one. We refer the reader to [6] for more details but to ease the presentation we provide in the paragraph below a simplified version of the interpolation problem specified by (5.8).

Given three positive integers k, l, m such that $k = l + m$, a function $f(z)$, and a collection of distinct points $\{z_j\}_1^k$ we are seeking two polynomials $a(z) = 1 + \sum_{n=1}^{l} a_n z^n$ and $b(z) = \sum_{n=0}^{m} b_n z^n$ such that

$$a(z_j)f(z_j) + b(z_j) = 0, \qquad 1 \le j \le k.$$

This interpolation problem is equivalent to the matrix problem

$$\begin{bmatrix} z_1^1 f(z_1) & z_1^2 f(z_1) & \dots & z_1^l f(z_1) & 1 & z_1 & \dots & z_1^m \\ z_2^1 f(z_1) & z_2^2 f(z_2) & \dots & z_2^l f(z_2) & 1 & z_2 & \dots & z_2^m \\ \vdots & \vdots & \vdots & \vdots & \vdots & \vdots & \vdots & \vdots \\ z_k^1 f(z_k) & z_k^2 f(z_k) & \dots & z_k^l f(z_k) & 1 & z_k & \dots & z_k^m \end{bmatrix} \begin{bmatrix} a_1 \\ \vdots \\ \vdots \\ a_l \\ b_0 \\ \vdots \\ b_m \end{bmatrix} = - \begin{bmatrix} f(z_1) \\ \vdots \\ \vdots \\ \vdots \\ f(z_k) \end{bmatrix}.$$

Denoting the determinant of the matrix on the left by D_k and assuming for now that $D_k \neq 0$ we can succinctly write the solution

$$a(z) + z^{l+1} b(z) =$$

$$\frac{1}{D_k} \det \begin{bmatrix} 1 & z & z^2 & \dots & z^l & z^{l+1} & z^{l+2} & \dots & z^k \\ f(z_1) & z_1^1 f(z_1) & z_1^2 f(z_1) & \dots & z_1^l f(z_1) & 1 & z_1 & \dots & z_1^m \\ f(z_2) & z_2^1 f(z_1) & z_2^2 f(z_2) & \dots & z_2^l f(z_2) & 1 & z_2 & \dots & z_2^m \\ \vdots & \vdots & \vdots & \vdots & \vdots & \vdots & \vdots & \vdots & \vdots \\ f(z_k) & z_k^1 f(z_k) & z_k^2 f(z_k) & \dots & z_k^l f(z_k) & 1 & z_k & \dots & z_k^m \end{bmatrix}.$$

In the case of the function f being the Stieltjes transform of a measure the determinants in question are computable. Now we turn to spelling out the most important points in the solution of the inverse problem. We start with a definition.

Definition 3. Given a (strictly) positive vector $\mathbf{e} \in \mathbf{R}^k$, a non-negative number c, an index l such that $0 \le l \le k$, another index p such that $0 \le p$, $p+l-1 \le k-l$, and a positive measure ν with support in $\mathbf{R}_+$, a *Cauchy-Stieltjes-Vandermonde (CSV) matrix* is that of the form

$$CSV_k^{(l,p)}(\mathbf{e}, \nu, c) = \begin{pmatrix} e_1^p \hat{\nu}_c(e_1) & e_1^{p+1}\hat{\nu}_c(e_1) & \cdots & e_1^{p+l-1}\hat{\nu}_c(e_1) & 1 & e_1 & \cdots & e_1^{k-l-1} \\ e_2^p \hat{\nu}_c(e_2) & e_2^{p+1}\hat{\nu}_c(e_2) & \cdots & e_2^{p+l-1}\hat{\nu}_c(e_2) & 1 & e_2 & \cdots & e_2^{k-l-1} \\ \vdots & \vdots & \ddots & \vdots & \vdots & \vdots & \ddots & \vdots \\ e_k^p \hat{\nu}_c(e_k) & e_k^{p+1}\hat{\nu}_c(e_k) & \cdots & e_k^{p+l-1}\hat{\nu}_c(e_k) & 1 & e_k & \cdots & e_k^{k-l-1} \end{pmatrix},$$

where $\hat{\nu}_c(y) = c + \int \frac{d\nu(x)}{y+x}$ is the (shifted) classical Stieltjes transform of the measure ν.

Remark 5. In this section we use a slightly different definition of the Stieltjes transform then the one in the context of the Weyl function (see Section 4). Thus, in this section, it is $W(-z)$ which is the Stieltjes transform of the spectral measure; this notation being in fact more in line with Stieltjes' notation in [20].

The explicit formulas for the determinant of the CSV matrix can be readily obtained. To this end we need some notations to facilitate the presentation. Recall that the multi-index notation that was introduced earlier in the part leading up to the definition 1. Moreover, let us denote $[i,j] = \{i, i+1, \cdots, j\}$, $\binom{[1,K]}{k} = \{J = \{j_1, j_2, \cdots, j_k\} | j_1 < \cdots < j_k, j_i \in [1,K]\}$. Then for two ordered multi-index sets I, J we define

$$\mathbf{x}_J = \prod_{j\in J} x_j, \qquad \Delta_J(\mathbf{x}) = \prod_{i<j\in J} (x_j - x_i),$$

$$\Delta_{I,J}(\mathbf{x};\mathbf{y}) = \prod_{i\in I}\prod_{j\in J}(x_i - y_j), \qquad \Gamma_{I,J}(\mathbf{x};\mathbf{y}) = \prod_{i\in I}\prod_{j\in J}(x_i + y_j),$$

along with the conventions

$$\Delta_\emptyset(\mathbf{x}) = \Delta_{\{i\}}(\mathbf{x}) = \Delta_{\emptyset,J}(\mathbf{x};\mathbf{y}) = \Delta_{I,\emptyset}(\mathbf{x};\mathbf{y}) = \Gamma_{\emptyset,J}(\mathbf{x};\mathbf{y}) = \Gamma_{I,\emptyset}(\mathbf{x};\mathbf{y}) = 1,$$

$$\binom{[1,K]}{0} = 1; \qquad \binom{[1,K]}{k} = 0, \quad k > K.$$

Since we will eventually obtain expressions in terms of the ratios of determinants it is helpful to study the structure of these determinants. The results stated below will be important at two stages of our analysis: the existence of the solution to the inverse problem and the large time asymptotic analysis of solutions to (1.10). The detailed proofs can be found in [6].

Theorem 8. *Let ν be a positive measure with support in $\mathbf{R}_+$ and let $\mathbf{x}$ denote the vector $[x_1, x_2, \ldots, x_l] \in \mathbf{R}^l$ and $d\nu^p(y) = y^p d\nu(y)$, respectively. Then*

1. *if either* $c = 0$ *or* $p + l - 1 < k - l$ *then*

$$\det CSV_k^{(l,p)}(\mathbf{e}, \nu, c) = (-1)^{lp+\frac{l(l-1)}{2}}$$
$$\Delta_{[1,k]}(\mathbf{e}) \int\limits_{0<x_1<x_2<\cdots<x_l} \frac{\Delta_{[1,l]}(\mathbf{x})^2}{\Gamma_{[1,k],[1,l]}(\mathbf{e};\mathbf{x})} d\nu^p(x_1)d\nu^p(x_2)\dots d\nu^p(x_l); \tag{5.12}$$

2. *if* $c > 0$ *and* $p + l - 1 = k - l$ *then*

$$\det CSV_k^{(l,p)}(\mathbf{e}, \nu, c) = (-1)^{lp+\frac{l(l-1)}{2}} \Delta_{[1,k]}(\mathbf{e})$$
$$\cdot \Big(\int\limits_{0<x_1<x_2<\cdots<x_l} \frac{\Delta_{[1,l]}(\mathbf{x})^2}{\Gamma_{[1,k],[1,l]}(\mathbf{e};\mathbf{x})} d\nu^p(x_1)d\nu^p(x_2)\dots d\nu^p(x_l) \tag{5.13}$$
$$+ c \int\limits_{0<y_1<y_2<\cdots<y_{l-1}} \frac{\Delta_{[1,l-1]}(\mathbf{y})^2}{\Gamma_{[1,k],[1,l-1]}(\mathbf{e};\mathbf{y})} d\nu^p(y_1)d\nu^p(y_2)\dots d\nu^p(y_{l-1}) \Big).$$

Our next step is to give a complete solution to the inverse problem as stated in Definition 2 in terms of the determinants of the CSV matrices. To this end we set (see (5.1)):

$$e_j = \frac{1}{m_{j'}n_{j'}}, \quad \mathbf{e}_{[1,j]} = e_1 e_2 \cdots e_j, \qquad \nu = \mu, \quad 1 \le j \le N, \tag{5.14}$$

and

$$\mathcal{D}_k^{(l,p)} = \left|\det CSV_k^{(l,p)}(\mathbf{e}, \mu, c)\right|. \tag{5.15}$$

Now, with Theorem 8 in hand, the main theorem of this section follows from the solution of the interpolation problem (5.8) with normalization conditions of Lemma 8, and Theorem 7. For details of computations we refer to [6].

Theorem 9. *Suppose the Weyl function* $W(z)$ *is given by* (5.1) *along with positive constants (masses)* $m_1, m_2, \cdots, m_N, n_1, n_2, \cdots, n_N$ *such that the products* $m_j n_j$ *are distinct. Then, there exists a unique solution to the inverse problem specified in Definition 2:*

$$g_{k'} = \frac{\mathcal{D}_k^{(\frac{k-1}{2},1)} \mathcal{D}_{k-1}^{(\frac{k-1}{2},1)}}{\mathbf{e}_{[1,k]} \mathcal{D}_k^{(\frac{k+1}{2},0)} \mathcal{D}_{k-1}^{(\frac{k-1}{2},0)}}, \qquad \text{if } k \text{ is odd}, \tag{5.16a}$$

$$g_{k'} = \frac{\mathcal{D}_k^{(\frac{k}{2},1)} \mathcal{D}_{k-1}^{(\frac{k}{2}-1,1)}}{\mathbf{e}_{[1,k]} \mathcal{D}_k^{(\frac{k}{2},0)} \mathcal{D}_{k-1}^{(\frac{k}{2},0)}}, \qquad \text{if } k \text{ is even}. \tag{5.16b}$$

Likewise,

$$h_{k'} = \frac{\mathbf{e}_{[1,k-1]}\mathcal{D}_k^{(\frac{k+1}{2},0)}\mathcal{D}_{k-1}^{(\frac{k-1}{2},0)}}{\mathcal{D}_k^{(\frac{k-1}{2},1)}\mathcal{D}_{k-1}^{(\frac{k-1}{2},1)}}, \qquad \text{if } k \text{ is odd,} \tag{5.17a}$$

$$h_{k'} = \frac{\mathbf{e}_{[1,k-1]}\mathcal{D}_k^{(\frac{k}{2},0)}\mathcal{D}_{k-1}^{(\frac{k}{2},0)}}{\mathcal{D}_k^{(\frac{k}{2},1)}\mathcal{D}_{k-1}^{(\frac{k}{2}-1,1)}}, \qquad \text{if } k \text{ is even.} \tag{5.17b}$$

5.2 The second inverse problem

To proceed to the next step we recall that the original peakon problem (1.10) was formulated in the x space. To go back to the x space we use the relation $h_j = m_j e^{x_j}$ (see equation (2.3)) to arrive at the inverse formulae relating the spectral data and the positions of peakons given by x_j.

Theorem 10. *Given positive constants m_j, n_j with distinct products $m_j n_j$, let Φ be the solution to the boundary value problem 2.4 with associated spectral data $\{d\mu, c\}$. Then the positions x_j (of peakons) in the discrete measures $m = 2\sum_{j=1}^N m_j\delta_{x_j}$ and $n = 2\sum_{j=1}^N n_j\delta_{x_j}$ can be expressed in terms of the spectral data as:*

$$x_{k'} = \ln \frac{\mathbf{e}_{[1,k-1]}\mathcal{D}_k^{(\frac{k+1}{2},0)}\mathcal{D}_{k-1}^{(\frac{k-1}{2},0)}}{m_{k'}\mathcal{D}_k^{(\frac{k-1}{2},1)}\mathcal{D}_{k-1}^{(\frac{k-1}{2},1)}}, \qquad \text{if } k \text{ is odd,} \tag{5.18a}$$

$$x_{k'} = \ln \frac{\mathbf{e}_{[1,k-1]}\mathcal{D}_k^{(\frac{k}{2},0)}\mathcal{D}_{k-1}^{(\frac{k}{2},0)}}{m_{k'}\mathcal{D}_k^{(\frac{k}{2},1)}\mathcal{D}_{k-1}^{(\frac{k}{2}-1,1)}}, \qquad \text{if } k \text{ is even,} \tag{5.18b}$$

with $\mathcal{D}_k^{(l,p)}$ defined in (5.15), $k' = N-k+1$, $1 \le k \le N$ and the convention that $\mathcal{D}_0^{l,p} = 1$.

Finally, we can relax the condition that the products of masses m_j, n_j be distinct. Indeed, it suffices to observe that the Vandermonde determinants $\Delta_{[1,r]}(\mathbf{e})$, $r = k, k-1$, cancel out in all expressions of the type

$$\frac{\mathcal{D}_k^{(l_1,p_1)}\mathcal{D}_{k-1}^{(l_2,p_2)}}{\mathcal{D}_k^{(l_3,p_3)}\mathcal{D}_{k-1}^{(l_4,p_4)}},$$

as a result of Theorem 8 (see (5.12) and (5.13)).

In summary, we completed the full circle of starting with initial positions of peakons, mapping them to the spectral data $\{d\mu, c\}$, while in the last chapter we solved explicitly the inverse problem of mapping back the spectral data to the positions x_j of peakons. In all this the time was fixed. In the following sections, we concentrate on the time evolution of the multipeakons of the 2-mCH. In light of the difference of the value of c in the Stieltjes transform according to whether N is even or odd, the discussions will be presented separately for N even, N odd, respectively.

6 Multipeakons for $N = 2K$

For even N, $c = 0$ (see Theorem 6). This impacts the asymptotic behaviour of solutions. Further comments on differences between solutions for odd and even N are in Section 8.1.

6.1 Closed formulae for $N = 2K$

If we assume that $x_1(0) < x_2(0) < \cdots < x_{2K}(0)$ then by continuity, this condition will hold at least in a small interval containing $t = 0$. At $t = 0$ we solve the forward problem (see Section 4) and obtain the Weyl function $W(z)$ (see Theorem 6) hence the spectral measure $d\mu(0) = \sum_{j=1}^{K} b_j(0)\delta_{\zeta_j}$ supported on the set of ordered eigenvalues $0 < \zeta_1 < \cdots < \zeta_K$.

With the help of Theorem 10 we obtain the following result.

Theorem 11. *The 2-mCH (1.7) with the regularization of the singular term Qm given by $\langle Q\rangle m$ admits the multipeakon solution*

$$u(x,t) = \sum_{k=1}^{2K} m_{k'} \exp(-|x - x_{k'}(t)|), \qquad v(x,t) = \sum_{k=1}^{2K} n_{k'} \exp(-|x - x_{k'}(t)|), \tag{6.1}$$

where $m_{k'}, n_{k'}$ are arbitrary positive constants, while $x_{k'}(t)$ are given by equations (5.18a) *and* (5.18b) *corresponding to the peakon spectral measure*

$$d\mu(t) = \sum_{j=1}^{K} b_j(t)\delta_{\zeta_j}, \tag{6.2}$$

with $b_j(t) = b_j(0)e^{\frac{2t}{\zeta_j}}$, $0 < b_j(0)$, and $c = 0$ in (5.15).

Proof. The only outstanding issue is the time evolution of b_j or, more generally, the time evolution of the spectral measure. Recall the Weyl function $W(z)$ defined in (4.1). By employing the time evolution (2.10), one easily obtains

$$\dot{W} = \frac{2}{z}W - \frac{2u_+}{z},$$

which, in turn, implies $\dot{b}_j = \frac{2}{\zeta_j} b_j, 1 \leq j \leq K$, by use of Corollary 3. ■

In the following, we provide examples of multipeakons in the case of even N. Before this is done, it is useful to examine the explicit formulas for the CSV determinants following Theorem 8 (see equation (5.15) for notation). We remind the reader that the eigenvalues ζ_j are positive and ordered $0 < \zeta_1 < \cdots < \zeta_K$.

Theorem 12. *Let $N = 2K$, $0 \leq l \leq K$, $0 \leq p$, $p + l - 1 \leq k - l$, $1 \leq k \leq 2K$ and let the peakon spectral measure be given by* (6.2). *Then*

1.

$$\mathcal{D}_k^{(l,p)} = \left|\Delta_{[1,k]}(\mathbf{e})\right| \sum_{I \in \binom{[1,K]}{l}} \frac{\Delta_I^2(\zeta)\mathbf{b}_I\zeta_I^p}{\Gamma_{[1,k],I}(\mathbf{e};\zeta)}; \tag{6.3}$$

2. in the asymptotic region $t \to +\infty$

$$\mathcal{D}_k^{(l,p)} = \left|\Delta_{[1,k]}(\mathbf{e})\right| \frac{\Delta_{[1,l]}^2(\zeta)\mathbf{b}_{[1,l]}\zeta_{[1,l]}^p}{\Gamma_{[1,k],[1,l]}(\mathbf{e};\zeta)}\left[1+\mathcal{O}(e^{-\alpha t})\right], \quad \textit{for some } \alpha > 0; \tag{6.4}$$

3. in the asymptotic region $t \to -\infty$

$$\mathcal{D}_k^{(l,p)} = \left|\Delta_{[1,k]}(\mathbf{e})\right| \frac{\Delta_{[1,l]^*}^2(\zeta)\mathbf{b}_{[1,l]^*}\zeta_{[1,l]^*}^p}{\Gamma_{[1,k],[1,l]^*}(\mathbf{e};\zeta)}\left[1+\mathcal{O}(e^{\beta t})\right], \quad 0 < \beta, \tag{6.5}$$

where $[1,l]^* = [l^* = K - l + 1, 1^* = K]$ (reflection of the interval $[1,K]$).

Now we are ready to present examples of expressions for positions $x_1, \cdots, x_{2K}$ of multipeakons based on formulas (5.18a), (5.18b), using (6.3) and $e_j = \frac{1}{m_{j'}n_{j'}}$, $j' = 2K - j + 1$.

Example 2 (2-peakon solution; K=1).

$$x_1 = \ln\left(\frac{b_1}{\zeta_1 m_1(1+\zeta_1 m_2 n_2)}\right), \qquad x_2 = \ln\left(\frac{b_1 n_2}{1+\zeta_1 m_2 n_2}\right).$$

Example 3 (4-peakon solution; K=2).

$$x_1 =$$

$$\ln\left(\frac{1}{m_1}\cdot\frac{b_1 b_2(\zeta_2-\zeta_1)^2}{\zeta_1\zeta_2\left(b_1\zeta_1(1+\zeta_2 m_2 n_2)(1+\zeta_2 m_3 n_3)(1+\zeta_2 m_4 n_4)+b_2\zeta_2(1+\zeta_1 m_2 n_2)(1+\zeta_1 m_3 n_3)(1+\zeta_1 m_4 n_4)\right)}\right),$$

$$x_2 = \ln\left(n_2\cdot\frac{b_1 b_2(\zeta_2-\zeta_1)^2\left(b_1(1+\zeta_2 m_3 n_3)(1+\zeta_2 m_4 n_4)+b_2(1+\zeta_1 m_3 n_3)(1+\zeta_1 m_4 n_4)\right)}{\left(b_1\zeta_1(1+\zeta_2 m_3 n_3)(1+\zeta_2 m_4 n_4)+b_2\zeta_2(1+\zeta_1 m_3 n_3)(1+\zeta_1 m_4 n_4)\right)}\right.$$

$$\left.\cdot\frac{1}{\left(b_1\zeta_1(1+\zeta_2 m_2 n_2)(1+\zeta_2 m_3 n_3)(1+\zeta_2 m_4 n_4)+b_2\zeta_2(1+\zeta_1 m_2 n_2)(1+\zeta_1 m_3 n_3)(1+\zeta_1 m_4 n_4)\right)}\right),$$

$$x_3 =$$

$$\ln\left(\frac{1}{m_3}\cdot\frac{\left(b_1(1+\zeta_2 m_4 n_4)+b_2(1+\zeta_1 m_4 n_4)\right)\left(b_1(1+\zeta_2 m_3 n_3)(1+\zeta_2 m_4 n_4)+b_2(1+\zeta_1 m_3 n_3)(1+\zeta_1 m_4 n_4)\right)}{(1+\zeta_1 m_4 n_4)(1+\zeta_2 m_4 n_4)\left(b_1\zeta_1(1+\zeta_2 m_3 n_3)(1+\zeta_2 m_4 n_4)+b_2\zeta_2(1+\zeta_1 m_3 n_3)(1+\zeta_1 m_4 n_4)\right)}\right),$$

$$x_4 = \ln\left(n_4\cdot\frac{b_1(1+\zeta_2 m_4 n_4)+b_2(1+\zeta_1 m_4 n_4)}{(1+\zeta_1 m_4 n_4)(1+\zeta_2 m_4 n_4)}\right).$$

6.2 Global existence for $N = 2K$

Recall that our solution in Theorem 11 was obtained under the assumption that $x_1 < x_2 < \cdots < x_{2K}$. However, even if we start with the initial positions satisfying $x_1(0) < x_2(0) < \cdots < x_{2K}(0)$, the order might cease to hold for sufficiently large times, in other words some of the peakons might collide. It is an interesting question to understand the nature of collisions but we leave this topic for future work. In this subsection we give sufficient conditions in terms of the spectrum and constant masses m_j, n_j which ensure that no collisions occur and thus the peakon solutions are global in t. The readers might want to consult Appendix B for a detailed proof. Granted global existence, one can talk sensibly about the large time behaviour of peakons.

Theorem 13. *Given arbitrary spectral data*

$$\{b_j > 0,\, 0 < \zeta_1 < \zeta_2 < \cdots < \zeta_K : 1 \le j \le K\},$$

suppose the masses m_k, n_k satisfy

$$\frac{\zeta_K^{\frac{k-1}{2}}}{\zeta_1^{\frac{k+1}{2}}} < m_{(k+1)'} n_{k'}, \qquad \textit{for all odd } k, \quad 1 \le k \le 2K-1, \quad (6.6a)$$

$$\frac{m_{(k+1)'} n_{(k+2)'}}{(1 + m_{(k+1)'} n_{(k+1)'} \zeta_1)(1 + m_{(k+2)'} n_{(k+2)'} \zeta_1)} < \frac{\zeta_1^{\frac{k+1}{2}}}{\zeta_K^{\frac{k-1}{2}}} \frac{2 \min_j (\zeta_{j+1} - \zeta_j)^{k-1}}{(k+1)(\zeta_K - \zeta_1)^{k+1}},$$

$$\textit{for all odd } k, \quad 1 \le k \le 2K-3. \quad (6.6b)$$

Then the positions obtained from inverse formulas (5.18a), (5.18b) *are ordered $x_1 < x_2 < \cdots < x_{2K}$ and the multipeakon solutions* (7.1) *exist for arbitrary $t \in R$.*

6.3 Large time peakon asymptotics for $n = 2K$

Once the global existence of solutions is guaranteed, for example by imposing sufficient conditions of Theorem 13, one can study the asymptotic behaviour of multipeakon solutions for large (positive and negative) time by employing Theorems 11 and 12. More precisely, by using the formulae for positions (5.18a), (5.18b), as well as asymptotic evaluations of determinants (6.4) and (6.5), one arrives at

Theorem 14. *Suppose the masses m_j, n_j satisfy the conditions of Theorem 13. Then the asymptotic position of a k-th (counting from the right) peakon as $t \to +\infty$ is given by*

$$x_{k'} = \frac{2t}{\zeta_{\frac{k+1}{2}}} + \ln \frac{b_{\frac{k+1}{2}}(0) \mathbf{e}_{[1,k-1]} \Delta^2_{[1,\frac{k-1}{2}],\{\frac{k+1}{2}\}}(\zeta)}{m_{k'} \Gamma_{[1,k],\{\frac{k+1}{2}\}}(\mathbf{e};\zeta) \zeta^2_{[1,\frac{k-1}{2}]}} + \mathcal{O}(e^{-\alpha_k t}),$$

$$\textit{for some positive } \alpha_k \textit{ and odd } k,$$

$$x_{k'} = \frac{2t}{\zeta_{\frac{k}{2}}} + \ln \frac{b_{\frac{k}{2}}(0) \mathbf{e}_{[1,k-1]} \Delta^2_{[1,\frac{k}{2}-1],\{\frac{k}{2}\}}(\zeta)}{m_{k'} \Gamma_{[1,k-1],\{\frac{k}{2}\}}(\mathbf{e};\zeta) \zeta^2_{[1,\frac{k}{2}-1]} \zeta_{\frac{k}{2}}} + \mathcal{O}(e^{-\alpha_k t}),$$

$$\textit{for some positive } \alpha_k \textit{ and even } k,$$

$$x_{k'} - x_{(k+1)'} = \ln m_{(k+1)'} n_{k'} \zeta_{\frac{k+1}{2}} + \mathcal{O}(e^{-\alpha_k t}),$$

$$\textit{for some positive } \alpha_k \textit{ and odd } k.$$

Likewise, as $t \to -\infty$, using the notation of Theorem 12, the asymptotic position *of the k-th peakon is given by*

$$x_{k'} = \frac{2t}{\zeta_{(\frac{k+1}{2})^*}} + \ln \frac{b_{(\frac{k+1}{2})^*}(0)\mathbf{e}_{[1,k-1]}\Delta^2_{([1,\frac{k-1}{2}])^*,\{(\frac{k+1}{2})^*\}}(\zeta)}{m_{k'}\Gamma_{[1,k],\{(\frac{k+1}{2})^*\}}(\mathbf{e};\zeta)\zeta^2_{[1,\frac{k-1}{2}]^*}} + \mathcal{O}(e^{\beta_k t}),$$

for some positive β_k and odd k,

$$x_{k'} = \frac{2t}{\zeta_{(\frac{k}{2})^*}} + \ln \frac{b_{(\frac{k}{2})^*}(0)\mathbf{e}_{[1,k-1]}\Delta^2_{[1,\frac{k}{2}-1]^*,\{(\frac{k}{2})^*\}}(\zeta)}{m_{k'}\Gamma_{[1,k-1],\{(\frac{k}{2})^*\}}(\mathbf{e};\zeta)\zeta^2_{[1,\frac{k}{2}-1]^*}\zeta_{(\frac{k}{2})^*}} + \mathcal{O}(e^{\beta_k t}),$$

for some positive β_k and even k,

$$x_{k'} - x_{(k+1)'} = \ln m_{(k+1)'} n_{k'} \zeta_{(\frac{k+1}{2})^*} + \mathcal{O}(e^{\beta_k t}),$$

for some positive β_k and odd k.

Remark 6. It follows from the above theorem that multipeakons of the 2-mCH equation exhibit *Toda-like sorting properties* of asymptotic speeds and *an asymptotic pairing.* The latter can be partially explained by the fact that there are K available eigenvalues to match $2K$ asymptotic speeds. Similar features were also observed in the mCH equation [6], as well as the interlacing cases of the 2-mCH equation [4]. It is clear now that these two features extend to the non-interlacing cases as well.

We end this section by providing graphs of a concrete 4-peakon solution. Let $K = 2$, and $b_1(0) = 10$, $b_2(0) = 1$, $\zeta_1 = 0.3$, $\zeta_2 = 3$, $m_1 = 8$, $m_2 = 24$, $m_3 = 5$, $m_4 = 10$, $n_1 = 12$, $n_2 = 10$, $n_3 = 24$, $n_4 = 16$. It is easy to check that the condition in Theorem 13 is satisfied. Hence, the order of $\{x_k, k = 1, 2, 3, 4\}$ will be preserved and one can use the explicit formulae for the 4-peakon solution, resulting in the following sequence of graphs (Figure 1), illustrating the asymptotic pairing of peakons.

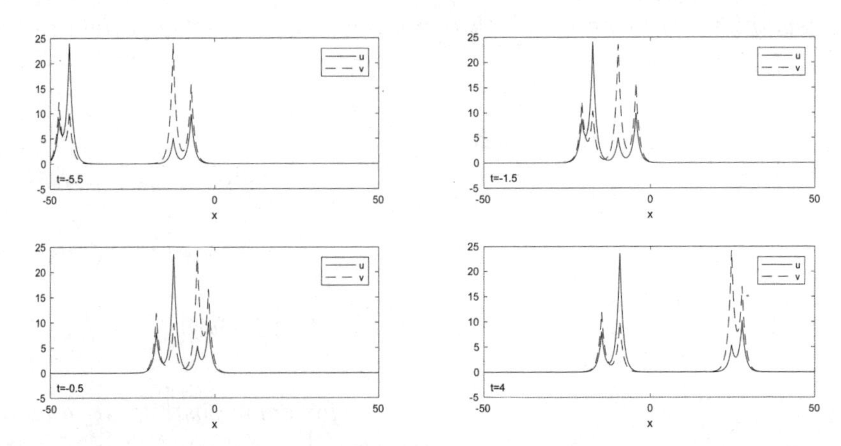

Figure 1. Graphs of 4-peakon at times $t = -5.5,\ -1.5,\ -0.5,\ 4$ in the case of $b_1(0) = 10$, $b_2(0) = 1$, $\zeta_1 = 0.3$, $\zeta_2 = 3$, $m_1 = 8$, $m_2 = 24$, $m_3 = 5$, $m_4 = 10$, $n_1 = 12$, $n_2 = 10$, $n_3 = 24$, $n_4 = 16$.

7 Multipeakons for $N = 2K + 1$

This section is devoted to the corresponding result for $N = 2K + 1$, which is presented in a way parallel to the previous section on the even case. The main source of difference between the two cases is of course the presence of the positive shift c which impacts the evaluations of the CSV determinants as illustrated by Theorem 8, in particular formula (5.13). Nevertheless, we will present a comparison in the form of a correspondence between the odd case and the even case in Section 8.1.

7.1 Closed formulae for $N = 2K + 1$

As before we assume that $x_1(0) < x_2(0) < \cdots < x_{2K+1}(0)$. Then this condition will hold at least in a small interval containing $t = 0$. The following *local existence* result follows from Theorem 10.

Theorem 15. *The 2-mCH equation* (1.7) *with the regularization of the singular term Qm given by $\langle Q\rangle m$ admits the multipeakon solution*

$$u(x,t) = \sum_{k=1}^{2K+1} m_{k'} \exp(-|x-x_{k'}(t)|), \qquad v(x,t) = \sum_{k=1}^{2K+1} n_{k'} \exp(-|x-x_{k'}(t)|), \tag{7.1}$$

where $m_{k'}, n_{k'}$ are arbitrary positive constants, while $x_{k'}(t)$ are given by equations (5.18a) *and* (5.18b) *corresponding to the peakon spectral measure*

$$d\mu = \sum_{j=1}^{K} b_j(t)\delta_{\zeta_j}, \tag{7.2}$$

$b_j(t) = b_j(0)e^{\frac{2t}{\zeta_j}}$, $0 < b_j(0)$, *with ordered eigenvalues* $0 < \zeta_1 < \cdots < \zeta_K$ *and* $c(t) = c(0) > 0$ *in* (5.15).

Proof. Similar to the even case, it is clear that the Weyl function $W(z)$ is defined in (4.1), undergoes the time evolution obtained earlier in the proof of Theorem 11, namely,

$$\dot{W} = \frac{2}{z}W - \frac{2u_+}{z},$$

which, in turn, implies $\dot{b}_j = \frac{2}{\zeta_j}b_j, 1 \leq j \leq K$ as well as $\dot{c} = 0$ by virtue of Corollary 3. The rest of the argument is the same as for the even case. ■

By using the above theorem, it is not hard to work out two of the most simple examples of solutions. Before we do that, however, we will examine the evaluation of CSV determinants presented in Theorem 8 (see equation (5.15) for notation), with due care to two facts: $N = 2K + 1$ and $c > 0$. The proof follows from the same steps as in Theorem 12 and we omit it.

Theorem 16. *Let $N = 2K+1$, $1 \leq k \leq 2K+1$, $0 \leq l \leq K+1$, $0 \leq p$, $p+l-1 \leq k-l$, and let the peakon spectral measure be given by (7.2) and a shift $c > 0$. Then*

1.

$$\mathcal{D}_k^{(l,p)} = |\Delta_{[1,k]}(\mathbf{e})| \sum_{I \in \binom{[1,K]}{l}} \frac{\Delta_I^2(\zeta)\mathbf{b}_I\zeta_I^p}{\Gamma_{[1,k],I}(\mathbf{e};\zeta)},$$

$$\text{if} \quad p+l-1 < k-l, \quad k \leq 2K+1; \tag{7.3a}$$

$$\mathcal{D}_k^{(l,p)} = |\Delta_{[1,k]}(\mathbf{e})| \Big(\sum_{I \in \binom{[1,K]}{l}} \frac{\Delta_I^2(\zeta)\mathbf{b}_I\zeta_I^p}{\Gamma_{[1,k],I}(\mathbf{e};\zeta)} + c \sum_{I \in \binom{[1,K]}{l-1}} \frac{\Delta_I^2(\zeta)\mathbf{b}_I\zeta_I^p}{\Gamma_{[1,k],I}(\mathbf{e};\zeta)} \Big)$$

$$\text{if} \quad p+l-1 = k-l, \quad k \leq 2K+1; \tag{7.3b}$$

with the proviso that the first term inside the bracket is set to zero if $l = K+1$, which only happens when $k = 2K+1, p = 0$.

2. *In the asymptotic region $t \to +\infty$*

$$\mathcal{D}_k^{(l,p)} = |\Delta_{[1,k]}(\mathbf{e})| \frac{\Delta_{[1,l]}^2(\zeta)\mathbf{b}_{[1,l]}\zeta_{[1,l]}^p}{\Gamma_{[1,k],[1,l]}(\mathbf{e};\zeta)} \Big[1 + \mathcal{O}(e^{-\alpha t})\Big], \quad 0 < \alpha,$$

$$\text{if} \quad 0 \leq l \leq K; \tag{7.4a}$$

$$\mathcal{D}_{2K+1}^{(K+1,0)} = c\,|\Delta_{[1,2K+1]}(\mathbf{e})| \frac{\Delta_{[1,K]}^2(\zeta)\mathbf{b}_{[1,K]}}{\Gamma_{[1,2K+1],[1,K]}(\mathbf{e};\zeta)},$$

$$\text{if} \quad k = 2K+1, l = K+1, p = 0. \tag{7.4b}$$

3. *In the asymptotic region $t \to -\infty$*

$$\mathcal{D}_k^{(l,p)} = |\Delta_{[1,k]}(\mathbf{e})| \frac{\Delta_{[1,l]^*}^2(\zeta)\mathbf{b}_{[1,l]^*}\zeta_{[1,l]^*}^p}{\Gamma_{[1,k],[1,l]^*}(\mathbf{e};\zeta)} \Big[1 + \mathcal{O}(e^{\beta t})\Big], \quad 0 < \beta,$$

$$\text{if} \quad p+l-1 < k-l, \quad k \leq 2K+1; \tag{7.5a}$$

$$\mathcal{D}_k^{(l,p)} = c\,|\Delta_{[1,k]}(\mathbf{e})| \frac{\Delta_{[1,l-1]^*}^2(\zeta)\mathbf{b}_{[1,l-1]^*}\zeta_{[1,l-1]^*}^p}{\Gamma_{[1,k],[1,l-1]^*}(\mathbf{e};\zeta)} \Big[1 + \mathcal{O}(e^{\beta t})\Big], \quad 0 < \beta,$$

$$\text{if} \quad p+l-1 = k-l, \quad k < 2K+1; \tag{7.5b}$$

$$\mathcal{D}_{2K+1}^{(K+1,0)} = c\,|\Delta_{[1,2K+1]}(\mathbf{e})| \frac{\Delta_{[1,K]}^2(\zeta)\mathbf{b}_{[1,K]}}{\Gamma_{[1,2K+1],[1,K]}(\mathbf{e};\zeta)},$$

$$\text{if} \quad k = 2K+1, l = K+1, p = 0, \tag{7.5c}$$

where, as before, $[1,l]^* = [l^* = K-l+1, 1^* = K]$.

There exists a relation between formulae with $c > 0$ and $c = 0$. Indeed, by comparing formulas (7.3a) and (7.3b) with (6.3), we arrive at the detailed dependence on c.

Corollary 4. Let $N = 2K+1$, $1 \leq k \leq 2K+1$, $0 \leq l \leq K+1$, $0 \leq p$, $p+l-1 \leq k-l$, and let the peakon spectral measure be given by (7.2) and a shift $c > 0$. Then

$$\mathcal{D}_k^{(l,p)}(c) = \mathcal{D}_k^{(l,p)}(0), \quad \text{if } p+l-1 < k-l,\ k \leq 2K+1; \tag{7.6a}$$
$$\mathcal{D}_k^{(l,p)}(c) = \mathcal{D}_k^{(l,p)}(0) + c\mathcal{D}_k^{(l-1,p)}(0), \quad \text{if } p+l-1 = k-l,\ k \leq 2K+1; \tag{7.6b}$$

with the convention that the first term in (7.6b) is set to zero if $l = K+1, k = 2K+1, p = 0$.

Below the reader will find two examples of explicit peakon solutions for odd $N = 2K+1$.

Example 4 (1-peakon solution; $K = 0$).

$$x_1 = \ln\left(\frac{c}{m_1}\right).$$

Example 5 (3-peakon solution; $K = 1$).

$$x_1 = \ln\left(\frac{b_1 c}{\zeta_1 m_1 \left(b_1\zeta_1 m_2 n_2 m_3 n_3 + c(1+\zeta_1 m_2 n_2)(1+\zeta_1 m_3 n_3)\right)}\right),$$
$$x_2 = \ln\left(\frac{b_1 n_2}{b_1\zeta_1 m_2 n_2 m_3 n_3 + c(1+\zeta_1 m_2 n_2)(1+\zeta_1 m_3 n_3)}\left(\frac{b_1 m_3 n_3}{1+\zeta_1 m_3 n_3} + c\right)\right),$$
$$x_3 = \ln\left(\frac{1}{m_3}\left(\frac{b_1 m_3 n_3}{1+\zeta_1 m_3 n_3} + c\right)\right).$$

7.2 Global existence for $N = 2K+1$

Similar to the even case, we can also provide a sufficient condition to ensure the global existence of peakon solutions when $N = 2K+1$. The main result is stated below while its proof is relegated to Appendix C.

Theorem 17. *Given arbitrary spectral data*

$$\{b_j > 0,\ 0 < \zeta_1 < \zeta_2 < \cdots < \zeta_K,\ c > 0 : 1 \leq j \leq K\},$$

suppose the masses m_k, n_k *satisfy*

$$\frac{1}{m_{(k+1)'} n_{k'}} < \frac{\zeta_1^{\frac{k+1}{2}}}{\zeta_K^{\frac{k-1}{2}}} min\{1, \hat{\beta}\}, \qquad \text{for all odd } k, \qquad 1 \leq k \leq 2K-1,$$

$$\frac{1}{m_{(k+2)'} n_{(k+1)'}} < \frac{\zeta_1^{\frac{k+1}{2}}}{\zeta_K^{\frac{k-1}{2}}} min\{1, \hat{\beta}_1\}, \qquad \text{for all odd } k, \qquad 1 \leq k \leq 2K-1,$$

where

$$\hat{\beta}=\begin{cases}\frac{2\zeta_K\, min_j(\zeta_{j+1}-\zeta_j)^{k-3}}{\zeta_1(k-1)(\zeta_K-\zeta_1)^{k-1}}\frac{(1+m_{k'}n_{k'}\zeta_1)(1+m_{(k+1)'}n_{(k+1)'}\zeta_1)}{m_{k'}n_{k'}m_{(k+1)'}n_{(k+1)'}}, & \text{for all odd } k,\ 3\leq k\leq 2K-1,\\ +\infty, & \text{for } k=1,\end{cases}$$

$$\hat{\beta}_1=\frac{2\, min_j(\zeta_{j+1}-\zeta_j)^{k-1}}{(k+1)(\zeta_K-\zeta_1)^{k+1}}\frac{(1+m_{(k+1)'}n_{(k+1)'}\zeta_1)(1+m_{(k+2)'}n_{(k+2)'}\zeta_1)}{m_{(k+1)'}n_{(k+1)'}m_{(k+2)'}n_{(k+2)'}}.$$

Then the positions obtained from inverse formulas (5.18a), (5.18b) *are ordered* $x_1 < x_2 < \cdots < x_{2K+1}$ *and the multipeakon solutions* (7.1) *exist for arbitrary* $t \in \mathbf{R}$.

7.3 Large time peakon asymptotics for $N = 2K + 1$

Again, based on the global existence of multipeakons guaranteed by Theorem 17, one can investigate the long time asymptotics of global multipeakon solutions. After a straightforward, but tedious, computation using the formulas for positions (5.18a), (5.18b) (for $N = 2K + 1$), as well as asymptotic evaluations of determinants presented in Theorem 16, we obtain

Theorem 18. *Suppose the masses* m_j, n_j *satisfy the conditions of Theorem 17. Then the asymptotic position of a k-th (counting from the right) peakon as* $t \to +\infty$ *is given by*

$$x_{k'}=\frac{2t}{\zeta_{\frac{k+1}{2}}}+\ln\frac{b_{\frac{k+1}{2}}(0)\mathbf{e}_{[1,k-1]}\Delta^2_{[1,\frac{k-1}{2}],\{\frac{k+1}{2}\}}(\zeta)}{m_{k'}\Gamma_{[1,k],\{\frac{k+1}{2}\}}(\mathbf{e};\zeta)\zeta^2_{[1,\frac{k-1}{2}]}}+\mathcal{O}(e^{-\alpha_k t}),\qquad \alpha_k>0$$

and odd $k \leq 2K-1$;

$$x_{(2K+1)'}=\ln\frac{c\mathbf{e}_{[1,2K]}}{m_{(2K+1)'}\zeta^2_{[1,K]}}+\mathcal{O}(e^{-\alpha t}),\qquad \alpha>0;$$

$$x_{k'}=\frac{2t}{\zeta_{\frac{k}{2}}}+\ln\frac{b_{\frac{k}{2}}(0)\mathbf{e}_{[1,k-1]}\Delta^2_{[1,\frac{k}{2}-1],\{\frac{k}{2}\}}(\zeta)}{m_{k'}\Gamma_{[1,k-1],\{\frac{k}{2}\}}(\mathbf{e};\zeta)\zeta^2_{[1,\frac{k}{2}-1]}\zeta_{\frac{k}{2}}}+\mathcal{O}(e^{-\alpha_k t}),\qquad \alpha_k>0$$

and even $k \leq 2K$;

$$x_{k'}-x_{(k+1)'}=\ln m_{(k+1)'}n_{k'}\zeta_{\frac{k+1}{2}}+\mathcal{O}(e^{-\alpha_k t}),\qquad \alpha_k>0$$

and odd $k \leq 2K-1$.

Likewise, as $t \to -\infty$, *using the notation of Theorem 12, the asymptotic position of the k-th peakon is given by*

$$x_{k'} = \frac{2t}{\zeta_{(\frac{k-1}{2})^*}} + \ln \frac{b_{(\frac{k-1}{2})^*}(0)\mathbf{e}_{[1,k-1]}\Delta^2_{[1,\frac{k-1}{2}-1]^*,\{(\frac{k-1}{2})^*\}}(\zeta)}{m_{k'}\Gamma_{[1,k-1],\{(\frac{k-1}{2})^*\}}(\mathbf{e};\zeta)\zeta^2_{[1,\frac{k-1}{2}-1]^*}\zeta_{(\frac{k-1}{2})^*}} + \mathcal{O}(e^{\beta_k t}),$$

$$\beta_k > 0 \text{ and odd } 1 < k \leq 2K+1;$$

$$x_{1'} = \ln \frac{c}{m_{1'}} + \mathcal{O}(e^{\beta_k t}), \qquad \beta_k > 0;$$

$$x_{k'} = \frac{2t}{\zeta_{(\frac{k}{2})^*}} + \ln \frac{b_{(\frac{k}{2})^*}(0)\mathbf{e}_{[1,k-1]}\Delta^2_{([1,\frac{k}{2}-1])^*,\{(\frac{k}{2})^*\}}(\zeta)}{m_{k'}\Gamma_{[1,k],\{(\frac{k}{2})^*\}}(\mathbf{e};\zeta)\zeta^2_{[1,\frac{k}{2}-1]^*}} + \mathcal{O}(e^{\beta_k t}),$$

$$\beta_k > 0 \text{ and even } k \leq 2K;$$

$$x_{k'} - x_{(k+1)'} = \ln m_{(k+1)'} n_{k'} \zeta_{(\frac{k}{2})^*} + \mathcal{O}(e^{\beta_k t}),$$

$$\beta_k > 0 \text{ and even } k \leq 2K.$$

Remark 7. Similar to the even case, the Toda-like sorting property can also be observed in this case. It is perhaps interesting to examine the role of the constant c. The constant c is playing the role of an additional eigenvalue $\zeta_{K+1} = \infty$ resulting in the formal asymptotic speed 0. We observe that for large positive times the first particle counting from the left comes to a stop, while the remaining $2K$ peakons form pairs of bound states akin to what is occurring for even N, effectively sharing in pairs the remaining K speeds. By contrast, for large negative times, the first particle counting from the right slows to a halt, while the remaining peakons form pairs. Note that a similar phenomenon also occurs in the mCH.

At the end of this section, we present graphs for a concrete 3-peakon solution, which confirm our theoretical predictions. Let $K = 1$, and $b_1(0) = 1,\ c = 3,\ \zeta_1 = 5,\ m_1 = 3,\ m_2 = 2,\ m_3 = 1.6,\ n_1 = 1.8,\ n_2 = 3,\ n_3 = 2.2$. Then the sufficient conditions in Theorem 17 are satisfied. Hence the order of $\{x_k, k = 1, 2, 3\}$ will be preserved and one can use the explicit formulae for the 3-peakon solution, resulting in the following sequence of graphs (Figure 2).

8 Reductions of multipeakons

We recall that the 2-mCH (1.7) is a two-component integrable generalization of the mCH (1.5). In the present contribution we constructed explicit formulae for generic multipeakons of the 2-mCH by using the inverse spectral method. Yet, in our past work we constructed the so-called interlacing multipeakons of the 2-mCH [4], while in [6] we gave explicit formulae for multipeakons of the mCH. In this section, as if to close the circle, we would like to show how the formulae obtained in previous sections reduce to those special cases.

8.1 From odd case to even case

In Section 6 and 7, we presented the multipeakon formulae according to the parity of the number N of masses. In this subsection, we will show how the multipeakon

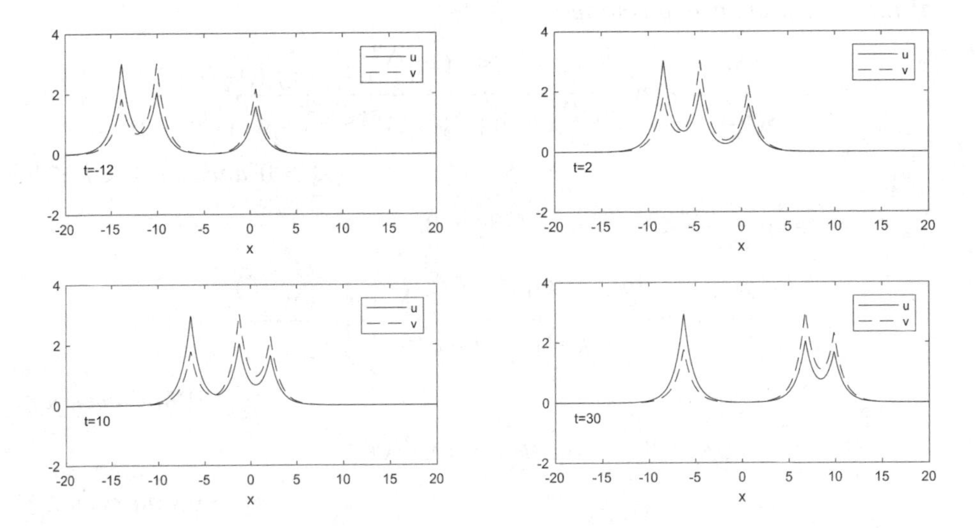

Figure 2. Graphs of 3-peakon at time $t=-12,\ 2,\ 10,\ 30$ in the case of $b_1(0)=1,\ c=3,\ \zeta_1=5,\ m_1=3,\ m_2=2,\ m_3=1.6,\ n_1=1.8,\ n_2=3,\ n_3=2.2$.

formula in the odd case can be used to derive the multipeakon formula for the even case. Consider the multipeakons

$$u=\sum_{j=1}^{2K+1} m_j(t)e^{-|x-x_j(t)|},\qquad v=\sum_{j=1}^{2K+1} n_j(t)e^{-|x-x_j(t)|}.$$

Suppose $m_1\to 0$, $n_1\to 0$, then $h_1\to 0$, $g_1\to 0$, since $h_j=m_je^{x_j}$, $g_j=n_je^{-x_j}$. Moreover, by continuity, it is not hard to see that $c\to 0$ in view of (3.9) and results from Section 4.

Now, with the help of multipeakon formulae (5.18a)-(5.18b), we also see that when $c\to 0$, then $\lim_{c\to 0} m_1e^{-|x-x_1|}=0$. Moreover, we can rewrite the multipeakon formulae (5.18a)-(5.18b) in the odd case as

$$x_{2K-2k+3}=\ln\frac{\mathbf{e}_{[1,2k-2]}\mathcal{D}_{2k-1}^{(k,0)}(c)\,\mathcal{D}_{2k-2}^{(k-1,0)}(c)}{m_{2K-2k+3}\mathcal{D}_{2k-1}^{(k-1,1)}(c)\,\mathcal{D}_{2k-2}^{(k-1,1)}(c)},\qquad 1\le k\le K+1,\tag{8.1a}$$

$$x_{2K-2k+2}=\ln\frac{\mathbf{e}_{[1,2k-1]}\mathcal{D}_{2k}^{(k,0)}(c)\,\mathcal{D}_{2k-1}^{(k,0)}(c)}{m_{2K-2k+2}\mathcal{D}_{2k}^{(k,1)}(c)\,\mathcal{D}_{2k-1}^{(k-1,1)}(c)},\qquad 1\le k\le K,\tag{8.1b}$$

where $e_j=\frac{1}{m_{2K+2-j}n_{2K+2-j}}$. Note, however, that neither m_1 nor n_1 appear in the expressions for $x_2,x_3,\cdots,x_{2K+1}$. Thus, we have

$$\lim_{c\to 0} x_{2K-2k+3}=\ln\frac{\mathbf{e}_{[1,2k-2]}\mathcal{D}_{2k-1}^{(k,0)}(0)\,\mathcal{D}_{2k-2}^{(k-1,0)}(0)}{m_{2K-2k+3}\mathcal{D}_{2k-1}^{(k-1,1)}(0)\,\mathcal{D}_{2k-2}^{(k-1,1)}(0)},\qquad 1\le k\le K,\tag{8.2a}$$

$$\lim_{c\to 0} x_{2K-2k+2}=\ln\frac{\mathbf{e}_{[1,2k-1]}\mathcal{D}_{2k}^{(k,0)}(0)\,\mathcal{D}_{2k-1}^{(k,0)}(0)}{m_{2K-2k+2}\mathcal{D}_{2k}^{(k,1)}(0)\,\mathcal{D}_{2k-1}^{(k-1,1)}(0)},\qquad 1\le k\le K.\tag{8.2b}$$

With the notation

$$\tilde{x}_j = \lim_{c\to 0} x_{j+1}, \qquad \tilde{m}_j = m_{j+1}, \qquad \tilde{n}_j = n_{j+1},$$

as well as

$$\tilde{e}_j = \frac{1}{\tilde{m}_{2K+1-j}\tilde{n}_{2K+1-j}}, \qquad \tilde{\mathbf{e}}_{[1,j]} = \tilde{e}_1 \cdots \tilde{e}_j, \qquad \tilde{\mathcal{D}}_k^{l,p} = \mathcal{D}_k^{l,p}$$

where $\tilde{\mathcal{D}}_k^{l,p}$ is obtained by replacing m_{j+1}, n_{j+1} by $\tilde{m}_j, \tilde{n}_j$ in the corresponding CSV matrix, we obtain

$$\tilde{x}_{2K-2k+2} = \ln \frac{\tilde{\mathbf{e}}_{[1,2k-2]}\tilde{\mathcal{D}}_{2k-1}^{(k,0)}(0)\,\tilde{\mathcal{D}}_{2k-2}^{(k-1,0)}(0)}{\tilde{m}_{2K-2k+2}\tilde{\mathcal{D}}_{2k-1}^{(k-1,1)}(0)\,\tilde{\mathcal{D}}_{2k-2}^{(k-1,1)}(0)}, \qquad 1 \le k \le K, \tag{8.3a}$$

$$\tilde{x}_{2K-2k+1} = \ln \frac{\tilde{\mathbf{e}}_{[1,2k-1]}\tilde{\mathcal{D}}_{2k}^{(k,0)}(0)\,\tilde{\mathcal{D}}_{2k-1}^{(k,0)}(0)}{\tilde{m}_{2K-2k+1}\tilde{\mathcal{D}}_{2k}^{k,1)}(0)\,\tilde{\mathcal{D}}_{2k-1}^{(k-1,1)}(0)}, \qquad 1 \le k \le K. \tag{8.3b}$$

Comparing the formulae (5.18a)-(5.18b) in the even case, these are nothing but the formulae for the positions $\tilde{x}_j$ corresponding the multipeakon ansatz:

$$u = \sum_{j=1}^{2K} \tilde{m}_j(t)e^{-|x-\tilde{x}_j(t)|}, \qquad v = \sum_{j=1}^{2K+1} \tilde{n}_j(t)e^{-|x-\tilde{x}_j(t)|}.$$

So far, we have shown how the formulae for the odd case reduce to those for the even case. It is not hard to see that a comparison between Examples 2 and 5 supports our claim above.

8.2 From the even case to the interlacing case

Let us consider the multipeakon ansatz with the even number of masses:

$$u = \sum_{j=1}^{2K} m_j(t)e^{-|x-x_j(t)|}, \qquad v = \sum_{j=1}^{2K} n_j(t)e^{-|x-x_j(t)|}.$$

Then the reduction $m_{2j} \to 0$, $n_{2j-1} \to 0$ gives the interlacing ansatz

$$u = \sum_{j=1}^{K} m_{2j-1}(t)e^{-|x-x_{2j-1}(t)|}, \qquad v = \sum_{j=1}^{K} n_{2j}(t)e^{-|x-x_{2j}(t)|},$$

which was considered in [4].

Note that this reduction does not change the form of the Weyl function, that is, the Weyl function still possesses the form:

$$W(z) = \int \frac{d\mu(x)}{x-z}, \quad d\mu = \sum_{i=1}^{K} b_j\delta_{\zeta_j}, \quad 0 < \zeta_1 < \cdots < \zeta_K, \quad 0 < b_j, \quad 1 \le j \le K.$$

Let us consider the multipeakon formulae (5.18a)-(5.18b) for the case of the even number of masses. It is not hard to verify that

$$(\mathbf{e}_{[1,k]})^l \frac{\mathcal{D}_k^{(l,p)}}{|\Delta_{[1,k]}(\mathbf{e})|} \to \sum_{J \in \binom{[1,K]}{l}} b_J \zeta_J^p \Delta_J^2 \triangleq H_l^p, \qquad \text{as} \qquad m_{2j},\ n_{2j-1} \to 0,$$

which leads to

$$x_{2K+1-2k} \to \ln\left(\frac{1}{m_{2K+1-2k}} \cdot \frac{(H_k^0)^2}{H_k^1 H_{k-1}^1}\right), \quad 1 \le k \le K,$$

$$x_{2K+2-2k} \to \ln\left(n_{2K+2-2k} \cdot \frac{H_k^0 H_{k-1}^0}{(H_{k-1}^1)^2}\right), \quad 1 \le k \le K.$$

These formulae are identical to those obtained in [4] for the interlacing multipeakons of 2-mCH equation. It is useful to compare examples 2 and 3 in the present paper and examples 5.4, 5.5 in [4] to see a concrete manifestation of the transition from a general configuration to an interlacing one. It is also helpful to observe that the inverse problem for the interlacing case was solved using diagonal Padé approximations at ∞ while in the present paper we are formulating our interpolation problem at e_j (see (5.14) all of which get moved to ∞ in the limit $m_{2j} \to 0, n_{2j-1} \to 0$.

8.3 From 2-mCH to 1-mCH

As is known, when $v = u$, the 2-mCH (1.7) reduces to the mCH (1.5). As for the formulae (5.18a)-(5.18b) in Theorem 10, the identical mass setting

$$n_j = m_j,$$

leads to

$$v = u = \sum_{j=1}^{N} m_j(t) e^{-|x - x_j(t)|}, \qquad n = m = 2\sum_{j=1}^{N} m_j \delta_{x_j}.$$

In this case we have

$$e_j = \frac{1}{m_{j'}^2}$$

and the formulae (5.18a)-(5.18b) in Theorem 10 reduce to the multipeakon formulae for the mCH equation (see [6, Theorem 4.21]). Moreover, it is not hard to see that the global existence condition and the long time asymptotics in Section 6 and 7 also cover those for the mCH equation in [6].

Acknowledgments

The first author was supported in part by the National Natural Science Foundation of China (#11688101, 11731014, 11701550) and the Youth Innovation Promotion Association CAS. The second author was supported in part by NSERC #163953.

A Lax pair for the 2-mCH peakon ODEs

The purpose of this appendix is to give a rigorous interpretation for the distributional Lax pair of the 2-mCH equation in the generic case, i.e. the Lax pair of the 2-mCH peakon ODEs (1.10). We note that the transition from the smooth sector to the distribution sector is not canonical. The argument presented below is closer to that in Appendix A of [6] than to the one used for the interlacing case of the 2-mCH [4]. Let us begin by reviewing some notation needed to present the argument.

Let Ω_k denote the region $x_k(t) < x < x_{k+1}(t)$, where x_k are smooth functions such that $-\infty = x_0(t) < x_1(t) < \cdots < x_N(t) < x_{N+1}(t) = +\infty$.

Let the function space PC^∞ consist of all piecewise smooth functions $f(x,t)$ such that the restriction of f to each region Ω_k is a smooth function $f_k(x,t)$ defined on Ω_k. Then, for each fixed t, $f(x,t)$ defines a regular distribution $T_f(t)$ [18] in the class $\mathcal{D}'(R)$ (for simplicity we will write f instead of T_f). Since distributions do not in general have values at individual points, we do not require $f(x_k)$ to be defined. However, the left and right limits are defined. Let us denote them $f_x(x_k-,t)$ and $f_x(x_k+,t)$ and set

$$[f](x_k) = f(x_k+,t) - f(x_k-,t), \qquad \langle f\rangle(x_k) = \frac{f(x_k+,t)+f(x_k-,t)}{2},$$

to denote the jump and the average, respectively. Denote by f_x (or f_t) the ordinary (classical) partial derivative with respect to x (or t), and by $\frac{\partial f_k}{\partial x}$ (or $\frac{\partial f_k}{\partial t}$) their restrictions to Ω_k. Let $D_x f$ denote the distributional derivative with respect to x. Then we have a well-known identity

$$D_x f = f_x + \sum_{k=1}^{N}[f](x_k)\delta_{x_k}.$$

Likewise, we can define $D_t f$ as a distributional limit

$$D_t f(t) = \lim_{a\to 0}\frac{f(t+a)-f(t)}{a},$$

which can be computed explicitly to be

$$D_t f = f_t - \sum_{k=1}^{N}\dot{x}_k[f(x_k)]\delta_{x_k},$$

where $\dot{x}_k = dx_k/dt$.

The following formulas will be useful in further analysis:

$$[fg] = \langle f\rangle[g] + [f]\langle g\rangle, \qquad \langle fg\rangle = \langle f\rangle\langle g\rangle + \frac{1}{4}[f][g] \tag{A.1}$$

$$\frac{d}{dt}[f](x_k) = [f_x](x_k)\dot{x}_k + [f_t](x_k), \tag{A.2}$$

$$\frac{d}{dt}\langle f\rangle(x_k)=\langle f_x\rangle(x_k)\dot{x}_k+\langle f_t\rangle(x_k), \tag{A.3}$$

for any $f, g \in PC^\infty$.

It is easy to see that the peakon solution $u(x,t), v(x,t)$ and the corresponding functions $\Psi_1,\ \Psi_2$ in the Lax pair (2.1) are piecewise smooth functions of class PC^∞. Moreover, u, v are continuous throughout $\mathbf{R}$ but u_x, v_x, Ψ_1, Ψ_2 have a jump at each x_k.

Let us now set $\Psi=(\Psi_1,\Psi_2)^T$, and consider an overdetermined system (see (2.1))

$$D_x\Psi=\frac{1}{2}\hat{L}\Psi, \qquad D_t\Psi=\frac{1}{2}\hat{A}\Psi, \tag{A.4}$$

where

$$\hat{L}=L+2\lambda\left(\sum_{k=1}^{N}M_k\delta_{x_k}\right), \tag{A.5}$$

$$\hat{A}=A-2\lambda\left(\sum_{k=1}^{N}M_kQ(x_k)\delta_{x_k}\right) \tag{A.6}$$

with

$$L=\begin{pmatrix} -1 & 0\\ 0 & 1\end{pmatrix}, \qquad M_k=\begin{pmatrix} 0 & m_k\\ -n_k & 0\end{pmatrix},$$

$$A=\begin{pmatrix} 4\lambda^{-2}+Q & -2\lambda^{-1}(u-u_x)\\ 2\lambda^{-1}(v+v_x) & -Q\end{pmatrix}$$

and $Q=(u-u_x)(v+v_x)$. We observe that the above splitting of the Lax pair corresponds to the distributional splitting into a regular distribution and a distribution with a singular support. In particular the singular part of ((A.5)) requires that we multiply $M_k\Psi=(m_k\Psi_2,-n_k\Psi_1)$ by δ_{x_k}. This is not defined, and to define this multiplication we need to assign some values to Ψ_1, Ψ_2 at x_k. This, in principle, is an arbitrary procedure but we want it to reflect the local behaviour of Ψ around x_k and postulate that our choice of values of Ψ depends linearly on the right and left hand limits.

Likewise, for the t-Lax equation (A.6) to be defined as a distribution equation, $QM_k\Psi=(Qm_k\Psi_2,-Qn_k\Psi_1)$ needs to be a multiplier of δ_{x_k}. Thus, in the same sense as above, the values of $Q(x_k)$ need to be assigned as well.

Henceforth, we will refer to these assignments as *regularizations*, even though this name in the theory of distributions refers usually to re-defining divergent integrals.

We now specify more concretely a family of regularizations we consider. We will refer to them as *invariant regularizations*. The reason for that name is explained in [6, Appendix A].

Definition 4. An invariant regularization of the Lax pair (A.4) consists of specifying the values of $\alpha, \beta \in \mathbf{R}$ and $Q(x_k) = (u - u_x)(v + v_x)(x_k)$ in the formulas

$$
\begin{aligned}
\Psi(x)\delta_{x_k} &= \Psi(x_k)\delta_{x_k},\\
\Psi(x_k) &= \alpha\big[\Psi\big](x_k) + \beta\big\langle\Psi\big\rangle(x_k),\\
Q(x)\delta_{x_k} &= Q(x_k)\delta_{x_k}.
\end{aligned}
$$

Theorem 19. *Let m and n be the discrete measures associated to u and v defined by (1.8). Given an invariant regularization in the sense of Definition 4 the distributional Lax pair (A.4) is compatible, i.e. $D_tD_x\Psi = D_xD_t\Psi$, if and only if the following conditions hold:*

$$
\beta = 1, \qquad \alpha = \pm\tfrac{1}{2}. \tag{A.7}
$$

Then

$$
\begin{aligned}
\dot m_k &= -m_k(Q(x_k) - \big\langle Q\big\rangle(x_k)), &\text{(A.8a)}\\
\dot n_k &= n_k(Q(x_k) - \big\langle Q\big\rangle(x_k)), &\text{(A.8b)}\\
\dot x_k &= Q(x_k). &\text{(A.8c)}
\end{aligned}
$$

Proof. The proof follows almost *verbatim* [6, Theorem A.2] (also see [4, 10]). We omit the details. ■

Now we can spell out the connection of the regularization problem to the content of the present paper.

Corollary 5. Let m and n be the discrete measures associated to u and v defined by (1.8). Suppose that $Q(x_k), \Psi(x_k)$ are assigned values

$$
Q(x_k) = \big\langle Q\big\rangle(x_k), \quad \Psi(x_k) = \Psi(x_k+),
$$

or

$$
Q(x_k) = \big\langle Q\big\rangle(x_k), \quad \Psi(x_k) = \Psi(x_k-).
$$

For either case, the compatibility condition $D_tD_x\Psi = D_xD_t\Psi$ for the distributional Lax pair (A.4) reads

$$
\dot m_k = 0, \quad \dot n_k = 0, \quad \dot x_k = \big\langle Q\big\rangle(x_k).
$$

B Proof of Theorem 13

Proof. It suffices to ensure that the ordering conditions $x_1 < x_2 < \cdots < x_{2K}$ hold for all time t. We write these conditions as:

$$
\begin{aligned}
&x_{(k+1)'} < x_{k'}, &&\text{for all odd } k, &&1 \le k \le 2K-1, &\text{(B.1a)}\\
&x_{(k+2)'} < x_{(k+1)'}, &&\text{for all odd } k, &&1 \le k \le 2K-3, &\text{(B.1b)}
\end{aligned}
$$

and use equations (5.18a), (5.18b) to obtain equivalent conditions

$$\frac{1}{m_{(k+1)'}n_{k'}} < \frac{\mathcal{D}_{k+1}^{(\frac{k+1}{2},1)}\mathcal{D}_{k-1}^{(\frac{k-1}{2},0)}}{\mathcal{D}_{k+1}^{(\frac{k+1}{2},0)}\mathcal{D}_{k-1}^{(\frac{k-1}{2},1)}}, \quad \text{for all odd } k,\ 1 \le k \le 2K-1, \tag{B.2a}$$

$$\frac{1}{m_{(k+2)'}n_{(k+1)'}} < \frac{\mathcal{D}_{k+2}^{(\frac{k+1}{2},1)}\mathcal{D}_{k}^{(\frac{k+1}{2},0)}}{\mathcal{D}_{k+2}^{(\frac{k+3}{2},0)}\mathcal{D}_{k}^{(\frac{k-1}{2},1)}}, \quad \text{for all odd } k,\ 1 \le k \le 2K-3. \tag{B.2b}$$

However, (6.3) implies that the inequality

$$\frac{\mathcal{D}_{k+1}^{(\frac{k+1}{2},1)}\mathcal{D}_{k-1}^{(\frac{k-1}{2},0)}}{\mathcal{D}_{k+1}^{(\frac{k+1}{2},0)}\mathcal{D}_{k-1}^{(\frac{k-1}{2},1)}} > \frac{\zeta_1^{\frac{k+1}{2}}}{\zeta_K^{\frac{k-1}{2}}} \tag{B.3}$$

holds uniformly in t and if we impose

$$\frac{1}{m_{(k+1)'}n_{k'}} < \frac{\zeta_1^{\frac{k+1}{2}}}{\zeta_K^{\frac{k-1}{2}}} \text{ for all odd } k, \qquad k \le 2K-1,$$

then (B.1a) holds.

Now we focus on the second inequality, namely (B.2b), which is valid whenever $K \ge 2$. It is convenient to consider a slightly more general expression, namely,

$$\frac{\mathcal{D}_{k+2}^{(l,1)}\mathcal{D}_{k}^{(l,0)}}{\mathcal{D}_{k+2}^{(l+1,0)}\mathcal{D}_{k}^{(l-1,1)}}, \qquad 1 \le l \le K-1,$$

for which, using similar steps to those to in [6], we can prove the bound

$$\begin{aligned}\frac{\mathcal{D}_{k+2}^{(l,1)}\mathcal{D}_{k}^{(l,0)}}{\mathcal{D}_{k+2}^{(l+1,0)}\mathcal{D}_{k}^{(l-1,1)}} &> \frac{\zeta_1^l}{\zeta_K^{l-1}}\frac{\min_j(\zeta_{j+1}-\zeta_j)^{2(l-1)}}{l(\zeta_K-\zeta_1)^{2l}}\cdot(e_{k+1}+\zeta_1)(e_{k+2}+\zeta_1)\\ &= \frac{\zeta_1^l}{\zeta_K^{l-1}}\frac{\min_j(\zeta_{j+1}-\zeta_j)^{2(l-1)}}{l(\zeta_K-\zeta_1)^{2l}}\cdot\frac{(1+m_{(k+1)'}n_{(k+1)'}\zeta_1)(1+m_{(k+2)'}n_{(k+2)'}\zeta_1)}{m_{(k+1)'}n_{(k+1)'}m_{(k+2)'}n_{(k+2)'}}.\end{aligned} \tag{B.4}$$

Hence, specializing to $l = \frac{k+1}{2}$ and assuming

$$\frac{1}{m_{(k+2)'}n_{(k+1)'}} < \frac{\zeta_1^{\frac{k+1}{2}}}{\zeta_K^{\frac{k-1}{2}}}\frac{2\min_j(\zeta_{j+1}-\zeta_j)^{k-1}}{(k+1)(\zeta_K-\zeta_1)^{k+1}}\cdot\frac{(1+m_{(k+1)'}n_{(k+1)'}\zeta_1)(1+m_{(k+2)'}n_{(k+2)'}\zeta_1)}{m_{(k+1)'}n_{(k+1)'}m_{(k+2)'}n_{(k+2)'}},$$

we obtain that (B.2b) and thus (B.1b) hold.

Finally, after rewriting the last condition as:

$$\frac{n_{(k+2)'}m_{(k+1)'}}{(1+m_{(k+1)'}n_{(k+1)'}\zeta_1)(1+m_{(k+2)'}n_{(k+2)'}\zeta_1)} < \frac{\zeta_1^{\frac{k+1}{2}}}{\zeta_K^{\frac{k-1}{2}}}\frac{2\min_j(\zeta_{j+1}-\zeta_j)^{k-1}}{(k+1)(\zeta_K-\zeta_1)^{k+1}},$$
$$\text{for all odd } k,\ k \le 2K-3$$

we obtain the second condition (6.6b). ■

C Proof of Theorem 17

Proof. Again, it suffices to ensure that the ordering conditions $x_1 < x_2 < \cdots < x_{2K+1}$ hold for all time t. We write these conditions in an equivalent form:

$$x_{(k+1)'} < x_{k'}, \qquad \text{for all odd } k, \qquad 1 \le k \le 2K-1, \tag{C.1a}$$

$$x_{(k+2)'} < x_{(k+1)'}, \qquad \text{for all odd } k, \qquad 1 \le k \le 2K-1, \tag{C.1b}$$

and use equations (5.18a), (5.18b) to obtain

$$\frac{1}{m_{(k+1)'}n_{k'}} < \frac{\mathcal{D}_{k+1}^{(\frac{k+1}{2},1)}(c)\mathcal{D}_{k-1}^{(\frac{k-1}{2},0)}(c)}{\mathcal{D}_{k+1}^{(\frac{k+1}{2},0)}(c)\mathcal{D}_{k-1}^{(\frac{k-1}{2},1)}(c)}, \text{ for all odd } k, 1 \le k \le 2K-1, \tag{C.2a}$$

$$\frac{1}{m_{(k+2)'}n_{(k+1)'}} < \frac{\mathcal{D}_{k+2}^{(\frac{k+1}{2},1)}(c)\mathcal{D}_{k}^{(\frac{k+1}{2},0)}(c)}{\mathcal{D}_{k+2}^{(\frac{k+3}{2},0)}(c)\mathcal{D}_{k}^{(\frac{k-1}{2},1)}(c)}, \text{ for all odd } k, 1 \le k \le 2K-1. \tag{C.2b}$$

From (7.6a) and (7.6b) we obtain

$$\begin{aligned}
\frac{\mathcal{D}_{k+1}^{(\frac{k+1}{2},1)}(c)\mathcal{D}_{k-1}^{(\frac{k-1}{2},0)}(c)}{\mathcal{D}_{k+1}^{(\frac{k+1}{2},0)}(c)\mathcal{D}_{k-1}^{(\frac{k-1}{2},1)}(c)} &= \frac{\big(\mathcal{D}_{k+1}^{(\frac{k+1}{2},1)}(0) + c\mathcal{D}_{k+1}^{(\frac{k-1}{2},1)}(0)\big)\mathcal{D}_{k-1}^{(\frac{k-1}{2},0)}(0)}{\mathcal{D}_{k+1}^{(\frac{k+1}{2},0)}(0)\big(\mathcal{D}_{k-1}^{(\frac{k-1}{2},1)}(0) + c\mathcal{D}_{k-1}^{(\frac{k-3}{2},1)}(0)\big)} \\
&= \frac{\mathcal{D}_{k+1}^{(\frac{k+1}{2},1)}(0)\mathcal{D}_{k-1}^{(\frac{k-1}{2},0)}(0) + c\mathcal{D}_{k+1}^{(\frac{k-1}{2},1)}(0)\mathcal{D}_{k-1}^{(\frac{k-1}{2},0)}(0)}{\mathcal{D}_{k+1}^{(\frac{k+1}{2},0)}(0)\mathcal{D}_{k-1}^{(\frac{k-1}{2},1)}(0) + c\mathcal{D}_{k+1}^{(\frac{k+1}{2},0)}(0)\mathcal{D}_{k-1}^{(\frac{k-3}{2},1)}(0)} \stackrel{def}{=} \frac{\mathcal{A}_1 + \mathcal{B}_1}{\mathcal{A}_2 + \mathcal{B}_2},
\end{aligned}$$

where $\mathcal{B}_2 = 0$ for $k = 1$. Hence the ratios $\frac{\mathcal{A}_1}{\mathcal{A}_2}, \frac{\mathcal{B}_1}{\mathcal{B}_2}$ satisfy (uniform in t) bounds

$$\frac{\mathcal{A}_1}{\mathcal{A}_2} > \frac{\zeta_1^{\frac{k+1}{2}}}{\zeta_K^{\frac{k-1}{2}}} \stackrel{def}{=} \alpha,$$

$$\frac{\mathcal{B}_1}{\mathcal{B}_2} > \frac{\zeta_1^{\frac{k-1}{2}}}{\zeta_K^{\frac{k-3}{2}}} \frac{2\min_j(\zeta_{j+1}-\zeta_j)^{k-3}}{(k-1)(\zeta_K-\zeta_1)^{k-1}} \frac{(1+m_{k'}n_{k'}\zeta_1)(1+m_{(k+1)'}n_{(k+1)'}\zeta_1)}{m_{k'}n_{k'}m_{(k+1)'}n_{(k+1)'}} \stackrel{def}{=} \beta,$$

by equations (B.3) and (B.4), respectively, with the convention that $\beta = \infty$ for the special case $k = 1$. Thus

$$\min\{\alpha, \beta\} < \frac{\mathcal{D}_{k+1}^{(\frac{k+1}{2},1)}(c)\mathcal{D}_{k-1}^{(\frac{k-1}{2},0)}(c)}{\mathcal{D}_{k+1}^{(\frac{k+1}{2},0)}(c)\mathcal{D}_{k-1}^{(\frac{k-1}{2},1)}(c)}$$

holds uniformly in t and if we impose

$$\frac{1}{m_{(k+1)'}n_{k'}} < \min\{\alpha, \beta\} \qquad \text{for all odd } k, \qquad 1 \le k \le 2K-1,$$

then equations (C.1a) will hold automatically.

Now we turn to the second inequality (C.2b). From Corollary 4 we obtain

$$\frac{\mathcal{D}_{k+2}^{(\frac{k+1}{2},1)}(c)\mathcal{D}_{k}^{(\frac{k+1}{2},0)}(c)}{\mathcal{D}_{k+2}^{(\frac{k+3}{2},0)}(c)\mathcal{D}_{k}^{(\frac{k-1}{2},1)}(c)} > \frac{\mathcal{A}_1+\mathcal{B}_1}{\mathcal{A}_2+\mathcal{B}_2}.$$

Since

$$\frac{\mathcal{A}_1}{\mathcal{A}_2} > \frac{\zeta_1^{\frac{k+1}{2}}}{\zeta_K^{\frac{k-1}{2}}} = \alpha,$$

$$\begin{aligned}\frac{\mathcal{B}_1}{\mathcal{B}_2} &> \frac{\zeta_1^{\frac{k+1}{2}}}{\zeta_K^{\frac{k-1}{2}}}\frac{2\min_j(\zeta_{j+1}-\zeta_j)^{k-1}}{(k+1)(\zeta_K-\zeta_1)^{k+1}}\frac{(1+m_{(k+1)'}n_{(k+1)'}\zeta_1)(1+m_{(k+2)'}n_{(k+2)'}\zeta_1)}{m_{(k+1)'}n_{(k+1)'}m_{(k+2)'}n_{(k+2)'}}\\ &\overset{def}{=} \beta_1,\end{aligned}$$

then

$$\min\{\alpha,\beta_1\} < \frac{\mathcal{D}_{k+2}^{(\frac{k+1}{2},1)}(c)\mathcal{D}_{k}^{(\frac{k+1}{2},0)}(c)}{\mathcal{D}_{k+2}^{(\frac{k+3}{2},0)}(c)\mathcal{D}_{k}^{(\frac{k-1}{2},1)}(c)}$$

is satisfied. Thus inequality

$$\frac{1}{m_{(k+2)'}n_{(k+1)'}} < \min\{\alpha,\beta_1\}, \qquad \text{for all odd } k, \qquad 1 \le k \le 2K-1,$$

implies (C.2b) and, consequently, (C.1b), thereby proving the claim. ■

References

[1] Aki K and Richards P G, *Quantitative Seismology*, University Science Books, Sausalito, California, second edition, 2002.

[2] Birkhoff G and Rota G C, *Ordinary differential equations*, John Wiley & Sons, Inc., New York, fourth edition, 1989.

[3] Camassa R and Holm D, An integrable shallow water equation with peaked solitons, *Phys. Rev. Lett.*, **71**, 1661–1664, 1993.

[4] Chang X, Hu X, and Szmigielski J, Multipeakons of a two-component modified Camassa-Holm equation and the relation with the finite Kac-van Moerbeke lattice, *Adv. Math.*, **299**,1–35, 2016.

[5] Chang X and Szmigielski J, Liouville integrability of conservative peakons for a modified CH equation, *J. Nonlinear Math. Phys.*, **24**, 584–595, 2017.

[6] Chang X and Szmigielski J, Lax integrability and the peakon problem for the modified Camassa-Holm equation, *Comm. Math. Phys.*, **358**, 295–341, 2018.

[7] Fokas A, The Korteweg-de Vries equation and beyond, *Acta Appl. Math.*, **39**, 295–305, 1995.

[8] Fuchssteiner B, Some tricks from the symmetry-toolbox for nonlinear equations: generalizations of the Camassa-Holm equation, *Physica D: Nonlinear Phenomena*, **95**, 229–243, 1996.

[9] Gasca M, Martinez J, and Mühlbach G, Computation of rational interpolants with prescribed poles, *J. Comput. Appl. Math.*, **26**, 297–309, 1989.

[10] Hone A N W, Lundmark H, and Szmigielski J, Explicit multipeakon solutions of Novikov's cubically nonlinear integrable Camassa-Holm type equation, *Dyn. Partial Differ. Equ.*, **6**, 253–289, 2009.

[11] Lundmark H and Szmigielski J, An inverse spectral problem related to the Geng-Xue two-component peakon equation, *Mem. Amer. Math. Soc.*, **244**, no. 1155, vii+87, 2016.

[12] Martínez J and Peña J, Factorizations of Cauchy-Vandermonde matrices, *Linear Algebra Appl.*, **284**, 229–237, 1998.

[13] Martínez J and Peña J, Fast algorithms of Björck-Pereyra type for solving Cauchy-Vandermonde linear systems, *Appl. Numer. Math.*, **26**, 343–352, 1998.

[14] Moser J, Three integrable systems connected with isospectral deformations, *Advances in Mathematics*, **16**, 197–220, 1975.

[15] Olver P and Rosenau P, Tri-Hamiltonian duality between solitons and solitary-wave solutions having compact support, *Phys. Rev. E*, **53**, 1900, 1996.

[16] Qiao Z, A new integrable equation with cuspons and W/M-shape-peaks solitons, *J. Math. Phys.*, **47**, 112701–112900, 2006.

[17] Qiao Z, New integrable hierarchy, its parametric solutions, cuspons, one-peak solitons, and M/W-shape peak solitons, *J. Math. Phys.*, **48**, 082701, 2007.

[18] Schwartz L, *Théorie des distributions. Tome I*, Actualités Sci. Ind., no. 1091 = Publ. Inst. Math. Univ. Strasbourg 9. Hermann & Cie., Paris, 1950.

[19] Song J, Qu C, and Qiao Z, A new integrable two-component system with cubic nonlinearity, *J. Math. Phys.*, **52**, 013503, 2011.

[20] Stieltjes T J, *Œuvres complètes/Collected papers. Vol. I, II*, Springer-Verlag, Berlin, 1993.

[21] Tian K and Liu Q, Tri-Hamiltonian duality between the Wadati-Konno-Ichikawa hierarchy and the Song-Qu-Qiao hierarchy, *J. Math. Phys.*, **54**, 043513, 2013.

C3. Exactly solvable (discrete) quantum mechanics and new orthogonal polynomials

Ryu Sasaki

Faculty of Science, Shinshu University,
Matsumoto 390-8621, Japan

Abstract

Orthogonal polynomials satisfying second order differential or difference equations as well as three term recurrence relations are called *classical* orthogonal polynomials. They are 40+ (q-)hypergeometric polynomials catalogued in the Askey scheme. New types of *infinitely many* orthogonal polynomials *not satisfying* three term recurrence relations are obtained as the main parts of the complete sets of eigenfunctions of exactly solvable one-dimensional quantum mechanics and their difference versions. Their special feature is the 'holes' in the degrees, which allow us to evade the constraints by Bochner's theorem. These new polynomials are rational deformations of the classical orthogonal polynomials generated by multiple Darboux transformations. They are called *multi indexed* orthogonal polynomials in which the degrees of the seed polynomials constitute the multi index. The discrete symmetries of the original solvable quantum mechanical systems provide the seed polynomials from the original eigenpolynomials.

1 Introduction

It is well known as Routh-Bochner Theorem [40, 5] that the orthogonal polynomials satisfying three term recurrence relations and *second order differential equations* are quite limited. There are four of them, the Hermite, Laguerre, Jacobi and Bessel polynomials, which constitute the minority of the *classical orthogonal polynomials*. The other, about 40, classical orthogonal polynomials consisting of the Akey-Wilson, (q-)Racah polynomials, etc. satisfy *second order difference equations* and they are catalogued by the Askey scheme and its q-version [2, 15, 17].

It is also well known that the Hermite, Laguerre and Jacobi polynomials constitute the main parts of the eigenfunctions of the exactly solvable Quantum Mechanics (QM) in one dimension [14, 6]. The eigenvalue problem of self-adjoint operators is the natural setting of the orthogonality of these polynomials and their orthogonality weight are simply the square of the *ground state eigenfunctions*. This reasoning has led to quantum mechanical reformulation of all the classical orthogonal polynomials in the Askey scheme for which the Schrödinger equations are second order difference equations with pure imaginary or real shifts [24, 38, 25, 33].

The reformulation has turned out quite fruitful, providing universal expressions for various formulas, *e.g.* the Rodrigues formulas as well as the Heisenberg operator solutions, the creation/annihilation operators and coherent states, etc. Many new properties and formulas have been uncovered for the orthogonal polynomials of a *discrete variable* [23], *i.e.* those belonging to discrete QM with real shifts, *e.g.* the (dual) (q-)Hahn, (q-)Racah polynomials etc. [24, 38]. For this group, the corresponding quantum mechanical setting is simply the eigenvalue problem of *tri-diagonal real symmetric* (Jacobi) matrices of finite or infinite dimensions [24, 38]. The *duality* [21, 43, 44] is recognised as a universal property of the polynomials of this group and the *dual polynomial* [3] is identified for each polynomial [24, 38] except for those having the Jackson integral measures. All the polynomials belonging to this group are shown to provide *exactly solvable Birth and Death processes* [41], which are well-known stochastic (Markov) processes [16, 15]. An even more interesting finding is that the Jackson integral measures arise naturally through the *self-adjoint extension processes* of certain infinite matrix formulations, including the big q-Jacobi (Laguerre), Al-Salam-Carlitz I, discrete q-Hermite polynomials, [24, 38]. It is rather remarkable that such interesting and explicit examples of self-adjoint extension have never been mentioned in most standard textbooks or monographs (Stone, Kolmogorov, Dunford-Schwartz, Yoshida, Simon) of Functional Analysis.

The quantum mechanical formulation of orthogonal polynomials has provided not only the *unified theory classical orthogonal polynomials* but also opened new dimensions for *non-classical orthogonal polynomials*. They satisfy second order differential or difference equations but not three term recurrence relations. This means that there are '*holes*' in their degrees but they still form complete sets in proper Hilbert spaces. The constraints due to Routh-Bochner [40, 5] do not apply due to the holes. In a sense they are deformations of the classical orthogonal polynomials. They are obtained from the eigenfunctions (vectors) of the original systems through Darboux-Crum transformations [8, 7, 26, 31] by using certain seed solutions. Depending on the nature of the seed solutions, there are two types of non-classical orthogonal polynomials. When a part of the eigenfunctions (vectors) is used, the obtained polynomials are called *Krein-Adler polynomials* [20],[1]. By applying certain discrete transformations of the original systems to the eigen functions (vectors), seed solutions called *virtual state solutions (vectors)* are obtained. The non-classical orthogonal polynomials obtained by using virtual state solutions are called *multi-indexed orthogonal polynomials* [11, 4, 39, 27, 34]. Of course, using both eigenfunctions and virtual state solutions as seed solutions for multiple Darboux-Crum transformations is possible and it leads to mixed type non-classical orthogonal polynomials.

Another type of non-classical orthogonal polynomials, called *Krall type polynomials* [18, 19], have been known for many years. They satisfy fourth order or higher differential or difference equations and three term recurrence relations. They will not be discussed in this chapter.

To provide a concise and comprehensive review of the non-classical orthogonal polynomials is the main aim of this chapter. We emphasise the strategy and main results and omit details. The contents of this chapter are as follows. In section 2, the new orthogonal polynomials satisfying second order differential equations are surveyed. Starting from QM formulation of the Hermite, Laguerre and Jacobi polynomials, their discrete symmetries and the virtual state solutions are explicitly presented. Based on the Darboux-Crum transformations, the multi-indexed

Laguerre and Jacobi polynomials are derived. The general forms of the Krein-Adler polynomials for the Hermite, Laguerre and Jacobi are presented. Explicit examples of the multi-indexed Laguerre and Jacobi polynomials are displayed. In section 3, the non-classical orthogonal polynomials satisfying second order difference equations with *real shifts* are presented. Starting from the QM formulation of the classical ones *i.e.* the Racah and q-Racah systems, which are simply the eigenvalue problems of certain tri-diagonal real symmetric matrices, the Hamiltonians, eigenvectors, etc. and their discrete symmetry transformations and the corresponding virtual state vectors are presented. The discrete QM analogue of the Darboux transformation is introduced by using one seed 'solution'. General construction of multi-indexed (q-)Racah polynomials through multiple Darboux transformations is presented. Explicit examples of multi-indexed (q-)Racah polynomials are exhibited. The final section is for a summary and comments.

2 New orthogonal polynomials in ordinary QM

For the construction of the non-classical orthogonal polynomials, the fundamental knowledge of the corresponding classical counterparts is essential; the complete sets of the eigenvalues, eigenfunctions, discrete symmetries and forms of Darboux transformations, etc. The passage from the classical to non-classical orthogonal polynomials is basically the same for the ordinary and discrete QM. Here we show the derivation and some explicit examples of the non-classical orthogonal polynomials for the familiar examples from the ordinary QM, *i.e.* the Hermite, Laguerre and Jacobi.

2.1 Classical polynomials: the Hermite, Laguerre and Jacobi

We begin with the summary of fundamental properties of the classical orthogonal polynomials in the one-dimensional quantum mechanical formulation. The eigenvalue problem is defined in a finite or (semi-)infinite interval $x_1 < x < x_2$,

$$\mathcal{H}\phi_n(x) = \mathcal{E}(n)\phi_n(x), \quad (n \in \mathbb{Z}_{\geq 0}), \quad (\phi_n, \phi_m) \stackrel{\text{def}}{=} \int_{x_1}^{x_2} \phi_n^*(x)\phi_m(x)dx = h_n\delta_{n\,m}, \quad (2.1)$$

with the Hamiltonian or the Schrödinger operator

$$\mathcal{H} = -\frac{d^2}{dx^2} + V(x), \quad V(x) \in \mathbb{R}, \quad V(x) \in C^{\infty}. \quad (2.2)$$

Throughout this section we consider the real eigenfunctions and wavefunctions, $\phi_n^*(x) = \phi_n(x)$. The properties of the Hermite (H), Laguerre (L) and Jacobi (J) systems are summarised as follows. The first is the Hermite system:

$$\text{H:} \qquad V(x) = x^2 - 1, \quad -\infty < x < \infty, \quad \mathcal{E}(n) = 2n \quad (n \in \mathbb{Z}_{\geq 0}), \quad (2.3)$$

$$\phi_n(x) = \phi_0(x)H_n(x), \quad \phi_0(x) = e^{-x^2/2}, \quad H_n(x) : \text{Hermite poly.} \quad (2.4)$$

The square of the ground state eigenfunction $\phi_0^2(x) = e^{-x^2}$ provides the orthogonality weight function for the Hermite polynomials:

$$((H_n, H_m)) = \int_{-\infty}^{\infty} e^{-x^2} H_n(x) H_m(x) dx = h_n \delta_{n\,m}, \quad h_n = 2^n n! \sqrt{\pi}. \tag{2.5}$$

The Hamiltonian without the constant term $\mathcal{H}' \stackrel{\text{def}}{=} \mathcal{H} + 1$ has a discrete symmetry

$$x \to ix \Rightarrow \mathcal{H}' \to -\mathcal{H}'. \tag{2.6}$$

By acting on the eigenfunctions, it generates square non-integrable seed solutions of *negative energy*

$$\varphi_n(x) \stackrel{\text{def}}{=} i^n \phi_n(ix), \quad \mathcal{H}\varphi_n(x) = \widetilde{\mathcal{E}}_n \varphi_n(x), \quad \widetilde{\mathcal{E}}_n = -2(n+1), \ (n \in \mathbb{Z}_{\geq 0}). \tag{2.7}$$

The second is the Laguerre system with $g > 1/2$:

$$\text{L: } V(x) = x^2 + \frac{g(g-1)}{x^2} - (1+2g), \ 0 < x < \infty, \ \mathcal{E}(n) = 4n \ (n \in \mathbb{Z}_{\geq 0}), \tag{2.8}$$

$$\phi_n(x; g) = \phi_0(x; g) L_n^{(g-1/2)}(x^2), \quad \phi_0(x; g) = e^{-x^2/2} x^g, \tag{2.9}$$

in which $L_n^{(\alpha)}(\eta)$ is the Laguerre polynomial and its weight function is given by $\phi_0^2(x; g)$ with $\alpha \stackrel{\text{def}}{=} g - 1/2$,

$$((L_n^{(\alpha)}, L_m^{(\alpha)})) = \int_0^{\infty} e^{-x^2} x^{2g} L_n^{(\alpha)}(x^2) L_m^{(\alpha)}(x^2) dx = h_n \delta_{n\,m}, \tag{2.10}$$

$$= \frac{1}{2} \int_0^{\infty} e^{-\eta} \eta^{\alpha} L_n^{(\alpha)}(\eta) L_m^{(\alpha)}(\eta) d\eta, \quad h_n = \frac{\Gamma(n+g+1/2)}{2n!}. \tag{2.11}$$

The lower end point $x = 0$ is a regular singular point with the characteristic exponents g and $1-g$. The Hamiltonian without the constant term $\mathcal{H}' \stackrel{\text{def}}{=} \mathcal{H} + 1 + 2g$ has the same discrete symmetry as the Hermite system

$$\text{L1:} \qquad x \to ix \Rightarrow \mathcal{H}' \to -\mathcal{H}',$$

called L1 symmetry transformation. By acting on the eigenfunctions it generates square non-integrable seed solutions of *negative energy*, called type I *virtual state solutions*,

$$\varphi_n^{\mathrm{I}}(x) \stackrel{\text{def}}{=} |\phi_n(ix; g)| = e^{x^2/2} x^g L_n^{(g-1/2)}(-x^2), \tag{2.12}$$

$$\mathcal{H}\varphi_n^{\mathrm{I}}(x) = \widetilde{\mathcal{E}}_n^{\mathrm{I}} \varphi_n^{\mathrm{I}}(x), \quad \widetilde{\mathcal{E}}_n^{\mathrm{I}} = -4(n+g+1/2), \quad (n \in \mathbb{Z}_{\geq 0}). \tag{2.13}$$

It is obvious that $\varphi_n^{\mathrm{I}}(x)$ has no zero in $0 < x < \infty$. It is easy to see that the Hamiltonian $\mathcal{H}' \stackrel{\text{def}}{=} \mathcal{H} + 1 + 2g$ is invariant under the exchange of the characteristic exponents at $x = 0$,

$$\text{L2: } \ g \leftrightarrow 1 - g \Rightarrow \mathcal{H}' \Leftrightarrow \mathcal{H}', \tag{2.14}$$

which is called L2 symmetry transformation. By acting on the lower lying eigenfunctions it generates square non-integrable seed solutions of *negative energy* called type II virtual state solutions,

$$\varphi_n^{\mathrm{II}}(x) \stackrel{\text{def}}{=} \phi_n(x;1-g) = e^{-x^2/2}x^{1-g}L_n^{(1/2-g)}(x^2), \tag{2.15}$$
$$\mathcal{H}\varphi_n^{\mathrm{II}}(x) = \tilde{\mathcal{E}}_n^{\mathrm{II}}\varphi_n^{\mathrm{II}}(x),\ \tilde{\mathcal{E}}_n^{\mathrm{II}} = -4(n-g-1/2),\ n=0,1,\ldots,[g-1/2]'. \tag{2.16}$$

For the above range of n, $\varphi_n^{\mathrm{II}}(x)$ has no zero in $0<x<\infty$, [28]. Here $[a]'$ denotes the greatest integer less than a. By applying type I and II transformations on the eigenfunctions, type III virtual state solutions are obtained:

$$\varphi_n^{\mathrm{III}}(x) \stackrel{\text{def}}{=} |\phi_n(ix;1-g)| = e^{x^2/2}x^{1-g}L_n^{(1/2-g)}(-x^2), \tag{2.17}$$
$$\mathcal{H}\varphi_n^{\mathrm{III}}(x) = \tilde{\mathcal{E}}_n^{\mathrm{III}}\varphi_n^{\mathrm{III}}(x),\ \tilde{\mathcal{E}}_n^{\mathrm{III}} = -4(n+1),\ n\in\mathbb{Z}_{\geq 0}. \tag{2.18}$$

The third is the Jacobi system with $g>1/2$, $h>1/2$:

$$\text{J:}\qquad V(x) = \frac{g(g-1)}{\sin^2 x} + \frac{h(h-1)}{\cos^2 x} - (g+h)^2,\quad 0<x<\pi/2, \tag{2.19}$$
$$\mathcal{E}(n) = 4n(n+g+h)\ (n\in\mathbb{Z}_{\geq 0}), \tag{2.20}$$
$$\phi_n(x;g,h) = \phi_0(x;g,h)P_n^{(\alpha,\beta)}(\cos 2x),\quad \alpha \stackrel{\text{def}}{=} g-1/2,\ \beta \stackrel{\text{def}}{=} h-1/2, \tag{2.21}$$
$$\phi_0(x;g,h) = (\sin x)^g(\cos x)^h, \tag{2.22}$$

in which $P_n^{(\alpha,\beta)}(\eta)$ is the Jacobi polynomial and its weight function is $\phi_0^2(x;g,h)$,

$$((P_n^{(\alpha,\beta)},P_m^{(\alpha,\beta)})) = \int_0^{\pi/2}(\sin x)^{2g}(\cos x)^{2h}P_n^{(\alpha,\beta)}(\cos 2x)P_m^{(\alpha,\beta)}(\cos 2x)dx \tag{2.23}$$
$$= \frac{1}{2^{(2+\alpha+\beta)}}\int_{-1}^{1}(1-\eta)^\alpha(1+\eta)^\beta P_n^{(\alpha,\beta)}(\eta)P_m^{(\alpha,\beta)}(\eta)d\eta \tag{2.24}$$
$$= h_n\delta_{nm},\quad h_n = \frac{\Gamma(n+g+\frac{1}{2})\Gamma(n+h+\frac{1}{2})}{2n!(2n+g+h)\Gamma(n+g+h)}. \tag{2.25}$$

The two end points $x=\pi/2$ and $x=0$ are regular singular points with the characteristic exponents $\{h,1-h\}$ and $\{g,1-g\}$, respectively. The Hamiltonian without the constant term $\mathcal{H}' \stackrel{\text{def}}{=} \mathcal{H}+(g+h)^2$ has two discrete symmetry transformations, exchanging the characteristic exponents at $x=\pi/2$ and $x=0$, respectively:

$$\text{J1: } h\leftrightarrow 1-h,\qquad \text{J2: } g\leftrightarrow 1-g\quad \Longrightarrow \mathcal{H}'\Leftrightarrow\mathcal{H}', \tag{2.26}$$

which are called J1 and J2 symmetry transformations, respectively. By acting on the lower lying eigenfunctions they generate square non-integrable seed solutions of *negative energy* called type I and type II virtual state solutions, respectively:

$$\varphi_n^{\mathrm{I}}(x) \stackrel{\text{def}}{=} \phi_n(x;g,1-h) = (\sin x)^g(\cos x)^{1-h}P_n^{(g-1/2,1/2-h)}(\cos 2x), \quad (2.27)$$

$$\mathcal{H}\varphi_n^{\mathrm{I}}(x) = \widetilde{\mathcal{E}}_n^{\mathrm{I}}\varphi_n^{\mathrm{I}}(x),\ \ \widetilde{\mathcal{E}}_n^{\mathrm{I}} = -4(n+g+1/2)(h-n-1/2),$$
$$n = 0,1,\ldots,[h-1/2]', \quad (2.28)$$

$$\varphi_n^{\mathrm{II}}(x) \stackrel{\text{def}}{=} \phi_n(x;1-g,h) = (\sin x)^{1-g}(\cos x)^h P_n^{(1/2-g,h-1/2)}(\cos 2x), \quad (2.29)$$

$$\mathcal{H}\varphi_n^{\mathrm{II}}(x) = \widetilde{\mathcal{E}}_n^{\mathrm{II}}\varphi_n^{\mathrm{II}}(x),\ \ \widetilde{\mathcal{E}}_n^{\mathrm{II}} = -4(g-n-1/2)(n+h+1/2),$$
$$n = 0,1,\ldots,[g-1/2]'. \quad (2.30)$$

These virtual state solutions have no nodes in $0 < x < \pi/2$ [27, 28]. By applying type I and II transformations on the eigenfunctions, type III virtual state solutions are obtained:

$$\varphi_n^{\mathrm{III}}(x) \stackrel{\text{def}}{=} \phi_n(x;1-g,1-h) = (\sin x)^{1-g}(\cos x)^{1-h}P_n^{(1/2-g,1/2-h)}(\cos 2x),$$
$$\mathcal{H}\varphi_n^{\mathrm{III}}(x) = \widetilde{\mathcal{E}}_n^{\mathrm{III}}\varphi_n^{\mathrm{III}}(x),\ \ \widetilde{\mathcal{E}}_n^{\mathrm{III}} = -4(n+1)(g+h-n-1).$$

In the above three systems, H, L, and J, the constant parts of the potential function $V(x)$ are so chosen as to achieve the *zero ground state energy* $\mathcal{E}(0) = 0$. That is, the Hamiltonians or the Schrödinger operators $\mathcal{H}$ are *positive semi-definite* and they have a factorised expression, reminiscent of the well-known theorem in Linear Algebra:

$$\mathcal{H} = \mathcal{A}^\dagger\mathcal{A}, \quad \mathcal{A} \stackrel{\text{def}}{=} \frac{d}{dx} - \partial_x \log|\phi_0(x)|, \quad \mathcal{A}^\dagger = -\frac{d}{dx} - \partial_x \log|\phi_0(x)|. \quad (2.31)$$

This expression is valid for any one-dimensional ordinary QM system having zero ground state energy $\mathcal{E}(0) = 0$, as the ground state wavefunction $\phi_0(x)$ has no node. By similarity transforming $\mathcal{H}$ in terms of the ground state eigenfunction $\phi_0(x)$, we obtain the second order differential operator $\widetilde{\mathcal{H}}$

$$\widetilde{\mathcal{H}} \stackrel{\text{def}}{=} \phi_0(x)^{-1} \circ \mathcal{H} \circ \phi_0(x) = -\frac{d^2}{dx^2} - 2\partial_x \log|\phi_0(x)| \cdot \frac{d}{dx}, \quad (2.32)$$

which governs the classical orthogonal polynomials H, L and J.
For H, it is $-H_n''(x) + 2xH_n'(x) = 2nH_n(x)$. For L and J they read

$$-\eta\partial_\eta^2 L_n^{(\alpha)}(\eta) - (\alpha+1-\eta)\partial_\eta L_n^{(\alpha)}(\eta) = n\,L_n^{(\alpha)}(\eta), \quad (2.33)$$

$$-(1-\eta^2)\partial_\eta^2 P_n^{(\alpha,\beta)}(\eta) - [\beta-\alpha+(\alpha+\beta+2)\eta]\partial_\eta P_n^{(\alpha,\beta)}(\eta)$$
$$= n(n+\alpha+\beta+1)P_n^{(\alpha,\beta)}(\eta), \quad (2.34)$$

after the change of the independent variables $\eta(x) = x^2$ and $\eta(x) = \cos 2x$, respectively.

2.2 Darboux transformation

Darboux transformations are essential for the derivation of non-classical orthogonal polynomials. In its original form, a Darboux transformation maps one solution $\psi(x)$ of a Schrödinger operator $\mathcal{H}$, to another $\psi^{(1)}(x)$ with the same energy $\mathcal{E}$ of a deformed Schrödinger operator $\mathcal{H}^{(1)}$ in terms of a seed solution $\varphi(x)$. Explicitly the transformation reads

$$\mathcal{H}\psi(x)=\mathcal{E}\psi(x),\quad \mathcal{H}\varphi(x)=\widetilde{\mathcal{E}}\varphi(x),\quad \psi(x),\varphi(x),\mathcal{E},\widetilde{\mathcal{E}}\in\mathbb{C}, \tag{2.35}$$

$$\mathcal{H}^{(1)}\psi^{(1)}(x)=\mathcal{E}\psi^{(1)}(x), \tag{2.36}$$

$$\psi^{(1)}(x)\stackrel{\text{def}}{=}\frac{\varphi(x)\psi'(x)-\varphi'(x)\psi(x)}{\varphi(x)},\quad \mathcal{H}^{(1)}\stackrel{\text{def}}{=}\mathcal{H}-2\partial_x^2\log|\varphi(x)|, \tag{2.37}$$

which is easy to verify. Let us prepare one solution $\psi(x)$ and M distinct seed solutions $\{\varphi_j(x),\tilde{\mathcal{E}}_j\}$ of $\mathcal{H}$:

$$\mathcal{H}\psi(x)=\mathcal{E}\psi(x),\quad \mathcal{H}\varphi_j(x)=\tilde{\mathcal{E}}_j\varphi_j(x),\ \mathcal{E},\tilde{\mathcal{E}}_j\in\mathbb{C},\ j=1,\ldots,M. \tag{2.38}$$

Applying Darboux transformations to $\psi(x)$ successively by using $\{\varphi_j(x)\}$ in turn, we arrive at:

Theorem 1 (Darboux [8]). *An isospectral solution $\psi^{(M)}(x)$ of a deformed Hamiltonian $\mathcal{H}^{(M)}$,*

$$\mathcal{H}^{(M)}\psi^{(M)}(x)=\mathcal{E}\psi^{(M)}(x),\quad \psi^{(M)}(x)\stackrel{def}{=}\frac{\mathrm{W}[\varphi_1,\ldots,\varphi_M,\psi](x)}{\mathrm{W}[\varphi_1,\ldots,\varphi_M](x)}, \tag{2.39}$$

$$\mathcal{H}^{(M)}\stackrel{def}{=}\mathcal{H}-2\frac{d^2\log|\mathrm{W}[\varphi_1,\ldots,\varphi_M](x)|}{dx^2}. \tag{2.40}$$

Here the Wronskian of n functions $\{f_1(x),\ldots,f_n(x)\}$ is defined by

$$\mathrm{W}\,[f_1,\ldots,f_n](x)\stackrel{\text{def}}{=}\det\Big(\frac{d^{j-1}f_k(x)}{dx^{j-1}}\Big)_{1\le j,k\le n}. \tag{2.41}$$

The theorem applies to any Schrödinger equation even with a complex potential.

2.3 Krein-Adler polynomials

The Krein-Adler polynomials [20, 1] are obtained as the main parts of the eigenfunctions of the system obtained by deforming the complete set of eigenfunctions $\{\phi_n(x)\}$ (H, L or J) through Darboux transformations by choosing a part of the eigenfunctions as the seed solutions. Let $\mathcal{D}$ be a set of distinct non-negative integers, $\mathcal{D}=\{d_1,d_2,\ldots,d_M\}$ which specifies the set of seed eigenfunctions $\{\phi_{d_1}(x),\ldots,\phi_{d_M}(x)\}$. The deformed system reads:

$$\mathcal{H}_{\mathcal{D}}\stackrel{\text{def}}{=}\mathcal{H}-2\partial_x^2\Big(\log|\mathrm{W}\,[\phi_{d_1},\phi_{d_2},\ldots,\phi_{d_M}](x)|\Big), \tag{2.42}$$

$$\mathcal{H}_{\mathcal{D}}\phi_{\mathcal{D};n}(x)=\mathcal{E}(n)\phi_{\mathcal{D};n}(x), \tag{2.43}$$

$$\phi_{\mathcal{D};n}(x)\stackrel{\text{def}}{=}\frac{\mathrm{W}\,[\phi_{d_1},\phi_{d_2},\ldots,\phi_{d_M},\phi_n](x)}{\mathrm{W}\,[\phi_{d_1},\phi_{d_2},\ldots,\phi_{d_M}](x)},\quad (n\in\mathbb{Z}_{\ge 0}\backslash\mathcal{D}). \tag{2.44}$$

The norms of the deformed eigenfunctions are

$$(\phi_{\mathcal{D};n}, \phi_{\mathcal{D};m}) = \prod_{j=1}^{M} \big(\mathcal{E}(n) - \mathcal{E}(d_j)\big) \cdot (\phi_n, \phi_m) \quad (n, m \in \mathbb{Z}_{\geq 0} \backslash \mathcal{D}). \tag{2.45}$$

It is necessary and sufficient for the set $\mathcal{D}$ to satisfy the conditions

$$\prod_{j=1}^{M} (m - d_j) \geq 0, \quad \forall m \in \mathbb{Z}_{\geq 0}, \tag{2.46}$$

for the positivity of the norms and non-singularity of the potential [1]. The Darboux-Crum [7] transformation is the case when $\mathcal{D}$ consists of consecutive integers from zero, $\mathcal{D} = \{0, 1, \ldots, M-1\}$, then the above conditions (2.46) are simply satisfied.

Let us denote the eigenfunctions of the (H, L, J) system generically as

$$\phi_n(x) = \phi_0(x) P_n\big(\eta(x)\big), \quad \eta(x) = x,\ x^2,\ \cos 2x \text{ for H, L, J.}$$

By using the following properties of the Wronskian

$$\mathrm{W}[gf_1, gf_2, \ldots, gf_n](x) = g(x)^n \mathrm{W}[f_1, f_2, \ldots, f_n](x), \tag{2.47}$$

$$\begin{aligned} &\mathrm{W}[f_1(y), f_2(y), \ldots, f_n(y)](x) \\ &\qquad = y'(x)^{n(n-1)/2} \mathrm{W}[f_1, f_2, \ldots, f_n](y), \end{aligned} \tag{2.48}$$

we arrive at

$$\phi_{\mathcal{D};n}(x) = \frac{\phi_0(x) \eta'(x)^M}{\mathrm{W}[P_{d_1}, P_{d_2}, \ldots, P_{d_M}](\eta)} \mathrm{W}[P_{d_1}, P_{d_2}, \ldots, P_{d_M}, P_n](\eta),$$

which gives the explicit expressions of the Krein-Adler polynomials

$$\mathrm{W}[P_{d_1}, P_{d_2}, \ldots, P_{d_M}, P_n](\eta), \qquad (n \in \mathbb{Z}_{\geq 0} \backslash \mathcal{D}), \tag{2.49}$$

with the degree $\ell_{\mathcal{D}} + n$ and the orthogonality weight function

$$\ell_{\mathcal{D}} = \sum_{j=1}^{M} d_j - M(M+1)/2, \qquad \frac{\phi_0^2(x) \eta'(x)^{2M}}{\big(\mathrm{W}[P_{d_1}, P_{d_2}, \ldots, P_{d_M}](\eta)\big)^2}. \tag{2.50}$$

This means that infinitely many families of *complete set of orthogonal rational functions having the same orthogonality weight functions with shifted parameters* can be generated.

Let us consider a simple explicit example of H and $\mathcal{D} = \{1, 2\}$ [9, 1]. The deformed Hamiltonian

$$\mathcal{H}_{\mathcal{D}} = -\frac{d^2}{dx^2} + x^2 + 3 + \frac{32x^2}{(2x^2+1)^2} - \frac{8}{2x^2+1}, \quad -\infty < x < \infty, \tag{2.51}$$

has an eigenpolynomial of degree n, $\mathrm{W}[H_1, H_2, H_n](x)$, $n \in \mathbb{Z}_{\geq 0}\backslash\{1,2\}$, with the orthogonality weight function $e^{-x^2}/(2x^2+1)^2$. Obviously three term recurrence relations do not hold but they form a complete set. Essentially the same deformation is achieved by using the seed solution $\varphi_2(x) = e^{x^2/2}H_2(ix)$ obtained by the discrete symmetry transformation (2.6) acting on $\phi_2(x)$. More generally, using an appropriate set of these seed solutions for H has the same effects as certain Krein-Adler transformations [35]. A similar theorem holds for L/J system when the type III virtual state solutions are used [35].

2.4 Multi-indexed Laguerre and Jacobi polynomials

The multi-indexed Laguerre/Jacobi polynomials are obtained by deforming the L/J system by multiple Darboux transformations in terms of type I and II virtual state solutions. In this subsection we assume that the couplings g and h are generic real numbers. Here we emphasise the logical structure [34] rather than following the historical developments. Let us denote, for simplicity of presentation, the type I/II virtual state polynomials as $\xi^{\mathrm{I}}_n(\eta)$ and $\xi^{\mathrm{II}}_n(\eta)$:

$$\begin{aligned}
&\mathrm{L}:\ \xi^{\mathrm{I}}_n(\eta) \stackrel{\text{def}}{=} L_n^{(g-1/2)}(-\eta), \qquad \xi^{\mathrm{II}}_n(\eta) \stackrel{\text{def}}{=} L_n^{(1/2-g)}(\eta),\\
&\mathrm{J}:\ \xi^{\mathrm{I}}_n(\eta) \stackrel{\text{def}}{=} P_n^{(g-1/2,1/2-h)}(\eta),\ \xi^{\mathrm{II}}_n(\eta) \stackrel{\text{def}}{=} P_n^{(1/2-g,h-1/2)}(\eta).
\end{aligned}$$

Let us prepare two sets of distinct positive integers which specify the degrees of type I and II virtual state solutions:

$$\mathcal{D} = \mathcal{D}^{\mathrm{I}} \cup \mathcal{D}^{\mathrm{II}}, \qquad \mathcal{D}^{\mathrm{I}} = \{d^{\mathrm{I}}_1, d^{\mathrm{I}}_2, \ldots, d^{\mathrm{I}}_M\},\quad \mathcal{D}^{\mathrm{II}} = \{d^{\mathrm{II}}_1, d^{\mathrm{II}}_2, \ldots, d^{\mathrm{II}}_N\}. \tag{2.52}$$

The multiple Darboux transformations using these seed solutions produce a deformed system:

$$\mathcal{H}_{\mathcal{D}} \stackrel{\text{def}}{=} \mathcal{H} - 2\partial_x^2\Big(\log\big|\mathrm{W}\,[\varphi^{\mathrm{I}}_{d_1}, \ldots, \varphi^{\mathrm{I}}_{d_M}, \varphi^{\mathrm{II}}_{d_1}, \ldots, \varphi^{\mathrm{II}}_{d_N}](x)\big|\Big), \tag{2.53}$$

$$\mathcal{H}_{\mathcal{D}}\phi_{\mathcal{D},n}(x;\boldsymbol{\lambda}) = \mathcal{E}(n)\phi_{\mathcal{D},n}(x;\boldsymbol{\lambda}) \quad (n \in \mathbb{Z}_{\geq 0}\backslash\mathcal{D}), \tag{2.54}$$

$$\phi_{\mathcal{D},n}(x;\boldsymbol{\lambda}) \stackrel{\text{def}}{=} \frac{\mathrm{W}\,[\varphi^{\mathrm{I}}_{d_1}, \ldots, \varphi^{\mathrm{I}}_{d_M}, \varphi^{\mathrm{II}}_{d_1}, \ldots, \varphi^{\mathrm{II}}_{d_N}, \phi_n](x)}{\mathrm{W}\,[\varphi^{\mathrm{I}}_{d_1}, \ldots, \varphi^{\mathrm{I}}_{d_M}, \varphi^{\mathrm{II}}_{d_1}, \ldots, \varphi^{\mathrm{II}}_{d_N}](x)}, \tag{2.55}$$

in which $\boldsymbol{\lambda}$ specify the coupling(s) dependence, $\boldsymbol{\lambda} = g$ for L and $\boldsymbol{\lambda} = \{g, h\}$ for J. The nodeless property of each seed solution $\varphi^{\mathrm{I,II}}_{d_j}$ conspires to achieve the nodelessness of the denominator function of $\phi_{\mathcal{D},n}(x;\boldsymbol{\lambda})$. Similar to the corresponding expressions of the Krein-Adler polynomials (2.45), we obtain the norms of the deformed eigenfunctions $\phi_{\mathcal{D},n}(x)$:

$$(\phi_{\mathcal{D},n}, \phi_{\mathcal{D},m}) = \prod_{j=1}^{M}\big(\mathcal{E}(n) - \tilde{\mathcal{E}}^{\mathrm{I}}_{d_j}\big)\prod_{j=1}^{N}\big(\mathcal{E}(n) - \tilde{\mathcal{E}}^{\mathrm{II}}_{d_j}\big)\cdot(\phi_n, \phi_m) \quad (n, m \in \mathbb{Z}_{\geq 0}). \tag{2.56}$$

This clearly shows the necessity of the *negative* virtual state energies $\{\tilde{\mathcal{E}}^{\mathrm{I,II}}_{d_j} < 0\}$.

The multi-indexed Laguerre/Jacobi polynomials are extracted by factorising the deformed eigenfunction $\phi_{\mathcal{D},n}(x;\boldsymbol{\lambda})$ into the ground state eigenfunction and the polynomial part by using the Wronskian formulas (2.47), (2.48):

$$\phi_{\mathcal{D},n}(x;\boldsymbol{\lambda}) = c_{\mathcal{F}}^{M+N}\psi_{\mathcal{D}}(x;\boldsymbol{\lambda})P_{\mathcal{D},n}(\eta(x);\boldsymbol{\lambda}), \quad \psi_{\mathcal{D}}(x;\boldsymbol{\lambda}) \stackrel{\text{def}}{=} \frac{\phi_0(x;\boldsymbol{\lambda}^{[M,N]})}{\Xi_{\mathcal{D}}(\eta(x);\boldsymbol{\lambda})}, \tag{2.57}$$

in which $c_{\mathcal{F}} = 2$ for L and $c_{\mathcal{F}} = -4$ for J. The coupling $\boldsymbol{\lambda}$ is shifted to $\boldsymbol{\lambda}^{[M,N]}$:

$$\boldsymbol{\lambda}^{[M,N]} = g + M - N \;\text{ for L}, \quad \boldsymbol{\lambda}^{[M,N]} = \{g+M-N, h-M+N\} \;\text{ for J}. \tag{2.58}$$

The general formulas for the multi-indexed polynomial $P_{\mathcal{D},n}(\eta(x);\boldsymbol{\lambda})$ and the denominator polynomial $\Xi_{\mathcal{D}}(\eta(x);\boldsymbol{\lambda})$ are:

$$\begin{aligned} P_{\mathcal{D},n}(\eta;\boldsymbol{\lambda}) &\stackrel{\text{def}}{=} \mathrm{W}[\mu_1,\ldots,\mu_M,\nu_1,\ldots,\nu_N,P_n](\eta) \\ &\quad\times\begin{cases} e^{-M\eta}\,\eta^{(M+g+\frac{1}{2})N} & : \mathrm{L} \\ \left(\frac{1-\eta}{2}\right)^{(M+g+\frac{1}{2})N}\left(\frac{1+\eta}{2}\right)^{(N+h+\frac{1}{2})M} & : \mathrm{J} \end{cases}, \end{aligned} \tag{2.59}$$

$$\begin{aligned} \Xi_{\mathcal{D}}(\eta;\boldsymbol{\lambda}) &\stackrel{\text{def}}{=} \mathrm{W}[\mu_1,\ldots,\mu_M,\nu_1,\ldots,\nu_N](\eta) \\ &\quad\times\begin{cases} e^{-M\eta}\,\eta^{(M+g-\frac{1}{2})N} & : \mathrm{L} \\ \left(\frac{1-\eta}{2}\right)^{(M+g-\frac{1}{2})N}\left(\frac{1+\eta}{2}\right)^{(N+h-\frac{1}{2})M} & : \mathrm{J} \end{cases}, \end{aligned} \tag{2.60}$$

$$\mu_j \stackrel{\text{def}}{=} \begin{cases} e^{\eta}\xi^{\mathrm{I}}_{d_j}(\eta) & : \mathrm{L} \\ \left(\frac{1+\eta}{2}\right)^{\frac{1}{2}-h}\xi^{\mathrm{I}}_{d_j}(\eta) & : \mathrm{J} \end{cases}, \quad \nu_j \stackrel{\text{def}}{=} \begin{cases} \eta^{\frac{1}{2}-g}\xi^{\mathrm{II}}_{d_j}(\eta) & : \mathrm{L} \\ \left(\frac{1-\eta}{2}\right)^{\frac{1}{2}-g}\xi^{\mathrm{II}}_{d_j}(\eta) & : \mathrm{J} \end{cases}, \tag{2.61}$$

in which P_n in (2.59) denotes the original polynomial for L/J. The multi-indexed polynomial $P_{\mathcal{D},n}$ is of degree $\ell_{\mathcal{D}}+n$ and the denominator polynomial $\Xi_{\mathcal{D}}$ is of degree $\ell_{\mathcal{D}}$ in η, in which $\ell_{\mathcal{D}}$ is given by

$$\ell_{\mathcal{D}} \stackrel{\text{def}}{=} \sum_{j=1}^{M} d^{\mathrm{I}}_j + \sum_{j=1}^{N} d^{\mathrm{II}}_j - \frac{1}{2}M(M-1) - \frac{1}{2}N(N-1) + MN \geq 1. \tag{2.62}$$

Here the label n specifies the energy eigenvalue $\mathcal{E}(n)$ of $\phi_{\mathcal{D},n}$ and it also counts the nodes due to the oscillation theorem. Since the degrees $0,1,\ldots,\ell_{\mathcal{D}}-1$ are missing, the multi-indexed polynomials $\{P_{\mathcal{D},n}\}$ do not satisfy three term recurrence relations but they form a complete set in L^2 with the orthogonality relations:

$$\begin{aligned} &\int d\eta\,\frac{\mathrm{W}(\eta;\boldsymbol{\lambda}^{[M,N]})}{\Xi_{\mathcal{D}}(\eta;\boldsymbol{\lambda})^2}P_{\mathcal{D},n}(\eta;\boldsymbol{\lambda})P_{\mathcal{D},m}(\eta;\boldsymbol{\lambda}) \\ &= h_n\delta_{nm}\times\begin{cases} \prod_{j=1}^{M}(n+g+d^{\mathrm{I}}_j+\frac{1}{2})\cdot\prod_{j=1}^{N}(n+g-d^{\mathrm{II}}_j-\frac{1}{2}) & : \mathrm{L} \\ 4^{-M-N}\prod_{j=1}^{M}(n+g+d^{\mathrm{I}}_j+\frac{1}{2})(n+h-d^{\mathrm{I}}_j-\frac{1}{2}) & \\ \qquad\times\prod_{j=1}^{N}(n+g-d^{\mathrm{II}}_j-\frac{1}{2})(n+h+d^{\mathrm{II}}_j+\frac{1}{2}) & : \mathrm{J} \end{cases}. \end{aligned} \tag{2.63}$$

The weight function of the original polynomials $\mathrm{W}(\eta;\boldsymbol{\lambda})d\eta = \phi_0(x;\boldsymbol{\lambda})^2dx$ reads explicitly, see (2.11) and (2.24):

$$\mathrm{W}(\eta;\boldsymbol{\lambda}) \stackrel{\text{def}}{=} \begin{cases} \frac{1}{2}e^{-\eta}\eta^{g-\frac{1}{2}} & :\mathrm{L} \\ \frac{1}{2^{g+h+1}}(1-\eta)^{g-\frac{1}{2}}(1+\eta)^{h-\frac{1}{2}} & :\mathrm{J} \end{cases}. \tag{2.64}$$

The very form of the deformed eigenfunctions (2.57) suggests that the system can be regarded as *orthogonal rational functions system* $\{P_{\mathcal{D},n}(\eta)/\Xi_{\mathcal{D}}(\eta)\}$, $n \in \mathbb{Z}_{\geq 0}$ with the same orthogonality weight function as the L/J system but parameter(s) are shifted $\mathrm{W}(\eta,\boldsymbol{\lambda}^{[M,N]})$. It is important to note that the lowest degree polynomial $P_{\mathcal{D},0}(\eta;\boldsymbol{\lambda})$ is related to the denominator polynomial $\Xi_{\mathcal{D}}(\eta;\boldsymbol{\lambda})$ by the parameter shift $\boldsymbol{\lambda}\to\boldsymbol{\lambda}+\boldsymbol{\delta}$ ($\boldsymbol{\delta}=1$ (L), $\boldsymbol{\delta}=\{1,1\}$ (J), *i.e.* $g\to g+1$ for L and $\{g,h\}\to\{g+1,h+1\}$ for J):

$$P_{\mathcal{D},0}(\eta;\boldsymbol{\lambda}) \propto \Xi_{\mathcal{D}}(\eta;\boldsymbol{\lambda}+\boldsymbol{\delta}). \tag{2.65}$$

The above formulas (2.59) and (2.60) are drastically simplified when type I (II) seed solutions only are used.

By similarity transforming the deformed Hamiltonian $\mathcal{H}_{\mathcal{D}}$ in terms of the ground state eigenfunction $\psi_{\mathcal{D}}(x;\boldsymbol{\lambda})$, we obtain the second order differential operator $\widetilde{\mathcal{H}}_{\mathcal{D}}$ governing the multi-indexed polynomials:

$$\begin{aligned} \widetilde{\mathcal{H}}_{\mathcal{D}} &\stackrel{\text{def}}{=} \psi_{\mathcal{D}}(x;\boldsymbol{\lambda})^{-1}\circ\mathcal{H}_{\mathcal{D}}\circ\psi_{\mathcal{D}}(x;\boldsymbol{\lambda}) \\ &= -4\biggl(c_2(\eta)\frac{d^2}{d\eta^2} + \Bigl(c_1(\eta,\boldsymbol{\lambda}^{[M,N]}) - 2c_2(\eta)\frac{\partial_\eta\Xi_{\mathcal{D}}(\eta;\boldsymbol{\lambda})}{\Xi_{\mathcal{D}}(\eta;\boldsymbol{\lambda})}\Bigr)\frac{d}{d\eta} \\ &\qquad + c_2(\eta)\frac{\partial_\eta^2\Xi_{\mathcal{D}}(\eta;\boldsymbol{\lambda})}{\Xi_{\mathcal{D}}(\eta;\boldsymbol{\lambda})} - c_1(\eta,\boldsymbol{\lambda}^{[M,N]}-\boldsymbol{\delta})\frac{\partial_\eta\Xi_{\mathcal{D}}(\eta;\boldsymbol{\lambda})}{\Xi_{\mathcal{D}}(\eta;\boldsymbol{\lambda})}\biggr), \end{aligned} \tag{2.66}$$

$$\widetilde{\mathcal{H}}_{\mathcal{D}}P_{\mathcal{D},n}(\eta;\boldsymbol{\lambda}) = \mathcal{E}(n)P_{\mathcal{D},n}(\eta;\boldsymbol{\lambda}), \tag{2.67}$$

in which

$$c_1(\eta,\boldsymbol{\lambda}) \stackrel{\text{def}}{=} \begin{cases} g+\frac{1}{2}-\eta & :\mathrm{L} \\ h-g-(g+h+1)\eta & :\mathrm{J} \end{cases}, \quad c_2(\eta) \stackrel{\text{def}}{=} \begin{cases} \eta & :\mathrm{L} \\ 1-\eta^2 & :\mathrm{J} \end{cases}. \tag{2.68}$$

The above differential equation (2.67) is Fuchsian for the J case, since the $\ell_{\mathcal{D}}$ zeros of $\Xi_{\mathcal{D}}(\eta;\boldsymbol{\lambda})$ are all simple with the same characteristic exponents, 0 and 3. The L case is obtained as a confluent limit.

The first examples of multi-indexed polynomials were the simplest ones with $\mathcal{D}=\{1\}$ for L and J by Gomez-Ullate et al. [11] and Quesne [39] and they were called exceptional orthogonal polynomials. In this case type I seed solutions define the same polynomials as those of type II seed solutions by a shift of parameters. The general $\mathcal{D}=\{\ell\}$, $\forall\ell\geq 1$ cases for type I and II were derived in [27], [29]. The present derivation of $P_{\mathcal{D},n}(\eta;\boldsymbol{\lambda})$ is due to [34].

2.4.1 Explicit examples of multi-indexed Laguerre polynomials

We present two explicit examples of multi-indexed Laguerre polynomials specified by (i) $\mathcal{D}=\{\ell^{\mathrm{I}}\}$, (ii) $\mathcal{D}=\{\ell^{\mathrm{II}}\}$.

(i) $\mathcal{D}=\{\ell^{\mathrm{I}}\}$ In this case $\xi^{\mathrm{I}}_\ell(\eta)=L^{(g-1/2)}_\ell(-\eta)$, $\eta=x^2$ and the potential and the eigenfunctions of the deformed QM system are:

$$\begin{aligned}
V^{\mathrm{I}}_\ell(x)&=x^2+\frac{g(g+1)}{x^2}-(3+2g)-2\partial_x^2\log|\xi^{\mathrm{I}}_\ell(\eta)|,\\
\phi^{\mathrm{I}}_{\ell,n}(x)&=2\psi^{\mathrm{I}}_\ell(x)P^{\mathrm{I}}_{\ell,n}(\eta),\quad \psi^{\mathrm{I}}_\ell(x)=\frac{e^{-x^2/2}x^{g+1}}{\xi^{\mathrm{I}}_\ell(\eta)},\\
P^{\mathrm{I}}_{\ell,n}(\eta)&=-\left(\partial_\eta\xi^{\mathrm{I}}_\ell(\eta)+\xi^{\mathrm{I}}_\ell(\eta)\right)L^{(g-1/2)}_n(\eta)+\xi^{\mathrm{I}}_\ell(\eta)\partial_\eta L^{(g-1/2)}_n(\eta).
\end{aligned}$$

The second order differential operator $\widetilde{\mathcal{H}}^{\mathrm{I}}_\ell$ governing $P^{\mathrm{I}}_{\ell,n}(\eta)$ is

$$\begin{aligned}
\widetilde{\mathcal{H}}^{\mathrm{I}}_\ell P^{\mathrm{I}}_{\ell,n}(\eta)&=nP^{\mathrm{I}}_{\ell,n}(\eta),\\
\widetilde{\mathcal{H}}^{\mathrm{I}}_\ell&=-\left[\eta\frac{d^2}{d\eta^2}+\left(L^{(g+1/2)}_1(\eta)-2\eta\frac{\partial_\eta\xi^{\mathrm{I}}_\ell(\eta)}{\xi^{\mathrm{I}}_\ell(\eta)}\right)\frac{d}{d\eta}\right.\\
&\qquad\left.+\eta\frac{\partial_\eta^2\xi^{\mathrm{I}}_\ell(\eta)}{\xi^{\mathrm{I}}_\ell(\eta)}-L^{(g-1/2)}_1(\eta)\frac{\partial_\eta\xi^{\mathrm{I}}_\ell(\eta)}{\xi^{\mathrm{I}}_\ell(\eta)}\right].
\end{aligned}$$

(ii) $\mathcal{D}=\{\ell^{\mathrm{II}}\}$ In this case $\xi^{\mathrm{II}}_\ell(\eta)=L^{(1/2-g)}_\ell(\eta)$, $\eta=x^2$ and the potential and the eigenfunctions of the deformed QM system are:

$$\begin{aligned}
V^{\mathrm{I}}_\ell(x)&=x^2+\frac{(g-1)(g-2)}{x^2}+(1-2g)-2\partial_x^2\log|\xi^{\mathrm{II}}_\ell(\eta)|,\\
\phi^{\mathrm{II}}_{\ell,n}(x)&=2\psi^{\mathrm{II}}_\ell(x)P^{\mathrm{II}}_{\ell,n}(\eta),\quad \psi^{\mathrm{II}}_\ell(x)=\frac{e^{-x^2/2}x^{g-1}}{\xi^{\mathrm{II}}_\ell(\eta)},\\
P^{\mathrm{II}}_{\ell,n}(\eta)&=\left(-\eta\partial_\eta\xi^{\mathrm{II}}_\ell(\eta)+(g-1/2)\xi^{\mathrm{II}}_\ell(\eta)\right)L^{(g-1/2)}_n(\eta)+\eta\xi^{\mathrm{II}}_\ell(\eta)\partial_\eta L^{(g-1/2)}_n(\eta).
\end{aligned}$$

The second order differential operator $\widetilde{\mathcal{H}}^{\mathrm{II}}_\ell$ governing $P^{\mathrm{II}}_{\ell,n}(\eta)$ is

$$\begin{aligned}
\widetilde{\mathcal{H}}^{\mathrm{II}}_\ell P^{\mathrm{II}}_{\ell,n}(\eta)&=nP^{\mathrm{II}}_{\ell,n}(\eta),\\
\widetilde{\mathcal{H}}^{\mathrm{II}}_\ell&=-\left[\eta\frac{d^2}{d\eta^2}+\left(L^{(g-3/2)}_1(\eta)-2\eta\frac{\partial_\eta\xi^{\mathrm{II}}_\ell(\eta)}{\xi^{\mathrm{II}}_\ell(\eta)}\right)\frac{d}{d\eta}\right.\\
&\qquad\left.+\eta\frac{\partial_\eta^2\xi^{\mathrm{II}}_\ell(\eta)}{\xi^{\mathrm{II}}_\ell(\eta)}-L^{(g-5/2)}_1(\eta)\frac{\partial_\eta\xi^{\mathrm{II}}_\ell(\eta)}{\xi^{\mathrm{II}}_\ell(\eta)}\right].
\end{aligned}$$

2.4.2 Explicit examples of multi-indexed Jacobi polynomials

We present two explicit examples of multi-indexed Jacobi polynomials, (i) $\mathcal{D}=\{\ell^{\mathrm{I}}\}$, (ii) $\mathcal{D}=\{\ell^{\mathrm{II}}\}$. In this subsection we use the abbreviation $P_n(\eta)\equiv P^{(g-1/2,h-1/2)}_n(\eta)$.

(i) $\mathcal{D}=\{\ell^{\mathrm{I}}\}$ In this case $\xi^{\mathrm{I}}_\ell(\eta)=P^{(g-1/2,1/2-h)}_\ell(\eta)$, $\eta=\cos 2x$ and the potential and the eigenfunctions of the deformed QM system are:

$$
\begin{aligned}
V^{\mathrm{I}}_\ell(x)&=\frac{g(g+1)}{\sin^2 x}+\frac{(h-1)(h-2)}{\cos^2 x}-2\partial_x^2\log|\xi^{\mathrm{I}}_\ell(\eta)|-(g+h)^2,\\
\phi^{\mathrm{I}}_{\ell,n}(x)&=-4\psi^{\mathrm{I}}_\ell(x)P^{\mathrm{I}}_{\ell,n}(\eta),\quad \psi^{\mathrm{I}}_\ell(x)=\frac{(\sin x)^{g+1}(\cos x)^{h-1}}{\xi^{\mathrm{I}}_\ell(\eta)},\\
P^{\mathrm{I}}_{\ell,n}(\eta)&=-\tfrac14\big(2(1+\eta)\partial_\eta\xi^{\mathrm{I}}_\ell(\eta)+(1-2h)\xi^{\mathrm{I}}_\ell(\eta)\big)P_n(\eta)+\tfrac12(1+\eta)\xi^{\mathrm{I}}_\ell(\eta)\partial_\eta P_n(\eta).
\end{aligned}
$$

The second order differential operator $\widetilde{\mathcal{H}}^{\mathrm{I}}_\ell$ governing $P^{\mathrm{I}}_{\ell,n}(\eta)$ is

$$
\begin{aligned}
&\widetilde{\mathcal{H}}^{\mathrm{I}}_\ell P^{\mathrm{I}}_{\ell,n}(\eta)=n(n+g+h)P^{\mathrm{I}}_{\ell,n}(\eta),\\
&\widetilde{\mathcal{H}}^{\mathrm{I}}_\ell=-\bigg[(1-\eta^2)\frac{d^2}{d\eta^2}+\bigg(h-g-2-(g+h+1)\eta-2(1-\eta^2)\frac{\partial_\eta\xi^{\mathrm{I}}_\ell(\eta)}{\xi^{\mathrm{I}}_\ell(\eta)}\bigg)\frac{d}{d\eta}\\
&\qquad\qquad+(1-\eta^2)\frac{\partial_\eta^2\xi^{\mathrm{I}}_\ell(\eta)}{\xi^{\mathrm{I}}_\ell(\eta)}-(h-g-2-(g+h-1)\eta)\frac{\partial_\eta\xi^{\mathrm{I}}_\ell(\eta)}{\xi^{\mathrm{I}}_\ell(\eta)}\bigg].
\end{aligned}
$$

(ii) $\mathcal{D}=\{\ell^{\mathrm{II}}\}$ In this case $\xi^{\mathrm{II}}_\ell(\eta)=P^{(1/2-g,h-1/2)}_\ell(\eta)$, $\eta=\cos 2x$ and the potential and the eigenfunctions of the deformed QM system are

$$
\begin{aligned}
V^{\mathrm{II}}_\ell(x)&=\frac{(g-1)(g-2)}{\sin^2 x}+\frac{h(h+1)}{\cos^2 x}-2\partial_x^2\log|\xi^{\mathrm{II}}_\ell(\eta)|-(g+h)^2,\\
\phi^{\mathrm{II}}_{\ell,n}(x)&=-4\psi^{\mathrm{II}}_\ell(x)P^{\mathrm{II}}_{\ell,n}(\eta),\quad \psi^{\mathrm{II}}_\ell(x)=\frac{(\sin x)^{g-1}(\cos x)^{h+1}}{\xi^{\mathrm{II}}_\ell(\eta)},\\
P^{\mathrm{II}}_{\ell,n}(\eta)&=-\tfrac14\left(2(1-\eta)\partial_\eta\xi^{\mathrm{II}}_\ell(\eta)+(2g-1)\xi^{\mathrm{II}}_\ell(\eta)\right)P_n(\eta)+\tfrac12(1-\eta)\xi^{\mathrm{II}}_\ell(\eta)\partial_\eta P_n(\eta).
\end{aligned}
$$

The second order differential operator $\widetilde{\mathcal{H}}^{\mathrm{II}}_\ell$ governing $P^{\mathrm{II}}_{\ell,n}(\eta)$ is

$$
\begin{aligned}
&\widetilde{\mathcal{H}}^{\mathrm{II}}_\ell P^{\mathrm{II}}_{\ell,n}(\eta)=n(n+g+h)P^{\mathrm{II}}_{\ell,n}(\eta),\\
&\widetilde{\mathcal{H}}^{\mathrm{I}}_\ell=-\bigg[(1-\eta^2)\frac{d^2}{d\eta^2}+\bigg(h-g+2-(g+h+1)\eta-2(1-\eta^2)\frac{\partial_\eta\xi^{\mathrm{II}}_\ell(\eta)}{\xi^{\mathrm{II}}_\ell(\eta)}\bigg)\frac{d}{d\eta}\\
&\qquad\qquad+(1-\eta^2)\frac{\partial_\eta^2\xi^{\mathrm{II}}_\ell(\eta)}{\xi^{\mathrm{II}}_\ell(\eta)}-(h-g+2-(g+h-1)\eta)\frac{\partial_\eta\xi^{\mathrm{II}}_\ell(\eta)}{\xi^{\mathrm{II}}_\ell(\eta)}\bigg].
\end{aligned}
$$

3 New orthogonal polynomials in discrete QM with real shifts

The discrete QM with real shifts is the simplest QM, consisting of eigenvalue problems of certain hermitian matrices, *i.e.* the *tri-diagonal real symmetric* (Jacobi) matrices of finite or infinite dimensions [24, 38]. The orthogonal polynomials of a discrete variable [23], that is, those having purely discrete orthogonality measures, belonging to the Askey scheme, *e.g.* the Hahn, Meixner, Racah, etc. and their

q-versions all constitute the main parts of the eigenvectors of these exactly solvable simplest QM systems. We will present new orthogonal polynomials obtained by deforming the Racah and q-Racah polynomials [32] which are the most generic members in this group. For the others, *e.g.* the multi-indexed Meixner or little q-Jacobi, etc., see [37].

3.1 Classical polynomials: the Racah and q-Racah

In order to adopt 1-d QM-like notation, let us denote the indices of the matrices by $x, y = 0, 1, \dots, x_{\max}$ and the vector component by v_x or $v(x)$. The matrix eigenvalue equation is rewritten as

$$\sum_{y=0}^{x_{\max}} \mathcal{H}_{x,y} v_y = \lambda v_x, \quad \text{or} \quad \mathcal{H} v(x) = \lambda v(x), \quad \sum_{y=0}^{x_{\max}} \mathcal{H}_{x,y} v(y) = \lambda v(x),$$

in which $x_{\max} = N$, a positive integer, for finite dimensional matrices and $x_{\max} = \infty$ for infinite matrices. In this section we discuss the Racah (R) and q-Racah (qR) systems only, which are finite dimensional. We assume that the Hamiltonian $\mathcal{H} = (\mathcal{H}_{x,y})$ is an irreducible tri-diagonal real symmetric (Jacobi) matrix, that is, not the direct sum of two or more such matrices. We also assume, without loss of generality, that the lowest eigenvalue is zero, $\mathcal{E}(0) = 0$. Note that the spectra of a Jacobi matrix is simple. Then the Hamiltonians $\mathcal{H}$ is *positive semi-definite* and it can be expressed in a *factorised form*:

$$\mathcal{H} = \mathcal{A}^\dagger \mathcal{A}, \qquad \mathcal{A} = (\mathcal{A}_{x,y}), \quad \mathcal{A}^\dagger = ((\mathcal{A}^\dagger)_{x,y}) = (\mathcal{A}_{y,x}), \tag{3.1}$$

$$\mathcal{A}_{x,y} \stackrel{\text{def}}{=} \sqrt{B(x)}\, \delta_{x,y} - \sqrt{D(x+1)}\, \delta_{x+1,y}, \tag{3.2}$$

$$(\mathcal{A}^\dagger)_{x,y} = \sqrt{B(x)}\, \delta_{x,y} - \sqrt{D(x)}\, \delta_{x-1,y}, \tag{3.3}$$

$$\mathcal{H}_{x,y} \stackrel{\text{def}}{=} -\sqrt{B(x)D(x+1)}\, \delta_{x+1,y} - \sqrt{B(x-1)D(x)}\, \delta_{x-1,y} + \big(B(x) + D(x)\big)\delta_{x,y}, \tag{3.4}$$

in which the potential functions $B(x)$ and $D(x)$ are real and positive but vanish at the boundary

$$B(x) > 0 \ \ (x \geq 0), \quad D(x) > 0 \ \ (x \geq 1), \quad D(0) = 0, \quad B(N) = 0. \tag{3.5}$$

The Schrödinger equation is the eigenvalue problem for the hermitian matrix $\mathcal{H}$,

$$\mathcal{H}\phi_n(x) = \mathcal{E}(n)\phi_n(x) \quad (n = 0, \dots, n_{\max} = N), \quad 0 = \mathcal{E}(0) < \cdots < \mathcal{E}(n_{\max}). \tag{3.6}$$

We write $\mathcal{H}$, $\mathcal{A}$ and $\mathcal{A}^\dagger$ as follows by using the simple shift operators $e^{\pm\partial}$:

$$e^{\pm\partial} = ((e^{\pm\partial})_{x,y}), \quad (e^{\pm\partial})_{x,y} \stackrel{\text{def}}{=} \delta_{x\pm1,y}, \quad (e^{\partial})^\dagger = e^{-\partial}, \tag{3.7}$$

$$\begin{aligned}\mathcal{H} &= -\sqrt{B(x)D(x+1)}\, e^{\partial} - \sqrt{B(x-1)D(x)}\, e^{-\partial} + B(x) + D(x) \\ &= -\sqrt{B(x)}\, e^{\partial} \sqrt{D(x)} - \sqrt{D(x)}\, e^{-\partial} \sqrt{B(x)} + B(x) + D(x),\end{aligned} \tag{3.8}$$

$$\mathcal{A}=\sqrt{B(x)}-e^{\partial}\sqrt{D(x)},\quad \mathcal{A}^{\dagger}=\sqrt{B(x)}-\sqrt{D(x)}\,e^{-\partial}. \tag{3.9}$$

The groundstate eigenvector is easily obtained. The zero mode equation, $\mathcal{A}\phi_0=0$, is

$$\sqrt{B(x)}\,\phi_0(x)-\sqrt{D(x+1)}\,\phi_0(x+1)=0\quad (x=0,1,\ldots,x_{\max}-1), \tag{3.10}$$

$$\phi_0(x)=\sqrt{\prod_{y=0}^{x-1}\frac{B(y)}{D(y+1)}}\quad (x=0,1,\ldots,x_{\max}), \tag{3.11}$$

with the normalisation $\phi_0(0)=1$. The eigenvectors are mutually orthogonal:

$$(\phi_n,\phi_m)\stackrel{\text{def}}{=}\sum_{x=0}^{x_{\max}}\phi_n(x)\phi_m(x)=\frac{1}{d_n^2}\delta_{n\,m}\quad (n,m=0,1,\ldots,n_{\max}). \tag{3.12}$$

The systems contain four real parameters a, b, c, d and q for qR, which are symbolically denoted by $\boldsymbol{\lambda}$:

$$\text{R}:\ \boldsymbol{\lambda}=(a,b,c,d),\quad \boldsymbol{\delta}=(1,1,1,1),\quad \kappa=1, \tag{3.13}$$

$$\text{qR}:\ q^{\boldsymbol{\lambda}}=(a,b,c,d),\quad \boldsymbol{\delta}=(1,1,1,1),\quad \kappa=q^{-1},\quad 0<q<1, \tag{3.14}$$

where $q^{\boldsymbol{\lambda}}$ stands for $q^{(\lambda_1,\lambda_2,\ldots)}=(q^{\lambda_1},q^{\lambda_2},\ldots)$ and $\boldsymbol{\delta}$ and κ will be used later. The potential functions $B(x)$ and $D(x)$ are:

$$B(x;\boldsymbol{\lambda})=\begin{cases}-\dfrac{(x+a)(x+b)(x+c)(x+d)}{(2x+d)(2x+1+d)} & :\text{R}\\[2ex] -\dfrac{(1-aq^x)(1-bq^x)(1-cq^x)(1-dq^x)}{(1-dq^{2x})(1-dq^{2x+1})} & :\text{qR}\end{cases}, \tag{3.15}$$

$$D(x;\boldsymbol{\lambda})=\begin{cases}-\dfrac{(x+d-a)(x+d-b)(x+d-c)x}{(2x-1+d)(2x+d)} & :\text{R}\\[2ex] -\tilde{d}\,\dfrac{(1-a^{-1}dq^x)(1-b^{-1}dq^x)(1-c^{-1}dq^x)(1-q^x)}{(1-dq^{2x-1})(1-dq^{2x})} & :\text{qR},\end{cases} \tag{3.16}$$

in which a new parameter $\tilde{d}$ is defined by

$$\tilde{d}\stackrel{\text{def}}{=}\begin{cases}a+b+c-d-1 & :\text{R}\\ abcd^{-1}q^{-1} & :\text{qR}\end{cases}. \tag{3.17}$$

We adopt the following choice of the parameter ranges:

$$\text{R}:\quad a=-N,\quad 0<d<a+b,\quad 0<c<1+d, \tag{3.18}$$

$$\text{qR}:\quad a=q^{-N},\quad 0<ab<d<1,\quad qd<c<1. \tag{3.19}$$

The eigenvalue problem and its factorised eigenvectors are:

$$\mathcal{H}(\boldsymbol{\lambda})\phi_n(x;\boldsymbol{\lambda})=\mathcal{E}(n;\boldsymbol{\lambda})\phi_n(x;\boldsymbol{\lambda})\quad(x=0,1,\ldots,N;n=0,1,\ldots,N),\tag{3.20}$$

$$\phi_n(x;\boldsymbol{\lambda})=\phi_0(x;\boldsymbol{\lambda})\check{P}_n(x;\boldsymbol{\lambda}),\tag{3.21}$$

$$\mathcal{E}(n;\boldsymbol{\lambda})=\begin{cases}n(n+\tilde{d}) & :\mathrm{R}\\ (q^{-n}-1)(1-\tilde{d}q^n) & :q\mathrm{R}\end{cases}\quad \eta(x;\boldsymbol{\lambda})=\begin{cases}x(x+d) & :\mathrm{R}\\ (q^{-x}-1)(1-dq^x) & :q\mathrm{R}\end{cases}\tag{3.22}$$

$$\check{P}_n(x;\boldsymbol{\lambda})=P_n\big(\eta(x;\boldsymbol{\lambda});\boldsymbol{\lambda}\big)=\begin{cases}{}_4F_3\Big(\begin{matrix}-n,\,n+\tilde{d},\,-x,\,x+d\\ a,\,b,\,c\end{matrix}\Big|\,1\Big) & :\mathrm{R}\\ {}_4\phi_3\Big(\begin{matrix}q^{-n},\,\tilde{d}q^n,\,q^{-x},\,dq^x\\ a,\,b,\,c\end{matrix}\Big|\,q\,;q\Big) & :q\mathrm{R}\end{cases}\tag{3.23}$$

$$\phi_0(x;\boldsymbol{\lambda})^2=\begin{cases}\dfrac{(a,b,c,d)_x}{(1+d-a,1+d-b,1+d-c,1)_x}\,\dfrac{2x+d}{d} & :\mathrm{R}\\ \dfrac{(a,b,c,d\,;q)_x}{(a^{-1}dq,b^{-1}dq,c^{-1}dq,q\,;q)_x\,\tilde{d}^x}\,\dfrac{1-dq^{2x}}{1-d} & :q\mathrm{R}\end{cases},\tag{3.24}$$

$$d_n(\boldsymbol{\lambda})^2=\begin{cases}\dfrac{(a,b,c,\tilde{d})_n}{(1+\tilde{d}-a,1+\tilde{d}-b,1+\tilde{d}-c,1)_n}\,\dfrac{2n+\tilde{d}}{\tilde{d}}\\ \quad\times\dfrac{(-1)^N(1+d-a,1+d-b,1+d-c)_N}{(\tilde{d}+1)_N(d+1)_{2N}} & :\mathrm{R}\\ \dfrac{(a,b,c,\tilde{d}\,;q)_n}{(a^{-1}\tilde{d}q,b^{-1}\tilde{d}q,c^{-1}\tilde{d}q,q\,;q)_n\,d^n}\,\dfrac{1-\tilde{d}q^{2n}}{1-\tilde{d}}\\ \quad\times\dfrac{(-1)^N(a^{-1}dq,b^{-1}dq,c^{-1}dq\,;q)_N\,\tilde{d}^N q^{\frac{1}{2}N(N+1)}}{(\tilde{d}q\,;q)_N(dq\,;q)_{2N}} & :q\mathrm{R}\end{cases}.\tag{3.25}$$

Here $(a)_n$ $((a;q)_n)$ is the $(q$-)shifted factorial ($(q$-)Pochhammer symbol):

$$(a)_n\stackrel{\text{def}}{=}\prod_{k=1}^{n}(a+k-1),\quad (a_1,\ldots,a_l)_n\stackrel{\text{def}}{=}\prod_{j=1}^{l}(a_j)_n,\tag{3.26}$$

$$(a\,;q)_n\stackrel{\text{def}}{=}\prod_{k=1}^{n}(1-aq^{k-1}),\quad (a_1,\ldots,a_l;q)_n\stackrel{\text{def}}{=}\prod_{j=1}^{l}(a_j;q)_n.\tag{3.27}$$

The $(q$-)Racah polynomials are expressed by truncated (basic)-hypergeometric series (3.23) and they are polynomials in the sinusoidal coordinates $\eta(x;\boldsymbol{\lambda})$ (3.22). In fact, if a (Laurent) polynomial $\check{f}$ in x (q^x) is invariant under the involution $\mathcal{I}$

$$\mathcal{I}(x)=-x-d\quad:\mathrm{R},\qquad \mathcal{I}(q^x)=q^{-x}d^{-1}\quad:q\mathrm{R},\quad \mathcal{I}^2=\mathrm{id},\tag{3.28}$$

$$\mathcal{I}\big(\check{f}(x)\big)=\check{f}(x)\ \Leftrightarrow\ \check{f}(x)=f\big(\eta(x;\boldsymbol{\lambda})\big),\tag{3.29}$$

it is a polynomial in the sinusoidal coordinate $\eta(x;\boldsymbol{\lambda})$. The parameter dependence of $\eta(x;\boldsymbol{\lambda})$ is important, since parameters are shifted by the multi-indexed deformation.

By similarity transforming $\mathcal{H}$ in terms of the ground state eigenfunction $\phi_0(x)$, we obtain the second order difference operator $\widetilde{\mathcal{H}}$ governing the classical polynomial $\check{P}_n(x)$:

$$\widetilde{\mathcal{H}} \stackrel{\text{def}}{=} \phi_0(x) \circ \mathcal{H} \circ \phi_0(x) = B(x)(1-e^{\partial}) + D(x)(1-e^{-\partial}), \tag{3.30}$$

$$B(x)\big(\check{P}_n(x) - \check{P}_n(x+1)\big) + D(x)\big(\check{P}_n(x) - \check{P}_n(x-1)\big) = \mathcal{E}(n)\check{P}_n(x). \tag{3.31}$$

3.2 Discrete symmetry and virtual state vectors

Here we introduce discrete symmetry and virtual state vectors of the (q-)Racah system. In ordinary QM, the virtual state solutions are square non-integrable solutions of the Schrödinger equation. In the (q-)Racah case, the Hamiltonians are finite-dimensional real symmetric tri-diagonal matrices. The eigenvalue equation for a given Hamiltonian matrix cannot have any extra solution other than the genuine eigenvectors. Thus we will use the term *virtual state vectors*. As will be shown shortly, virtual state vectors are the 'solutions' of the eigenvalue problem for a *virtual* Hamiltonian $\mathcal{H}'$, except for one of the boundaries, $x = x_{\max}$ (3.46). For the polynomials corresponding to infinite dimensional matrix eigenvalue problems, another type of virtual state vectors is possible [37].

By the twist operation $\mathfrak{t}$ of the parameters:

$$\mathfrak{t}(\boldsymbol{\lambda}) \stackrel{\text{def}}{=} (\lambda_4 - \lambda_1 + 1, \lambda_4 - \lambda_2 + 1, \lambda_3, \lambda_4), \quad \mathfrak{t}^2 = \text{id}, \tag{3.32}$$

we introduce two functions $B'(x)$ and $D'(x)$ by

$$B'(x;\boldsymbol{\lambda}) \stackrel{\text{def}}{=} B\big(x;\mathfrak{t}(\boldsymbol{\lambda})\big), \quad D'(x;\boldsymbol{\lambda}) \stackrel{\text{def}}{=} D\big(x;\mathfrak{t}(\boldsymbol{\lambda})\big), \tag{3.33}$$

namely,

$$B'(x;\boldsymbol{\lambda}) = \begin{cases} -\dfrac{(x+d-a+1)(x+d-b+1)(x+c)(x+d)}{(2x+d)(2x+1+d)} & : \mathrm{R} \\ -\dfrac{(1-a^{-1}dq^{x+1})(1-b^{-1}dq^{x+1})(1-cq^x)(1-dq^x)}{(1-dq^{2x})(1-dq^{2x+1})} & : q\mathrm{R} \end{cases}, \tag{3.34}$$

$$D'(x;\boldsymbol{\lambda}) = \begin{cases} -\dfrac{(x+a-1)(x+b-1)(x+d-c)x}{(2x-1+d)(2x+d)} & : \mathrm{R} \\ -\dfrac{cdq}{ab}\dfrac{(1-aq^{x-1})(1-bq^{x-1})(1-c^{-1}dq^x)(1-q^x)}{(1-dq^{2x-1})(1-dq^{2x})} & : q\mathrm{R} \end{cases}. \tag{3.35}$$

We restrict the parameter range to

$$\mathrm{R}: \;\; d + M < a + b, \qquad q\mathrm{R}: \;\; ab < dq^M, \tag{3.36}$$

in which M is a positive integer and later it will be identified with the possible maximal number of repeated Darboux transformations. It is easy to verify

$$B(x;\boldsymbol{\lambda})D(x+1;\boldsymbol{\lambda}) = \alpha(\boldsymbol{\lambda})^2 B'(x;\boldsymbol{\lambda})D'(x+1;\boldsymbol{\lambda}), \tag{3.37}$$

$$B(x;\boldsymbol{\lambda}) + D(x;\boldsymbol{\lambda}) = \alpha(\boldsymbol{\lambda})\big(B'(x;\boldsymbol{\lambda}) + D'(x;\boldsymbol{\lambda})\big) + \alpha'(\boldsymbol{\lambda}), \tag{3.38}$$

$$B'(x;\boldsymbol{\lambda}) > 0 \;\; (x = 0, 1, \ldots, x_{\max} + M - 1), \tag{3.39}$$

$$D'(x;\boldsymbol{\lambda}) > 0 \;\; (x = 1, 2, \ldots, x_{\max}), \;\; D'(0;\boldsymbol{\lambda}) = D'(x_{\max}+1;\boldsymbol{\lambda}) = 0. \tag{3.40}$$

Here the constant $\alpha(\boldsymbol{\lambda})$ is positive and $\alpha'(\boldsymbol{\lambda})$ is negative:

$$0<\alpha(\boldsymbol{\lambda})=\begin{cases}1 & :\mathrm{R}\\ abd^{-1}q^{-1} & :q\mathrm{R},\end{cases}\quad 0>\alpha'(\boldsymbol{\lambda})=\begin{cases}-c(a+b-d-1) & :\mathrm{R}\\ -(1-c)(1-abd^{-1}q^{-1}) & :q\mathrm{R}.\end{cases}\tag{3.41}$$

The above relations (3.37)–(3.40) imply a linear relation between the original Hamiltonian $\mathcal{H}$ and virtual Hamiltonian $\mathcal{H}'$

$$\mathcal{H}(\boldsymbol{\lambda})=\alpha(\boldsymbol{\lambda})\mathcal{H}'+\alpha'(\boldsymbol{\lambda}),\quad \mathcal{H}'\stackrel{\text{def}}{=}\mathcal{H}\big(\mathfrak{t}(\boldsymbol{\lambda})\big),\tag{3.42}$$

which is defined by the twisted parameters (the $\boldsymbol{\lambda}$ dependence is suppressed for simplicity):

$$\mathcal{H}\big(\mathfrak{t}(\boldsymbol{\lambda})\big)=-\sqrt{B'(x)}\,e^{\partial}\sqrt{D'(x)}-\sqrt{D'(x)}\,e^{-\partial}\sqrt{B'(x)}+B'(x)+D'(x).\tag{3.43}$$

This means that $\mathcal{H}(\mathfrak{t}(\boldsymbol{\lambda}))$ is *positive definite* and it has *no zero-mode.* In other words, the two term recurrence relation determining the 'zero-mode' of $\mathcal{H}(\mathfrak{t}(\boldsymbol{\lambda}))$

$$\begin{aligned}&\mathcal{A}\big(\mathfrak{t}(\boldsymbol{\lambda})\big)=\sqrt{B'(x;\boldsymbol{\lambda})}-e^{\partial}\sqrt{D'(x;\boldsymbol{\lambda})},\\ &\mathcal{A}\big(\mathfrak{t}(\boldsymbol{\lambda})\big)\tilde{\phi}_0(x;\boldsymbol{\lambda})=0\quad(x=0,1,\ldots,x_{\max}-1),\end{aligned}\tag{3.44}$$

can be 'solved' from $x=0$ to $x=x_{\max}-1$ to determine all the components

$$\tilde{\phi}_0(x;\boldsymbol{\lambda})\stackrel{\text{def}}{=}\sqrt{\prod_{y=0}^{x-1}\frac{B'(y;\boldsymbol{\lambda})}{D'(y+1;\boldsymbol{\lambda})}}\qquad(x=0,1,\ldots,x_{\max}).\tag{3.45}$$

But at the end point $x=x_{\max}$, the 'zero-mode' equation (3.44) is not satisfied. The eigenvalue problem for the virtual Hamiltonian

$$\mathcal{H}'(\boldsymbol{\lambda})\tilde{\varphi}_{\mathrm{v}}(x;\boldsymbol{\lambda})=\mathcal{E}'_{\mathrm{v}}(\boldsymbol{\lambda})\tilde{\varphi}_{\mathrm{v}}(x;\boldsymbol{\lambda}),\tag{3.46}$$

can be *solved except for the end point* $x=x_{\max}$ by the factorisation ansatz

$$\tilde{\varphi}_{\mathrm{v}}(x;\boldsymbol{\lambda})\stackrel{\text{def}}{=}\tilde{\phi}_0(x;\boldsymbol{\lambda})\check{\xi}_{\mathrm{v}}(x;\boldsymbol{\lambda}),$$

as in the original (q-)Racah system. By using the explicit form of $\tilde{\phi}_0(x;\boldsymbol{\lambda})$ (3.45), the new Schrödinger equation for $x=0,\ldots,x_{\max}-1$ is rewritten as

$$B'(x;\boldsymbol{\lambda})\big(\check{\xi}_{\mathrm{v}}(x;\boldsymbol{\lambda})-\check{\xi}_{\mathrm{v}}(x+1;\boldsymbol{\lambda})\big)+D'(x;\boldsymbol{\lambda})\big(\check{\xi}_{\mathrm{v}}(x;\boldsymbol{\lambda})-\check{\xi}_{\mathrm{v}}(x-1;\boldsymbol{\lambda})\big)=\mathcal{E}'_{\mathrm{v}}(\boldsymbol{\lambda})\check{\xi}_{\mathrm{v}}(x;\boldsymbol{\lambda}).\tag{3.47}$$

This is the same form of equation as that for the (q-)Racah polynomials. So its solution for $x\in\mathbb{C}$ is given by the (q-)Racah polynomial (3.23) with the twisted parameters:

$$\check{\xi}_{\mathrm{v}}(x;\boldsymbol{\lambda})=\check{P}_{\mathrm{v}}\big(x;\mathfrak{t}(\boldsymbol{\lambda})\big),\qquad \mathcal{E}'_{\mathrm{v}}(\boldsymbol{\lambda})=\mathcal{E}_{\mathrm{v}}\big(\mathfrak{t}(\boldsymbol{\lambda})\big).\tag{3.48}$$

Among such 'solutions' of (3.46), those with the negative energy and having definite sign

$$\check{\xi}_{\rm v}(x;\boldsymbol{\lambda})>0\quad (x=0,1,\ldots,x_{\max},x_{\max}+1;{\rm v}\in\mathcal{V}), \tag{3.49}$$

$$\tilde{\mathcal{E}}_{\rm v}(\boldsymbol{\lambda})<0\quad ({\rm v}\in\mathcal{V}), \tag{3.50}$$

are called the *virtual state vectors*: $\{\tilde{\varphi}_{\rm v}(x)\}$, ${\rm v}\in\mathcal{V}$. The index set of the virtual state vectors is

$$\mathcal{V}=\{1,2,\ldots,{\rm v}_{\max}\},\quad {\rm v}_{\max}=\min\bigl\{[\lambda_1+\lambda_2-\lambda_4-1]',[\tfrac{1}{2}(\lambda_1+\lambda_2-\lambda_3-\lambda_4)]\bigr\}, \tag{3.51}$$

where $[x]$ denotes the greatest integer not exceeding x and $[x]'$ denotes the greatest integer not equal or exceeding x. The negative virtual state energy conditions (3.50) is met by ${\rm v}_{\max}\leq[\lambda_1+\lambda_2-\lambda_4-1]'$. For the positivity of $\check{\xi}_{\rm v}(x;\boldsymbol{\lambda})$ (3.49), we write them down explicitly:

$$\check{\xi}_{\rm v}(x;\boldsymbol{\lambda})=\begin{cases}{}_4F_3\Bigl(\begin{matrix}-{\rm v},\,{\rm v}-a-b+c+d+1,\,-x,\,x+d\\ d-a+1,\,d-b+1,\,c\end{matrix}\Bigm|1\Bigr) & :{\rm R}\\ {}_4\phi_3\Bigl(\begin{matrix}q^{-{\rm v}},\,a^{-1}b^{-1}cdq^{{\rm v}+1},\,q^{-x},\,dq^x\\ a^{-1}dq,\,b^{-1}dq,\,c\end{matrix}\Bigm|q\,;q\Bigr) & :q{\rm R}\end{cases} \tag{3.52}$$

$$=\begin{cases}\displaystyle\sum_{k=0}^{\rm v}\frac{(-{\rm v},{\rm v}-a-b+c+d+1,-x,x+d)_k}{(d-a+1,d-b+1,c)_k}\frac{1}{k!} & :{\rm R}\\ \displaystyle\sum_{k=0}^{\rm v}\frac{(q^{-{\rm v}},a^{-1}b^{-1}cdq^{{\rm v}+1},q^{-x},dq^x;q)_k}{(a^{-1}dq,b^{-1}dq,c;q)_k}\frac{q^k}{(q;q)_k} & :q{\rm R}\end{cases}. \tag{3.53}$$

Each k-th term in the sum is non-negative for $2{\rm v}_{\max}\leq\lambda_1+\lambda_2-\lambda_3-\lambda_4$. As shown shortly (3.60), it is the virtual state polynomial $\check{\xi}_{\rm v}(x)$ rather than the full virtual state vector $\tilde{\varphi}_{\rm v}(x)$ that plays an important role in the Darboux transformations.

For later use, we summarise the properties of the virtual state vectors:

$$\tilde{\phi}_0(x;\boldsymbol{\lambda})\stackrel{\rm def}{=}\phi_0\bigl(x;\mathfrak{t}(\boldsymbol{\lambda})\bigr),\quad \tilde{\varphi}_{\rm v}(x;\boldsymbol{\lambda})\stackrel{\rm def}{=}\phi_{\rm v}\bigl(x;\mathfrak{t}(\boldsymbol{\lambda})\bigr)=\tilde{\phi}_0(x;\boldsymbol{\lambda})\check{\xi}_{\rm v}(x;\boldsymbol{\lambda}), \tag{3.54}$$

$$\check{\xi}_{\rm v}(x;\boldsymbol{\lambda})\stackrel{\rm def}{=}\check{P}_{\rm v}\bigl(x;\mathfrak{t}(\boldsymbol{\lambda})\bigr)=P_{\rm v}\bigl(\eta(x;\mathfrak{t}(\boldsymbol{\lambda}));\mathfrak{t}(\boldsymbol{\lambda})\bigr), \tag{3.55}$$

$$\mathcal{H}(\boldsymbol{\lambda})\tilde{\varphi}_{\rm v}(x;\boldsymbol{\lambda})=\tilde{\mathcal{E}}_{\rm v}(\boldsymbol{\lambda})\tilde{\varphi}_{\rm v}(x;\boldsymbol{\lambda})\quad (x=0,1,\ldots,x_{\max}-1),$$

$$\mathcal{H}(\boldsymbol{\lambda})\tilde{\varphi}_{\rm v}(x_{\max};\boldsymbol{\lambda})\neq\tilde{\mathcal{E}}_{\rm v}(\boldsymbol{\lambda})\tilde{\varphi}_{\rm v}(x_{\max};\boldsymbol{\lambda}),\qquad \mathcal{E}'_{\rm v}(\boldsymbol{\lambda})=\mathcal{E}_{\rm v}\bigl(\mathfrak{t}(\boldsymbol{\lambda})\bigr), \tag{3.56}$$

$$\tilde{\mathcal{E}}_{\rm v}(\boldsymbol{\lambda})=\alpha(\boldsymbol{\lambda})\mathcal{E}'_{\rm v}(\boldsymbol{\lambda})+\alpha'(\boldsymbol{\lambda})=\begin{cases}-(c+{\rm v})(a+b-d-1-{\rm v}) & :{\rm R}\\ -(1-cq^{\rm v})(1-abd^{-1}q^{-1-{\rm v}}) & :q{\rm R}\end{cases}, \tag{3.57}$$

$$\nu(x;\boldsymbol{\lambda})\stackrel{\rm def}{=}\frac{\phi_0(x;\boldsymbol{\lambda})}{\tilde{\phi}_0(x;\boldsymbol{\lambda})}=\begin{cases}\dfrac{\Gamma(1-a)\Gamma(x+b)\Gamma(d-a+1)\Gamma(b-d-x)}{\Gamma(1-a-x)\Gamma(b)\Gamma(x+d-a+1)\Gamma(b-d)} & :{\rm R}\\ \dfrac{(a^{-1}q^{1-x},b,a^{-1}dq^{x+1},bd^{-1};q)_\infty}{(a^{-1}q,bq^x,a^{-1}dq,bd^{-1}q^{-x};q)_\infty} & :q{\rm R}\end{cases}. \tag{3.58}$$

Note that $\alpha'(\boldsymbol{\lambda})=\tilde{\mathcal{E}}_0(\boldsymbol{\lambda})<0$. The function $\nu(x;\boldsymbol{\lambda})$ can be analytically continued into a meromorphic function of x or q^x through the functional relations:

$$\nu(x+1;\boldsymbol{\lambda})=\frac{B(x;\boldsymbol{\lambda})}{\alpha B'(x;\boldsymbol{\lambda})}\nu(x;\boldsymbol{\lambda}),\quad \nu(x-1;\boldsymbol{\lambda})=\frac{D(x;\boldsymbol{\lambda})}{\alpha D'(x;\boldsymbol{\lambda})}\nu(x;\boldsymbol{\lambda}). \tag{3.59}$$

By $B(x_{\max};\boldsymbol{\lambda})=0$, it vanishes for integer $x_{\max}+1\le x\le x_{\max}+M$, $\nu(x;\boldsymbol{\lambda})=0$, and at negative integer points it takes nonzero finite values in general.

3.3 Darboux transformations

Darboux transformations for the simplest QM have been introduced in [31], in which Wronskian's roles are played by Casoratians. The Casorati determinant of a set of n functions $\{f_j(x)\}$ is defined by

$$\mathrm{W}_C[f_1,\ldots,f_n](x)\stackrel{\text{def}}{=}\det\Big(f_k(x+j-1)\Big)_{1\le j,k\le n},$$

(for $n=0$, we set $\mathrm{W}[\cdot](x)=1$), which satisfies identities

$$\begin{aligned}
&\mathrm{W}_C[gf_1,gf_2,\ldots,gf_n](x)=\prod_{k=0}^{n-1}g(x+k)\cdot\mathrm{W}_C[f_1,f_2,\ldots,f_n](x),\\
&\mathrm{W}_C\big[\mathrm{W}_C[f_1,f_2,\ldots,f_n,g],\mathrm{W}_C[f_1,f_2,\ldots,f_n,h]\big](x)\\
&=\mathrm{W}_C[f_1,f_2,\ldots,f_n](x+1)\,\mathrm{W}_C[f_1,f_2,\ldots,f_n,g,h](x)\quad(n\ge 0).
\end{aligned}$$

For simplicity of presentation the parameter ($\boldsymbol{\lambda}$) dependence of various quantities is suppressed in this subsection. Let us deform the (q-)Racah system by a Darboux transformation with a seed 'solution' which is one of the virtual state vectors $\tilde{\varphi}_{d_1}(x)$ ($1\le d_1\in\mathcal{V}$) (3.54).

We rewrite the original Hamiltonian in such a factorisation in which the virtual state vectors $\tilde{\varphi}_{d_1}(x)$ are almost annihilated except for the upper end point:

$$\begin{aligned}
&\mathcal{H}=\hat{\mathcal{A}}_{d_1}^{\dagger}\hat{\mathcal{A}}_{d_1}+\tilde{\mathcal{E}}_{d_1},\\
&\hat{\mathcal{A}}_{d_1}\stackrel{\text{def}}{=}\sqrt{\hat{B}_{d_1}(x)}-e^{\partial}\sqrt{\hat{D}_{d_1}(x)},\quad \hat{\mathcal{A}}_{d_1}^{\dagger}=\sqrt{\hat{B}_{d_1}(x)}-\sqrt{\hat{D}_{d_1}(x)}\,e^{-\partial},\\
&\hat{\mathcal{A}}_{d_1}\tilde{\varphi}_{d_1}(x)=0\ \ (x=0,1,\ldots,x_{\max}-1),\quad \hat{\mathcal{A}}_{d_1}\tilde{\varphi}_{d_1}(x_{\max})\neq 0.
\end{aligned}$$

This is achieved by introducing potential functions $\hat{B}_{d_1}(x)$ and $\hat{D}_{d_1}(x)$ determined by the virtual state polynomial $\check{\xi}_{d_1}(x)$:

$$\hat{B}_{d_1}(x)\stackrel{\text{def}}{=}\alpha B'(x)\frac{\check{\xi}_{d_1}(x+1)}{\check{\xi}_{d_1}(x)},\quad \hat{D}_{d_1}(x)\stackrel{\text{def}}{=}\alpha D'(x)\frac{\check{\xi}_{d_1}(x-1)}{\check{\xi}_{d_1}(x)}.\tag{3.60}$$

We have $\hat{B}_{d_1}(x)>0$ ($x=0,1,\ldots,x_{\max}$), $\hat{D}_{d_1}(0)=\hat{D}_{d_1}(x_{\max}+1)=0$, $\hat{D}_{d_1}(x)>0$ ($x=1,2,\ldots,x_{\max}$) and

$$B(x)D(x+1)=\hat{B}_{d_1}(x)\hat{D}_{d_1}(x+1),\quad B(x)+D(x)=\hat{B}_{d_1}(x)+\hat{D}_{d_1}(x)+\tilde{\mathcal{E}}_{d_1},$$

where use is made of (3.47) in the second equation.

A new deformed Hamiltonian system with $\mathcal{H}_{d_1}$, $\phi_{d_1n}(x)$, $\tilde{\varphi}_{d_1\mathrm{v}}(x)$, is obtained by changing the order of the two matrices $\hat{\mathcal{A}}_{d_1}^{\dagger}$ and $\hat{\mathcal{A}}_{d_1}$:

$$\mathcal{H}_{d_1} \stackrel{\text{def}}{=} \hat{\mathcal{A}}_{d_1}\hat{\mathcal{A}}_{d_1}^{\dagger} + \tilde{\mathcal{E}}_{d_1}, \tag{3.61}$$

$$\phi_{d_1n}(x) \stackrel{\text{def}}{=} \hat{\mathcal{A}}_{d_1}\phi_n(x) \quad (x=0,1,\ldots,x_{\max}; n=0,1,\ldots,n_{\max}), \tag{3.62}$$

$$\tilde{\varphi}_{d_1\mathrm{v}}(x) \stackrel{\text{def}}{=} \hat{\mathcal{A}}_{d_1}\tilde{\varphi}_{\mathrm{v}}(x) + \delta_{x,x_{\max}}\rho_{d_1\mathrm{v}} \quad (x=0,1,\ldots,x_{\max}; \mathrm{v}\in\mathcal{V}\backslash\{d_1\}), \tag{3.63}$$

$$\rho_{d_1\mathrm{v}} \stackrel{\text{def}}{=} -\frac{\sqrt{\alpha B'(x_{\max})}\,\tilde{\phi}_0(x_{\max})}{\sqrt{\check{\xi}_{d_1}(x_{\max})\check{\xi}_{d_1}(x_{\max}+1)}}\,\check{\xi}_{d_1}(x_{\max})\check{\xi}_{\mathrm{v}}(x_{\max}+1), \tag{3.64}$$

$$\mathcal{H}_{d_1}\phi_{d_1n}(x) = \mathcal{E}(n)\phi_{d_1n}(x) \quad (x=0,1,\ldots,x_{\max}; n=0,1,\ldots,n_{\max}), \tag{3.65}$$

$$\begin{aligned}\mathcal{H}_{d_1}\tilde{\varphi}_{d_1\mathrm{v}}(x) &= \tilde{\mathcal{E}}_{\mathrm{v}}\tilde{\varphi}_{d_1\mathrm{v}}(x) \quad (x=0,1,\ldots,x_{\max}-1; \mathrm{v}\in\mathcal{V}\backslash\{d_1\}),\\ \mathcal{H}_{d_1}\tilde{\varphi}_{d_1\mathrm{v}}(x_{\max}) &\neq \tilde{\mathcal{E}}_{\mathrm{v}}\tilde{\varphi}_{d_1\mathrm{v}}(x_{\max}),\end{aligned} \tag{3.66}$$

$$\phi_{d_1n}(x) = \frac{-\sqrt{\alpha B'(x)}\,\tilde{\phi}_0(x)}{\sqrt{\check{\xi}_{d_1}(x)\check{\xi}_{d_1}(x+1)}}\,\mathrm{W}_C\big[\check{\xi}_{d_1}, \nu\check{P}_n\big](x), \tag{3.67}$$

$$\tilde{\varphi}_{d_1\mathrm{v}}(x) = \frac{-\sqrt{\alpha B'(x)}\,\tilde{\phi}_0(x)}{\sqrt{\check{\xi}_{d_1}(x)\check{\xi}_{d_1}(x+1)}}\,\mathrm{W}_C[\check{\xi}_{d_1}, \check{\xi}_{\mathrm{v}}](x). \tag{3.68}$$

The $\rho_{d_1\mathrm{v}}$ term is necessary for the Casoratian expression for $\tilde{\varphi}_{d_1\mathrm{v}}(x)$ in (3.68) to hold at $x = x_{\max}$.

The non-classical orthogonal polynomial is extracted from the above Casoratian $\mathrm{W}_C\big[\check{\xi}_{d_1}, \nu\check{P}_n\big](x)$ by separating certain kinematical factors as will be shown in the next subsection, see (3.78). As in ordinary QM, the positivity of each virtual state vector is inherited by the next generation $\tilde{\varphi}_{d_1\mathrm{v}}$.

3.4 Multi-indexed (q-)Racah polynomials

Let us prepare a set of distinct positive integers which specify the degrees of virtual state vectors:

$$\mathcal{D} = \{d_1, d_2, \ldots, d_M\} \subseteq \mathcal{V}. \tag{3.69}$$

After M-step Darboux transformations, we arrive at the deformed Hamiltonian $\mathcal{H}_{\mathcal{D}}$, $\mathcal{A}_{\mathcal{D}}^{\dagger}$, $\mathcal{A}_{\mathcal{D}}$, and its eigenvectors $\{\phi_{\mathcal{D},n}(x)\}$:

$$\mathcal{H}_{\mathcal{D}} = \mathcal{A}_{\mathcal{D}}^{\dagger}\mathcal{A}_{\mathcal{D}}, \quad \mathcal{H}_{\mathcal{D}}\phi_{\mathcal{D}n}(x) = \mathcal{E}(n)\phi_{\mathcal{D}n}(x), \tag{3.70}$$

$$\phi_{\mathcal{D}n}(x) = \frac{(-1)^M\sqrt{\prod_{j=1}^M \alpha B'(x+j-1)}\,\tilde{\phi}_0(x)\,\mathrm{W}_C[\check{\xi}_{d_1}, \ldots, \check{\xi}_{d_M}, \nu\check{P}_n](x)}{\sqrt{\mathrm{W}_C[\check{\xi}_{d_1}, \ldots, \check{\xi}_{d_M}](x)\,\mathrm{W}_C[\check{\xi}_{d_1}, \ldots, \check{\xi}_{d_M}](x+1)}}, \tag{3.71}$$

$$\mathcal{A}_{\mathcal{D}} \stackrel{\text{def}}{=} \sqrt{B_{\mathcal{D}}(x)} - e^{\partial}\sqrt{D_{\mathcal{D}}(x)}, \quad \mathcal{A}_{\mathcal{D}}^{\dagger} = \sqrt{B_{\mathcal{D}}(x)} - \sqrt{D_{\mathcal{D}}(x)}\,e^{-\partial}, \tag{3.72}$$

with

$$\mathcal{A}_{\mathcal{D}}\phi_{\mathcal{D}\,0}(x)=0 \quad (x=0,1,\ldots,x_{\max}).$$

The potential functions $B_{\mathcal{D}}(x)$ and $D_{\mathcal{D}}(x)$ are:

$$B_{\mathcal{D}}(x) \stackrel{\text{def}}{=} \alpha B'(x+M)\frac{\mathrm{W}_C[\check{\xi}_{d_1},\ldots,\check{\xi}_{d_M}](x)}{\mathrm{W}_C[\check{\xi}_{d_1},\ldots,\check{\xi}_{d_M}](x+1)}\,\frac{\mathrm{W}_C[\check{\xi}_{d_1},\ldots,\check{\xi}_{d_M},\nu](x+1)}{\mathrm{W}_C[\check{\xi}_{d_1},\ldots,\check{\xi}_{d_M},\nu](x)}, \tag{3.73}$$

$$D_{\mathcal{D}}(x) \stackrel{\text{def}}{=} \alpha D'(x)\frac{\mathrm{W}_C[\check{\xi}_{d_1},\ldots,\check{\xi}_{d_M}](x+1)}{\mathrm{W}_C[\check{\xi}_{d_1},\ldots,\check{\xi}_{d_M}](x)}\,\frac{\mathrm{W}_C[\check{\xi}_{d_1},\ldots,\check{\xi}_{d_M},\nu](x-1)}{\mathrm{W}_C[\check{\xi}_{d_1},\ldots,\check{\xi}_{d_M},\nu](x)}. \tag{3.74}$$

In order to extract the multi-indexed (q-)Racah polynomials, we need some auxiliary functions $\varphi(x;\boldsymbol{\lambda})$ and $\varphi_M(x;\boldsymbol{\lambda})$, $M\in\mathbb{Z}_{\geq 0}$:

$$\varphi(x;\boldsymbol{\lambda}) \stackrel{\text{def}}{=} \frac{\eta(x+1;\boldsymbol{\lambda})-\eta(x;\boldsymbol{\lambda})}{\eta(1;\boldsymbol{\lambda})}, \tag{3.75}$$

$$\begin{aligned}\varphi_M(x;\boldsymbol{\lambda}) &\stackrel{\text{def}}{=} \prod_{1\leq j<k\leq M}\frac{\eta(x+k-1;\boldsymbol{\lambda})-\eta(x+j-1;\boldsymbol{\lambda})}{\eta(k-j;\boldsymbol{\lambda})}\\ &= \prod_{1\leq j<k\leq M}\varphi\bigl(x+j-1;\boldsymbol{\lambda}+(k-j-1)\boldsymbol{\delta}\bigr),\end{aligned} \tag{3.76}$$

and $\varphi_0(x;\boldsymbol{\lambda})=\varphi_1(x;\boldsymbol{\lambda})=1$. Two polynomials $\check{\Xi}_{\mathcal{D}}(x;\boldsymbol{\lambda})$ and $\check{P}_{\mathcal{D},n}(x;\boldsymbol{\lambda})$, to be called the *denominator polynomial* and the *multi-indexed orthogonal polynomial*, respectively, are extracted from the Casoratians as follows:

$$\mathrm{W}_C[\check{\xi}_{d_1},\ldots,\check{\xi}_{d_M}](x;\boldsymbol{\lambda})=\mathcal{C}_{\mathcal{D}}(\boldsymbol{\lambda})\varphi_M(x;\boldsymbol{\lambda})\check{\Xi}_{\mathcal{D}}(x;\boldsymbol{\lambda}), \tag{3.77}$$

$$\mathrm{W}_C[\check{\xi}_{d_1},\ldots,\check{\xi}_{d_M},\nu\check{P}_n](x;\boldsymbol{\lambda})=\mathcal{C}_{\mathcal{D},n}(\boldsymbol{\lambda})\varphi_{M+1}(x;\boldsymbol{\lambda})\check{P}_{\mathcal{D},n}(x;\boldsymbol{\lambda})\nu(x;\boldsymbol{\lambda}+M\tilde{\boldsymbol{\delta}}), \tag{3.78}$$

$$\tilde{\boldsymbol{\delta}} \stackrel{\text{def}}{=} (0,0,1,1),\qquad \mathfrak{t}(\boldsymbol{\lambda})+\beta\boldsymbol{\delta}=\mathfrak{t}(\boldsymbol{\lambda}+\beta\tilde{\boldsymbol{\delta}})\quad(\forall\beta\in\mathbb{R}). \tag{3.79}$$

The constants $\mathcal{C}_{\mathcal{D}}(\boldsymbol{\lambda})$ and $\mathcal{C}_{\mathcal{D},n}(\boldsymbol{\lambda})$ are specified later. The eigenvector (3.71) is rewritten as

$$\begin{aligned}\phi^{\text{gen}}_{\mathcal{D}\,n}(x;\boldsymbol{\lambda}) &= (-1)^M\kappa^{\frac{1}{4}M(M-1)}\frac{\mathcal{C}_{\mathcal{D},n}(\boldsymbol{\lambda})}{\mathcal{C}_{\mathcal{D}}(\boldsymbol{\lambda})}\sqrt{\prod_{j=1}^{M}\alpha(\boldsymbol{\lambda})B'\bigl(0;\boldsymbol{\lambda}+(j-1)\tilde{\boldsymbol{\delta}}\bigr)}\\ &\quad\times\frac{\phi_0(x;\boldsymbol{\lambda}+M\tilde{\boldsymbol{\delta}})}{\sqrt{\check{\Xi}_{\mathcal{D}}(x;\boldsymbol{\lambda})\check{\Xi}_{\mathcal{D}}(x+1;\boldsymbol{\lambda})}}\check{P}_{\mathcal{D},n}(x;\boldsymbol{\lambda}).\end{aligned} \tag{3.80}$$

The multi-indexed orthogonal polynomial $\check{P}_{\mathcal{D},n}(x;\boldsymbol{\lambda})$ (3.78) has an expression

$$\check{P}_{\mathcal{D},n}(x;\boldsymbol{\lambda})=\mathcal{C}_{\mathcal{D},n}(\boldsymbol{\lambda})^{-1}\varphi_{M+1}(x;\boldsymbol{\lambda})^{-1}\times\begin{vmatrix}\check{\xi}_{d_1}(x_1)&\cdots&\check{\xi}_{d_M}(x_1)&r_1(x_1)\check{P}_n(x_1)\\ \check{\xi}_{d_1}(x_2)&\cdots&\check{\xi}_{d_M}(x_2)&r_2(x_2)\check{P}_n(x_2)\\ \vdots&\cdots&\vdots&\vdots\\ \check{\xi}_{d_1}(x_{M+1})&\cdots&\check{\xi}_{d_M}(x_{M+1})&r_{M+1}(x_{M+1})\check{P}_n(x_{M+1})\end{vmatrix},\tag{3.81}$$

where $x_j\stackrel{\text{def}}{=}x+j-1$ and $r_j(x)=r_j(x;\boldsymbol{\lambda},M)$ $(1\le j\le M+1)$ are given by

$$r_j(x+j-1;\boldsymbol{\lambda},M)\stackrel{\text{def}}{=}\begin{cases}\dfrac{(x+a,x+b)_{j-1}(x+d-a+j,x+d-b+j)_{M+1-j}}{(d-a+1,d-b+1)_M}&:\text{R}\\[2ex] \dfrac{(aq^x,bq^x;q)_{j-1}(a^{-1}dq^{x+j},b^{-1}dq^{x+j};q)_{M+1-j}}{(abd^{-1}q^{-1})^{j-1}q^{Mx}(a^{-1}dq,b^{-1}dq;q)_M}&:q\text{R}\end{cases}.\tag{3.82}$$

One can show that $\check{\Xi}_{\mathcal{D}}$ (3.77) and $\check{P}_{\mathcal{D},n}$ (3.81) are indeed polynomials in η:

$$\check{\Xi}_{\mathcal{D}}(x;\boldsymbol{\lambda})\stackrel{\text{def}}{=}\Xi_{\mathcal{D}}\big(\eta(x;\boldsymbol{\lambda}+(M-1)\tilde{\boldsymbol{\delta}});\boldsymbol{\lambda}\big),\quad \check{P}_{\mathcal{D},n}(x;\boldsymbol{\lambda})\stackrel{\text{def}}{=}P_{\mathcal{D},n}\big(\eta(x;\boldsymbol{\lambda}+M\tilde{\boldsymbol{\delta}});\boldsymbol{\lambda}\big),\tag{3.83}$$

and their degrees are generically $\ell_{\mathcal{D}}$ and $\ell_{\mathcal{D}}+n$, respectively, with (see (2.62)):

$$\ell_{\mathcal{D}}\stackrel{\text{def}}{=}\sum_{j=1}^{M}d_j-\tfrac{1}{2}M(M-1).\tag{3.84}$$

The involution properties (3.29) of these polynomials are the consequence of those of the basic polynomials $\check{P}_n(x)$ and $\check{\xi}_{d_j}(x)$. We adopt the standard normalisation for $\check{\Xi}_{\mathcal{D}}$ and $\check{P}_{\mathcal{D},n}$: $\check{\Xi}_{\mathcal{D}}(0;\boldsymbol{\lambda})=1$, $\check{P}_{\mathcal{D},n}(0;\boldsymbol{\lambda})=1$, which determine the constants $\mathcal{C}_{\mathcal{D}}(\boldsymbol{\lambda})$ and $\mathcal{C}_{\mathcal{D},n}(\boldsymbol{\lambda})$,

$$\mathcal{C}_{\mathcal{D}}(\boldsymbol{\lambda})\stackrel{\text{def}}{=}\frac{1}{\varphi_M(0;\boldsymbol{\lambda})}\prod_{1\le j<k\le M}\frac{\tilde{\mathcal{E}}_{d_j}(\boldsymbol{\lambda})-\tilde{\mathcal{E}}_{d_k}(\boldsymbol{\lambda})}{\alpha(\boldsymbol{\lambda})B'(j-1;\boldsymbol{\lambda})},\tag{3.85}$$

$$\mathcal{C}_{\mathcal{D},n}(\boldsymbol{\lambda})\stackrel{\text{def}}{=}(-1)^M\mathcal{C}_{\mathcal{D}}(\boldsymbol{\lambda})\tilde{d}_{\mathcal{D},n}(\boldsymbol{\lambda})^2,\ \tilde{d}_{\mathcal{D},n}(\boldsymbol{\lambda})^2\stackrel{\text{def}}{=}\frac{\varphi_M(0;\boldsymbol{\lambda})}{\varphi_{M+1}(0;\boldsymbol{\lambda})}\prod_{j=1}^{M}\frac{\mathcal{E}(n;\boldsymbol{\lambda})-\tilde{\mathcal{E}}_{d_j}(\boldsymbol{\lambda})}{\alpha(\boldsymbol{\lambda})B'(j-1;\boldsymbol{\lambda})}.\tag{3.86}$$

The expression for $\tilde{d}^2_{\mathcal{D},n}$ shows the necessity of the negative virtual state energies $\{\tilde{\mathcal{E}}_{d_j}<0\}$. The lowest degree multi-indexed orthogonal polynomial $\check{P}_{\mathcal{D},0}(x;\boldsymbol{\lambda})$ is related to $\check{\Xi}_{\mathcal{D}}(x;\boldsymbol{\lambda})$ by the parameter shift $\boldsymbol{\lambda}\to\boldsymbol{\lambda}+\boldsymbol{\delta}$:

$$\check{P}_{\mathcal{D},0}(x;\boldsymbol{\lambda})=\check{\Xi}_{\mathcal{D}}(x;\boldsymbol{\lambda}+\boldsymbol{\delta}).\tag{3.87}$$

The potential functions $B_{\mathcal{D}}$ and $D_{\mathcal{D}}$ (3.73)–(3.74) can be expressed neatly in terms of the denominator polynomial:

$$B_{\mathcal{D}}(x;\boldsymbol{\lambda})=B(x;\boldsymbol{\lambda}+M\tilde{\boldsymbol{\delta}})\,\frac{\check{\Xi}_{\mathcal{D}}(x;\boldsymbol{\lambda})}{\check{\Xi}_{\mathcal{D}}(x+1;\boldsymbol{\lambda})}\,\frac{\check{\Xi}_{\mathcal{D}}(x+1;\boldsymbol{\lambda}+\boldsymbol{\delta})}{\check{\Xi}_{\mathcal{D}}(x;\boldsymbol{\lambda}+\boldsymbol{\delta})},\tag{3.88}$$

$$D_{\mathcal{D}}(x;\boldsymbol{\lambda})=D(x;\boldsymbol{\lambda}+M\tilde{\boldsymbol{\delta}})\,\frac{\check{\Xi}_{\mathcal{D}}(x+1;\boldsymbol{\lambda})}{\check{\Xi}_{\mathcal{D}}(x;\boldsymbol{\lambda})}\,\frac{\check{\Xi}_{\mathcal{D}}(x-1;\boldsymbol{\lambda}+\boldsymbol{\delta})}{\check{\Xi}_{\mathcal{D}}(x;\boldsymbol{\lambda}+\boldsymbol{\delta})}.\tag{3.89}$$

The groundstate eigenvector $\phi_{\mathcal{D}0}$ is expressed by $\phi_0(x)$ (3.11) and $\check{\Xi}_{\mathcal{D}}(x;\boldsymbol{\lambda})$:

$$\begin{aligned}\phi_{\mathcal{D}0}(x;\boldsymbol{\lambda})&=\sqrt{\prod_{y=0}^{x-1}\frac{B_{\mathcal{D}}(y)}{D_{\mathcal{D}}(y+1)}}\\&=\phi_0(x;\boldsymbol{\lambda}+M\tilde{\boldsymbol{\delta}})\sqrt{\frac{\check{\Xi}_{\mathcal{D}}(1;\boldsymbol{\lambda})}{\check{\Xi}_{\mathcal{D}}(x;\boldsymbol{\lambda})\check{\Xi}_{\mathcal{D}}(x+1;\boldsymbol{\lambda})}}\,\check{\Xi}_{\mathcal{D}}(x;\boldsymbol{\lambda}+\boldsymbol{\delta})\\&=\psi_{\mathcal{D}}(x;\boldsymbol{\lambda})\check{P}_{\mathcal{D},0}(x;\boldsymbol{\lambda})\propto\phi^{\text{gen}}_{\mathcal{D}0}(x;\boldsymbol{\lambda}),\end{aligned}\tag{3.90}$$

$$\psi_{\mathcal{D}}(x;\boldsymbol{\lambda})\stackrel{\text{def}}{=}\sqrt{\check{\Xi}_{\mathcal{D}}(1;\boldsymbol{\lambda})}\,\frac{\phi_0(x;\boldsymbol{\lambda}+M\tilde{\boldsymbol{\delta}})}{\sqrt{\check{\Xi}_{\mathcal{D}}(x;\boldsymbol{\lambda})\,\check{\Xi}_{\mathcal{D}}(x+1;\boldsymbol{\lambda})}},\quad \psi_{\mathcal{D}}(0;\boldsymbol{\lambda})=1.\tag{3.91}$$

We arrive at the normalised eigenvector $\phi_{\mathcal{D}n}(x;\boldsymbol{\lambda})$ with the orthogonality relation,

$$\phi_{\mathcal{D}n}(x;\boldsymbol{\lambda})\stackrel{\text{def}}{=}\psi_{\mathcal{D}}(x;\boldsymbol{\lambda})\check{P}_{\mathcal{D},n}(x;\boldsymbol{\lambda})\propto\phi^{\text{gen}}_{\mathcal{D}n}(x;\boldsymbol{\lambda}),\quad \phi_{\mathcal{D}n}(0;\boldsymbol{\lambda})=1,\tag{3.92}$$

$$\sum_{x=0}^{x_{\max}}\frac{\psi_{\mathcal{D}}(x;\boldsymbol{\lambda})^2}{\check{\Xi}_{\mathcal{D}}(1;\boldsymbol{\lambda})}\check{P}_{\mathcal{D},n}(x;\boldsymbol{\lambda})\check{P}_{\mathcal{D},m}(x;\boldsymbol{\lambda})=\frac{\delta_{nm}}{d_n(\boldsymbol{\lambda})^2\tilde{d}_{\mathcal{D},n}(\boldsymbol{\lambda})^2}\quad(n,m=0,\ldots,n_{\max}).\tag{3.93}$$

The similarity transformed Hamiltonian is

$$\begin{aligned}\widetilde{\mathcal{H}}_{\mathcal{D}}(\boldsymbol{\lambda})&\stackrel{\text{def}}{=}\psi_{\mathcal{D}}(x;\boldsymbol{\lambda})^{-1}\circ\mathcal{H}_{\mathcal{D}}(\boldsymbol{\lambda})\circ\psi_{\mathcal{D}}(x;\boldsymbol{\lambda})\\&=B(x;\boldsymbol{\lambda}+M\tilde{\boldsymbol{\delta}})\,\frac{\check{\Xi}_{\mathcal{D}}(x;\boldsymbol{\lambda})}{\check{\Xi}_{\mathcal{D}}(x+1;\boldsymbol{\lambda})}\biggl(\frac{\check{\Xi}_{\mathcal{D}}(x+1;\boldsymbol{\lambda}+\boldsymbol{\delta})}{\check{\Xi}_{\mathcal{D}}(x;\boldsymbol{\lambda}+\boldsymbol{\delta})}-e^{\partial}\biggr)\\&\quad+D(x;\boldsymbol{\lambda}+M\tilde{\boldsymbol{\delta}})\,\frac{\check{\Xi}_{\mathcal{D}}(x+1;\boldsymbol{\lambda})}{\check{\Xi}_{\mathcal{D}}(x;\boldsymbol{\lambda})}\biggl(\frac{\check{\Xi}_{\mathcal{D}}(x-1;\boldsymbol{\lambda}+\boldsymbol{\delta})}{\check{\Xi}_{\mathcal{D}}(x;\boldsymbol{\lambda}+\boldsymbol{\delta})}-e^{-\partial}\biggr),\end{aligned}\tag{3.94}$$

and the multi-indexed orthogonal polynomials $\check{P}_{\mathcal{D},n}(x;\boldsymbol{\lambda})$ are its eigenpolynomials:

$$\widetilde{\mathcal{H}}_{\mathcal{D}}(\boldsymbol{\lambda})\check{P}_{\mathcal{D},n}(x;\boldsymbol{\lambda})=\mathcal{E}(n;\boldsymbol{\lambda})\check{P}_{\mathcal{D},n}(x;\boldsymbol{\lambda}).\tag{3.95}$$

It should be stressed that these multi-indexed orthogonal polynomials in simple QM provide infinitely many examples of exactly solvable birth and death processes [15, 16, 41].

3.4.1 Explicit examples of multi-indexed (q-)Racah polynomials

We present a simple and explicit example of multi-indexed orthogonal polynomials of the Racah and q-Racah systems corresponding to $\mathcal{D} = \{\ell\}$, ($\ell \geq 1$, $M = 1$). In this case $\check{\Xi}_\ell(x;\boldsymbol{\lambda}) \equiv \check{\xi}_\ell(x;\boldsymbol{\lambda})$ (3.52). The 1-indexed (q-)Racah polynomials are obtained by evaluating (3.81) at $M = 1$. They are a degree $\ell + n$ polynomial in $\eta = x(x+d+1)$ (R) and $\eta = (q^{-x}-1)(1-dq^{x+1})$ (qR):

$$\begin{aligned}
\text{R: } \check{P}_{\ell,n}(x;\boldsymbol{\lambda}) &= \frac{c}{(2x+d+1)\big(\mathcal{E}(n;\boldsymbol{\lambda}) - \tilde{\mathcal{E}}_\ell(\boldsymbol{\lambda})\big)} \\
&\quad \times \Big(\check{\xi}_\ell(x;\boldsymbol{\lambda})(x+a)(x+b)\check{P}_n(x+1;\boldsymbol{\lambda}) \\
&\qquad - \check{\xi}_\ell(x+1;\boldsymbol{\lambda})(x+d-a+1)(x+d-b+1)\check{P}_n(x;\boldsymbol{\lambda})\Big), \\
q\text{R: } \check{P}_{\ell,n}(x;\boldsymbol{\lambda}) &= \frac{1-c}{(1-dq^{2x+1})\big(\mathcal{E}(n;\boldsymbol{\lambda}) - \tilde{\mathcal{E}}_\ell(\boldsymbol{\lambda})\big)} \\
&\quad \times \Big(\check{\xi}_\ell(x;\boldsymbol{\lambda})(aq^x, bq^x; q)\check{P}_n(x+1;\boldsymbol{\lambda}) \\
&\qquad - \check{\xi}_\ell(x+1;\boldsymbol{\lambda})abd^{-1}q^{-1}(a^{-1}dq^{x+1}, b^{-1}dq^{x+1}; q)\check{P}_n(x;\boldsymbol{\lambda})\Big).
\end{aligned}$$

They satisfy a second order difference equation:

$$\begin{aligned}
&\widetilde{\mathcal{H}}_\ell(\boldsymbol{\lambda})\check{P}_{\ell,n}(x;\boldsymbol{\lambda}) = \mathcal{E}(n;\boldsymbol{\lambda})\check{P}_{\ell,n}(x;\boldsymbol{\lambda}), \\
&\widetilde{\mathcal{H}}_\ell(\boldsymbol{\lambda}) = B(x;\boldsymbol{\lambda}+\tilde{\boldsymbol{\delta}})\,\frac{\check{\xi}_\ell(x;\boldsymbol{\lambda})}{\check{\xi}_\ell(x+1;\boldsymbol{\lambda})}\left(\frac{\check{\xi}_\ell(x+1;\boldsymbol{\lambda}+\boldsymbol{\delta})}{\check{\xi}_\ell(x;\boldsymbol{\lambda}+\boldsymbol{\delta})} - e^{\partial}\right) \\
&\qquad + D(x;\boldsymbol{\lambda}+\tilde{\boldsymbol{\delta}})\,\frac{\check{\xi}_\ell(x+1;\boldsymbol{\lambda})}{\check{\xi}_\ell(x;\boldsymbol{\lambda})}\left(\frac{\check{\xi}_\ell(x-1;\boldsymbol{\lambda}+\boldsymbol{\delta})}{\check{\xi}_\ell(x;\boldsymbol{\lambda}+\boldsymbol{\delta})} - e^{-\partial}\right), \\
&\mathcal{E}(n;\boldsymbol{\lambda}) = \begin{cases} n(n+\tilde{d}), & \text{R}, \\ (q^{-n}-1)(1-\tilde{d}q^n), & q\text{R}. \end{cases}
\end{aligned}$$

3.5 Krein-Adler polynomials

The Krein-Adler polynomials for the systems in simplest QM or discrete QM with real shifts have been presented in [33]. Most formulas look quite similar to those of the multi-indexed polynomials in §3.4, (3.69)–(3.74), (3.80)–(3.95). Now $\mathcal{D}$ (3.69) denotes the degrees of eigenpolynomials to be used as seed solutions and it has to satisfy the same conditions as (2.46). The virtual state polynomial $\check{\xi}_{d_j}(x)$ is replaced by the eigenpolynomial $\check{P}_{d_j}(x)$.

4 Summary and comments

Amongst many achievements of the quantum mechanical reformulation of the theory of classical orthogonal polynomials, we have reported on the subject of non-classical orthogonal polynomials, which satisfy second order differential or difference equations. They form a complete set of orthogonal vectors in a certain Hilbert space in spite of a 'hole' in their degrees, which is essential to evade the constraints due to Routh and Bochner. Due to length limitation, the parallel results for the group of polynomials including the Wilson and Askey-Wilson polynomials [30] have been omitted. For readability and consideration of length we have skipped various interesting and important arguments. Among them an intriguing problem is what would replace the three term recurrence relations, or the *bi-spectrality* [13] in a wider context, for the non-classical orthogonal polynomials [42, 36, 10, 12, 22]. For more details of the multi-indexed Laguerre and Jacobi polynomials, see [34] and [32] for (q-) Racah systems.

We apologise to various authors whose many important ideas and contributions cannot be reported or cited due to the constraint on length.

References

[1] Adler V É, A modification of Crum's method, *Theor. Math. Phys.* **101** (1994) 1381-1386.

[2] Andrews G E, Askey R and Roy R, *Special Functions, Encyclopedia of mathematics and its applications*, vol. 71 Cambridge Univ. Press, Cambridge, (1999).

[3] Atakishiyev N M and Klimyk A U, Duality of q-polynomials, orthogonal on countable sets of points, *Electr. Trans. Numer. Anal.* **24** (2006) 108-180, `arXiv:math/0411249[math.CA]`.

[4] Bagchi B, Quesne C and Roychoudhury R, Isospectrality of conventional and new extended potentials, second-order supersymmetry and role of PT symmetry, *Pramana J. Phys.* **73** (2009) 337-347, `arXiv:0812.1488[quant-ph]`.

[5] Bochner S, Über Sturm-Liouvillesche Polynomsysteme, *Math. Zeit.* **29** (1929) 730-736.

[6] Cooper F, Khare A and Sukhatme U, Supersymmetry and quantum mechanics, *Phys. Rep.* **251** (1995) 267-385.

[7] Crum M M, Associated Sturm-Liouville systems, *Quart. J. Math. Oxford Ser.* (2) **6** (1955) 121-127, `arXiv:physics/9908019[physics.hist-ph]`.

[8] Darboux G, Sur une proposition relative aux équations linéaires, *C. R. Acad. Paris* **94** (1882) 1456-1459.

[9] Dubov S Yu, Eleonskiĭ V M and Kulagin N E, Equidistant spectra of anharmonic oscillators, *Soviet Phys. JETP* **75** (1992) 446-451.

[10] Durán A J, Higher order recurrence relation for exceptional Charlier, Meixner, Hermite and Laguerre orthogonal polynomials, *Integral Transforms Spec. Funct.* **26** (2015) 357-376, `arXiv:1409.4697[math.CA]`.

[11] Gómez-Ullate D, Kamran N and Milson R, An extension of Bochner's problem: exceptional invariant subspaces, *J. Approx. Theory* **162** (2010) 987-1006, `arXiv:0805.3376[math-ph]`; An extended class of orthogonal polynomials defined by a Sturm-Liouville problem, *J. Math. Anal. Appl.* **359** (2009) 352-367, `arXiv:0807.3939[math-ph]`.

[12] Gómez-Ullate D, Kasman A, Kuijlaars A B J and R. Milson, Recurrence relations for exceptional Hermite Polynomials, *J. Approx. Theory* **204** (2016) 1-16, `arXiv:1506.03651[math.CA]`.

[13] Grünbaum F A and Haine L, Bispectral Darboux transformations: an extension of the Krall polynomials, *IMRN* (1997), No. 8, 359–392.

[14] Infeld L and Hull T E, The factorization method, *Rev. Mod. Phys.* **23** (1951) 21-68.

[15] Ismail M E H, *Classical and quantum orthogonal polynomials in one variable, Encyclopedia of mathematics and its applications*, vol. 98, Cambridge Univ. Press, Cambridge, (2005).

[16] Karlin S and McGregor J L, The differential equations of birth-and-death processes, *Trans. Amer. Math. Soc.* **85** (1957) 489-546.

[17] Koekoek R, Lesky P A and Swarttouw R F, *Hypergeometric orthogonal polynomials and their q-analogues,* Springer Monographs in Mathematics, Springer-Verlag, Berlin, (2010).

[18] Krall H L, On higher derivatives of orthogonal polynomials, *Bull. Amer. Math. Soc.* **42** (1936) 867-870; Certain differential equations for Tchebyshev polynomials, *Duke Math. J.* **4** (1938) 705-719;
Krall H L and Frink O, A new class of orthogonal polynomials. *Trans. Amer. Math. Soc.* **65** (1949) 100-115.

[19] Krall A M, *Hilbert space, boundary value problems and orthogonal polynomials*, Operator Theory Advances and Applications **133**, Birkhäuser, Basel (2002).

[20] Krein M G, On continuous analogue of Christoffel's formula in orthogonal polynomial theory (Russian), *Doklady Acad. Nauk.* CCCP, **113** (1957) 970-973.

[21] Leonard D, Orthogonal polynomials, duality, and association schemes, *SIAM J. Math. Anal.* **13** (1982) 656-663.

[22] Miki H and Tsujimoto S, A new recurrence formula for generic exceptional orthogonal polynomials, *J. Math. Phys.* **56** (2015) 033502 (13pp), `arXiv:1410.0183[math.CA]`.

[23] Nikiforov A F, Suslov S K and Uvarov V B, *Classical Orthogonal Polynomials of a Discrete Variable*, Springer-Verlag, Berlin (1991).

[24] Odake S and Sasaki R, Orthogonal Polynomials from Hermitian Matrices, *J. Math. Phys.* **49** (2008) 053503 (43 pp), `arXiv:0712.4106[math.CA]`.

[25] Odake S and Sasaki R, Exactly solvable 'discrete' quantum mechanics; shape invariance, Heisenberg solutions, annihilation-creation operators and coherent states, *Prog. Theor. Phys.* **119** (2008) 663-700, `arXiv:0802.1075[quant-ph]`.

[26] Odake S and Sasaki R, Crum's theorem for 'discrete' quantum mechanics, *Prog. Theor. Phys.* **122** (2009) 1067-1079, `arXiv:0902.2593[math-ph]`.

[27] Odake S and Sasaki R, Infinitely many shape invariant potentials and new orthogonal polynomials, *Phys. Lett.* **B679** (2009) 414-417, `arXiv:0906.0142 [math-ph]`.

[28] Odake S and Sasaki R, Infinitely many shape invariant potentials and cubic identities of the Laguerre and Jacobi polynomials, *J. Math. Phys.* **51** (2010) 053513 (9 pp), `arXiv:0911.1585[math-ph]`.

[29] Odake S and Sasaki R, Another set of infinitely many exceptional (X_ℓ) Laguerre polynomials, *Phys. Lett.* **B684** (2010) 173-176, `arXiv:0911.3442[math-ph]`.

[30] Odake S and Sasaki R, Exceptional Askey-Wilson type polynomials through Darboux-Crum transformations, *J. Phys.* **A43** (2010) 335201 (18pp), `arXiv: 1004.0544[math-ph]`; Multi-indexed Wilson and Askey-Wilson polynomials, *J. Phys.* **A46** (2013) 045204 (22 pp), `arXiv:1207.5584[math-ph]`.

[31] Odake S and Sasaki R, Dual Christoffel transformations, *Prog. Theor. Phys.* **126** (2011) 1-34, `arXiv:1101.5468[math-ph]`.

[32] Odake S and Sasaki R, Exceptional (X_ℓ) (q)-Racah polynomials, *Prog. Theor. Phys.* **125** (2011) 851-870, `arXiv:1102.0812[math-ph]`; Multi-indexed (q)-Racah polynomials, *J. Phys.* **A45** (2012) 385201 (21 pp), `arXiv:1203.5868 [math-ph]`.

[33] Odake S and Sasaki R, Discrete quantum mechanics, (Topical Review) *J. Phys.* **A44** (2011) 353001 (47 pp), `arXiv:1104.0473[math-ph]`.

[34] Odake S and Sasaki R, Exactly solvable quantum mechanics and infinite families of multi-indexed orthogonal polynomials, *Phys. Lett.* **B702** (2011) 164-170, `arXiv:1105.0508[math-ph]`.

[35] Odake S and Sasaki R, Krein-Adler transformations for shape-invariant potentials and pseudo virtual states, *J. Phys.* **A46** (2013) 245201 (24pp), `arXiv:1212.6595[math-ph]`.

[36] Odake S, Recurrence Relations of the Multi-Indexed Orthogonal Polynomials, *J. Math. Phys.* **54** (2013) 083506 (18pp), `arXiv:1303.5820[math-ph]`; Recurrence Relations of the Multi-Indexed Orthogonal Polynomials : II, *J.*

Math. Phys. **56** (2015) 053506 (18pp), `arXiv:1410.8236[math-ph]`; Recurrence Relations of the Multi-Indexed Orthogonal Polynomials : III, *J. Math. Phys.* **57** (2016) 023514 (24pp), `arXiv:1509.08213[math-ph]`; Recurrence Relations of the Multi-Indexed Orthogonal Polynomials: IV: closure relations and creation/annihilation operators, *J. Math. Phys.* **57** (2016) 113503 (22pp), `arXiv:1606.0283[math-ph]`.

[37] Odake S and Sasaki R, Multi-indexed Meixner and little q-Jacobi (Laguerre) Polynomials, *J. Phys.* **A50** (2017) 165204 (23pp), `arXiv:1610.09854 [math-ph]`.

[38] Odake S and Sasaki R, Orthogonal polynomials from hermitian matrices II, *J. Math. Phys.* **59** (2018) No.1 013504 (42pp), `arXiv:1604.00714[math.CA]`.

[39] Quesne C, Exceptional orthogonal polynomials, exactly solvable potentials and supersymmetry, *J. Phys.* **A41** (2008) 392001 (6 pp), `arXiv:0807.4087 [quant-ph]`.

[40] Routh E, On some properties of certain solutions of a differential equation of the second order, *Proc. London Math. Soc.* **16** (1884) 245-261.

[41] Sasaki R, Exactly solvable birth and death processes, *J. Math. Phys.* **50** (2009) 103509 (18pp), `arXiv:0903.3097[math-ph]`.

[42] Sasaki R, Tsujimoto S and Zhedanov A, Exceptional Laguerre and Jacobi polynomials and the corresponding potentials through Darboux-Crum transformations, *J. Phys.* **A43** (2010) 315204, `arXiv:1004.4711[math-ph]`.

[43] Terwilliger P, Leonard pairs and the q-Racah polynomials, *Linear Algebra Appl.* **387** (2004) 235–276, `arXiv:math/0306301[math.QA]`;
Terwilliger P and Vidunas R. Leonard pairs and the Askey-Wilson relations, *J. Algebra Appl.* **3** (2004) no. 4, 411–426, `arXiv:math/0305356[math.QA]`;

[44] Terwilliger P, Two linear transformations each tridiagonal with respect to an eigenbasis of the other, *Linear Algebra Appl.* **330** (2001) no. 1-3, 149–203; Two linear transformations each tridiagonal with respect to an eigenbasis of the other; an algebraic approach to the Askey scheme of orthogonal polynomials, `arXiv:math/0408390[math.QA]`.

◎ 编辑手记

世界著名数学家 E. M. 劳埃德(E. M. Llyod)曾指出:

> 创造一个有活力的数学模型是一项需要了不起的技能和判断力的迷人的工作,必须在过细的描述与过简的描述之间做出合理的权衡. 过细就会造成一组无法解决的数学问题,过简则可能掩盖有兴趣的细节,致使无法发现所要寻求的最优预报. 一个在这种意义上合理地接近所考虑的过程的模型可以用来设计、预言和发现最优的工作条件,使得所花代价最小,效率最大.

本书就是这样一本成功的创造了一个优秀数学模型的英文专著,中文书名或可译为《非线性系统及其绝妙的数学结构:第 2 卷》.

本书的主编共有二位:诺伯特·欧拉(Norbert Euler)和玛丽亚·克拉拉·努奇(Maria Clara Nucci).

下面分别介绍一下这两位主编:

诺伯特·欧拉,墨西哥人,目前是国际科学中心 A. C. (墨西哥库埃纳瓦卡)的客座教授. 25 年来,他一直在全球多所大学教授各种本科和研究生水平的数学课程. 他是一位活跃的研究人员,迄今为止已发表了 80 多篇关于非线性系统主题的同行评审研究文章,也是多本书的合著者,还参与了一些国际期刊的编辑工作.

玛丽亚·克拉拉·努奇,意大利人,意大利佩鲁贾大学(University of Perugia)数学物理副教授,她曾以优异的成绩毕业于该大学. 1986 ~ 1991 年间,她在美国亚特兰大佐治亚理工学院(Georgia Institute of Technology)担任客座助理教授. 她还曾被澳大利亚、加拿大、法国、德国的大学邀请. 她在许多国际会议和研讨会上展示过她的研究成果. 从 1995 ~ 2009 年,她担任了《数学分析与应用》杂志的副主编,自 2005 年起担任《非线性数学物理学》杂志的编委. 她是 100 多篇出版物的作者或合著者. 她兴趣广泛,研究领域包括流体力学、刚体力学、流行病学、天体物理学、数学史和量子力学.

正如本书前言中所述:

非线性系统是由于其自身的优势而被发现和研究的,因为它描述了从大的波浪的运动到最小粒子的振动的世界.事实上,人们发现它们的数学结构是如此的丰富多样,以至于人们需要用非常广泛的多重视角才能理解它们.在本书中读者能够通过不同的权威作者的文章来找到这些多重视角,这些文章补充了《非线性系统及其绝妙的数学结构:第1卷》中的作者在构建非线性问题方面的工作,这是一项真正的至关重要的任务.本书包含17篇由非线性系统的不同方面的顶尖专家撰写的特邀论文,包括常微分方程和偏微分方程、差分方程、离散或格点方程、非交换方程和矩阵方程以及时滞方程.本书内容被分为三个主要的部分:Part A:可积性,Lax对与对称性.Part B:代数与几何方法.Part C:应用.下面我们对每部分给出一个简短的描述.

Part A包含7篇论文,分别是A1到A7.在这部分中,作者主要解决了如何检测可积系统的基本问题,以及非线性系统对称方法的使用问题.在A1中,作者P.阿尔巴雷斯(P. Albres)、P. G.埃斯特韦斯(P. G. Estévez)和C.萨尔东(C. Sardón)讨论了与不同的可积方程、带有可积方程的不可积方程,以及构造Lax对相关的反向变换的作用.在A2中,作者M.布拉萨克(M. Blaszak)和A.谢尔盖耶夫(A. Sergyeyev)回顾了构建(3+1)维可积无散射(也被称为流体力学型)偏微分系统的现代方法,该系统的基础是Lax对,包括相关的R-矩阵理论,并通过几个例子进行了说明.在A3中,作者T. J.布里奇曼(T. J. Bridgman)和W.赫尔曼(W. Hereman)构建了非线性偏差分系统的Lax对,因为可积非线性偏差分方程的基本特征之一实际上是存在Lax对.在A4中,作者V. A.多罗德尼岑(V. A. Dorodnitsyn)、R.科兹洛夫(R. Kozlov)、S. V.梅列什科(S. V. Meleshko)和P.温特尼茨(P. Winternitz)考虑了单时滞标量一阶时滞常微分方程,讨论了它们的对称性质,展现了某些不变量解法.在A5中,作者R.赫尔南德斯·赫雷德罗(R. Hernández Heredero)和V.索科洛夫(V. Sokolov)通过更高的对称性回顾了可积偏微分方程的分类.在A6中,作者C.穆丽尔(C. Muriel)和J. L.罗梅罗(J. L. Romero)描述了常微分方程的λ对称性的主要应用,以及它在偏微分方程中的扩展.在A7中,作者M. C.努齐(M. C. Nucci)给出了对heir方程方法的回顾,以及它在确定偏微分方程的额外对称解、非经典对称性和广义对称性中的作用.

Part B包含了7篇论文,分别是B1到B7.在这部分中作者描述了获得非线性系统显式解和(或)描述系统解的结构的不同方法.在B1中,作者A.德加斯佩里斯(A. Degasperis)、S.隆巴尔多(S. Lombardo)和M.索马卡尔(M. Sommacal)在可积性框架下研究了两个耦合非线性薛定谔(Schrödinger)方程的连续波的解的线性稳定性问题.在B2中,作者D.戈麦斯-乌利亚特(D. Gómez-Ullate)、Y.格兰蒂(Y. Grandati)和R.米尔森(R. Milson)关注了潘勒韦(Painlevé,旧译"班乐卫")方程和系统的有理解的分类,并在用经典正交多项式描述了行列式表示后,明确构造了潘勒韦IV方程、潘勒韦V方程和A_4潘勒韦系统的埃尔米特(Hermite)型解.在B3中,作者A.N.W.霍恩(A.H.W.Hone)、P.兰普(P.Lampe)和T.E.库卢卡斯(T.E.Kouloukas)给出了对集群代数的介绍,并解释了离散可积系统是如何出现在聚类突变中的.在B4中,作者N.乔西(N.Joshi)和中佐野(Nakazono)通过使用境正一郎(Sakai Shoichiro)几何方法回顾了椭圆差分潘勒韦方程的构造,并给出了更多的例子.在B5中,作者S.加治(S.Kaji)、K.梶原(K.Kajiwara)和H.帕克(H.Park)研究了由用铰链连接起来的恒等连接的副本组成的一系列连接(四面体旋转环),其特征是具有恒定速度和恒定扭力的离散曲线,并通过半离散改进的科尔泰沃赫-德弗里斯(Korteweg-de Vries)(KdV)方程在四面体旋转环的构形空间中来定义流动.在B6中,作者J-P马格诺特(J-P.Magnot)和E.G.雷耶斯(E.G.Reyes)通过使用弗洛利歇尔(Frölicher)广义微分几何和微分空间展示了卡多姆采夫-彼得维亚什维利(Kadomtsev-Petviashvili)分层的柯西(Cauchy)问题的适定性的证明.在B7中,作者D.J.张(D.J.Zhang)以朗斯基(Wronskian)形式对朗斯基方法和解决方案进行了回顾,并添加了几个示例,包括科尔泰沃赫-德弗里斯、阿布罗维茨-考普-纽厄尔-塞居尔(Ablowitz-Kaup-Newell-Segu)分层和归约,以及格点势科尔泰沃赫-德弗里斯方程.

Part C包含了三篇论文,分别是C1到C3.作者为了解决特殊的非线性问题应用了不同的方法.在C1中,作者R.卡马萨(R.Camassa)讨论了艾里(Airy)浅水系统的某些精确解的性质,揭示了当层厚度消失时动力学的特殊特征.在C2中,作者X.常(X.Chang)和J.斯米盖尔斯基(J.Szmigielski)对施图姆-刘维尔(Sturm-Liouville)系统的等光谱变形进行了一般性的讨论,该讨论根据双组分修正的卡马萨-霍尔姆(Camassa-Holm)方程进行了解释,并针对离散测量(多峰)的情况进行了求解.在C3中,作者R.佐佐木(R.Sasaki)对满足二阶微分或差分方程的正交多项式进行了精确的和综合的回顾,该内容可以从经典正交多项式理论的量子力学重构中推导出来.

本书的目录为:

前言
作者
Part A:可积性,Lax对与对称性
A1:互逆变换及其在可积性中的作用与PDEs的分类
A2:将Lax对与相伴(3+1)维可积无散射系统联系起来
A3:偏差分方程的边约束布辛奈斯克(Boussinesq)系统的Lax对
A4:时滞常微分方程的李(Lie)点对称
A5:可积性的对称方法:研究进展
A6:λ-对称概念的演变与主要应用
A7:偏微分方程的heir方程:25年的回顾
Part B:代数与几何方法

B1:耦合非线性薛定谔方程:平面波的光谱与不稳定性

B2:潘勒韦系统的有理解

B3:簇代数与离散可积性

B4:椭圆差分潘勒韦方程的回顾

B5:由离散空间曲线的可积变形控制的连接机制

B6:卡多姆采夫-彼得维亚什维利分层的柯西问题与无限维群

B7:可积系统的朗斯基解

Part C:应用

C1:浅水模型中的全球梯度突变:通过拉伸坐标进行进化

C2:弹性杆的振动,等谱变形与改善的卡马萨-霍尔姆方程

C3:完全可解的(离散)量子力学与新的正交多项式

特别值得一提的是本书的 Part B 中的 B2 和 B4 都提到了潘勒韦这个人,他是法国前总理,更是一位被遗忘的数学奇人①.

潘勒韦一生在出世的数学家和入世的政治家两个不同的角色之间切换,尽管他在政治生涯中取得了最高职位,但他作为数学家的贡献更为耀眼.

17 世纪以来,法国历史上出现了众多一流数学家:笛卡儿(Descartes)、韦达(Viète)、帕斯卡(Pascal)、费马(Fermat)、拉格朗日(Lagrange)、拉普拉斯(Laplace)、达朗贝尔(d'Alembert)、勒让德(Legendre)、蒙日(Monge)、彭赛列(Poncelet)、柯西(Cauchy)、傅里叶(Fourier)、伽罗瓦(Galois)、庞加莱(Poincaré)、阿达马(Hadamard)、格罗滕迪克(Grothendieck)等大师级人物,如天空中的群星般璀璨,不可胜数. 与人们印象中对于数学家的刻板印象不同,很多法国数学家热心于社会政治活动,在法国还有数学家从政的传统. 例如:1799 年,拉普拉斯曾给数学爱好者拿破仑(Napoléon)当过六个星期的内政部长;1831 年,伽罗瓦两度因政治原因入狱;2010 年,菲尔兹奖得主 C. 维拉尼(C. Villani)出任法国国民议会议员. 数学家、政治家和航空赞助人保罗·潘勒韦(1863—1933)也是这样一位奇人.

1. 所处时代里最著名的数学家之一

潘勒韦出生在巴黎的一个工匠家庭,他的童年正值法国的动荡年代,从小在文学和科学方面都具有天赋. 直到中学毕业前,潘勒韦都尚未决定自己的人生方向,在政治和工程之间举棋不定,但他最终选择了科学生涯. 1883 年,潘勒韦进入巴黎高等师范学校(cole normale supérieure,简称巴黎高师)学习数学,在P. 阿佩尔(P. Appel)、G. 达布(G. Darboux)、C. 埃尔米特、J. 皮卡(J. Picard)、庞加莱和 J. 塔内里(J. Tannery)等教授的影响下被数学深深吸引. 他在他的博士论

① 引自《科普中国》,2022 年 09 月 23 日.

文导师——当时法国最杰出的数学家之一皮卡的建议下,1886 年前往德国哥廷根大学(Universität Göttingen)跟随 F. 克莱因(F. Klein)和 H. A. 施瓦茨(H. A. Schwarz)深造,次年以题为《关于解析函数的奇异线》的论文获得了巴黎西岱大学(Université Paris Cité)博士学位.

当时法国领先学者的标准职业道路是在外省获得第一个教职,然后尝试返回巴黎. 潘勒韦博士毕业后被聘为里尔大学(Université de Lille)数学和应用力学讲师,1892 年回到巴黎,先后在巴黎西岱大学、巴黎综合理工学院(École Polytechnique)和法兰西公学院(Le Collège de France)任教,1903 年成为巴黎西岱大学数学教授,1905 年成为巴黎综合理工学院力学教授. 潘勒韦的主要研究领域涉及微分方程及分析力学,他对数学最早的兴趣是代数曲线和曲面的有理变换,提出了双均匀变换的概念,并对非线性分析理论表现出了极大兴趣. 潘勒韦具有敏锐的数学直觉,他有一句名言:"在实数域的两个真理之间,最简单和最短的路径通常穿过复数域."

在瑞典现代数学之父 G. 米塔-列夫勒(G. Mittag-Leffler)的斡旋下,1888 年庞加莱因对三体问题的研究获得了瑞典和挪威君合国国王奥斯卡二世颁发的数学奖. 1895 年 9 月至 1895 年 11 月,同样对三体问题感兴趣的潘勒韦应国王之邀前往斯德哥尔摩大学(Stockholm University)讲学. 潘勒韦的讲义《微分方程的解析理论教程》两年后出版,其中包括对 n 体问题奇异性的第一次系统研究. 例如他证明了"三体问题的奇点都是碰撞奇点",并提出了著名的"潘勒韦猜想":当 $n>3$ 时,n 体问题存在非碰撞奇点. 用通俗的话来说,就是如果系统中有三颗以上的星球,就可以将其中一颗甩到无穷远处. 留美中国数学家夏志宏(1992 年)和薛金鑫(2014 年)分别证明了当 $n\geqslant5$ 和 $n=4$ 时潘勒韦猜想成立. 潘勒韦取得的最重要的成就之一是发现了后来以他的名字命名的非线性常微分方程以及新的超越函数. 众所周知,线性常微分方程可以使用初等函数或经典的特殊函数求解,而求解非线性微分方程比求解线性方程困难得多. 19 世纪发现的椭圆函数扩展了特殊函数族,可用来求解一类二阶非线性常微分方程. 潘勒韦利用 W. 魏尔斯特拉斯(K. Weierstrass)、L. 富克斯(L. Fuchs)和 S. 柯瓦列夫斯卡娅(S. V. Kovalevskaya)的想法,研究了一类二阶非线性常微分方程,其通解的二阶导数是自身和一阶导数的有理函数,该函数在复平面上局部解析,并且通解没有可移动临界奇点. 这类方程被称为具有"潘勒韦特性",有些文章中(如维基百科)"唯一可移动的奇点是极点"的定义是错误的.

潘勒韦与 B. O. 冈比埃(B. O. Gambier)、R. 富克斯(R. Fuchs)等发现,具有潘勒韦特性的非线性常微分方程总是可以转化为 50 种规范形式之一,其中 44 个方程可以经约化后使用已知函数求解,只有 6 个方程需要引进"新的"超越函数. 这 6 个常微分方程被称为"潘勒韦方程",而其解被称为"潘勒韦超越函数",具有与经典特殊函数非常不同的性质. 某些潘勒韦方程的不可约性一直是一个有争议的话题,20 世纪 80 年代末,日本数学家 K. 西冈(K. Nishioka)和 H. 梅村(H. Umemura)证明了所有潘勒韦方程都不可约化为线性方程或利用椭圆函数求解. 由于在现代几何、量子场论、可积系统和统计力学中的应用,近年来潘勒韦超越函数重新引起了数学界的兴趣,并且被推广到了高阶非线性常微分方程以及非线性偏微分方程的研究.

潘勒韦系统地分析了刚体系统的运动,其中涉及在滑动过程中的干(库仑(Coulomb))摩擦力.他给出了此类系统的一般运动方程,并指出使用库仑摩擦定律可能导致的自相矛盾的情况,提出了摩擦系统动力学中的"潘勒韦悖论".后来潘勒韦还曾尝试创建力学公理,他相信力学公理允许定义仅适用于直线和匀速平移运动的绝对运动坐标系.类似于潘勒韦方程,由于近几十年来非线性动力学方法的发展,潘勒韦悖论再次回到公众视线.米塔-列夫勒对潘勒韦的评价是:"他不惧怕最困难的问题,是一位真正的发明家."与潘勒韦师出同门的阿达马说:"潘勒韦继承了庞加莱的工作,达到了人类力量的极限."

潘勒韦的数学才华很快得到国际公认,成为那个时代最著名的数学家之一.他曾获得法国科学院数学科学大奖(1890年)、波尔丁大奖(1894年)和蓬塞莱大奖(1896年),1900年当选为法兰西科学院(Académie des sciences)院士.在同年于巴黎举办的国际数学家大会(ICM)上,潘勒韦担任分析分会主席.1904年,他在海德堡ICM上做了题为《积分微分方程的现代问题》的大会报告.潘勒韦指导的一名博士是1907年毕业于巴黎高师的法图(P. Fatou),因勒贝格积分中的法图引理和复变动态系统中的法图集而知名.

2. 航空先行者及执掌多个部门的政治家

如果潘勒韦继续从事数学研究,前途未可限量.然而19世纪末著名的"德雷福斯(Dreyfus)事件"改变了他的人生,使他迈出了政治生涯的第一步.德雷福斯是一位法国犹太裔军官,1894年12月被反犹联盟指控叛国罪并被判处无期徒刑.

1898年初,以著名作家左拉(Zola)投书支持德雷福斯的清白为开端,掀起了为期十多年、天翻地覆的法国社会大改造运动.1899年,潘勒韦在新的军事法庭上作证,持续为德雷福斯争取正义,直到1906年他被无罪释放,正式成为国家英雄.与潘勒韦亦师亦友的庞加莱和阿达马均为德雷福斯平反奔走呼号.

1901年,潘勒韦与J. Petit de Villeneuve结婚,他们的儿子让(Jean Painlevé)于次年出生.不幸的是,潘勒韦的妻子在生产六周后死于产褥热.让由其父孀居的姐姐抚养成人,后来成为著名的纪录片导演和制片人,执导了200多部科学和自然电影.潘勒韦是一位理想主义、人道主义及和平主义者,他于1910年停止了所有教学和研究工作,成为一名全职政治家.潘勒韦一直担任法国众议院议员,第一次世界大战开始后,他主持了多个军事方面的委员会.潘勒韦于1915年加入内阁,历任法国公共教育部长、国防发明部长、战争部长、航空部长、财政部长等职位.

潘勒韦自幼就对探索科学的奥秘感兴趣,对前卫技术充满好奇和激情,他在1903年利用流体力学理论证明了飞行的可能性.1908年,美国航空先驱莱特兄弟(Wright brothers)在几乎没有政府支持的情况下降落法国,展示他们的飞机并与法国方面谈判专利.1908年10月10日,潘勒韦登上了威尔伯·莱特

(Wilbur Wright)的飞机,成为第一位飞上天空的法国人.这架飞机携带了45升汽油,在10米高的空中飞行了55公里,历时1小时9分钟后成功着陆.这位狂热的航空科学家亲身体验了自己的计算结果,顺利完成了征服天空的壮举.

潘勒韦非常清楚飞机的重要性,他认为这是一种具有广阔前景的新型交通工具,他游说法国众议院,建议成立一个涉及航空的军事部门并获得了成功,为法国航空业奠定了政治基础.1909年潘勒韦成为法国第一位航空动力学教授,致力于航空科学的理论研究,担任多个空中航行委员会的主席,并率先在大学开设了空气动力学课程.

1910年,潘勒韦与他的好友——法国著名数学家博雷尔(Borel)合作撰写了《航空学》一书.博雷尔是20世纪初测度论的开拓者之一,拓扑学中的博雷尔集以他的名字命名.博雷尔也是一位政治家,于1925年担任海洋部长.

1917年,作为战争部长的潘勒韦在一次讲话中说:"科学是为人类社会保证公平合理的法律和组织,它将通过增加工业力量和对自然的控制来解决社会问题,不断创造新的财富,但不会从任何人手中夺走它们.科学将通过博爱的驯化和智慧的发展最终软化人类行为,其本质上的集体努力已经使我们从心底和思想上深刻地感受到了高度团结所赋予生命的教诲."1924~1925年间,潘勒韦当选为众议院议长,他还两次出任法国第三共和内阁总理,第一次是在第一次世界大战关键时期的1917年9月12日至1917年11月13日,第二次是在1925年4月17日至1925年11月22日的金融危机期间,因其改革计划未得到众议院批准而辞职,博雷尔就是潘勒韦第二次担任总理期间的内阁成员.

3. 为中法文化科学交流搭桥

由于从事科学研究的关系,潘勒韦对于古老神秘的中华文明十分好奇.早在1914年他就结识了蔡元培.1919年巴黎和会期间,北洋政府交通总长叶恭绰前往欧洲部分国家和美、日、朝鲜诸国考察.在叶恭绰、潘勒韦和韩汝甲等人的努力下,巴黎西岱大学中国学院于1920年3月17日成立,潘勒韦为首任院长,后来赴巴黎西岱大学勤工俭学的中国学生大多入读该学院.潘勒韦曾向叶恭绰提出,法国政府愿用退还的部分庚款印行四库全书,为此他于1919年9月专程到上海商议此事,但由于资金缺口及时局动荡未能成功.

1920年6月22日至1920年9月11日,应北洋政府之邀,潘勒韦率领由法国文化界、知识界著名人士组成的代表团前往中国访问,他的随员中有法国文学家博纳尔(Bonnard)、巴黎西岱大学经济学教授马丹、铁路工程师纳达尔以及数学家博雷尔.潘勒韦特别强调此行是一次文化之旅,代表团与中国学术文化界进行了广泛交流.1920年7月1日,潘勒韦参观了北京大学,并在北京大学理科大讲堂演讲,蔡元培校长致欢迎词.1920年6月29日至1920年7月1日,《北京大学日刊》连续三天进行宣传.1920年7月4日的《申报》以《北大欢迎班乐卫》为题做了报道,并刊登了蔡元培的欢迎词和潘勒韦的演讲.潘勒韦说:"三四千年之前,欧洲文明各国尚未形成,而中国之天文学、数学,竟能预测日、月食,实足钦佩."鉴于潘勒韦对中法文化交流的热心,以及在数学领域的贡献,

1920年8月31日蔡元培在北京大学主持仪式,聘请潘勒韦担任北京大学名誉教授.北京大学教务会议还决定授予潘勒韦,美国外交家、远东事务权威P.S.芮恩施(P.S. Reinsch),法国教育家和外交家、里昂中法大学校长P.儒班(P. Joubin),美国著名哲学家、教育家、心理学家J.杜威(J. Dewey)"理学荣誉博士"称号,开了国内大学授予外国学者荣誉博士称号的先河.授予仪式那天只有潘勒韦一人在北京,蔡元培在致辞中说:"北京大学第一次授予学位,而受者为班乐卫先生,可为特别纪念者有两点:第一,大学宗旨,凡治哲学文学及应用科学者,都要从纯粹科学入手.治纯粹科学者,都要从数学入手.所以各系次序,列数学为第一系.班乐卫先生为世界数学大家,可以代表此义.第二,科学为公,各大学自然有共通研究之对象.但大学所在地,对于其地之社会、历史,不得不有特别注重之任务,就是分工之理.北京大学既设在中国于世界学者共通研究之对象外,对于中国特有之对象,尤负特别责任.班乐卫先生最提倡中国学问的研究,又可以代表此义.所以我以为本校第一次授予学位属于班乐卫先生,不但是北京大学至重要之纪念,实可为我国教育界之大纪念."

1920年底蔡元培抵达法国考察,遍访当地名流.1921年1月和2月,蔡元培两次拜访老友潘勒韦,请他推荐几位法国学者访华.潘勒韦推荐的第一位科学家就是享誉世界的玛丽·居里(Marie Curie),另外三位是物理学家J.B.佩林(J.B. Perrin)、P.朗之万(P. Langevin)及数学家阿达马.为此蔡元培专程前往玛丽·居里的实验室,邀请她到中国访问,遗憾的是一直未能成行.1931年,朗之万参加了国联组织的中国教育与科学发展考察团来华访问,与中国物理学家进行了广泛接触和交流,做了多次学术演讲.1936年,阿达马前往上海交通大学、浙江大学演讲,后应清华大学之邀赴京讲学3个多月.四位大师分别培养了施士元、李书华、汪德昭、熊庆来、吴新谋等中国弟子,对中国近代数学和物理学的发展产生了重要影响.

4.政治生涯里的"广义相对论"插曲

1921~1922年间,潘勒韦的注意力转向广义相对论.1915年11月,爱因斯坦(Einstein)提出了广义相对论的核心——场方程,不久后德国物理学家K.施瓦茨席尔德(K. Schwarzschild)证明了被称为"施瓦茨席尔德度规"的球对称真空解,其重要特征是施瓦茨席尔德半径和奇点.潘勒韦与A.古尔斯特兰德(A. Gullstrand)先后独立推导出在施瓦茨席尔德半径处没有奇点的爱因斯坦方程解,这个解后来被命名为古尔斯特兰德-潘勒韦坐标.古尔斯特兰德是瑞典乌普萨拉大学(Uppsala University)眼科与光学教授、1910年诺贝尔生理或医学奖得主、诺贝尔物理学奖评委,曾极力反对爱因斯坦因相对论获奖.1921年10月和11月,潘勒韦在法国科学院发表了两篇笔记,其中考虑广义相对论的数学形式,直接从问题的对物性推导出上述爱因斯坦场方程解.

1921年底潘勒韦写信给爱因斯坦,介绍自己的解决方案,并邀请爱因斯坦前往巴黎探讨.1922年3月底爱因斯坦接受法国物理学会之邀访问巴黎,成为第一次世界大战后首位在法国公开露面的德国人,因此引起轰动.爱因斯坦在法兰西公学院做了公开演讲,并与潘勒韦、H. 贝克勒尔(H. Becquerel)、L. 布里尤安(L. Brillouin)、嘉当(Cartan)、阿达马、朗之万等人进行了激烈辩论. 爱因斯坦对于潘勒韦方案中线性元素的非二次交叉项感到困惑,因此否定了他的想法. 在这场辩论后,潘勒韦发表了第三篇笔记,将他在牛顿(Newton)理论中使用的几何形式扩展到了广义相对论.

法国科学院是一个相当保守的学术机构,包括其中最活跃的一些成员,直到 1921 年都对广义相对论持敌对态度,认为它破坏了牛顿经典力学. 在经过某些科学院成员对于广义相对论的一场恶毒攻击之后,潘勒韦的工作旨在"缓和"这场辩论,引导那些对爱因斯坦新理论感到困惑的同事们对两种理论进行比较性的研究. 根据潘勒韦的科学背景,当时很难做到完全客观,他还没有准备好放弃整个经典力学的大厦. 然而他的尝试具有高度建设性,为随后在科学院进行的富有启发性的辩论做出了贡献,使得爱因斯坦的巴黎之行富有成效. 潘勒韦最早构造出了在施瓦茨席尔德半径处没有奇点的爱因斯坦方程解,虽然后来他对其有效性也表示怀疑,但是作为数学家,潘勒韦确信这一有争议的方案的形式推导是正确的. 潘勒韦对于广义相对论的兴趣持续了六个月后即重返政坛,而他的一些超前思想却被遗忘了几十年. 尽管当时包括爱因斯坦在内的许多著名物理学家都认为在施瓦茨席尔德半径上的物理奇点是实际存在的,而 1933 年 G. 勒梅特(G. Lemaitre)发现了潘勒韦的解实际上是施瓦茨席尔德度规的一个坐标变换,人们才得知坐标系的变换揭示了施瓦茨席尔德半径仅仅是一个坐标奇点,更深远的意义是它代表了黑洞的事件视界. 直到 20 世纪 60 年代,一些如微分几何等更高级的数学工具进入了广义相对论的研究,物理学家们才普遍认可这一点.

5. 富有理性和活力的一生

潘勒韦天生淳朴、精力充沛、充满活力,身上散发着一种即使在其对手中也很少有人能够抗拒的人格魅力. 他一生在出世的数学家和入世的政治家两个不同的角色之间切换,撰写出版了 144 部/篇学术著作、教科书及论文,最后一部著作是 1930 年出版的《无粘性流体阻力教程》. 1925 年,潘勒韦辞去法国总理职位后,继续在政府中担任高官. 1932 年,他被推举为法国总统候选人,却在大选前退出. 潘勒韦毕生享受理性思维和科学精神带来的愉悦,成为"数学治国"的典范. 潘勒韦是沿法国东部边界的军事防御工事——马其诺防线的主要设计者之一,他还提议制定一项禁止制造轰炸机的国际公约,并建立一支国际空军以维护全球和平,但由于 1933 年 1 月法国政府倒台而付之东流.

一些历史学家认为,尽管潘勒韦在他的政治生涯中取得了最高职位,但他作为数学家的贡献更为显著. 潘勒韦在其生命的黄昏又回到了自己最喜爱的研究领域,他曾说:"如果我必须离去,我会尽量优雅地做到这一点!"1933 年 10 月 29 日,潘勒韦因心力衰竭在巴黎家中去世,预言成为现实. 1933 年 11 月 4 日举行国葬,潘勒韦长眠于先贤祠,法国失去了其最优秀的一个儿子. 巴黎拉丁区的一个广场及里尔大学的一个数学实验室以潘勒韦冠名,太阳系小行星 953 被命名为“Painlevé”,有一艘法国航母也被命名为“潘勒韦号”,但仅存在于图纸上.就像他的许多愿景一样,未必都能付诸现实或者被人长时间遗忘,然而潘勒韦终生为之奋斗,乐此不疲.

国内许多科研人员都是想得多、读得少.达尔文曾说:

无知要比博学的人更容易产生自信.

所以多读是对学问保持敬畏的有效途径.

刘培杰
2023年12月29日
于哈工大

刘培杰数学工作室
已出版(即将出版)图书目录——原版影印

书　　名	出版时间	定　价	编号
数学物理大百科全书.第1卷(英文)	2016—01	418.00	508
数学物理大百科全书.第2卷(英文)	2016—01	408.00	509
数学物理大百科全书.第3卷(英文)	2016—01	396.00	510
数学物理大百科全书.第4卷(英文)	2016—01	408.00	511
数学物理大百科全书.第5卷(英文)	2016—01	368.00	512
zeta函数,q-zeta函数,相伴级数与积分(英文)	2015—08	88.00	513
微分形式:理论与练习(英文)	2015—08	58.00	514
离散与微分包含的逼近和优化(英文)	2015—08	58.00	515
艾伦·图灵:他的工作与影响(英文)	2016—01	98.00	560
测度理论概率导论,第2版(英文)	2016—01	88.00	561
带有潜在故障恢复系统的半马尔柯夫模型控制(英文)	2016—01	98.00	562
数学分析原理(英文)	2016—01	88.00	563
随机偏微分方程的有效动力学(英文)	2016—01	88.00	564
图的谱半径(英文)	2016—01	58.00	565
量子机器学习中数据挖掘的量子计算方法(英文)	2016—01	98.00	566
量子物理的非常规方法(英文)	2016—01	118.00	567
运输过程的统一非局部理论:广义波尔兹曼物理动力学,第2版(英文)	2016—01	198.00	568
量子力学与经典力学之间的联系在原子、分子及电动力学系统建模中的应用(英文)	2016—01	58.00	569
算术域(英文)	2018—01	158.00	821
高等数学竞赛:1962—1991年的米洛克斯·史怀哲竞赛(英文)	2018—01	128.00	822
用数学奥林匹克精神解决数论问题(英文)	2018—01	108.00	823
代数几何(德文)	2018—04	68.00	824
丢番图逼近论(英文)	2018—01	78.00	825
代数几何学基础教程(英文)	2018—01	98.00	826
解析数论入门课程(英文)	2018—01	78.00	827
数论中的丢番图问题(英文)	2018—01	78.00	829
数论(梦幻之旅):第五届中日数论研讨会演讲集(英文)	2018—01	68.00	830
数论新应用(英文)	2018—01	68.00	831
数论(英文)	2018—01	78.00	832

刘培杰数学工作室
已出版(即将出版)图书目录——原版影印

书　　名	出版时间	定　价	编号
湍流十讲(英文)	2018－04	108.00	886
无穷维李代数:第3版(英文)	2018－04	98.00	887
等值、不变量和对称性(英文)	2018－04	78.00	888
解析数论(英文)	2018－09	78.00	889
《数学原理》的演化:伯特兰·罗素撰写第二版时的手稿与笔记(英文)	2018－04	108.00	890
哈密尔顿数学论文集(第4卷):几何学、分析学、天文学、概率和有限差分等(英文)	2019－05	108.00	891
偏微分方程全局吸引子的特性(英文)	2018－09	108.00	979
整函数与下调和函数(英文)	2018－09	118.00	980
幂等分析(英文)	2018－09	118.00	981
李群,离散子群与不变量理论(英文)	2018－09	108.00	982
动力系统与统计力学(英文)	2018－09	118.00	983
表示论与动力系统(英文)	2018－09	118.00	984
分析学练习.第1部分(英文)	2021－01	88.00	1247
分析学练习.第2部分,非线性分析(英文)	2021－01	88.00	1248
初级统计学:循序渐进的方法:第10版(英文)	2019－05	68.00	1067
工程师与科学家微分方程用书:第4版(英文)	2019－07	58.00	1068
大学代数与三角学(英文)	2019－06	78.00	1069
培养数学能力的途径(英文)	2019－07	38.00	1070
工程师与科学家统计学:第4版(英文)	2019－06	58.00	1071
贸易与经济中的应用统计学:第6版(英文)	2019－06	58.00	1072
傅立叶级数和边值问题:第8版(英文)	2019－05	48.00	1073
通往天文学的途径:第5版(英文)	2019－05	58.00	1074
拉马努金笔记.第1卷(英文)	2019－06	165.00	1078
拉马努金笔记.第2卷(英文)	2019－06	165.00	1079
拉马努金笔记.第3卷(英文)	2019－06	165.00	1080
拉马努金笔记.第4卷(英文)	2019－06	165.00	1081
拉马努金笔记.第5卷(英文)	2019－06	165.00	1082
拉马努金遗失笔记.第1卷(英文)	2019－06	109.00	1083
拉马努金遗失笔记.第2卷(英文)	2019－06	109.00	1084
拉马努金遗失笔记.第3卷(英文)	2019－06	109.00	1085
拉马努金遗失笔记.第4卷(英文)	2019－06	109.00	1086
数论:1976年纽约洛克菲勒大学数论会议记录(英文)	2020－06	68.00	1145
数论:卡本代尔1979:1979年在南伊利诺伊卡本代尔大学举行的数论会议记录(英文)	2020－06	78.00	1146
数论:诺德韦克豪特1983:1983年在诺德韦克豪特举行的Journees Arithmetiques数论大会会议记录(英文)	2020－06	68.00	1147
数论:1985－1988年在纽约城市大学研究生院和大学中心举办的研讨会(英文)	2020－06	68.00	1148

刘培杰数学工作室
已出版(即将出版)图书目录——原版影印

书　　名	出版时间	定　价	编号
数论:1987 年在乌尔姆举行的 Journees Arithmetiques 数论大会会议记录(英文)	2020—06	68.00	1149
数论:马德拉斯 1987:1987 年在马德拉斯安娜大学举行的国际拉马努金百年纪念大会会议记录(英文)	2020—06	68.00	1150
解析数论:1988 年在东京举行的日法研讨会会议记录(英文)	2020—06	68.00	1151
解析数论:2002 年在意大利切特拉罗举行的 C. I. M. E. 暑期班演讲集(英文)	2020—06	68.00	1152
量子世界中的蝴蝶:最迷人的量子分形故事(英文)	2020—06	118.00	1157
走进量子力学(英文)	2020—06	118.00	1158
计算物理学概论(英文)	2020—06	48.00	1159
物质,空间和时间的理论:量子理论(英文)	2020—10	48.00	1160
物质,空间和时间的理论:经典理论(英文)	2020—10	48.00	1161
量子场理论:解释世界的神秘背景(英文)	2020—07	38.00	1162
计算物理学概论(英文)	2020—06	48.00	1163
行星状星云(英文)	2020—10	38.00	1164
基本宇宙学:从亚里士多德的宇宙到大爆炸(英文)	2020—08	58.00	1165
数学磁流体力学(英文)	2020—07	58.00	1166
计算科学:第 1 卷,计算的科学(日文)	2020—07	88.00	1167
计算科学:第 2 卷,计算与宇宙(日文)	2020—07	88.00	1168
计算科学:第 3 卷,计算与物质(日文)	2020—07	88.00	1169
计算科学:第 4 卷,计算与生命(日文)	2020—07	88.00	1170
计算科学:第 5 卷,计算与地球环境(日文)	2020—07	88.00	1171
计算科学:第 6 卷,计算与社会(日文)	2020—07	88.00	1172
计算科学. 别卷,超级计算机(日文)	2020—07	88.00	1173
多复变函数论(日文)	2022—06	78.00	1518
复变函数入门(日文)	2022—06	78.00	1523
代数与数论:综合方法(英文)	2020—10	78.00	1185
复分析:现代函数理论第一课(英文)	2020—07	58.00	1186
斐波那契数列和卡特兰数:导论(英文)	2020—10	68.00	1187
组合推理:计数艺术介绍(英文)	2020—07	88.00	1188
二次互反律的傅里叶分析证明(英文)	2020—07	48.00	1189
旋瓦兹分布的希尔伯特变换与应用(英文)	2020—07	58.00	1190
泛函分析:巴拿赫空间理论入门(英文)	2020—07	48.00	1191
卡塔兰数入门(英文)	2019—05	68.00	1060
测度与积分(英文)	2019—04	68.00	1059
组合学手册. 第一卷(英文)	2020—06	128.00	1153
*一代数、局部紧群和巴拿赫*一代数丛的表示. 第一卷,群和代数的基本表示理论(英文)	2020—05	148.00	1154
电磁理论(英文)	2020—08	48.00	1193
连续介质力学中的非线性问题(英文)	2020—09	78.00	1195
多变量数学入门(英文)	2021—05	68.00	1317
偏微分方程入门(英文)	2021—05	88.00	1318
若尔当典范性:理论与实践(英文)	2021—07	68.00	1366
伽罗瓦理论. 第 4 版(英文)	2021—08	88.00	1408
R 统计学概论	2023—03	88.00	1614
基于不确定静态和动态问题解的仿射算术(英文)	2023—03	38.00	1618

刘培杰数学工作室
已出版(即将出版)图书目录——原版影印

书　　名	出版时间	定　价	编号
典型群,错排与素数(英文)	2020—11	58.00	1204
李代数的表示:通过 gln 进行介绍(英文)	2020—10	38.00	1205
实分析演讲集(英文)	2020—10	38.00	1206
现代分析及其应用的课程(英文)	2020—10	58.00	1207
运动中的抛射物数学(英文)	2020—10	38.00	1208
2—纽结与它们的群(英文)	2020—10	38.00	1209
概率,策略和选择:博弈与选举中的数学(英文)	2020—11	58.00	1210
分析学引论(英文)	2020—11	58.00	1211
量子群:通往流代数的路径(英文)	2020—11	38.00	1212
集合论入门(英文)	2020—10	48.00	1213
酉反射群(英文)	2020—11	58.00	1214
探索数学:吸引人的证明方式(英文)	2020—11	58.00	1215
微分拓扑短期课程(英文)	2020—10	48.00	1216
抽象凸分析(英文)	2020—11	68.00	1222
费马大定理笔记(英文)	2021—03	48.00	1223
高斯与雅可比和(英文)	2021—03	78.00	1224
π与算术几何平均:关于解析数论和计算复杂性的研究(英文)	2021—01	58.00	1225
复分析入门(英文)	2021—03	48.00	1226
爱德华·卢卡斯与素性测定(英文)	2021—03	78.00	1227
通往凸分析及其应用的简单路径(英文)	2021—01	68.00	1229
微分几何的各个方面.第一卷(英文)	2021—01	58.00	1230
微分几何的各个方面.第二卷(英文)	2020—12	58.00	1231
微分几何的各个方面.第三卷(英文)	2020—12	58.00	1232
沃克流形几何学(英文)	2020—11	58.00	1233
仿射和韦尔几何应用(英文)	2020—12	58.00	1234
双曲几何学的旋转向量空间方法(英文)	2021—02	58.00	1235
积分:分析学的关键(英文)	2020—12	48.00	1236
为有天分的新生准备的分析学基础教材(英文)	2020—11	48.00	1237
数学不等式.第一卷.对称多项式不等式(英文)	2021—03	108.00	1273
数学不等式.第二卷.对称有理不等式与对称无理不等式(英文)	2021—03	108.00	1274
数学不等式.第三卷.循环不等式与非循环不等式(英文)	2021—03	108.00	1275
数学不等式.第四卷.Jensen 不等式的扩展与加细(英文)	2021—03	108.00	1276
数学不等式.第五卷.创建不等式与解不等式的其他方法(英文)	2021—04	108.00	1277

刘培杰数学工作室
已出版(即将出版)图书目录——原版影印

书　名	出版时间	定　价	编号
冯・诺依曼代数中的谱位移函数:半有限冯・诺依曼代数中的谱位移函数与谱流(英文)	2021－06	98.00	1308
链接结构:关于嵌入完全图的直线中链接单形的组合结构(英文)	2021－05	58.00	1309
代数几何方法. 第1卷(英文)	2021－06	68.00	1310
代数几何方法. 第2卷(英文)	2021－06	68.00	1311
代数几何方法. 第3卷(英文)	2021－06	58.00	1312

书　名	出版时间	定　价	编号
代数、生物信息和机器人技术的算法问题. 第四卷,独立恒等式系统(俄文)	2020－08	118.00	1199
代数、生物信息和机器人技术的算法问题. 第五卷,相对覆盖性和独立可拆分恒等式系统(俄文)	2020－08	118.00	1200
代数、生物信息和机器人技术的算法问题. 第六卷,恒等式和准恒等式的相等 问题、可推导性和可实现性(俄文)	2020－08	128.00	1201
分数阶微积分的应用:非局部动态过程,分数阶导热系数(俄文)	2021－01	68.00	1241
泛函分析问题与练习:第2版(俄文)	2021－01	98.00	1242
集合论、数学逻辑和算法论问题:第5版(俄文)	2021－01	98.00	1243
微分几何和拓扑短期课程(俄文)	2021－01	98.00	1244
素数规律(俄文)	2021－01	88.00	1245
无穷边值问题解的递减:无界域中的拟线性椭圆和抛物方程(俄文)	2021－01	48.00	1246
微分几何讲义(俄文)	2020－12	98.00	1253
二次型和矩阵(俄文)	2021－01	98.00	1255
积分和级数. 第2卷,特殊函数(俄文)	2021－01	168.00	1258
积分和级数. 第3卷,特殊函数补充:第2版(俄文)	2021－01	178.00	1264
几何图上的微分方程(俄文)	2021－01	138.00	1259
数论教程:第2版(俄文)	2021－01	98.00	1260
非阿基米德分析及其应用(俄文)	2021－03	98.00	1261
古典群和量子群的压缩(俄文)	2021－03	98.00	1263
数学分析习题集. 第3卷,多元函数:第3版(俄文)	2021－03	98.00	1266
数学习题:乌拉尔国立大学数学力学系大学生奥林匹克(俄文)	2021－03	98.00	1267
柯西定理和微分方程的特解(俄文)	2021－03	98.00	1268
组合极值问题及其应用:第3版(俄文)	2021－03	98.00	1269
数学词典(俄文)	2021－01	98.00	1271
确定性混沌分析模型(俄文)	2021－06	168.00	1307
精选初等数学习题和定理. 立体几何. 第3版(俄文)	2021－03	68.00	1316
微分几何习题:第3版(俄文)	2021－05	98.00	1336
精选初等数学习题和定理. 平面几何. 第4版(俄文)	2021－05	68.00	1335
曲面理论在欧氏空间 *En* 中的直接表示(俄文)	2022－01	68.00	1444
维纳一霍普夫离散算子和托普利兹算子:某些可数赋范空间中的诺特性和可逆性(俄文)	2022－03	108.00	1496
Maple 中的数论:数论中的计算机计算(俄文)	2022－03	88.00	1497
贝尔曼和克努特问题及其概括:加法运算的复杂性(俄文)	2022－03	138.00	1498

刘培杰数学工作室

已出版(即将出版)图书目录——原版影印

书　名	出版时间	定　价	编号
复分析:共形映射(俄文)	2022－07	48.00	1542
微积分代数样条和多项式及其在数值方法中的应用(俄文)	2022－08	128.00	1543
蒙特卡罗方法中的随机过程和场模型:算法和应用(俄文)	2022－08	88.00	1544
线性椭圆型方程组:论二阶椭圆型方程的迪利克雷问题(俄文)	2022－08	98.00	1561
动态系统解的增长特性:估值、稳定性、应用(俄文)	2022－08	118.00	1565
群的自由积分解:建立和应用(俄文)	2022－08	78.00	1570
混合方程和偏差自变数方程问题:解的存在和唯一性(俄文)	2023－01	78.00	1582
拟度量空间分析:存在和逼近定理(俄文)	2023－01	108.00	1583
二维和三维流形上函数的拓扑性质:函数的拓扑分类(俄文)	2023－03	68.00	1584
齐次马尔科夫过程建模的矩阵方法:此类方法能够用于不同目上的的复杂系统研究、设计和完善(俄文)	2023－03	68.00	1594
周期函数的近似方法和特性:特殊课程(俄文)	2023－04	158.00	1622
扩散方程解的矩函数:变分法(俄文)	2023－03	58.00	1623
多赋范空间和广义函数:理论及应用(俄文)	2023－03	98.00	1632
分析中的多值映射:部分应用(俄文)	2023－06	98.00	1634
数学物理问题(俄文)	2023－03	78.00	1636
函数的幂级数与三角级数分解(俄文)	2024－01	58.00	1695
星体理论的数学基础:原子三元组(俄文)	2024－01	98.00	1696
素数规律:专著(俄文)	2024－01	118.00	1697
狭义相对论与广义相对论:时空与引力导论(英文)	2021－07	88.00	1319
束流物理学和粒子加速器的实践介绍:第2版(英文)	2021－07	88.00	1320
凝聚态物理中的拓扑和微分几何简介(英文)	2021－05	88.00	1321
混沌映射:动力学、分形学和快速涨落(英文)	2021－05	128.00	1322
广义相对论:黑洞、引力波和宇宙学介绍(英文)	2021－06	68.00	1323
现代分析电磁均质化(英文)	2021－06	68.00	1324
为科学家提供的基本流体动力学(英文)	2021－06	88.00	1325
视觉天文学:理解夜空的指南(英文)	2021－06	68.00	1326
物理学中的计算方法(英文)	2021－06	68.00	1327
单星的结构与演化:导论(英文)	2021－06	108.00	1328
超越居里:1903年至1963年物理界四位女性及其著名发现(英文)	2021－06	68.00	1329
范德瓦尔斯流体热力学的进展(英文)	2021－06	68.00	1330
先进的托卡马克稳定性理论(英文)	2021－06	88.00	1331
经典场论导论:基本相互作用的过程(英文)	2021－07	88.00	1332
光致电离量子动力学方法原理(英文)	2021－07	108.00	1333
经典域论和应力:能量张量(英文)	2021－05	88.00	1334
非线性太赫兹光谱的概念与应用(英文)	2021－06	68.00	1337
电磁学中的无穷空间并矢格林函数(英文)	2021－06	88.00	1338
物理科学基础数学.第1卷,齐次边值问题、傅里叶方法和特殊函数(英文)	2021－07	108.00	1339
离散量子力学(英文)	2021－07	68.00	1340
核磁共振的物理学和数学(英文)	2021－07	108.00	1341
分子水平的静电学(英文)	2021－08	68.00	1342
非线性波:理论、计算机模拟、实验(英文)	2021－06	108.00	1343
石墨烯光学:经典问题的电解解决方案(英文)	2021－06	68.00	1344
超材料多元宇宙(英文)	2021－07	68.00	1345
银河系外的天体物理学(英文)	2021－07	68.00	1346
原子物理学(英文)	2021－07	68.00	1347
将光打结:将拓扑学应用于光学(英文)	2021－07	68.00	1348
电磁学:问题与解法(英文)	2021－07	88.00	1364
海浪的原理:介绍量子力学的技巧与应用(英文)	2021－07	108.00	1365

刘培杰数学工作室
已出版(即将出版)图书目录——原版影印

书　　名	出版时间	定　价	编号
多孔介质中的流体:输运与相变(英文)	2021—07	68.00	1372
洛伦兹群的物理学(英文)	2021—08	68.00	1373
物理导论的数学方法和解决方法手册(英文)	2021—08	68.00	1374
非线性波数学物理学入门(英文)	2021—08	88.00	1376
波:基本原理和动力学(英文)	2021—07	68.00	1377
光电子量子计量学.第1卷,基础(英文)	2021—07	88.00	1383
光电子量子计量学.第2卷,应用与进展(英文)	2021—07	68.00	1384
复杂流的格子玻尔兹曼建模的工程应用(英文)	2021—08	68.00	1393
电偶极矩挑战(英文)	2021—08	108.00	1394
电动力学:问题与解法(英文)	2021—09	68.00	1395
自由电子激光的经典理论(英文)	2021—08	68.00	1397
曼哈顿计划——核武器物理学简介(英文)	2021—09	68.00	1401
粒子物理学(英文)	2021—09	68.00	1402
引力场中的量子信息(英文)	2021—09	128.00	1403
器件物理学的基本经典力学(英文)	2021—09	68.00	1404
等离子体物理及其空间应用导论.第1卷,基本原理和初步过程(英文)	2021—09	68.00	1405
磁约束聚变等离子体物理:理想MHD理论(英文)	2023—03	68.00	1613
相对论量子场论.第1卷,典范形式体系(英文)	2023—03	38.00	1615
相对论量子场论.第2卷,路径积分形式(英文)	2023—06	38.00	1616
相对论量子场论.第3卷,量子场论的应用(英文)	2023—06	38.00	1617
涌现的物理学(英文)	2023—05	58.00	1619
量子化旋涡:一本拓扑激发手册(英文)	2023—04	68.00	1620
非线性动力学:实践的介绍性调查(英文)	2023—05	68.00	1621
静电加速器:一个多功能工具(英文)	2023—06	58.00	1625
相对论多体理论与统计力学(英文)	2023—06	58.00	1626
经典力学.第1卷,工具与向量(英文)	2023—04	38.00	1627
经典力学.第2卷,运动学和匀加速运动(英文)	2023—04	58.00	1628
经典力学.第3卷,牛顿定律和匀速圆周运动(英文)	2023—04	58.00	1629
经典力学.第4卷,万有引力定律(英文)	2023—04	38.00	1630
经典力学.第5卷,守恒定律与旋转运动(英文)	2023—04	38.00	1631
对称问题:纳维尔一斯托克斯问题(英文)	2023—04	38.00	1638
摄影的物理和艺术.第1卷,几何与光的本质(英文)	2023—04	78.00	1639
摄影的物理和艺术.第2卷,能量与色彩(英文)	2023—04	78.00	1640
摄影的物理和艺术.第3卷,探测器与数码的意义(英文)	2023—04	78.00	1641
拓扑与超弦理论焦点问题(英文)	2021—07	58.00	1349
应用数学:理论、方法与实践(英文)	2021—07	78.00	1350
非线性特征值问题:牛顿型方法与非线性瑞利函数(英文)	2021—07	58.00	1351
广义膨胀和齐性:利用齐性构造齐次系统的李雅普诺夫函数和控制律(英文)	2021—06	48.00	1352
解析数论焦点问题(英文)	2021—07	58.00	1353
随机微分方程:动态系统方法(英文)	2021—07	58.00	1354
经典力学与微分几何(英文)	2021—07	58.00	1355
负定相交形式流形上的瞬子模空间几何(英文)	2021—07	68.00	1356
广义卡塔兰轨道分析:广义卡塔兰轨道计算数字的方法(英文)	2021—07	48.00	1367
洛伦兹方法的变分:二维与三维洛伦兹方法(英文)	2021—08	38.00	1378
几何、分析和数论精编(英文)	2021—08	68.00	1380
从一个新角度看数论:通过遗传方法引入现实的概念(英文)	2021—07	58.00	1387
动力系统:短期课程(英文)	2021—08	68.00	1382
几何路径:理论与实践(英文)	2021—08	48.00	1385

刘培杰数学工作室
已出版(即将出版)图书目录——原版影印

书　　名	出版时间	定　价	编号
论天体力学中某些问题的不可积性(英文)	2021－07	88.00	1396
广义斐波那契数列及其性质(英文)	2021－08	38.00	1386
对称函数和麦克唐纳多项式:余代数结构与 Kawanaka 恒等式(英文)	2021－09	38.00	1400
杰弗里·英格拉姆·泰勒科学论文集:第 1 卷.固体力学(英文)	2021－05	78.00	1360
杰弗里·英格拉姆·泰勒科学论文集:第 2 卷.气象学、海洋学和湍流(英文)	2021－05	68.00	1361
杰弗里·英格拉姆·泰勒科学论文集:第 3 卷.空气动力学以及落弹数和爆炸的力学(英文)	2021－05	68.00	1362
杰弗里·英格拉姆·泰勒科学论文集:第 4 卷.有关流体力学(英文)	2021－05	58.00	1363
非局域泛函演化方程:积分与分数阶(英文)	2021－08	48.00	1390
理论工作者的高等微分几何:纤维丛、射流流形和拉格朗日理论(英文)	2021－08	68.00	1391
半线性退化椭圆微分方程:局部定理与整体定理(英文)	2021－07	48.00	1392
非交换几何、规范理论和重整化:一般简介与非交换量子场论的重整化(英文)	2021－09	78.00	1406
数论论文集:拉普拉斯变换和带有数论系数的幂级数(俄文)	2021－09	48.00	1407
挠理论专题:相对极大值,单射与扩充模(英文)	2021－09	88.00	1410
强正则图与欧几里得若尔当代数:非通常关系中的启示(英文)	2021－10	48.00	1411
拉格朗日几何和哈密顿几何:力学的应用(英文)	2021－10	48.00	1412
时滞微分方程与差分方程的振动理论:二阶与三阶(英文)	2021－10	98.00	1417
卷积结构与几何函数理论:用以研究特定几何函数理论方向的分数阶微积分算子与卷积结构(英文)	2021－10	48.00	1418
经典数学物理的历史发展(英文)	2021－10	78.00	1419
扩展线性丢番图问题(英文)	2021－10	38.00	1420
一类混沌动力系统的分歧分析与控制:分歧分析与控制(英文)	2021－11	38.00	1421
伽利略空间和伪伽利略空间中一些特殊曲线的几何性质(英文)	2022－01	68.00	1422
一阶偏微分方程:哈密尔顿—雅可比理论(英文)	2021－11	48.00	1424
各向异性黎曼多面体的反问题:分段光滑的各向异性黎曼多面体反边界谱问题:唯一性(英文)	2021－11	38.00	1425
项目反应理论手册.第一卷,模型(英文)	2021－11	138.00	1431
项目反应理论手册.第二卷,统计工具(英文)	2021－11	118.00	1432
项目反应理论手册.第三卷,应用(英文)	2021－11	138.00	1433
二次无理数:经典数论入门(英文)	2022－05	138.00	1434

刘培杰数学工作室
已出版(即将出版)图书目录——原版影印

书　　名	出版时间	定　价	编号
数,形与对称性:数论,几何和群论导论(英文)	2022－05	128.00	1435
有限域手册(英文)	2021－11	178.00	1436
计算数论(英文)	2021－11	148.00	1437
拟群与其表示简介(英文)	2021－11	88.00	1438
数论与密码学导论:第二版(英文)	2022－01	148.00	1423
几何分析中的柯西变换与黎兹变换:解析调和容量和李普希兹调和容量、变化和振荡以及一致可求长性(英文)	2021－12	38.00	1465
近似不动点定理及其应用(英文)	2022－05	28.00	1466
局部域的相关内容解析:对局部域的扩展及其伽罗瓦群的研究(英文)	2022－01	38.00	1467
反问题的二进制恢复方法(英文)	2022－03	28.00	1468
对几何函数中某些类的各个方面的研究:复变量理论(英文)	2022－01	38.00	1469
覆盖、对应和非交换几何(英文)	2022－01	28.00	1470
最优控制理论中的随机线性调节器问题:随机最优线性调节器问题(英文)	2022－01	38.00	1473
正交分解法:涡流流体动力学应用的正交分解法(英文)	2022－01	38.00	1475
芬斯勒几何的某些问题(英文)	2022－03	38.00	1476
受限三体问题(英文)	2022－05	38.00	1477
利用马利亚万微积分进行 Greeks 的计算:连续过程、跳跃过程中的马利亚万微积分和金融领域中的 Greeks(英文)	2022－05	48.00	1478
经典分析和泛函分析的应用:分析学的应用(英文)	2022－03	38.00	1479
特殊芬斯勒空间的探究(英文)	2022－03	48.00	1480
某些图形的施泰纳距离的细谷多项式:细谷多项式与图的维纳指数(英文)	2022－05	38.00	1481
图论问题的遗传算法:在新鲜与模糊的环境中(英文)	2022－05	48.00	1482
多项式映射的渐近簇(英文)	2022－05	38.00	1483
一维系统中的混沌:符号动力学,映射序列,一致收敛和沙可夫斯基定理(英文)	2022－05	38.00	1509
多维边界层流动与传热分析:粘性流体流动的数学建模与分析(英文)	2022－05	38.00	1510
演绎理论物理学的原理:一种基于量子力学波函数的逐次置信估计的一般理论的提议(英文)	2022－05	38.00	1511
R^2 和 R^3 中的仿射弹性曲线:概念和方法(英文)	2022－08	38.00	1512
算术数列中除数函数的分布:基本内容、调查、方法、第二矩、新结果(英文)	2022－05	28.00	1513
抛物型狄拉克算子和薛定谔方程:不定常薛定谔方程的抛物型狄拉克算子及其应用(英文)	2022－07	28.00	1514
黎曼-希尔伯特问题与量子场论:可积重正化、戴森-施温格方程(英文)	2022－08	38.00	1515
代数结构和几何结构的形变理论(英文)	2022－08	48.00	1516
概率结构和模糊结构上的不动点:概率结构和直觉模糊度量空间的不动点定理(英文)	2022－08	38.00	1517

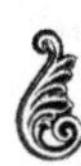

刘培杰数学工作室
已出版(即将出版)图书目录——原版影印

书　　名	出版时间	定　价	编号
反若尔当对:简单反若尔当对的自同构(英文)	2022—07	28.00	1533
对某些黎曼—芬斯勒空间变换的研究:芬斯勒几何中的某些变换(英文)	2022—07	38.00	1534
内诣零流形映射的尼尔森数的阿诺索夫关系(英文)	2023—01	38.00	1535
与广义积分变换有关的分数次演算:对分数次演算的研究(英文)	2023—01	48.00	1536
强子的芬斯勒几何和吕拉几何(宇宙学方面):强子结构的芬斯勒几何和吕拉几何(拓扑缺陷)(英文)	2022—08	38.00	1537
一种基于混沌的非线性最优化问题:作业调度问题(英文)	2023—03	38.00	1538
广义概率论发展前景:关于趣味数学与置信函数实际应用的一些原创观点(英文)	2023—03	48.00	1539
纽结与物理学:第二版(英文)	2022—09	118.00	1547
正交多项式和 q—级数的前沿(英文)	2022—09	98.00	1548
算子理论问题集(英文)	2022—09	108.00	1549
抽象代数:群、环与域的应用导论:第二版(英文)	2023—01	98.00	1550
菲尔兹奖得主演讲集:第三版(英文)	2023—01	138.00	1551
多元实函数教程(英文)	2022—09	118.00	1552
球面空间形式群的几何学:第二版(英文)	2022—09	98.00	1566
对称群的表示论(英文)	2023—01	98.00	1585
纽结理论:第二版(英文)	2023—01	88.00	1586
拟群理论的基础与应用(英文)	2023—01	88.00	1587
组合学:第二版(英文)	2023—01	98.00	1588
加性组合学:研究问题手册(英文)	2023—01	68.00	1589
扭曲、平铺与镶嵌:几何折纸中的数学方法(英文)	2023—01	98.00	1590
离散与计算几何手册:第三版(英文)	2023—01	248.00	1591
离散与组合数学手册:第二版(英文)	2023—01	248.00	1592
分析学教程. 第 1 卷,一元实变量函数的微积分分析学介绍(英文)	2023—01	118.00	1595
分析学教程. 第 2 卷,多元函数的微分和积分,向量微积分(英文)	2023—01	118.00	1596
分析学教程. 第 3 卷,测度与积分理论,复变量的复值函数(英文)	2023—01	118.00	1597
分析学教程. 第 4 卷,傅里叶分析,常微分方程,变分法(英文)	2023—01	118.00	1598

刘培杰数学工作室
已出版(即将出版)图书目录——原版影印

书　　名	出版时间	定　价	编号
共形映射及其应用手册(英文)	2024-01	158.00	1674
广义三角函数与双曲函数(英文)	2024-01	78.00	1675
振动与波:概论:第二版(英文)	2024-01	88.00	1676
几何约束系统原理手册(英文)	2024-01	120.00	1677
微分方程与包含的拓扑方法(英文)	2024-01	98.00	1678
数学分析中的前沿话题(英文)	2024-01	198.00	1679
流体力学建模:不稳定性与湍流(英文)	2024-03	88.00	1680
动力系统:理论与应用(英文)	2024-03	108.00	1711
空间统计学理论:概述(英文)	2024-03	68.00	1712
梅林变换手册(英文)	2024-03	128.00	1713
非线性系统及其绝妙的数学结构.第1卷(英文)	2024-03	88.00	1714
非线性系统及其绝妙的数学结构.第2卷(英文)	2024-03	108.00	1715
Chip-firing 中的数学(英文)	2024-04	88.00	1716
阿贝尔群的可确定性:问题、研究、概述(俄文)	2024-05	716.00(全7册)	1727
素数规律:专著(俄文)	2024-05	716.00(全7册)	1728
函数的幂级数与三角级数分解(俄文)	2024-05	716.00(全7册)	1729
星体理论的数学基础:原子三元组(俄文)	2024-05	716.00(全7册)	1730
技术问题中的数学物理微分方程(俄文)	2024-05	716.00(全7册)	1731
概率论边界问题:随机过程边界穿越问题(俄文)	2024-05	716.00(全7册)	1732
代数和幂等配置的正交分解:不可交换组合(俄文)	2024-05	716.00(全7册)	1733

联系地址:哈尔滨市南岗区复华四道街10号　哈尔滨工业大学出版社刘培杰数学工作室
邮　　编:150006
联系电话:0451-86281378　　13904613167
E-mail:lpj1378@163.com